THE CIT[illegible] Fifth edition

The fifth edition of the highly successful *The City Reader* juxtaposes the best classic and contemporary writings on the city. It contains fifty-seven selections including seventeen new selections by Elijah Anderson, Robert Bruegmann, Michael Dear, Jan Gehl, Harvey Molotch, Clarence Perry, Daphne Spain, Nigel Taylor, Sam Bass Warner, and others, five of which have been newly written exclusively for *The City Reader.* Classic writings from Ebenezer Howard, Ernest W. Burgess, Le Corbusier, Lewis Mumford, Jane Jacobs, and Louis Wirth meet the best contemporary writings of Sir Peter Hall, Manuel Castells, David Harvey, Kenneth Jackson and others.

The City Reader: Fifth edition is now integrated with all nine other titles in the series. *The City Reader: Fifth edition* has been extensively updated and expanded to reflect the latest thinking in each of the disciplinary areas included and in topical areas such as sustainable urban development, climate change, globalization, and the impact of technology on cities. The plate sections have been extensively revised and expanded and a new plate section on global cities has been added.

The anthology features general and part introductions and introductions to the selected articles.

Richard T. LeGates is Professor Emeritus of Urban Studies and Planning at San Francisco State University.

Frederic Stout is Lecturer in Urban Studies at Stanford University.

THE ROUTLEDGE URBAN READER SERIES

Series editors

Richard T. LeGates
Professor emeritus of Urban Studies and Planning, San Francisco State University.

Frederic Stout
Lecturer in Urban Studies, Stanford University

The Routledge Urban Reader Series responds to the need for comprehensive coverage of the classic and essential texts that form the basis of intellectual work in the various academic disciplines and professional fields concerned with cities and city planning.

The readers focus on the key topics encountered by undergraduates, graduate students, and scholars in urban studies, geography, sociology, political science, anthropology, economics, culture studies and professional fields such as city and regional planning, urban design, architecture, environmental studies, international relations and landscape architecture. They discuss the contributions of major theoreticians and practitioners and other individuals, groups, and organizations that study the city or practice in a field that directly affects the city.

As well as drawing together the best of classic and contemporary writings on the city, each reader features extensive introductions to the book, sections, and individual selections prepared by the volume editors to place the selections in context, illustrate relations among topics, provide information on the author and point readers towards additional related bibliographic material.

Each reader contains:

- Between thirty-five and sixty *selections* divided into six, seven or eight sections. Almost all of the selections are previously published works that have appeared as journal articles or portions of books.
- A *general introduction* describing the nature and purpose of the reader.
- *Section introductions* for each section of the reader to place the readings in context.
- *Selection introductions* for each selection describing the author, the intellectual background and context of the selection, competing views of the subject matter of the selection and bibliographic references to other readings by the same author and other readings related to the topic.
- One or more plate sections and illustrations at the beginning of each section.
- An index.

The series consists of the following titles:

THE CITY READER

The City Reader: Fifth edition – an interdisciplinary urban reader aimed at urban studies, urban planning, urban geography and urban sociology courses – is the *anchor urban reader*. Routledge published the first edition of *The City Reader* in 1996, a second edition in 2000, a third edition in 2003, and a fourth edition in 2007. *The City Reader* has become one of the most widely used anthologies in urban studies, urban geography, urban sociology and urban planning courses in the world.

URBAN DISCIPLINARY READERS

The series contains *urban disciplinary readers* organized around social science disciplines and professorial fields: urban sociology, urban geography, urban politics, urban and regional planning, and urban design. The urban disciplinary readers include both classic writings and recent, cutting-edge contributions to the respective disciplines. They are lively, high-quality, competitively priced readers which faculty can adopt as course texts and which also appeal to a wider audience.

TOPICAL URBAN ANTHOLOGIES

The urban series includes *topical urban readers* intended both as primary and supplemental course texts and for the trade and professional market. The topical titles include readers related to sustainable urban development, global cities, cybercities, and city cultures.

INTERDISCIPLINARY ANCHOR TITLE

The City Reader: Fifth edition
Richard T. LeGates and Frederic Stout (eds)

URBAN DISCIPLINARY READERS

The Urban Geography Reader
Nick Fyfe and Judith Kenny (eds)

The Urban Sociology Reader
Jan Lin and Christopher Mele (eds)

The Urban Politics Reader
Elizabeth Strom and John Mollenkopf (eds)

The Urban and Regional Planning Reader
Eugenie Birch (ed.)

The Urban Design Reader
Michael Larice and Elizabeth Macdonald (eds)

TOPICAL URBAN READERS

The City Cultures Reader: Second edition
Malcolm Miles, Tim Hall with Iain Borden (eds)

The Cybercities Reader
Stephen Graham (ed.)

The Sustainable Urban Development Reader: Second edition
Stephen M. Wheeler and Timothy Beatley (eds)

The Global Cities Reader
Neil Brenner and Roger Keil (eds)

For further information on the Routledge Urban Reader Series please visit our website:
www.geographyarena.com/geographyarena/urbanreaderseries
or contact

Andrew Mould
Routledge
2 Park Square
Milton Park
Abingdon
Oxon OX14 4RN
andrew.mould@routledge.co.uk

Richard T. LeGates
Department of Urban Studies and Planning
San Francisco State University
1600 Holloway Avenue
San Francisco, CA 94132
(510) 642-3256
dlegates@sfsu.edu

Frederic Stout
Urban Studies Program
Stanford University
Stanford, CA 94305-2048
fstout@stanford.edu

The City Reader
Fifth edition

Edited by

Richard T. LeGates

and

Frederic Stout

LONDON AND NEW YORK

First published 1996
by Routledge
2 Park Square, Milton Park, Abingdon, Oxon OX14 4RN

Simultaneously published in the USA and Canada
by Routledge
270 Madison Avenue, New York, NY 10016

Second edition 2000
Third edition 2003
Fourth edition 2007
Fifth edition 2011

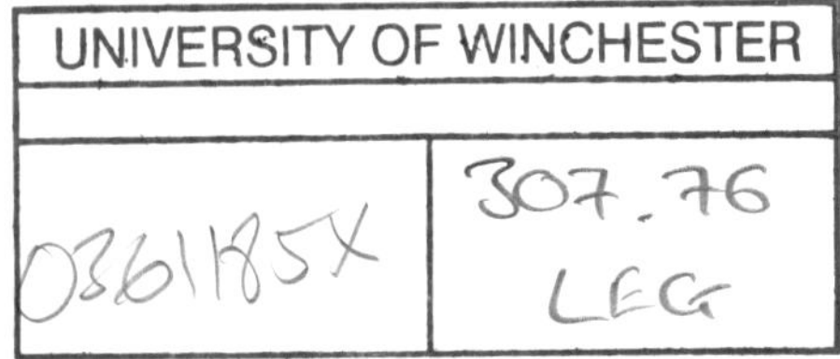

Routledge is an imprint of the Taylor & Francis Group, an informa business

Designed and typeset in Amasis MT Lt and Akzidenz Grotesk
by Keystroke, Station Road, Codsall, Wolverhampton

Printed and bound
by the MPG Books Group Ltd., in the UK

British Library Cataloguing in Publication Data
A catalogue record for this book is available from the British Library

Library of Congress Cataloging in Publication Data
A catalogue record for this book has been requested.

ISBN 13: 978–0–415–55664–4 (hbk)
ISBN 13: 978–0–415–55665–1 (pbk)
ISBN 13: 978–0–203–86926–0 (ebk)

To Joanne Fraser
and Lisa Ryan,
significant others

In Memoriam
Robert Elliot Stout
1938–2010

COMMENTS ON PREVIOUS EDITIONS

"Comprehensive and deep, this collection embodies the grand tradition, both classical and contemporary, of the urban field. It is a course itself; or a great lode for reference."

Robert J.S. Ross, Ph.D., *Professor of Sociology and Director of International Studies Stream, Clark University*

"This is the most useful reader on the market for students of cities. LeGates and Stout have refined the selections with each edition. My students tell me that the introductory notes and references make the readings more meaningful."

Ben Kohl, *Assistant Professor of Geography and Urban Studies, Temple University, Philadelphia*

"*The City Reader* brings together key works on the urban experience, problems, and policy alternatives in an engaging, accessibly structured and informative way. It draws together classic works and recent scholarship, capturing the dynamism of cities, urban processes and our interpretations of urban life. This is an impressive, comprehensive resource."

Dr Niall Majury, *School of Geography, Queen's University Belfast*

"*The City Reader* is a survival kit for practitioners who seek a better understanding of the towns and cities in which they work."

Steve Crocker, *Senior Skills Advisor, Neighbourhood Renewal Unit, Office of the Deputy Prime Minister, London*

"*The City Reader* weaves urban studies classics and modern writings in a masterful anthology. Editors' introductions to each section and piece make it an effective and accessible classroom tool."

Verrdie A. Craig, *Department of Geography, Rutgers University*

". . . remarkably well-chosen articles and excerpts . . . provides a solid introduction to the main currents of thought about urban form and processes in the twentieth century."

Environment and Planning

"An excellent, wide-ranging, stimulating reader; attractively presented and easy to read."

Brian Whalley, *Department of Built Environment, De Montfort University*

"A comprehensive mapping of the terrain of Urban Studies, old and new."

Jamie Peck, *Department of Geography, University of Manchester*

"An excellent overview, real breadth of coverage. Particularly valuable as a collection of key contributions which give a real flavour for the temporal development of Urban Studies."

David Valler, *Department of Town and Regional Planning, University of Sheffield*

"An excellent comprehensive overview of urban development and source material."

Allan Bryce, *School of Architecture, University of Dundee*

"This volume is a most welcome collection, without precedent in range and quality."

Alan Simpson, *Urban Design Associates*

"An excellent anthology bringing together of the most important papers and ideas that are relevant to the study of the urban environment."

K.J. Bussey, *Department of Land Economy, Paisley University*

"A real achievement. This book brings together 99 percent of the prominent names in Urban Studies."

Ian Robert Douglas, *Watson Institute for International Studies, Brown University, USA*

"An excellent range of texts. *The City Reader* gathers together some central classics of urban theory, with a few surprises and a number of other pieces, which can be difficult to acquire. Editors' comments are consistently illuminating."

Nick Freeman, *Department of English, University of Bristol*

"This is an essential reader for teaching about the cities and Urban Planning in developing countries."

Horng-Chang Hsieh, *Urban Planning Department, Taiwan University*

"I think this is a splendid selection of writings which illustrate the development of modern thinking on urban problems. This is by far the best book of its type."

Dr Tom Begg, *Queen Margaret College, Edinburgh*

"Provides an international overview of urban design issues and a historical perspective on visionary planners who have shaped thinking about development."

Andrew McCafferty, *Department of Built Environment, Northumbria University*

Contents

Plates

URBAN PLANNING AND URBAN DESIGN (BETWEEN PAGES 540 AND 541)

CITIES IN A GLOBAL SOCIETY (BETWEEN PAGES 608 AND 609)

Contributors

Elijah Anderson is the William K. Lanman, Jr., Professor of Sociology at Yale University.

Donald Appleyard (1928–1982) was a professor of urban design in the Department of City and Regional Planning at the University of California, Berkeley.

Sherry Arnstein (1930–1997) was the chief adviser on citizen participation in the Model Cities Program at the United States Department of Housing and Urban Development in the late 1960s and early 1970s.

Timothy Beatley is Teresa Heinz Professor of Sustainable Communities in the Department of Urban and Environmental Planning at the University of Virginia. He is coeditor (with Steven Wheeler) of *The Sustainable Urban Development Reader*, 2nd ed. in the Routledge Urban Reader Series.

Jonathan V. Beaverstock is a professor of economic geography at the University of Loughborough, England. He co-directs the Globalization and World Cities (GaWC) Research Network with Peter Taylor.

Neil Brenner is a professor of sociology and metropolitan studies at New York University. He is coeditor (with Roger Keil) of *The Global Cities Reader* in the Routledge Urban Reader Series.

Robert Bruegmann is a professor of architecture and art history at the University of Illinois at Chicago.

Ernest W. Burgess (1886–1966) was a sociology professor at the University of Chicago, and a core member of the first generation of Chicago school sociologists.

Peter Calthorpe is a California-based architect, urban designer, author, and co-founder of the Congress of the New Urbanism.

Manuel Castells holds the Wallis Annenberg Chair in Communication Technology and Society at the University of Southern California and is a research professor at the Open University of Catalonia in Barcelona, and is a professor emeritus of city and regional planning and sociology at the University of California, Berkeley.

V. Gordon Childe (1892–1957) was a professor of archaeology at the University of Edinburgh and director of the Institute of Archaeology at the University of London.

Paul Davidoff (1930–1984) was a lawyer, urban planner, professor, and civil rights activist. As the director of the Suburban Action Institute he worked to integrate housing. He proposed the advocacy planning model of urban planning.

Kingsley Davis (1908–1996) was a Distinguished Professor of Sociology at the University of Southern California, Ford Professor of Sociology and Comparative Studies Emeritus at the University of California, Berkeley, and a senior research fellow at the Hoover Institution at Stanford University. He pioneered the field of historical urban demography.

Mike Davis is a distinguished professor of creative writing at the University of California, Riverside, and a MacArthur fellow.

Michael Dear is a professor of city and regional planning at the University of California, Berkeley.

W.E.B. (William Edward Burghardt) Du Bois (1868–1963) was a professor, editor, author, novelist, playwright, and political activist. Du Bois was the first African American to receive a PhD degree from Harvard and one of the preeminent intellectuals of his generation.

Friedrich Engels (1820–1895) was a friend, partner and supporter of Karl Marx and one of the founders of the international Communist movement.

Robert Fishman is a historian and professor of architecture and urban planning at the University of Michigan.

Richard Florida is a professor at the Rotman School of Management, University of Toronto.

John Forester is a professor of city and regional planning at Cornell University.

William Fulton is an urban planner and the president of the Solimar Research Group in Ventura, California.

Jan Gehl is a Danish architect and urban planner who specializes in how to design streets, public spaces, and private outdoor space to increase social interaction.

David R. Godschalk is an emeritus professor of urban and regional planning at the University of North Carolina.

Sir Peter Hall is a professor of planning at the Bartlett School of Architecture and Planning, University College, London. Previously he taught city and regional planning at the University of California, Berkeley.

David Harvey is a professor at the New York University Graduate Center.

Ebenezer Howard (1850–1928) was a British social reformer and the founder of the Garden City movement.

J.B. (John Brinckerhoff) Jackson (1909–1997) was the founder and editor of the journal *Landscape* and taught courses on the American landscape at Harvard and the University of California, Berkeley.

Kenneth T. Jackson is the Jacques Barzun Professor of History and the Social Sciences at Columbia University.

Allan Jacobs is an emeritus professor of city and regional planning at the University of California, Berkeley. He served as San Francisco's city planning director from 1976 to 1984.

Jane Jacobs (1916–2006) was a community activist and best-selling author.

Edward J. Kaiser is an emeritus professor of urban and regional planning at the University of North Carolina.

Roger Keil is a professor of environmental studies at York University in Toronto and the director of the Canadian Centre for European Studies. He is co-editor (with Neil Brenner) of *The Global Cities Reader* in the Routledge Urban Reader Series.

George L. Kelling is a senior fellow at the Manhattan Institute and a professor of criminal justice at Rutgers University.

H.D.F. (Humphrey Davey Findley) Kitto (1897–1982) was a professor of classics at the University of Bristol, England.

Le Corbusier (Charles-Edouard Jeanneret) (1887–1965) was an architect, urban visionary, and an important force in the modernist movement.

Richard T. LeGates is a professor emeritus of urban studies and planning at San Francisco State University. He has also taught at the University of California, Berkeley, and Stanford University.

Kevin Lynch (1918–1994) was a professor of urban planning at the Massachusetts Institute of Technology.

Ali Madanipour is a professor of urban design at the University of Newcastle, England.

Harvey Molotch is a professor of sociology and metropolitan studies at New York University.

Lewis Mumford (1895–1990) was a distinguished urbanist, cultural historian, biographer, architectural critic, occasional academic, and public intellectual.

Frederick Law Olmsted (1822–1903) was a social reformer, landscape architect, and founder of the parks movement in America.

Myron Orfield is a law professor at the University of Minnesota, former Minnesota State Congressman and Senator, and geographical information systems (GIS) expert.

Clarence Perry (1872–1943) was an architect and education expert whose work with the Russell Sage Foundation on residential neighborhood design has had a large influence on urban planning.

Michael Porter is the Bishop William Lawrence University Professor of Business Administration at Harvard Business School and director of Harvard's Institute for Strategy and Competitiveness.

Robert D. Putnam is the Peter and Isabel Malkin Professor of Public Policy at Harvard University and a former Dean of Harvard's Kennedy School of Government.

Saskia Sassen is Robert S. Lynd Professor of Sociology at Columbia University and chair of Columbia's Committee on Global Thought.

Camillo Sitte (1843–1903) was an Austrian architect whose careful on-site studies of European public spaces and theory of urban planning according to artistic principles inspired an urban design movement in the early twentieth century.

Richard G. Smith is a senior lecturer in urban theory and globalization at Swansea University, Wales.

Daphne Spain is a professor of urban and environmental planning in the School of Architecture at the University of Virginia.

Frederic Stout is a lecturer at Stanford University's Program on Urban Studies. He has also taught at the University of California, Davis, San Francisco State University, and New College of California.

Nigel Taylor is a principal lecturer in planning and architecture at the Centre for Sustainable Planning and Environments at the University of the West of England and a visiting professor in the Department of Architecture and Planning at the University of Bologna, Italy.

Peter J. Taylor is a professor of geography at the University of Loughborough, England and the director of the Globalization and World Cities (GaWC) Research Network.

Wilbur Thompson is a professor emeritus of economics at Wayne State University.

Sam Bass Warner is a visiting professor of urban history at the Massachusetts Institute of Technology.

Melvin M. Webber (1920–2006) was a professor of city and regional planning at the University of California, Berkeley, and director of Berkeley's Institute of Urban and Regional Development.

Stephen Wheeler is an associate professor in the Landscape Architecture Program at the University of California, Davis, and an authority on sustainable urban development. He is co-editor (with Timothy Beatley) of *The Sustainable Urban Development Reader* in the Routledge Urban Reader Series.

William H. Whyte (1918–1999) was a sociologist whose perceptive studies of the way in which people use parks, plazas, and other public spaces in cities have influenced urban design practice.

James Q. Wilson is the Ronald Reagan Professor of Public Policy at Pepperdine University.

William Julius Wilson is the Lewis P. and Linda L. Geyser University Professor at Harvard University.

Louis Wirth (1897–1952) was a professor of sociology at the University of Chicago and a leading figure in the first generation of Chicago school sociologists.

Frank Lloyd Wright (1867–1959) is widely regarded as the greatest American architect of his time.

Tingwei Zhang is professor of urban planning and policy at the University of Illinois, Chicago (UIC), and director of UIC's Great Cities Institute's Center for Southeast Asian Studies.

Acknowledgements

We received constant encouragement and many valuable suggestions from our colleagues, both for selections to include and approaches to critical commentary. We wish particularly to thank Andrew Mould, our editor at Routledge, for his support, encouragement, and helpful suggestions. Michael Jones and Faye Leerink at Routledge provided invaluable help in securing permissions. Christine Firth did a first-rate job of copy-editing the manuscript. Senior Production Editor Emma Hart ably managed the book production process.

Editors of the Routledge Urban Reader Series provided inspiration, advice, and assistance in selecting and commenting on selections within their domains of expertise: Timothy Beatley (University of Virginia), Eugenie Birch (University of Pennsylvania), Iain Borden (University College London), Neil Brenner (New York University), Nicholas Fyfe (University of Dundee), Stephen Graham (Newcastle University), Tim Hall (University College London), Roger Keil (York University), Judith Kenny (University of Minnesota), Michael Larice (University of Pennsylvania), Jan Lin (Occidental College), Elizabeth Macdonald (University of California, Berkeley), Chris Mele (University of Buffalo), Malcolm Miles (University College London), John Mollenkopf (City University of New York), Elizabeth Strom (University of South Florida), Stephen Wheeler (University of California, Davis).

Ayse Pamuk, Raquel Pinderhughes, Jasper Rubin, and Ashok Das at San Francisco State University; Paul Turner, Leonard Ortolano, Linda Darling-Hammond, Doug McAdam, and Gerry Gast at Stanford University; Dan Lewis of Northwestern University, and Chester Hartman at the Poverty and Race Research and Action Council – all gave us many valuable suggestions. Alexander Garvin of Yale University, Peter Calthorpe of Calthorpe Associates, and Chris McGee of San Francisco State University were generous in sharing their insights about what visual images to include and contributed their own copies of images they had assembled over the years. Lisa Ryan contributed her artistic talents to creating visual images that appear at the beginning of the Prologue and Part Three. Many others too numerous to mention made helpful suggestions. All errors and infelicities are, of course, ours.

INTRODUCTION

The fifth edition of *The City Reader* comes as each of us has officially retired from a combined total of eighty years' teaching students about cities and the twentieth anniversary of our proposal to Routledge to prepare the first edition of what was to become *The City Reader* and the Routledge Urban Reader Series.

During the first half of our teaching careers, our students in urban studies and city and regional planning courses at San Francisco State University, Stanford University, and the University of California, Berkeley, where we are based (and at Tongji, Tsinghua, Northeastern, Renmin and Xiamen Universities in China, Charles University in the Czech Republic, the American University of Sharjah in the United Arab Emirates, the University of Dundee in Scotland, Cardiff University in Wales, and the Universities of Leicester, Nottingham, Plymouth, and University College London in England where we have visited and lectured) have often asked us what is the best writing on a given topic or what one single piece of new writing captures current thinking about an important topic in urban studies or urban planning right now. Since there was no one source to which we could refer them, each of us accumulated photocopies of what we consider to be essential writings and bibliographic references to many more. As time passed our colleagues began to come to us for suggested course readings, and we in turn added other selections they have found most useful to our list. We realized that a systematic organization of the best writings we use to meet both requests would make a good anthology to introduce students of urban studies and city planning to the field and to supplement course texts used in these and other courses concerned with cities. Accordingly we set to work in 1991 to produce *The City Reader*. The contents of the first edition of *The City Reader* were further enriched by suggestions from members of a distinguished review panel, who added their own suggestions to our list of selections to include. The first edition of *The City Reader*, published by Routledge in 1996, contained a generous selection of both kinds of essential readings – enduring writings and the exciting new writings that we, our students, colleagues, and expert reviewers considered to best introduce students to cities.

The first edition was well received and we learned a great deal more about what readings students and faculty find most helpful from using the first edition in our own courses and receiving feedback from faculty colleagues about what selections their students found most useful. Our only regret was that space limitations made it impossible to include as many of the writings we had accumulated, and that reviewers suggested, as we would have liked.

In 2000 Routledge published an expanded and improved second edition of *The City Reader* that quickly established itself as required reading in courses in urban studies, urban and regional planning, urban geography, and urban sociology worldwide. Based on the success of the second edition Routledge suggested that we act as general editors for a series of urban readers modeled on *The City Reader*. We saw this as a way to draw on the expertise of scholars that went far beyond our own and to make many of the excellent selections we could not fit in *The City Reader* accessible to students worldwide. We enthusiastically agreed to oversee a series of urban readers organized around disciplinary perspectives (such as *The Urban Sociology Reader* and *The Urban Geography Reader*), applied fields (such as *The Urban and Regional Planning Reader* and *The Urban Design Reader*) and important substantive themes (such as *The Sustainable Urban Development Reader* and *The Global Cities Reader*). Between 2001 and 2003 we assembled a talented team of scholars to edit nine additional readers in what has now become the Routledge Urban Reader Series. Between 2004 and 2008 all of the nine volumes were published. *The City Cultures Reader* and *The Sustainable Urban Development Reader* are already in their second editions and other readers will be soon.

The fifth edition of *The City Reader* continues and expands the tradition established in the earlier editions. Faculty and students familiar with earlier editions will find the classic and contemporary selections which have proven most useful in the past as well as exciting new material on urban history, metropolitan fiscal equity, global climate change, globalization, urban design, information technology and much more. The fifth edition places greater emphasis on globalization and cities worldwide than prior editions.

The fifth edition of *The City Reader* serves as the anchor text in the Routledge Urban Reader Series and articulates with the other nine volumes in the series, all of which are now in print. Editors expert in the academic disciplines most related to the study of cities and in topical areas of greatest interest to urbanists have now published readers under our direction with a format similar to *The City Reader*. These volumes provide in-depth coverage of material in their respective disciplines and subject matter areas. Routledge published a third edition of *The City Reader* in 2003. In 2004 three additional titles joined the series: a second edition of *The City Cultures Reader* edited by Malcolm Miles and Tim Hall with Iain Borden, *The Sustainable Urban Development Reader* edited by Stephen Wheeler and Timothy Beatley, and *The Cybercities Reader* edited by Stephen Graham. In 2005 *The Urban Geography Reader* edited by Nick Fyfe and Judith Kenny, *The Urban Sociology Reader* edited by Jan Lin and Christopher Mele, and *The Global Cities Reader* edited by Neil Brenner and Roger Keil were published. *The Urban Politics Reader* edited by Elizabeth Strom and John Mollenkopf was published in 2006, *The Urban Design Reader* edited by Elizabeth Macdonald and Michael Larice in 2007, and *The City and Regional Planning Reader* edited by Eugenie Birch in 2008.

It is a great satisfaction that the reader series provides space to include many more selections, covering topics introduced in *The City Reader* in much greater depth, and selections covering many additional topics beyond our subject matter expertise. Our talented team of seventeen editors has vastly leveraged our original concept and created a comprehensive compendium for understanding cities. From our first edition of *The City Reader* with fifty-five selections, the urban reader concept has expanded to include almost 400 selections in ten volumes.

Completion of all the readers in the series made it possible both to draw upon the accumulated expertise of our colleagues and to use this edition of *The City Reader* to anchor the entire series. We called upon the expertise of the editors of other volumes in the series in deciding which selections to include in this edition. As we revised the book, the part introduction and selection introductions, we referred constantly to material in the other nine volumes. Readers will see many references to material in the entire series in this edition of *The City Reader*. Most of the audience for *The City Reader* is in North America and Europe, but the fifth edition expands coverage of cities in othe parts of the world and includes a new selection by University of Illinois, Chicago, Professor Tingwei Zhang on Chinese cities.

The City Reader and the other readings in the series focus on essential writings. We and the other editors picked enduring issues in urban studies and planning across different cultures and times. In our courses, we have found that H.D.F. Kitto's "The Polis" raises fundamental questions about individuals' relations to their communities which are as relevant today as they were 2,400 years ago; that Louis Wirth's seventy-five-year-old essay on "Urbanism as a Way of Life" speaks to our students trying to understand contemporary urban violence, economic dislocation, homelessness, and anomie. *The City Reader* and other volumes in the series also include the best contemporary writing on cities. We find that our students are excited by William Julius Wilson's theories on the Black underclass, Robert Putnam's ideas about declining social capital from *Bowling Alone*, and Manuel Castells' reflections on the space of flows. Most writings in this edition of *The City Reader* are from twentieth-century writers and more than half were written very recently.

This is an international anthology. In an increasingly global world, students must learn from writers beyond the borders of their country of origin. In addition to writers from the United States, the fifth edition contains writings by scholars from Australia, Austria, Canada, China, Denmark, England, France, Germany, Iran, Norway, Spain, and Switzerland. Many of the writers included are truly world citizens.

The City Reader is an interdisciplinary anthology. The disciplines and professional fields represented in *The City Reader* include anthropology, archaeology, architecture, city planning, classics, culture studies, demography, economics, geography, history, landscape architecture, law, photography, political science, sociology, and urban design. Many of the writers blend insights from more than one discipline. Some of the best writing in *The City Reader* does not fit in conventional disciplinary boxes at all.

Cities can be studied to good advantage from both interdisciplinary and disciplinary perspectives. The disciplinary Routledge Urban Readers contain writings by scholars from the respective academic disciplines – geography, political science, and sociology that bring to bear their disciplinary expertise and provide depth in the literature of the specific discipline beyond what is possible in *The City Reader.* Pairing *The City Reader* and one of the Routledge urban disciplinary readers will provide students in courses in urban geography, urban politics, or urban sociology both the broad interdisciplinary perspective of *The City Reader* and the closer perspective of the disciplinary reader. Thus, for example, using both *The City Reader* and *The Urban Sociology Reader* will give students in urban sociology courses both an interdisciplinary understanding of cities and in-depth coverage of urban sociology topics written primarily by urban sociologists. Using both *The City Reader* and *The Urban Geography Reader* as course texts will give students in urban geography courses both a broad interdisciplinary introduction to scholarship on cities and in-depth coverage of the best scholarship by geographers on topics of particular interest to urban geographers. Using both *The City Reader* and *The Urban Politics Reader* will give students in urban politics courses both an interdisciplinary understanding of cities and in-depth coverage of urban politics topics written by political scientists.

The City Reader emphasizes the connection between the built environment of cities and the natural environment. As the world's population soars and urbanization continues, the imperative to design sustainable cities becomes ever more important. Readings by the Bruntland Commission, Timothy Beatley, Stephen Wheeler, Peter Calthorpe, and the Congress of the New Urbanism introduce students to sustainable urban development, green urbanism, ecological design, and the New Urbanism. Courses in environmental studies, environmental planning, sustainable urban development, and related topics may benefit from pairing *The City Reader* with the second edition of *The Sustainable Urban Development Reader.*

Similarly, pairing *The City Reader* with either of the readers organized around an applied field will provide a similar balance between broad interdisciplinary understanding and more focused knowledge from the field. Courses on urban and regional planning and urban design would benefit from pairing *The City Reader* with *The Urban and Regional Planning Reader, The Urban Design Reader*, or both.

An anthology of essential writings on cities should have a flexible organization. There is no one best way to organize material on cities. The content of urban studies and city planning courses varies widely and courses are organized in as many different ways as there are courses. This dictates a flexible structure for *The City Reader.* Readings are grouped into eight broad categories: The Evolution of Cities; Urban Culture and Society; Urban Space; Urban Politics, Governance, and Economics; Urban Planning History and Visions; Urban Planning Theory and Practice; Perspectives on Urban Design; and Cities in a Global Society. Professors may choose to assign readings in whatever order best fits the logic of their own course. If *The City Reader* is paired with one or more of the other Routledge Urban Reader Series volumes the possibilities for mixing and matching selections greatly increase.

Two other goals in picking the selections were to expose students to examples of great thinking and excellent writing. H.D.F. Kitto, Jane Jacobs, Robert Putnam, Mike Davis, J.B. Jackson, Ebenezer Howard, Lewis Mumford, and William Whyte are fine stylists as well as seminal thinkers. Reading their work is a lesson in how to communicate in a clear and engaging style. They are excellent models for how to write. Similarly intellectual sparks fly from virtually everything that Manuel Castells, Lewis Mumford, David Harvey, Kevin Lynch, Peter Hall, and other great thinkers represented in *The City Reader* write. Beyond the rich substantive content of the selections, we picked selections that will stimulate readers to think and write.

In the fifth edition of *The City Reader*, we have said a good deal about the role of visions in urban studies and planning. We close with our own vision of how this anthology will be used. *The City Reader* is aimed primarily at students who will encounter many of the writers and writings for the first time. *The City Reader* will work well for students in general education courses who do not pursue urban studies further. We hope the writings touch responsive chords and will inspire all the students who use *The City Reader* to think more deeply and read more widely about cities. To that end for each selection we point the way to other related writings by the same authors and other writers on the same subject matter. For students who will take additional coursework in urban studies, city and regional planning, geography, sociology, political science, history, or other academic disciplines and professional fields, we have designed *The City Reader* to provide an interdisciplinary overview of cities. As described above, *The City Reader* is designed to work well if it is paired with another volume in the Routledge Urban Reader Series. This fifth edition of *The City Reader* provides clear links to material in Routledge Urban Reader Series volumes so that readers can pursue material they find interesting in greater depth. As a reference work *The City Reader* and some or all of the other readers will provide a lifelong resource.

We hope *The City Reader* will prove to be a book that students, professors, and practitioners will keep and periodically reread. One test to which we put each of the essential writings included is that it should still be relevant to reread and enjoy for many years to come.

Richard T. LeGates
Frederic Stout
San Francisco, July 2010

Prologue

Prologue

How to Study Cities

Richard T. LeGates

Studying cities is a vast and never-ending enterprise. There is too much material for any one individual to master and always more to learn. Fortunately many fine scholars, past and present, have focused their attention on cities. We now know a great deal about how cities evolved, their social structures, urban culture, their internal spatial organization and relationships to other cities in systems of cities, what economic functions they perform, how they are governed, how they are (and might be) planned and designed, and their possible futures. One premise of *The City Reader* is that much of the classic writing about cities over the past hundred years remains remarkably relevant today.

DISCIPLINARY AND INTERDISCIPLINARY TEACHING ABOUT CITIES

While academic teaching about cities occurs in courses as different as English literature and civil engineering, most urban scholarship can be grouped under the heading of "urban studies," as an urban specialization within one of the social science disciplines such as "urban geography," "urban sociology," "urban politics," "urban economics," or "urban anthropology," in applied professional courses in urban planning, urban design, architecture, and landscape architecture, or in interdisciplinary culture studies programs. A description of these fields and disciplines and how they fit into universities is helpful to students encountering this material for the first time.

Almost all modern universities organize teaching and research into academic units called schools or colleges such as a college of social science or a school of architecture and urban planning. Schools and colleges in turn are generally organized into academic departments around a single discipline such as a department of geography and perhaps interdisciplinary programs, such as an urban studies program, that require students to take courses in a variety of different academic disciplines and fields. Professors educated in different academic disciplines are located within the departments and programs. City and regional planning and urban studies departments are generally interdisciplinary with faculty educated in a variety of academic disciplines and professional fields. A ten-person faculty in a mid-sized US city and regional planning department, for example, might have a core faculty of ten professors: three who received PhDs in city and regional planning, two with degrees in urban design or architecture, one with a law degree, and one each with a PhD in economics, geography, statistics, and political science.

Most universities encourage research and teaching that cross disciplinary boundaries. For example, a university may encourage a historian to teach a course that serves students in an urban studies department or the urban studies department may include the economics department's urban economics course as a required or elective course for the urban studies major. While professors from many different academic disciplines as well as interdisciplinary scholars study cities, most of the academic literature about cities – and most of the readings in *The City Reader* and the Routledge Urban Reader Series – has been written by social scientists: faculty trained to systematically study different aspects of human society. Some writing in *The City Reader*, and many of the selections in *The Urban and Regional Planning Reader* and the *The Urban Design Reader*, were written by scholars in applied fields related to urban planning and design – city and regional planning, urban design, architecture, and landscape architecture.

Most universities have a school or college of social science. Schools of social science contain social science departments where professors educated in the social science disciplines of geography, sociology, economics, political science and anthropology teach. History departments are sometimes located within schools of social science; sometimes within schools of humanities. Within these social science departments professors interested in cities teach urban courses from the point of view of their disciplines: courses on urban geography, urban sociology, urban politics, etc. Professors in these discipline-based courses may include material written by scholars from other academic disciplines in their courses. For example a geography professor may use content and methods developed by economists and sociologists in her urban geography course.

Departments of city and regional planning (often town and country planning in the UK) are often located within professional colleges that group architecture, planning, landscape architecture and sometimes other departments related to the built environment.

The borderlines between substantive content, methods, and theory in the different academic disciplines and professional fields related to the study of cities are fuzzy. Urban economists tend to study the economics of cities using quantitative methods and urban sociologists tend to study the social aspects of cities using qualitative methods. But some urban economists use qualitative methods and some urban sociologists are very quantitative. Most of the academic literature about urban politics has been written by political scientists based on theory and methods political scientists use. But law professor Myron Orfield (p. 296) mapped metropolitan areas using geographical information systems (GIS) software to develop his theory of metropolitics, which has become influential among political scientists. Sociologist Saskia Sassen's writings on the global system of cities (p. 554) are widely read by political scientists as well as planners, economists, and scholars from other disciplines.

Disciplines have the advantage that they are based on more or less agreed-upon methods for acquiring knowledge and a more or less agreed-upon body of knowledge shared by the discipline. All history professors, for example, in order to get their history PhD degree must study methods of historical research that historians use. All history professors will have taken enough different history courses that they have a broad overall knowledge of history in addition to their specialties in one or more specific time periods, issue areas, or methods of historical inquiry.

A disadvantage of disciplines is that they encourage rigid thinking within the four corners of the discipline itself. There is a danger that professors who are rigorously trained in economics, for example, will see only economic factors as important when they study or teach about an issue like urban sprawl. Because they have been trained in the importance of economics they may neglect political, social, and spatial aspects of sprawl. Of course understanding urban sprawl as an economic problem related to differential land costs, changing job locations, infrastructure finance, and other economic issues is important. But understanding the sociology of suburbanites, the relationship of single-family suburban home design to sprawl, spatial aspects of ethnic clustering in suburbs and a host of other issues that bear only indirectly on economics will further enrich understanding of suburbs and can help inform good urban planning and policy. In sum, the strength of interdisciplinary approaches is that, done properly, they provide for a richer, more holistic, more varied understanding of multiple dimensions of the phenomena being studied than a study from a single disciplinary perspective.

The danger of interdisciplinary approaches is that they may become so loose and standardless that they lack intellectual rigor. Well-trained and specialized disciplinary scholars are often justifiably critical of colleagues who do wide but shallow interdisciplinary teaching, research, and writing.

THEORY AND PRACTICE

Academic writing about cities is guided by theory – logically coherent bodies of principles advanced to explain phenomena. Theory in the social sciences is intended to provide a framework for understanding. Manuel Castells' theory about the "space of flows" (p. 572), for example, provides rich insights that help explain how digital information flows affect the global system of cities.

Some professors value only basic research and theory-building and look down on applied research and writing intended to produce solutions to actual urban problems. They see applied research as derivative and inferior – a kind of vocational education that is not worthy of true scholars. This is silly. Cities lend themselves well to applied research. Good scholarship directed at problem solving can be just as theoretically subtle and methodologically sophisticated as pure academic research. Willam Whyte's thoughtful prescriptions for park and plaza design based on his observations of New York City parks and plazas (p. 510), James Q. Wilson and George L. Kelling's "broken windows" theory of community policing based on their observations of police work in Newark, New Jersey (p. 263), and John Forester's theories about mediating urban planning conflicts (p. 421) based on dozens of interviews with practicing planners are as intellectually rigorous as any of the less applied selections in *The City Reader*.

Peter Hall (p. 373) and others deplore the lack of connection between urban theory and urban practice. We agree. Theory and practice should be linked in studying cities. Theory can inform practice and practice can inform theory. John Forester's approach is a good example of how to do this. Forester (p. 421) developed theories about how urban planners manage conflict by talking to practitioners. The theory he developed is in turn helpful to practitioners.

METHODS FOR STUDYING CITIES

Scholars who study cities use both quantitative and qualitative research methods. Both approaches can contribute to understanding cities. The best urban research designs often combine both quantitative and qualitative research and triangulate on problems using multiple methods.

Quantitative methods involve analyzing data using statistical methods. Virtually all quantitative analysis is now assisted by computers. A professor of urban politics doing statistical analysis of city voting data to see if recent immigrants feel differently about immigration than longer-term non-immigrant residents would be doing quantitative urban research. Applied statistics is a regular part of most urban studies and urban planning curricula and sometimes required in the various social science disciplines. Students learn to use computerized statistical packages such as the Statistical Package for the Social Sciences (SPSS) to do quantitative analysis.

Time is an important dimension in much urban research. Researchers may choose to look at an issue at one specific point in time. Imagine scientists studying a hundred foot long cylindrical sample of polar ice that had built up over a thousand years to determine the amount of carbon from the atmosphere that settled on the ice at different times in the past. Cutting a small slice of the cylinder where carbon froze into the cylinder in the year 1682 and analyzing it for carbon content would be an example of what is called cross-sectional research from the natural sciences. Similarly social science research that focuses only on a single time period is cross-sectional research. Friedrich Engels' study of the deplorable living and working conditions of factory workers in Manchester, England, in 1844 (p. 46) is a good example of cross-sectional urban research. The conditions he described in that year were different from earlier conditions and they would change in the future. But his snapshot of what conditions were like in Manchester during that one year provides a devastating picture of what early industrial cities were like.

A research design that chooses to look at how conditions change over time is called a longitudinal research design. Kingsley Davis' study of the urbanization of the human population from the early Middle Ages through the latter part of the twentieth century (p. 20) is an example of longitudinal research. By looking at population data for European cities over a thousand year period, Davis was able to describe changes that would not have been possible from a cross-sectional study.

Geographical space is an important aspect in much urban research. Most statistical analysis of urban phenomena is aspatial (does not include geographical space as a variable). But because many urban phenomena have a spatial dimension, geographical information systems software that permits users to map data is very important in studying cities and preparing city and regional plans. GIS is taught in geography, urban planning, and other departments. Myron Orfield used GIS to map attributes of cities in metropolitan regions of the United States (p. 296) to identify common needs and political interests. This is an excellent example of urban spatial analysis.

Qualitative research need not involve numbers or statistical analysis. William Whyte's use of observation (including time lapse photography) to find out how people use urban parks and plazas (p. 510) is a good example of effective use of one qualitative urban research method. Urban sociologists William Julius Wilson (p. 117) and Elijah Anderson (p. 127) conducted exhaustive qualitative field research in black ghetto areas of Chicago and Philadelphia by spending time there observing what was going on and talking to the residents. Andersen's descriptions of what residents told him paints a complex and subtle portrait that would be impossible to capture with quantitative methods. The interviews urban designer Kevin Lynch and his students conducted with Boston residents to understand how they perceived the city image (p. 499) is another excellent example of effective qualitative research.

There is never only one "right" way to do urban research. Multiple methods help researchers triangulate on a problem. Thus a researcher might choose to do both cross-sectional and longitudinal, qualitative and quantitative research on urban sprawl. The quantitative research might involve both aspatial analysis using a computerized statistical package and mapping and spatial statistical analysis using GIS. Within this broad research design the researcher could choose a variety of methods depending on his or her skills, the time available, and costs. A literature search, observation, interviewing, depth interviews, web-based research, a phone or mail survey, focus group(s), case studies, and many other methods are widely used in urban research.

While this is not a book about urban research methods, some of the selection introductions comment on the research methods used in the selection. For all the other selections, it is always important to pay attention to the research methods used as well as the substance.

ORGANIZATIONS AND JOURNALS DEVOTED TO THE STUDY OF CITIES

A number of academic associations organize conferences, set standards, publish academic journals, and work to advance scholarship related to understanding cities. In North America the academic association most directly concerned with urban studies is the Urban Affairs Association (UAA). The European counterpart organization is the European Urban Research Association (EURA). Both the UAA and EURA include faculty and students from a variety of social sciences, urban planning, and other backgrounds among their members.

In North America planning schools are members of the Association of Collegiate Schools of Planning (ACSP); in Europe they belong to the Association of European Schools of Planning (AESOP). There are organizations of planning schools for Latin America, Australia and New Zealand, Canada, Brazil and other Portuguese-speaking countries, and France and other francophone countries. The Global Planning Education Association Network (GPEAN) maintains a website with links to all of the above associations of planning schools. Each of the member associations' sites has links to their members' sites.

Members of the ACSP who are interested in international planning education have formed the Global Planning Educators Interest Group (GPEIG) which maintains a lively and informative website with a worldwide urban planning focus.

World congresses of planning schools are held every five years. The first World Planning Schools Congress occurred in Shanghai in 2001; the second in Mexico City in 2006. At the time of this writing the third World Planning Schools Congress will take place in July, 2011 in Perth, Australia.

Disciplinary academic organizations like the American Sociological Association (ASA), the American Political Science Association (APSA), and the Association of American Geographers (AAG) have specialized member groups dealing with urban concerns. Urban "tracks" at these conferences bring urbanists with similar interests together to present and discuss scholarly papers and otherwise share information.

The main professional association of city planners in the United States is the American Planning Association (APA). In the United Kingdom an equivalent organization is the Royal Town Planning Institute (RTPI). Practicing planners in these organizations meet to discuss their professional interests. Both associations publish excellent journals and have informative websites.

Journals like the *Journal of Urban Affairs, Urban Studies*, and *International Journal of Urban and Regional Research* specialize in publishing scholarly articles related to cities. The *Journal of Planning Education and Research* (published by the ACSP) is North America's leading academic urban planning journal. The *American Planning Association Journal* (published by the APA) is an excellent scholarly journal with a more applied focus. The *Town Planning Review* is the leading UK scholarly urban planning journal. *Urban Forum* published in Chinese by Tongji University is China's leading urban planning journal.

The evolution of cities

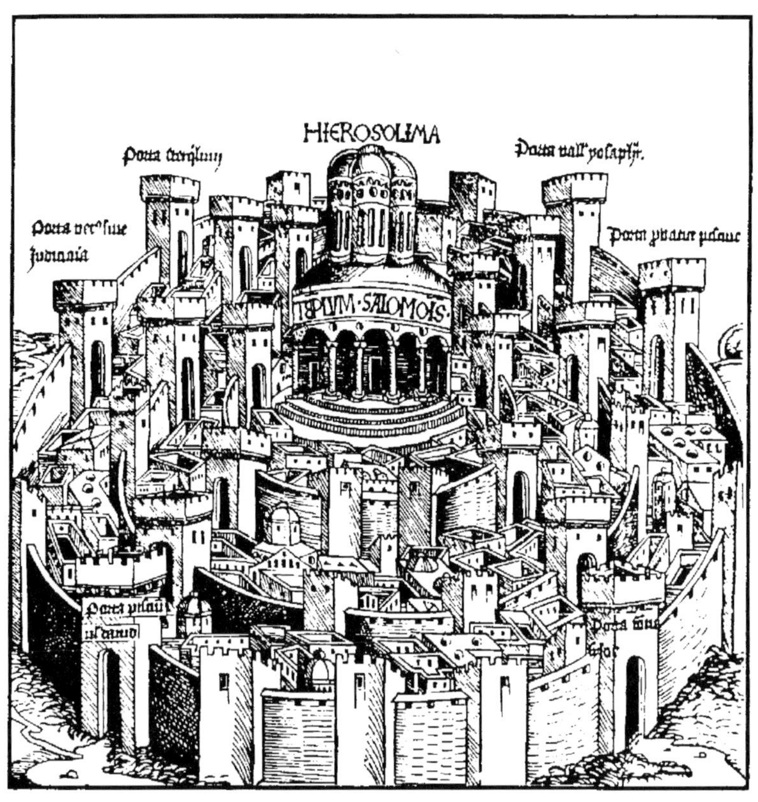

INTRODUCTION TO PART ONE

Cities are civilization. Humankind's rise to civilization took tens of thousands of years, but ever since the first true cities arose in Mesopotamia and in the Indus and Nile valleys sometime around 3500 BCE, the influence of city-based cultures and the steady spread and increase of urban populations around the world have been the central facts of human history.

As demographer Kingsley Davis (p. 20) points out, "urbanization" and "the growth of cities" are not the same thing. "Urbanization," as Davis defines it, is the *increase in the proportion* of a population that is urban as opposed to rural. That such an increase could take place without the growth of cities *per se* (for example, by the death of vast numbers of the rural population) or that city populations could grow without an increase in urbanization (as when the total population, urban and rural, increases at a similar rate) are important concepts that underlie the history of urban life. Most importantly, this definition of urbanization helps to explain how immigration from the countryside to the city has repeatedly been the key factor in the history of urban development, as it continues to be nowadays.

The history of cities is characterized by both continuities – the slowly evolving pattern of urban functions common to all cities – and discontinuities or periods of dramatic change in urban structure and purpose. The first great discontinuity of urban history is what the Australian archaeologist V. Gordon Childe (p. 31) called the "Urban Revolution," the momentous shift from simple, pre-urban tribal communities and village-based agricultural production to the complex social, economic, and political systems that characterized the earliest cities of Mesopotamia, Egypt, and the Indus Valley. True, the earliest cities, in the ancient Near East and elsewhere, grew out of accumulated neolithic knowledge, and certain extensive neolithic communities such as Çatal Hüyük in Anatolia predate the Mesopotamian cities by several millennia and may be regarded as at least proto-urban. For Childe, however, the development of writing was one of the crucial cultural elements of true urbanism, and the emergence of the cities of the ancient Near East, where writing and record-keeping began, constituted only the second of a series of massive transformations (the first being the Neolithic Revolution that established settled agriculture) that gave shape to the whole of human evolutionary development. Although the successive stages overlapped, each of Childe's three "revolutions" (the agricultural, the urban, and the industrial) totally changed the world as it had been before.

In certain important respects, many of the ancient cities are remarkably similar. They are frequently walled – except in Egypt, where the surrounding deserts may have been regarded as sufficient defenses – and in places like Peru, where secure empires surrounded individual cities, making urban defense unnecessary. In addition, almost all contain a distinct citadel precinct, often separately walled, encompassing a temple, a palace, and the central granary. Many of the earliest cities also boasted some sort of pyramid or ziggurat. And, as Karl Wittfogel pointed out in *Oriental Despotism* (1957), almost all were located along major rivers and based their power (and that of their rulers) on the control of massive irrigation systems serving the surrounding countryside. In addition, most of these earliest cities were dominated by what Lewis Mumford (p. 91) calls the "monologue of power" by all-powerful religious and military rulers.

Thus, both the physical structure and socio-economic complexity of the earliest cities are unlike anything that had come before. Whereas the neolithic village had been ruled by a council of elders, the cities were mostly ruled by totalitarian god-kings and their attendant priests who formed a class totally apart from the rest of the citizenry. And whereas neolithic communities may have built earthen enclosures as ceremonial centers for ritual pageantry and hill forts for refuge and defense, the ancient cities – from Uruk and Babylon on the Euphrates to Teotihuacan and Tenochtitlan in the Valley of Mexico – transformed these institutions into elaborate structures so massive that their remains are still visible today.

Many of the ancient cities elsewhere – in China, sub-Saharan Africa, Southeast Asia, Mexico, Mesoamerica, and the Andes – arose quite independently of the cities of the ancient Near East. Still, what is remarkable is how similar ancient cities everywhere were in terms of social structure, economic function, political order, and architectural monumentality. Even today – and although all cities are in some ways unique – the basic urban functions of citadel (mainly associated with government and the ruling order), marketplace (where the economic functions of production and exchange take place), and community (the place of homes, families, and the local culture of neighborhoods) continue to define cities and urban life.

Still, the cities of ancient Greece made a sharp break with the past and developed on a very different model from the citadel-dominated cities of the ancient Near East. Perhaps because they arose in narrow mountain valleys rather than on broad alluvial plains, the Greek cities that emerged around 1200 BCE and developed into an astonishing cultural efflorescence by 500 BCE were small (sometimes with a population of only a few thousand), economically self-contained, and almost village-like in their social and political institutions. It was the concept of urban citizenship and democratic self-government that was the distinctive contribution of the Greeks to the evolution of urban civilization. Greek democracy was by no means perfect and hardly inclusive. Women, slaves, and foreigners were all excluded from the rights and responsibilities of citizenship. But the cultural, artistic, and intellectual consequences of the democratic principle were extraordinary. "Within a couple of centuries," writes Lewis Mumford in *The City in History* (1961), "the Greeks discovered more about the nature and potentialities of man than the Egyptians or the Sumerians seem to have discovered in as many millennia."

If cities are civilization, they are also the cultural instrumentality by which humanity has attempted, since neolithic times, to achieve a higher, more inclusive concept of community. At the core of the Greek contribution to the history of urban civilization was the concept of the "polis." Sometimes translated as "city-state," at other times identified as the collective citizenry of a Greek city, the polis, as described so brilliantly by H.D.F. Kitto (p. 40) in *The Greeks*, was both a community and a *sense* of community that helped to define Greek city-dwellers' relationship to their city and their fellow citizens, to the world at large, and to themselves. In *The Politics*, Aristotle called human beings the "*zoon politikon*" (the "political animal" or, more properly, "the animal that belongs to a polis") and described the ideal city-state as one small enough so that a single citizen's voice could be heard by all the assembled fellow citizens. The Greek cities had citadels such as the Acropolis of Athens, to be sure, but for the Greek citizen, all aspects of public life were lived in the agora or public marketplace, and contact with rural nature was within a short walk. In that sense, the polis was a reincarnation, in an urban context, of the face-to-face human relationships that characterized the pre-urban community of the neolithic village.

Marking another discontinuity or sharp break in the history of urban life, the city of Rome began as a cluster of villages along the Tiber in central Italy, emerged as a powerful republic similar to the earlier Greek cities, but then exploded into a giant metropolis and a city of world empire – indeed, a citadel city for a far-flung urban community that presages, in a sense, the worldwide system of cities that dominates global society today. Rome's contributions to civilization were considerable. Its roads, aqueducts, and sewers set new standards of engineering excellence. Its systems of military and colonial administration spread a common law, and established a common peace, throughout a large and populous area that extended from Persia to the borders of Scotland. Roman imperial expansion also spread Roman literature, philosophy, and art, establishing the basis for a widespread cultural hegemony. And Rome planted colonial towns wherever its legions marched, often leaving traces of an original *castrum* (or military camp) laid out along the cardinal points of the compass at the center of later medieval cities.

But if the administrative and infrastructural accomplishments of the Romans were impressive, their record in the field of social development is more problematic. In the place of the Greek conception of community and participation in the life of the polis, the Romans erected a citizenship of imperial privilege rooted in a rigid social hierarchy of patricians, clients, and plebeians. Beginning with Augustus, the Roman emperors proclaimed themselves gods, staged extravagant spectacles to awe the cowed populace, and, it has been said, ruled by the provision of "bread and circuses" to the vast Roman populace. In the end, Rome, with a population of one million, came to be seen as a kind of imperial parasite on the entire Mediterranean world, and both city and empire eventually fell of their own weight.

For much of the medieval period that followed the fall of the Roman Empire in the West, Europe was a cultural backwater. In the early Middle Ages, self-contained monastery communities kept the larger world at bay, some provincial towns retreated inside the walls of the Roman amphitheaters, and the population of Rome itself dwindled to a few thousand. During this same period, some cities in China and India grew in influence and became the centers of imperial systems of their own. Raided by Vikings from the North and invaded by North African Arabs on its southern flank, much of Europe reverted to rural conditions, and serfdom became widespread under a system of warlord feudalism.

Meanwhile, the cities of Islam – Cairo and Baghdad and Moorish Córdoba in present-day Spain – were the real centers of power in what had been parts of the Roman Empire. And other urban centers – the Khmer civilization at Angkor, successive capitals of China, Great Zimbabwe in Africa – often rivaled and sometimes surpassed Europe's cities in wealth and power. After about the year 1000 CE, however, Europe began to revive, and the late medieval cities became true centers of commerce, culture, and community. As Henri Pirenne argued in *Medieval Cities* (1925), it was the economic function of the great trading towns that led inevitably to their growing power and political independence. Having used their wealth to win from the barons the right to self-government, the medieval towns became islands of freedom in a sea of feudal obligation.

The defensive walls of the medieval city in Europe provided a clear demarcation line between the urban and the rural, and the small size of most towns allowed for an easy reciprocity between urban industry and commerce, on the one hand, and agricultural pursuits on the other. Within the town walls, the guilds provided for the organization of economic and social life, while the church saw to the citizens' spiritual needs and established a framework for social ritual and communal unity. Cathedrals, guildhouses, charitable institutions, universities, and colorful marketplaces were all characteristic medieval institutions. Together, they established the perfect stage for what Lewis Mumford called "the urban drama" (p. 91) but as soon as "the unity of this social order was broken [with the advent of nation-statism and capitalist industrialization] everything about it was set in confusion . . . and the city became a battleground for conflicting cultures, dissonant ways of life."

The cities of the period between the ancient world and the modern world were extraordinarily diverse, and the urban civilizations of Asia and the Americas often rendered European visitors awestruck. In Europe itself, the slow decay of medieval urban unity was hastened by the forces of the Renaissance and the rise of absolutist monarchies. The powerful new national rulers built their royal palaces, such as Louis XIV's Versailles, outside of the traditional urban centers. Their interventions into the existing urban fabric included building broad boulevards and open squares fit for the display of baroque pomp and power. The Enlightenment and the Age of Revolutions brought down the divine right of kings and reestablished the political power of the urban commercial interests, but in a new socio-political context. In the end, it was market capitalism and a new industrial economic order based on powerful new productive technologies that destroyed the last vestiges of the medieval city by separating the church from its social role and reducing the marketplace to its purely economic functions while at the same time extending the economy worldwide. Thus, the capitalist city, especially the city of the Industrial Revolution, made yet another sharp break in the history of cities. Capitalism created an entirely new urban paradigm and established the physical, social, economic, and political preconditions of all that was to follow. With the Industrial Revolution, we see the emergence of urban modernism.

While the political and economic consequences of the Renaissance had helped to spread European domination worldwide through extensive projects of exploration, discovery, and imperialist expansion, the forces of industrialization helped to complete that process of world domination by dividing the world between the advanced industrialized nations (originally Europe and North America) and the underdeveloped, non-industrialized nations. Industrial modernism also created a new social order based on property-owning capitalists and propertyless proletarians. And the cities, especially the new industrial centers, became dismal conurbations of factories and slums such as the world had never before seen.

One of the earliest and most acute observers of the new urban-industrial order was Friedrich Engels (p. 40), himself the son of a major German industrialist. In *The Condition of the Working Class in England in 1844* (1845), Engels detailed the unrelenting squalor and misery that characterized the working-class districts of Manchester and the strategies employed by the capitalist bourgeoisie to protect themselves from the physical and social horror that was the source of their wealth. There were many responses to these horrifying conditions – the introduction of urban parks, systems of water supply and public hygiene, agitation for poor relief and model housing, even utopian visions of perfect societies – and all of these contributed to the development of modern urban planning (see Part Five, Urban Planning History and Visions, pp. 315–365).

The "shock cities" of the Industrial Revolution – the Manchester of Engels or the Chicago of Upton Sinclair's *The Jungle* (1906) – are often considered as though they were unique and discrete phenomena, but the first phase of industrial urbanism proved to be merely the beginning of a long process of urban adaptation and transformation. The noted historian Sam Bass Warner (p. 55) reviews the century of change that followed the Industrial Revolution in "Evolution and Transformation: The American Industrial Metropolis, 1840–1940," an essay written specially for this edition of *The City Reader*. "A parade of surprises characterized the century," he writes, and among those surprises were wholly new and unexpected developments in technology, in the economy, in urban social life, and in the very shape of the city. In the 1840s, industrial workers lived in squalid and environmentally polluted conditions close to the industrial enterprises where they worked. These were what Lewis Mumford called the "factory camps" of the early industrial period, and life expectancies, even in the most technologically advanced nations, were barely more than 40 years. But gradually water and steam power gave way to electricity, and municipalized transit systems allowed workers to live at some distance from the smoke and soot of their workplaces. Railways allowed the factories themselves to locate outside the center cities, and a new kind of central-business-district downtown began to take shape as the city became a metropolitan system and took on the structure described by Ernest W. Burgess in "The Growth of the City" (p. 161). In the United States, slavery was abolished and the long march of the struggle to secure full civil rights for African Americans began. Labor unions emancipated industrial workers, and women emancipated themselves. And especially after 1920, new forms of finance capital and commercial mass marketing increased the standard of living for the urban middle class and even the industrial workers. The story of a century of urban-industrial development was not entirely sunny. Along with material progress came population growth, cultural accommodations demanded by massive immigration, and the kind of social and psychological alienation described by Louis Wirth in "Urbanism as a Way of Life" (p. 96). But cleaner air and water, greater overall wealth, and advances in public sanitation and medical technology meant that by 1940 the average life expectancy of a resident of a modern industrial city was close to 65 or 70 years.

As Warner and other urban historians have noted, one compelling strategy for coping with the challenges and complexities of the new urban reality was middle-class flight to the suburbs. Suburbanization, with its consequent segregation by social class, became one of the continuing features of the modern city and one of the sources of its ongoing social disharmony and class conflict. Throughout the twentieth century, particularly in North America, the model of middle-class suburbia has grown in size and influence to the point where it is no longer just an appendage to the central city. Instead, suburbia now defines many cities, leaving the old inner cores to the poorest elements of the urban population and in need of massive efforts at renewal and redevelopment. Although, as Warner noted, at first there were "streetcar suburbs" built along inter-urban railroad lines, the newer suburbs, especially those developments built after World War II, were automobile-based and created the "sprawl" that characterizes more and more cities worldwide that Robert Bruegmann (p. 211) describes as an almost inevitable process in response to population growth and increased social wealth. The new tract-home developments have spawned a vast literature, much of it criticizing suburbia as a cultural wasteland and a segregated sanctuary of class privilege. In *The Levittowners* (1967), Herbert Gans presented a rather sympathetic view of the community of tract-homes built by developer Arthur Levitt on Long Island, New York. He described a family-oriented community of skilled workers and mid-level managers – that is, a true *middle* class, not an upper-middle-class elite. But the more general view of automobile-dominated suburbia, a view that subjected sprawl to cultural as well as design criticism, is ably summarized in Kenneth T. Jackson's "The Drive-in Culture of Contemporary America" from *Crabgrass Frontier* (p. 65).

Beginning as sprawl suburbia but quickly transcending suburbia's initial limitations, a new city type arose in California during the early decades of the twentieth century that signaled a new phase in the history of urbanism worldwide. Sometimes dismissed as a mere conurbation of suburbs "in search of a city" and frequently derided as the ultimate in mindless, post-urban chaos, Los Angeles did indeed break all the existing rules and natural boundaries of urban development but emerged finally as a new, radically decentralized urban paradigm: the contemporary multi-nucleated metropolis poised on the edge of postmodernity. The essential characteristics of Los Angeles – a city that grew from less than 600,000 in 1920 to about 4 million today (about 10 million in the county of Los Angeles) – were present almost from the beginning, particularly its sense of "spatial freedom" and its preference for the middle-class single-family dwelling as an "expression of its design for living." For good or ill, these characteristics were further emphasized by a grid pattern of freeways and a reliance – many would say over-reliance – on the automobile that replaced a once-extensive network of streetcars and created a metropolis without a single downtown. Today, Los Angeles is a true world city, and its products – both industrial and cultural – are influential around the globe.

In the nineteenth century, middle-class suburbs developed outside major urban centers, spaced along commuter rail lines. In the twentieth century, the influence of the automobile turned once-attractive small-scale suburbs into an endless, congested sprawl. These first two stages in the development of suburbia depended on the existence of a vital central city, both as a center for production and employment and for cultural amenities. With the emergence of Los Angeles, however, that pattern began to change, and today the new "Edge City" suburban ring is clearly different from the earlier suburban developments in size, complexity, and even function. This is where most of the new houses, most of the new jobs, and even most of the new cultural centers are located. Increasingly, the major commute pattern is not from suburb to central city, but from suburb to suburb. Indeed, as Robert Fishman (p. 75) argues in *Bourgeois Utopias: The Rise and Fall of Suburbia* (1987), the new "Edge City" suburbs are not suburbs at all, but a fundamentally new kind of decentralized city that he calls "technoburbs."

What the future holds for urban civilization is infinitely debatable. Will central cities disappear? Will "Edge Cities" take over the primary urban functions? Will the urbanization process itself reverse direction, as Melvin Webber predicted in 1968 (p. 549), and lead to counter-urbanization and a general dispersal of the human population? Or will certain urban regions become worldwide command and control centers – new citadels of global power, internally characterized by the uneasy side-by-side coexistence of corporate power and service-sector marginality, as Saskia Sassen (p. 554) argues in "The Impact of the New Technologies and Globalization on Cities"? No one knows for certain, of course, but it increasingly appears that urban history is well into the early stages of a major new sharp break with the past. Perhaps a new paradigm is looming on the horizon, a new discontinuity in urban history variously designated as postmodernism or post-urbanism or the restructuring of global capitalism that will almost certainly be characterized by telecommunications networking, techno-virtuality, global systems of economic exchange, and new ecological constraints demanding an increased concern with issues of sustainability. But cities and urban society themselves will continue to be central to the history of humanity, and certain immemorial features of urban life – the side-by-side coexistence of rich and poor, the ongoing search for social justice and meaningful community, the myriad opportunities for interchange and innovation – will surely continue despite fundamental shifts and changes in the very definition of what a city is. The new urban paradigm that seems to be emerging now – and that is discussed at greater length in the selections in Part Eight, Cities in a Global Society – will perhaps be part-regional sprawl, part-technoburb, part-virtual metropolis – or what Manuel Castells calls "the space of flows, the space of place" (p. 572). However it evolves, the new urban world – the world of cities in a global society – promises to be a major new stage in the history of citadels, marketplaces, and communities . . . and in the ongoing evolution of the city as a human institution.

"The Urbanization of the Human Population"

Scientific American (1965)

Kingsley Davis

Editors' Introduction

The demographics of the urbanization process are the foundation of all urban history. Demography – from the Greek *demos*, "people" – is the study of human populations. Kingsley Davis (1908–1996) pioneered the study of historical urban demography and was particularly fascinated by the history of world urbanization: that is, the increase over time of the proportion of the total human population that is urban as opposed to rural.

This selection synthesizes Davis's conclusions about how urbanization has occurred throughout the world during all of human history. He raises fundamental issues and lays out a clear framework for understanding population dynamics and urban growth. Davis's careful distinctions of possible sources of urbanization are fundamental. He concludes that urbanization is caused by rural–urban migration, not because of other possible factors such as differential birth and mortality rates.

Davis's extraordinary data on how tiny European urban settlements were after the fall of Rome, and how slowly they grew throughout the Middle Ages and Early Modern period, provides the demographic backdrop for the historic growth of European urbanization. During the long period of medieval urbanization the proportion of the population that was urban as opposed to rural changed very slowly. In sharp contrast, urbanization increases very rapidly around the year 1800, and Davis concludes that as the Industrial Revolution in England, along with rapid population growth, combined with rural–urban shifts to change both the proportion of the population living in cities and absolute city size very quickly. Friedrich Engels (p. 46) describes in horrifying detail what this revolution in urban demography meant to the impoverished urban proletariat of Manchester and other nineteenth-century industrial cities. His analysis is extremely relevant in assessing prospects for the twenty-first century as the advanced industrial societies and eventually the world reach what some environmental analysts regard as the full "carrying capacity" of the globe.

Davis argues that urbanization follows an attenuated S curve in which pre-industrial cities urbanize very slowly at the long bottom of the S, shoot up at the middle of the S as they industrialize, and then level off at the top of the S (see Plate 1). He observes that advanced industrialized countries are now reaching the top part of an S curve, many rapidly urbanizing but less developed countries are at the steep middle of the S, and other emerging countries are still moving along the long slowly rising bottom of the S. It is helpful to visualize Kingsley Davis's theory of urbanization as a "family of S curves" on a horizontal *x*-axis representing time and a vertical *y*-axis representing the percentage of a country's population that is urban. England (which was only about 5 percent urban in 1300 CE, but is now about 93 percent urban) is at the top of the family of S curves. The steep part of England's S starts about 1750 with the start of the industrial revolution. A long nearly level top to the S shows that England became nearly fully urbanized several decades ago. The long attenuated parts of the S curves for Germany and France begin later and the time at which the S starts to rise rapidly also occurs later than in England. The long bottom part of China's S curve extends until about 1980 and only then begins to rise rapidly.

The S curves for a few very poor African countries still have only the bottom part of the S because they have not yet begun to urbanize rapidly.

The developing countries of Asia, South America, and Africa already have many huge and rapidly growing cities. As the twenty-first century progresses, it appears likely the human population will increasingly live in "megacities" of 10 million inhabitants and more, often flowing together in vast urban conurbations sometimes called "mega-urban regions."

Davis concludes that there will be an end to urbanization – but not necessarily to absolute population growth, the physical size of cities or the absolute number of people cities contain. He found that the rural population in less developed countries continues to grow as these countries urbanize, unlike European cities in the nineteenth century, where industrialization led to depopulation of rural areas. His vision of developing societies unable to sustain their populations helps to explain Saskia Sassen's description of growing poverty and inequality worldwide and the growth of large, poorly paid immigrant labor forces in the largest cities in the developed world (p. 554). Research and scholarly debate continues on the nature and causes of world urbanization. The majority view, established by numerous United Nations projections, holds that total world population will increase from nearly 7 billion in 2010 to about 9 billion in 2050, and that urbanization will continue everywhere but especially in the developing world.

Historians continue to shed light on the growth of cities, but because the records from which they work are often fragmentary and incomplete not everyone agrees with Davis or any other standard account. Debate continues on the relative importance of war, plague, medical advances, trade, technology, religion, and ideology on urban growth. And debate is even more intense in the normative area – about what, if anything, governments should do about population growth and urbanization. Davis stressed the impact of overall population growth (which he saw as a real danger) on world urbanization and implies that family planning is essential if cities are to meet human needs. But many governments reject family planning on religious or policy grounds, and some European countries now face declining populations and are currently debating the desirability of enacting family-friendly policies to reward child-bearing.

Davis's other writings include many articles and studies on demographics and natural resources as well as two anthologies: *Cities: Their Origin, Growth and Human Impact* (San Francisco, CA: W.H. Freeman, 1973) and *Resources, Environment and Population: Present Knowledge, Future Options*, with Mikhail S. Bernstram, (New York: Population Council, Oxford University Press, 1991). For more on Davis and his writings, see David Hoer, *Kingsley Davis: A Biography and Selections from his Writings* (Edison, NJ: Transaction, 2004).

Data on world urbanization are contained in Tertius Chandler and Gerald Fox, *3000 Years of Urban Growth* (New York: Academic Press, 1974) and Tertius Chandler, *Four Thousand Years of Urban Growth: An Historical Census* (Lewiston, NY: Edwin Mellen Press, 1987). These books assemble estimates of the population size of individual cities everywhere in the world over a 4,000-year period. The sources of the estimates range from contemporary accounts to scholarly estimates completed just before the second of the two books was published in 1987. Footnotes explain where the estimates come from. This is a valuable compendium for source information. The authors' attempts to synthesize the source material so that they provide longitudinal data on the population of different cities over time, however, are problematic. Since the sources are so varied and conflicting and not based on consistent definitions or methods, Chandler and Fox's estimates – particularly for the earliest time periods and for cities where the records are least complete and reliable – must be judged with extreme caution. Much more reliable data on the population of all Western European cities that achieved a population of 10,000 or more at any time between 1500 and 1800 are reported at fifty-year increments starting in 1500 in Jan DeVries, *European Urbanization, 1500–1800* (Cambridge, MA: Harvard University Press, 1984). Further insight on demography and urbanization can be found in World Bank, *World Development Indicators, 2005* (Washington, DC: World Bank, 2005), Ad van der Woude, Akira Hayami, and Jan de Vries (eds), *Urbanization in History: A Process of Dynamic Interaction* (Oxford: Oxford University Press, 1990), and the frequent revisions of *World Population Prospects* published by the Population Division of the United Nations. Also useful is Paul Knox and Linda McCarthy, *Urbanization: An Introduction to Urban Geography*, 2nd edn (Upper Saddle River, NJ: Prentice-Hall, 2005).

For developments in urbanization in less developed countries, see Alan Gilbert (ed.), *The Mega-City in Latin America* (New York: United Nations University Press, 1996), Carole Rakodi (ed.), *The Urban Challenge in Africa*

(New York: United Nations University Press, 1997), and Fu-chen Lo and Yue-man Yeung (eds), *Emerging World Cities in Pacific Asia* (New York: United Nations University Press, 1996).

For an environmental view of world urbanization, consult Cedric Pugh (ed.), *Sustainability, the Environment, and Urbanization* (London: Earthscan, 1996). Lester R. Brown and Jodi L. Jacobson provide a somewhat alarming summary of world population studies and reflections on the future in *The Future of Urbanization: Facing the Ecological and Economic Constraints* (New York: Worldwatch Paper 77, 1987) and George Martine, Gordon McGranahan, Mark Montgomery, and Rogelio Fernández-Castilla (eds), *The New Global Frontier: Urbanization, Poverty and Environment in the 21st Century* (London: Earthscan, 2008).

The future of urbanization is of course an important issue for policy planners. For a fascinating review of population-related policy issues, including the possibility of a "world population implosion," see Nicholas Eberstadt, *Prosperous Paupers and Other Population Problems* (New Brunswick, NJ: Transaction, 2000). For more on the possibility of declining populations in the future, consult Phillip Longman, *The Empty Cradle: How Falling Birthrates Threaten World Prosperity and What To Do About It* (New York: Basic Books, 2004), Ben J. Wattenberg, *Fewer: How the New Demography of Depopulation Will Shape Our Future* (Chicago, IL: Ivan R. Dee, 2004), and, for the special case of the effects of China's "One Child" policy, Valerie Hudson, *Bare Essentials: The Security Implications of Asia's Surplus Male Population* (Cambridge, MA: MIT Press, 2004).

Urbanized societies, in which a majority of the people live crowded together in towns and cities, represent a new and fundamental step in man's social evolution. Although cities themselves first appeared some 5,500 years ago, they were small and surrounded by an overwhelming majority of rural people; moreover, they relapsed easily to village or small-town status. The urbanized societies of today, in contrast, not only have urban agglomerations of a size never before attained but also have a high proportion of their population concentrated in such agglomerations. In 1960, for example, nearly 52 million Americans lived in only 16 urbanized areas. Together these areas covered less land than one of the smaller counties (Cochise) of Arizona. According to one definition used by the U.S. Bureau of the Census, 96 million people – 53 percent of the nation's population – were concentrated in 213 urbanized areas that together occupied only 0.7 percent of the nation's land. Another definition used by the bureau puts the urban population at about 70 percent. The large and dense agglomerations comprising the urban population involve a degree of human contact and of social complexity never before known. They exceed in size the communities of any other large animal; they suggest the behavior of communal insects rather than of mammals.

Neither the recency nor the speed of this evolutionary development is widely appreciated. Before 1850 no society could be described as predominantly urbanized, and by 1900 only one – Great Britain – could be so regarded. Today, only 65 years later, all industrial nations are highly urbanized, and in the world as a whole the process of urbanization is accelerating rapidly.

Some years ago my associates and I at Columbia University undertook to document the progress of urbanization by compiling data on the world's cities and the proportion of human beings living in them; in recent years the work has been continued in our center – International Population and Urban Research – at the University of California at Berkeley. The data obtained in these investigations . . . show the historical trend in terms of one index of urbanization: the proportion of the population living in cities of 100,000 or larger. Statistics of this kind are only approximations of reality, but they are accurate enough to demonstrate how urbanization has accelerated. Between 1850 and 1950 the index changed at a much higher rate than from 1800 to 1850, but the rate of change from 1950 to 1960 was twice that of the preceding 50 years! If the pace of increase that obtained between 1950 and 1960 were to remain the same, by 1990 the fraction of the world's people living in cities of 100,000 or larger would be more than half. Using another index of urbanization – the proportion of the world's population living in urban places of all sizes – we found that by 1960 the figure had already reached 33 percent.

Clearly the world as a whole is not fully urbanized, but it soon will be. This change in human life is so recent that even the most urbanized countries still exhibit the rural origins of their institutions. Its full implications for man's organic and social evolution can only be surmised.

In discussing the trend – and its implications insofar as they can be perceived – I shall use the term "urbanization" in a particular way. It refers here to the proportion of the total population concentrated in urban settlements, or else to a rise in this proportion. A common mistake is to think of urbanization as simply the growth of cities. Since the total population is composed of both the urban population and the rural, however, the "proportion urban" is a function of both of them. Accordingly, cities can grow without any urbanization, provided that the rural population grows at an equal or a greater rate.

Historically, urbanization and the growth of cities have occurred together, which accounts for the confusion. As the reader will soon see, it is necessary to distinguish the two trends. In the most advanced countries today, for example, urban populations are still growing, but their proportion of the total population is tending to remain stable or to diminish. In other words, the process of urbanization – the switch from a spread-out pattern of human settlement to one of concentration in urban centers – is a change that has a beginning and an end, but the growth of cities has no inherent limit. Such growth could continue even after everyone was living in cities, through sheer excess of births over deaths.

The difference between a rural village and an urban community is of course one of degree; a precise operational distinction is somewhat arbitrary, and it varies from one nation to another. Since data are available for communities of various sizes, a dividing line can be chosen at will. One convenient index of urbanization, for example, is the proportion of people living in places of 100,000 or more. In the following analysis I shall depend on two indexes: the one just mentioned and the proportion of population classed as "urban" in the official statistics of each country. In practice the two indexes are highly correlated; therefore either one can be used as an index of urbanization.

Actually the hardest problem is not that of determining the "floor" of the urban category but of ascertaining the boundary of places that are clearly urban by any definition. How far east is the boundary of Los Angeles? Where along the Hooghly River does Calcutta leave off and the countryside begin? In the past the population of cities and towns has usually been given as the number of people living within the political boundaries. Thus the population of New York is frequently given as around eight million, this being the population of the city proper. The error in such a figure was not large before World War I, but since then, particularly in the advanced countries, urban populations have been spilling over the narrow political boundaries at a tremendous rate. In 1960 the New York–Northeastern New Jersey urbanized area, as delineated by the Bureau of the Census, had more than 14 million people. That delineation showed it to be the largest city in the world and nearly twice as large as New York City proper.

As a result of the outward spread of urbanites, counts made on the basis of political boundaries alone underestimate the city populations and exaggerate the rural. For this reason our office delineated the metropolitan areas of as many countries as possible for dates around 1950. These areas included the central, or political, cities and the zones around them that are receiving the spillover.

This reassessment raised the estimated proportion of the world's population in cities of 100,000 or larger from 15.1 percent to 16.7 percent. As of 1960 we have used wherever possible the "urban agglomeration" data now furnished to the United Nations by many countries. The U.S., for example, provides data for "urbanized areas," meaning cities of 50,000 or larger and the built-up agglomerations around them.

. . . My concern is with the degree of urbanization in whole societies. It is curious that thousands of years elapsed between the first appearance of small cities and the emergence of urbanized societies in the nineteenth century. It is also curious that the region where urbanized societies arose – northwestern Europe – was not the one that had given rise to the major cities of the past; on the contrary, it was a region where urbanization had been at an extremely low ebb. Indeed, the societies of northwestern Europe in medieval times were so rural that it is hard for modern minds to comprehend them. Perhaps it was the non-urban character of these societies that erased the parasitic nature of towns and eventually provided a new basis for a revolutionary degree of urbanization.

At any rate, two seemingly adverse conditions may have presaged the age to come: one the low productivity of medieval agriculture in both per-acre and per-man terms, the other the feudal social system. The first meant that towns could not prosper on the basis of local agriculture alone but had to trade and to manufacture something to trade. The second meant that they could not gain political dominance over their hinterlands and thus become warring city-states. Hence they specialized in commerce and manufacture and evolved local institutions suited to this role. Craftsmen were housed in the towns, because there the merchants could regulate quality and cost. Competition among towns stimulated specialization and technological innovation. The need for literacy, accounting skills and geographical knowledge caused the towns to invest in secular education.

Although the medieval towns remained small and never embraced more than a minor fraction of each region's population, the close connection between industry and commerce that they fostered, together with their emphasis on technique, set the stage for the ultimate breakthrough in urbanization. This breakthrough came only with the enormous growth in productivity caused by the use of inanimate energy and machinery. How difficult it was to achieve the transition is agonizingly apparent from statistics showing that even with the conquest of the New World the growth of urbanization during three postmedieval centuries in Europe was barely perceptible. I have assembled population estimates at two or more dates for 33 towns and cities in the sixteenth century, 46 in the seventeenth and 61 in the eighteenth. The average rate of growth during the three centuries was less than 0.6 percent per year. Estimates of the growth of Europe's population as a whole between 1650 and 1800 work out to slightly more than 0.4 percent. The advantage of the towns was evidently very slight. Taking only the cities of 100,000 or more inhabitants, one finds that in 1600 their combined population was 1.6 percent of the estimated population of Europe; in 1700, 1.9 percent; and in 1800, 2.2 percent. On the eve of the industrial revolution Europe was still an overwhelmingly agrarian region.

With industrialization, however, the transformation was striking. By 1801 nearly a tenth of the people of England and Wales were living in cities of 100,000 or larger. This proportion doubled in 40 years and doubled again in another 60 years. By 1900 Britain was an urbanized society. In general, the later each country became industrialized, the faster was its urbanization. The change from a population with 10 percent of its members in cities of 100,000 or larger to one in which 30 percent lived in such cities took about 79 years in England and Wales, 66 in the U.S., 48 in Germany, 36 in Japan and 26 in Australia. The close association between economic development and urbanization has persisted: . . . in 199 countries around 1960 the proportion of the population living in cities varied sharply with per capita income.

Clearly, modern urbanization is best understood in terms of its connection with economic growth, and its implications are best perceived in its latest manifestations in advanced countries. What becomes apparent as one examines the trend in these countries is that urbanization is a finite process, a cycle through which nations go in their transition from agrarian to industrial society. The intensive urbanization of most of the advanced countries began within the past hundred years; in the underdeveloped countries it got under way more recently. In some of the advanced countries its end is now in sight. The fact that it will end, however, does not mean that either economic development or the growth of cities will necessarily end.

The typical cycle of urbanization can be represented by a curve in the shape of an attenuated S. Starting from the bottom of the S, the first bend tends to come early and to be followed by a long attenuation. In the United Kingdom, for instance, the swiftest rise in the proportion of people living in cities of 100,000 or larger occurred from 1811 to 1851. In the U.S. it occurred from 1820 to 1890, in Greece from 1879 to 1921. As the proportion climbs above 50 percent the curve begins to flatten out; it falters, or even declines, when the proportion urban has reached about 75 percent. In the United Kingdom, one of the world's most urban countries, the proportion was slightly higher in 1926 (78.7 percent) than in 1961 (78.3 percent).

At the end of the curve some ambiguity appears. As a society becomes advanced enough to be highly urbanized it can also afford considerable suburbanization and fringe development. In a sense the slowing down of urbanization is thus more apparent than real: an increasing proportion of urbanites simply live in the country and are classified as rural. Many countries now try to compensate for this ambiguity by enlarging the boundaries of urban places; they did so in numerous censuses taken around 1960. Whether in these cases the old classification of urban or the new one is erroneous depends on how one looks at it; at a very advanced stage the entire concept of urbanization becomes ambiguous.

The end of urbanization cannot be unraveled without going into the ways in which economic development governs urbanization. Here the first question is: where do the urbanites come from? The possible answers are few: the proportion of people in cities can rise because rural settlements grow larger and are reclassified as towns or cities; because the excess of births over deaths is greater in the city than in the country, or because people move from the country to the city.

The first factor has usually had only slight influence. The second has apparently never been the case. Indeed, a chief obstacle to the growth of cities in the past has been their excessive mortality. London's water in the middle of the nineteenth century came mainly from wells and rivers that drained cesspools, graveyards and tidal areas. The city was regularly ravaged by cholera. Tables for 1841 show an expectation of life of about 36 years for London and 26 for Liverpool and Manchester, as compared to 41 for England and Wales as a whole. After 1850, mainly as a result of sanitary measures and some improvement in nutrition and housing, city health improved, but as late as the period 1901–1910 the death rate of the urban counties in England and Wales, as modified to make the age structure comparable, was 33 percent higher than the death rate of the rural counties. As Bernard Benjamin, a chief statistician of the British General Register Office, has remarked: "Living in the town involved not only a higher risk of epidemic and crowd diseases . . . but also a higher risk of degenerative disease – the harder wear and tear of factory employment and urban discomfort." By 1950, however, virtually the entire differential had been wiped out.

As for birth rates, during rapid urbanization in the past they were notably lower in cities than in rural areas. In fact, the gap tended to widen somewhat as urbanization proceeded in the latter half of the nineteenth century and the first quarter of the twentieth. In 1800 urban women in the U.S. had 36 percent fewer children than rural women did; in 1840, 38 percent and in 1930, 41 percent. Thereafter the difference diminished.

With mortality in the cities higher and birth rates lower, and with reclassification a minor factor, the only real source for the growth in the proportion of people in urban areas during the industrial transition was rural–urban migration. This source had to be plentiful enough not only to overcome the substantial disadvantage of the cities in natural increase but also, above that, to furnish a big margin of growth in their populations. If, for example, the cities had a death rate a third higher and a birth rate a third lower than the rural rates (as was typical in the latter half of the nineteenth century), they would require each year perhaps 40 to 45 migrants from elsewhere per 1,000 of their population to maintain a growth rate of 3 percent per year. Such a rate of migration could easily be maintained as long as the rural portion of the population was large, but when this condition ceased to obtain, the maintenance of the same urban rate meant an increasing drain on the countryside.

Why did the rural–urban migration occur? The reason was that the rise in technological enhancement of human productivity, together with certain constant factors, rewarded urban concentration. One of the constant factors was that agriculture uses land as its prime instrument of production and hence spreads out people who are engaged in it, whereas manufacturing, commerce and services use land only as a site. Moreover, the demand for agricultural products is less elastic than the demand for services and manufactures. As productivity grows, services and manufactures can absorb more manpower by paying higher wages. Since nonagricultural activities can use land simply as a site, they can locate near one another (in towns and cities) and thus minimize the fraction of space inevitably involved in the division of labor. At the same time, as agricultural technology is improved, capital costs in farming rise and manpower becomes not only less needed but also economically more burdensome. A substantial portion of the agricultural population is therefore sufficiently disadvantaged, in relative terms, to be attracted by higher wages in other sectors.

In this light one sees why a large *flow* of people from farms to cities was generated in every country that passed through the industrial revolution. One also sees why, with an even higher proportion of people already in cities and with the inability of city people to replace themselves by reproduction, the drain eventually became so heavy that in many nations the rural population began to decline in absolute as well as relative terms. In Sweden it declined after 1920, in England and Wales after 1861, in Belgium after 1910.

Realizing that urbanization is transitional and finite, one comes on another fact – a fact that throws light on

covered by the urban agglomeration is growing faster than the population: it grew by 51 percent from 1950 to 1960, whereas the population rose by 15 percent.

If, then, one projects the rise in population and the rise in territory for the New York urbanized region one finds the density problem solved. It is not solved for long, though, because New York is not the only city in the region that is expanding. So are Philadelphia, Trenton, Hartford, New Haven and so on. By 1960 a huge stretch of territory about 600 miles long and 30 to 100 miles wide along the eastern seaboard contained some 37 million people. (I am speaking of a longer section of the seaboard than the Boston-to-Washington conurbation referred to by some other authors.) Since the whole area is becoming one big polynucleated city, its population cannot long expand without a rise in density. Thus persistent human multiplication promises to frustrate the ceaseless search for space – for ample residential lots, wide-open suburban school grounds, sprawling shopping centers, one-floor factories, broad freeways.

How people feel about giant agglomerations is best indicated by their headlong effort to escape them. The bigger the city, the higher the cost of space; yet the more the level of living rises, the more people are willing to pay for low-density living. Nevertheless, as urbanized areas expand and collide, it seems probable that life in low-density surroundings will become too dear for the great majority.

One can of course imagine that cities may cease to grow and may even shrink in size while the population in general continues to multiply. Even this dream, however, would not permanently solve the problem of space. It would eventually obliterate the distinction between urban and rural, but at the expense of the rural.

It seems plain that the only way to stop urban crowding and to solve most of the urban problems besetting both the developed and the underdeveloped nations is to reduce the overall rate of population growth. Policies designed to do this have as yet little intelligence and power behind them. Urban planners continue to treat population growth as something to be planned for, not something to be itself planned. Any talk about applying brakes to city growth is therefore purely speculative, overshadowed as it is by the reality of uncontrolled population increase.

"The Urban Revolution"

Town Planning Review (1950)

V. Gordon Childe

Editors' Introduction

The study of the earliest cities belongs to the fields of prehistory, anthropology, and archaeology, and more is being learned every day about the first emergence of urban civilization. V. Gordon Childe (1892–1957) is arguably the single most influential archaeologist of the twentieth century. Born in Australia, Childe won a scholarship to Queen's College, Oxford, returned to Australia, where he briefly pursued a career in left-wing politics, then returned to the UK as Professor of Archaeology at the University of Edinburgh and, later, Director of the Institute of Archaeology at the University of London.

Childe's most important book, the one that revolutionized the world of archaeological research by laying out an entirely new theoretical framework for understanding the phases of human development throughout history and prehistory, was *Man Makes Himself* (1936). In that pioneering work, Childe threw out the "three age system" (Stone Age, Bronze Age, Iron Age) that had been left over from nineteenth-century conceptions of human historical development. In its place he proposed a series of four stages (paleolithic, neolithic, urban, industrial) punctuated by three "revolutions" or fundamental shifts in cultural development.

According to Childe, the first revolution – from old Stone Age hunter-gatherer cultures to settled agriculture – was the Neolithic Revolution. The second – the movement from neolithic agriculture to complex, hierarchical systems of manufacturing and trade that began during the fourth and third millennia BCE – was the Urban Revolution. And the third major shift in the record of human cultural and historical development – the only truly new development since the rise of cities – was the Industrial Revolution of the eighteenth and nineteenth centuries. It is important to bear in mind that Childe lived before the computer age and developed his typology before current debates about how information technology is revolutionizing urban society. Manuel Castells (p. 572) considers "the informational city" to be a radically different type of city and "the rise of the network society" to be a transformation as profound as the earlier revolutions that Childe identified.

Childe is best known for his writings on the first cities, which arose in Mesopotamia (present-day Iraq) beginning about 4000 BCE. These cities sprang up in the area bounded by the Tigris and Euphrates Rivers – often referred to as "the Fertile Crescent." Plate 2, "A View of Ancient Babylon," illustrates the form these first cities took. Monumental gates, massive mud-brick walls, courtyards, residences for priest-kings, and a ziggurat.

Childe's work continues to figure prominently in ongoing debates about when, where, and why the first cities arose and in the antecedent debate about what a city is. Not everyone has accepted Childe's notion that the shift from neolithic to urban was a total break with the past. Evidence of ancient earthworks, wells, irrigation systems, and even continental trade networks have been traced back as far as 10,000 years in a number of areas in both the Old World and the New. Archaeologist James Mellaart has argued that evidence from the great neolithic communities of Çatal Hüyük and Hacilar in ancient Turkey, which predate the earliest Mesopotamian cities by some thousands of years, calls the entire Childe theory into question. Still, it is clear that in most locations agriculture generally predated the rise of the first cities by thousands of years, and that the full elaboration of those cultural institutions we associate with urban life emerged only with the rise of the Mesopotamian cities.

Not everyone agrees with Childe's definition of a city. More recent archaeologists excavating older, smaller, less culturally advanced settlements than the Mesopotamian cities Childe studied often argue that these settlements were urban enough to qualify as true cities. Scholars working in South and Central America point out that many of the cultural features Childe believed essential to the definition of a city (including the wheel, writing, and the plow) did not exist in large and culturally advanced Amerindian settlements that appear truly urban in other respects.

In the selection from *Town Planning Review* reprinted here, Childe details the constituent elements of the Urban Revolution that accompanied the initial rise of complex civilizations in Mesopotamia and elsewhere in the ancient Near East. Childe felt that the major factors motivating the transformation were rooted in the material base of the society: its means of production and its available physical and technological resources. Thus, the economic division of labor, the elaboration of socio-political hierarchies, and even the emergence of basic religious and intellectual patterns of thought characteristic of urban civilizations all rested on the underlying need to increase food production through massive irrigation systems and to protect the communities themselves through the erection of massive walls and fortifications.

Many modern scholars question the deterministic Marxist categories Childe employed. Although he stresses the importance of writing as an element of any truly urban society, Childe has been faulted for his apparent disregard of the primacy of non-material aspects of culture. His system has very little room for what Lewis Mumford (p. 91) called "the urban drama" or what Jane Jacobs (p. 105) called the "street ballet." Still, no one has ever called Childe's vision limited or ideologically cramped. On the contrary, he provided an expansive macro-historical foundation upon which generations of others have built.

A tireless researcher and writer, Childe produced a veritable stream of books, many of which are still classics. Among the most notable are *The Dawn of European Civilization* (London: Routledge & Kegan Paul, 1925), *The Most Ancient East* (New York: Grove, 1928), *What Happened in History* (Harmondsworth: Penguin, 1942), and *Social Evolution* (London: Watts, 1951). Other books on Mesopotamian cities include Nicholas Postgate and J.N. Postgate, *Early Mesopotamia: Society and Economy at the Dawn of History* (London: Routledge, 1994), Georges Roux, *Ancient Iraq* (New York: Penguin, 1993), and C. Leonard Wooley's two classics, *The Sumerians* (Oxford: Oxford University Press, 1928) and *Ur of the Chaldees* (Oxford: Oxford University Press, 1929). For more on Childe and his contributions to the field, consult Sally Green, *Prehistorian: A Biography of V. Gordon Childe* (Dana Point, CA: Moonraker, 1981), and David R. Harris, *The Archaeology of V. Gordon Childe: Contemporary Perspectives*, 2nd edn (Chicago, IL: University of Chicago Press, 1994).

For surveys of the current state of research into cities in the ancient world, see Gwendolyn Leick, *Mesopotamia* (London: Penguin, 2003) and Charles Gates, *Ancient Cities: The Archaeology of Urban Life in the Ancient Near East and Egypt, Greece, and Rome* (London: Routledge, 2003). Of special interest is Joyce Marcus and Jeremy Sabloff, *The Ancient City: New Perspectives on Urbanism in the Old and New World* (Santa Fe, NM: School for Advanced Research Press, 2009).

Earlier studies on the rise of the earliest cities elsewhere in the world that are still of interest include Mortimer Wheeler, *Civilizations of the Indus Valley and Beyond* (London: Thames & Hudson, 1966), Karl Wittfogel, *Oriental Despotism* (New Haven, CT: Yale University Press, 1957), Basil Davidson, *The Lost Cities of Africa* (Boston, MA: Little, Brown, 1959), Richard E.W. Adams, *Prehistoric Mesoamerica*, 3rd edn (Norman, OK: University of Oklahoma Press, 2005), Sylvanus G. Morley and George W. Brainerd, *The Ancient Maya* (Stanford, CA: Stanford University Press, 1956), Jacques Soustelle, *The Daily Life of the Aztecs* (New York: Macmillan, 1962), James Mellaart, *Earliest Civilizations of the Near East* (New York: McGraw-Hill, 1965) and *Çatal Hüyük* (New York: McGraw-Hill, 1967), and Paul Wheatley, *The Pivot of the Four Quarters: A Preliminary Inquiry into the Origins and Character of the Ancient Chinese City* (Chicago, IL: Aldine, 1971).

The concept of 'city' is notoriously hard to define. The aim of the present essay is to present the city historically – or rather prehistorically – as the resultant and symbol of a 'revolution' that initiated a new economic stage in the evolution of society. The word 'revolution' must not of course be taken as denoting a sudden violent catastrophe; it is here used for the culmination of a progressive change in the economic structure and social organization of communities that caused, or was accompanied by, a dramatic increase in the population affected – an increase that would appear as an obvious bend in the population graph were vital statistics available. Just such a bend is observable at the time of the Industrial Revolution in England. Though not demonstrable statistically, comparable changes of direction must have occurred at two earlier points in the demographic history of Britain and other regions. Though perhaps less sharp and less durable, these too should indicate equally revolutionary changes in economy. They may then be regarded likewise as marking transitions between stages in economic and social development.

Sociologists and ethnographers last century classified existing pre-industrial societies in a hierarchy of three evolutionary stages, denominated respectively 'savagery,' 'barbarism' and 'civilization.' If they be defined by suitably selected criteria, the logical hierarchy of stages can be transformed into a temporal sequence of ages, proved archaeologically to follow one another in the same order wherever they occur. Savagery and barbarism are conveniently recognized and appropriately defined by the methods adopted for procuring food. Savages live exclusively on wild food obtained by collecting, hunting or fishing. Barbarians on the contrary at least supplement these natural resources by cultivating edible plants and – in the Old World north of the Tropics – also by breeding animals for food.

Throughout the Pleistocene Period – the Palaeolithic Age of archaeologists – all known human societies were savage in the foregoing sense, and a few savage tribes have survived in out of the way parts to the present day. In the archaeological record barbarism began less than ten thousand years ago with the Neolithic Age of archaeologists. It thus represents a later, as well as a higher stage, than savagery. Civilization cannot be defined in quite such simple terms. Etymologically the word is connected with 'city,' and sure enough life in cities begins with this stage. But 'city' is itself ambiguous so archaeologists like to use 'writing' as a criterion of civilization; it should be easily recognizable and proves to be a reliable index to more profound characters. Note, however, that, because a people is said to be civilized or literate, it does not follow that all its members can read and write, nor that they all lived in cities. Now there is no recorded instance of a community of savages civilizing themselves, adopting urban life or inventing a script. Wherever cities have been built, villages of preliterate farmers existed previously (save perhaps where an already civilized people have colonized uninhabited tracts). So civilization, wherever and whenever it arose, succeeded barbarism.

We have seen that a revolution as here defined should be reflected in the population statistics. In the case of the Urban Revolution the increase was mainly accounted for by the multiplication of the numbers of persons living together, i.e., in a single built-up area. The first cities represented settlement units of hitherto unprecedented size. Of course it was not just their size that constituted their distinctive character. We shall find that by modern standards they appeared ridiculously small and we might meet agglomerations of population today to which the name city would have to be refused. Yet a certain size of settlement and density of population is an essential feature of civilization.

Now the density of population is determined by the food supply which in turn is limited by natural resources, the techniques for their exploitation and the means of transport and food-preservation available. The last factors have proved to be variables in the course of human history, and the technique of obtaining food has already been used to distinguish the consecutive stages termed savagery and barbarism. Under the gathering economy of savagery population was always exceedingly sparse. In aboriginal America the carrying capacity of normal unimproved land seems to have been from .05 to .10 per square mile. Only under exceptionally favourable conditions did the fishing tribes of the Northwest Pacific coast attain densities of over one human to the square mile. As far as we can guess from the extant remains, population densities in Palaeolithic and pre-neolithic Europe were less than the normal American. Moreover such hunters and collectors usually live in small roving bands. At best several bands may come together for quite brief periods on ceremonial occasions such as the Australian corroborees. Only in exceptionally favoured regions can fishing tribes establish anything like villages. Some settlements on the Pacific coasts comprised thirty or so substantial and durable houses, accommodating groups of several hundred persons. But even these villages were only occupied during the winter; for the rest of the year their inhabitants disposed in smaller groups. Nothing comparable has been found in pre-neolithic times in the Old World.

The Neolithic Revolution certainly allowed an expansion of population and enormously increased the carrying capacity of suitable land. On the Pacific Islands neolithic societies today attain a density of 30 or more persons to the square mile. In pre-Columbian North America, however, where the land is not obviously restricted by surrounding seas, the maximum density recorded is just under 2 to the square mile.

Neolithic farmers could of course, and certainly did, live together in permanent villages, though, owing to the extravagant rural economy generally practised, unless the crops were watered by irrigation, the villages had to be shifted at least every twenty years. But on the whole the growth of population was not reflected so much in the enlargement of the settlement unit as in a multiplication of settlements. In ethnography neolithic villages can boast only a few hundred inhabitants (a couple of 'pueblos' in New Mexico house over a thousand, but perhaps they cannot be regarded as neolithic). In prehistoric Europe the largest neolithic village yet known, Barkaer in Jutland, comprised 52 small, one-roomed dwellings, but 16 to 30 houses was a more normal figure; so the average, local group in neolithic times would average 200 to 400 members.

These low figures are of course the result of technical limitations. In the absence of wheeled vehicles and roads for the transport of bulky crops men had to live within easy walking distance of their cultivations. At the same time the normal rural economy of the Neolithic Age, what is now termed slash-and-burn or *jhumming*, condemns much more than half the arable land to lie fallow so that large areas were required. As soon as the population of a settlement rose above the numbers that could be supported from the accessible land, the excess had to hive off and found a new settlement.

The Neolithic Revolution had other consequences beside increasing the population, and their exploitation might in the end help to provide for the surplus increase. The new economy allowed, and indeed required, the farmer to produce every year more food than was needed to keep him and his family alive. In other words it made possible the regular production of a social surplus. Owing to the low efficiency of neolithic technique, the surplus produced was insignificant at first, but it could be increased till it demanded a reorganization of society.

Now in any Stone Age society, palaeolithic or neolithic, savage or barbarian, everybody can at least in theory make at home the few indispensable tools, the modest cloths and the simple ornaments everyone requires. But every member of the local community, not disqualified by age, must contribute actively to the communal food supply by personally collecting, hunting, fishing, gardening or herding. As long as this holds good, there can be no full-time specialists, no persons nor class of persons who depend for their livelihood on food produced by others and secured in exchange for material or immaterial goods or services.

We find indeed today among Stone Age barbarians and even savages expert craftsmen (for instance flint-knappers among the Ona of Tierra del Fuego), men who claim to be experts in magic, and even chiefs. In Palaeolithic Europe too there is some evidence for magicians and indications of chieftainship in pre-neolithic times. But on closer observation we discover that today these experts are not full-time specialists. The Ona flintworker must spend most of his time hunting; he only adds to his diet and his prestige by making arrowheads for clients who reward him with presents. Similarly a pre-Columbian chief, though entitled to customary gifts and services from his followers, must still personally lead hunting and fishing expeditions and indeed could only maintain his authority by his industry and prowess in these pursuits. The same holds good of barbarian societies that are still in the neolithic stage, like the Polynesians where industry in gardening takes the place of prowess in hunting. The reason is that there simply will not be enough food to go round unless every member of the group contributes to the supply. The social surplus is not big enough to feed idle mouths.

Social division of labour, save those rudiments imposed by age and sex, is thus impossible. On the contrary community of employment, the common absorption in obtaining food by similar devices guarantees a certain solidarity to the group. For co-operation is essential to secure food and shelter and for defence against foes, human and subhuman. This identity of economic interests and pursuits is echoed and magnified by identity of language, custom and belief; rigid conformity is enforced as effectively as industry in the common quest for food. But conformity and industrious co-operation need no State organization to maintain them. The local group usually consists either of a single clan (persons who believe themselves descended from a common ancestor or who have earned a mystical claim to such descent by ceremonial adoption) or a group of clans related by habitual intermarriage. And the sentiment of kinship is reinforced or supplemented by common rites focused on some ancestral shrine or sacred place. Archaeology can provide no evidence for kinship organization, but shrines occupied the central place in preliterate villages in Mesopotamia, and the long barrow, a collective tomb that overlooks the presumed site of most neolithic villages in Britain, may well have been also the ancestral shrine on which converged the emotions and ceremonial activities of the villagers below. However, the solidarity thus idealized and concretely symbolized, is really based on the same principles as that of a pack of wolves or a herd of sheep; Durkheim has called it 'mechanical.'

Now among some advanced barbarians (for instance tattooers or woodcarvers among the Maori) still technologically neolithic we find expert craftsmen tending towards the status of full-time professionals, but only at the cost of breaking away from the local community. If no single village can produce a surplus large enough to feed a full-time specialist all the year round, each should produce enough to keep him a week or so. By going round from village to village an expert might thus live entirely from his craft. Such itinerants will lose their membership of the sedentary kinship group. They may in the end form an analogous organization of their own – a craft clan, which, if it remain hereditary, may become a caste, or, if it recruit its members mainly by adoption (apprenticeship throughout Antiquity and the Middle Ages was just temporary adoption), may turn into a guild. But such specialists by emancipation from kinship ties, have also forfeited the protection of the kinship organization which alone under barbarism, guaranteed to its members security of person and property. Society must be reorganized to accommodate and protect them.

In pre-history specialization of labour presumably began with similar itinerant experts. Archaeological proof is hardly to be expected, but in ethnography metal-workers are nearly always full-time specialists. And in Europe at the beginning of the Bronze Age metal seems to have been worked and purveyed by perambulating smiths who seem to have functioned like tinkers and other itinerants of much more recent times. Though there is no such positive evidence, the same probably happened in Asia at the beginning of metallurgy. There must of course have been in addition other specialist craftsmen whom, as the Polynesian example warns us, archaeologists could not recognize because they worked in perishable materials. One result of the Urban Revolution will be to rescue such specialists from nomadism and to guarantee them security in a new social organization.

About 5,000 years ago irrigation cultivation (combined with stockbreeding and fishing) in the valleys of the Nile, the Tigris-Euphrates and the Indus had begun to yield a social surplus, large enough to support a number of resident specialists who were themselves released from food-production. Water-transport, supplemented in Mesopotamia and the Indus valley by wheeled vehicles and even in Egypt by pack animals, made it easy to gather food stuffs at a few centres. At the same time dependence on river water for the irrigation of the crops restricted the cultivable areas while the necessity of canalizing the waters and protecting habitations against annual floods encouraged the aggregation of population. Thus arose the first cities – units of settlement ten times as great as any known neolithic village. It can be argued that all cities in the old world are offshoots of those of Egypt, Mesopotamia and the Indus basin. So the latter need not be taken into account if a minimum definition of civilization is to be inferred from a comparison of its independent manifestations.

But some three millennia later cities arose in Central America, and it is impossible to prove that the Mayas owed anything directly to the urban civilizations of the Old World. Their achievements must therefore be taken into account in our comparison, and their inclusion seriously complicates the task of defining the essential preconditions for the Urban Revolution. In the Old World the rural economy which yielded the surplus was based on the cultivation of cereals combined with stock-breeding. But this economy had been made more efficient as a result of the adoption of irrigation (allowing cultivation without prolonged fallow periods) and of important inventions and discoveries – metallurgy, the plough, the sailing boat and the wheel. None of these devices was known to the Maya; they bred no animals for milk or meat; though they cultivated the cereal maize, they used the same sort of slash-and-burn method as neolithic farmers in prehistoric Europe or in the Pacific Islands today. Hence the minimum definition of a city, the

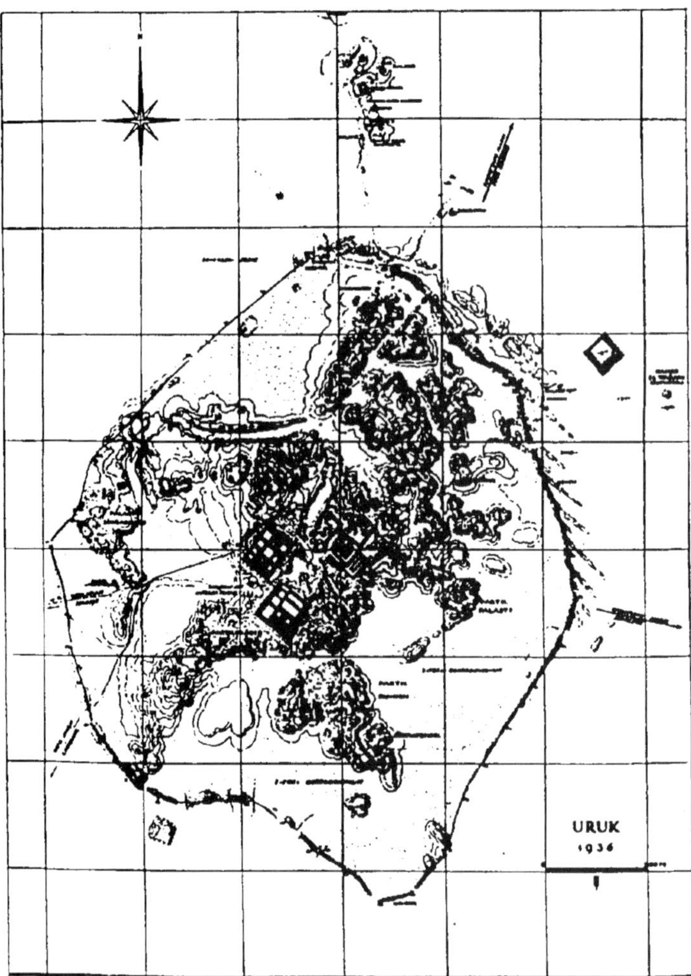

Figure 1 Plan of the city of Erek (Uruk)

greatest factor common to the Old World and the New will be substantially reduced and impoverished by the inclusion of the Maya. Nevertheless ten rather abstract criteria, all deducible from archaeological data, serve to distinguish even the earliest cities from any older or contemporary village.

(1) In point of size the first cities must have been more extensive and more densely populated than any previous settlements, although considerably smaller than many villages today. It is indeed only in Mesopotamia and India that the first urban populations can be estimated with any confidence or precision. There excavation has been sufficiently extensive and intensive to reveal both the total area and the density of building in sample quarters and in both respects has disclosed significant agreement with the less industrialized Oriental cities today. The population of Sumerian cities, thus calculated, ranged between 7,000 and 20,000; Harappa and Mohenjo-daro in the Indus valley must have approximated to the higher figure. We can only infer that Egyptian and Maya cities were of comparable magnitude from the scale of public works, presumably executed by urban populations.

(2) In composition and function the urban population already differed from that of any village. Very likely indeed most citizens were still also peasants, harvesting the lands and waters adjacent to the city. But all cities must have accommodated in addition classes who did not themselves procure their own food by agriculture, stock-breeding, fishing or collecting – full-time specialist craftsmen, transport workers, merchants, officials and priests. All these were of

course supported by the surplus produced by the peasants living in the city and in dependent villages, but they did not secure their share directly by exchanging their products or services for grains or fish with individual peasants.

(3) Each primary producer paid over the tiny surplus he could wring from the soil with his still very limited technical equipment as tithe or tax to an imaginary deity or a divine king who thus concentrated the surplus. Without this concentration, owing to the low productivity of the rural economy, no effective capital would have been available.

(4) Truly monumental public buildings not only distinguish each known city from any village but also symbolize the concentration of the social surplus. Every Sumerian city was from the first dominated by one or more stately temples, centrally situated on a brick platform raised above the surrounding dwellings and usually connected with an artificial mountain, the staged tower or ziggurat. But attached to the temples were workshops and magazines, and an important appurtenance of each principal temple was a great granary. Harappa, in the Indus basin, was dominated by an artificial citadel, girt with a massive rampart of kiln-baked bricks, containing presumably a palace and immediately overlooking an enormous granary and the barracks of artisans. No early temples nor palaces have been excavated in Egypt, but the whole Nile valley was dominated by the gigantic tombs of the divine pharaohs while royal granaries are attested from the literary record. Finally the Maya cities are known almost exclusively from the temples and pyramids of sculptured stone round which they grew up.

Hence in Sumer the social surplus was first effectively concentrated in the hands of a god and stored in his granary. That was probably true in Central America while in Egypt the pharaoh (king) was himself a god. But of course the imaginary deities were served by quite real priests who, besides celebrating elaborate and often sanguinary rites in their honour, administered their divine masters' earthly estates. In Sumer indeed the god very soon, if not even before the revolution, shared his wealth and power with a mortal viceregent, the 'City-King,' who acted as civil ruler and leader in war. The divine pharaoh was naturally assisted by a whole hierarchy of officials.

(5) All those not engaged in food-production were of course supported in the first instance by the surplus accumulated in temple or royal granaries and were thus dependent on temple or court. But naturally priests, civil and military leaders and officials absorbed a major share of the concentrated surplus and thus formed a 'ruling class.' Unlike a palaeolithic magician or a neolithic chief, they were, as an Egyptian scribe actually put it, 'exempt from all manual tasks.' On the other hand, the lower classes were not only guaranteed peace and security, but were relieved from intellectual tasks which many find more irksome than any physical labour. Besides reassuring the masses that the sun was going to rise next day and the river would flood again next year (people who have not five thousand years of recorded experience of natural uniformities behind them are really worried about such matters!), the ruling classes did confer substantial benefits upon their subjects in the way of planning and organization.

(6) They were in fact compelled to invent systems of recording and exact, but practically useful, sciences. The mere administration of the vast revenues of a Sumerian temple or an Egyptian pharaoh by a perpetual corporation of priests or officials obliged its members to devise conventional methods of recording that should be intelligible to all their colleagues and successors, that is, to invent systems of writing and numeral notation. Writing is thus a significant, as well as a convenient, mark of civilization. But while writing is a trait common to Egypt, Mesopotamia, the Indus valley and Central America, the characters themselves were different in each region and so were the normal writing materials – papyrus in Egypt, clay in Mesopotamia. The engraved seals or stelae that provide the sole extant evidence for early Indus and Maya writing no more represent the normal vehicles for the scripts than do the comparable documents from Egypt and Sumer.

(7) The invention of writing – or shall we say the inventions of scripts – enabled the leisured clerks to proceed to the elaboration of exact and predictive sciences – arithmetic, geometry and astronomy. Obviously beneficial and explicitly attested by the Egyptian and Maya documents was the correct determination of the tropic year and the creation of a calendar. For it enabled the rulers to regulate successfully the cycle of agricultural operations. But once more the Egyptian, Maya and Babylonian calendars were as different as any systems based on a single natural unit could be. Calendrical and mathematical sciences are common features of the earliest civilizations and they too are corollaries of the archaeologists' criterion, writing.

(8) Other specialists, supported by the concentrated social surplus, gave a new direction to artistic expression. Savages even in palaeolithic times had tried, sometimes with astonishing success, to depict animals and even men as they saw them – concretely and naturalistically. Neolithic peasants never did that; they hardly ever tried to represent natural objects, but preferred to symbolize them by abstract geometrical patterns which at most may suggest by a few traits a fantastical man or beast or plant. But Egyptian, Sumerian, Indus and Maya artist-craftsmen – full-time sculptors, painters, or seal-engravers – began once more to carve, model or draw likenesses of persons or things, but no longer with the naive naturalism of the hunter, but according to conceptualized and sophisticated styles which differ in each of the four urban centres.

(9) A further part of the concentrated social surplus was used to pay for the importation of raw materials, needed for industry or cult and not available locally. Regular 'foreign' trade over quite long distances was a feature of all early civilizations and, though common enough among barbarians later, is not certainly attested in the Old World before 3000 B.C. nor in the New before the Maya 'empire.' Thereafter regular trade extended from Egypt at least as far as Byblos on the Syrian coast while Mesopotamia was related by commerce with the Indus valley. While the objects of international trade were at first mainly 'luxuries,' they already included industrial materials, in the Old World notably metal, the place of which in the New was perhaps taken by obsidian. To this extent the first cities were dependent for vital materials on long distance trade as no neolithic village ever was.

(10) So in the city, specialist craftsmen were both provided with raw materials needed for the employment of their skill and also guaranteed security in a State organization based now on residence rather than kinship. Itinerancy was no longer obligatory. The city was a community to which a craftsman could belong politically as well as economically.

Yet in return for security they became dependent on temple or court and were relegated to the lower classes. The peasant masses gained even less material advantages; in Egypt for instance metal did not replace the old stone and wood tools for agricultural work. Yet, however imperfectly, even the earliest urban communities must have been held together by a sort of solidarity missing from any neolithic village. Peasants, craftsmen, priests and rulers form a community, not only by reason of identity of language and belief, but also because each performs mutually complementary functions, needed for the well-being (as redefined under civilization) of the whole. In fact the earliest cities illustrate a first approximation to an organic solidarity based upon a functional complementarity and interdependence between all its members such as subsist between the constituent cells of an organism. Of course this was only a very distant approximation. However necessary the concentration of the surplus really was with the existing forces of production, there seemed a glaring conflict on economic interests between the tiny ruling class, who annexed the bulk of the social surplus, and the vast majority who were left with a bare subsistence and effectively excluded from the spiritual benefits of civilization. So solidarity had still to be maintained by the ideological devices appropriate to the mechanical solidarity of barbarism as expressed in the pre-eminence of the temple or the sepulchral shrine, and now supplemented by the force of the new State organization. There could be no room for sceptics or sectaries in the oldest cities.

These ten traits exhaust the factors common to the oldest cities that archaeology, at best helped out with fragmentary and often ambiguous written sources, can detect. No specific elements of town planning for example can be proved characteristic of all such cities; for on the one hand the Egyptian and Maya cities have not yet been excavated; on the other neolithic villages were often walled, an elaborate system of sewers drained the Orcadian hamlet of Skara Brae; two-storeyed houses were built in pre-Columbian pueblos, and so on.

The common factors are quite abstract. Concretely Egyptian, Sumerian, Indus and Maya civilizations were as different as the plans of their temples, the signs of their scripts and their artistic conventions. In view of this divergence and because there is so far no evidence for a temporal priority of one Old World centre (for instance, Egypt) over the rest nor yet for contact between Central America and any other urban centre, the four revolutions just considered may be regarded as mutually independent. On the contrary, all later civilizations in the Old World may in a sense be regarded as lineal descendants of those of Egypt, Mesopotamia or the Indus.

But this was not a case of like producing like. The maritime civilizations of Bronze Age Crete or classical Greece for example, to say nothing of our own, differ more from their reputed ancestors than these did among themselves. But the urban revolutions that gave them birth did not start from scratch. They could and probably did draw upon the capital accumulated in the three allegedly primary centres. That is most obvious in the case of cultural capital. Even today we use the Egyptians' calendar and the Sumerians' divisions of the day and the hour. Our European ancestors did not have to invent for themselves these divisions of time nor repeat the observations on which they are based; they took over – and very slightly improved – systems elaborated 5,000 years ago! But the same is in a sense true of material capital as well. The Egyptians, the Sumerians and the Indus people had accumulated vast reserves of surplus food. At the same time they had to import from abroad necessary raw materials like metals and building timber as well as 'luxuries.' Communities controlling these natural resources could in exchange claim a slice of the urban surplus. They could use it as capital to support full-time specialists – craftsmen or rulers – until the latters' achievement in technique and organization had so enriched barbarian economics that they too could produce a substantial surplus in their turn.

"The Polis"

from *The Greeks* (1951)

H.D.F. Kitto

Editors' Introduction

At its peak ancient Athens had only about as many residents as Peoria, Illinois – a little over 100,000 – not a city that leaps out as a great center of world civilization. But British classicist Humphrey Davy Findley Kitto (1897–1982) reminds us not to commit the vulgar error of confusing size with significance. During its golden age, Athens and the 700 or so other tiny settlements of ancient Greece made a monumental contribution to human culture. What the Greeks achieved in philosophy, literature, drama, poetry, art, logic, mathematics, sculpture, and architecture has exercised a profound influence on Western civilization.

A Greek invention of enduring interest to urbanists is the concept of the polis. Since we have not got the thing, which the Greeks called "the polis," Kitto notes, we do not possess an equivalent word. "City-state" or, perhaps, "self-governing community" come closest.

The classical Greek polis came of age in the fifth century BCE, about halfway between the emergence of the great Mesopotamian cities Childe describes and the present time. The physical form of the polis stressed public space. Private houses were low and turned away from the street. In contrast the Greeks emphasized public temples, stadiums, and the agora (a combined marketplace and public forum), as illustrated in Plate 3. In a larger polis, like Athens, these public buildings were spacious and often beautifully constructed of marble. Even in the smaller ones the community devoted many of its resources to them.

If the physical form of the polis was often stunning, it was the social organization of the polis that remains of particular fascination. The polis represents a form of community, which has exerted a powerful fascination for more than two millennia. One of the enduring questions of urban history is whether the ideals of the polis can be applied to, for example, the class-polarized cities of the Industrial Revolution described by Friedrich Engels (p. 46), or the sprawl suburbia analyzed by Kenneth T. Jackson (p. 65) and Robert Bruegmann (p. 211), or the cities of the present-day global society described in Part Eight of this book.

In this selection Kitto describes how the Greek polis made it possible for all citizens to realize their spiritual, moral, and intellectual capacities. The polis was a living community; almost an extended family. While the Greeks were very private in many ways, Kitto notes that their public life was essentially communistic. The polis as a social institution defined the very nature of being human for its citizens.

Not that the polis supported development of every resident: women and slaves were not citizens and did not participate in much of the life of the polis. Foreigners could attend plays in the Greek theater, but were barred from many institutions reserved to the (free, non-foreign, male) citizens. Sir Peter Hall in *Cities in Civilization* further questions the extent to which many citizens actually participated in public affairs. He hypothesizes that only a small percentage of those eligible to participate in public decision-making actually did so. He also notes that while farmers and other of the least educated and least articulate citizens of the Greek polis may have been physically present and possessed the same voting rights as educated upper-class Athenians, it is unlikely that they participated very effectively compared to the higher classes. For the most part, Hall believes, they were passive spectators rather than active participants in public affairs.

While a balanced view of the polis must acknowledge the existence of slavery, exclusion of women from civic life, limitations on the rights of foreigners, and the influence of education and class on social relations, Athens and the other Greek poleis were astonishingly democratic compared to any other urban civilization that preceded them. It is easy to dismiss Kitto as a hopeless romantic and his description of the Greek polis as an ivory tower depiction of a Camelot that never was. But that may be too harsh. The Greek polis as a social institution clearly represented a remarkable advance over social relations in any previous society. And the values it represented for its citizens are of enduring importance in an imperfect world.

In the debate about why the polis arose in Greece when it did, Kitto rejects deterministic answers such as the argument by geographical and economic determinists that the mountainous terrain required little, separate city-states. Rather, Kitto attributes the rise of the polis to the *character* of the Greeks themselves. He also expresses nostalgia for human qualities of life in the polis that appear threatened nowadays. Compare the vision of polis as a supportive, humanistic, structure for human fulfillment with the vision of large modern cities as centers of alienation and anomie depicted by Louis Wirth (p. 96), or ghettos housing the Black underclass as described by William Julius Wilson (p. 117) and Elijah Anderson (p. 127). Note the connections between humanistic values Kitto felt that the polis nurtured and Robert Putnam's concept of "social capital" growing out of civic engagement (p. 134), idealized life in J.B. Jackson's "almost perfect town" (p. 202), the return to human-scale community Peter Calthorpe and William Fulton advocate in *The Regional City* (p. 360), and the values expressed in *The Charter of the New Urbanism* (p. 360).

Other books helpful in understanding the polis and its significance are Christian Meier, *Athens: A Portrait of the City in its Golden Age* (New York: Metropolitan Books, 1998), Cecil Maurice Bowra, *The Greek Experience* (London: Weidenfeld & Nicolson, 1957), and a new edition of classic writings by a great classicist, Jacob Burckhardt, *The Greeks and Greek Civilization* (New York: St. Martin's, 1998). Also of interest are Lisa Nevett, *House and Society in the Ancient Greek World* (Cambridge: Cambridge University Press, 1999), and Nicholas Cahill, *Household and City Organization at Olynthus* (New Haven, CT: Yale University Press, 2002). Two excellent studies of Greek democracy are James O'Neil, *The Origins and Development of Ancient Greek Democracy* (Lanham, MD: Rowman & Littlefield, 1995), and Josiah Ober, *Political Dissent in Democratic Athens* (Princeton, NJ: Princeton University Press, 1998).

For accounts of Greek city planning see Richard Ernest Wycherley, *How the Greeks Built Cities*, 2nd edn (London: Macmillan, 1963) and *The Stones of Athens* (Princeton, NJ: Princeton University Press, l978). Dora Crouch, *Water Management in Ancient Greek Cities* (Oxford: Oxford University Press, 1993) is a gem, and Spiro Kostof, "Polis and Akropolis," Chapter 7 in *A History of Architecture* (Oxford: Oxford University Press, 1980) provides insight on classical Greek architecture.

Two masterful accounts of the role of cities in civilization give particular emphasis to the contribution of the Greek polis. See Lewis Mumford's chapters on "The Emergence of the Polis" and "Citizen Versus Ideal City" in *The City in History* (New York: Harcourt Brace Jovanovich, 1961) and Sir Peter Hall, "The Fountainhead," in Hall's *Cities in Civilization* (New York: Pantheon, 1998).

■

"Polis" is the Greek word which we translate as "city-state". It is a bad translation, because the normal polis was not much like a city, and was very much more than a state. But translation, like politics, is the art of the possible; since we have not got the thing which the Greeks called "the polis", we do not possess an equivalent word. From now on, we will avoid the misleading term "city-state", and use the Greek word instead . . . We will first inquire how this political system arose, then we will try to reconstitute the word "polis" and recover its real meaning by watching it in action. It may be a long task, but all the time we shall be improving our acquaintance with the Greeks. Without a clear conception what the polis was, and what it meant to the Greek, it is quite impossible to understand properly Greek history, the Greek mind, or the Greek achievement.

First then, what was the polis? . . .

. . . In Crete . . . we find over fifty quite independent poleis, fifty small "states" . . . What is true of Crete is true of Greece in general, or at least of those parts which play any considerable part in Greek history . . .

It is important to realize their size. The modern reader picks up a translation of Plato's *Republic* or Aristotle's *Politics*; he finds Plato ordaining that his ideal city shall have 5,000 citizens, and Aristotle that each citizen should be able to know all the others by sight; and he smiles, perhaps, at such philosophic fantasies. But Plato and Aristotle are not fantasts. Plato is imagining a polis on the normal Hellenic scale; indeed he implies that many existing Greek poleis are too small – for many had less than 5,000 citizens. Aristotle says, in his amusing way . . . that a polis of ten citizens would be impossible, because it could not be self-sufficient, and that a polis of a hundred thousand would be absurd, because it could not govern itself properly . . . Aristotle speaks of a hundred thousand citizens; if we allow each to have a wife and four children, and then add a liberal number of slaves and resident aliens, we shall arrive at something like a million – the population of Birmingham; and to Aristotle an independent "state" as populous as Birmingham is a lecture-room joke . . .

In fact, only three poleis had more than 20,000 citizens: Syracuse and Acragas (Girgenti) in Sicily, and Athens. At the outbreak of the Peloponnesian War the population of Attica was probably about 350,000, half Athenian (men, women and children), about a tenth resident aliens, and the rest slaves. Sparta, or Lacedaemon, had a much smaller citizen-body, though it was larger in area. The Spartans had conquered and annexed Messenia, and possessed 3,200 square miles of territory. By Greek standards this was an enormous area: it would take a good walker two days to cross it. The important commercial city of Corinth had a territory of 330 square miles . . . The island of Ceos, which is about as big as Bute, was divided into four poleis. It had therefore four armies, four governments, possibly four different calendars, and, it may be, four different currencies and systems of measures – though this is less likely. Mycenae was in historical times a shrunken relic of Agamemnon's capital, but still independent. She sent an army to help the Greek cause against Persia at the battle of Plataea; the army consisted of eighty men. Even by Greek standards this was small, but we do not hear that any jokes were made about an army sharing a cab.

To think on this scale is difficult for us, who regard a state of ten million as small, and are accustomed to states which, like the U.S.A. and the U.S.S.R., are so big that they have to be referred to by their initials; but when the adjustable reader has become accustomed to the scale, he will not commit the vulgar error of confusing size with significance . . .

But before we deal with the nature of the polis, the reader might like to know how it happened that the relatively spacious pattern of pre-Dorian Greece became such a mosaic of small fragments. The Classical scholar too would like to know; there are no records, so that all we can do is to suggest plausible reasons. There are historical, geographical and economic reasons; and when these have been duly set forth, we may conclude perhaps that the most important reason of all is simply that this is the way in which the Greeks preferred to live.

[Here Kitto describes the evolution of the Greek acropolis from a fortified hilltop strong-point built for protection against Dorian invaders to a place of assembly, religion, and commerce.]

At this point we may invoke the very sociable habits of the Greeks, ancient or modern. The English farmer likes to build his house on his land, and to come into town when he has to. What little leisure he has he likes to spend on the very satisfying occupation of looking over a gate. The Greek prefers to live in the town or village, to walk out to his work, and to spend his rather ampler leisure talking in the town or village square. Therefore the market becomes a market-town, naturally beneath the acropolis. This became the center of the communal life of the people – and we shall see presently how important that was.

But why did not such towns form larger units? This is the important question.

There is an economic point. The physical barriers which Greece has so abundantly made the transport of goods difficult, except by sea, and the sea was not yet used with any confidence. Moreover, the variety of which we spoke earlier enabled quite a small area to be reasonably self-sufficient for a people who made such small material demands on life as the Greek. Both of these facts tend in the same direction; there was in Greece no great economic interdependence, no reciprocal pull between the different parts of the country, strong enough to counteract the desire of the Greek to live in small communities.

There is a geographical point. It is sometimes asserted that this system of independent poleis was imposed on Greece by the physical character of the country. The theory is attractive, especially to those who like to have one majestic explanation of any phenomenon, but it does not seem to be true. It is of course obvious that the physical subdivision of the country helped; the system could not have existed, for example, in Egypt, a country which depends entirely on the proper management of the Nile flood, and therefore must have a central government. But there are countries cut up quite as much as Greece – Scotland, for instance – which have never developed the polis-system; and conversely there were in Greece many neighbouring poleis, such as Corinth and Sicyon, which remained independent of each other although between them there was no physical barrier that would seriously incommode a modern cyclist. Moreover, it was precisely the most mountainous parts of Greece that never developed poleis, or not until later days – Arcadia and Aetolia, for example, which had something like a canton-system. The polis flourished in those parts where communications were relatively easy. So that we are still looking for our explanation.

Economics and geography helped, but the real explanation is the character of the Greeks . . . As it will take some time to deal with this, we may first clear out of the way an important historical point. How did it come about that so preposterous a system was able to last for more than twenty minutes?

The ironies of history are many and bitter, but at least this must be put to the credit of the gods, that they arranged for the Greeks to have the Eastern Mediterranean almost to themselves long enough to work out what was almost a laboratory-experiment to test how far, and in what conditions, human nature is capable of creating and sustaining a civilization . . . this lively and intelligent Greek people was for some centuries allowed to live under the apparently absurd system which suited and developed its genius instead of becoming absorbed in the dull mass of a large empire, which would have smothered its spiritual growth . . . no history of Greece can be intelligible until one has understood what the polis meant to the Greek; and when we have understood that, we shall also understand why the Greeks developed it, and so obstinately tried to maintain it. Let us then examine the word in action.

It meant at first that which was later called the Acropolis, the stronghold of the whole community and the centre of its public life . . . "polis" very soon meant either the citadel or the whole people which, as it were, "used" this citadel. So we read in Thucydides, "Epidamnus is a polis on the right as you sail into the Ionian gulf." This is not like saying "Bristol is a city on the right as you sail up the Bristol Channel", for Bristol is not an independent state which might be at war with Gloucester, but only an urban area with a purely local administration. Thucydides' words imply that there is a town – though possibly a very small one – called Epidamnus, which is the political centre of the Epidamnians, who live in the territory of which the town is the centre – not the "capital" – and are Epidamnians whether they live in the town or in one of the villages in this territory.

Sometimes the territory and the town have different names. Thus, Attica is the territory occupied by the Athenian people; it comprised Athens – the "polis" in the narrower sense – the Piraeus, and many villages; but the people collectively were Athenians, not Attics, and a citizen was an Athenian in whatever part of Attica he might live.

In this sense "polis" is our "state" . . . The actual business of governing might be entrusted to a monarch, acting in the name of all according to traditional usages, or to the heads of certain noble families, or to a council of citizens owning so much property, or to all the citizens. All these and many modifications of them, were natural forms of "polity"; all were sharply distinguished by the Greek from Oriental monarchy, in which the monarch is irresponsible, not holding his powers in trust by the grace of god, but being himself a god. If there were irresponsible government there was no polis . . .

. . . [T]he size of the polis made it possible for a member to appeal to all his fellow citizens in person, and this he naturally did if he thought that another member of the polis had injured him. It was the common assumption of the Greeks that the polis took its origin in the desire for Justice. Individuals are lawless, but the polis will see to it that wrongs are redressed. But not by an elaborate machinery of state-justice, for such a machine could not be operated except by individuals, who may be as unjust as the original wrongdoer. The injured party will be sure of obtaining Justice only if he can declare his wrongs to the whole polis. The word therefore now means "people" in actual distinction from state.

[. . .]

. . . Demosthenes the orator talks of a man who, literally, "avoids the city" – a translation which might lead the unwary to suppose that he lived in something corresponding to the Lake District, or Purley. But the phrase "avoids the polis" tells us nothing about his domicile; it means that he took no part in public life – and was therefore something of an oddity. The affairs of the community did not interest him.

We have now learned enough about the word polis to realize that there is no possible English rendering of such a common phrase as, "It is everyone's duty to help the polis." We cannot say "help the state", for that arouses no enthusiasm; it is "the state" that takes half our incomes from us. Not "the community", for with us "the community" is too big and too various to be grasped except theoretically. One's village, one's trade union, one's class, are entities that mean something to us at once, but "work for the community", though an admirable sentiment, is to most of us vague and flabby. In the years before the war, what did most parts of Great Britain know about the depressed areas? How much do bankers, miners and farmworkers understand each other? But the "polis" every Greek knew; there it was, complete, before his eyes. He could see the fields which gave it its sustenance – or did not, if the crops failed; he could see how agriculture, trade and industry dovetailed into one another; he knew the frontiers, where they were strong and where weak; if any malcontents were planning a *coup*, it was difficult for them to conceal the fact. The entire life of the polis, and the relation between its parts, were much easier to grasp, because of the small scale of things. Therefore to say "It is everyone's duty to help the polis" was not to express a fine sentiment but to speak the plainest and most urgent common sense. Public affairs had an immediacy and a concreteness which they cannot possibly have for us.

[. . .]

Pericles' Funeral Speech, recorded or recreated by Thucydides, will illustrate this immediacy, and will also take our conception of the polis a little further. Each year, Thucydides tells us, if citizens had died in war – and they had, more often than not – a funeral oration was delivered by "a man chosen by the polis". Today, that would be someone nominated by the Prime Minister, or the British Academy, or the BBC [British Broadcasting Corporation]. In Athens it meant that someone was chosen by the Assembly who had often spoken to that Assembly; and on this occasion Pericles spoke from a specially high platform, that his voice might reach as many as possible. Let us consider two phrases that Pericles used in that speech.

He is comparing the Athenian polis with the Spartan, and makes the point that the Spartans admit foreign visitors only grudgingly, and from time to time expel all strangers, "while we make our polis common to all". "Polis" here is not the political unit; there is no question of naturalizing foreigners – which the Greeks did rarely, simply because the polis was so intimate a union. Pericles means here: "We throw open to all our common cultural life", as is shown by the words that follow, difficult though they are to translate: "nor do we deny them any instruction or spectacle" – words that are almost meaningless until we realize that the drama, tragic and comic, the performance of choral hymns, public recitals of Homer, games, were all necessary and normal parts of "political" life. This is the sort of thing Pericles has in mind when he speaks of "instruction and spectacle", and of "making the polis open to all".

But we must go further than this. A perusal of the speech will show that in praising the Athenian polis Pericles is praising more than a state, a nation, or a people: he is praising a way of life; he means no less when, a little later, he calls Athens the "school of Hellas". – And what of that? Do not we praise "the English way of life"? The difference is this; we expect our State to be quite indifferent to "the English way of life" – indeed, the idea that the State should actively try to promote it would fill most of us with alarm. The Greeks thought of the polis as an active, formative thing, training the minds and characters of the citizens; we think of it as a piece of machinery for the production of safety and convenience. The training in virtue, which the medieval state left to the Church, and the polis made its own concern, the modern state leaves to God knows what.

"Polis", then, originally "citadel", may mean as much as "the whole communal life of the people, political, cultural, moral" – even "economic", for how else are we to understand another phrase in this same speech, "the produce of the whole world comes to us, because of the magnitude of our polis"? This must mean "our national wealth".

Religion too was bound up with the polis – though not every form of religion. The Olympian gods were indeed worshipped by Greeks everywhere, but each polis had, if not its own gods, at least its own particular cults of these gods . . . But beyond these Olympians, each polis had its minor local deities, "heroes" and nymphs, each worshipped with his immemorial rite, and scarcely imagined to exist outside the particular locality where the rite was performed. So . . . there is a sense in which it is true to say that the polis is an independent religious, as well as political, unit . . .

[. . .]

. . . Aristotle made a remark which we most inadequately translate "Man is a political animal." What

Aristotle really said is "Man is a creature who lives in a polis"; and what he goes on to demonstrate, in his *Politics*, is that the polis is the only framework within which man can fully realize his spiritual, moral and intellectual capacities.

Such are some of the implications of this word . . . The polis was a living community, based on kinship, real or assumed – a kind of extended family, turning as much as possible of life into family life, and of course having its family quarrels, which were the more bitter because they were family quarrels.

This it is that explains not only the polis but also much of what the Greek made and thought, that he was essentially social. In the winning of his livelihood he was essentially individualist: in the filling of his life he was essentially "communist". Religion, art, games, the discussion of things – all these were needs of life that could be fully satisfied only through the polis – not, as with us, through voluntary associations of like-minded people, or through entrepreneurs appealing to individuals. (This partly explains the difference between Greek drama and the modern cinema.) Moreover, he wanted to play his own part in running the affairs of the community. When we realize how many of the necessary, interesting and exciting activities of life the Greek enjoyed through the polis, all of them in the open air, within sight of the same acropolis, with the same ring of mountains or of sea visibly enclosing the life of every member of the state – then it becomes possible to understand Greek history, to understand that in spite of the promptings of common sense the Greek could not bring himself to sacrifice the polis, with its vivid and comprehensive life, to a wider but less interesting unity . . .

[. . .]

"The Great Towns"

from *The Condition of the Working Class in England in 1844* (1845)

Friedrich Engels

Editors' Introduction

It was the peculiar fate of Friedrich Engels (1820–1895) to live most of his adult life in the shadow of his better known friend and colleague Karl Marx and to be remembered as a fiercely bearded icon of International Communism. It was, however, a more humanly accessible Engels who, full of youthful idealism at the age of only 24, came face to face with the social horrors of the Industrial Revolution. Young Engels was sent by his industrialist father to learn business management in the factories of Manchester in the North of England. The unintended consequences of that particular paternal decision was *The Condition of the Working Class in England in 1844* (1845), a book that ranks as one of the earliest masterpieces of urban socio-politics.

By the 1840s, the Industrial Revolution had transformed conditions in many English cities, particularly in the Midlands and North. Manchester, which Engels observed in detail, was emblematic of what the new industrial cities were like. Plate 5 – Augustus Welby Pugin's *Contrasts* (1841) – compares the skyline of a fifteenth-century city dominated by church steeples to the same town in 1840. In the second view mills, factories, and a huge prison dominate the scene.

In the selection "The Great Towns" reprinted here (and note that "great" means large, not excellent!), Engels employs a peripatetic method of observation and analysis. Although he summarizes the socialist theory of the origin and historic role of the industrial working class, and although he quotes from many contemporaneous sources to bolster his analysis, Engels constructs the bulk of his argument by merely walking around the city and reporting what he sees. Quickly growing impatient with *telling* his readers about the social misery of working-class life, Engels begins *showing* them the horrors of industrial urbanism by conducting them on a tour of Manchester's working-class districts. As in Dante's *Inferno*, the tour descends deeper and deeper into the filth, misery, and despair that constitute the greater part of the Manchester conurbation. Engels was thought by some to have exaggerated slum conditions in support of his radical ideology, but subsequent mainstream British academic researchers like Charles Booth and several prestigious royal commissions produced a series of reports documenting conditions in British cities every bit as terrible as those that Engels described.

Engels wrote just as the first photographs of cities were being produced, but unfortunately he did not illustrate his book with actual pictures of the conditions that he described. Later in the nineteenth century Jacob Riis, Lewis Hine, and other photographers were to document slum conditions. The response of these and other photographers to the new reality of the nineteenth-century city is discussed in the selection by Frederic Stout, "Visions of a New Reality" (p. 150).

No one can read *The Condition of the Working Class* without acknowledging that Engels had come to know the various neighborhoods of proletarian Manchester – Old Town, Irish Town, Long Millgate, and Salford – intimately and that his observations were acute and objective. Of particular interest are his descriptions of the public health consequences (in terms of air and water pollution) of unrestrained overbuilding. In this, Engels anticipated many of the points made by environmental reformers like Frederick Law Olmsted (p. 321) and utopian planners like Ebenezer Howard (p. 328). He may even be said to lay the groundwork for the arguments of the sustainable planning advocates like Stephen Wheeler (p. 458), the World Commission on Environment and Development (p. 351), and Timothy Beatley (p. 446).

Responding to the spatial arrangements of the spatial segregation of urban-industrialism, Engels observed that the façades of the main thoroughfares mask the horrors that lie beyond from the eyes of the factory owners and the middle-class managers who commute into the city from outlying suburbs. This became a common theme in urban analysis. It was reiterated by Michael Harrington, *The Other America* (New York: Macmillan, 1962), Mike Davis in *City of Quartz* (p. 195), and many other observers of urban socio-spatial inequality.

The entire tradition of twentieth-century urban planning, capitalist and socialist alike, owes an enormous debt to Engels. The connection he draws between the physical decrepitude of the urban infrastructure and the alienation and despair of the urban poor remains valid to the present day. The urban parks movement and the construction of ideal company towns – Saltaire and Port Sunlight in the UK, Lowell and Pullman in the United States – as well as more recent attempts at inner-city redevelopment, all address issues first identified by Engels.

The conditions described by Engels form the basis for the social realist tradition in literature, a tradition that begins with Charles Dickens and Mrs. Gaskell in England and is continued in the works of Upton Sinclair and Theodore Dreiser in the United States. One can wonder whether the cultural impact of the industrial working class that Engels and the social realists describe will in any way be paralleled by the cultural impact of the post-industrial "creative class" that Richard Florida describes (p. 143).

For more on early Manchester, see the chapter on "Manchester, 1760–1840," in Peter Hall, *Cities in Civilization* (New York: Pantheon, 1998). Other significant investigations of urban poverty in England include Henry Mayhew, *London Labour and the London Poor* (four volumes, 1851–1862; selections reprinted, Ware, UK: Wordsworth, 2008), Charles Booth, *Conditions and Occupation of the People in East London and Hackney* (*Journal of the Royal Historical Society*, 1887), Jack London, *People of the Abyss* (New York: Macmillan, 1903), and George Orwell, *Down and Out in Paris and London* (London: Secker & Warburg, 1933).

In the United States, important studies of slum conditions include Jacob Riis, *How the Other Half Lives* (New York: Scribners, 1903), Upton Sinclair, *The Jungle* (New York: Doubleday, 1906), and a whole series of reports on conditions in the African American ghettos such as W.E.B. Du Bois, *The Philadelphia Negro* (p. 110), St. Clair Drake and Horace Cayton, *Black Metropolis* (Chicago, IL: University of Chicago Press, 1945), William Julius Wilson, *When Work Disappears* (p. 117), and Elijah Anderson, *Code of the Street* (p. 127).

There are two recent and excellent biographies of Engels: Tristram Hunt, *Marx's General: The Revolutionary Life of Friedrich Engels* (New York: Metropolitan Books, 2009), and John Green, *Engels: A Revolutionary Life* (London: Artery, 2008). For a sampling of Engels's most important writings as well as those of his lifelong friend Karl Marx, see Robert C. Tucker, *The Marx–Engels Reader* (New York: Norton, 1978). For an excellent summary of nineteenth-century urban poverty conditions and the broader socio-political context, consult Peter Hall, "The City of Dreadful Night," in Hall, *Cities of Tomorrow* (London: Blackwell, 1988), Eric Hobsbawm, *The Age of Revolution, 1789–1848* (New York: Vintage, 1996), and Kenneth Morgan, *The Birth of Industrial Britain, 1750–1850* (London: Longman, 1999). Robert C. Allen, *The British Industrial Revolution in Global Perspective* (Cambridge: Cambridge University Press, 2009) is a brilliant new analysis of the economic history. For further information on Engels's Manchester, and its connections to an emerging social realism in fiction, consult literary historian Steven Marcus's magisterial *Engels, Manchester, and the Working Class* (New York: Random House, 1974). Also of interest is Robert Roberts's extraordinary *The Classic Slum* (Manchester: University of Manchester Press, 1971), a first-person account of growing up in Salford during the early years of the twentieth century.

A town, such as London, where a man may wander for hours together without reaching the beginning of the end, without meeting the slightest hint which could lead to the inference that there is open country within reach, is a strange thing. This colossal centralization, this heaping together of two and a half millions of human beings at one point, has multiplied the power of this two and a half millions a hundred-fold; has raised London to the commercial capital of the world, created the giant docks and assembled the thousand vessels that continually cover the Thames. I know nothing more imposing than the view which the Thames offers during the ascent from the sea to London Bridge. The masses of buildings, the wharves on both sides, especially from Woolwich upwards, the countless ships along both shores, crowding ever closer and closer together, until, at last, only a narrow passage remains in the middle of the river, a passage through which hundreds of steamers shoot by one another; all this is so vast, so impressive, that a man cannot collect himself, but is lost in the marvel of England's greatness before he sets foot upon English soil.

But the sacrifices which all this has cost become apparent later. After roaming the streets of the capital a day or two, making headway with difficulty through the human turmoil and the endless lines of vehicles, after visiting the slums of the metropolis, one realizes for the first time that these Londoners have been forced to sacrifice the best qualities of their human nature, to bring to pass all the marvels of civilization which crowd their city; that a hundred powers which slumbered within them have remained inactive, have been suppressed in order that a few might be developed more fully and multiply through union with those of others. The very turmoil of the streets has something repulsive, something against which human nature rebels. The hundreds of thousands of all classes and ranks crowding past each other, are they not all human beings with the same qualities and powers, and with the same interest in being happy? And have they not, in the end, to seek happiness in the same way, by the same means? And still they crowd by one another as though they had nothing in common, nothing to do with one another, and their only agreement is the tacit one, that each keep to his own side of the pavement, so as not to delay the opposing streams of the crowd, while it occurs to no man to honour another with so much as a glance. The brutal indifference, the unfeeling isolation of each in his private interest becomes the more repellent and offensive, the more these individuals are crowded together, within a limited space. And, however much one may be aware that this isolation of the individual, this narrow self-seeking is the fundamental principle of our society everywhere, it is nowhere so shamelessly barefaced, so self-conscious as just here in the crowding of the great city. The dissolution of mankind into monads, of which each one has a separate principle and a separate purpose, the world of atoms, is here carried out to its utmost extreme.

Hence it comes, too, that the social war, the war of each against all, is here openly declared. . . . , people regard each other only as useful objects; each exploits the other, and the end of it all is, that the stronger treads the weaker under foot, and that the powerful few, the capitalists, seize everything for themselves, while to the weak many, the poor, scarcely a bare existence remains.

What is true of London, is true of Manchester, Birmingham, Leeds, is true of all great towns. Everywhere barbarous indifference, hard egotism on one hand, and nameless misery on the other, everywhere social warfare, every man's house in a state of siege, everywhere reciprocal plundering under the protection of the law, and all so shameless, so openly avowed that one shrinks before the consequences of our social state as they manifest themselves here undisguised, and can only wonder that the whole crazy fabric still hangs together.

Since capital, the direct or indirect control of the means of subsistence and production, is the weapon with which this social warfare is carried on, it is clear that all the disadvantages of such a state must fall upon the poor. For him no man has the slightest concern. Cast into the whirlpool, he must struggle through as well as he can. If he is so happy as to find work, i.e. if the bourgeoisie does him the favour to enrich itself by means of him, wages await him which scarcely suffice to keep body and soul together; if he can get no work he may steal, if he is not afraid of the police, or starve, in which case the police will take care that he does so in a quiet and inoffensive manner. During my residence in England, at least twenty or thirty persons have died of simple starvation under the most revolting circumstances, and a jury has rarely been found possessed of the courage to speak the plain truth in the matter. Let the testimony of the witnesses be never so clear and unequivocal, the bourgeoisie, from which the jury is selected, always finds some backdoor through which to escape the frightful verdict, death from starvation. The bourgeoisie dare not speak the truth in these cases, for it would speak its own condemnation. But indirectly, far more than directly, many have died of starvation, where long continued want of proper nourishment has called forth fatal illness, when it has produced such debility that causes which might otherwise have remained inoperative, brought on severe illness and death. The English working-men call this social murder, and accuse our whole society of perpetrating this crime perpetually. Are they wrong?

True, it is only individuals who starve, but what security has the working-man that it may not be his turn tomorrow? Who assures him employment, who vouches for it that, if for any reason or no reason his lord and master discharges him tomorrow, he can struggle along with those dependent upon him, until he may find some one else "to give him bread"? Who guarantees that willingness to work shall suffice to obtain work, that uprightness, industry, thrift, and the rest of the virtues recommended by the bourgeoisie, are really his road to happiness? No one. He knows that he has something today, and that it does not depend upon himself whether he shall have something tomorrow. He knows that every breeze that blows, every whim of his employer, every bad turn of trade may hurl him back into the fierce whirlpool from which he has temporarily saved himself, and in which it is hard and often impossible to keep his head above water. He knows that, though he may have the means of living today, it is very uncertain whether he shall tomorrow.

[. . .]

Manchester lies at the foot of the southern slope of a range of hills, which stretch hither from Oldham, their last peak, Kersallmoor, being at once the race-course and the Mons Sacer of Manchester. Manchester proper lies on the left bank of the Irwell, between that stream and the two smaller ones, the Irk and the Medlock, which here empty into the Irwell. On the right bank of the Irwell, bounded by a sharp curve of the river, lies Salford, and farther westward Pendleton; northward from the Irwell lie Upper and Lower Broughton; northward of the Irk, Cheetham Hill; south of the Medlock lies Hulme; farther east Chorlton on Medlock; still farther, pretty well to the east of Manchester, Ardwick. The whole assemblage of buildings is commonly called Manchester, and contains about four hundred thousand inhabitants, rather more than less. The town itself is peculiarly built, so that a person may live in it for years, and go in and out daily without coming into contact with a working-people's quarter or even with workers; that is, so long as he confines himself to his business or to pleasure walks. This arises chiefly from the fact, that by unconscious tacit agreement, as well as with outspoken conscious determination, the working-people's quarters are sharply separated from the sections of the city reserved for the middle class; or, if this does not succeed, they are concealed with the cloak of charity. Manchester contains, at its heart, a rather extended commercial district, perhaps half a mile long and about as broad, and consisting almost wholly of offices and warehouses. Nearly the whole district is abandoned by dwellers, and is lonely and deserted at night; only watchmen and policemen traverse its narrow lanes with their dark lanterns. This district is cut through by certain main thoroughfares upon which the vast traffic concentrates, and in which the ground level is lined with brilliant shops. In these streets the upper floors are occupied, here and there, and there is a good deal of life upon them until late at night. With the exception of this commercial district, all Manchester proper, all Salford and Hulme, a great part of Pendleton and Chorlton, two-thirds of Ardwick, and single stretches of Cheetham Hill and Broughton are all unmixed working-people's quarters, stretching like a girdle, averaging a mile and a half in breadth, around the commercial district. Outside, beyond this girdle, lives the upper and middle bourgeoisie, the middle bourgeoisie in regularly laid out streets in the vicinity of the working quarters, especially in Chorlton and the lower-lying portions of Cheetham Hill; the upper bourgeoisie in remoter villas with gardens in Chorlton and Ardwick or on the breezy heights of Cheetham Hill, Broughton and Pendleton, in free, wholesome country air, in fine, comfortable homes, passed once every half or quarter hour by omnibuses going into the city. And the finest part of the arrangement is this, that the members of this money aristocracy can take the shortest road through the middle of all the labouring districts to their places of business, without ever seeing that they are in the midst of the grimy misery that lurks to the right and the left. For the thoroughfares leading from the Exchange in all directions out of the city are lined, on both sides, with an almost unbroken series of shops, and are so kept in the hands of the middle and lower bourgeoisie, which, out of self-interest, cares for a decent and cleanly external appearance and *can* care for it. True, these shops bear some relation to the districts which lie behind them, and are more elegant in the commercial and residential quarters than when they hide grimy working-men's dwellings; but they suffice to conceal from the eyes of the wealthy men and women of strong stomachs and weak nerves the misery and grime which form the complement of their wealth. So, for instance, Deansgate, which leads from the Old Church directly southward, is lined first with mills and warehouses, then with second-rate shops and alehouses; farther south, when it leaves the commercial district, with less inviting shops, which grow dirtier and more interrupted by beerhouses and gin palaces the farther one goes, until at the southern end the appearance of the shops leaves no doubt that workers and workers only are their customers. So Market Street running south east from the Exchange; at first brilliant shops of the best sort, with counting-houses or warehouses above; in the continuation, Piccadilly, immense hotels and warehouses; in the farther continuation, London Road, in the neighbourhood of the Medlock, factories, beerhouses, shops for the humbler bourgeoisie and the working population; and from this point onward, large gardens and villas of the wealthier merchants and manufacturers. In this way any one who knows Manchester can infer the adjoining districts, from the appearance of the thoroughfare, but one is seldom in a position to catch from the street a glimpse of the real labouring districts. I know very well that this hypocritical plan is more or less common to all great cities; I know, too, that the retail dealers are forced by the nature of their business to take possession of the great highways; I know that there are more good buildings than bad ones upon such streets everywhere, and that the value of land is greater near them than in remoter districts; but at the same time I have never seen so systematic a shutting out of the working-class from the thoroughfares, so tender a concealment of everything which might affront the eye and the nerves of the bourgeoisie, as in Manchester. And yet, in other respects, Manchester is less built according to a plan, after official regulations, is more an outgrowth of accident, than any other city; and when I consider in this connection the eager assurances of the middle-class, that the working-class is doing famously, I cannot help feeling that the liberal manufacturers, the "Big Wigs" of Manchester, are not so innocent after all, in the matter of this sensitive method of construction.

I may mention just here that the mills almost all adjoin the rivers or the different canals that ramify throughout the city, before I proceed at once to describe the labouring quarters. First of all, there is the Old Town of Manchester [Figure 1], which lies between the northern boundary of the commercial district and the Irk. Here the streets, even the better ones, are narrow and winding, as Todd Street, Long Millgate, Withy Grove, and Shude Hill, the houses dirty, old, and tumble-down, and the construction of the side streets utterly horrible. Going from the Old Church to Long Millgate, the stroller has at once a row of old-fashioned houses at the right, of which not one has kept its original level; these are remnants of the old pre-manufacturing Manchester, whose former inhabitants have removed with their descendants into better-built districts, and have left the houses, which were not good enough for them, to a working-class population strongly mixed with Irish blood. Here one is in an almost undisguised working-men's quarter, for even the shops and beer-houses hardly take the trouble to exhibit a trifling degree of cleanliness. But all this is nothing in comparison with the courts and lanes which lie behind, to which access can be gained only through covered passages in which no two human beings can pass at the same time. Of the irregular cramming together of dwellings in ways which defy all rational plan, of the tangle in which they are crowded literally one upon the other, it is impossible to convey an idea. And it is not the buildings surviving from the old times of Manchester which are to blame for this; the confusion has only recently reached its height when every scrap of space left by the old way of building has been filled up and patched over until not a foot of land is left to be further occupied.

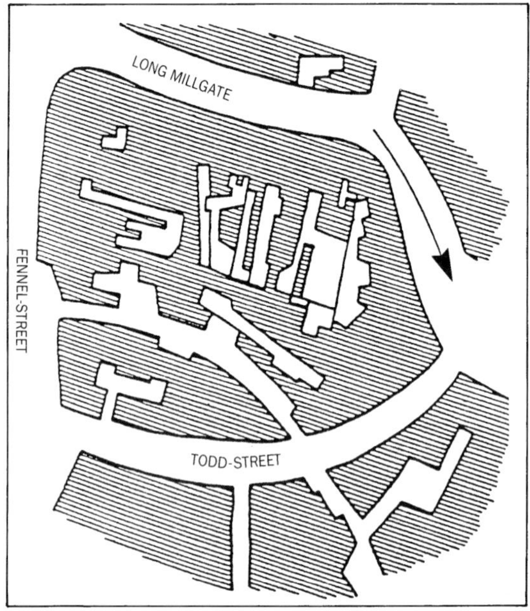

Figure 1

[. . .]

The south bank of the Irk is here very steep and between fifteen and thirty feet high. On this declivitous hillside there are planted three rows of houses, of which the lowest rise directly out of the river, while the front walls of the highest stand on the crest of the hill in Long Millgate. Among them are mills on the river; in short, the method of construction is as crowded and disorderly here as in the lower part of Long Millgate. Right and left a multitude of covered passages lead from the main street into numerous courts, and he who turns in thither gets into a filth and disgusting grime the equal of which is not to be found – especially in the courts which lead down to the Irk and which contain unqualifiedly the most horrible dwellings which I have yet beheld. In one of these courts there stands directly at the entrance, at the end of the covered passage, a privy without a door, so dirty that the inhabitants can pass into and out of the court only by passing through foul pools of stagnant urine and excrement. This is the first court on the Irk above Ducie Bridge – in case any one should care to look into it. Below it on the river there are several tanneries which fill the whole neighbourhood with the stench of animal putrefaction. Below Ducie Bridge the only entrance to most of the houses is by means of narrow dirty stairs and over heaps of refuse and filth. The first court below Ducie Bridge, known as Allen's Court, was in such a state at the time of the cholera that the sanitary police ordered it evacuated, swept, and disinfected with chloride of lime. Dr. Kay gives a terrible description of the state of this court at that time. Since then it seems to have been partially torn away and rebuilt; at least looking down from Ducie Bridge, the passer-by sees several ruined walls and heaps of debris with some newer houses. The view from this bridge, mercifully concealed from mortals of small stature by a parapet as high as a man, is characteristic for the whole district. At the bottom flows, or rather stagnates, the Irk, a narrow, coal-black, foul-smelling stream full of debris and refuse, which it deposits on the shallower right bank. In dry weather, a long string of the most disgusting, blackish-green, slime pools are left standing on this bank, from the depths of which bubbles of miasmatic gas constantly arise and give forth a stench unendurable even on the bridge forty or fifty feet above the surface of the stream. But besides this, the stream itself is checked every few paces by high weirs, behind which slime and refuse accumulate and rot in thick masses. Above the bridge are tanneries, bonemills, and gasworks, from which all drains and refuse find their way into the Irk, which receives further the contents of all the neighbouring sewers and privies. It may be easily imagined, therefore, what sort of residue the stream deposits. Below the bridge you look upon the piles of debris, the refuse, filth, and offal from the courts on the steep left bank; here each house is packed close behind its neighbour and a piece of each is visible, all black, smoky, crumbling, ancient, with broken panes and window-frames. The background is furnished by old barrack-like factory buildings. On the lower right bank stands a long row of houses and mills; the second house being a ruin without a roof, piled with debris; the third stands so low that the lowest floor is uninhabitable, and therefore without windows or doors. Here the background embraces the pauper burial-ground, the station of the Liverpool and Leeds railway, and, in the rear of this, the Workhouse, the "Poor-Law-Bastille" of Manchester, which, like a citadel, looks threateningly down from behind its high walls and parapets on the hilltop, upon the working-people's quarter below.

Above Ducie Bridge, the left bank grows more flat and the right bank steeper, but the condition of the dwellings on both banks grows worse rather than better. He who turns to the left here from the main street, Long Millgate, is lost; he wanders from one court to another, turns countless corners, passes nothing but narrow, filthy nooks and alleys, until after a few minutes he has lost all clue, and knows not whither to turn. Everywhere half or wholly ruined buildings, some of them actually uninhabited, which means a great deal here; rarely a wooden or stone floor to be seen in the houses, almost uniformly broken, ill-fitting windows and doors, and a state of filth! Everywhere heaps of debris, refuse, and offal; standing pools for gutters, and a stench which alone would make it impossible for a human being in any degree civilized to live in such a district. The newly-built extension of the Leeds railway, which crosses the Irk here, has swept away some of these courts and lanes, laying others completely open to view. Immediately under the railway bridge there stands a court, the filth and horrors of which surpass all the others by far, just because it was hitherto so shut off, so secluded that the way to it could not be found without a good deal of trouble. I should never have discovered it myself, without the breaks made by the railway, though I thought I knew this whole region thoroughly. Passing along a rough bank, among stakes and washing-lines, one penetrates into this chaos of small one-storeyed, one-roomed huts, in most of which there is no artificial floor; kitchen, living and sleeping room all in one. In such a hole, scarcely five feet long by six broad, I found two beds – and such bedsteads and beds! – which, with a staircase and chimney-place, exactly filled the room. In several others I found absolutely nothing, while the door stood open, and the inhabitants leaned against it. Everywhere before the doors refuse and offal; that any sort of pavement lay underneath could not be seen but only felt, here and there, with the feet. This whole collection of cattle-sheds for human beings was surrounded on two sides by houses and a factory, and on the third by the river, and besides the narrow stair up the bank, a narrow doorway alone led out into another almost equally ill-built, ill-kept labyrinth of dwellings.

Enough! The whole side of the Irk is built in this way, a planless, knotted chaos of houses, more or less on the verge of uninhabitableness, whose unclean interiors fully correspond with their filthy external surroundings. And how could the people be clean with no proper opportunity for satisfying the most natural and ordinary wants? Privies are so rare here that they are either filled up every day, or are too remote for most of the inhabitants to use. How can people wash when they have only the dirty Irk water at hand, while pumps and water pipes can be found in decent parts of the city alone? In truth, it cannot be charged to the account of these helots of modern society if their dwellings are not more clean than the pig-sties which are here and there to be seen among them. The landlords are not ashamed to let dwellings like the six or seven cellars on the quay directly below Scotland Bridge, the floors of which stand at least two feet below the low-water level of the Irk that flows not six feet away from them; or like the upper floor of the corner-house on the opposite shore directly above the bridge, where the ground-floor, utterly uninhabitable, stands deprived of all fittings for doors and windows, a case by no means rare in this region, when this open ground-floor is used as a privy by the whole neighbourhood for want of other facilities!

If we leave the Irk and penetrate once more on the opposite side from Long Millgate into the midst of the working-men's dwellings, we shall come into a somewhat newer quarter, which stretches from St. Michael's Church to Withy Grove and Shude Hill. Here there is somewhat better order. In place of the chaos of buildings, we find at least long straight lanes and alleys or courts, built according to a plan and usually square. But if, in the former case, every house was built according to caprice, here each lane and court is so built, without reference to the situation of the adjoining ones. The lanes run now in this direction, now in that, while every two minutes the wanderer gets into a blind alley, or, on turning a corner, finds himself back where he started from; certainly no one who has not lived a considerable time in this labyrinth can find his way through it.

If I may use the word at all in speaking of this district, the ventilation of these streets and courts is, in consequence of this confusion, quite as imperfect as in the Irk region; and if this quarter may, nevertheless, be said to have some advantage over that of the Irk, the houses being newer and the streets occasionally having gutters, nearly every house has, on the other hand, a cellar dwelling, which is rarely found in the Irk district, by reason of the greater age and more careless construction of the houses. As for the rest the filth, debris, and offal heaps, and the pools in the streets are common to both quarters, and in the district now under discussion, another feature most injurious to the cleanliness of the inhabitants, is the multitude of pigs walking about in all the alleys, rooting into the offal heaps, or kept imprisoned in small pens. Here, as in most of the working-men's quarters of Manchester, the pork-raisers rent the courts and build pigpens in them. In almost every court one or even several such pens may be found into which the inhabitants of the court throw all refuse and offal, whence the swine grow fat; and the atmosphere, confined on all four sides, is utterly corrupted by putrefying animal and vegetable substances. Through this quarter, a broad and measurably decent street has been cut, Millers Street, and the background has been pretty successfully concealed. But if any one should be led by curiosity to pass through one of the numerous passages which lead into the courts, he will find this piggery repeated at every twenty paces.

Such is the Old Town of Manchester, and on re-reading my description, I am forced to admit that instead of being exaggerated, it is far from black enough to convey a true impression of the filth, ruin, and uninhabitableness, the defiance of all considerations of cleanliness, ventilation, and health which characterize the construction of this single district, containing at least twenty to thirty thousand inhabitants. And such a district exists in the heart of the second city of England, the first manufacturing city of the world. If any one wishes to see in how little space a human being can move, how little air – and *such* air! – he can breathe, how little of civilization he may share and yet live, it is only necessary to travel hither. True, this is the *Old* Town, and the people of Manchester emphasize the fact whenever any one mentions to them the frightful condition of this Hell upon Earth; but what does that prove? Everything which here arouses horror and indignation is of recent origin, belongs to the *industrial* epoch. The couple of hundred houses, which belong to old Manchester, have been long since abandoned by their original inhabitants; the industrial epoch alone has crammed into them the swarms of workers whom they now shelter; the industrial epoch alone has built up every spot between these old houses to win a covering for the masses whom it has conjured hither from the agricultural districts and from Ireland; the industrial epoch alone enables the owners of these cattlesheds to rent them for high prices to human beings, to plunder the poverty of the workers, to undermine the health of thousands, in order that they *alone*, the owners, may grow rich. In the industrial epoch alone has it become possible that the worker scarcely freed from feudal servitude could be used as mere material, a mere chattel; that he must let himself be crowded into a dwelling too bad for every other, which he for his hard-earned wages buys the right to let go utterly to ruin. This manufacture has achieved, which, without these workers, this poverty, this slavery could not have lived. True, the original construction of this quarter was bad, little good could have been made out of it; but, have the landowners, has the municipality done anything to improve it when rebuilding? On the contrary, wherever a nook or corner was free, a house has been run up; where a superfluous passage remained, it has been built up; the value of land rose with the blossoming out of manufacture, and the more it rose, the more madly was the work of building up carried on, without reference to the health or comfort of the inhabitants, with sole reference to the highest possible profit on the principle that *no hole is so bad but that some poor creature must take it who can pay for nothing better.*

[. . .]

It may not be out of place to make some general observations just here as to the customary construction of working-men's quarters in Manchester. We have seen how in the Old Town pure accident determined the grouping of the houses in general. Every house is built without reference to any other, and the scraps of space between them are called courts for want of another name. In the somewhat newer portions of the same quarter, and in other working-men's quarters, dating from the early days of industrial activity, a somewhat more orderly arrangement may be found. The space between two streets is divided into more regular, usually square courts.

These courts were built in this way from the beginning, and communicate with the streets by means of covered passages. If the totally planless construction is injurious to the health of the workers by preventing ventilation, this method of shutting them up in courts surrounded on all sides by buildings is far more so. The air simply cannot escape; the chimneys of the houses are the sole drains for the imprisoned atmosphere of the courts, and they serve the purpose only so long as fire is kept burning. Moreover, the houses surrounding such courts are usually built back to back, having the rear wall in common; and this alone suffices to prevent any sufficient through ventilation. And, as the police charged with care of the streets does not trouble itself about the condition of these courts, as everything quietly lies where it is thrown, there is no cause for wonder at the filth and heaps of ashes and offal to be found here. I have been in courts, in Millers Street, at least half a foot below the level of the thoroughfare, and without the slightest drainage for the water that accumulates in them in rainy weather! More recently another different method of building was adopted, and has now become general. Working-men's cottages are almost never built singly, but always by the dozen or score; a single contractor building up one or two streets at a time. These are then arranged as follows: One front is formed of cottages of the best class, so fortunate as to possess a back door and small court, and these command the highest rent. In the rear of these cottages runs a narrow alley, the back street, built up at both ends, into which either a narrow roadway or a covered passage leads from one side. The cottages which face this back street command least rent, and are most neglected. These have their rear walls in common with the third row of cottages which face a second street, and command less rent than the first row and more than the second. The streets are laid out somewhat as in [Figure 2].

By this method of construction, comparatively good ventilation can be obtained for the first row of cottages, and the third row is no worse off than in the former method. The middle row, on the other hand, is at least as badly ventilated as the houses in the courts, and the back street is always in the same filthy, disgusting condition as they. The contractors prefer this method because it saves them space, and furnishes the means of fleecing better-paid workers through the higher rents of the cottages in the first and third rows. These three different forms of cottage building are found all over Manchester and throughout Lancashire and Yorkshire,

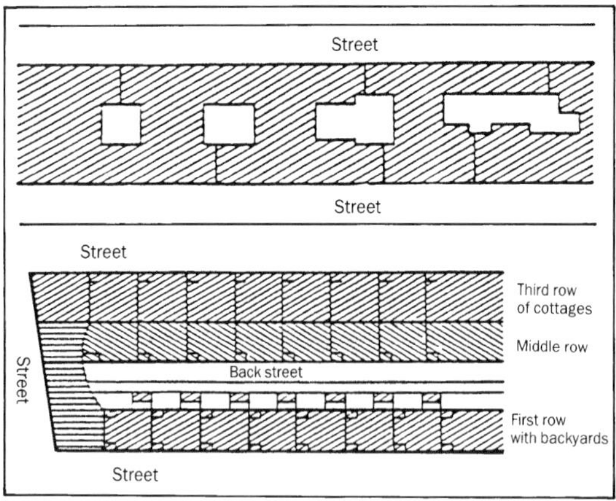

Figure 2

often mixed up together, but usually separate enough to indicate the relative age of parts of towns. The third system, that of the back alleys, prevails largely in the great working-men's district east of St. George's Road and Ancoats Street, and is the one most often found in the other working-men's quarters of Manchester and its suburbs.

[. . .]

Such are the various working-people's quarters of Manchester as I had occasion to observe them personally during twenty months. If we briefly formulate the result of our wanderings, we must admit that 350,000 working-people of Manchester and its environs live, almost all of them, in wretched, damp, filthy cottages, that the streets which surround them are usually in the most miserable and filthy condition, laid out without the slightest reference to ventilation, with reference solely to the profit secured by the contractor. In a word, we must confess that in the working-men's dwellings of Manchester, no cleanliness, no convenience, and consequently no comfortable family life is possible; that in such dwellings only a physically degenerate race, robbed of all humanity, degraded, reduced morally and physically to bestiality, could feel comfortable and at home.

[. . .]

To sum up briefly the facts thus far cited. The great towns are chiefly inhabited by working-people, since in the best case there is one bourgeois for two workers, often for three, here and there for four; these workers have no property whatsoever of their own, and live wholly upon wages, which usually go from hand to mouth. Society, composed wholly of atoms, does not trouble itself about them; leaves them to care for themselves and their families, yet supplies them no means of doing this in an efficient and permanent manner. Every working-man, even the best, is therefore constantly exposed to loss of work and food, that is, to death by starvation, and many perish in this way. The dwellings of the workers are everywhere badly planned, badly built, and kept in the worst condition, badly ventilated, damp, and unwholesome. The inhabitants are confined to the smallest possible space, and at least one family usually sleeps in each room. The interior arrangement of the dwellings is poverty-stricken in various degrees, down to the utter absence of even the most necessary furniture. The clothing of the workers, too, is generally scanty, and that of great multitudes is in rags. The food is, in general, bad; often almost unfit for use, and in many cases, at least at times, insufficient in quantity, so that, in extreme cases, death by starvation results. Thus the working-class of the great cities offers a graduated scale of conditions in life, in the best cases a temporarily endurable existence for hard work and good wages, good and endurable, that is, from the worker's standpoint; in the worst cases, bitter want, reaching even homelessness and death by starvation. The average is much nearer the worst case than the best. And this series does not fall into fixed classes, so that one can say, this fraction of the working-class is well off, has always been so, and remains so. If that is the case here and there, if single branches of work have in general an advantage over others, yet the condition of the workers in each branch is subject to such great fluctuations that a single working-man may be so placed as to pass through the whole range from comparative comfort to the extremest need, even to death by starvation, while almost every English working-man can tell a tale of marked changes of fortune.

"Evolution and Transformation: The American Industrial Metropolis, 1840–1940"

Sam Bass Warner

Editors' Introduction

Perhaps no place and time of urban history has been more exhaustively studied than the cities of the United States during the industrial transformation and after – that is, during the period from the mid-nineteenth century when industrial enterprises, which had originated in Europe, began to secure a foothold in North America to the period in the mid-twentieth century when the United States became a world power and its cities were regarded as the paradigm examples of advanced, "modern" urbanism. It was an extraordinary period of rapid transformation, both socially and technologically, and it is the subject of Sam Bass Warner's "Evolution and Transformation: The American Industrial Metropolis, 1840–1940," an essay specially prepared for this edition of *The City Reader*.

Warner summarizes the broad range of areas in which the very terms of urban life changed during the century after the onset of the Industrial Revolution. Among these were economic cycles of boom and bust culminating in the Great Depression of the 1930s, new types of power (from the muscle power of men and animals to water, steam, and electricity) and new types of transportation (from walking and horse-drawn wagons to railroads, interurban tramways, and the first appearance of those extraordinary new machines, the truck and the personal automobile). He notes that these new technologies and economic changes led the way toward new urban spatial arrangements – inner-city neighborhoods, suburbs, specialized industrial districts, and commercial downtowns with their department stores and soaring skyscrapers. He also notes the extraordinary, and often unexpected, social transformations that accompanied and intertwined with these other developments. Among these were the cultural accommodation of massive waves of foreign immigration, the growing power of labor unions, voting and employment rights for women, and the beginnings of the more gradual process of full enfranchisement for African Americans that began with emancipation after the Civil War and continued through decades of Jim Crow segregation, discrimination, and inequality.

Born and raised in Boston, Sam Bass Warner graduated from Harvard in 1950, took a master's degree in journalism from Boston University, and received his doctorate in history from Harvard in 1959. What followed was a career as a teacher and scholar that would include professorships at Washington University in St. Louis, the University of Michigan, Boston University, Brandeis University and, since 1994, the Department of Urban Studies and Planning at Massachusetts Institute of Technology. Along the way, he has been awarded fellowships at Harvard's Charles Warren Center for Studies in American History, the Guggenheim Foundation, and the Rockefeller Foundation. He has also served as a member of the Advisory Council of the United States National Archives, the National Research Council, and the Inter-University Consortium for Political Research and has been a member of the Executive Committee of the Organization of American Historians and President of the Urban History Association.

Warner's exceptionally distinguished career has been based on his skills as a teacher and mentor and on a series of books that place him in the first rank of historians of the city as a unique social, technological, and organizational complex central to the human experience. His first book, *Streetcar Suburbs: The Process of Growth in Boston, 1870–1900* (Cambridge, MA: Harvard University Press, 1962), broke new ground by radically expanding the common understanding of suburban growth away from a narrow post-World War II, automobile-based phenomenon toward a more historically accurate view of the role that streetcar systems played, as early as the 1870s, in decentralizing the dense central cities, creating new residential options for the urban middle class, and thereby contributing to increased class segregation in American society. *Streetcar Suburbs* made a strong plea for better, more socially conscious planning in urban America, and Warner's next book was an influential edited volume, *Planning for a Nation of Cities* (Cambridge, MA: MIT Press, 1966). That book was followed in 1968 by *The Private City: Philadelphia in Three Periods of its Growth* (Philadelphia, PA: University of Pennsylvania Press, 1968), a masterful study of both the successes and failures of private enterprise and private philanthropy as the primary forces behind nineteenth-century American urban development. Four years later, Warner published *The Urban Wilderness: A History of the American City* (Berkeley, CA: University of California Press, 1972), a sweeping overview that returned to his heartfelt themes: change as the one constant in the process of urban growth and the need for socially conscious planning. One reviewer called the book "a domestic policy brief," and urban historian Richard C. Wade wrote that *The Urban Wilderness* "is not a history of American cities, but rather a discussion of the need to transform them."

Other books by Sam Bass Warner include *The Way We Really Live: Social Change in Metropolitan Boston since 1920* (Boston, MA: Trustees of the Public Library of the City of Boston: 1977), *To Dwell is to Garden: A History of Boston's Community Gardens*, with Hansi Durlach (Boston, MA: Northeastern University Press, 1987), *Greater Boston: Adapting Regional Traditions to the Present* (Philadelphia, PA: University of Pennsylvania Press, 2001), and *Imaging the City: Continuing Struggles and New Directions*, with Lawrence J. Vale (New Brunswick, NJ: Center for Urban Research, Policy, 2001).

Appended to Warner's "Evolution and Transformation: The American Industrial Metropolis, 1840–1940," is an annotated list of the major sources that the author relied on in preparing this essay for *The City Reader*. Needless to say, the literature on nineteenth- and twentieth-century urban development is vast, and no brief list can do justice to its breadth and depth. For a world perspective, the best place to start is Lewis Mumford, *The City in History: Its Origins, its Transformations, and its Prospects* (New York: Harcourt Brace & World, 1961) and Peter Hall, *Cities in Civilization* (New York: Pantheon, 1998). Also of interest is Joel Kotkin's brief but stimulating *The City: A Global History* (New York: Random House, 2005). For the American perspective, consult Howard Chudacoff, Peter Baldwin, and Thomas Paterson, *Major Problems in American Urban and Suburban History*, 2nd edn (Belmont, CA: Wadsworth, 2004), Raymond Mohl, *The New City: Urban America in the Industrial Age, 1860–1920* (Wheeling, IL: Harlan Davidson, 1985), and David R. Goldfield (ed.), *Encyclopedia of American Urban History* (two vols, New York: Sage, 2006).

■

Never before in human history had people built a society based upon steam-powered machines. Friedrich Engels saw the beginnings of this process at its early stages. His was the moment when the gathering of mechanized factories established large cities. He could not have known in 1844 that as the process unfolded, all of England, and all modern nations, would come to be organized by huge industrial metropolises of more than a million inhabitants.

This essay will pick up the story of the American industrial metropolis in the years after 1840. It will follow its path of development in two stages: the years to 1920 and the years from 1920 to 1940. Just as Engels picked up the story of the merchant economy turning to an industrial one, so during the 1920s a wholly new urban society began to emerge.

A parade of surprises characterized the century. Basic changes in the energy available to the city, first steam and then electricity, transformed the form of the urban settlement and the life within it. How might anyone have imagined a city of lights, a downtown and the crowds on its streets, a city of skyscrapers and miles and miles of small houses? These obvious surprises merely record the surface of change because technology, social and cultural invention, politics, and historical inheritances interact in complicated ways. A water pipe and a light bulb and a steel building frame are but the tools of complex social and cultural invention. The railroad and the steamship made the American industrial metropolis into a crossroads for the young people of Europe. This abundance of youthful hands revived a system of garment manufacture as old as Shakespeare's tailor. Manufacturers gave out materials for families to sew and assemble in their own rooms. In this way international transportation fostered the tenement sweatshop, but the newcomers' unions gave the city new voices and fresh expectations. These same machines, the railroad and the steamship, connected and interconnected with all manner of business institutions of finance, manufacture and marketing to press down upon American society and its cities a completely unstable economy that alternated booms with dangerous economic collapse.

THE AMERICAN INDUSTRIAL CITY EMERGES

Commonly a focus on factories and machines leads a historian to neglect the essential role of animals and humans in carrying out industrialization. In fact, coal and biopower together fueled the growth of the modern American metropolis. The coal-fired steam engines of the railroads and steamships tied the city to the world's resources and markets, while coal-fired furnaces and engines enabled factories to escape their water-powered sites so that they might locate near the city's port or alongside the railroad lines. Later, subsequent to the 1890s, coal powered the generation of electricity for streetcars, subways, elevated railroads, electric lights, elevators, and numberless small motors. In the twentieth century, petroleum joined coal as the fuel for transportation.

Steam-powered machines have captured our historical imagination, but they all depended on biopower to make them work. The city of the nineteenth and early twentieth centuries depended on horses to pull the streetcars, wagons, and taxis, and to power many construction tasks. Further, both the horses and the machines depended upon the city's army of men, women and children to tend them. Even the few automatic machines of the time, like rotary presses, looms and screw machines, required constant attention. Most, like sewing machines, steam pressers, lathes, and saws, required a person to feed them. In addition, every object that moved in the city was first lifted, carried, or carted by a laborer. Indeed, historians have been surprised to notice that the increase in heavy manufactures – like steel and pipes and shipyards for iron and steel ships – brought with it an increase in the proportion of unskilled laborers as opposed to craftsmen.

Thus from 1840 to about 1920 during this major surge of industrialization, every newcomer to the city represented a gain, a fresh addition of a little biopower to the human settlement. Yet, until the twentieth century, the city quite literally consumed young people and children. It injured and killed young men and women workers with industrial accidents and pollutants, malnutrition, overcrowded housing and disease. It killed children faster than they could be born. No large nineteenth-century city in the world could sustain itself by the natural reproduction of its resident population.

Because large cities drew their human resources from national and international trading networks, changes at a distance impacted local conditions. The 1840s famine in Ireland and the appearance of crowded Irish slums and shanties in New York and Boston is such a well-known case. The change in urban survival also followed upon distance events.

Fresh farmland in the western United States and Europe, when tilled with new farm machinery and new agricultural practices, produced a flood of inexpensive grain, hay and meat. The refrigeration of meat and dairy products, the canning of meat, fruits and vegetables, and the pasteurization of milk allowed the nation's railroad network to fill the markets of the city with safe products. Here in the city a chain of very long-term incremental municipal investments in water and sewer systems and the introduction of public health inspection created an environment of plentiful and clean foodstuffs, even for the poor. To be sure, every man, woman and child dwelling in large American cities at any given moment during the twentieth century did not have enough to eat, but the level of starvation and malnutrition ceased to be a population control. Thereby, from the early decades of the twentieth century onwards, the large American city and its equivalents in modern Europe became habitable places for humans.

IMMIGRATION EXPERIENCES

Overseas immigrants to American cities followed the pathways of food, cotton and coal: steamships to and from Europe, railroads among cities. The immigrant parade itself depended upon the sequential destruction of local peasant economies by the railroad and its delivery of cheap grain and inexpensive manufactured goods to the villages and towns. First, modern agricultural methods and land enclosures destroyed the old rural economies of Great Britain, Ireland and Germany, sending these country folks to America and Britain's new factories. Then as the transportation network spread, Norwegians, Swedes, Danes, Poles, Hungarians, Italians, Russians and Greeks, and peoples of the Mediterranean followed.

The degree of survival of these immigrant peoples revealed itself only slowly in the cities' census statistics. Nevertheless, parallel and unconnected events accompanied this remarkable demographic change. By 1920 the progress of mechanization had proceeded so far that the addition of a new pair of willing hands and a strong back did not add to the city's wealth. The city continued to seek skilled workers, but the unskilled proved redundant. During this same post-World War I decade, fear of foreigners, fears of European socialist and communist ideas, anti-Semitism and labor unions' desire to cut back on competition among workers combined to pass federal legislation severely rationing overseas immigration. The rationing itself favored the early comers – Germans, Irish, and Britons. In consequence, the history of the metropolis from 1920 to 1940 differs markedly from its earlier experiences. During the 1930s, the major urban migrants were native American farmers driven off their farms by low prices and drought. Those already settled in the city subsequently spent many of the following decades learning to master their ethnic, religious, and racial prejudices.

In a new continent rich in untapped natural resources and prolific inventions, the returns to capital far exceeded those to labor. Consequently, extremes of wealth and sharp class divisions emerged as the century wore on. For those with capital or access to funds, hard work, luck, and leverage enabled business people to amass substantial profits from new resources, new inventions, new business methods, and the appreciation of land values in the rapidly growing city. The migrant, however, whether from overseas or an American farm, who arrived in the city without money had somehow to find a way to advance from unskilled labor jobs. Machine tending in a factory or construction work offered many a working-class income. Also, white native Americans often found jobs as clerks in offices or as sales people.

SOCIAL TRANSFORMATIONS

The women traditionally took up domestic service, and many immigrant girls worked in factories before marriage. During the 1840s and subsequently, married women often supported their families by leasing a house and running it as a boarding house. The housekeeper offered single rooms and two meals a day. Toward the end of the century, women found fresh opportunities selling in stores, nursing, elementary school teaching and staffing telephone switchboards. In all these roles, the customs and prejudices of men kept the women's wages well below those of the men.

Children helped out their families in home manufactures and in family stores. Many worked as office and errand boys. One future governor of New York, Al Smith, upon the death of his father, ran about Lower Manhattan carrying messages to teamsters telling them where their wagons should go for the next load.

Social connections have always been valuable, and the help of family, friends and neighbors proved the most reliable source of jobs. There were private employment offices, to be sure, but they were expensive, unreliable and often fraudulent. Once the job had been found, the fate of the worker depended upon the patronage of the owner and the tyranny of the foreman. There were few remedies for mistreatment, harassment, short wages, false clocks and extra hours.

THE ROLE OF LABOR

Since the eighteenth century, carpenters and masons had organized themselves into unions, and later, others – especially printers, shoe and textile workers – continued the union movement. At mid-century, however, many courts held strikes and workers' boycotts of offending firms to be criminal offenses. Yet despite the legal obstacles in boom years, the craft unions and working people's politics made progress. But when economic depressions set in and thousands of men and women became unemployed, organized labor and its programs collapsed.

So the movement to establish a ten-hour workday fell in the depression of the 1850s, and its revival was later squelched by the deep depression of 1873 that persisted in the form of low wages and prices until the mid-1890s. Yet, bit by bit, craft unions of skilled workers and some factory operatives made headway. Secure unions often could be built upon ethnic solidarity, like those of the German cigar makers and the Philadelphia English cotton workers.

Because factories grew ever larger, and because they located within and next to large cities, every American metropolis experienced massive labor actions. The railway workers strike of 1877 broke out in many cities, especially Pittsburgh where workers burned the railroad cars and fired rifles and a cannon at the troops sent to quell the strike. Later the Homestead strike in Pittsburgh (1892), the Pullman strike in Chicago (1894), the garment workers' strike in New York City (1910), and the silk workers in Patterson, New Jersey, (1913) proved major events dividing middle-class voters from their working-class fellow citizens. These strikes often turned violent because employers hired private police to attack the workers.

In 1914, the Clayton Anti-Trust Act established the legality of unions in the United States, but until business people and the general public regarded unions as a regular element in American business, labor relations remained an urban battleground. Frightened politicians built armories next to elite neighborhoods to protect them from the possible dangers of mobs of workers and established state police corps to keep order during strikes. The demands of some workers for socialism unleashed a long-standing media attack that persists to this day. It consciously muddles revolutionary socialism and anarchism with sensible social democratic reform proposals. Indeed, during the early twentieth century, "gas and water socialism" – the call for the municipal ownership of streetcar and utility monopolies – found acceptance in a number of cities.

THE PHYSICAL CITY AND THE NEW DOWNTOWN

A mix of transportation change and changes in the organization of work set the geography of the new metropolis. In the city of 1840, most people walked. Only steam ferries carried any volume of passengers. Cumbersome large coaches called omnibuses ran on the main streets but they were expensive, as were the horse-drawn taxis. Only the rich could afford to keep a private carriage. The new railroads of the 1840s and 1850s laid out what later would become important commuter lines, and the introduction of horse-drawn streetcars during these same years in time enabled the city to double its settlement radius.

Even before these transportation changes had taken effect, more business and new business began to create a new urban element, the modern downtown. The former merchant and warehouse area split into parts as it grew. The warehouse area expanded and began to attach itself to the new railroad yards. The former all-purpose merchant counting houses divided and subdivided into offices of wholesale and commission merchants, importers, commodity traders, bankers, insurance and real estate offices, stock brokers, lawyers and surveyors. From this multiplication arose a downtown office concentration of four- and five-story office buildings. Large hotels settled next this concentration, and dry goods merchants who began to expand their offerings for the well-to-do settled their stores on the downtown fringe. Nearby, there was often a street of inexpensive stores that catered to the downtown clerks and working class customers.

Altogether, this downtown base of the mid-nineteenth century grew and elaborated into a gathering of newspapers, hotels, skyscraper offices buildings, and downtown department stores: Macy's and Gimbel's in New York, Wannamaker's in Philadelphia, Jordan Marsh in Boston, Marshall Field's in Chicago. Only railroads, shipyards, foundries, coal yards and gas works required large spaces. Everything else fitted into the small lots along the city's streets. Whether old like Boston and New York, or new like Chicago, small shops, two- and three-family houses, boarding houses, factories, workrooms and livery stables mixed in together. This pepper and salt mix of work and residence created a city of multiethnic neighborhoods as immigrants and natives alike settled near their workplaces. A local mortgage market financed much of this mixed urban fabric. City residents with a little capital lent small sums for short terms of five to ten years through the agency of downtown brokers to builders and homeowners. By such a union of small savings, local builders and local brokers the beginning of the industrial metropolis was built.

THE SEGREGATED CITY

In these mid-nineteenth-century decades, only three groups of residents lived separated from everyone else: the poor (mostly new immigrants), free African

Americans, and the wealthy. These years may have been the meanest and nastiest time for housing for the poor. Landlords cut up old houses into single rooms, some in the basement, others without an outside window; a tap for cold running water appeared only in some buildings, and privies sat in the back yard. Where there had been an open yard, owners filled the rear spaces. Since these slums grew in the oldest sections of town, they stood near the port and in the shadow of the new downtown. Free African Americans settled along a few poor streets, poor because they were confined by white prejudice to low-paying servant, peddler, and wharf-side jobs. A few had managed to become doctors, lawyers and owners of small businesses, but they too were not welcome to live among whites. Successful old merchant families and the newly rich set themselves apart from the general run of mixed neighborhoods. Developers laid out small sections designed in the latest fashions to cater to these buyers: Boston's Beacon Hill and Back Bay, New York's Gramercy Park and Fifth Avenue, Chicago's Gold Coast, and San Francisco's Nob Hill.

From these partially differentiated beginnings, the functionally specialized and socially segregated modern industrial metropolis emerged. Changes in work and transportation again set the frame. Office work, retailing, business services and financial institutions multiplied and expanded to build the crowded downtown of large buildings and skyscrapers. The railroad lines out from the city center became industrial corridors for meat packing, railroad equipment building, automobile assembly, lumber yards, printing, piano factories, and shoe and textile mills. Some of the largest enterprises, like steel mills and oil refining, established satellite cities of their own at the outer edges of the metropolis: Gary, Indiana; Oakland, California; Quincy, Massachusetts; Newark and Bayonne, New Jersey.

The railroads that supported these concentrations also facilitated the expansion of the earlier elite settlements, while the high cost of rail commuting kept most residents out. Luxury suburbs extended outwards from their old inner bases: Boston's Brookline and western suburbs, Philadelphia's Main Line, New York's Long Island, and Chicago's North Shore communities.

Most residents in the late nineteenth century and the early decades of the twentieth depended initially on the horse car and then, after 1890, on its improvement, the electric streetcar. Boston, New York, Philadelphia, and Chicago also constructed subways and elevated railroads to deal with inner city crowding and to carry the downtown workers and shoppers to their homes. These public carriers allowed the industrial metropolis to spread outwards in a rough social geometry of poor and African Americans in the old inner sections of the city, the working class in multiple housing along the transit lines, and the middle class in single and double homes beyond. The residential hallmarks of the twentieth-century metropolis were such places as New York's Brooklyn and the Bronx, Philadelphia's West Philadelphia, Boston's Roxbury and Dorchester, Chicago's South Side, San Francisco's West Portal and Sunset districts, and the ever-widening towns and suburbs that were to become the vast metropolis of Greater Los Angeles.

YET ANOTHER STAGE OF URBAN EVOLUTION

Because the depression of the 1930s slowed the pace of change, from World War I to 1940 the American industrial metropolis continued along the pathways laid down earlier. Yet, aided by hindsight, it is possible to observe the beginnings of economic and social changes that would transform these urban regions once again. Just as it required half a century after 1840 for the industrial metropolis to realize its mature organization, so it took another fifty years for the metropolis of the 1920s to fully assume its later patterns of a controlled and government-supported market economy, the dispersed geography of the automobile and the open social forms of full citizenship for African Americans and women.

During the 1920s, new marketing techniques addressed the failings of industrial production. At the heart of the problem lay the manufacturers' ability to make more goods than the wages they paid would allow people to purchase. In consequence, the industrial economy fluctuated between full production at robust prices and over-supply, distress prices, lay-offs, unemployment and the inadequate aid of municipal soup kitchens and wood yards.

A partial solution lay in understandings among firms to keep prices uniform and steady while competing in marketing. Two steps needed to be taken. Wholesale distributors needed to be eliminated because they manipulated prices to their advantage when a glut occurred. Manufacturers, therefore, must undertake their own distribution. Second, manufacturers must set aside large budgets for advertising, packaging, and brand promotion. Nationally advertised and distributed products multiplied during these years: soft drinks, cigars and cigarettes, automobiles, radios, packaged flour and cereals. Investment bankers began to appreciate the possibilities of large-scale retailing and they invested in chains of stores like Woolworth's and A & P groceries. The banking firm, Lehman Brothers, even assembled a chain of downtown department stores. The city's downtown now attained a new level of fantasy with elegant store windows, advertising billboards and bright theater lights.

Purchases on time contracts completed the marketing revolution. With sales of goods "on time," the manufacturer and storekeeper literally give their wares to a customer in return for the patron's promise to pay for them sometime in the future. John Wannamaker introduced charge cards for his department store customers, others soon followed. Until General Motors began selling cars on time payments in 1919, automobiles had been bought with cash. Soon, alongside these retail innovations, consumer installment loan companies sprang up to lend small sums at high interest rates.

Urban housing underwent a similar transformation to planned marketing. Developers of large suburban properties and owners of expensive center city real estate both wanted security for their investments. In the suburbs, the enemies were the cheap house, store, gas station, bar or apartment house that might settle next to an area planned for medium to high priced homes. Downtown, fear took form of factory buildings impinging on a retail street, or a monster building, like the Equitable Life Insurance Company offices overpowering their neighbors. In consequence, these interests promoted zoning laws that controlled land according to categories of use like residential, industrial and retail, and also set forth some limits on building types. New York City assumed the innovator's role in 1915, and soon zoning spread from state to state urged on by federal encouragement. The uniformities of metropolitan suburbs have their origins in these investment planning and marketing strategies of the 1920s.

A post-World War I housing boom filled out suburbs that railway commuters and streetcar and subway and elevated riders had begun, but the automobile did not yet alter the shape of the metropolis. Suburbanization in the 1920s did draw many families of modest incomes outward, and in consequence the long process of reducing the density of inner-city neighborhoods began during these years. The popularity of the new machine and its rapid diffusion brought daily traffic jams to the downtown. Cities spent enormous sums to pave streets and make traffic improvements since in these years neither state nor federal funds were available for use on city roadways. By 1940, glimpses of the future automobile metropolis appeared. The immensely popular General Motors Futurama exhibit at the New York World's Fair in 1939 demonstrated a region of continuous automobile flow. Even then, some new highways had been built or were building: the Pasadena Freeway in Los Angeles, the Pennsylvania Turnpike, the Merritt Parkway in Connecticut, and the first miles of Boston's circumferential highway, Route 128.

THE MODERN URBAN ECONOMY

Neither the federal government nor the states restrained the banking system during the 1920s so that developers over-borrowed to build city apartment houses and suburban homes. Some developers even established small banks to furnish themselves with capital. During the boom that peaked in 1926, the local mortgage market became more and more one of banks and insurance companies and less and less one of local lenders and their brokers. In consequence, the overextension of mortgages and consumer credit on time purchases joined the uncontrolled international banking, stock and bond markets in a disastrous collapse. By contrast, during the 1970s the new metropolis was sustained in its building and prosperity by federal regulation of banks and mortgages. Also the federal Cold War budget financed extensive government purchases that in turn made up for any shortfall in consumers' ability to buy.

The American public and American business people have never been able to settle upon an institution that would counterbalance the power of employers. During World War I, in order to maintain full production, the federal government oversaw labor relations and even nationalized the railroads to see that they were managed efficiently. The experience offered some precedents that might have been adapted to peacetime. Instead, the Russian Revolution and fears of socialism fostered a violent attack on unions as they launched strikes in 1919–1922 to offset wartime price inflation. The union movement did not recover from these attacks until the 1930s and the massive industrial strikes in the automobile and steel industries. A reformed federal administration also established the rules for organizing and collective bargaining. There followed the World War II unionization of a large segment of workers that proved an integral element in the later reorganization of the metropolis. Because union wages served as a yardstick for all workers, large numbers of families were able to purchase small new homes in the automobile suburbs.

THE ONGOING SOCIAL STRUGGLE

For African Americans, the long road from slavery to full citizenship took an encouraging turn during the 1920s. The largest metropolises, New York and Chicago, had attracted many African American migrants north to meet their demands for more workers. In New York because of a local collapse in the real estate market, and in Chicago because of a fierce race riot in 1919, African Americans came to be confined to the concentrated ghettos of Harlem and the South Side. The communities proved large enough to support newspapers, theaters, and special African American stores and services. In New York, they also drew black talent from across the nation who began a literary and cultural movement now known as the Harlem Renaissance. Black writers captured few white readers, but the jazz musicians and song-writers found ever growing national audiences when they brought their music before white listeners. Finally, after years of being belittled, African American men and women came to be widely admired. The civil rights movement and the desegregation of the northern metropolises would be years in the future, but the "Jazz Age" marked the beginnings of respect.

A romantic haze now obscures the image of another 1920s figure, "the flapper," a then-novel young woman with short hair and short skirts. Her dancing and partying made her a volunteer in the culture wars of this Prohibition era. The attempt to halt the sale of alcoholic beverages in 1920, the Volstead Act, was part of a backlash that fed upon the fears of rural Americans and their city cousins. The pace of material change frightened many, and the waves of new ideas and immigrants frightened others. Like its companion legislation to ration foreign immigration, Prohibition was an exercise in social control.

In contrast to "flappers," working-class women had long shocked public sensibilities with labor activism. Even the Lowell cotton spinning and weaving girls, subjects of a much-admired paternalistic management, went out on strike in 1834 and 1836. Also, middle-class and wealthy women living in the American metropolises had a long and distinguished history of charity and reform. They established settlement houses, informal schools and aid stations, in the poorest sections of town. They supported housing and factory investigations and child labor laws, and promoted coal smoke abatement. Schools, playgrounds and gardens were also central concerns. Some women had worked for Prohibition, others for the nineteenth Amendment to the Constitution permitting women to vote, and still others campaigned against the United States' entry into World War I. When the right to vote arrived in August 1919, it could have been regarded as a recognition of women's leadership. The reform initiatives continued with the establishment of the League of Women Voters, who worked for informed and open politics. In the suburbs, where government units were small and the politics personal, women had a strong effect on behalf of the new ideas of zoning and land planning as well as support for taxes to fund public schools. If you happen to visit an especially attractive suburb nowadays, chances are that some of its charm stems from the work of women in the town.

Yet the flapper's cheeky behavior and the right to vote proved but small beginnings. The flapper was essentially a girl, not a woman, and during the 1920s and many years thereafter, American women were to be housewives and mothers. If they held jobs, their wages were to be well below those of men, and their occupations restricted to a short list. Only with the shift of the urban economy toward the multiplication of white-collar work and the development of easy forms of birth control did women move forward to take a position as full citizens.

AN ERA OF CHANGE

Overall, during the century from 1840 to 1940, the industrial metropolis fostered patterns that never could have been anticipated. Settlements of such magnitude had never existed in the extended fashion afforded by the railroad and the streetcar. Never had such a large city been a fit place for its human population. Almost all the technology and almost all the business methods were new, and in the American case the politics rested on inventions of the late eighteenth century. In fact the factories and mean streets and slum houses of the mid-nineteenth century gave few clues for what lay ahead.

Although automobiles sold by the millions before 1940, it was only with the building of the interstate highway system that some of the possibilities of an automobile-dominated metropolis began to reveal themselves. Once again the machine and its roadways disguised the complex processes of economic, social and cultural adaptation: the destruction of the inherited downtowns, the collapse of urban public education, the heightened segregation of race and class, the Civil Rights and Feminist movements, the isolation of families and the rush of women into the workplace.

SUGGESTED FURTHER READINGS

For a study of the parallel changes in two surprise cities, see Harold L. Platt, *Shock Cities: The Environmental Transformation and Reform of Manchester and Chicago* (Chicago, IL: University of Chicago Press, 2005).

A wonderful novel about the Russian Jewish immigrants' move into the garment industry in New York City is Abraham Cahan, *The Rise of David Levinsky* (1917; New York: Harper & Row, 1960).

On biopower and the use of animals, see Clay McShane and Joel A. Tarr, *The Horse and the City* (Baltimore, MD: Johns Hopkins University Press, 2007).

For a detailed portrait of the ways of a fully industrialized city at the turn of the nineteenth century with excellent illustrations, two volumes are especially relevant: Paul Underwood Kellogg, *Pittsburgh Survey*, six vols. (New York: Charities Publication Committee, 1909–1914): Crystal Eastman, *Work-Accidents and the Law*, vol. 2; Margaret F. Byington, *Homestead: The Households of a Mill Town*, vol. 4.

The full report on Fogel's urban demographic research is in Robert William Fogel, *The Conquest of High Mortality and Hunger in Europe and America, NBER Working Paper Series on Historical Factors in Long Run Rates of Growth, Historical Paper 16* (Cambridge, MA: National Bureau of Economic Research, 1990); a short summary appears in Robert William Fogel, *The Escape from Hunger and Premature Death, 1700–2100: Europe, America and the Third World* (New York: Cambridge University Press, 2004).

An account by a woman who went to the villages and saw the rural destruction process in action is Emily Greened Balch, *Our Slavic Fellow Citizens* (1910; reprint, New York: Arno Press, 1969).

On the role of boarding houses, see Thomas Butler Gunn, *The Physiology of New York Boarding-Houses* (1857; reprint, New Brunswick, NJ: Rutgers University Press, 2009).

On the role of labor, see David Brody, *In Labor's Cause: Main Themes on the History of the American Worker* (New York: Oxford University Press, 1993). For the impact of labor strife on the urban landscape, see Robert M. Fogelson, *America's Armories: Architecture, Society and Public Order* (Cambridge, MA: Harvard University Press, 1989).

On the new downtowns, see Mona Domosh, *Invented Cities: The Creation of Landscape in Nineteenth Century New York and Boston* (New Haven, CT: Yale University Press, 1996), and Robert M. Fogelson, *Downtown: Its Rise and Fall 1880–1950* (New Haven, CT: Yale University Press, 2001).

Kenneth A. Scherzer provides a very careful and convincing qualitative study of urban settlement patterns. The reader will be helped by accompanying it with some picture histories in his *The Unbounded Community: Neighborhood Life and Social Structure in New York City, 1830–1875* (Durham, NC: Duke University Press, 1992); also see John A. Kouwenhoven, *The Columbia Historical Portrait of New York* (Garden City, NY: Doubleday, 1953) and Harold M. Mayer and Richard C. Wade, *Chicago: Growth of a Metropolis* (Chicago, IL: University of Chicago Press, 1969).

On the methods whereby the nineteenth-century metropolis was built, see Elizabeth Blackmar, *Manhattan for Rent 1785–1850* (Ithaca, NY: Cornell University Press, 1989) and Sam Bass Warner, *Streetcar Suburbs* (Cambridge, MA: Harvard University Press, 1962).

For descriptions of mid-nineteenth-century slums, see John H. Griscom, *The Sanitary Condition of the Laboring Population of New York with Suggestions for its Improvement* (1848; reprint, New York: Arno Press, 1970) and Oscar Handlin, *Boston's Immigrants*, 2nd edn (Cambridge, MA: Harvard University Press, 1967).

On the African American Big City Experience, see Roger Lane, *William Dorsey's Philadelphia and Ours* (New York: Oxford University Press, 1991).

On metropolitan growth patterns by sectors and with satellites, see Homer Hoyt, *One Hundred Years of Land Values in Chicago* (Chicago, IL: University of Chicago Press, 1933) and Graham R. Taylor, *Satellite Cities: A Study of Industrial Suburbs* (1915; reprint, New York: Arno Press, 1970). Also see Robert M. Fogelson, *The Fragmented Metropolis: Los Angeles, 1850–1930* (Cambridge, MA: Harvard University Press, 1967).

On the first appearances of the new economy, see Eric E. Lampard, "Introductory Essay," and William Leach, "Brokers and the New Corporate Industrial Order," both in William R. Taylor (ed.), *Inventing Times Square* (New York: Russell Sage Foundation, 1991), pp. 16–35, 99–117.

On the new housing patterns, see Marc A. Weiss, *The Rise of the Community Builders: The American Real Estate Industry and Urban Land Planning* (New York: Columbia University Press, 1987).

On the emptying out of dense inner city areas, see Clay McShane, *Down the Asphalt Path: The Automobile and the American City* (New York: Columbia University Press, 1994) and Sam Bass Warner, *Greater Boston* (Philadelphia, PA: University of Pennsylvania Press, 2001).

On the process of banking and real estate collapse, see Miles Colean, *American Housing, Problems and Perspectives* (New York: Twentieth Century Fund, 1944).

On labor, see David Montgomery, *The Fall of the House of Labor* (New York: Cambridge University Press, 1987) and David Brody, *Workers in Industrial America: Essays on the 20th Century Struggle*, 2nd edn (New York: Oxford University Press, 1993).

On African Americans, see James Weldon Johnson, *Black Manhattan* (1930; reprint, New York: Athenaeum, 1968) and George Hutchinson (ed.), *The Cambridge Companion to the Harlem Renaissance* (New York: Cambridge University Press, 2007).

"The Drive-in Culture of Contemporary America"

from *Crabgrass Frontier: The Suburbanization of the United States* (1985)

Kenneth T. Jackson

Editors' Introduction

As Friedrich Engels (p. 46), Sam Bass Warner (p. 55), Robert Bruegmann (p. 211), Robert Fishman (p. 75), and others have shown, suburbia has a long history, extending back at least as far as the European and American railway suburbs that arose as retreats from the polluted industrial cities for the comfortable middle class. But suburbanization took on new form and historical significance in the 1920s and in the years following World War II. The initial locus was the United States, and the catalyzing technology was the automobile. In *Crabgrass Frontier*, Kenneth T. Jackson, sometimes called the dean of American urban historians, provides a sweeping overview of the "suburban revolution" in the United States. In the chapter entitled "The Drive-in Culture of Contemporary America," he lays out a devastating critique of the mostly negative social and cultural effects that the private automobile has had on urban society.

Kenneth Jackson did not originate the critique of suburbia. Indeed, the suburban developments of the 1940s, 1950s, and 1960s in North America and elsewhere gave birth to a massive literature, most of it highly critical. Damned as culturally dead and socially and racially segregated, the post-World War II suburbs were called "sprawl" and stigmatized as "anti-cities" (to use Lewis Mumford's term to describe Los Angeles). Titles such as John Keats's *The Crack in the Picture Window* (1956), Richard Gordon's *The Split-level Trap* (1961), Mark Baldassare's *Trouble in Paradise* (1986), Robert Fogelson's *Bourgeois Nightmares* (2005), David Goetz's *Death by Suburb: How to Keep the Suburb from Killing Your Soul* (2007), and Saralee Rosenberg's *Dear Neighbor, Drop Dead* (2008) capture the tone of much of the commentary. Indeed, James Howard Kunstler in *The Geography of Nowhere* (1993) calls the automobile suburbs "the evil empire," Joel S. Hirschhorn titles his analysis *Sprawl Kills* (2005), and another radical analysis screams *Bomb the Suburbs* (2001)! Nevertheless, Jackson's well-documented analysis of the artifacts of suburban culture – everything from three-car garages to drive-in churches – stands as the definitive statement on how the automobile transformed both the structure and social life of modern cities.

Kenneth T. Jackson (b. 1939) is the Jacques Barzun Professor of History and the Social Sciences at Columbia University. He earned his PhD at the University of Chicago and is a past president of the Urban History Association and the Organization of American Historians. He is the editor of the *Encyclopedia of New York City*, 2nd edn (New Haven, CT: Yale University Press, 2010) and the author of several influential books including *The Ku Klux Klan in the City, 1915–1930* (Chicago, IL: I. R. Dee, 1992) and *Cities in American History*, with Stanley Schultz (New York: Knopf, 1972). Jackson has been called an "urban pessimist" because of his dark view of suburbanization – a current work in progress on American transportation policy is entitled *The Road to Hell* – but in a recent interview he noted that he now sees "a ray of hope" for cities "after a long decline."

Classic works on suburbia include Herbert Gans, *The Levittowners* (New York: Pantheon, 1967), remarkable for its overall positive view of middle-class tract-home life, Sam Bass Warner, *Streetcar Suburbs: The Process of Growth in Boston, 1870–1900* (Cambridge, MA: Harvard University Press, 1962), and Robert Fishman, *Bourgeois Utopias: The Rise and Fall of Suburbia* (New York: Basic Books, 1987). For a representative collection of analytical essays on the subject, see Becky Nicolaides and Andrew Wieze (eds), *The Suburb Reader* (London: Routledge, 2006).

Other studies of suburbia include J. Eric Oliver, *Democracy in Suburbia* (Princeton, NJ: Princeton University Press, 2001), Mark Salzman, *Lost in Place: Growing Up Absurd in Suburbia* (New York: Vintage, 1995), Valerie C. Johnson, *Black Power in the Suburbs* (New York: SUNY Press, 2002), Becky Nicolaides, *My Blue Heaven: Life and Politics in the Working-class Suburbs of Los Angeles* (Chicago, IL: University of Chicago Press, 2002), Adam Rome, *The Bulldozer and the Countryside: Suburban Sprawl and the Rise of American Environmentalism* (Cambridge: Cambridge University Press, 2001), Oliver Gillham, *The Limitless City: A Primer on the Urban Sprawl Debate* (Washington, DC: Island Press, 2002), and Dolores Hayden, *Building Suburbia: Green Fields and Urban Growth, 1820–2000* (New York: Pantheon, 2003). For a different view – spirited defenses of "sprawl" as an age-old natural process of urban expansion – see Robert Bruegmann, *Sprawl: A Compact History* (Chicago, IL: University of Chicago Press, 2005) (p. 211) and Paul Barker, *The Freedoms of Suburbia* (London: Frances Lincoln, 2009).

■

The postwar years brought unprecedented prosperity to the United States, as color televisions, stereo systems, frost-free freezers, electric blenders, and automatic garbage disposals became basic equipment in the middle-class American home. But the best symbol of individual success and identity was a sleek, air-conditioned, high-powered, personal statement on wheels. Between 1950 and 1980, when the American population increased by 50 percent, the number of their automobiles increased by 200 percent. In high school the most important rite of passage came to be the earning of a driver's license and the freedom to press an accelerator to the floor. Educational administrators across the country had to make parking space for hundreds of student vehicles. A car became one's identity, and the important question was: "What does he drive?" Not only teenagers, but also millions of older persons literally defined themselves in terms of the number, cost, style, and horsepower of their vehicles. "Escape," thinks a character in a novel by Joyce Carol Oates. "As long as he had his own car he was an American and could not die."

Unfortunately, Americans did die, often behind the wheel. On September 9, 1899, as he was stepping off a streetcar at 74th Street and Central Park West in New York, Henry H. Bliss was struck and killed by a motor vehicle, thus becoming the first fatality in the long war between flesh and steel. Thereafter, the carnage increased almost annually until Americans were sustaining about 50,000 traffic deaths and about 2 million nonfatal injuries per year. Automobility proved to be far more deadly than war for the United States. It was as if a Pearl Harbor attack took place on the highways every two weeks, with crashes becoming so commonplace that an entire industry sprang up to provide medical, legal, and insurance services for the victims.

The environmental cost was almost as high as the human toll. In 1984 the 159 million cars, trucks, and buses on the nation's roads were guzzling millions of barrels of oil every day, causing traffic jams that shattered nerves and clogged the cities they were supposed to open up and turning much of the countryside to pavement. Not surprisingly, when gasoline shortages created long lines at the pumps in 1974 and 1979, behavioral scientists noted that many people experienced anger, depression, frustration, and insecurity, as well as a formidable sense of loss.

Such reactions were possible because the automobile and the suburb have combined to create a drive-in culture that is part of the daily experience of most Americans . . . Moreover, the American people have proven to be no more prone to motor vehicle purchases than the citizens of other lands. After World War II, the Europeans and the Japanese began to catch up, and by 1980 both had achieved the same level of automobile ownership that the United States had reached in 1950. In automotive technology, American dominance slipped away in the postwar years as German, Swedish, and Japanese engineers pioneered the development of diesel engines, front-wheel drives, disc brakes, fuel-injection, and rotary engines.

Although it is not accurate to speak of a uniquely American love affair with the automobile, and although John B. Rae claimed too much when he wrote in 1971 that "modern suburbia is a creature of the automobile and could not exist without it," the motor vehicle has fundamentally restructured the pattern of everyday life in the United States. As a young man, Lewis Mumford advised his countrymen to "forget the damned motor car and build cities for lovers and friends." As it was, of course, the nation followed a different pattern. Writing in the *American Builder* in 1929, the critic Willard Morgan noted that the building of drive-in structures to serve a motor-driven population had ushered in "a completely new architectural form."

THE INTERSTATE HIGHWAY

The most popular exhibit at the New York World's Fair in 1939 was General Motors' "Futurama." Looking twenty-five years ahead, it offered a "magic Aladdin-like flight through time and space." Fair-goers stood in hour-long lines, waiting to travel on a moving side-walk above a huge model created by designer Norman Bel Geddes. Miniature superhighways with 50,000 automated cars wove past model farms en route to model cities . . . The message of "Futurama" was as impressive as its millions of model parts: "The job of building the future is one which will demand our best energies, our most fruitful imagination; and that with it will come greater opportunities for all."

The promise of a national system of impressive roadways attracted a diverse group of lobbyists, including the Automobile Manufacturers Association, state-highway administrators, motor-bus operators, the American Trucking Association, and even the American Parking Association – for the more cars on the road, the more cars would be parked at the end of the journey. Truck companies, for example, promoted legislation to spend state gasoline taxes on highways, rather than on schools, hospitals, welfare, or public transit. In 1943 these groups came together as the American Road Builders Association, with General Motors as the largest contributor, to form a lobbying enterprise second only to that of the munitions industry. By the mid-1950s, it had become one of the most broad-based of all pressure groups, consisting of the oil, rubber, asphalt, and construction industries; the car dealers and renters; the trucking and bus concerns; the banks and advertising agencies that depended upon the companies involved; and the labor unions. On the local level, professional real estate groups and home-builders' associations joined the movement in the hope that highways would cause a spurt in housing turnover and a jump in prices. They envisaged no mere widening of existing roads, but the creation of an entirely new superhighway system and the initiation of the largest peacetime construction project in history.

[. . .]

Sensitive to mounting political pressure, President Dwight Eisenhower appointed a committee in 1954 to "study" the nation's highway requirements. Its conclusions were foregone, in part because the chairman was Lucius D. Clay, a member of the board of directors of General Motors. The committee considered no alternative to a massive highway system, and it suggested a major redirection of national policy to benefit the car and the truck. The Interstate Highway Act became law in 1956, when the Congress provided for a 41,000-mile (eventually expanded to a 42,500-mile) system, with the federal government paying 90 percent of the cost. President Eisenhower gave four reasons for signing the measure: current highways were unsafe; cars too often became snarled in traffic jams; poor roads saddled business with high costs for transportation; and modern highways were needed because "in case of atomic attack on our key cities, the road net must permit quick evacuation of target areas." Not a single word was said about the impact of highways on cities and suburbs, although the concrete thoroughfares and the thirty-five-ton tractor-trailers which used them encouraged the continued outward movement of industries toward the beltways and interchanges. Moreover, the interstate system helped continue the downward spiral of public transportation and virtually guaranteed that future urban growth would perpetuate a centerless sprawl . . .

[. . .]

The inevitable result of the bias in American transport funding, a bias that existed for a generation before the Interstate Highway program was initiated, is that the United States now has the world's best road system and very nearly its worst public transit offerings. Los Angeles, in particular, provides the nation's most dramatic example of urban sprawl tailored to the mobility of the automobile. Its vast, amorphous conglomeration of housing tracts, shopping centers, industrial parks, freeways, and independent towns blend into each other in a seamless fabric of concrete and asphalt, and nothing over the years has succeeded in gluing this automobile-oriented civilization into any kind of cohesion – save that of individual routine. Los Angeles's basic shape comes from three factors, all of which long preceded the freeway system. The first was cheap land (in the 1920s rather than 1970s) and the desire for single-family houses. In 1950, for example, nearly two-thirds of all the dwelling units in the Los Angeles area were fully detached, a much higher percentage than in Chicago (28 percent), New York City (20 percent), or Philadelphia (15 percent), and its residential density was the lowest of major cities. The second was the dispersed-location of its oil fields and refineries, which led to the creation of industrial suburbs like Whittier and Fullerton and of residential suburbs like La Habra, which housed oil workers and their families. The third was its once excellent mass transit system, which at its peak included more than 1,100 miles of track and constituted the largest electric interurban railway in the world.

[. . .]

THE GARAGE

The drive-in structure that is closest to the hearts, bodies, and cars of the American family is the garage. It is the link between the home and the outside world. The word is French, meaning storage space, but its transformation into a multipurpose enclosure internally integrated with the dwelling is distinctively American.

[. .]

After World War I, house plans of the expensive variety began to include garages, and by the mid-1920s driveways were commonplace and garages had become important selling points. The popular 1928 *Home Builders* pattern book offered designs for fifty garages in wood, Tudor, and brick varieties. In affluent sections, such large and efficiently planned structures included housing above for the family chauffeur. In less pretentious neighborhoods, the small, single-purpose garages were scarcely larger than the vehicles themselves . . . Although there was a tendency to move garages closer to the house, they typically remained at the rear of the property before 1925, often with access via an alley which ran parallel to the street. The car was still thought of as something similar to a horse – dependable and important, but not something that one needed to be close to in the evening.

By 1935, however, the garage was beginning to merge into the house itself, and in 1937 the *Architectural Record* noted that "the garage has become a very essential part of the residence." The tendency accelerated after World War II, as alleys went the way of the horse-drawn wagon, as property widths more often exceeded fifty feet, and as the car became not only a status symbol, but almost a member of the family, to be cared for and sheltered. The introduction of a canopied and unenclosed structure called a "car port" represented an inexpensive solution to the problem, particularly in mild climates, but in the 1950s the enclosed garage was back in favor and a necessity even in a tract house. Easy access to the automobile became a key aspect of residential design, and not only for the well-to-do. By the 1960s garages often occupied about 400 square feet (about one-third that of the house itself) and usually contained space for two automobiles and a variety of lawn and woodworking tools. Offering direct access to the house (a conveniently placed door usually led directly into the kitchen), the garage had become an integrated part of the dwelling, and it dominated the front facades of new houses. In California garages and driveways were often so prominent that the house could almost be described as accessory to the garage. Few people, however, went to the extremes common in England, where the automobile was often so precious that living rooms were often converted to garages.

THE MOTEL

As the United States became a rubber-tire civilization, a new kind of roadside architecture was created to convey an instantly recognizable image to the fast-moving traveler. Criticized as tasteless, cheap, forgettable, and flimsy by most commentators, drive-in structures did attract the attention of some talented architects, most notably Los Angeles's Richard Neutra. For him, the automobile symbolized modernity, and its design paralleled his own ideals of precision and efficiency. This correlation between the structure and the car began to be celebrated in the late 1960s and 1970s when architects Robert Venturi, Denise Scott Brown, and Steven Izenour developed such concepts as "architecture as symbol" and the "architecture of communication." Their book, *Learning From Las Vegas*, was instrumental in encouraging a shift in taste from general condemnation to appreciation of the commercial strip and especially of the huge and garish signs which were easily recognized by passing motorists.

A ubiquitous example of the drive-in culture is the motel. In the middle of the nineteenth century, every city, every county seat, every aspiring mining town, every wide place in the road with aspirations to larger size, had to have a hotel. Whether such structures were grand palaces on the order of Boston's Tremont House or New York's Fifth Avenue Hotel, or whether they were jerry-built shacks, they were typically located at the center of the business district, at the focal point of community activities. To a considerable extent, the hotel was the place for informal social interaction and business, and the very heart and soul of the city.

Between 1910 and 1920, however, increasing numbers of traveling motorists created a market for overnight accommodation along the highways. The first tourists simply camped wherever they chose along the road. By 1924, several thousand municipal campgrounds were opened which offered cold water spigots and outdoor privies. Next came the "cabin camps," which consisted of tiny, white clapboard cottages arranged in a semicircle and often set in a grove of trees. Initially called "tourist courts," these establishments were cheap, convenient, and informal, and by 1926 there were an estimated two thousand of them, mostly in the West and in Florida.

[. . .]

It was not until 1952 that Kemmons Wilson and Wallace E. Johnson opened their first "Holiday Inn" on Summer Avenue in Memphis. But long before that, in 1926, a San Luis Obispo, California, proprietor had coined a new word, "motel," to describe an establishment that allowed a guest to park his car just outside his room . . .

Motels began to thrive after World War II, when the typical establishment was larger and more expensive than the earlier cabins. Major chains set standards for prices, services, and respectability that the traveling public could depend on. As early as 1948, there were 26,000 self-styled motels in the United States. Hard-won respectability attracted more middle-class families, and by 1960 there were 60,000 such places, a figure that doubled again by 1972. By that time an old hotel was closing somewhere in downtown America every thirty hours. And somewhere in suburban America, a plastic and glass Shangri-La was rising to take its place.

[. . .]

THE DRIVE-IN THEATER

The downtown movie theaters and old vaudeville houses faced a similar challenge from the automobile. In 1933 Richard M. Hollinshead set up a 16-mm projector in front of his garage in Riverton, New Jersey, and then settled down to watch a movie. Recognizing a nation addicted to the motorcar when he saw one, Hollinshead and Willis Smith opened the world's first drive-in movie in a forty-car parking lot in Camden on June 6, 1933. Hollinshead profited only slightly from his brainchild, however, because in 1938 the United States Supreme Court refused to hear his appeal against Loew's Theaters, thus accepting the argument that the drive-in movie was not a patentable item. The idea never caught on in Europe, but by 1958 more than four thousand outdoor screens dotted the American landscape. Because drive-ins offered bargain-basement prices and double or triple bills, the theaters tended to favor movies that were either second-run or second-rate. Horror films and teenage romance were the order of the night . . . Pundits often commented that there was a better show in the cars than on the screen.

In the 1960s and 1970s the drive-in movie began to slip in popularity. Rising fuel costs and a season that lasted only six months contributed to the problem, but skyrocketing land values were the main factor. When drive-ins were originally opened, they were typically out in the hinterlands. When subdivisions and shopping malls came closer, the drive-ins could not match the potential returns from other forms of investments. According to the National Association of Theater Owners, only 2,935 open-air theaters still operated in the United States in 1983, even though the total number of commercial movie screens in the nation, 18,772, was at a thirty-five-year high. The increase picked up not by the downtown and the neighborhood theaters, but by new multiscreen cinemas in shopping centers. Realizing that the large parking lots of indoor malls were relatively empty in the evening, shopping center moguls came to regard theaters as an important part of a successful retailing mix.

THE GASOLINE SERVICE STATION

The purchase of gasoline in the United States has thus far passed through five distinct epochs. The first stage was clearly the worst for the motorist, who had to buy fuel by the bucketful at a livery stable, repair shop, or dry goods store. Occasionally, vendors sold gasoline from small tank cars which they pushed up and down the streets. In any event, the automobile owner had to pour gasoline from a bucket through a funnel into his tank. The entire procedure was inefficient, smelly, wasteful, and occasionally dangerous.

The second stage began about 1905, when C.H. Laessig of St. Louis equipped a hot-water heater with a glass gauge and a garden hose and turned the whole thing on its end. With this simple maneuver, he invented an easy way to transfer gasoline from a storage tank to an automobile without using a bucket. Later in the same year, Sylvanus F. Bowser invented a gasoline pump which automatically measured the outflow. The entire assembly was labeled a "filling station." At this stage, which lasted until about 1920, such an apparatus consisted of a single pump outside a retail store which was primarily engaged in other businesses and which provided precious few services for the motorist . . .

Between 1920 and 1950, service stations entered into a third phase and became, as a group, one of the most widespread kinds of commercial buildings in the United States. Providing under one roof all the functions of gasoline distribution and normal automotive maintenance, these full-service structures were often built in the form of little colonial houses, Greek temples, Chinese pagodas, and Art Deco palaces. Many were local landmarks and a source of community pride . . .

After 1935 the gasoline station evolved again, this time into a more homogeneous entity that was standardized across the entire country and that reflected the mass-marketing techniques of billion-dollar oil companies. Some of the more familiar designs were innovative or memorable, such as the drum-like Mobile station by New York architect Frederick Frost, which featured a dramatically curving facade while conveying the corporate identity. Another popular service station style was the Texaco design of Walter Dorwin Teaguea, smooth white exterior with elegant trim and the familiar red star and bold red lettering. Whatever the product or design, the stations tended to be operated by a single entrepreneur and represented an important part of small business in American life.

The fifth stage of gasoline-station development began in the 1970s, with the slow demise of the traditional service-station businessman. New gasoline outlets were of two types. The first was the super station, often owned and operated by the oil companies themselves. Most featured a combination of self-service and full-service pumping consoles, as well as fully equipped "car care centers." Service areas were separated from the pumping sections so that the two functions would not interfere with each other. Mechanics never broke off work to sell gas.

The more pervasive second type might be termed the "mini-mart station." The operators of such establishments have now gone full circle since the early twentieth century. Typically, they know nothing about automobiles and expect the customers themselves to pump the gasoline. Thus, "the man who wears the star" has given way to the teenager who sells six-packs, bags of ice, and pre-prepared sandwiches.

THE SHOPPING CENTER

Large-scale retailing, long associated with central business districts, began moving away from the urban cores between the world wars. The first experiments to capture the growing suburban retail markets were made by major department stores in New York and Chicago in the 1920s . . .

Another threat to the primacy of the central business district was the "string street" or "shopping strip," which emerged in the 1920s and which was designed to serve vehicular rather than pedestrian traffic. These bypass roads encouraged city dwellers with cars to patronize businesses on the outskirts of town. Short parades of shops could already have been found near the streetcar and rapid transit stops, but, as has been noted, these new retailing thoroughfares generally radiated out from the city business district toward low-density, residential areas, functionally dominating the urban street system. They were the prototypes for the familiar highway strips of the 1980s which stretch far into the countryside.

[. . .]

The concept of the enclosed, climate-controlled mall, first introduced at the Southdale Shopping Center near Minneapolis in 1956, added to the suburban advantage . . .

During the 1970s, a new phenomenon – the super regional mall – added a more elaborate twist to suburban shopping. Prototypical of the new breed was Tyson's Corner, on the Washington Beltway in Fairfax County, Virginia. Anchored by Bloomingdale's, it did over $165 million in business in 1983 and provided employment to more than 14,000 persons. Even larger was Long Island's Roosevelt Field, a 180-store, 2.2 million square foot megamall that attracted 275,000 visitors a week and did $230 million in business in 1980. Most elaborate of all was Houston's Galleria, a world-famed setting for 240 prestigious boutiques, a quartet of cinemas, 26 restaurants, an Olympic-sized ice-skating pavilion, and two luxury hotels. There were few windows in these mausoleums of merchandising, and clocks were rarely seen – just as in gambling casinos.

Boosters of such megamalls argue that they are taking the place of the old central business districts and becoming the identifiable collecting points for the rootless families of the newer areas. As weekend and afternoon attractions, they have a special lure for teenagers, who often go there on shopping dates or to see the opposite sex. As one official noted in 1971: "These malls are now their street corners. The new shopping centers have killed the little merchant, closed most movies, and are now supplanting the older shopping centers in the suburbs." They are also especially attractive to mothers with young children and to the elderly, many of whom visit regularly to get out of the house without having to worry about crime or inclement weather.

[. . .]

THE HOUSE TRAILER AND MOBILE HOME

The phenomenon of a nation on wheels is perhaps best symbolized by the uniquely American development of the mobile home. "Trailers are here to stay," predicted the writer Howard O'Brien in 1936. Although in its infancy at that time, the mobile-home industry has flourished in the United States. The house trailer itself came into existence in the teens of this century as an individually designed variation on a truck or a car, and it began to be produced commercially in the 1920s. Originally, trailers were designed to travel, and they were used primarily for vacation purposes. During the Great Depression of the 1930s, however, many people, especially salesmen, entertainers, construction workers, and farm laborers, were forced into a nomadic way of life as they searched for work, any work. They found that these temporary trailers on rubber tires provided the necessary shelter while also meeting their economic and migratory requirements. Meanwhile, Wally Byam and other designers were streamlining the mobile home into the classic tear-drop form made famous by Airstream.

During World War II, the United States government got into the act by purchasing tens of thousands of trailers for war workers and by forbidding their sale to the general public. By 1943 the National Housing Agency alone owned 35,000 of the aluminum boxes, and more than 60 percent of the nation's 200,000 mobile homes were in defense areas . . .

Not until the mid-1950s did the term "mobile home" begin to refer to a place where respectable people could marry, mature, and die. By then it was less a "mobile" than a "manufactured" home. No longer a trailer, it became a modern industrialized residence with almost all the accoutrements of a normal house. By the late 1950s, widths were increased to ten feet, the Federal Housing Administration (FHA) began to recognize the mobile home as a type of housing suitable for mortgage insurance, and the maturities on sales contracts were increased from three to five years.

In the 1960s, twelve-foot widths were introduced, and then fourteen, and manufacturers began to add fireplaces, skylights, and cathedral ceilings. In 1967 two trailers were attached side by side to form the first "double wide." These new dimensions allowed for a greater variety of room arrangement and became particularly attractive to retired persons with fixed incomes. They also made the homes less mobile. By 1979 even the single-width "trailer" could be seventeen feet wide (by about sixty feet long), and according to the Manufactured Housing Institute, fewer than 2 percent were ever being moved from their original site. Partly as a result of this increasing permanence, individual communities and the courts began to define the structures as real property and thus subject to real estate taxes rather than as motor vehicles subject only to license fees.

Although it continued to be popularly perceived as a shabby substitute for "stick" housing (a derogatory word used to describe the ordinary American balloon-frame dwelling), the residence on wheels reflected American values and industrial practices. Built with easily machined and processed materials, such as sheet metal and plastic, it represented a total consumer package, complete with interior furnishings, carpets, and appliances. More importantly, it provided a suburban-type alternative to the inner-city housing that would otherwise have been available to blue-collar workers, newly married couples, and retired persons . . .

A DRIVE-IN SOCIETY

Drive-in motels, drive-in movies, and drive-in shopping facilities were only a few of the many new institutions that followed in the exhaust of the internal-combustion engine. By 1984 mom-and-pop grocery stores had given way almost everywhere to supermarkets, most banks had drive-in windows, and a few funeral homes were making it possible for mourners to view the deceased, sign the register, and pay their respects without emerging from their cars. Odessa Community College in Texas even opened a drive-through registration window.

Particularly pervasive were fast-food franchises, which not only decimated the family-style restaurants but cut deeply into grocery store sales. In 1915, James G. Huneker, a raconteur whose tales of early twentieth-century American life were compiled as *New Cosmopolis*, complained of the infusion of cheap, quick-fire "food hells," and of the replacement of relaxed dining with "canned music and automatic lunch taverns." With the automobile came the notion of "grabbing" something to eat. The first drive-in restaurant, Royce Hailey's Pig Stand, opened in Dallas in 1921, and later in the decade, the first fast-food franchise, "White Tower," decided that families touring in motorcars needed convenient meals along the way. The places had to look clean, so they were painted white. They had to be familiar, so a minimal menu was standardized at every outlet. To catch the eye, they were built like little castles, replete with fake ramparts and turrets. And to forestall any problem with a land lease, the little white castles were built to be moveable.

The biggest restaurant operation of all began in 1954, when Ray A. Kroc, a Chicago area milkshake-machine salesman, joined forces with Richard and Maurice McDonald, the owners of a fast-food emporium in San Bernardino, California. In 1955 the first of Mr. Kroc's "McDonald's" outlets was opened in Des Plaines, a Chicago suburb long famous as the site of an annual Methodist encampment. The second and third, both in California, opened later in 1955 . . . [T]he McDonald's enterprise is based on free parking and drive-in access, and its methods have been copied by dozens of imitators. Late in 1984, on an interstate highway north of Minneapolis, McDonald's began construction of the most complete drive-in complex in the world. To be called McStop, it will feature a motel, gas station, convenience store, and, of course, a McDonald's restaurant.

[. . .]

THE CENTERLESS CITY

More than anyplace else, California became the symbol of the postwar suburban culture. It pioneered the booms in sports cars, foreign cars, vans, and motor homes, and by 1984 its 26 million citizens owned almost 19 million motor vehicles and had access to the world's most extensive freeway system. The result has been a new type of centerless city, best exemplified by once sleepy and out-of-the-way Orange County, just south and east of Los Angeles. After Walt Disney came down from Hollywood, bought out the ranchers, and opened Disneyland in 1955, Orange County began to evolve from a rural backwater into a suburb and then into a collection of medium and small towns. It had never had a true urban focus, in large part because its oil-producing sections each spawned independent suburban centers, none of which was particularly dominant over the others. The tradition continued when the area became a subdivider's dream in the 1960s and 1970s. By 1980 there were twenty-six Orange County cities, none with more than 225,000 residents. Like the begats of the Book of Genesis, they merged and multiplied into a huge agglomeration of two million people with its own Census Bureau metropolitan area designation – Anaheim, Santa Ana, Garden Grove. Unlike the traditional American metropolitan region, however, Orange County lacked a commutation focus, a place that could obviously be accepted as the center of local life. Instead, the experience of a local resident was typical: "I live in Garden Grove, work in Irvine, shop in Santa Ana, go to the dentist in Anaheim, my husband works in Long Beach, and I used to be the president of the League of Women Voters in Fullerton."

A centerless city also developed in Santa Clara County, which lies forty-five miles south of San Francisco and which is best known as the home of

Silicon Valley. Stretching from Palo Alto on the north to the garlic and lettuce fields of Gilroy to the south, Santa Clara County has the world's most extensive concentration of electronics concerns. In 1940, however, it was best known for prunes and apricots, and it was not until after World War II that its largest city, San Jose, also became the nation's largest suburb. With fewer than 70,000 residents in 1940, San Jose exploded to 636,000 by 1980, superseding San Francisco as the region's largest municipality . . .

The numbers were larger in California, but the pattern was the same on the edges of every American city, from Buffalo Grove and Schaumburg near Chicago, to Germantown and Collierville near Memphis, to Creve Coeur and Ladue near St. Louis. And perhaps more important than the growing number of people living outside of city boundaries was the sheer physical sprawl of metropolitan areas. Between 1950 and 1970, the urbanized area of Washington, DC, grew from 181 to 523 square miles, of Miami from 116 to 429, while in the larger megalopolises of New York, Chicago, and Los Angeles, the region of settlement was measured in the thousands of square miles.

THE DECENTRALIZATION OF FACTORIES AND OFFICES

The deconcentration of post-World War II American cities was not simply a matter of split-level homes and neighborhood schools. It involved almost every facet of national life, from manufacturing to shopping to professional services. Most importantly, it involved the location of the workplace, and the erosion of the concept of suburb as a place from which wage-earners commuted daily to jobs in the center. So far had the trend progressed by 1970 that in nine of the fifteen largest metropolitan areas suburbs were the principal sources of employment, and in some cities, like San Francisco, almost three-fourths of all work trips were by people who neither lived nor worked in the core city. In Wilmington, Delaware, 66 percent of area jobs in 1940 were in the core city; by 1970, the figure had fallen below one-quarter. And despite the fact that Manhattan contained the world's highest concentration of office space and business activity, in 1970, about 78 percent of the residents in the New York suburbs also worked in the suburbs. Many outlying communities thus achieved a kind of autonomy from the older downtown areas . . .

Manufacturing is now among the most dispersed of nonresidential activities. As the proportion of industrial jobs in the United States work force fell from 29 percent to 23 percent of the total in the 1970s, those manufacturing enterprises that survived often relocated either to the suburbs or to the lower-cost South and West . . .

Office functions, once thought to be securely anchored to the streets of big cities, have followed the suburban trend. In the nineteenth century, businesses tried to keep all their operations under one centralized roof. It was the most efficient way to run a company when the mails were slow and uncertain and communication among employees was limited to the distance that a human voice could carry. More recently, the economics of real estate and a revolution in communications have changed these circumstances, and many companies are now balkanizing their accounting departments, data-processing divisions, and billing departments. Just as insurance companies, branch banks, regional sales staffs, and doctors' offices have reduced their costs and presumably increased their accessibility by moving to suburban locations, so also have back-office functions been splitting away from front offices and moving away from central business districts.

[. . .]

Since World War II, the American people have experienced a transformation of the manmade environment around them. Commercial, residential, and industrial structures have been redesigned to fit the needs of the motorist rather than the pedestrian. Garish signs, large parking lots, one-way streets, drive-in windows, and throw-away fast-food buildings – all associated with the world of suburbia – have replaced the slower-paced, neighborhood-oriented institutions of an earlier generation. Some observers of the automobile revolution have argued that the car has created a new and better urban environment and that the change in spatial scale, based upon swift transportation, has formed a new kind of organic entity, speeding up personal communication and rendering obsolete the older urban settings. Lewis Mumford, writing from his small-town retreat in Amenia, New York, has emphatically disagreed. His prize-winning book, *The City in History*, was a celebration of the medieval community and an excoriation of "the formless urban exudation" that he saw American cities becoming. He noted that the automobile megalopolis was not a final stage in city development but an anti-city which "annihilates the city whenever it collides with it."

[. . .]

There are some signs that the halcyon days of the drive-in culture and automobile are behind us. More than one hundred thousand gasoline stations, or about one-third of the American total, have been eliminated in the last decade. Empty tourist courts and boarded-up motels are reminders that the fast pace of change can make commercial structures obsolete within a quarter-century of their erection. Even that suburban bellwether, the shopping center, which revolutionized merchandising after World War II, has come to seem small and out-of-date as newer covered malls attract both the trendy and the family trade. Some older centers have been recycled as bowling alleys or industrial buildings, and some have been remodeled to appeal to larger tenants and better-heeled customers. But others stand forlorn and boarded up. Similarly, the characteristic fast-food emporiums of the 1950s, with uniformed "car hops" who took orders at the automobile window, are now relics of the past. One of the survivors, Delores Drive-in, which opened in Beverly Hills in 1946, was recently proposed as an historic landmark, a sure sign that the species is in danger.

"Beyond Suburbia: The Rise of the Technoburb"

from *Bourgeois Utopias: The Rise and Fall of Suburbia* (1987)

Robert Fishman

Editors' Introduction

Robert Fishman (b. 1946) is a professor of history at the University of Michigan who established his academic reputation with his first book, the magisterial *Urban Utopias in the Twentieth Century* (1977), a study of the work of Ebenezer Howard, Le Corbusier, and Frank Lloyd Wright (see Part Five of this volume). For his second book, Fishman decided to address a totally prosaic, nonvisionary subject – the history of suburbia – only to discover that "the suburban ideal" was, in the final analysis, yet another form of utopia, the utopia of the middle class.

The real focus of *Bourgeois Utopias* is the suburban ideal, more than suburbia itself and the logic of Fishman's analysis leads him to many surprising insights and conclusions. In the medieval period and up through the eighteenth century, suburbs were clusters of houses inhabited by poor and/or disreputable people on the outskirts of towns. When suburbs were first established for the upper and middle classes – a phenomenon that has thrived more in North America than in Europe where working-class suburbs and *banlieus* often predominate – the ideal was to create a perfect synthesis of urban sophistication and rural virtue. Here was a conception as utopian as that of any visionary social reformer but with an important difference: "Where other modern utopias have been collectivist," writes Fishman, "suburbia has built its vision of community on the primacy of private property and the individual family."

What suburbia has evolved into today is "technoburbia," a dominant new urban reality that can no longer be considered suburbia in the traditional sense. In Redmond, Washington, and in Cupertino, California, the Microsoft and Apple corporate headquarters mix with residential neighborhoods, retail centers, and even bands of open space to make up a new urban form where city and suburb – urbanized and un-urbanized areas, high-tech and conventional development – flow seamlessly together.

To describe this new reality Fishman has coined two new terms "technoburb" and "techno-city." Fishman defines technoburbs as peripheral zones, perhaps as large as a county, that have emerged as viable socioeconomic units. The new technoburbs are spread out along highway growth corridors. Along the highways of metropolitan regions shopping malls, industrial parks, campus-like office complexes, hospitals, schools, and a whole range of housing types succeed each other.

By "techno-city" Fishman means the whole metropolitan region, which has been transformed by the coming of the technoburb. In Fishman's view we may still refer to the New York Metropolitan region as "New York City," but increasingly by "New York City" we mean the entire New York City region. And much of the economic and cultural life of the region no longer resides just in the core city. The old central cities have become increasingly marginal, while the technoburb has emerged as the focus of American life. In Fishman's view, the new technoburbs surrounding the old urban cores do not represent "the suburbanization of the United States," as Kenneth Jackson (p. 65) would have it, but "the end of suburbia in its traditional sense and the creation of a new kind of decentralized city." That suburbia has become the city itself is, perhaps, the final irony of modern urbanism.

Fishman lays out a strong indictment of what is wrong with technoburbs. They consist of an unplanned jumble of discordant elements – housing, industry, commerce, even agriculture – with little coherent pattern or structure. They waste land. Technoburbs are dependent on highway systems, yet their highway systems are in a state of chronic chaos. They have no proper boundaries, but consist of a crazy quilt of separate and overlapping political jurisdictions that make meaningful region-wide planning virtually impossible and, as Myron Orfield observes (p. 296), access to revenue to pay for local government services becomes highly inequitable.

Yet Fishman notes that all new urban forms appeared chaotic in their early stages. Even the most "organic" cityscapes of the past evolved slowly after much chaos and trial and error. For example, it took planners of genius like Frederick Law Olmsted (p. 321) and Ebenezer Howard (p. 328) to create orderly parks and garden suburbs (like Olmsted's Riverside, a romantic suburb on the outskirts of Chicago) out of the chaos of the nineteenth-century city or to imagine and actually build Garden Cities. Fishman acknowledges that there is a functional logic to sprawl. Perhaps, he speculates, if sprawl is better understood and better managed it might prove to be a positive rather than a negative development. Fishman looks to Frank Lloyd Wrights's Broadacre City vision (p. 345) as an example of how inspired planners may yet devise an aesthetic to tame techoburbia, while Robert Bruegmann's selection from *Sprawl* (p. 211) suggests that building ever-outward is merely a logical, indeed "natural" response to increased population pressures and the desire of the middle class to avoid the disagreeable aspects of inner-city life.

The technocity, Fishman concludes, is still under construction both physically and culturally. How it will evolve is unclear, although Richard Florida (p. 143) offers persuasive insights into who will live in the new techno-communities and how they will work and socially interact. The jury is still out on whether technoburbia will ultimately be judged as an advance over earlier urban forms. Another question is how technoburbia relates to the cities of the emerging global society discussed in Part Eight of this book.

This selection is from Fishman's *Bourgeois Utopias: The Rise and Fall of Surburbia* (New York: Basic Books, 1987). His other major books on cities are *Urban Utopias in the Twentieth Century* (New York: Basic Books, 1977) and *The American Planning Tradition: Culture and Policy* (Washington, DC: Woodrow Wilson Center Press, 2000).

For other views of emerging postmodern suburbia, see journalist Joel Garreau, *Edge City* (New York: Anchor, 1992), Edward Soja, "Taking Los Angeles Apart," in Soja, *Postmodern Geographies: The Reassertion of Space in Critical Social Theory* (New York: Verso, 1989), Michael Dear, "The Los Angeles School of Urbanism: An Intellectual History" (p. 170), and Peter Calthorpe and William Fulton's description of the emerging "Regional City" (p. 360). For an excellent collection of articles on the history of suburbia, see Becky Nicolaides and Andrew Wiese (eds), *The Suburb Reader* (London: Routledge, 2006). Also of interest are two books that call for a reconfiguration of the city–suburb relationship: Myron Orfield, *American Metropolitics: The New Suburban Reality* (Washington, DC: Brookings Institution, 2002) and David Rusk, *Cities without Suburbs: A Census 2000 Update* (Washington, DC: Woodrow Wilson Center Press, 2003).

■

If the nineteenth century could be called the Age of Great Cities, post-1945 America would appear to be the Age of Great Suburbs. As central cities stagnated or declined in both population and industry, growth was channeled almost exclusively to the peripheries. Between 1950 and 1970 American central cities grew by 10 million people, their suburbs by 85 million. Suburbs, moreover, accounted for at least three-quarters of all new manufacturing and retail jobs generated during that period. By 1970 the percentage of Americans living in suburbs was almost exactly double what it had been in 1940, and more Americans lived in suburban areas (37.6 percent) than in central cities (31.4 percent) or in rural areas (31 percent). In the 1970s central cities experienced a net out-migration of 13 million people, combined with an unprecedented deindustrialization, increasing poverty levels, and housing decay.

[. . .]

From its origins in eighteenth-century London, suburbia has served as a specialized portion of the expanding metropolis. Whether it was inside or outside the political borders of the central city, it was always functionally dependent on the urban core. Conversely, the growth of suburbia always meant a strengthening of the specialized services at the core.

In my view, the most important feature of postwar American development has been the almost simultaneous decentralization of housing, industry, specialized services, and office jobs; the consequent breakaway of the urban periphery from a central city it no longer needs; and the creation of a decentralized environment that nevertheless possesses all the economic and technological dynamism we associate with the city. This phenomenon, as remarkable as it is unique, is not suburbanization but a new city.

Unfortunately, we lack a convenient name for this new city, which has taken shape on the outskirts of all our major urban centers. Some have used the terms "exurbia" or "outer city." I suggest (with apologies) two neologisms: the "technoburb" and the "techno-city." By "technoburb" I mean a peripheral zone, perhaps as large as a county, that has emerged as a viable socioeconomic unit. Spread out along its highway growth corridors are shopping malls, industrial parks, campus-like office complexes, hospitals, schools, and a full range of housing types. Its residents look to their immediate surroundings rather than to the city for their jobs and other needs; and its industries find not only the employees they need but also the specialized services.

The new city is a technoburb not only because high tech industries have found their most congenial homes in such archetypal technoburbs as Silicon Valley in northern California and Route 128 in Massachusetts. In most technoburbs such industries make up only a small minority of jobs, but the very existence of the decentralized city is made possible only through the advanced communications technology which has so completely superseded the face-to-face contact of the traditional city. The technoburb has generated urban diversity without traditional urban concentration.

By "techno-city" I mean the whole metropolitan region that has been transformed by the coming of the technoburb. The techno-city usually still bears the name of its principal city, for example, "the New York metropolitan area"; its sports teams bear that city's name (even if they no longer play within the boundaries of the central city); and its television stations appear to broadcast from the central city. But the economic and social life of the region increasingly bypasses its supposed core. The techno-city is truly multicentered, along the pattern that Los Angeles first created. The technoburbs, which might stretch over seventy miles from the core in all directions, are often in more direct communication with one another – or with other techno-cities across the country – than they are with the core. The techno-city's real structure is aptly expressed by the circular superhighways or beltways that serve so well to define the perimeters of the new city. The beltways put every part of the urban periphery in contact with every other part without passing through the central city at all.

[. . .]

The old central cities have become increasingly marginal, while the technoburb has emerged as the focus of American life. The traditional suburbanite – commuting at ever-increasing cost to a center where the available resources barely duplicate those available much closer to home – becomes increasingly rare. In this transformed urban ecology the history of suburbia comes to an end.

PROPHETS OF THE TECHNO-CITY

Like all new urban forms, the techno-city and its technoburbs emerged not only unpredicted but unobserved. We are still seeing this new city through the intellectual categories of the old metropolis. Only two prophets, I believe, perceived the underlying forces that would lead to the techno-city at the time of their first emergence. Their thoughts are therefore particularly valuable in understanding the new city.

At the turn of the twentieth century, when the power and attraction of the great city was at its peak, H.G. Wells daringly asserted that the technological forces that had created the industrial metropolis were now moving to destroy it. In his 1900 essay "The Probable Diffusion of Great Cities," Wells argued that the seemingly inexorable concentration of people and resources in the largest cities would soon be reversed. In the course of the twentieth century, he prophesied, the metropolis would see its own resources drain away to decentralized "urban regions" so vast that the very concept of "the city" would become, in his phrase, "as obsolete as 'mailcoach.'"

Wells based his prediction on a penetrating analysis of the emerging networks of transportation and communication. Throughout the nineteenth century, rail transportation had been a relatively simple system favoring direct access to large centers. With the spread of branchlines and electric tramways, however, a complex rail network had been created that could serve as the basis for a decentralized region. (As Wells wrote, Henry E. Huntington was proving the truth of his propositions for the Los Angeles region.)

Wells pictured the "urban region" of the year 2000 as a series of villages with small homes and factories set in the open fields, yet connected by high speed rail transportation to any other point in the region. (It was a vision not very different from those who saw Los Angeles developing into just such a network of villages.) The old cities would not completely disappear, but they would lose both their financial and their industrial functions, surviving simply because of an inherent human love of crowds. The "post-urban" city, Wells predicted, will be "essentially a bazaar, a great gallery of shops and places of concourse and rendezvous, a pedestrian place, its pathways reinforced by lifts and moving platforms, and shielded from the weather, and altogether a very spacious, brilliant, and entertaining agglomeration." In short, the great metropolis will dwindle to what we would today call a massive shopping mall, while the productive life of the society would take place in the decentralized urban region.

Wells's prediction was taken up in the late 1920s and early 1930s by Frank Lloyd Wright, who moved from similar assumptions to an even more radical view. Wright had actually seen the beginnings of the automobile and truck era; he was, perhaps not coincidentally, living mostly in Los Angeles in the late 1910s and early 1920s. Wright, like Wells, argued that "the great city was no longer modern" and that it was destined to be replaced by a decentralized society.

He called this new society Broadacre City. It has often been confused with a kind of universal suburbanization, but for Wright "Broadacres" was the exact opposite of the suburbia he despised. He saw correctly that suburbia represented the essential extension of the city into the countryside, whereas Broadacres represented the disappearance of all previously existing cities.

As Wright envisioned it, Broadacres was based on universal automobile ownership combined with a network of superhighways, which removed the need for population to cluster in a particular spot. Indeed, any such clustering was necessarily inefficient, a point of congestion rather than of communication. The city would thus spread out over the countryside at densities low enough to permit each family to have its own homestead and even to engage in part-time agriculture. Yet these homesteads would not be isolated; their access to the superhighway grid would put them within easy reach of as many jobs and specialized services as any nineteenth-century urbanite. Traveling at more than sixty miles an hour, each citizen would create his own city within the hundreds of square miles he could reach in an hour's drive.

Like Wells, Wright saw industrial production inevitably leaving the cities for the space and convenience of rural sites. But Wright went one step further in his attempt to envision the way that a radically decentralized environment could generate that diversity and excitement which only cities had possessed.

He saw that even in the most scattered environment, the crossing of major highways would possess a certain special status. These intersections would be the natural sites of what he called the roadside market, a remarkable anticipation of the shopping center: "great spacious roadside pleasure places these markets, rising high and handsome like some flexible form of pavilion – designed as places of cooperative exchange, not only of commodities but of cultural facilities." To the roadside markets he added a range of highly civilized yet small scale institutions: schools, a modern cathedral, a center for festivities, and the like. In such an environment, even the entertainment functions of the city would disappear. Soon, Wright devoutly wished, the centralized city itself would disappear.

Taken together, Wells's and Wright's prophecies constitute a remarkable insight into the decentralizing tendencies of modern technology and society. Both were presented in utopian form, an image of the future presented as somehow "inevitable" yet without any sustained attention to how it would actually be achieved. Nevertheless, something like the transformation that Wells and Wright foresaw has taken place in the United States, a transformation all the more remarkable in that it occurred without a clear recognition that it was happening. While diverse groups were engaged in what they believed was "the suburbanization" of America, they were in fact creating a new city.

[. . .]

TECHNOBURB/TECHNO-CITY: THE STRUCTURE OF THE NEW METROPOLIS

To claim that there is a pattern or structure in the new American city is to contradict what appears to be overwhelming evidence. One might sum up the structure of the technoburb by saying that it goes against every rule of planning. It is based on two extravagances that have always aroused the ire of planners: the waste of land inherent in a single family house with its own yard, and the waste of energy inherent in the use of the personal automobile. The new city is absolutely dependent on its road system, yet that system is almost always in a state of chaos and congestion. The landscape of the technoburb is a hopeless jumble of housing, industry, commerce, and even agricultural uses. Finally, the technoburb has no proper boundaries; however defined, it is divided into a crazy quilt of separate and overlapping political jurisdictions, which make any kind of coordinated planning virtually impossible.

Yet the technoburb has become the real locus of growth and innovation in our society. And there is a real structure in what appears to be wasteful sprawl, which provides enough logic and efficiency for the technoburb to fulfill at least some of its promises.

If there is a single basic principle in the structure of the technoburb, it is the renewed linkage of work and residence. The suburb had separated the two into distinct environments; its logic was that of the massive commute, in which workers from the periphery traveled each morning to a single core and then dispersed each evening. The technoburb, however, contains both work and residence within a single decentralized environment.

By the standards of a preindustrial city where people often lived and worked under the same roof, or even of the turn of the century industrial zones where factories were an integral part of working class neighborhoods, the linkage between work and residence in the technoburb is hardly close. A recent study of New Jersey shows that most workers along the state's growth corridors now live in the same county in which they work. But this relative dispersion must be contrasted to the former pattern of commuting into urban cores like Newark or New York. In most cases traveling time to work diminishes, even when the distances traveled are still substantial; as the 1980 census indicates, the average journey to work appears to be diminishing both in distance and, more importantly, in time.

For commuting within the technoburb is multidirectional, following the great grid of highways and secondary roads that, as Frank Lloyd Wright understood, defines the community. This multiplicity of destinations makes public transportation highly inefficient, but it does remove that terrible bottleneck which necessarily occurred when work was concentrated at a single core within the region. Each house in a technoburb is within a reasonable driving time of a truly "urban" array of jobs and services, just as each workplace along the highways can draw upon an "urban" pool of workers.

Those who believed that the energy crisis of the 1970s would cripple the technoburb failed to realize that the new city had evolved its own pattern of transportation in which a multitude of relatively short automobile journeys in a multitude of different directions substitutes for that great tidal wash in and out of a single urban core which had previously defined commuting. With housing, jobs, and services all on the periphery, this sprawl develops its own form of relative efficiency. The truly inefficient form would be any attempted revival of the former pattern of long distance mass transit commuting into a core area. To account for the new linkage of work and residence in the technoburb, we must first confront this paradox: the new city required a massive and coordinated relocation of housing, industry, and other "core" functions to the periphery; yet there were no coordinators directing the process. Indeed, the technoburb emerged in spite of, not because of, the conscious purposes motivating the main actors. The postwar housing boom was an attempt to escape from urban conditions; the new highways sought to channel traffic into the cities; planners attempted to limit peripheral growth; the government programs that did the most to destroy the hegemony of the old industrial metropolis were precisely those designed to save it.

This paradox can be seen clearly in the area of transportation policy. Wright had grasped the basic point in his Broadacre City plan: a fully developed highway grid eliminates the primacy of a central business district. It creates a whole series of highway crossings, which can serve as business centers while promoting the multidirectional travel that prevents any single center from attaining unique importance. Yet, from the time of Robert Moses to the present, highway planners have imagined that the new roads, like the older rail transportation, would enhance the importance of the old centers by funneling cars and trucks into the downtown area and the surrounding industrial belt. At most, the highways were to serve traditional suburbanization; in other words, the movement from the periphery to the core during consuming time and space, destroying the natural landscape. The wealth that postindustrial America has generated has been used to create an ugly and wasteful pseudocity, too spread out to be efficient, too superficial to create a true culture. The truth of both indictments is impossible to deny, yet it must be rescued from the polemical overstatements that seem to afflict anyone who deals with these topics. The first charge is the more fundamental, for it points to a genuine structural discontinuity in post-1945 decentralization. By detaching itself physically, socially, and economically from the city, the technoburb is profoundly antiurban as suburbia never had been. Suburbanization strengthened the central core as the cultural and economic heart of an expanding region; by excluding industry, suburbia left intact and even augmented the urban factory districts.

Technoburb development, however, completely undermines the factory district and potentially threatens even the commercial core. The competition from new sites on the outskirts renders obsolete the whole complex of housing and factory sites that had been built up in the years 1890 to 1930 and provides alternatives to the core for even the most specialized shopping and administrative services.

This competition, moreover, has occurred in the context of a massive migration of southern blacks to northern cities. Blacks, Hispanics, and other recent migrants could afford housing only in the old factory districts, which were being abandoned by both employers and the white working class. The result was a twentieth century version of Disraeli's "two nations." Now, however, the outer reaches of affluence include both the middle class and the better-off working class – a majority of the population; while the largely black and Hispanic minority are forced into decaying neighborhoods, which lack not only decent housing but jobs.

This bleak picture has been modified somewhat by the continued ability of the traditional urban cores to retain certain key areas of white collar and professional employment; and by the choice of some highly paid core workers to live in high-rise or recently renovated housing around the core. Compared both to the decaying factory zones and to peripheral expansion, the "gentrification" phenomenon has been highly visible yet statistically insignificant. It has done as much to displace low income city dwellers as to benefit them. The late twentieth century American environment thus shows all the signs of the two nations syndrome: one caught in an environment of poverty, cut off from the majority culture, speaking its own languages and dialects; the other an increasingly homogenized culture of affluence, more and more remote from an urban environment it finds dangerous.

[. . .]

The case against the technoburb can easily be summarized. Compared even to the traditional suburb, it at first appears impossible to comprehend. It has no clear boundaries; it includes discordant rural, urban, and suburban elements; and it can best be measured in counties rather than in city blocks. Consequently the new city lacks any recognizable center to give meaning to the whole. Major civic institutions seem scattered at random over an undifferentiated landscape.

Even planned developments – however harmonious they might appear from the inside – can be no more than fragments in a fragmented environment. A single house, a single street, even a cluster of streets and houses can be and frequently are well designed. But true public space is lacking or totally commercialized. Only the remaining pockets of undeveloped farmland maintain real openness, and these pockets are inevitably developed, precipitating further flight and further sprawl.

The case for the techno-city can only be made hesitantly and conditionally. Nevertheless, we can hope that its deficiencies are in large part the early awkwardness of a new urban type. All new city forms appear in their early stages to be chaotic. "There were a hundred thousand shapes and substances of incompleteness, wildly mingled out of their places, upside down, burrowing in the earth, aspiring in the earth, moldering in the water, and unintelligible as any dream." This was Charles Dickens describing London in 1848, in his novel *Dombey and Son* (Chapter 6). As I have indicated, sprawl has a functional logic that may not be apparent to those accustomed to more traditional cities. If that logic is understood imaginatively, as Wells and especially Wright attempted to do, then perhaps a matching aesthetic can be devised.

We must remember that even the most "organic" cityscapes of the past evolved slowly after much chaos and trial and error. The classic late nineteenth century railroad suburb – the standard against which critics judge today's sprawl – evolved out of the disorder of nineteenth century metropolitan growth. First, planners of genius like John Nash and Frederick Law Olmsted comprehended the process and devised aesthetic formulas to guide it. These formulas were then communicated – slowly and incompletely – to speculative builders, who nevertheless managed to capture the basic idea. Finally, individual property owners constantly upgraded their holdings to eliminate discordant elements and bring their community closer to the ideal.

We might hope that a similar process is now at work in the postsuburban outer city. As a starting point for a technoburb aesthetic, there are Wright's Broadacre City plans and drawings, which still repay study for anyone seeking a vision of a modern yet organic American landscape. More useful still is the American New Town tradition, starting from Radburn, New Jersey, with its careful designs intended to reconcile decentralization with older ideas of community. Already, New Town designs have been adopted by speculative builders, not only in a highly publicized project like James Rouse's Columbia, Maryland, but in hundreds of smaller planned communities, which are beginning to leave their mark on the landscape.

At the level of civic architecture there is Wright's Marin County Civic Center to serve as a model for public monuments in a decentralized environment. The multilevel, enclosed shopping mall has attained a spaciousness not unworthy of the great urban shopping districts of the past, while newly built college campuses and campuslike office complexes and research centers contribute significantly to the environment. Some commercial highway strips have been rescued from cacophony and have managed to achieve a liveliness that is not tawdry. (This evolution parallels the evolution of the nineteenth-century urban core, originally a remarkably ugly cluster of small buildings and large signs, which was transformed into a reasonably dignified center for commerce by the turn of the century.)

Most importantly, there is a growing sense that open land must be preserved as an integral part of the landscape, through regional land use plans, purchases for parklands, and tax abatements for working farms. These governmental measures, combined with thousands of small scale efforts by individuals, could create a fitting environment for the new city. These efforts, moreover, could provide the starting point for a more profound diversification of the outer city. An increased understanding and respect for the landscape of each region could lead to a growing rejection of a mass culture that erases all such distinctions.

The techno-city, therefore, is still under construction, both physically and culturally. Its economic and social successes are undeniable, as are its costs. Most importantly, the new pattern of decentralization has fundamentally altered the urban form on which suburbia had depended for its function and meaning. Whatever the fate of the new city, suburbia in its traditional sense now belongs to the past.

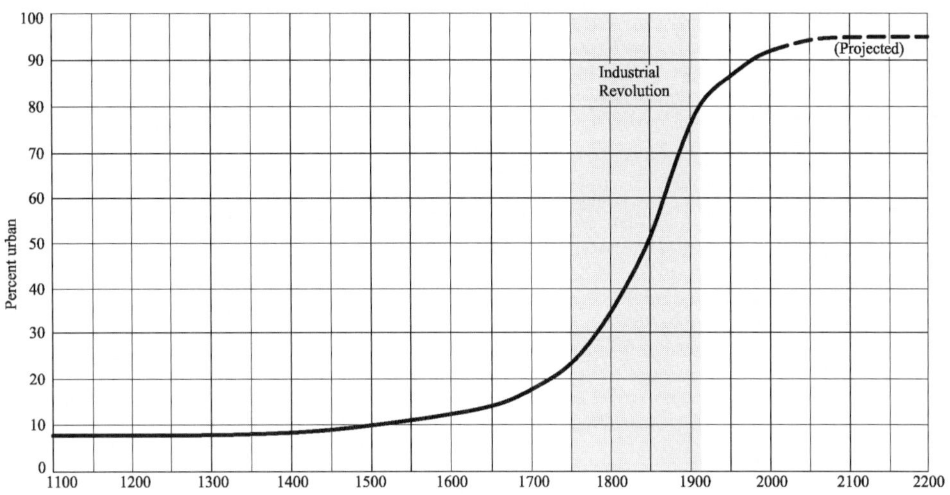

Percent of England urban 1100 ADE to 2200 ADE

Urbanization refers to the *percent* of the population of a geographical area such as a nation state or region of the world that lives in urban as opposed to non-urban places (such as farms and small towns). Urbanization often follows a pattern that historical urban demographer Kingsley Davis describes as an attenuated S curve with a long left tail as the percentage of the population in a region slowly becomes more urban, a steep middle portion of the S as the region urbanizes rapidly, and then a nearly flat upper part of the S once the region is essentially fully urban. England urbanized very slowly until the beginning of the industrial revolution (about 1750), then urbanized rapidly, and today -- with 92% of its population urban -- is barely urbanizing if at all.

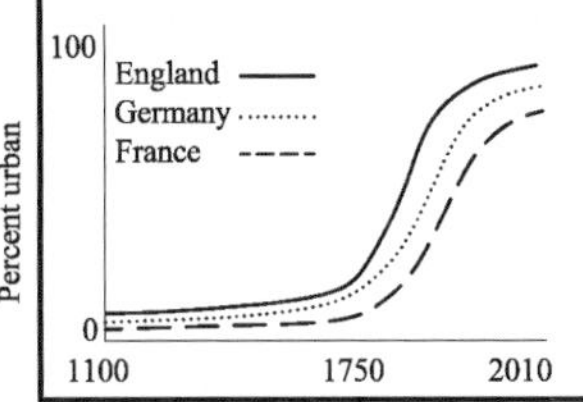

Davis argues that there are often "families" of similar S curves in a region. The S curves of Germany and France began increasing later and proceeded less rapidly than England's, but the overall shape is similar.

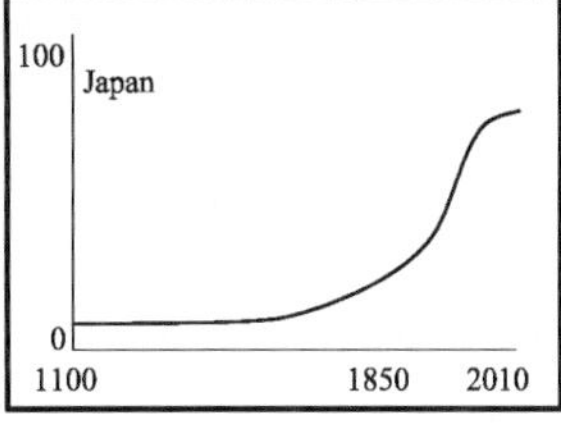

History, politics, economics, and culture affect urbanization. Japan's S curve began to increase rapidly only after 1850 when Japan opened up to the West politically and began to industrialize.

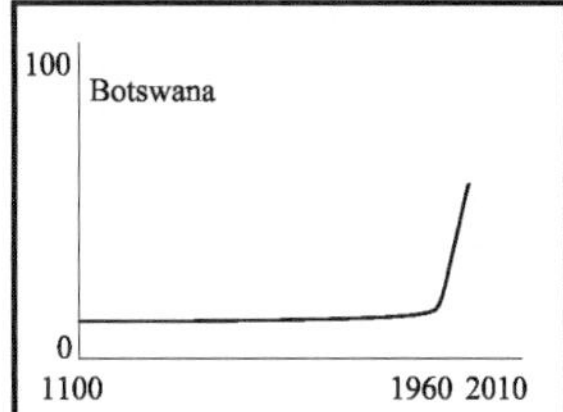

Some developing countries today are urbanizing very rapidly. Only about 2% of Botswana's population was urban as late as 1960, but today Botswana is more than 50% urban.

Data graphic by Michael Brestel based on Kingsley Davis, "The Urbanization of the Human Population," *Scientific American* (September 1965).

Plate 1 **The demographic S curves of urbanization**

Plate 2 **A view of the city of Babylon.** The first cities arose about 4000 BCE in Mesopotamia between the Tigris and Euphrates rivers during what V. Gordon Childe termed "the urban revolution." The first Mesopotamian city is thought to be Uruk, but the most famous is surely Babylon, home to the legendary Tower of Babel, the Hanging Gardens, and powerful kings like Hammurabi and Nebuchadnezzar. Note the magnificent outer walls enclosing and protecting the whole community as well as the monumental interior citadel with its ziggurat, temple, and royal palace. This painting of Babylon as it may have appeared ca. 2500 BCE is by Maurice Bardin and is in the collection of the Oriental Institute at the University of Chicago.

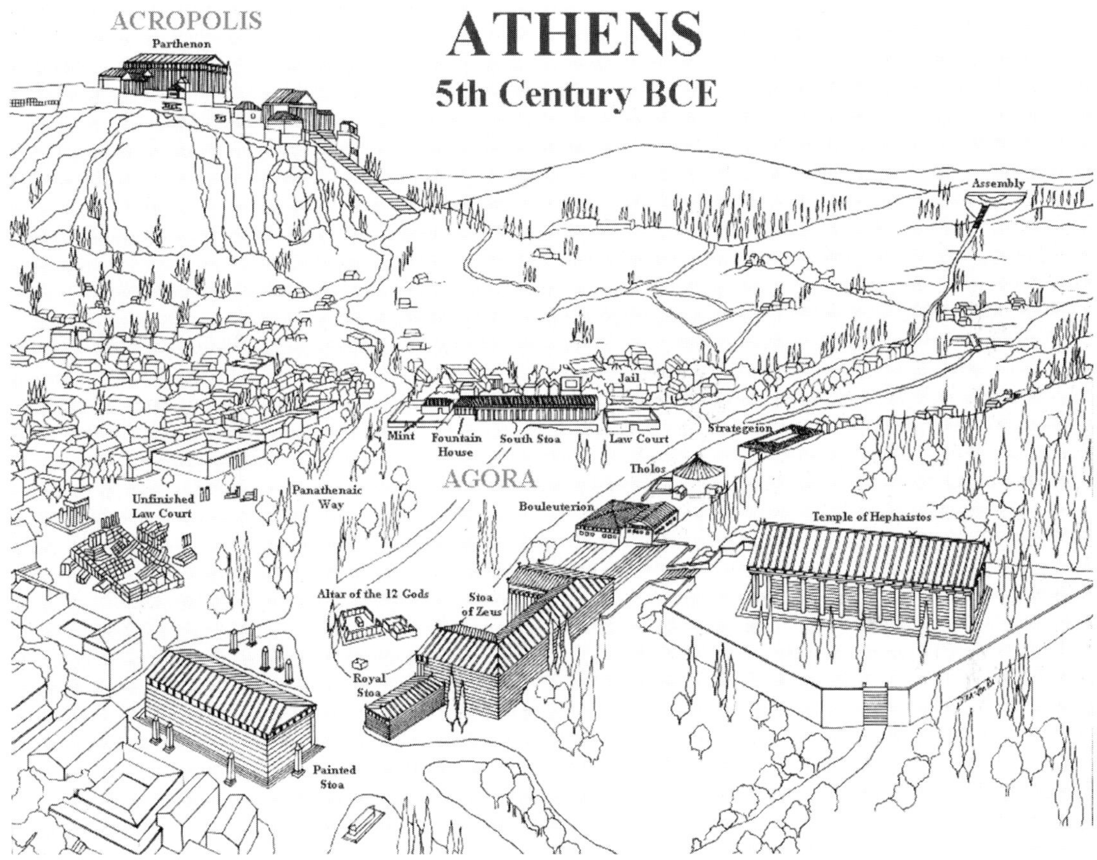

Plate 3 **A view of ancient Athens.** The Greek *polis* stressed public over private life. The Athenian acropolis (literally "high city") with its Parthenon and associated temple structures retained its importance as a symbolic citadel, but citizens normally conversed, shopped, and settled disputes in the low city below the acropolis – especially in the public agora or marketplace. Greek citizens exercised and competed in public stadiums and gymnasia and participated in the cultural life of the community in large open-air theaters.

Plate 4 **A walled medieval city: Carcassonne, France.** As Europe began to revive after the period of disorganization and strife that followed the fall of the Roman Empire in the West, small cities like Carcassonne, France, fostered trade, economic expansion, and self-governing institutions like guilds. Note the density of the city, how the walls define and protect its limits, and how the farms and orchards are near at hand. This clear and sensible layout inspired Garden City and New Urbanist planners in the modern period.

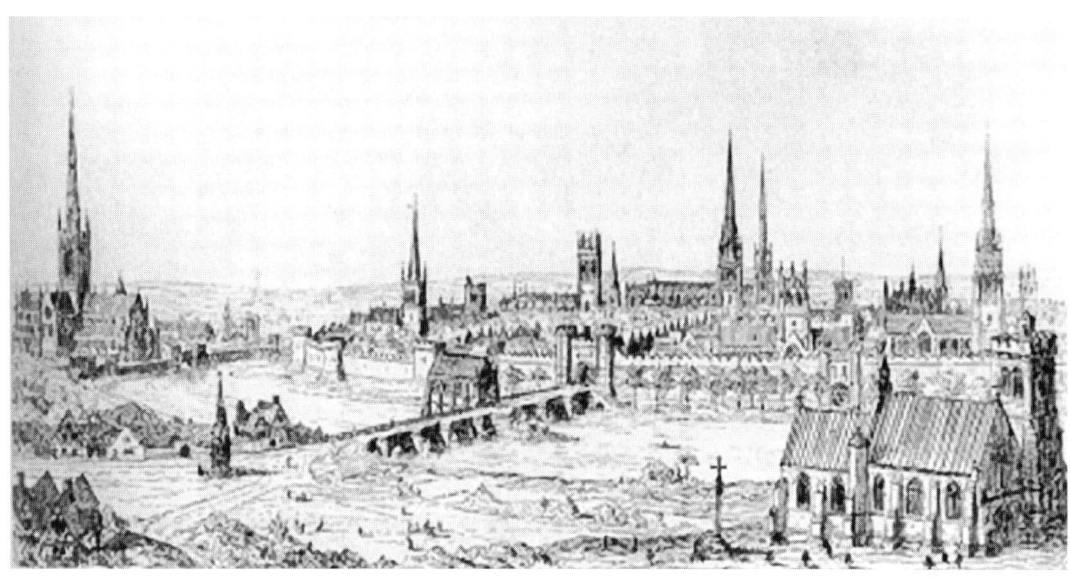

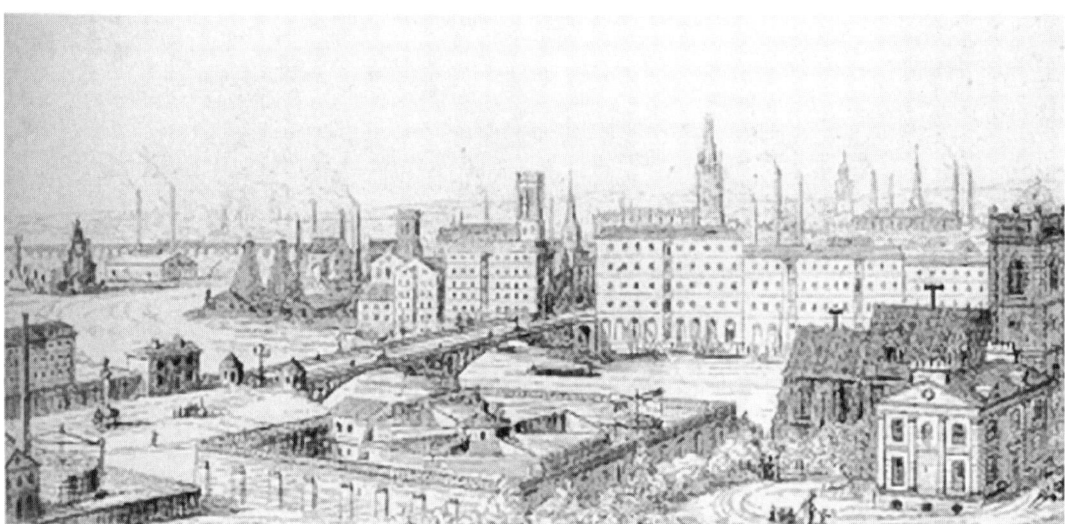

Plate 5 **The nineteenth-century industrial city.** During the first part of the nineteenth century, new industrial cities based on steam-powered machinery sprang up in Europe. In these prints titled "A Catholic Town in 1440" and "The Same Town in 1840," Augustus Welby Pugin, a contemporary observer, contrasts the same city before and after the Industrial Revolution. The first print shows a city where church spires are the dominant architectural element, the land surrounding the medieval city walls is largely empty, and the air and water are clean. In the second print, factory smokestacks have largely replaced steeples, the air is filled with smoke, development has sprawled to the once-empty land, and the foreground is dominated by a massive panopticon prison. Pugin subtitled his work: *A Parallel Between the Noble Edifices of the Middle Ages and the Corresponding Buildings of the Present Day Showing the Present Decay of Taste.*

Plate 6 **A modern downtown of the 1920s.** One of the urban features that Friedrich Engels noted in mid-nineteenth-century Manchester, England, was the appearance of large streets where the upper classes could travel without coming into contact with the squalid living conditions of the residential slums that comprised the greater part of the city. This image of bustling Market Street in San Francisco in about 1925 is typical of the new downtown of the modernist period. Note the way streetcars, buses, automobiles, and pedestrians all share the public space.

Plate 7 **Levittown, New York, 1947.** While suburbs have a long and varied history, it was during the period after World War II that many of the suburbs surrounding US cities arose. Levittown, New York (and its counterpart Levittown, Pennsylvania) provided entirely new communities of affordable, cute, single-family houses on individual lots to returning GIs and other first-time (white) homebuyers.

Plate 8 **The auto-centered metropolis, 1922.** Developers show off the location of a proposed new "50 Foot Boulevard" in the Westwood Village area of Santa Monica, California in 1922. Capitalizing on the mass production of Henry Ford's Model-T, they and thousands of their counterparts built communities in Southern California that required a car.

Plate 9 **Sprawl suburbia.** Robert Fishman uses the term Technoburbia to describe the form of urban development that jumbles business, residential, commercial, and other uses together in the area surrounding older core cities. Journalist Joel Garreau uses the term Edge City. Others just call it suburban sprawl, and the houses form a repeated pattern – best seen from the air – that many regard as a cultural wasteland.

Urban culture and society

INTRODUCTION TO PART TWO

As Shakespeare wrote, and as urbanists ever since have never tired of quoting, "the people are the city." Cities comprise a series of institutions – the government citadel, the economic market, and the community – and, of those three, it is the community that is the least well defined but most authentically expressive of the people of the city at large in their families, neighborhoods, and local cultures. Urban form and design describe the physical appearance and infrastructural layout of cities, and the efficiently planned city often evolves in response to the needs of citadel and market forces. But it is the people of the city in their communities – their individual aspirations and collective struggles, their day-to-day lives and their moments of heightened awareness – that constitute the core subject of urban studies and the final purpose of city planning.

In turning to the people of the cities themselves, we move to a consideration of the subtle and ever-shifting interplay between society, community, and culture. Part Two addresses how urban society affects urban culture and how culture affects the daily lives and the life prospects of city dwellers. It asks what culture is in an urban context and how it expresses itself in different social contexts – either as high culture or popular culture. Finally, it analyzes how community operates in the urban context and speculates about what it could be. These and other aspects of urban society and culture are explored in Jan Lin and Christopher Mele (eds), *The Urban Sociology Reader*, (London: Routledge, 2005) and Malcolm Miles, Tim Hall, and Iain Borden (eds), *The City Cultures Reader*, 2nd edn (London: Routledge 2004).

In studying the people of the city, the key discipline is sociology. That "science of society" arose alongside the emergence of the modern industrial city itself. Aristotle may well have been practicing sociology when he observed that all humans are *zoon politikon* – that is, animals that live in the *polis* (see p. 42) – but modern sociology is an academic discipline that developed in parallel with the profound urban transformations of the Industrial Revolution. In recent years, new analytical frameworks – variously called social studies, social theory, social relations, culture studies, and urban anthropology – have joined the discipline either as adjuncts or rivals. But the basic sociological vision remains central to all investigations of people in cities.

The subject matter of sociology is broad and inclusive. August Comte (1798–1857), one of the founders of the field, hoped to combine all of history, psychology, and economics into a single discipline that could help solve all the problems of modern society. Nowadays, the field includes social interactions between individuals and groups, patterns of social stratification (by class or socio-economic status), social deviance (for example, crime), and issues of race, ethnicity, and gender. Not all of the great pioneering sociologists – Karl Marx, Ferdinand Tonnies, Emile Durkheim, Georg Simmel, and Max Weber, to list just a few – were as optimistic as Comte, and one approach, the sociobiology of Edward O. Wilson and others that stresses the genetic roots of human social behavior, is regarded by some as distinctly pessimistic. "Each person," Wilson writes, "is molded by an interaction of his environment, especially his cultural environment, with the genes that affect social behavior." Mainstream sociology downplays genetic inheritance and emphasizes an understanding of diverse cultural-environmental influences that can be celebrated as elements of a rich human mosaic and that can be modified, if necessary or desirable, through the application of education policies, social action, and urban planning initiatives.

There is no better person with whom to begin a discussion of "Urban Culture and Society" than Lewis Mumford (p. 91). Not a trained sociologist but rather a self-educated social philosopher of the first rank, Mumford was one of the great public intellectuals of the twentieth century and certainly the foremost American urbanist. Mumford never lost sight of the human dimension of cities. For over sixty years he sparred with those who argued that cities arose and prospered for purely economic reasons or that cities were best defined in terms of size and density. Quite to the contrary, Mumford argued that cities are expressions of the human spirit and that they exist to nurture the ever-evolving human personality. This perspective comes through loud and clear in a talk he gave to a group of urban planners in 1937 entitled "What is a City?" To Mumford, defining a city only in terms of population size, or density, or attributes of the built environment is grossly inadequate. Rather, the human side of cities is their very essence, and city streets are a stage on which life's drama is played out. The idea of the "urban drama" was central to Mumford's vision of the city and he returned to it in major works such as *The Culture of Cities* (1938) and *The City in History* (1961). Like William Whyte (p.510) and Jane Jacobs (p.105), Mumford takes real delight in city life. For him, cities reflect and enlarge the human spirit, and he argues that creating better, more human cities will enrich civilization itself.

Mumford was not alone in focusing on the connections between urban life and the human personality. In "Urbanism as a Way of Life," Louis Wirth (p.96) asked the fundamental question "What does it mean to be urban?" and concluded that an urban "way of life" resulted in an "urban type" of character and personality. Wirth was one of a gifted group of sociologists at the University of Chicago who, in the 1920s and 1930s, developed a pioneering body of urban sociological theory that still shapes the field of urban sociology today. Studying rural migrants to Chicago from the peasant societies of Southern and Eastern Europe, Wirth perceived that the whole way of life in modern cities was fundamentally different from the way of life in rural cultures. In "Urbanism as a Way of Life," he attempts to abstract the essential characteristics of urban as opposed to rural life and to find the sources of the widely perceived urban characteristics of brusqueness and impersonality. As the face-to-face transactions of static rural village life are replaced by the distanced and mediated transactions of a large city, human personalities are transformed, and the new urbanites respond to each other and to society as a whole in entirely different ways than they did in their rural folk communities.

While Wirth's is a *theoretical* study, it is important to remember that his theories were generated by empirical observations that he and his colleagues conducted in Chicago during the early decades of the twentieth century. Architectural critic and urban community activist Jane Jacobs (p.105) did her own kind of street level social observation in writing *The Death and Life of Great American Cities* (1961), a book that shook the complacent world of establishment planning by reintroducing the values of community to the design of urban spaces. Architecture and urban design may not *determine* human behavior, but bad design can numb the human spirit and good design can have powerful, positive influences on human beings. Of the many values designers seek to build into their designs perhaps none is more important than fostering community and human interaction. To Jacobs, traffic engineering should be only one consideration in designing a street. In "The Uses of Sidewalks: Safety," she argues that a street designed so that people can see their children from house windows and will want to congregate on the front door stoops – one very much like her own Hudson Street in Greenwich Village – will be much more user-friendly. It will also be much safer than one which moves traffic efficiently, but is inhospitable to neighborhood life and insensitive to the potential for street life to reduce urban crime. Jacobs stresses the importance of designing streets to promote safety, particularly for women. A safer environment, she argues, is essential to the creation and preservation of community.

The African-American ghettoes of the United States have been the subject of an enormous body of sociological research that both celebrates their distinctive culture and analyzes the social pathology of segregated, poverty-ridden inner-city communities. In multiethnic societies worldwide, racial divisions tend to compound class distinctions to create an even further crisis of community in the form of racially segregated neighborhoods that have remained as symbols of inequality and oppression. *The Philadelphia Negro* (1899) by W.E.B. Du Bois (p. 110) specifically describes the African-American district of Philadelphia, Pennsylvania, as it developed in the years following the American Civil War, but the social and cultural dynamic of housing segregation and racial discrimination in the workplace that Du Bois describes can be applied to ghetto and barrio experiences throughout the United States and to the "social exclusion" experienced by residents of developing immigrant communities worldwide. In the years since Du Bois first surveyed the life of the racially segregated ghetto community, conditions have in some ways grown worse: so much so that the persistence of racial segregation, and the emergence of an "underclass" population radically disconnected from the rest of the urban community, threaten the social stability of some of the largest, wealthiest cities in the world, even in the age of urban globalism discussed in Part Eight (see p. 541).

Beginning with Du Bois's pioneering studies and continuing through the work of E. Franklin Frazier (*The Negro Family in the United States*, 1939), St. Clair Drake and Horace Cayton (*Black Metropolis*, 1945), and Kenneth B. Clark (*Dark Ghetto*, 1965), African-American scholars have taken the lead in examining the social and cultural dynamics of ghetto communities in the United States' Northern cities. More recently, an important debate, called the "underclass" debate, arose concerning the plight of the mostly Black residents of American inner-city ghettos. One of the principals in that debate was William Julius Wilson (p.117), an African-American sociologist.

In "From Institutional to Jobless Ghettos" from *When Work Disappears* (1996), Wilson illustrates the role of ideology in shaping urban theory. As a leading liberal voice, he argues that the situation of the poorest urban Blacks in the United States has grown worse during the last generation and that at the present time, poor ghetto Blacks, especially youth, are in deep trouble. But why is this so, and what can policy-makers do about it? Wilson stresses the loss of jobs accessible to unskilled inner-city youth. It is this loss of jobs, he argues that has destroyed ghetto family structure and lies at the root of crime, substance abuse, and other ghetto ills. As Wilson sees it, many young Black males are not "marriageable" because they lack minimum job skills, have substance abuse problems, or are in prison. Without "marriageable" males, many young Black women cannot form two-parent nuclear families, and teen pregnancies, out-of-wedlock births, and female-headed, welfare-dependent, single-parent households result. These peculiarities of the inner-city African-American community are analyzed by Yale University sociologist Elijah Anderson (p. 127) as the difference between "decent families" – meaning those that attempt to maintain respectable, middle-class values – and "street families" – meaning those that embrace the worst kind of violent and offensive behavior that he calls "the code of the street." Brilliantly employing the participant-observer methodology in *Code of the Street: Decency, Violence, and the Moral Life of the Inner City* (1999), Anderson details the deeply embedded value systems that make American ghetto life – and the life of immigrant and minority communities in cities worldwide – so stressful and frustrating, especially for youth and for families struggling for survival in bleak urban environments.

Perhaps the biggest difference between liberals and conservatives in the underclass debate lies in their views on the role of government in solving social problems. Many liberals would like to see government intervene with a universal, not necessarily race-specific, full employment program. If the poor of all races are employed, they reason, family stability will return, and substance abuse and criminal behavior will drop. Conservatives argue that the last thing the poor need is more government assistance. For example, Charles Murray, author of *Losing Ground* (1984) and *The Bell Curve* (1994), argues that patronizing government programs sapped initiative from the poor and created perverse incentives to stay out of the labor market. And Michael Porter (p. 282) argues that programs that do not address the real competitive advantages of inner cities are counterproductive. Murray's remedy: a form of low-level guaranteed national income in place of the existing liberal programs of Social Security and welfare. Porter's remedy: redirect government aid and corporate philanthropy to economic development programs that will employ low-skilled urban residents in jobs that modern economies need.

In the United States, the "underclass debate" changed direction when, in the mid-1990s, the American Congress and President Clinton changed "welfare as we know it" by radically cutting back on many welfare support programs in favor of decentralized "workfare." On balance, the results have seemed promising, with welfare rolls declining and employment trending upward. At the same time, however, another concern emerged – or rather re-emerged – about the fundamental quality of civic culture in contemporary urban society. The leading voice raising this concern was that of Robert Putnam (p.134), author of *Bowling Alone*

(2000). Drawing on evidence from both Europe and the United States, and without downplaying the importance of the issues of race and poverty, Putnam asked urban leaders and members of the urban middle class to confront the evident social reality that people are no longer as connected to the basic institutions of their communities – the neighborhood groups, the fraternal organizations, even the political parties – as they once were. Putnam attributes the growing lack of community participation – what he calls the decline in "social capital" and loss of civic engagement – to many factors: the movement of women into the workforce, increased social and geographical mobility, and "the technological transformation of leisure." Another explanation might be that contemporary urban society is deeply divided and that every major city has now become culturally contested terrain. Culture used to mean "high culture" – the world of the symphony, opera, and ballet. But culture, of course, is not just the product of an identifiable artistic or intellectual class. Working-class neighborhoods and inner-city ghettos also produce formal poetry and the linguistic inventiveness of street talk, rhapsodies and jazz music, paintings and graffiti. These are surely elements of "the urban drama," but who will dominate a city's culture? Whose "social capital" will be hegemonic in the socially "contested city"? And what exactly constitutes "civic engagement"?

Richard Florida (p. 143) addresses another aspect of the debate on the future of urban economies – the role of the "creative class." In *The Rise of the Creative Class* (2002), Florida argues that a creative environment – or at least one that is compatible with creativity and creative people – is essential to urban life and a city's economic success, especially in the postmodern Information Age. Consisting of "information managers" and "symbolic analysts" such as engineers, artists, software programmers, writers, and corporate strategists involved in the new post-industrial economy, these new urban dwellers are knowledge workers who add value to their enterprises, and to society as a whole, by exercising their creative imaginations. More than just an educated class of high-end service workers, members of the creative class bring a new vitality to the city and transform urban culture through their commitment to the values of individuality, meritocracy, and diversity.

Sometimes, of course, cultures clash within an urban environment. As socio-economic classes – or racial, ethnic, or immigrant groups – compete for benefits and status within the urban order, short-term confrontations and long-term accommodations tend to take place as urban society slowly evolves. In the new globalized urban world, however, some culture clashes can take on a more desperate character, and this is nowhere more in evidence than in the conflict between the broadly "liberal" and tolerant values of the modern Western city and the more narrowly prescribed moral values of radical fundamentalist Islam that forms the basis of the current War on Terrorism. When the events of September 11, 2001, took place in New York, many cultural analysts attempted to explain what had happened. Some posited a "clash of civilizations." Others explored how many in the non-dominant, non-Western world conceived of the West, especially the modern Western city, as the locus and source of ungodliness and moral depravity. With the Islamic world extending from Indonesia to Morocco, and with growing Islamic populations in the cities of Europe and North America, the culture clash defined by the War on Terrorism has already had deep effects on city life almost everywhere and has introduced elements of fear, uncertainty, and heightened security measures into the daily round of urban life. Elsewhere in the globalizing urban world, accelerated rural-to-urban migrations have brought new populations into existing cities, causing new conflicts and social transformations. These new developments are discussed in greater detail in Part Eight, Cities in a Global Society (p. 541).

Finally, in exploring urban society and culture, academic analysts recognize that urban culture has other dimensions that do not lend themselves so easily to purely sociological analysis or narrative description. Among these are the expressions of the creative arts and the other cultural productions that emerge from urban communities everywhere, indeed from the very conditions of modern urban life. In "Visions of a New Reality: The City and the Emergence of Modern Visual Culture," the image-essay that ends Part Two, Frederic Stout argues that popular illustrated journalism, photography, and cinema are among the most characteristic of urban cultural genres and that visual culture generally is an artifact of modern urban society.

"What is a City?"

Architectural Record (1937)

Lewis Mumford

Editors' Introduction

Lewis Mumford (1895–1990) has been called the United States' last great public intellectual – that is, a scholar not based in academia who writes for an educated popular audience. Beginning with the publication of his first book *The Story of Utopias* in 1922 and continuing throughout a career that saw the publication of some twenty-five influential volumes, Mumford made signal contributions to social philosophy, American literary and cultural history, the history of technology and, preeminently, the history of cities and urban planning practice.

Born in Brooklyn and coming of age at a time when the modern city was reaching a new peak in the history of urban civilization, Mumford saw the urban experience as an essential component in the development of human culture and the human personality. He consistently argued that the physical design of cities and their economic functions were secondary to their relationship to the natural environment and to the spiritual values of human community. Mumford applied these principles to his architectural criticism for *The New Yorker* magazine and his work with the Regional Planning Association of America in the 1920s and 1930s, his campaign against plans to build a highway through Washington Square in New York's Greenwich Village in the 1950s, and his lifelong championing of the environmental theories of Patrick Geddes and the Garden City ideals of Ebenezer Howard.

In "What is a City?" – the text of a 1937 talk to an audience of urban planners – Mumford lays out his fundamental propositions about city planning and the human potential, both individual and social, of urban life. The city, he writes, is "a theater of social action," and everything else – art, politics, education, commerce – serve only to make the "social drama . . . more richly significant, as a stage-set, well-designed, intensifies and underlines the gestures of the actors and the action of the play." The city as a form of social drama expressed as much in daily life as in revolutionary moments – it was a theme and an image to which Mumford would return over and over again. In *The Culture of Cities* of 1938, he rhapsodized about the artist Albrecht Dürer witnessing a religious procession in Antwerp in 1519 that was a dramatic performance "where the spectators were also communicants." And in "The Urban Drama" from *The City in History* of 1961, he reflected on the ways that the social life of the ancient city established a kind of dramatic dialogue "in which common life itself takes on the features of a drama, heightened by every device of costume and scenery, for the setting itself magnifies the voice and increases the apparent stature of the actors." Mumford was quick to point out that the earliest urban dialogue was really a one-way "monologue of power" from the king to his cowering subjects. Such an absence of true dialogue, he wrote, was "bound to have a fatal last act." But real dialogue developed slowly but irresistibly in the forum, the agora, or the neighborhood. In the end, said Mumford, great moments of urban civilization often found expression in theatrical and literary dialogues – in everything from Plato's *Republic* to the plays of Shakespeare – that sum up the city's "total experience of life." It is an arresting insight and leads us to wonder what movies, television shows, popular websites and video games say about the quality of our present-day urban civilization.

Mumford's influence on the theory and practice of modern urban planning can hardly be overstated. His "urban drama" idea clearly resonates with an entire line of urban cultural analysts. Jane Jacobs, for example, talks about "street ballet" (p. 105). William Whyte (p. 510) says that a good urban plaza should function like a stage. Allan

Jacobs and Donald Appleyard (p. 518) urge planners to fulfill human needs for "fantasy and exoticism." The city, they write, "has always been a place of excitement; it is a theater, a stage upon which citizens can display themselves and be seen by others." And Mumford would no doubt have approved of economist Richard Florida (p. 143) and his argument for the importance to urban culture of a "creative class."

As a historian, Mumford's emphasis on community values and the city's role in enlarging the potential of the human personality connects him with a long line of urban theorists that includes Louis Wirth (p. 96) and many others. *The City in History* (1961) is undoubtedly Mumford's masterpiece, but an earlier version of the same material, *The Culture of Cities* (1938), is still of interest. *The Urban Prospect* (1968) is an outstanding collection of his essays on urban planning and culture, and *The Myth of the Machine* (1967) and *The Pentagon of Power* (1970) are excellent analyses of the influence of technology on human culture. The magisterial *The Transformations of Man* (1956) invites comparison with V. Gordon Childe's theory of the urban revolution (p. 31). And Mumford's ideas about urban regionalism and his advocacy of Ebenezer Howard's Garden City (p. 328) are foundational to the theories of Peter Calthorpe (p. 360) and other New Urbanists.

A sampling of Mumford's writings is included in Donald L. Miller (ed.), *The Lewis Mumford Reader* (Athens, GA: University of Georgia Press, 1995). Mumford's illuminating correspondence with Patrick Geddes is contained in Frank G. Novak, *Lewis Mumford and Patrick Geddes: The Correspondence* (London: Routledge, 1995). His correspondence with Frank Lloyd Wright is contained in Bruce Brooks Pfeiffer et al., *Frank Lloyd Wright and Lewis Mumford: Thirty Years of Correspondence* (New York: Princeton Architectural Press, 2001), and his writings for *The New Yorker* are contained in Robert Wojtowicz (ed.), *Sidewalk Critic: Lewis Mumford's Writings on New York* (New York: Princeton Architectural Press, 1998).

Mumford is now being rediscovered by a new generation of environmental planners. Examples of books applying his perspective to current ecological issues are Mark Luccarelli Lewis, *Mumford and the Ecological Region: The Politics of Planning* (New York: Guilford, 1997) and Robert Wojtowicz, *Lewis Mumford and American Modernism: Eutopian Theories for Architecture and Urban Planning* (Cambridge: Cambridge University Press, 1998).

Biographies of Lewis Mumford are Donald L. Miller, *Lewis Mumford: A Life* (New York: Weidenfeld & Nicolson, 1989), Thomas P. Hughes and Agatha C. Hughes (eds), *Lewis Mumford: Public Intellectual* (Oxford: Oxford University Press, 1990), and Frank G. Novak, *Lewis Mumford* (New York: Twayne, 1998). An excellent bibliography of Mumford's writings is Elmer S. Newman, *Lewis Mumford: A Bibliography, 1914–1970* (New York: Harcourt Brace Jovanovich, 1971).

■

Most of our housing and city planning has been handicapped because those who have undertaken the work have had no clear notion of the social functions of the city. They sought to derive these functions from a cursory survey of the activities and interests of the contemporary urban scene. And they did not, apparently, suspect that there might be gross deficiencies, misdirected efforts, mistaken expenditures here that would not be set straight by merely building sanitary tenements or straightening out and widening irregular streets.

The city as a purely physical fact has been subject to numerous investigations. But what is the city as a social institution? The earlier answers to these questions, in Aristotle, Plato, and the Utopian writers from Sir Thomas More to Robert Owen, have been on the whole more satisfactory than those of the more systematic sociologists: most contemporary treatises on "urban sociology" in America throw no important light upon the problem. One of the soundest definitions of the city was that framed by John Stow, an honest observer of Elizabethan London, who said:

> Men are congregated into cities and commonwealths for honesty and utility's sake, these shortly be the commodities that do come by cities, commonalties and corporations. First, men by this nearness of conversation are withdrawn from barbarous fixity and force, to certain mildness of manners, and to humanity and justice . . . Good behavior is yet called urbanitas because it is rather found in cities than elsewhere. In sum, by often hearing, men be better persuaded in religion, and for that they live in the eyes of others, they be by example the more easily trained to justice, and by shamefastness restrained from injury.

> And whereas commonwealths and kingdoms cannot have, next after God, any surer foundation than the love and good will of one man towards another, that also is closely bred and maintained in cities, where men by mutual society and companying together, do grow to alliances, commonalties, and corporations.

It is with no hope of adding much to the essential insight of this description of the urban process that I would sum up the sociological concept of the city in the following terms:

The city is a related collection of primary groups and purposive associations: the first, like family and neighborhood, are common to all communities, while the second are especially characteristic of city life. These varied groups support themselves through economic organizations that are likewise of a more or less corporate, or at least publicly regulated, character; and they are all housed in permanent structures, within a relatively limited area. The essential physical means of a city's existence are the fixed site, the durable shelter, the permanent facilities for assembly, interchange, and storage; the essential social means are the social division of labor, which serves not merely the economic life but the cultural processes. The city in its complete sense, then, is a geographic plexus, an economic organization, an institutional process, a theater of social action, and an aesthetic symbol of collective unity. The city fosters art and is art; the city creates the theater and is the theater. It is in the city, the city as theater, that man's more purposive activities are focused, and work out, through conflicting and cooperating personalities, events, groups, into more significant culminations.

Without the social drama that comes into existence through the focusing and intensification of group activity there is not a single function performed in the city that could not be performed – and has not in fact been performed – in the open country. The physical organization of the city may deflate this drama or make it frustrate; or it may, through the deliberate efforts of art, politics, and education, make the drama more richly significant, as a stage-set, well-designed, intensifies and underlines the gestures of the actors and the action of the play. It is not for nothing that men have dwelt so often on the beauty or the ugliness of cities: these attributes qualify men's social activities. And if there is a deep reluctance on the part of the true city dweller to leave his cramped quarters for the physically more benign environment of a suburb – even a model garden suburb! – his instincts are usually justified: in its various and many-sided life, in its very opportunities for social disharmony and conflict, the city creates drama; the suburb lacks it.

One may describe the city, in its social aspect, as a special framework directed toward the creation of differentiated opportunities for a common life and a significant collective drama. As indirect forms of association, with the aid of signs and symbols and specialized organizations, supplement direct face-to-face intercourse, the personalities of the citizens themselves become many-faceted: they reflect their specialized interests, their more intensively trained aptitudes, their finer discriminations and selections: the personality no longer presents a more or less unbroken traditional face to reality as a whole. Here lies the possibility of personal disintegration; and here lies the need for reintegration through wider participation in a concrete and visible collective whole. What men cannot imagine as a vague formless society, they can live through and experience as citizens in a city. Their unified plans and buildings become a symbol of their social relatedness; and when the physical environment itself becomes disordered and incoherent, the social functions that it harbors become more difficult to express.

One further conclusion follows from this concept of the city: social facts are primary, and the physical organization of a city, its industries and its markets, its lines of communication and traffic, must be subservient to its social needs. Whereas in the development of the city during the last century we expanded the physical plant recklessly and treated the essential social nucleus, the organs of government and education and social service, as mere afterthought, today we must treat the social nucleus as the essential element in every valid city plan: the spotting and inter-relationship of schools, libraries, theaters, community centers is the first task in defining the urban neighborhood and laying down the outlines of an integrated city.

In giving this sociological answer to the question: What is a City? one has likewise provided the clue to a number of important other questions. Above all, one has the criterion for a clear decision as to what is the desirable size of a city – or may a city perhaps continue to grow until a single continuous urban area might cover half the American continent, with the rest of the world tributary to this mass? From the standpoint of the purely physical organization of urban utilities

– which is almost the only matter upon which metropolitan planners in the past have concentrated – this latter process might indeed go on indefinitely. But if the city is a theater of social activity, and if its needs are defined by the opportunities it offers to differentiated social groups, acting through a specific nucleus of civic institutes and associations, definite limitations on size follow from this fact.

In one of Le Corbusier's early schemes for an ideal city, he chose three million as the number to be accommodated: the number was roughly the size of the urban aggregate of Paris, but that hardly explains why it should have been taken as a norm for a more rational type of city development. If the size of an urban unit, however, is a function of its productive organization and its opportunities for active social intercourse and culture, certain definite facts emerge as to adequate ratio of population to the process to be served. Thus, at the present level of culture in America, a million people are needed to support a university. Many factors may enter which will change the size of both the university and the population base; nevertheless one can say provisionally that if a million people are needed to provide a sufficient number of students for a university, then two million people should have two universities. One can also say that, other things being equal, five million people will not provide a more effective university than one million people would. The alternative to recognizing these ratios is to keep on overcrowding and overbuilding a few existing institutions, thereby limiting, rather than expanding, their genuine educational facilities.

What is important is not an absolute figure as to population or area: although in certain aspects of life, such as the size of city that is capable of reproducing itself through natural fertility, one can already lay down such figures. What is more important is to express size *always as a function of the social relationships to be served* . . . There is an optimum numerical size, beyond which each further increment of inhabitants creates difficulties out of all proportion to the benefits. There is also an optimum area of expansion, beyond which further urban growth tends to paralyze rather than to further important social relationships. Rapid means of transportation have given a regional area with a radius of from forty to a hundred miles, the unity that London and Hampstead had before the coming of the underground railroad. But the activities of small children are still bounded by a walking distance of about a quarter of a mile; and for men to congregate freely and frequently in neighborhoods the maximum distance means nothing, although it may properly define the area served for a selective minority by a university, a central reference library, or a completely equipped hospital. The area of potential urban settlement has been vastly increased by the motor car and the airplane; but the necessity for solid contiguous growth, for the purposes of intercourse, has in turn been lessened by the telephone and the radio. In the Middle Ages a distance of less than a half a mile from the city's center usually defined its utmost limits. The block-by-block accretion of the big city, along its corridor avenues, is in all important respects a denial of the vastly improved type of urban grouping that our fresh inventions have brought in. For all occasional types of intercourse, the region is the unit of social life but the region cannot function effectively, as a well-knit unit, if the entire area is densely filled with people – since their very presence will clog its arteries of traffic and congest its social facilities.

Limitations on size, density, and area are absolutely necessary to effective social intercourse; and they are therefore the most important instruments of rational economic and civic planning. The unwillingness in the past to establish such limits has been due mainly to two facts: the assumption that all upward changes in magnitude were signs of progress and automatically "good for business," and the belief that such limitations were essentially arbitrary, in that they proposed to "decrease economic opportunity" – that is, opportunity for profiting by congestion – and to halt the inevitable course of change. Both these objections are superstitious.

Limitations on height are now common in American cities; drastic limitations on density are the rule in all municipal housing estates in England: that which could not be done has been done. Such limitations do not obviously limit the population itself: they merely give the planner and administrator the opportunity to multiply the number of centers in which the population is housed, instead of permitting a few existing centers to aggrandize themselves on a monopolistic pattern. These limitations are necessary to break up the functionless, hypertrophied urban masses of the past. Under this mode of planning, the planner proposes to replace the "mononucleated city," as Professor Warren Thompson has called it, with a new type of "polynucleated city," in which a cluster of communities, adequately spaced and bounded, shall do duty for the badly organized mass city. Twenty such cities, in a region whose environment and whose resources were adequately planned, would have all the benefits of a metropolis that held a million people, without its ponderous disabilities: its capital frozen into unprofitable utilities, and its land values congcaled at levels that stand in the way of effective adaptation to new needs.

Mark the change that is in process today. The emerging sources of power, transport, and communication do not follow the old highway network at all. Giant power strides over the hills, ignoring the limitations of wheeled vehicles; the airplane, even more liberated, flies over swamps and mountains, and terminates its journey, not on an avenue, but in a field. Even the highway for fast motor transportation abandons the pattern of the horse-and-buggy era. The new highways, like those of New Jersey and Westchester, to mention only examples drawn locally, are based more or less on a system definitively formulated by Benton MacKaye in his various papers on the Townless Highway. The most complete plans form an independent highway network, isolated both from the adjacent countryside and the towns that they bypass: as free from communal encroachments as the railroad system. In such a network no single center will, like the metropolis of old, become the focal point of all regional advantages: on the contrary, the "whole region" becomes open for settlement.

Even without intelligent public control, the likelihood is that within the next generation this dissociation and decentralization of urban facilities will go even farther. The Townless Highway begets the Highwayless Town in which the needs of close and continuous human association on all levels will be uppermost. This is just the opposite of the earlier mechanocentric picture of Roadtown, as pictured by Edgar Chambless and the Spanish projectors of the Linear City. For the highwayless town is based upon the notion of effective zoning of functions through initial public design, rather than by blind legal ordinances. It is a town in which the various functional parts of the structure are isolated topographically as urban islands, appropriately designed for their specific use with no attempt to provide a uniform plan of the same general pattern for the industrial, the commercial, the domestic, and the civic parts.

The first systematic sketch of this type of town was made by Messrs. Wright and Stein in their design for Radburn in 1929; a new type of plan that was repeated on a limited scale – and apparently in complete independence – by planners in Köln and Hamburg at about the same time. Because of restrictions on design that favored a conventional type of suburban house and stale architectural forms, the implications of this new type of planning were not carried very far in Radburn. But in outline the main relationships are clear: the differentiation of foot traffic from wheeled traffic in independent systems, the insulation of residence quarters from through roads; the discontinuous street pattern; the polarization of social life in specially spotted civic nuclei, beginning in the neighborhood with the school and the playground and the swimming pool. This type of planning was carried to a logical conclusion in perhaps the most functional and most socially intelligent of all Le Corbusier's many urban plans: that for Nemours in North Africa, in 1934.

Through these convergent efforts, the principles of the polynucleated city have been well established. Such plans must result in a fuller opportunity for the primary group, with all its habits of frequent direct meeting and face-to-face intercourse: they must also result in a more complicated pattern and a more comprehensive life for the region, for this geographic area can only now, for the first time, be treated as an instantaneous whole for all the functions of social existence. Instead of trusting to the mere massing of population to produce the necessary social concentration and social drama, we must now seek these results through deliberate local nucleation and a finer regional articulation. The words are jargon; but the importance of their meaning should not be missed. To embody these new possibilities in city life, which come to us not merely through better technical organization but through acuter sociological understanding, and to dramatize the activities themselves in appropriate individual and urban structures, forms the task of the coming generation.

"Urbanism as a Way of Life"

American Journal of Sociology (1938)

Louis Wirth

Editors' Introduction

Louis Wirth (1897–1952) was a member of the famed Chicago school of urban sociology that included such academic luminaries as Ernest W. Burgess (author of "The Growth of the City," p. 161), Robert E. Park, and St. Clair Drake. Together, these scholars at the University of Chicago set out to reinvent modern sociology by taking academic research to the streets and by using the city of Chicago itself as a "living laboratory" for the study of urban problems and social processes.

Wirth was born in Germany, emigrated to the United States as a child, and rose within academia to become the president of the American Sociological Association. His major contribution to urban sociology was the formulation of nothing less fundamental than a meaningful and logically coherent "sociological definition" of urban life. As he lays it out in the magnificent synthesis that is his 1938 essay "Urbanism as a Way of Life," a "sociologically significant definition of the city" looks beyond the mere physical structure of the city, or its economic product, or its characteristic cultural institutions – however important all these may be – to discover those underlying "elements of urbanism which mark it as a distinctive mode of human group life."

Wirth argues that three key characteristics of cities – large population size, social heterogeneity, and population density – contribute to the development of a peculiarly "urban way of life" and, indeed, a distinct "urban personality." For centuries, at least as far back as Aesop's fable of the city mouse and the country mouse, casual observers have noted sharp personality differences between urban and rural people and between nature-based and machine-based styles of living. Wirth attempts to explain those differences in terms of the functional responses of urban dwellers to the characteristic environmental conditions of modern urban society. If, for example, city people are regarded as rather more socially tolerant that rural people – and, at the same time, more impersonal and seemingly less friendly – these are merely adaptations to the experience of living in large, dense, socially diverse urban environments. Wirth's analysis invites comparison with Georg Simmel's "The Metropolis and Mental Life," delivered as a lecture in 1903 and reprinted in Jan Lin and Christopher Mele (eds.), *The Urban Sociology Reader* (London: Routledge, 2005).

Although some see Wirth's explanation of the sociology of urban life as nothing more than the social scientific verification of the obvious, others have argued that there is actually no such thing as an "urban personality" or an "urban way of life." Sociologist Herbert Gans, for example, argues that both inner-city "urban villagers" and suburbanites tend to maintain their preexisting cultures and personalities, and Oscar Lewis's work on "the culture of poverty" – along with a whole body of Marxist analysis – suggests that culture and personality types differ widely with socioeconomic class, not merely being "urban." Wirth's work, however, led to the development of a whole school of urban social ecology, and Wirth's ideas about personality and adaptation to urban conditions – many of them quite pessimistic – inform the full range of more recent urban planning theories and the planning practitioners who attempt to create and nurture a sense of community in the urban environment.

Can physical design of the built environment improve people's sense of community, psychological well-being and adjustment to urban life? Many sociologists and psychologists as well as architects and urban designers have proposed ways to design cities to address the troubling concerns that Wirth, Simmel, and others have raised about the psychological disorientation that city life can bring. All of the selections on urban design in Part seven evidence some degree of environmental determinism – the idea that environment, including the built environment, will to affect human behavior. Kevin Lynch (p. 499), a great figure in twentieth-century urban design believed that improving the image of the city would increase residents' comfort level with their surroundings. Lynch identified elements of the city image that people perceive and proposed strategies to design the image of the city by improving the design of its various elements. Danish architect and planner Jan Gehl (p. 530) is convinced that if people spend more time outside enjoying the space between buildings there will be more social interaction and human happiness. Sociologist-turned-urban designer William Whyte (p. 510) lays out a set of a whole range of very practical design suggestions to increase use and enjoyment of parks and plazas. Building in more sittable space, making food available, and reducing the disconnect between the street and these important public spaces will, Whyte argues, improve people's life experience in cities. But the question remains: can the kind of profound alienation of big city life described by Wirth really be ameliorated just by good design?

Other books by Louis Wirth include *Contemporary Social Problems* (Chicago, IL: University of Chicago Press, 1940), *The Effect of War on American Minorities* (New York: Social Science Research Council, 1943), *Community Life and Social Policy* (Chicago, IL: University of Chicago Press, 1956), and *The Ghetto* (Chicago, IL: University of Chicago Press, 1956). *Louis Wirth on Cities and Social Life: Selected Papers* (Chicago, IL: University of Chicago Press, 1964) is a useful collection. Also of interest is Roger A. Salerno, *Louis Wirth: A Bio-Bibliography* (Westport, CT: Greenwood, 1987).

For other important analyses of the relationship between urban life and the human personality, see Erving Goffman, *The Presentation of Self in Everyday Life* (Garden City, NY: Doubleday, 1959) and Richard Sennett, *The Uses of Disorder: Personal Identity and City Life* (New York: Norton, 1970). Of related interest are Sylvia Fleis Fava, "Suburbanism as a Way of Life" (*American Sociological Review*, 21(1), 1956) and Fred Dewey, "Cyberurbanism as a Way of Life," from *Architecture of Fear* (Princeton, NJ: Princeton University Press, 1997), reprinted in Stephen Graham (ed.), *The Cybercities Reader* (London: Routledge, 2004).

■

THE CITY AND CONTEMPORARY CIVILIZATION

Just as the beginning of Western civilization is marked by the permanent settlement of formerly nomadic peoples in the Mediterranean basin, so the beginning of what is distinctively modern in our civilization is best signalized by the growth of great cities. Nowhere has mankind been farther removed from organic nature than under the conditions of life characteristic of great cities . . . The city and the country may be regarded as two poles in reference to one or the other of which all human settlements tend to arrange themselves. In viewing urban-industrial and rural-folk society as ideal types of communities, we may obtain a perspective for the analysis of the basic models of human association as they appear in contemporary civilization.

A SOCIOLOGICAL DEFINITION OF THE CITY

Despite the preponderant significance of the city in our civilization, however, our knowledge of the nature of urbanism and the process of urbanization is meager. Many attempts have indeed been made to isolate the distinguishing characteristics of urban life. Geographers, historians, economists, and political scientists have incorporated the points of view of their respective disciplines into diverse definitions of the city. While it is in no sense intended to supersede these, the formulation of a sociological approach to the city may incidentally serve to call attention to the interrelations between them by emphasizing the peculiar characteristics of the city as a particular form of human association. A sociologically significant definition of the city seeks to select those elements of urbanism which mark it as a distinctive mode of human group life.

[. . .]

While urbanism, or that complex of traits which makes up the characteristic mode of life in cities, and urbanization, which denotes the development and extensions of these factors, are thus not exclusively found in settlements which are cities in the physical and demographic sense, they do, nevertheless, find their most pronounced expression in such areas, especially in metropolitan cities. In formulating a definition of the city it is necessary to exercise caution in order to avoid identifying urbanism as a way of life with any specific locally or historically conditioned cultural influences which, while they may significantly affect the specific character of the community, are not the essential determinants of its character as a city.

It is particularly important to call attention to the danger of confusing urbanism with industrialism and modern capitalism. The rise of cities in the modern world is undoubtedly not independent of the emergence of modern power-driven machine technology, mass production, and capitalistic enterprise. But different as the cities of earlier epochs may have been by virtue of their development in a preindustrial and precapitalistic order from the great cities of today, they were, nevertheless, cities.

For sociological purposes a city may be defined as a relatively large, dense, and permanent settlement of socially heterogeneous individuals. On the basis of the postulates which this minimal definition suggests, a theory of urbanism may be formulated in the light of existing knowledge concerning social groups.

A THEORY OF URBANISM

In the rich literature on the city we look in vain for a theory of urbanism presenting in a systematic fashion the available knowledge concerning the city as a social entity. We do indeed have excellent formulations of theories on such special problems as the growth of the city viewed as a historical trend and as a recurrent process, and we have a wealth of literature presenting insights of sociological relevance and empirical studies offering detailed information on a variety of particular aspects of urban life. But despite the multiplication of research and textbooks on the city, we do not as yet have a comprehensive body of competent hypotheses which may be derived from a set of postulates implicitly contained in a sociological definition of the city, and from our general sociological knowledge which may be substantiated through empirical research. The closest approximations to a systematic theory of urbanism that we have are to be found in a penetrating essay, "Die Stadt," by Max Weber, and a memorable paper by Robert E. Park titled "The City: Suggestions for the Investigation of Human Behavior in the Urban Environment." But even these excellent contributions are far from constituting an ordered and coherent framework of theory upon which research might profitably proceed.

In the pages that follow, we shall seek to set forth a limited number of identifying characteristics of the city. Given these characteristics we shall then indicate what consequences or further characteristics follow from them in the light of general sociological theory and empirical research. We hope in this manner to arrive at the essential propositions comprising a theory of urbanism. Some of these propositions can be supported by a considerable body of already available research materials; others may be accepted as hypotheses for which a certain amount of presumptive evidence exists, but for which more ample and exact verification would be required. At least such a procedure will, it is hoped, show what in the way of systematic knowledge of the city we now have and what are the crucial and fruitful hypotheses for future research.

[. . .]

There are a number of sociological propositions concerning the relationship between (a) numbers of population, (b) density of settlement, (c) heterogeneity of inhabitants and group life, which can be formulated on the basis of observation and research.

SIZE OF THE POPULATION AGGREGATE

Ever since Aristotle's *Politics*, it has been recognized that increasing the number of inhabitants in a settlement beyond a certain limit will affect the relationships between them and the character of the city. Large numbers involve, as has been pointed out, a greater range of individual variation. Furthermore, the greater the number of individuals participating in a process of interaction, the greater is the potential differentiation between them. The personal traits, the occupations, the cultural life, and the ideas of the members of an urban community may, therefore, be expected to range between more widely separated poles than those of rural inhabitants.

That such variations should give rise to the spatial segregation of individuals according to color, ethnic heritage, economic and social status, tastes and preferences, may readily be inferred. The bonds of kinship, of neighborliness, and the sentiments arising out of living together for generations under a common folk tradition are likely to be absent or, at best, relatively weak in an aggregate the members of which have such diverse origins and backgrounds. Under such circumstances competition and formal control mechanisms furnish the substitutes for the bonds of solidarity that are relied upon to hold a folk society together.

[. . .]

The multiplication of persons in a state of interaction under conditions which make their contact as full personalities impossible produces that segmentalization of human relationships which has sometimes been seized upon by students of the mental life of the cities as an explanation for the "schizoid" character of urban personality. This is not to say that the urban inhabitants have fewer acquaintances than rural inhabitants, for the reverse may actually be true; it means rather that in relation to the number of people whom they see and with whom they rub elbows in the course of daily life, they know a smaller proportion, and of these they have less intensive knowledge.

Characteristically, urbanites meet one another in highly segmental roles. They are, to be sure, dependent upon more people for the satisfactions of their life-needs than are rural people and thus are associated with a greater number of organized groups, but they are less dependent upon particular persons, and their dependence upon others is confined to a highly fractionalized aspect of the other's round of activity. This is essentially what is meant by saying that the city is characterized by secondary rather than primary contacts. The contacts of the city may indeed be face-to-face, but they are nevertheless impersonal, superficial, transitory, and segmental. The reserve, the indifference, and the blasé outlook which urbanites manifest in their relationships may thus be regarded as devices for immunizing themselves against the personal claims and expectations of others.

The superficiality, the anonymity, and the transitory character of urban social relations make intelligible, also, the sophistication and the rationality generally ascribed to city-dwellers. Our acquaintances tend to stand in a relationship of utility to us in the sense that the role which each one plays in our life is overwhelmingly regarded as a means for the achievement of our own ends. Whereas, therefore, the individual gains, on the one hand, a certain degree of emancipation or freedom from the personal and emotional controls of intimate groups, he loses, on the other hand, the spontaneous self-expression, the morale, and the sense of participation that comes with living in an integrated society. This constitutes essentially the state of anomie or the social void to which Durkheim alludes in attempting to account for the various forms of social disorganization in technological society.

The segmental character and utilitarian accent of interpersonal relations in the city find their institutional expression in the proliferation of specialized tasks which we see in their most developed form in the professions. The operations of the pecuniary nexus lead to predatory relationships, which tend to obstruct the efficient functioning of the social order unless checked by professional codes and occupational etiquette. The premium put upon utility and efficiency suggests the adaptability of the corporate device for the organization of enterprises in which individuals can engage only in groups. The advantage that the corporation has over the individual entrepreneur and the partnership in the urban-industrial world derives not only from the possibility it affords of centralizing the resources of thousands of individuals or from the legal privilege of limited liability and perpetual succession, but from the fact that the corporation has no soul.

[. . .]

DENSITY

As in the case of numbers, so in the case of concentration in limited space certain consequences of relevance in sociological analysis of the city emerge. Of these only a few can be indicated.

As Darwin pointed out for flora and fauna and as Durkheim noted in the case of human societies, an increase in numbers when area is held constant (i.e. an increase in density) tends to produce differentiation and specialization, since only in this way can the area support increased numbers. Density thus reinforces the effect of numbers in diversifying men and their activities and in increasing the complexity of the social structure.

On the subjective side, as Simmel has suggested, the close physical contact of numerous individuals necessarily produces a shift in the mediums through which we orient ourselves to the urban milieu, especially to our fellow-men. Typically, our physical contacts are close but our social contacts are distant. The urban world puts a premium on visual recognition. We see the uniform which denotes the role of the functionaries and are oblivious to the personal eccentricities that are hidden behind the uniform. We tend to acquire and develop a sensitivity to a world of artifacts and become progressively farther removed from the world of nature.

We are exposed to glaring contrasts between splendor and squalor, between riches and poverty, intelligence and ignorance, order and chaos. The competition for space is great, so that each area generally tends to be put to the use which yields the greatest economic return. Place of work tends to become dissociated from place of residence, for the proximity of industrial and commercial establishments makes an area both economically and socially undesirable for residential purposes.

Density, land values, rentals, accessibility, healthfulness, prestige, aesthetic consideration, absence of nuisances such as noise, smoke, and dirt determine the desirability of various areas of the city as places of settlement for different sections of the population . . . The different parts of the city thus acquire specialized functions. The city consequently tends to resemble a mosaic of social worlds in which the transition from one to the other is abrupt. The juxtaposition of divergent personalities and modes of life tends to produce a relativistic perspective and a sense of toleration of differences which may be regarded as prerequisites for rationality and which lead toward the secularization of life.

The close living together and working together of individuals who have no sentimental and emotional ties foster a spirit of competition, aggrandizement, and mutual exploitation. To counteract irresponsibility and potential disorder, formal controls tend to be resorted to. Without rigid adherence to predictable routines a large, compact society would scarcely be able to maintain itself. The clock and the traffic signal are symbolic of the basis of our social order in the urban world. Frequent close physical contact, coupled with great social distance, accentuates the reserve of unattached individuals toward one another and, unless compensated for by other opportunities for response, gives rise to loneliness. The necessary frequent movement of great numbers of individuals in a congested habitat gives occasion to friction and irritation. Nervous tensions which derive from such personal frustrations are accentuated by the rapid tempo and the complicated technology under which life in dense areas must be lived.

HETEROGENEITY

The social interaction among such a variety of personality types in the urban milieu tends to break down the rigidity of caste lines and to complicate the class structure, and thus induces a more ramified and differentiated framework of social stratification than is found in more integrated societies. The heightened mobility of the individual, which brings him within the range of stimulation by a great number of diverse individuals and subjects him to fluctuating status in the differentiated social groups that compose the social structure of the city, tends toward the acceptance of instability and insecurity in the world at large as a norm. This fact helps to account, too, for the sophistication and cosmopolitanism of the urbanite. No single group has the undivided allegiance of the individual. The groups with which he is affiliated do not lend themselves readily to a simple hierarchical arrangement. By virtue of his different interests arising out of different aspects of social life, the individual acquires membership in widely divergent groups, each of which functions only with reference to a single segment of his personality. Nor do these groups easily permit of a concentric arrangement so that the narrower ones fall within the circumference of the more inclusive ones, as is more likely to be the case in the rural community or in primitive societies. Rather the groups with which the person typically is affiliated are tangential to each other or intersect in highly variable fashion.

Partly as a result of the physical footlooseness of the population and partly as a result of their social mobility, the turnover in group membership generally is rapid. Place of residence, place and character of employment, income and interests fluctuate, and the task of holding organizations together and maintaining and promoting intimate and lasting acquaintanceship between the members is difficult. This applies strikingly to the local areas within the city into which persons become segregated more by virtue of differences in race, language, income, and social status, than through choice or positive attraction to people like themselves. Overwhelmingly the city-dweller is not a home-owner, and since a transitory habitat does not generate binding traditions and sentiments, only rarely is he truly a neighbor. There is little opportunity for the individual to obtain a conception of the city as a whole or to survey his place in the total scheme. Consequently he finds it difficult to determine what is to his own "best interests" and to decide between the issues and leaders presented to him by the agencies of mass suggestion. Individuals who are thus detached from the organized bodies which integrate society comprise the fluid masses that make collective behavior in the urban community so unpredictable and hence so problematical.

Although the city, through the recruitment of variant types to perform its diverse tasks and the accentuation of their uniqueness through competition and the premium upon eccentricity, novelty, efficient performance, and inventiveness, produces a highly differentiated population, it also exercises a leveling influence. Wherever large numbers of differently constituted individuals congregate, the process of depersonalization also enters . . . Individuality under these circumstances must be replaced by categories. When large numbers have to make common use of facilities and institutions, an arrangement must be made to adjust the facilities and institutions to the needs of the average person rather than to those of particular individuals. The services of the public utilities, of the recreational, educational, and cultural institutions, must be adjusted to mass requirements. Similarly, the cultural institutions, such as the schools, the movies, the radio, and the newspapers, by virtue of their mass clientele, must necessarily operate as leveling influences. The political process as it appears in urban life could not be understood without taking account of the mass appeals made through modern propaganda techniques. If the individual would participate at all in the social, political, and economic life of the city, he must subordinate some of his individuality to the demands of the larger community and in that measure immerse himself in mass movements.

THE RELATION BETWEEN A THEORY OF URBANISM AND SOCIOLOGICAL RESEARCH

By means of a body of theory such as that illustratively sketched above, the complicated and many-sided phenomena of urbanism may be analyzed in terms of a limited number of basic categories. The sociological approach to the city thus acquires an essential unity and coherence enabling the empirical investigator not merely to focus more distinctly upon the problems and processes that properly fall in his province but also to treat his subject matter in a more integrated and systematic fashion. A few typical findings of empirical research in the field of urbanism, with special reference to the United States, may be indicated to substantiate the theoretical propositions set forth in the preceding pages, and some of the crucial problems for further study may be outlined.

On the basis of the three variables, number, density of settlement, and degree of heterogeneity, of the urban population, it appears possible to explain the characteristics of urban life and to account for the differences between cities of various sizes and types.

Urbanism as a characteristic mode of life may be approached empirically from three interrelated perspectives: (1) as a physical structure comprising a population base, a technology, and an ecological order; (2) as a system of social organization involving a characteristic social structure, a series of social institutions, and a typical pattern of social relationships; and (3) as a set of attitudes and ideas, and a constellation of personalities engaging in typical forms of collective behavior and subject to characteristic mechanisms of social control.

URBANISM IN ECOLOGICAL PERSPECTIVE

Since in the case of physical structure and ecological processes we are able to operate with fairly objective indices, it becomes possible to arrive at quite precise and generally quantitative results. The dominance of the city over its hinterland becomes explicable through the functional characteristics of the city which derive in large measure from the effect of numbers and density. Many of the technical facilities and the skills and organizations to which urban life gives rise can grow and prosper only in cities where the demand is sufficiently great. The nature and scope of the services rendered by these organizations and institutions and the advantage which they enjoy over the less developed facilities of smaller towns enhances the dominance of the city and the dependence of ever wider regions upon the central metropolis.

The urban population composition shows the operation of selective and differentiating factors. Cities contain a larger proportion of persons in the prime of life than rural areas which contain more old and very young people. In this, as in so many other respects, the larger the city the more this specific characteristic of urbanism is apparent. With the exception of the largest cities, which have attracted the bulk of the foreign-born males, and a few other special types of cities, women predominate numerically over men. The heterogeneity of the urban population is further indicated along racial and ethnic lines. The foreign born and their children constitute nearly two-thirds of all the inhabitants of cities of one million and over. Their proportion in the urban population declines as the size of the city decreases, until in the rural areas they comprise only about one-sixth of the total population. The larger cities similarly have attracted more Negroes and other racial groups than have the smaller communities. Considering that age, sex, race, and ethnic origin are associated with other factors such as occupation and interest, it becomes clear that one major characteristic of the urban-dweller is his dissimilarity from his fellows. Never before have such large masses of people of diverse traits as we find in our cities been thrown together into such close physical contact as in the great cities of America. Cities generally, and American cities in particular, comprise a motley of peoples and cultures, of highly differentiated modes of life between which there often is only the faintest communication, the greatest indifference and the broadest tolerance, occasionally bitter strife, but always the sharpest contrast.

The failure of the urban population to reproduce itself appears to be a biological consequence of a combination of factors in the complex of urban life, and the decline in the birth-rate generally may be regarded as one of the most significant signs of the urbanization of the Western world. While the proportion of deaths in cities is slightly greater than in the country, the outstanding difference between the failure of present-day cities to maintain their population and that of cities of the past is that in former times it was due to the exceedingly high death-rates in cities, whereas today, since cities have become more livable from a health standpoint, it is due to low birth-rates. These biological characteristics of the urban population are significant sociologically, not merely because they reflect the urban mode of existence but also because they condition the growth and future dominance of cities and their basic social organization. Since cities are the consumers rather than the producers of men, the value of human life and the social estimation of the personality will not be unaffected by the balance between births and deaths. The pattern of land use, of land values, rentals, and ownership, the nature and functioning of the physical structures, of housing, of transportation and communication facilities, of public utilities – these and many other phases of the physical mechanism of the city are not isolated phenomena unrelated to the city as a social entity, but are affected by and affect the urban mode of life.

URBANISM AS A FORM OF SOCIAL ORGANIZATION

The distinctive features of the urban mode of life have often been described sociologically as consisting of the substitution of secondary for primary contacts, the weakening of bonds of kinship, and the declining social significance of the family, the disappearance of the neighborhood, and the undermining of the traditional basis of social solidarity. All these phenomena can be substantially verified through objective indices. Thus, for instance, the low and declining urban reproduction rates suggest that the city is not conducive to the traditional type of family life, including the rearing of children and the maintenance of the home as the locus of a whole round of vital activities. The transfer of industrial, educational, and recreational activities to specialized institutions outside the home has deprived the family of some of its most characteristic historical functions. In cities mothers are more likely to be employed, lodgers are more frequently part of the household, marriage tends to be postponed, and the proportion of single and unattached people is greater. Families are smaller and more frequently without children than in the country. The family as a unit of social life is emancipated from the larger kinship group characteristic of the country, and the individual members pursue their own diverging interests in their vocational, educational, religious, recreational, and political life.

[. . .]

On the whole, the city discourages an economic life in which the individual in time of crisis has a basis of subsistence to fall back upon, and it discourages self-employment. While incomes of city people are on the average higher than those of country people, the cost of living seems to be higher in the larger cities. Home ownership involves greater burdens and is rarer. Rents are higher and absorb a large proportion of the income. Although the urban-dweller has the benefit of many communal services, he spends a large proportion of his income for such items as recreation and advancement and a smaller proportion for food. What the communal services do not furnish the urbanite must purchase, and there is virtually no human need which has remained unexploited by commercialism. Catering to thrills and furnishing means of escape from drudgery, monotony, and routine thus become one of the major functions of urban recreation, which at its best furnishes means for creative self-expression and spontaneous group association, but which more typically in the urban world results in passive spectatorism on the one hand, or sensational record-smashing feats on the other.

Being reduced to a stage of virtual impotence as an individual, the urbanite is bound to exert himself by joining with others of similar interest into organized groups to obtain his ends. This results in the enormous multiplication of voluntary organizations directed toward as great a variety of objectives as there are human needs and interests. While on the one hand the traditional ties of human association are weakened, urban existence involves a much greater degree of interdependence between man and man and a more complicated, fragile, and volatile form of mutual interrelations over many phases of which the individual as such can exert scarcely any control. Frequently there is only the most tenuous relationship between the economic position or other basic factors that determine the individual's existence in the urban world and the voluntary groups with which he is affiliated. While in a primitive and in a rural society it is generally possible to predict on the basis of a few known factors who will belong to what and who will associate with whom in almost every relationship of life, in the city we can only project the general pattern of group formation and affiliation, and this pattern will display many incongruities and contradictions.

URBAN PERSONALITY AND COLLECTIVE BEHAVIOR

It is largely through the activities of the voluntary groups, be their objectives economic, political, educational, religious, recreational, or cultural, that the urbanite expresses and develops his personality, acquires status, and is able to carry on the round of activities that constitute his life-career. It may easily be inferred, however, that the organizational framework which these highly differentiated functions call into being does not of itself insure the consistency and integrity of the personalities whose interests it enlists. Personal disorganization, mental breakdown, suicide, delinquency, crime, corruption, and disorder might be expected under these circumstances to be more prevalent in the urban than in the rural community. This has been confirmed insofar as comparable indices are available; but the mechanisms underlying these phenomena require further analysis.

Since for most group purposes it is impossible in the city to appeal individually to the large number of discrete and differentiated individuals, and since it is only through the organizations to which men belong that their interests and resources can be enlisted for a collective cause, it may be inferred that social control in the city should typically proceed through formally organized groups. It follows, too, that the masses of men in the city are subject to manipulation by symbols and stereotypes managed by individuals working from afar or operating invisibly behind the scenes through their control of the instruments of communication. Self-government either in the economic, the political, or the cultural realm is under these circumstances reduced to a mere figure of speech or, at best, is subject to the unstable equilibrium of pressure groups. In view of the ineffectiveness of actual kinship ties we create fictional kinship groups. In the face of the disappearance of the territorial unit as a basis of social solidarity we create interest units. Meanwhile the city as a community resolves itself into a series of tenuous segmental relationships superimposed upon a territorial base with a definite center but without a definite periphery and upon a division of labor which far transcends the immediate locality and is world-wide in scope. The larger the number of persons in a state of interaction with one another the lower is the level of communication and the greater is the tendency for communication to proceed on an elementary level, i.e. on the basis of those things which are assumed to be common or to be of interest to all.

It is obviously, therefore, to the emerging trends in the communication system and to the production and distribution technology that has come into existence with modern civilization that we must look for the symptoms which will indicate the probable future development of urbanism as a mode of social life. The direction of the ongoing changes in urbanism will for good or ill transform not only the city but the world. Some of the more basic of these factors and processes and the possibilities of their direction and control invite further detailed study.

It is only insofar as the sociologist has a clear conception of the city as a social entity and a workable theory of urbanism that he can hope to develop a unified body of reliable knowledge, which passes as "urban sociology" is certainly not at the present time. By taking his point of departure from a theory of urbanism such as that sketched in the foregoing pages to be elaborated, tested, and revised in the light of further analysis and empirical research, it is to be hoped that the criteria of relevance and validity of factual data can be determined. The miscellaneous assortment of disconnected information which has hitherto found its way into sociological treatises on the city may thus be sifted and incorporated into a coherent body of knowledge. Incidentally, only by means of some such theory will the sociologists escape the futile practice of voicing in the name of sociological science a variety of often unsupportable judgments concerning such problems as poverty, housing, city-planning, sanitation, municipal administration, policing, marketing, transportation, and other technical issues. While the sociologist cannot solve any of these practical problems – at least not by himself – he may, if he discovers his proper function, have an important contribution to make to their comprehension and solution. The prospects for doing this are brightest through a general, theoretical, rather than through an *ad hoc* approach.

"The Uses of Sidewalks: Safety"

from *The Death and Life of Great American Cities* (1961)

Jane Jacobs

Editors' Introduction

Jane Jacobs (1916–2006) started writing about city life and urban planning as a neighborhood activist and as associate editor of *Architectural Forum*, not as a trained planning professional. Dismissed as the original "little old lady in tennis shoes" and derided as a political amateur more concerned about personal safety issues than state-of-the-art planning techniques, she nonetheless struck a responsive chord with a 1960s public eager to believe the worst about arrogant city planning technocrats and just as eager to rally behind movements for neighborhood control and community resistance to bulldozer redevelopment.

The Death and Life of Great American Cities hit the world of city planning like an earthquake when it appeared in 1961. The book was a frontal attack on the planning establishment, especially on the massive urban renewal projects that were being carried out by powerful redevelopment bureaucrats like Robert Moses in New York. Jacobs derided urban renewal as a process that served only to create instant slums. She questioned universally accepted articles of faith – for example that parks were good and that crowding was bad. Indeed she suggested that parks were often dangerous and that crowded neighborhood sidewalks were the safest places for children to play. Jacobs ridiculed the planning establishment's most revered historical traditions as "the Radiant Garden City Beautiful" – an artful phrase that not only airily dismissed the contributions of Le Corbusier (p. 336), Ebenezer Howard (p. 328), and Daniel Burnham but lumped them together as well! Lewis Mumford, "Home Remedies for Urban Cancer" (1962), reprinted in both Elizabeth McDonald and Michael Larice (eds), *The Urban Design Reader* (London: Routledge, 2006) and Eugenie Birch (ed.), *The Urban and Regional Planning Reader* (London: Routledge, 2008), praises Jacobs' humanity and obvious love of city life but savages her attack on city planners like Ebenezer Howard and Patrick Geddes that Mumford had championed for decades.

The selection from *The Death and Life of Great American Cities* reprinted here presents Jane Jacobs at her very best. In "The Uses of Sidewalks: Safety," she outlines her basic notions of what makes a neighborhood a community and what makes a city livable. Safety – particularly for women and children – comes from "eyes on the street," the kind of involved neighborhood surveillance of public space that modern planning practice in the Corbusian tradition had destroyed with its insistence on superblocks and skyscraper developments. A sense of personal belonging and social cohesiveness comes from well-defined neighborhoods and narrow, crowded, multi-use streets. Finally, basic urban vitality comes from residents' participation in an intricate "street ballet," a diurnal pattern of observable and comprehensible human activity that is possible only in places like Jacobs' own Hudson Street in her beloved Greenwich Village.

It was this last quality, her unabashed love of cities and urban life, that is Jane Jacobs' most obvious and enduring characteristic. *The Death and Life of Great American Cities* was a scathing attack on the planning establishment – and, in many ways, it was a grassroots political call to arms – but it was also a loving invitation to experience the joys of city living that led many young, college-educated people to seek out neighborhoods like belonging to those we might call the natural proprietors of the street. The buildings on a street equipped to handle strangers and to insure the safety of both residents and strangers must be oriented to the street. They cannot turn their backs or blank sides on it and leave it blind.

And third, the sidewalk must have users on it fairly continuously, both to add to the number of effective eyes on the street and to induce the people in buildings along the street to watch the sidewalks in sufficient numbers. Nobody enjoys sitting on a stoop or looking out a window at an empty street. Almost nobody does such a thing. Large numbers of people entertain themselves, off and on, by watching street activity.

In settlements that are smaller and simpler than big cities, controls on acceptable public behavior, if not on crime, seem to operate with greater or lesser success through a web of reputation, gossip, approval, disapproval and sanctions, all of which are powerful if people know each other and word travels. But a city's streets, which must control the behavior not only of the people of the city but also of visitors from suburbs and towns who want to have a big time away from the gossip and sanctions at home, have to operate by more direct, straightforward methods. It is a wonder cities have solved such an inherently difficult problem at all. And yet in many streets they do it magnificently.

It is futile to try to evade the issue of unsafe city streets by attempting to make some other features of a locality, say interior courtyards, or sheltered play spaces, safe instead. By definition again, the streets of a city must do most of the job of handling strangers, for this is where strangers come and go. The streets must not only defend the city against predatory strangers, they must protect the many, many peaceable and well-meaning strangers who use them, insuring their safety too as they pass through. Moreover, no normal person can spend his life in some artificial haven, and this includes children. Everyone must use the streets.

On the surface, we seem to have here some simple aims: to try to secure streets where the public space is unequivocally public, physically unmixed with private or with nothing-at-all space, so that the area needing surveillance has clear and practicable limits; and to see that these public street spaces have eyes on them as continuously as possible.

But it is not so simple to achieve these objects, especially the latter. You can't make people use streets they have no reason to use. You can't make people watch streets they do not want to watch. Safety on the streets by surveillance and mutual policing of one another sounds grim, but in real life it is not grim. The safety of the street works best, most casually, and with least frequent taint of hostility or suspicion precisely where people are using and most enjoying the city streets voluntarily and are least conscious, normally, that they are policing.

The basic requisite for such surveillance is a substantial quantity of stores and other public places sprinkled along the sidewalks of a district; enterprises and public places that are used by evening and night must be among them especially. Stores, bars and restaurants, as the chief examples, work in several different and complex ways to abet sidewalk safety.

First, they give people – both residents and strangers – concrete reasons for using the sidewalks on which these enterprises face.

Second, they draw people along the sidewalks past places which have no attractions to public use in themselves but which become traveled and peopled as routes to somewhere else; this influence does not carry very far geographically, so enterprises must be frequent in a city district if they are to populate with walkers those other stretches of street that lack public places along the sidewalk. Moreover, there should be many different kinds of enterprises, to give people reasons for crisscrossing paths.

Third, storekeepers and other small businessmen are typically strong proponents of peace and order themselves; they hate broken windows and hold-ups; they hate having customers made nervous about safety. They are great street watchers and sidewalk guardians if present in sufficient numbers.

Fourth, the activity generated by people on errands, or people aiming for food or drink, is itself an attraction to still other people.

This last point, that the sight of people attracts still other people, is something that city planners and city architectural designers seem to find incomprehensible. They operate on the premise that city people seek the sight of emptiness, obvious order and quiet. Nothing could be less true. People's love of watching activity and other people is constantly evident in cities everywhere.

[. . .]

Under the seeming disorder of the old city, wherever the old city is working successfully, is a marvelous order for maintaining the safety of the streets and the freedom of the city. It is a complex order. Its essence is intricacy of sidewalk use, bringing with it a constant succession of eyes. This order is all composed of movement and change, and although it is life, not art, we may fancifully call it the art form of the city and liken it to the dance – not to a simple-minded precision dance with everyone kicking up at the same time, twirling in unison and bowing off en masse, but to an intricate ballet in which the individual dancers and ensembles all have distinctive parts which miraculously reinforce each other and compose an orderly whole. The ballet of the good city sidewalk never repeats itself from place to place, and in any one place is always replete with new improvisations.

The stretch of Hudson Street where I live is each day the scene of an intricate sidewalk ballet. I make my own first entrance into it a little after eight when I put out the garbage can, surely a prosaic occupation, but I enjoy my part, my little clang, as the droves of junior high school students walk by the center of the stage dropping candy wrappers. (How do they eat so much candy so early in the morning?)

While I sweep up the wrappers I watch the other rituals of morning: Mr. Halpert unlocking the laundry's handcart from its mooring to a cellar door, Joe Cornacchia's son-in-law stacking out the empty crates from the delicatessen, the barber bringing out his sidewalk folding chair, Mr. Goldstein arranging the coils of wire which proclaim the hardware store is open, the wife of the tenement's superintendent depositing her chunky 3-year-old with a toy mandolin on the stoop, the vantage point from which he is learning the English his mother cannot speak. Now the primary children, heading for St. Luke's, dribble through to the south; the children for St. Veronica's cross, heading to the west, and the children for P.S. 41, heading toward the east. Two new entrances are being made from the wings: well-dressed and even elegant women and men with briefcases emerge from doorways and side streets . . . Most of these are heading for the bus and subways, but some hover on the curbs, stopping taxis which have miraculously appeared at the right moment, for the taxis are part of a wider morning ritual: having dropped passengers from midtown in the downtown financial district, they are now bringing downtowners up to midtown. Simultaneously, numbers of women in housedresses have emerged and as they crisscross with one another they pause for quick conversations that sound with either laughter or joint indignation; never, it seems, anything between. It is time for me to hurry to work too, and I exchange my ritual farewell with Mr. Lofaro, the short, thick-bodied, white-aproned fruit man who stands outside his doorway a little up the street, his arms folded, his feet planted, looking solid as earth itself. We nod; we each glance quickly up and down the street then look back to each other and smile. We have done this many a morning for more than ten years, and we both know what it means: All is well.

[. . .]

I know the deep night ballet and its seasons best from waking; long after midnight to tend a baby and, sitting in the dark, seeing the shadows and hearing the sounds of the sidewalk. Mostly it is a sound like infinitely pattering snatches of party conversation and, about three in the morning, singing, very good singing. Sometimes there is sharpness and anger or sad, sad weeping, or a flurry of search for a string of beads broken. One night, a young man came roaring along, bellowing terrible language at two girls whom he had apparently picked up and who were disappointing him. Doors opened; a wary semicircle formed around him, not too close, until the police came. Out came the heads, too, along Hudson Street, offering opinion, "Drunk . . . Crazy . . . A wild kid from the suburbs." (He turned out to be a wild kid from the suburbs. Sometimes, on Hudson Street, we are tempted to believe the suburbs must be a difficult place to bring up children.)

I have made the daily ballet of Hudson Street sound more frenetic than it is, because writing it telescopes it. In real life, it is not that way. In real life, to be sure, something is always going on, the ballet is never at a halt, but the general effect is peaceful and the general tenor even leisurely. People who know well such animated city streets will know how it is. I am afraid people who do not will always have it a little wrong in their heads like the old prints of rhinoceroses made from travelers' descriptions of rhinoceroses. On Hudson Street, the same as in the North End of Boston or in any other animated neighborhoods of great cities, we are not innately more competent at keeping the sidewalks safe than are the people who try to live off the hostile truce of Turf in a blind-eyed city. We are the lucky possessors of a city order that makes it relatively simple to keep the peace because there are plenty of eyes on the street. But there is nothing simple about that order itself, or the bewildering number of components that go into it. Most of those components are specialized in one way or another. They unite in their joint effect upon the sidewalk, which is not specialized in the least. That is its strength.

"The Negro Problems of Philadelphia," "The Question of Earning a Living," and "Color Prejudice"

from *The Philadelphia Negro* (1899)

W.E.B. Du Bois

Editors' Introduction

William Edward Burghardt Du Bois (1868–1963) was one of the preeminent intellectuals of his generation. As a professor, editor, author, novelist, playwright, and politician, he made notable contributions in history, sociology, ethnic studies, literature, politics, and other fields. A brilliant student, Du Bois excelled at Fisk University in Nashville, Tennessee, the University of Berlin where he studied with the great sociologist Max Weber, and at Harvard University, where in 1895 he obtained the first PhD degree Harvard awarded to an African American. Du Bois defies easy classification. He was always an independent and critical thinker. During his long and varied career, he was a pan-Africanist who advocated solidarity among Black Africans and Blacks elsewhere in the world; a radical pacifist who was indicted, tried, and acquitted as an unregistered foreign agent during the McCarthy era for circulating the Stockholm peace plan; a humanist who wrote novels and plays and published many of the writers of the "Harlem Renaissance"; a civil rights leader who founded the National Association for the Advancement of Colored People (NAACP) publication *Crisis* in 1910 and served as its influential editor until 1934; a writer of children's books which taught Black pride; and a world political figure who urged United Nations protection for Black Americans as a nation within a nation. Du Bois joined the Communist Party at age 93 and became a Ghanaian citizen just before his death in 1963 at age 95.

At the time that Du Bois completed his education, Philadelphia had the largest and oldest settlement of African Americans in the northern United States. The settlement house movement was under way, and some well-intentioned Philadelphians were concerned to understand "the Negro problem" and to help the many poor Blacks in the city. Two wealthy leaders of Philadelphia society suggested a study of Negroes in the Seventh Ward, the city's Black ghetto.

Du Bois was given a one-year appointment as an assistant instructor in the Sociology Department at the University of Pennsylvania. Living with his bride of three months in one room over a cafeteria in the worst part of Philadelphia's worst Black ghetto, with no contact with students and little with faculty, Du Bois wrote *The Philadelphia Negro* from which the following selection is taken. He was only 31 when his monumental study was published.

While Du Bois found many problems in Philadelphia's segregated African American community in the 1890s (largely the result of pervasive race prejudice in the larger American society), there was work available for able-bodied laborers, no evidence of drug use, substantial homeownership, middle- and upper-income craftspeople, business people, and professionals to serve the community and act as role models, and little Black-on-Black violent crime. This is in marked contrast to William Julius Wilson's description of poor Black ghetto areas of Chicago in the 1980s (p. 117) and Elijah Anderson's descriptions of "street culture" in the Philadelphia of the 1990s (p. 127). Wilson, for example, describes "underclass" ghettos in Chicago consisting almost entirely of renters (many in public housing), with very few employed residents, extremely high concentrations of single-parent families, welfare dependency, drug use, and violent crime.

Ethnographic studies by sociologists and anthropologists often shed light on variations within communities, which are viewed as homogenous by outsiders. While white Philadelphians who never visited the Seventh Ward tended to view the area as homogenous and all African Americans as similar, Du Bois found a physical and social structure within the neighborhood – alleys peopled by criminals, loafers, and prostitutes separate from streets of the working poor and still other streets where an established group of Black middle-class homeowners lived.

In addition to *The Philadelphia Negro* (Philadelphia, PA: University of Pennsylvania Press, 1899) from which this selection is taken, Du Bois's writings include *Suppression of the Slave Trade to the United States of America* (New York: Longmans, Green, 1896), *Souls of Black Folk* (Chicago, IL: A.C. McClurg, 1903), *The Negro* (New York: Henry Holt, 1915), *Black Reconstruction* (New York: Harcourt, Brace, 1935), and *The World and Africa* (New York: Viking, 1947). There are many anthologies of Du Bois's writings and speeches. Perhaps the best is David Levering Lewis (ed.), *W.E.B. Du Bois: A Reader* (New York: Henry Holt, 1995). Also by Lewis, and of great interest, is *W.E.B. Du Bois: The Fight for Equality and the American Century, 1919–1963* (New York: Henry Holt, 1995).

For more by and about W.E.B. Du Bois see *The Autobiography of W.E.B. Du Bois* (New York: International Publishers, 1968), Francis L. Broderick, *W.E.B. Du Bois* (Standford, CA: Stanford University Press, 1959), Walter Wilson (ed.), *The Selected Writings of W.E.B. Du Bois* (New York: New American Library, 1970), Henry Lee Moon, *The Emerging Thought of W.E.B. Du Bois* (New York: Simon & Schuster, 1972), Marable Manning, *W.E.B. DuBois, Black Radical Democrat* (Boston, MA: Twayne, 1986; new edition published by Paradigm, 2005), and Patricia and Fredrick McKissack, *W.E.B. Du Bois* (New York: Franklin Watts, 1990). For readings on the current state of Black America, see the bibliographical references in the Editors' Introductions to William Julius Wilson's "From Institutional to Jobless Ghettos" (p. 118) and Elijah Anderson's *Code of the Street* (p. 128).

4. THE NEGRO PROBLEMS OF PHILADELPHIA

In Philadelphia, as elsewhere in the United States, the existence of certain peculiar social problems affecting the Negro people are plainly manifest. Here is a large group of people – perhaps forty-five thousand, a city within a city – who do not form an integral part of the larger social group. This in itself is not altogether unusual; there are other unassimilated groups: Jews, Italians, even Americans; and yet in the case of the Negroes the segregation is more conspicuous, more patent to the eye, and so intertwined with a long historic evolution, with peculiarly pressing social problems of poverty, ignorance, crime and labor, that the Negro problem far surpasses in scientific interest and social gravity most of the other race or class questions.

The student of these questions must first ask, What is the real condition of this group of human beings? Of whom is it composed, what sub-groups and classes exist, what sort of individuals are being considered? Further, the student must clearly recognize that a complete study must not confine itself to the group, but must specially notice the environment; the physical environment of city, sections and houses, the far mightier social environment – the surrounding world of custom, wish, whim and thought which envelops this group and powerfully influences its social development.

[. . .]

The Seventh Ward starts from the historic center of Negro settlement in the city, South Seventh street and Lombard, and includes the long narrow strip, beginning at South Seventh and extending west, with South and Spruce streets as boundaries, as far as the Schuylkill River. The colored population of this ward numbered 3,621 in 1860, 4,616 in 1870, and 8,861 in 1890. It is a thickly populated district of varying character; north of it is the residence and business section of the city; south of it a middle class and workingmen's residence section; at the east end it joins Negro, Italian and Jewish slums; at the west end, the wharves of the river and an industrial section separating it from the grounds of the University of Pennsylvania and the residence section of West Philadelphia.

Starting at Seventh street and walking along Lombard, let us glance at the general character of the ward. Pausing a moment at the corner of Seventh and Lombard, we can at a glance view the worst Negro slums of the city. The houses are mostly brick, some wood, not very old, and in general uncared for rather than dilapidated. The blocks between Eighth, Pine, Sixth, and South have for many decades been the center of Negro population. Here the riots of the thirties took place, and here once was a depth of poverty and degradation almost unbelievable. Even today there are many evidences of degradation . . . The alleys near, as Ratcliffe street, Middle alley, Brown's court, Barclay street, etc., are haunts of noted criminals, male and female, of gamblers and prostitutes, and at the same time of many poverty-stricken people, decent but not energetic. There is an abundance of political clubs, and nearly all the houses are practically lodging houses, with a miscellaneous and shifting population. The corners, night and day, are filled with Negro loafers – able-bodied young men and women, all cheerful, some with good natured, open faces, some with traces of crime and excess, a few pinched with poverty. They are mostly gamblers, thieves and prostitutes, and few have fixed and steady occupation of any kind. Some are stevedores, porters, laborers and laundresses. On its face this slum is noisy and dissipated, but not brutal, although now and then highway robberies and murderous assaults in other parts of the city are traced to its denizens. Nevertheless a stranger can usually walk about here day and night with little fear of being molested if he be not too inquisitive.

Passing up Lombard, beyond Eighth, the atmosphere suddenly changes, because these next two blocks have few alleys and the residences are good-sized and pleasant. Here some of the best Negro families of the ward live. Some are wealthy in a small way, nearly all are Philadelphia born, and they represent an early wave of emigration from the old slum section . . .

[. . .]

21. THE QUESTION OF EARNING A LIVING

For a group of freedmen the question of economic survival is the most pressing of all questions; the problem as to how, under the circumstances of modern life, any group of people can earn a decent living, so as to maintain their standard of life, is not always easy to answer. But when the question is complicated by the fact that the group has a low degree of efficiency on account of previous training; is in competition with well-trained, eager and often ruthless competitors; is more or less handicapped by a somewhat wide-reaching discrimination; and finally is seeking not merely to maintain a standard of living but steadily to raise it to a higher plane – such a situation presents baffling problems to the sociologist and philanthropist.

Of the men 21 years of age and over, there were in gainful occupations, the following:

In the learned professions		61	2.0 per cent
Conducting business on their own account		207	6.5
In the skilled trades		236	7.0
Clerks, etc.		159	5.0
Laborers, better class	602		
Laborers, common class	852	1454	45.0
Servants		1079	34.0
Miscellaneous		11	0.5
		3207	100 per cent
Total male population 21 and over		3850	

Taking the occupations of women 21 years of age and over, we have:

Domestic servants	1262	37.0 per cent
Housewives and day laborers	937	27.0
Housewives	568	17.0

Day laborers, maids, etc.	297	9.0
In skilled trades	221	6.0
Conducting businesses	63	2.0
Clerks, etc.	40	1.0
Learned professions	37	1.0
	3425	100 per cent
Total female population 21 and over	3740	

47. COLOR PREJUDICE

Incidentally throughout this study the prejudice against the Negro has been again and again mentioned. It is time now to reduce this somewhat indefinite term to something tangible. Everybody speaks of the matter, everybody knows that it exists, but in just what form it shows itself or how influential it is few agree. In the Negro's mind, color prejudice in Philadelphia is that widespread feeling of dislike for his blood, which keeps him and his children out of decent employment, from certain public conveniences and amusements, from hiring houses in many sections, and in general, from being recognized as a man. Negroes regard this prejudice as the chief cause of their present unfortunate condition. On the other hand most white people are quite unconscious of any such powerful and vindictive feeling; they regard color prejudice as the easily explicable feeling that intimate social intercourse with a lower race is not only undesirable but impractical if our present standards of culture are to be maintained, and although they are aware that some people feel the aversion more intensely than others, they cannot see how such a feeling has much influence on the real situation or alters the social condition of the mass of Negroes.

As a matter of fact, color prejudice in this city is something between these two extreme views: it is not today responsible for all, or perhaps the greater part of the Negro problems, or of the disabilities under which the race labors; on the other hand it is a far more powerful social force than most Philadelphians realize. The practical results of the attitude of most of the inhabitants of Philadelphia towards persons of Negro descent are as follows:

1. As to getting work:

   No matter how well trained a Negro may be, or how fitted for work of any kind, he cannot in the ordinary course of competition hope to be much more than a menial servant.

   He cannot get clerical or supervisory work to do save in exceptional cases.

   He cannot teach save in a few of the remaining Negro schools.

   He cannot become a mechanic except for small transient jobs, and cannot join a trades union.

   A Negro woman has but three careers open to her in this city: domestic service, sewing, or married life.

2. As to keeping work:

   The Negro suffers in competition more severely than white men.

   Change in fashion is causing him to be replaced by whites in the better-paid positions of domestic service.

   Whim and accident will cause him to lose a hard-earned place more quickly than the same things would affect a white man.

   Being few in number compared with the whites the crime or carelessness of a few of his race is easily imputed to all, and the reputation of the good, industrious, and reliable suffer thereby.

   Because Negro workmen may not often work side by side with white workmen, the individual black workman is rated not only by his own efficiency, but by the efficiency of a whole group of black fellow workmen which may often be low.

   Because of these difficulties which virtually increase competition in his case, he is forced to take lower wages for the same work than white workmen.

3. As to entering new lines of work:

   Men are used to seeing Negroes in inferior positions; when, therefore, by any chance a Negro gets in a better position, most men immediately conclude that he is not fitted for it, even before he has a chance to show his fitness.

   If, therefore, he set up a store, men will not patronize him.

   If he is put into public position men will complain.

If he gain a position in the commercial world, men will quietly secure his dismissal or see that a white man succeeds him.

4. As to his expenditure:

The comparative smallness of the patronage of the Negro, and the dislike of other customers, makes it usual to increase the charges or difficulties in certain directions in which a Negro must spend money.

He must pay more house-rent for worse houses than most white people pay.

He is sometimes liable to insult or reluctant service in some restaurants, hotels and stores, at public resorts, theaters and places of recreation; and at nearly all barber shops.

5. As to his children:

The Negro finds it extremely difficult to rear children in such an atmosphere and not have them either cringing or impudent: if he impresses upon them patience with their lot, they may grow up satisfied with their condition; if he inspires them with ambition to rise, they may grow to despise their own people, hate the whites, and become embittered with the world.

His children are discriminated against, often in public schools.

They are advised when seeking employment to become waiters and maids.

They are liable to species of insult and temptation peculiarly trying to children.

6. As to social intercourse:

In all walks of life the Negro is liable to meet some objection to his presence or some discourteous treatment; and the ties of friendship or memory seldom are strong enough to hold across the color line.

If an invitation is issued to the public for any occasion, the Negro can never know whether he would be welcomed or not; if he goes he is liable to have his feelings hurt and get into unpleasant altercation; if he stays away, he is blamed for indifference.

If he meet a lifelong white friend on the street, he is in a dilemma; if he does not greet the friend he is put down as boorish and impolite; if he does greet the friend he is liable to be flatly snubbed.

If by chance he is introduced to a white woman or man, he expects to be ignored on the next meeting, and usually is.

White friends may call on him, but he is scarcely expected to call on them, save for strictly business matters.

If he gain the affections of a white woman and marry her he may invariably expect that slurs will be thrown on her reputation and on his, and that both his and her race will shun their company. When he dies he cannot be buried beside white corpses.

7. The result:

Any one of these things happening now and then would not be remarkable or call for especial comment; but when one group of people suffer all these little differences of treatment and discriminations and insults continually, the result is either discouragement, or bitterness, or over-sensitiveness, or recklessness. And a people feeling thus cannot do their best.

Presumably the first impulse of the average Philadelphian would be emphatically to deny any such marked and blighting discrimination as the above against a group of citizens in this metropolis. Every one knows that in the past color prejudice in the city was deep and passionate; living men can remember when a Negro could not sit in a street car or walk many streets in peace. These times have passed, however, and many imagine discrimination against the Negro has passed with them. Careful inquiry will convince any such one of his error. To be sure a colored man to-day can walk the streets of Philadelphia without personal insult; he can go to theaters, parks and some places of amusement without meeting more than stares and discourtesy; he can be accommodated at most hotels and restaurants, although his treatment in some would not be pleasant. All this is a vast advance and augurs much for the future. And yet all that has been said of the remaining discrimination is but too true.

During the investigation of 1896 there was collected a number of actual cases, which may illustrate the discriminations spoken of. So far as possible these have been sifted and only those which seem undoubtedly true have been selected:

I. As to getting work

It is hardly necessary to dwell upon the situation of the Negro in regard to work in the higher walks of life:

the white boy may start in the lawyer's office and work himself into a lucrative practice; he may serve a physician as office boy or enter a hospital in a minor position, and have his talent alone between him and affluence and fame; if he is bright in school, he may make his mark in a university, become a tutor with some time and much inspiration for study, and eventually fill a professor's chair. All these careers are at the very outset closed to the Negro on account of his color; what lawyer would give even a minor case to a Negro assistant? What university would appoint a promising young Negro as tutor? Thus the young white man starts in life knowing that within some limits and barring accidents, talent and application will tell. The young Negro starts knowing that on all sides his advance is made difficult if not wholly shut off by his color. Let us come, however, to ordinary occupations which concern more nearly the mass of Negroes. Philadelphia is a great industrial and business center with thousands of foremen, managers and clerks – the lieutenants of industry who direct its progress. They are paid for thinking and for skill to direct, and naturally such positions are coveted because they are well paid, well thought-of and carry some authority. To such positions Negro boys and girls may not aspire no matter what their qualifications. Even as teachers and ordinary clerks and stenographers they find almost no openings. Let us note some actual instances:

A young woman who graduated with credit from the Girls Normal School in 1892 has taught in the kindergarten, acted as substitute, and waited in vain for a permanent position. Once she was allowed to substitute in a school with white teachers; the principal commended her work, but when the permanent appointment was made a white woman got it.

A girl who graduated from a Pennsylvania high school and from a business college sought work in the city as a stenographer and typewriter. A prominent lawyer undertook to find her a position; he went to friends and said, "Here is a girl that does excellent work and is of good character; can you not give her work?" Several immediately answered yes. "But," said the lawyer, "I will be perfectly frank with you and tell you she is colored"; and not in the whole city could he find a man willing to employ her. It happened, however, that the girl was so light in complexion that few not knowing would have suspected her descent. The lawyer therefore gave her temporary work in his own office until she found a position outside the city. "But," said he, "to this day I have not dared to tell my clerks that they worked beside a Negress." Another woman graduated from the high school and the Palmer College of Shorthand, but all over the city has met with nothing but refusal of work.

Several graduates in pharmacy have sought three years' required apprenticeship in the city and in only one case did one succeed, although they offered to work for nothing. One young pharmacist came from Massachusetts and for weeks sought in vain for work here at any price; "I wouldn't have a darky to clean out my store, much less to stand behind the counter," answered one druggist.

A colored man answered an advertisement for a clerk in the suburbs. "What do you suppose we'd want of a nigger?" was the plain answer. A graduate of the University of Pennsylvania in mechanical engineering, well recommended, obtained work in the city, through an advertisement, on account of his excellent record. He worked a few hours and then was discharged because he was found to be colored. He is now a waiter at the University Club, where his white fellow graduates dine. Another young man attended Spring Garden Institute and studied drawing for lithography. He had good references from the institute and elsewhere, but application at the five largest establishments in the city could secure him no work. A telegraph operator has hunted in vain for an opening, and two graduates of the Central High School have sunk to menial labor. "What's the use of an education?" asked one. Mr. A— has elsewhere been employed as a traveling salesman. He applied for a position here by letter and was told he could have one. When they saw him they had no work for him.

Such cases could be multiplied indefinitely. But that is not necessary; one has but to note that, notwithstanding the acknowledged ability of many colored men, the Negro is conspicuously absent from all places of honor, trust, emolument, as well as from those of respectable grade in commerce and industry.

Even in the world of skilled labor the Negro is largely excluded. Many would explain the absence of Negroes from higher vocations by saying that while a few may now and then be found competent, the great mass are not fitted for that sort of work and are destined for some time to form a laboring class. In the matter of the trades, however, there can be raised no serious question of ability; for years the Negroes filled satisfactorily the trades of the city, and to-day in many parts of the South they are still prominent. And yet in Philadelphia a determined prejudice, aided by public opinion, has succeeded nearly in driving them from the field:

A——, who works at a bookbinding establishment on Front street, has learned to bind books and often does so for his friends. He is not allowed to work at the trade in the shop, however, but must remain a porter at a porter's wages.

B——is a brushmaker; he has applied at several establishments, but they would not even examine his testimonials. They simply said: "We do not employ colored people."

C——is a shoemaker; he tried to get work in some of the large department stores. They "had no place" for him.

D——was a bricklayer, but experienced so much trouble in getting work that he is now a messenger.

E——is a painter, but has found it impossible to get work because he is colored.

F——is a telegraph line man, who formerly worked in Richmond, Va. When he applied here he was told that Negroes were not employed.

G——is an iron puddler, who belonged to a Pittsburgh union. Here he was not recognized as a union man and could not get work except as a stevedore.

H——was a cooper, but could get no work trials, and is now a common laborer.

I——is a candy-maker, but has never been able to find employment in the city; he was always told the white help would not work with him.

J——is a carpenter; he can only secure odd jobs or work where only Negroes are employed.

K——was an upholsterer, but could get no work save in the few colored shops which had workmen; he is now a waiter on a dining car.

L——was a first-class baker; he applied for work some time ago near Green street and was told shortly, "We don't work no niggers here."

[. . .]

"From Institutional to Jobless Ghettos"

from *When Work Disappears: The World of the New Urban Poor* (1996)

William Julius Wilson

Editors' Introduction

Harvard sociologist William Julius Wilson (b. 1935) spent much of his career at the University of Chicago. Like the earlier Chicago School sociologists Ernest W. Burgess (p. 161) and Louis Wirth (p. 96) writing in the 1920s and 1930s, and St. Clair Drake and Horace Cayton writing in the 1940s, Wilson uses careful empirical studies of Chicago to generate important urban theory. An African American, Wilson has been particularly concerned about the situation of poor Blacks in the United States' decaying central city neighborhoods.

Wilson is critical of timid liberals who avoid confronting tough questions about race and poverty because they are afraid that anything negative they say about Blacks will appear racist. He argues that there is an *urban underclass* (a term many liberals will not use) and that residents of poor Black ghettos today are socially isolated and caught in a tangle of pathology characterized by unemployment, crime (including violent Black-on-Black crime), teenage pregnancy, out-of-wedlock births, welfare dependency, and drug use. Wilson feels that the situation of Blacks, particularly poor urban Blacks, requires objective research and honest reportage. He is unwilling to let conservatives dominate theoretical discourse about the causes of and cures for Black poverty.

Wilson's research has convinced him that conditions for poor Black ghetto residents are far worse in many ways than a century ago when W.E.B. Du Bois studied *The Philadelphia Negro* (p. 110) or in 1945 when Drake and Cayton published *Black Metropolis*, their study of the African American community of Chicago at that time. Both Du Bois's and Drake and Cayton's studies found more Blacks employed and role models of upward mobility present, a higher proportion of nuclear families, less Black-on-Black violence, and much less drug use than exist in the poorest Black ghettos nowadays.

Wilson argues that the changed structure of the US economy is more responsible for the plight of poor Blacks today than is racism. A generation ago, Wilson argues, an able-bodied unskilled Black man could readily find work sufficient to support himself and a family – albeit often physically hard, racially segregated, and dirty work. But in the last generation unskilled manual urban jobs have largely disappeared from American cities as manufacturing has moved overseas and post-industrial jobs require more education. Without work, Black males cannot support a family. Hence, Wilson argues, Black ghettos now contain few "marriageable" Black males capable of supporting a family. A high incidence of out-of-wedlock births, family dissolution, and welfare-dependent female-headed households follows directly from that fact. With little sense of self-worth, Wilson argues, unemployed Blacks naturally turn to drug dependence and crime.

One of Wilson's most controversial contentions is that race and racism are declining in importance as causes of Black distress. Paradoxically, he argues, less racial discrimination has made matters in Black ghettos worse. According to Wilson, as upwardly mobile Blacks move out of Black ghetto areas, community leadership and positive role models disappear and pathology is concentrated.

Wilson is skeptical that *race-specific* policies like affirmative action will address problems as pervasive and profound as he describes. Rather he favors universal social policies aimed at improving the lot of all poor people regardless of race: new education, training, and particularly full employment policy.

This selection is from *When Work Disappears: The World of the New Urban Poor* (New York: Knopf, 1996). Wilson's two most influential prior books are *The Declining Significance of Race: Blacks and Changing American Institutions*, 2nd edn (Chicago, IL: University of Chicago Press, 1980) and *The Truly Disadvantaged* (Chicago, IL: University of Chicago Press, 1987). Also of interest is *The Bridge Over the Racial Divide: Rising Inequality and Coalition Politics* (Los Angeles, CA: University of California Press, 2001).

The literature on urban social inequality, welfarism, and the "underclass debate" is vast. Classic studies include Richard A. Cloward and Frances Fox Piven, *Regulating the Poor: The Functions of Public Welfare*, 2nd edn (New York: Vintage, 1993), Michael B. Katz, *In the Shadow of the Poorhouse: A Social History of Welfare in America* (New York: Basic Books, 1987), *The Undeserving Poor: From the War on Poverty to the War on Welfare* (New York: Random House, 1990), and Christopher Jencks, *Rethinking Social Policy: Race, Poverty, and the Underclass* (New York: Harpers, 1993). More recent studies of interest include Benjamin I. Page and James Roy Simmons, *What Government Can Do: Dealing with Poverty and Inequality* (Chicago, IL: University of Chicago Press, 2000), R. Kent Weaver and Michael H. Armacost, *Ending Welfare As We Know It* (Washington, DC: Brookings Institution Press, 2000), and Theda Skocpol and Richard C. Leone, *The Missing Middle: Working Families and the Future of American Social Policy* (New York: Norton, 2001). Two studies examine the continued poor economic performance of inner-city Blacks through the 1990s: Ronald Mincy (ed.), *Black Males Left Behind* (Washington, DC: Urban Institute Press, 2006) and Peter Edelman and Paul Offner, *Reconnecting Disadvantaged Young Men* (Washington, DC: Urban Institute Press, 2006). Also of interest, taking radically different positions, are Douglas Massey and Nancy Denton, *American Apartheid: Segregation and the Making of the Underclass* (Cambridge, MA: Harvard University Press, 1998) and John McWhorter, *Losing the Race: Self-Sabotage in Black America* (New York: Free Press, 2000). Two books by sociologist Elijah Anderson (see p. 127) break new ground in the area of inner-city ethnography: *Streetwise: Race, Class and Change in an Urban Community* (Chicago, IL: University of Chicago Press, 1992) and *Code of the Street: Decency, Violence, and the Moral Life of the Inner City* (New York: Norton, 2000).

For an overview of the literature on the underclass debate and writings by Latino scholars exploring the relevance and limitations of underclass theory for America's varied Latino communities, see Joan Moore and Raquel Pinderhughes, *In the Barrios: Latinos and the Underclass Debate* (New York: Russell Sage Foundation, 1993). Conservative explanations of inequality and suggested public policy regarding poverty and race include Edward Banfield, *The Unheavenly City Revisited* (Boston, MA: Little, Brown, 1974) and Charles Murray, *Losing Ground* (New York: Basic Books, 1984).

An elderly woman who has lived in one inner-city neighborhood on the South Side of Chicago for more than forty years reflected:

> I've been here since March 21, 1953. When I moved in, the neighborhood was intact. It was intact with homes, beautiful homes, mini mansions, with stores, laundromats, with cleaners, with Chinese [cleaners]. We had drugstores. We had hotels. We had doctors over on Thirty-ninth Street. We had doctors' offices in the neighborhood. We had the middle class and upper middle class. It has gone from affluent to where it is today. And I would like to see it come back, that we can have some of the things we had. Since I came in young, and I'm a senior citizen now, I would like to see some of the things come back so I can enjoy them like we did when we first came in.

[. . .]

A 91-year-old woman spoke of safety concerns: "It's not safe anymore because the streets aren't. When all the black businesses and shows closed down, the economy went to the dogs. The stores, the businesses, the shows, everywhere was lighted, the stores and businesses have disappeared."

The negative social forces triggered a decision by a concerned mother to send her son away.

I have a 13-year-old. I sent him away when he was nine because the gangs was at him so tough, because he wouldn't join – he's a basketball player. That's all he ever cared about. They took his gym shoes off his feet. They took his clothes. Made him walk home from school. Jumped on him every day. Took his jacket off his back in subzero weather. You know, and we only live two blocks from the school . . . A boy pulled a gun to his head and told him, "If you don't join, next week you won't be here." I had to send him out of town. His father stayed out of town. He came here last week for a week. He said, "Mom, I want to come home so bad," I said no!

The social deterioration of ghetto neighborhoods is the central concern expressed in the testimony of these residents. As a representative from the media put it, the ghetto has gone "from bad to worse." Few observers of the urban scene in the late 1960s anticipated the extensive breakdown of social institutions and the sharp rise in rates of social dislocation that have since swept the ghettos and spread to other neighborhoods that were once stable. For example, in the neighborhood of Woodlawn, located on the South Side of Chicago, there were over eight hundred commercial and industrial establishments in 1950. Today, it is estimated that only about a hundred are left, many of them represented by "tiny catering places, barber shops, and thrift stores with no more than one or two employees." As Löic Wacquant, a member of the Urban Poverty and Family Life Study research team, put it:

The once-lively streets – residents remember a time, not so long ago, when crowds were so dense at rush hour that one had to elbow one's way to the train station – now have the appearance of an empty, bombed-out war zone. The commercial strip has been reduced to a long tunnel of charred stores, vacant lots littered with broken glass and garbage, and dilapidated buildings left to rot in the shadow of the elevated train line. At the corner of Sixty-third Street and Cottage Grove Avenue, the handful of remaining establishments that struggle to survive are huddled behind wrought iron bars . . . The only enterprises that seem to be thriving are liquor stores and currency exchanges, these "banks of the poor" where one can cash checks, pay bills and buy money orders for a fee.

[. . .]

In 1950, almost two-thirds of Woodlawn's population was white; by 1960 the white population had declined to just 10 percent. Despite the sudden white exodus, the number of residents in the neighborhood increased slightly during this period. After 1960, however, a sizable exodus of black residents followed, including a significant number of working- and middle-class families. The population of the neighborhood declined from over 80,000 in 1960 to 53,814 in 1970; it further slipped to 36,323 in 1980 and finally to 24,473 in 1990. The loss of residents was accompanied by a substantial reduction in the economic, social, and political resources that make a community vibrant. Woodlawn is only one of a growing number of poor black neighborhoods in Chicago plagued by depopulation and social and economic deterioration.

When the black respondents in our large . . . survey were asked to rate their neighborhood as a place to live, only a third said that their area was a good or very good place to live and only 18 percent of those in the ghetto poverty census tracts felt that their neighborhood was a desirable place to live. (The Bureau of the Census defines a census tract as "a relatively homogeneous area with respect to population characteristics, economic status, and living conditions with an average population of 4,000." Poverty tracts are those in which at least 20 percent of the residents are poor, and ghetto poverty tracts are those in which at least 40 percent are poor.)

[. . .]

Many of the respondents described the negative effects of their neighborhood on their own personal outlook. An unmarried, employed clerical worker from a ghetto poverty census tract on the West Side stated:

There is a more positive outlook if you come from an upwardly mobile neighborhood than you would here. In this type of neighborhood, all you hear is negative [things] and that can kind of bring you down when you're trying to make it. So your neighborhood definitely has something to do with it.

This view was shared by a 17-year-old college student and part-time worker from an impoverished West Side neighborhood.

I'd say about 40 percent in my neighborhood . . . I'd say 40 percent are alcoholics . . . And . . . only 5 percent of the alcoholics have homes. Then you got

> the other 35 percent who are in the street . . . They probably live somewhere, but they in street, on the corner every day, same old thing, because they don't have no chance in life. They live based on today. [They say] "Oh, we gonna get high today." "Oh, whoopee!" "What you gonna do tomorrow, man?" "I don't know, man, I don't know." You can ask any of 'em: "What you gonna do tomorrow?" "I don't know, man. I know when it gets here." And I can really understand, you know, being in that state. If you around totally negative people, people who are not doing anything, that's the way you gonna be regardless.

The state of the inner-city public schools was another major concern expressed by our respondents. The complaints ranged from overcrowded conditions to unqualified and uncaring teachers. Sharply voicing her views on these subjects, a 25-year-old married mother of two children from a South Side census tract that just recently became poor stated: "My daughter ain't going to school here, she was going to a nursery school where I paid and of course they took the time and spent it with her, 'cause they was getting the money. But the public schools, no! They are overcrowded and the teachers don't care."

[. . .]

The respondents were also asked whether their neighborhoods had changed as a place to live over the years. Seventy-one percent of the African–American respondents felt that their neighborhoods had either stayed the same or had gotten worse.

An unemployed black man from a West Side housing project felt that the only thing that had changed in his neighborhood was that it was "going down instead of going back up." He further stated, "It ain't like it used to be. They laid off a lot of people. There used to be a time when you got a broken window, you call up housing and they send someone over to fix it, but it ain't like that no more."

Respondents frequently made statements about the increase in drug trafficking and drug consumption when discussing how their neighborhood had changed. "Well, OK, I realize there was drugs when I was growing up but they weren't as open as they are now," stated a divorced telephone dispatcher and mother of five children from a neighborhood that recently changed from a nonpoverty to a poverty area. "It's nothing to see a 10-year-old kid strung out or a 10-year-old kid selling drugs. I mean, when they were doing it back then they were sneaking around doing it. It's like an open thing now."

[. . .]

The feelings of many of the respondents in our study were summed up by a 33-year-old married mother of three from a very poor West Side neighborhood:

> If you live in an area in your neighborhood where you have people that don't work, don't have no means of support, you know, don't have no jobs, who're gonna break into your house to steal what you have, to sell to get them some money, then you can't live in a neighborhood and try to concentrate on tryin' to get ahead, then you get to work and you have to worry if somebody's breakin' into your house or not. So, you know, it's best to try to move in a decent area, to live in a community with people that works.

In 1959, less than one-third of the poverty population in the United States lived in metropolitan central cities. By 1991, the central cities included close to half of the nation's poor. Many of the most rapid increases in concentrated poverty have occurred in African–American neighborhoods. For example, in the ten community areas that represent the historic core of Chicago's Black Belt (see Figure 1), eight had rates of poverty in 1990 that exceeded 45 percent, including three with rates higher than 50 percent and three that surpassed 60 percent. Twenty-five years earlier, in 1970, only two of these neighborhoods had poverty rates above 40 percent.

In recent years, social scientists have paid particular attention to the increases in urban neighborhood poverty. "Defining an urban neighborhood for analytical purposes is no easy task." The community areas of Chicago referred to in Figure 1 include a number of adjacent census tracts. The seventy-seven community areas within the city of Chicago represent statistical units derived by urban sociologists at the University of Chicago for the 1930 census in their effort to analyze varying conditions within the city. These delineations were originally drawn up on the basis of settlement and history of the area, local identification and trade patterns, local institutions, and natural and artificial barriers. There have been major shifts in population and land use since then. But these units remain useful in tracing changes over time, and they continue to capture much of the contemporary reality of Chicago neighborhoods.

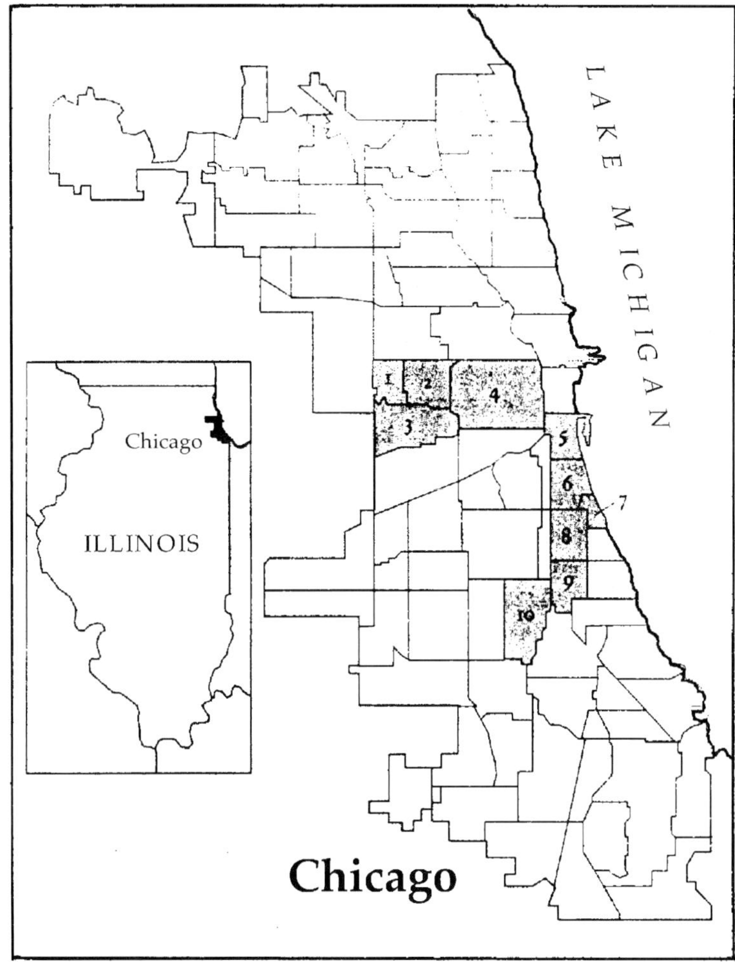

1. West Garfield Park
2. East Garfield Park
3. North Lawndale
4. Near West Side
5. Near South Side
6. Douglas
7. Oakland
8. Grand Boulevard
9. Washington Park
10. Englewood

Figure 1 Community areas in Chicago's Black Belt

Other cities, however, do not have such convenient classifications of neighborhoods, which means that comparison across cities cannot be drawn using community areas. The measurable unit considered most appropriate to represent urban neighborhoods is the census tract. In attempts to examine this problem of ghetto poverty across the nation empirically, social scientists have tended to define ghetto neighborhoods as those located in the *ghetto poverty* census tracts. As indicated earlier, ghetto poverty census tracts are those in which at least 40 percent of the residents are poor. For example, Paul Jargowsky and Mary Jo Bane state: "Visits to various cities confirmed that the 40 percent criterion came very close to identifying areas that looked like ghettos in terms of their housing conditions. Moreover, the areas selected by the 40 percent criterion corresponded closely with the neighborhoods that city officials and local Census Bureau officials considered ghettos." The ghetto poor in Jargowsky and Bane's study are therefore designated as those among the poor who live in these ghetto poverty areas. Three-quarters of all the ghetto poor in metropolitan areas reside in one hundred of the nation's largest central cities; however, it is important to remember that the ghetto areas in these central cities also include a good many families and individuals who are not poor.

In the nation's one hundred largest central cities, nearly one in seven census tracts is at least 40 percent poor. The number of such tracts has more than doubled since 1970 – indeed, it is alarming that 579 tracts fell to ghetto poverty level in these cities between 1970 and 1980, and 624 additional tracts joined these ranks in the following decade.

Paul Jargowsky's research reveals that a vast majority of people (almost seven out of eight) living in metropolitan-area ghettos in 1990 were minority group members. The number of African–Americans in these ghettos grew by more than one-third from 1980 to 1990, reaching nearly 6 million. Most of this growth involved poor people. The proportion of metropolitan blacks who live in ghetto areas climbed from more than a third (37 percent) to almost half (45 percent). Indeed, the metropolitan black poor are becoming increasingly isolated. The poverty rate among metropolitan blacks who reside in ghettos increased while the rate among those who live in nonghettos decreased.

The increase in the *number* of ghetto blacks is related to the *geographical spread* of the ghetto. Jargowsky and Bane found that in the cities they studied (Philadelphia, Cleveland, Milwaukee, and Memphis) areas that had become ghettos by 1980 had been mixed-income tracts in 1970 – but tracts that were contiguous to areas identified as ghettos. The exodus of the nonpoor from mixed-income areas was a major factor in the spread of ghettos in these cities in the 1970s. Since 1980, ghetto census tracts have increased in a substantial majority of the metropolitan areas in the country, including those with fewer people living in them. Nine new ghetto census tracts were added in Philadelphia, even though it experienced one of the largest declines in the proportion of people living in ghetto tracts. In a number of other cities, including Baltimore, Boston, and Washington, D.C., a smaller percentage of poor blacks live in a larger number of ghetto census tracts. Chicago had a 61.5 percent increase in the number of ghetto census tracts from 1980 to 1990, even though the number of poor residing in those areas increased only slightly.

Jargowsky reflects on the significance of the substantial spread of ghetto areas:

> The geographic size of a city's ghetto has a large effect on the perception of the magnitude of the problem associated with ghetto poverty. How big an area of the city do you consider off limits? How far out of your way will you drive not to go through a dangerous area? Indeed, the lower density exacerbated the problem. More abandoned buildings mean more places for crack dens and criminal enterprises. Police trying to protect a given number of citizens have to be stretched over a wider number of square miles, making it less likely that criminals will be caught. Lower density also makes it harder for a sense of community to develop, or for people to feel that they can find safety in numbers. From the point of view of local political officials, the increase in the size of the ghetto is a disaster. Many of those leaving the ghetto settle in non-ghetto areas outside the political jurisdiction of the central city. Thus, geographic size of the ghetto is expanding, cutting a wider swath through the hearts of our metropolitan areas.

In sum, the 1970s and 1980s witnessed a sharp growth in the number of census tracts classified as ghetto poverty areas, an increased concentration of the poor in these areas, and sharply divergent patterns of poverty concentration between racial minorities and whites. One of the legacies of historic racial and class subjugation in America is a unique and growing concentration of minority residents in the most impoverished areas of the nation's metropolises.

Some have argued that this concentration of poverty is not new but mirrors conditions prevalent in the 1930s. According to Douglas Massey and Nancy Denton, during the Depression poverty was just as concentrated in the ghettos of the 1930s as in those of the 1970s. The black communities of the 1930s and those of the 1970s shared a common experience: a high degree of racial segregation from the larger society. Massey and Denton argue that "concentrated poverty is created by a pernicious interaction between a group's overall rate of poverty and its degree of segregation in society. When a highly segregated group experiences a high or rising rate of poverty, geographically concentrated poverty is the inevitable result." However convincing the logic of that argument, it does not explain the following: In the ten neighborhoods that make up Chicago's Black Belt, the poverty rate increased almost 20 percent between 1970 and 1980 (from 32.5 to 50.4 percent) despite the fact that the overall black poverty rate for the city of Chicago increased only 7.5 percent during this same period (from 25.1 to 32.6 percent).

Concentrated poverty may be the inevitable result when a highly segregated group experiences an increase in its overall rate of poverty. But segregation does not explain why the concentration of poverty in *certain* neighborhoods of this segregated group should increase to nearly three times the group's *overall* rate of poverty increase. There is no doubt that the disproportionate concentration of poverty among African–Americans is one of the legacies of historic racial segregation. It is also true that segregation often compounds black vulnerability in the face of other changes in the society, including, as we shall soon see, economic changes. Nonetheless, to focus mainly on segregation to account for the growth of concentrated poverty is to overlook some of the dynamic aspects of the social and demographic changes occurring in cities like Chicago. Given the existence of segregation, we must consider the way in which other changes in society have interacted with segregation to produce the dramatic social transformation of inner-city neighborhoods, especially since 1970.

For example, the communities that make up the Black Belt in Chicago have been overwhelmingly black for the last four decades, yet they lost almost half their residents between 1970 and 1980. This rapid depopulation has had profound consequences for the social and economic deterioration of segregated Black Belt neighborhoods, including increases in concentrated poverty and joblessness. If comparisons are made strictly between the Depression years of the 1930s and the 1980s, rates of ghetto poverty and joblessness in these neighborhoods will indeed be similar. But such a comparison obscures significant changes that have occurred in these neighborhoods across the fifty-year span between those two points.

. . . Many of the gains made in inner-city neighborhoods following the Depression were wiped out after 1970. To maintain that concentrated black poverty in the 1970s or in the 1980s is equivalent in severity and pervasiveness to that which occurred during the Depression does not explain its dramatic rise since 1970; nor does it address a far more fundamental problem that is at the heart of the extraordinary increases in and spread of concentrated poverty – namely, the rapid growth of joblessness, which accelerated through these two decades. The problems reported by the residents of poor Chicago neighborhoods are not a consequence of poverty alone. Something far more devastating has happened that can only be attributed to the emergence of concentrated and persistent joblessness and its crippling effects on neighborhoods, families, and individuals. The city of Chicago epitomizes these changes.

[. . .]

The most fundamental difference between today's inner-city neighborhoods and those studied by Drake and Cayton [in the 1940s] is the much higher levels of joblessness. Indeed, there is a new poverty in our nation's metropolises that has consequences for a range of issues relating to the quality of life in urban areas, including race relations.

By "thc ncw urban poverty," I mean poor, segregated neighborhoods in which a substantial majority of individual adults are either unemployed or have dropped out of the labor force altogether. For example, in 1990 only one in three adults aged 16 and over in the twelve Chicago community areas with ghetto poverty rates held a job in a typical week of the year. Each of these community areas, located on the South and West Sides of the city, is overwhelmingly black. We can add to these twelve high-jobless areas three additional predominantly black community areas, with rates approaching ghetto poverty, in which only 42 percent of the adult population were working in a typical week in 1990. Thus, in these fifteen black community areas – comprising a total population of 425,125 – only 37 percent of all the adults were gainfully employed in a typical week in 1990. By contrast, 54 percent of the adults in the seventeen other predominantly black community areas in Chicago – a total population of 545,408 – worked in a typical week in 1990. This was close to the citywide employment figure of 57 percent for all adults. Finally, except for one Asian community area with an employment rate of 46 percent, and one Latino community area with an employment rate of 49 percent, a majority of the adults held a job in a typical week in each of the remaining forty-five community areas of Chicago.

But Chicago is by no means the only city that features new poverty neighborhoods. In the ghetto census tracts of the nation's one hundred largest central cities, there were only 65.5 employed persons for every hundred adults who did not hold a job in a typical week in 1990. In contrast, the nonpoverty areas contained 182.3 employed persons for every hundred of those not working. In other words, the ratio of employed to jobless persons was three times greater in census tracts not marked by poverty.

Looking at Drake and Cayton's Bronzeville, I can illustrate the magnitude of the changes that have occurred in many, inner-city ghetto neighborhoods in recent years. A majority of adults held jobs in the three Bronzeville areas in 1950, but by 1990 only four in ten in Douglas worked in a typical week, one in three in Washington Park, and one in four in Grand Boulevard. In 1950, 69 percent of all males 14 and over who lived in the Bronzeville neighborhoods worked in a typical week, and in 1960, 64 percent of this group were so employed. However, by 1990 only 37 percent of all males 16 and over held jobs in a typical week in these three neighborhoods.

Upon the publication of the first edition of *Black Metropolis* in 1945, there was much greater class integration within the black community. As Drake and Cayton pointed out, Bronzeville residents had limited success in "sorting themselves out into broad community areas designated as 'lower class' and 'middle class' . . . Instead of middle-class areas, Bronzeville tends to have middle-class buildings in all areas, or a few middle-class blocks here and there." Though they may have lived on different streets, blacks of all classes in inner-city areas such as Bronzeville lived in the same community and shopped at the same stores. Their children went to the same schools and played in the same parks. Although there was some class antagonism, their neighborhoods were more stable than the inner-city neighborhoods of today; in short, they featured higher levels of what social scientists call "social organization."

When I speak of social organization I am referring to the extent to which the residents of a neighborhood are able to maintain effective social control and realize their common goals. There are three major dimen-sions of neighborhood social organization: (1) the prevalence, strength, and interdependence of social networks; (2) the extent of collective supervision that the residents exercise and the degree of personal responsibility they assume in addressing neighbor-hood problems; and (3) the rate of resident participation in voluntary and formal organizations. Formal institutions (e.g., churches and political party organizations), voluntary associations (e.g., block clubs and parent–teacher organizations), and informal networks (e.g., neighborhood friends and acquaintances, coworkers, marital and parental ties) all reflect social organization.

Neighborhood social organization depends on the extent of local friendship ties, the degree of social cohesion, the level of resident participation in formal and informal voluntary associations, the density and stability of formal organizations, and the nature of informal social controls. Neighborhoods in which adults are able to interact in terms of obligations, expectations, and relationships are in a better position to supervise and control the activities and behavior of children. In neighborhoods with high levels of social organization, adults are empowered to act to improve the quality of neighborhood life – for example, by breaking up congregations of youths on street corners and by supervising the leisure activities of youngsters.

Neighborhoods plagued by high levels of joblessness are more likely to experience low levels of social organization: the two go hand in hand. High rates of joblessness trigger other neighborhood problems that undermine social organization, ranging from crime, gang violence, and drug trafficking to family breakups and problems in the organization of family life.

Consider, for example, the problems of drug trafficking and violent crime. As many studies have revealed, the decline in legitimate employment opportunities among inner-city residents has increased incentives to sell drugs. The distribution of crack in a neighborhood attracts individuals involved in violence and lawlessness. Between 1985 and 1992, there was a sharp increase in the murder rate among men under the age of 24; for men 18 years old and younger, murder rates doubled. Black males in particular have been involved in this upsurge in violence. For example, whereas the homicide rate for white males between 14 and 17 increased from 8 per 100,000 in 1984 to 14 in 1991, the rate for black males tripled during that time (from 32 per 100,000 to 112). This sharp rise in violent crime among younger males has accompanied the widespread outbreak of addiction to crack cocaine. The association is especially strong in inner-city ghetto neighborhoods plagued by joblessness and weak social organization.

Violent persons in the crack-cocaine marketplace have a powerful impact on the social organization of a neighborhood. Neighborhoods plagued by high levels of joblessness, insufficient economic opportunities, and high residential mobility are unable to control the volatile drug market and the violent crimes related to it. As informal controls weaken, the social processes that regulate behavior change.

As a result, the behavior and norms in the drug market are more likely to influence the action of others in the neighborhood, even those who are not involved in drug activity. Drug dealers cause the use and spread of guns in the neighborhood to escalate, which in turn raises the likelihood that others, particularly the youngsters, will come to view the possession of weapons as necessary or desirable for self-protection, settling disputes, and gaining respect from peers and other individuals.

Moreover, as Alfred Blumstein pointed out, the drug industry actively recruits teenagers in the neighborhood "partly because they will work more cheaply than adults, partly because they may be less vulnerable to the punishments imposed by the adult criminal justice system, partly because they tend to be daring and willing to take risks that more mature adults would eschew." Inner-city black youths with limited prospects for stable or attractive employment are easily lured into drug trafficking and therefore increasingly find themselves involved in the violent behavior that accompanies it.

A more direct relationship between joblessness and violent crime is revealed in recent research by Delbert Elliott of the University of Colorado, a study based on National Longitudinal Youth Survey data collected from 1976 to 1989, covering ages 11 to 30. As Elliott points out, the transition from adolescence to adulthood usually results in a sharp drop in most crimes as individuals take on new adult roles and responsibilities. "Participation in serious violent offending behavior (aggravated assault, forcible rape, and robbery) increases [for all males] from ages 11 and 12 to ages 15 and 16, then declines dramatically with advancing age." Although black and white males reveal similar age curves, "the negative slope of the age curve for blacks after age 20 is substantially less than that of whites."

The black–white differential in the proportion of males involved in serious violent crime, although almost even at age 11, increases to 3 : 2 over the remaining years of adolescence, and reaches a differential of nearly 4 : 1 during the late twenties. However, when Elliott compared only *employed* black and white males, he found no significant differences in violent behavior patterns among the two groups by age 21. Employed black males, like white males, experienced a precipitous decline in serious violent behavior following their adolescent period. Accordingly, a major reason for the racial gap in violent behavior after adolescence is joblessness; a large proportion of jobless black males do not assume adult roles and responsibilities, and their serious violent behavior is therefore more likely to extend into adulthood. The new poverty neighborhoods feature a high concentration of jobless males and, as a result, suffer rates of violent criminal behavior that exceed those in other urban neighborhoods.

. . . In 1990, 37 percent of Woodlawn's 27,473 adults were employed and only 23 percent of Oakland's 4,935 adults were working. When asked how much of a problem unemployment was in their neighborhood, 73 percent of the residents in Woodlawn and 76 percent in Oakland identified it as a major problem. The responses to the survey also revealed the residents' concerns about a series of related problems, such as crime and drug abuse, that are symptomatic of severe problems of social organization. Indeed, crime was identified as a major problem by 66 percent of the residents in each neighborhood. Drug abuse was cited as a major problem by as many as 86 percent of the adult residents in Oakland and 79 percent of those in Woodlawn.

Although high-jobless neighborhoods also feature concentrated poverty, high rates of neighborhood poverty are less likely to trigger problems of social organization if the residents are working. This was the case in previous years when the working poor stood out in areas like Bronzeville. Today, the nonworking poor predominate in the highly segregated and impoverished neighborhoods.

The rise of new poverty neighborhoods represents a movement away from what the historian Allan Spear has called an institutional ghetto – whose structure and activities parallel those of the larger society, as portrayed in Drake and Cayton's description of Bronzeville – toward a jobless ghetto, which features a severe lack of basic opportunities and resources, and inadequate social controls.

What can account for the growing proportion of jobless adults and the corresponding increase in problems of social organization in innercity communities such as Bronzeville? An easy answer is racial segregation. However, a race-specific argument is not sufficient to explain recent changes in neighborhoods like Bronzeville. After all, Bronzeville was just as *segregated by skin color in 1950* as it is today, yet the level of employment was much higher then.

Nonetheless, racial segregation does matter. If large segments of the African–American population had not been historically segregated in inner-city ghettos, we would not be talking about the new urban poverty. The segregated ghetto is not the result of voluntary or positive decisions on the part of the residents who live there. As Massey and Denton have carefully documented, the segregated ghetto is the product of systematic racial practices such as restrictive covenants, redlining by banks and insurance companies, zoning, panic peddling by real estate agents, and the creation of massive public housing projects in low-income areas.

Segregated ghettos are less conducive to employment and employment preparation than are other areas of the city. Segregation in ghettos exacerbates employment problems because it leads to weak informal employment networks and contributes to the social isolation of individuals and families, thereby reducing their chances of acquiring the human capital skills, including adequate educational training, that facilitate mobility in a society. Since no other group in society experiences the degree of segregation, isolation, and poverty concentration as do African–Americans, they are far more likely to be disadvantaged when they have to compete with other groups in society, including other despised groups, for resources and privileges.

To understand the new urban poverty, one has to account for the ways in which segregation interacts with other changes in society to produce the recent escalating rates of joblessness and problems of social organization in inner-city ghetto neighborhoods.

"The Code of the Street" and "Decent and Street Families"

from *Code of the Street: Decency, Violence, and the Moral Life of the Inner City* (1999)

Elijah Anderson

Editors' Introduction

Both W.E.B. Du Bois, writing as a historian, and William Julius Wilson, focusing on policy issues affecting inner-city poverty, employ the technique of ethnography – the close observation of people interacting in social situations – as an element of their overall analyses. Elijah Anderson (b. 1943) is an exceptionally accomplished ethnographer who looks closely at the minute details of African American urban experience and culture. Anderson is deeply concerned about urban policy issues, but he lets the reality of ghetto life speak for itself in ways that are sometimes startling and always brutally honest: in one chapter in *Code of the Street*, he describes the correct way to survive the experience of being robbed!

Code of the Street (New York: Norton, 1999) won the 2000 Komarovsky Award from the Eastern Sociological Association and is highly regarded as a signal contribution to urban ethnography and the study of urban inequality, race relations, and the policy issues of social control and cultural deviance. In the selection here, Anderson examines two kinds of cultures operating within the African American inner-city community: the "decent" life characterized by adherence to middle-class norms of behavior and the "street" life characterized by boisterousness, lawlessness, violence, and disregard of the rights of others.

The core problem of ghetto life, writes Anderson, is the pattern of "interpersonal violence and aggression" that wreaks havoc daily on the lives of community residents and increasingly spills over into downtown and residential middle-class areas." The source of this violence is "the circumstances of life among the ghetto poor – the lack of jobs . . . limited basic public services . . . the stigma of race . . . rampant drug use and drug trafficking . . . alienation and the absence of hope for the future." Young people in particular are the victims of this system of social pathology, and its effects can only be counteracted by "a strong, loving, 'decent' (as the inner-city residents put it) family that is committed to middle-class values." But standing against middle-class decency is "the code of the street" which Anderson describes as "a set of informal rules governing interpersonal public behavior, particularly violence."

For residents of inner-city ghettos, especially youth, the code of the street rules the way life is played. "At the heart of the code," writes Anderson, "is the issue of respect – loosely defined as being treated 'right' or granted one's 'props' . . . or the deference one deserves." Sociologically, the code is "a cultural adaptation to a profound lack of faith in the police and the judicial system" which is "viewed as representing the dominant white society" and an "oppositional culture . . . whose norms are often consciously opposed to those of mainstream society." Such oppositional cultures are completely understandable, given the racism and lack of opportunity that define the ghetto and may even be morally justifiable. But tragically, they are not useful to young people striving to rise within the larger society. And the code of the street cannot be ignored because "decent" and "street" systems of behavior coexist and constantly interact within the ghetto community. Thus, even children from solid and supportive "decent families" need to engage in "code-switching" in order "to handle themselves in a street-oriented environment."

Elijah Anderson was a Distinguished Professor of Social Sciences and Sociology at the University of Pennsylvania for many years before moving to Yale, where he became the William K. Lanman, Jr. Professor of Sociology in 2007. Regarded as a ground-breaking scholar of American urban life, Anderson published *A Place on the Corner: A Study of Black Street Corner Men* (Chicago, IL: University of Chicago Press, 1978; 2nd edn, 2003) and *Streetwise: Race, Class and Change in an Urban Community* (Chicago, IL: University of Chicago Press, 1990) for which he won the American Sociological Association's Robert E. Park Award. He also wrote the introduction to the 1990 reissue of W.E.B. Du Bois's *The Philadelphia Negro* (Philadelphia, PA: University of Pennsylvania Press, 1990; p. 110) and edited *Against the Wall: Poor, Young, Black and Male* (Philadelphia, PA: University of Pennsylvania Press, 2008).

The literature on African American ghetto culture is vast, but an excellent place to start is Alex Haley, *The Autobiography of Malcolm X* (New York: Grove, 1964). The most important, and ground-breaking, academic study is Elliott Liebow, *Tally's Corner: A Study of Negro Streetcorner Men* (Boston, MA: Little, Brown, 1967). Also of interest is Mitchell Duneier, *Slim's Table: Race, Respectability, and Masculinity* (Chicago, IL: University of Chicago Press, 1994). For a Latino perspective, see Philippe Bourgeois, *In Search of Respect: Selling Crack in El Barrio* (Cambridge: Cambridge University Press, 2nd edn, 2002). And for a youth and popular culture perspective, see Bakari Kitwana, *The Hip Hop Generation: Young Blacks and the Crisis in African American Culture* (New York: Basic Civitas Books, 2003).

THE CODE OF THE STREET

Of all the problems besetting the poor inner-city black community, none is more pressing than that of interpersonal violence and aggression. This phenomenon wreaks havoc daily on the lives of community residents and increasingly spills over into downtown and residential middle-class areas. Muggings, burglaries, carjackings, and drug-related shootings, all of which may leave their victims or innocent bystanders dead, are now common enough to concern all urban and many suburban residents.

The inclination to violence springs from the circumstances of life among the ghetto poor – the lack of jobs that pay a living wage, limited basic public services (police response in emergencies, building maintenance, trash pickup, lighting, and other services that middle-class neighborhoods take for granted), the stigma of race, the fallout from rampant drug use and drug trafficking, and the resulting alienation and absence of hope for the future. Simply living in such an environment places young people at special risk of falling victim to aggressive behavior. Although there are often forces in the community that can counteract the negative influences – by far the most powerful is a strong, loving, "decent" (as inner-city residents put it) family that is committed to middle-class values – the the despair is pervasive enough to have spawned an oppositional culture, that of "the street," whose norms are often consciously opposed to those of mainstream society. These two orientations – decent and street – organize the community socially, and the way they coexist and interact has important consequences for its residents, particularly for children growing up in the inner city. Above all, this environment means that even youngsters whose home lives reflect mainstream values – and most of the homes in the community do – must be able to handle themselves in a street-oriented environment.

This is because the street culture has evolved a "code of the street," which amounts to a set of informal rules governing interpersonal public behavior, particularly violence. The rules prescribe both proper comportment and the proper way to respond if challenged. They regulate the use of violence and so supply a rationale allowing those who are inclined to aggression to precipitate violent encounters in an approved way. The rules have been established and are enforced mainly by the street-oriented; but on the streets the distinction between street and decent is often irrelevant. Everybody knows that if the rules are violated, there are penalties. Knowledge of the code is thus largely defensive, and it is literally necessary for operating in public. Therefore, though families with a decency orientation are usually opposed to the values of the code, they often reluctantly encourage their children's familiarity with it in order to enable them to negotiate the inner-city environment.

At the heart of the code is the issue of respect – loosely defined as being treated "right" or being granted one's "props" (or proper due) or the deference one deserves. However, in the troublesome public environment of the inner city, as people increasingly feel buffeted by forces beyond their control, what one deserves in the way of respect becomes ever more problematic and uncertain. This situation in turn further opens up the issue of respect to sometimes intense interpersonal negotiation, at times resulting in altercations. In the street culture, especially among young people, respect is viewed as almost an external entity, one that is hard-won but easily lost – and so must constantly be guarded. The rules of the code in fact provide a framework for negotiating respect. With the right amount of respect, individuals can avoid being bothered in public. This security is important, for if they *are* bothered, not only may they face physical danger, but they will have been disgraced or "dissed" (disrespected). Many of the forms dissing can take may seem petty to middle-class people (maintaining eye contact for too long, for example), but to those invested in the street code, these actions, a virtual slap in the face, become serious indications of the other person's intentions. Consequently, such people become very sensitive to advances and slights, which could well serve as a warning of imminent physical attack or confrontation.

The hard reality of the world of the street can be traced to the profound sense of alienation from mainstream society and its institutions felt by many poor inner-city black people, particularly the young. The code of the street is actually a cultural adaptation to a profound lack of faith in the police and the judicial system – and in others who would champion one's personal security. The police, for instance, are most often viewed as representing the dominant white society and as not caring to protect inner-city residents. When called, they may not respond, which is one reason many residents feel they must be prepared to take extraordinary measures to defend themselves and their loved ones against those who are inclined to aggression. Lack of police accountability has in fact been incorporated into the local status system: the person who is believed capable of "taking care of himself" is accorded a certain deference and regard, which translates into a sense of physical and psychological control. The code of the street thus emerges where the influence of the police ends and where personal responsibility for one's safety is felt to begin. Exacerbated by the proliferation of drugs and easy access to guns, this volatile situation results in the ability of the street-oriented minority (or those who effectively "go for bad") to dominate the public spaces.

DECENT AND STREET FAMILIES

Almost everyone residing in poor inner-city neighborhoods is struggling financially and therefore feels a certain distance from the rest of America, but there are degrees of alienation, captured by the terms "decent" and "street" or "ghetto," suggesting social types. The decent family and the street family in a real sense represent two poles of value orientation, two contrasting conceptual categories. The labels "decent" and "street," which the residents themselves use, amount to evaluative judgments that confer status on local residents. The labeling is often the result of a social contest among individuals and families of the neighborhood. Individuals of either orientation may coexist in the same extended family. Moreover, decent residents may judge themselves to be so while judging others to be of the street, and street individuals often present themselves as decent, while drawing distinctions between themselves and still other people. There is also quite a bit of circumstantial behavior – that is, one person may at different times exhibit both decent and street orientations, depending on the circumstances. Although these designations result from much social jockeying, there do exist concrete features that define each conceptual category, forming a social typology.

The resulting labels are used by residents of inner-city communities to characterize themselves and one another, and understanding them is part of understanding life in the inner-city neighborhood. Most residents are decent or are trying to be. The same family is likely to have members who are strongly oriented toward decency and civility, whereas other members are oriented toward the street – and to all that it implies. There is also a great deal of "code-switching": a person may behave according to either set of rules, depending on the situation. Decent people, especially young people, often put a premium on the ability to code-switch. They share many of the middle-class values of the wider white society but know that the open display of such values carries little weight on the street: it doesn't provide the emblems that say, "I can take care of myself." Hence such people develop a repertoire of behaviors that do provide that security. Those strongly associated with the street, who have less exposure to the wider society, may have difficulty code-switching; imbued with the code of the street, they either don't know the rules for decent behavior or may see little value in displaying such knowledge.

At the extreme of the street-oriented group are those who make up the criminal element. People in this class are profound casualties of the social and economic system, and they tend to embrace the street code wholeheartedly. They tend to lack not only a decent education – though some are highly intelligent – but also an outlook that would allow them to see far beyond their immediate circumstances. Rather, many pride themselves on living the "thug life," actively defying not simply the wider social conventions but the law itself. They sometimes model themselves after successful local drug dealers and rap artists like Tupac Shakur and Snoop Doggy Dogg, and they take heart from professional athletes who confront the system and stand up for themselves. In their view, policemen, public officials, and corporate heads are unworthy of respect and hold little moral authority. Highly alienated and embittered, they exude generalized contempt for the wider scheme of things and for a system they are sure has nothing but contempt for them.

Members of this group are among the most desperate and most alienated people of the inner city. For them, people and situations are best approached both as objects of exploitation and as challenges possibly "having a trick to them," and in most situations their goal is to avoid being "caught up in the trick bag." Theirs is a cynical outlook, and trust of others is severely lacking, even trust of those they are close to. Consistently, they tend to approach all persons and situations as part of life's obstacles, as things to subdue or to "get over." To get over, individuals develop an effective "hustle" or "game plan," setting themselves up in a position to prevail by being "slick" and outsmarting others. In line with this, one must always be wary of one's counterparts, to assume that they are involved with you only for what they can get out of the situation.

Correspondingly, life in public often features an intense competition for scarce social goods in which "winners" totally dominate "losers" and in which losing can be a fate worse than death. So one must be on one's guard constantly. One is not always able to trust others fully, in part because so much is at stake socially, but also because everyone else is understood to be so deprived. In these circumstances, violence is quite prevalent – in families, in schools, and in the streets – becoming a way of public life that is effectively governed by the code of the street.

Decent and street families deal with the code of the street in various ways. An understanding of the dynamics of these families is thus critical to an understanding of the dynamics of the code. It is important to understand here that the family one emerges from is distinct from the "family" one finds in the streets. For street-oriented people especially, the family outside competes with blood relatives for an individual's loyalties and commitments. Nevertheless, blood relatives always come first. The folklore of the street says, in effect, that if I will fight and "take up for" my friend, then you know what I will do for my own brother, cousin, nephew, aunt, sister, or mother – and vice versa. Blood is thicker than mud.

Decent families

In decent families there is almost always a real concern with and a certain amount of hope for the future. Such attitudes are often expressed in a drive to work "to have something" or "to build a good life," while at the same time trying to "make do with what you have." This means working hard, saving money for material things, and raising children – any "child you touch" – to try to make something out of themselves. Decent families tend to accept mainstream values more fully than street families, and they attempt to instill them in their children. Probably the most meaningful description of the mission of the decent family, as seen by members and outsiders alike, is to instill "backbone" and a sense of responsibility in its younger members. In their efforts toward this goal, decent parents are much more able and willing than street-oriented ones to ally themselves with outside institutions such as schools and churches. They value hard work and self-reliance and are willing to sacrifice for their children: they harbor hopes for a better future for their children, if not for themselves. Rather than dwelling on the hardships and inequities facing them, many such decent people, particularly the increasing number of grandmothers raising grandchildren, often see their difficult situation as a test from God and derive great support from their faith and church community.

The role of the "man of the house" is significant. Working-class black families have traditionally placed a high value on male authority. Generally, the man is seen as the "head of household," with the woman as his partner and the children as their subjects. His role includes protecting the family from threats, at times literally putting his body in the line of fire on the street. In return he expects to rule his household and to get respect from the other members, and he encourages his sons to grow up with the same expectations. Being a breadwinner or good provider is often a moral issue, and a man unable to provide for a family invites disrespect from his partner. Many young men who lack the resources to do so often say, "I can't play house," and opt out of forming a family, perhaps leaving the woman and any children to fend for themselves. Intact nuclear families, although in the minority in the impoverished inner city, provide powerful role models. Typically, husband and wife work at low-paying jobs, sometimes juggling more than one such job each. They may be aided financially by the contributions of a teenage child who works part-time. Such families, along with other such local families, are often vigilant in their desire to keep the children away from the streets.

In public such an intact family makes a striking picture as the man may take pains to show he is in complete control – with the woman and the children following his lead. On the inner-city streets this appearance helps him play his role as protector, and he may exhibit exaggerated concern for his family, particularly when other males are near. His actions and words, including loud and deep-voiced assertions to get his small children in line, let strangers know: "This is my family, and I am in charge." He signals that he is capable of protecting them and that his family is not to be messed with.

I witnessed such a display one Saturday afternoon at the Gallery, an indoor shopping mall with a primarily black, Hispanic, and working- to middle-class white clientele. Rasheed Taylor, his wife, Iisha, and their children, Rhonda, Jimmy, and Malika, wandered about the crowded food court looking for a place to sit down to eat. They finally found a table next to mine. Before sitting down, Mr. Taylor asked me if the seats were available, to which I replied they were. He then summoned his family, and they walked forward promptly and in an orderly way to take the seats. The three children sat on one side and the parents on the other. Mr. Taylor took food requests and with a stern look in his eye told the children to stay seated until he and his wife returned with the food. The children nodded attentively. After the adults left, the children seemed to relax, talking more freely and playing with one another. When the parents returned, the kids straightened up again, received their food, and began to eat, displaying quiet and gracious manners all the while. It was very clear to everybody looking on that Mr. Taylor was in charge of this family, with everyone showing him utter deference and respect.

Extremely aware of the problematic and often dangerous environment in which they reside, decent parents tend to be strict in their child-rearing practices, encouraging children to respect authority and walk a straight moral line. They sometimes display an almost obsessive concern about trouble of any kind and encourage their children to avoid people and situations that might lead to it. But this is very difficult, since the decent and the street families live in such close proximity. . . .

As indicated above, people who define themselves as decent tend themselves to be polite and considerate of others and teach their children to be the same way. But this is sometimes difficult, mainly because of the social environment in which they reside, and they often perceive a need to "get ignorant" – to act aggressively, even to threaten violence. For whether a certain child gets picked on may well depend not just on the reputation of the child but, equally important, on how "bad" the child's family is known to be. How many people the child can gather together for the purposes of defense or revenge often emerges as a critical issue. Thus social relations can become practical matters of personal defense. Violence can come at any time, and many persons feel a great need to be ready to defend themselves.

At home, at work, and in church, decent parents strive to maintain a positive mental attitude and a spirit of cooperation. When disciplining their children, they tend to use corporal punishment, but unlike street parents, who can often be observed lashing out at their children, they may explain the reason for the spanking. These parents express their care and love for teenage children by guarding against the appearance of any kind of "loose" behavior (violence, drug use, staying out very late) that might be associated with the streets. In this regard, they are vigilant, observing children's peers as well and sometimes embarrassing their own children by voicing value judgments in front of friends.

These same parents are aware, however, that the right material things as well as a certain amount of cash are essential to young people's survival on the street. So they may purchase expensive things for their children, even when money is tight, in order that the children will be less tempted to turn to the drug trade or other aspects of the underground economy for money.

The street family

So-called street parents, unlike decent ones, often show a lack of consideration for other people and have a rather superficial sense of family and community. They may love their children but frequently find it difficult both to cope with the physical and emotional demands of parenthood and to reconcile their needs with those of their children. Members of these families, who are more fully invested in the code of the street than the decent people are, may aggressively socialize their children into it in a normative way. They more fully believe in the code and judge themselves and others according to its values.

In fact, the overwhelming majority of families in the inner-city community try to approximate the decent-family model, but many others clearly represent the decent families' worst fears. Not only are their financial resources extremely limited, but what little they have may easily be misused. The lives of the street-oriented are often marked by disorganization. In the most desperate circumstances, people frequently have a limited understanding of priorities and consequences, and so frustrations mount over bills, food, and, at times, liquor, cigarettes, and drugs. Some people tend toward self-destructive behavior; many street-oriented women are crack-addicted ("on the pipe"), alcoholic, or involved in complicated relationships with men who abuse them.

In addition, the seeming intractability of their situation, caused in large part by the lack of well-paying jobs and the persistence of racial discrimination, has engendered deep-seated bitterness and anger in many of the most desperate and poorest blacks, especially young people. The need both to exercise a measure of control and to lash out at somebody is often reflected in the adults' relations with their children. At the very least, the frustrations associated with persistent poverty shorten the fuse in such people, contributing to a lack of patience with anyone – child or adult – who irritates them.

People who fit the conception of street are often considered to be lowlife or "bad people," especially by the "decent people," and they are generally seen as incapable of being anything but a bad influence on the community and a bother to their neighbors. For example, on a relatively quiet block in West Oak Lane, on the edge of a racially integrated, predominantly middle-class neighborhood, there is a row of houses inhabited by impoverished people. One of them is Joe Dickens, a heavyset, thirty-two-year-old black man. Joe rents the house he lives in, and he shares it with his three children – two daughters (aged seven and five) and a three-year-old son. With patches on the brickwork, an irregular pillar holding up the porch roof, and an unpainted plywood front door, his house sticks out on the block. The front windows have bars; the small front yard is filled with trash and weeds; the garbage cans at the side of the house are continually overflowing.

Even more obtrusive is the lifestyle of the household. Dickens's wife has disappeared from the scene. It is rumored that her crack habit got completely out of control, and she gravitated to the streets and became a prostitute to support her habit. Dickens could not accept this behavior and let her go; he took over running the house and caring for the children as best he could. And to the extent that the children are fed, clothed, and housed under his roof, he might be considered a responsible parent.

But many of the neighbors do not view him as responsible. They see him yelling and cursing at the kids when he pays attention to them at all. Mostly, he allows them to "rip and run" unsupervised up and down the street at all hours, riding their Big Wheels and making a racket. They are joined by other neighborhood children playing on the streets and sidewalks without adult supervision. Dickens himself pays more attention to his buddies, who seem always to be hanging out at the house – on the porch in warm weather – playing loud rap music, drinking beer, and playing cards.

Dickens generally begins his day at about 11 a.m., when he may go out for cheesesteaks and videos for his visitors. In fact, one gets the impression that the house is the scene of an ongoing party. The noise constantly disturbs the neighbors, sometimes prompting them to call the police. But the police rarely respond to the complaints, leaving the neighbors frustrated and demoralized. Dickens seems almost completely indifferent to his neighbors and inconsiderate of their concerns, a defining trait of street-oriented people.

Dickens's decent neighbors are afraid to confront him because they fear getting into trouble with him and his buddies. They are sure that he believes in the principle that might makes right and that he is likely to try to harm anyone who annoys him. Furthermore, they suspect he is a crack dealer. The neighbors cannot confirm this, but some are convinced anyway, and activities around his house support this conclusion. People come and go at all hours of the day and night; they often leave their car engines running, dash into the house, and quickly emerge and drive off. Dickens's children, of course, see much of this activity. At times the children are made to stand outside on the porch while business is presumably being transacted inside. These children are learning by example the values of toughness and self-absorption: to be loud, boisterous, proudly crude, and uncouth – in short, street.

* * *

Street-oriented women tend to perform their motherly duties sporadically. The most irresponsible women can be found at local bars and crack houses, getting high and socializing with other adults. Reports of crack addicts abandoning their children have become common in drug-infested inner-city communities. Typically, neighbors or relatives discover the abandoned children, often hungry and distraught over the absence of their mother. After repeated absences a friend or relative, particularly a grandmother, will often step in to care for the children, sometimes petitioning the authorities to send her, as guardian of the children, the mother's welfare check, if she gets one. By this time, however, the children may well have learned the first lesson of the streets: you cannot take survival itself, let alone respect, for granted; you have to fight for your place in the world. Some of the children learn to fend for themselves, foraging for food and money any way they can. They are sometimes employed by drug dealers or become addicted themselves.

These children of the street, growing up with little supervision, are said to "come up hard." They often learn to fight at an early age, using short-tempered adults around them as role models. The street-oriented home may be fraught with anger, verbal disputes, physical aggression, even mayhem. The children are victimized by these goings-on and quickly learn to hit those who cross them.

The people who see themselves as decent refer to the general set of cultural deficits exhibited by people like Joe Dickens – a fundamental lack of social polish and commitment to norms of civility – as "ignorance." In their view ignorance lies behind the propensity to violence that makes relatively minor social transgressions snowball into more serious disagreements, and they believe that the street-oriented are quick to resort to violence in almost any dispute.

The fact that the decent people, as a rule civilly disposed, socially conscious, and self-reliant men and women, share the neighborhood streets and other public places with those associated with the street, the inconsiderate, the ignorant, and the desperate, places the "good" people at special risk. In order to live and function in the community, they must adapt to a street reality that is often dominated by people who at best are suffering severely in some way and who are apt to resort quickly to violence to settle disputes. This process of adapting means learning and observing the code of the street. Decent people may readily defer to people, especially strangers, who seem to be at all street-oriented. When they encounter such people at theaters and other public places talking loudly or making excessive noise, they are reluctant to correct them for fear of verbal abuse that could lead to violence. Similarly, they will often avoid confrontations over a parking space or traffic error for fear of a verbal or physical altercation. But under their breaths they may mutter "street niggers" to a black companion, drawing a sharp cultural distinction between themselves and such individuals.

"Bowling Alone: America's Declining Social Capital"

Journal of Democracy (1995)

Robert D. Putnam

Editors' Introduction

Robert Putnam has been called "the most influential academic in the world today," and his work has been praised by political leaders as varied as Bill Clinton, Tony Blair, and George W. Bush. As a social critic, he stands in a tradition that includes Alexis de Tocqueville (*Democracy in America*, 1835), Paul Goodman (*Growing Up Absurd*, 1960), and Philip Slater (*The Pursuit of Loneliness: American Culture at the Breaking Point*, 1970). In this article from the *Journal of Democracy*, and in a subsequent book of the same title, Putnam asserts the "bowling alone" phenomenon – that more and more people take up bowling as a form of recreation, but fewer and fewer belong to organized leagues – as a metaphor for what urban life has become in contemporary middle-class America and for millions of others worldwide living in an increasingly materialistic and solipsistic culture of corporate work, obsessive consumption, and overdetermined leisure.

Whereas cities once held out the promise of a wider, higher form of human community, Putnam argues that contemporary urbanites now follow a path of less, not more, civic engagement and that our collective stock of "social capital" – the meaningful human contacts of all kinds that characterize true communities – is so dangerously eroded that it verges on depletion. In a massive follow-up study seeking the causes of this social disengagement, Putnam discovered evidence of a negative correlation between racial and ethnic diversity and social capital formation. Although much of this study was completed as early as 2001, Putnam withheld full disclosure of the findings until 2008, leading to some criticism that questioned the ethics of suppressing information, even temporarily, that might be judged politically incorrect.

In some ways, Putnam's original critique is an updated version of Louis Wirth's 1938 essay "Urbanism as a Way of Life" (p. 96). Urban dwellers are not connected to one another through collective action as they or their forebears in small towns or rural areas once were. Instead, virtually all measures of social engagement – voting, participation in social organizations, active church membership, even friendships and family ties – seem to grow weaker every year. In part, this phenomenon arises from the well-known causes: social and geographic mobility, the decreasing importance of families as women join the corporate workforce, and the technological transformation of leisure. However understandable, Putnam argues, these forces have now risen to the level of social crisis and must be addressed by conscious policies to increase civic engagement of all sorts and to strengthen the connection between people in their roles as neighbors, co-workers, and fellow citizens.

Putnam's ideas about urban civic engagement emerge from a deep philosophical tradition that includes Peter Kropotkin's *Mutual Aid* (New York: Knopf, 1922, originally published in 1902) and John Gardner's "Building Community" (*Independent Sector*, 1991). They raise questions as to whether any modern urban society can hope to regain the intimacy of the ancient Greek polis as described by H.D.F. Kitto (p. 40) or if the planning practices of New Urbanists such as Peter Calthorpe and William Fulton (p. 360) and the design strategies of Jan Gehl (p. 530) can truly lead to the heightened degrees of "community" that Jane Jacobs describes in *The Death and*

Life of Great American Cities (p. 105). These issues take on increased importance when considering the social and technological futures of urban society discussed in Part Eight of this volume.

Robert Putnam (b. 1941) is the Peter and Isabel Malkin Professor of Public Policy at Harvard University, as well as the former dean of the Kennedy School of Government, and has written widely in the fields of politics, comparative politics, international relations, and public policy. He attended Swarthmore and Balliol College, Oxford, before taking his PhD at Yale in 1970. He is the author of *The Beliefs of Politicians: Ideology, Conflict, and Democracy in Britain and Italy* (New Haven, CT: Yale University Press, 1973) and, with Robert Leonardi and Rafaella Nanetti, *Making Democracy Work: Civic Traditions in Modern Italy* (Princeton, NJ: Princeton University Press, 1993). Following the wide popular success of *Bowling Alone* (New York: Simon & Schuster, 2000), Putnam founded the Saguaro Seminar on Civic Engagement in America where he brings together community leaders and policy makers to develop working plans for increasing civic connectedness in specific urban contexts. He is also the editor of *Democracies in Flux: The Evolution of Social Capital in Contemporary Society* (New York: Oxford University Press, 2002).

Other sources of information on the idea of urban social capital include James S. Coleman, "Social Capital in the Creation of Human Capital" (*American Journal of Sociology*, 94, 1988) and Robert Wuthnow, *Sharing the Journey: Support Groups and America's New Quest for Community* (New York: Free Press, 1994). A number of elaborations and critiques of the social capital idea can be found in Robert M. Silverman (ed.), *Community-Based Organizations: The Intersection of Social Capital and Local Context in Contemporary Urban Society* (Detroit, MI: Wayne State University Press, 2004). Among the more interesting, and sometimes critical, analyses are Ivan Light, "Social Capital for What?", Randy Stoeker, "The Mystery of the Missing Social Capital and the Ghost of Social Structure: Why Community Development Can't Win," and James DeFilippis, "The Myth of Social Capital in Community Development," *Housing Policy Debate*, 12(4) (2001).

Many students of the new democracies that have emerged over the past decade and a half have emphasized the importance of a strong and active civil society to the consolidation of democracy. Especially with regard to the postcommunist countries, scholars and democratic activists alike have lamented the absence or obliteration of traditions of independent civic engagement and a widespread tendency toward passive reliance on the state. To those concerned with the weakness of civil societies in the developing or postcommunist world, the advanced Western democracies and above all the United States have typically been taken as models to be emulated. There is striking evidence, however, that the vibrancy of American civil society has notably declined over the past several decades.

Ever since the publication of Alexis de Tocqueville's *Democracy in America* [1835], the United States has played a central role in systematic studies of the links between democracy and civil society. Although this is in part because trends in American life are often regarded as harbingers of social modernization, it is also because America has traditionally been considered unusually "civic" (a reputation that, as we shall later see, has not been entirely unjustified).

When Tocqueville visited the United States in the 1830s, it was the Americans' propensity for civic association that most impressed him as the key to their unprecedented ability to make democracy work. "Americans of all ages, all stations in life, and all types of disposition," he observed, "are forever forming associations. There are not only commercial and industrial associations in which all take part, but others of a thousand different types – religious, moral, serious, futile, very general and very limited, immensely large and very minute. Nothing, in my view, deserves more attention than the intellectual and moral associations in America."

Recently, American social scientists of a neo-Tocquevillean bent have unearthed a wide range of empirical evidence that the quality of public life and the performance of social institutions (and not only

The argument in this essay was amplified and to some extent modified in Robert D. Putnam, *Bowling Alone: The Collapse and Revival of American Community* (New York: Simon & Schuster, 2000)

in America) are indeed powerfully influenced by norms and networks of civic engagement. Researchers in such fields as education, urban poverty, unemployment, the control of crime and drug abuse, and even health have discovered that successful outcomes are more likely in civically engaged communities. Similarly, research on the varying economic attainments of different ethnic groups in the United States has demonstrated the importance of social bonds within each group. These results are consistent with research in a wide range of settings that demonstrates the vital importance of social networks for job placement and many other economic outcomes.

Meanwhile, a seemingly unrelated body of research on the sociology of economic development has also focused attention on the role of social networks. Some of this work is situated in the developing countries, and some of it elucidates the peculiarly successful "network capitalism" of East Asia. Even in less exotic Western economies, however, researchers have discovered highly efficient, highly flexible "industrial districts" based on networks of collaboration among workers and small entrepreneurs. Far from being paleoindustrial anachronisms, these dense interpersonal and interorganizational networks undergird ultramodern industries, from the high tech of Silicon Valley to the high fashion of Benetton.

The norms and networks of civic engagement also powerfully affect the performance of representative government. That, at least, was the central conclusion of my own 20-year, quasi-experimental study of subnational governments in different regions of Italy. Although all these regional governments seemed identical on paper, their levels of effectiveness varied dramatically. Systematic inquiry showed that the quality of governance was determined by longstanding traditions of civic engagement (or its absence). Voter turnout, newspaper readership, membership in choral societies and football clubs – these were the hallmarks of a successful region. In fact, historical analysis suggested that these networks of organized reciprocity and civic solidarity, far from being an epiphenomenon of socioeconomic modernization, were a precondition for it.

No doubt the mechanisms through which civic engagement and social connectedness produce such results – better schools, faster economic development, lower crime, and more effective government – are multiple and complex. While these briefly recounted findings require further confirmation and perhaps qualification, the parallels across hundreds of empirical studies in a dozen disparate disciplines and subfields are striking. Social scientists in several fields have recently suggested a common framework for understanding these phenomena, a framework that rests on the concept of social capital. By analogy with notions of physical capital and human capital – tools and training that enhance individual productivity – "social capital" refers to features of social organization such as networks, norms, and social trust that facilitate coordination and cooperation for mutual benefit.

For a variety of reasons, life is easier in a community blessed with a substantial stock of social capital. In the first place, networks of civic engagement foster sturdy norms of generalized reciprocity and encourage the emergence of social trust. Such networks facilitate coordination and communication, amplify reputations, and thus allow dilemmas of collective action to be resolved. When economic and political negotiation is embedded in dense networks of social interaction, incentives for opportunism are reduced. At the same time, networks of civic engagement embody past success at collaboration, which can serve as a cultural template for future collaboration. Finally, dense networks of interaction probably broaden the participants' sense of self, developing the "I" into the "we," or (in the language of rational-choice theorists) enhancing the participants' "taste" for collective benefits.

I do not intend here to survey (much less contribute to) the development of the theory of social capital. Instead, I use the central premise of that rapidly growing body of work – that social connections and civic engagement pervasively influence our public life, as well as our private prospects – as the starting point for an empirical survey of trends in social capital in contemporary America. I concentrate here entirely on the American case, although the developments I portray may in some measure characterize many contemporary societies.

WHATEVER HAPPENED TO CIVIC ENGAGEMENT?

We begin with familiar evidence on changing patterns of political participation, not least because it is immediately relevant to issues of democracy in the narrow sense. Consider the well-known decline in turnout in national elections over the last three decades. From a relative high point in the early 1960s, voter turnout had by 1990 declined by nearly a quarter; tens of millions of Americans had forsaken their parents' habitual readiness to engage in the simplest act of citizenship. Broadly similar trends also characterize participation in state and local elections.

It is not just the voting booth that has been increasingly deserted by Americans. A series of identical questions posed by the Roper Organization to national samples ten times each year over the last two decades reveals that since 1973 the number of Americans who report that "in the past year" they have "attended a public meeting on town or school affairs" has fallen by more than a third (from 22 percent in 1973 to 13 percent in 1993). Similar (or even greater) relative declines are evident in responses to questions about attending a political rally or speech, serving on a committee of some local organization, and working for a political party. By almost every measure, Americans' direct engagement in politics and government has fallen steadily and sharply over the last generation, despite the fact that average levels of education – the best individual-level predictor of political participation – have risen sharply throughout this period. Every year over the last decade or two, millions more have withdrawn from the affairs of their communities.

Not coincidentally, Americans have also disengaged psychologically from politics and government over this era. The proportion of Americans who reply that they "trust the government in Washington" only "some of the time" or "almost never" has risen steadily from 30 percent in 1966 to 75 percent in 1992.

These trends are well known, of course, and taken by themselves would seem amenable to a strictly political explanation. Perhaps the long litany of political tragedies and scandals since the 1960s (assassinations, Vietnam, Watergate, Irangate, and so on) has triggered an understandable disgust for politics and government among Americans, and that in turn has motivated their withdrawal. I do not doubt that this common interpretation has some merit, but its limitations become plain when we examine trends in civic engagement of a wider sort.

Our survey of organizational membership among Americans can usefully begin with a glance at the aggregate results of the General Social Survey, a scientifically conducted, national-sample survey that has been repeated 14 times over the last two decades. Church-related groups constitute the most common type of organization joined by Americans; they are especially popular with women. Other types of organizations frequently joined by women include school-service groups (mostly parent–teacher associations), sports groups, professional societies, and literary societies. Among men, sports clubs, labor unions, professional societies, fraternal groups, veterans' groups, and service clubs are all relatively popular.

Religious affiliation is by far the most common associational membership among Americans. Indeed, by many measures America continues to be (even more than in Tocqueville's time) an astonishingly "churched" society. For example, the United States has more houses of worship per capita than any other nation on Earth. Yet religious sentiment in America seems to be becoming somewhat less tied to institutions and more self-defined.

How have these complex crosscurrents played out over the last three or four decades in terms of Americans' engagement with organized religion? The general pattern is clear: the 1960s witnessed a significant drop in reported weekly churchgoing – from roughly 48 percent in the late 1950s to roughly 41 percent in the early 1970s. Since then, it has stagnated or (according to some surveys) declined still further. Meanwhile, data from the General Social Survey show a modest decline in membership in all "church-related groups" over the last 20 years. It would seem, then, that net participation by Americans, both in religious services and in church-related groups, has declined modestly (by perhaps a sixth) since the 1960s.

For many years, labor unions provided one of the most common organizational affiliations among American workers. Yet union membership has been falling for nearly four decades, with the steepest decline occurring between 1975 and 1985. Since the mid-1950s, when union membership peaked, the unionized portion of the nonagricultural workforce in America has dropped by more than half, falling from 32.5 percent in 1953 to 15.8 percent in 1992. By now, virtually all of the explosive growth in union membership that was associated with the New Deal has been erased. The solidarity of union halls is now mostly a fading memory of aging men.

The parent–teacher association (PTA) has been an especially important form of civic engagement in twentieth-century America because parental involvement in the educational process represents a particularly productive form of social capital. It is, therefore, dismaying to discover that participation in parent–teacher organizations has dropped drastically over the last generation, from more than 12 million in 1964 to barely 5 million in 1982 before recovering to approximately 7 million now.

Next, we turn to evidence on membership in (and volunteering for) civic and fraternal organizations. These data show some striking patterns. First, membership in traditional women's groups has declined more or less steadily since the mid-1960s. For example, membership in the national Federation of Women's Clubs is down by more than half (59 percent) since 1964, while membership in the League of Women Voters (LWV) is off 42 percent since 1969.

Similar reductions are apparent in the numbers of volunteers for mainline civic organizations, such as the Boy Scouts (off by 26 percent since 1970) and the Red Cross (off by 61 percent since 1970). But what about the possibility that volunteers have simply switched their loyalties to other organizations? Evidence on "regular" (as opposed to occasional or "drop-by") volunteering is available from the Labor Department's Current Population Surveys of 1974 and 1989. These estimates suggest that serious volunteering declined by roughly one-sixth over these 15 years, from 24 percent of adults in 1974 to 20 percent in 1989. The multitudes of Red Cross aides and Boy Scout troop leaders now missing in action have apparently not been offset by equal numbers of new recruits elsewhere.

Fraternal organizations have also witnessed a substantial drop in membership during the 1980s and 1990s. Membership is down significantly in such groups as the Lions (off 12 percent since 1983), the Elks (off 18 percent since 1979), the Shriners (off 27 percent since 1979), the Jaycees (off 44 percent since 1979), and the Masons (down 39 percent since 1959). In sum, after expanding steadily throughout most of this century, many major civic organizations have experienced a sudden, substantial, and nearly simultaneous decline in membership over the last decade or two.

The most whimsical yet discomfiting bit of evidence of social disengagement in contemporary America that I have discovered is this: more Americans are bowling today than ever before, but bowling in organized leagues has plummeted in the last decade or so. Between 1980 and 1993 the total number of bowlers in America increased by 10 percent, while league bowling decreased by 40 percent. (Lest this be thought a wholly trivial example, I should note that nearly 80 million Americans went bowling at least once during 1993, nearly a third more than voted in the 1994 congressional elections and roughly the same number as claim to attend church regularly. Even after the 1980s' plunge in league bowling, nearly 3 percent of American adults regularly bowl in leagues.) The rise of solo bowling threatens the livelihood of bowling-lane proprietors because those who bowl as members of leagues consume three times as much beer and pizza as solo bowlers, and the money in bowling is in the beer and pizza, not the balls and shoes. The broader social significance, however, lies in the social interaction and even occasionally civic conversations over beer and pizza that solo bowlers forgo. Whether or not bowling beats balloting in the eyes of most Americans, bowling teams illustrate yet another vanishing form of social capital.

COUNTERTRENDS

At this point, however, we must confront a serious counterargument. Perhaps the traditional forms of civic organization whose decay we have been tracing have been replaced by vibrant new organizations. For example, national environmental organizations (like the Sierra Club) and feminist groups (like the National Organization for Women) grew rapidly during the 1970s and 1980s and now count hundreds of thousands of dues-paying members. An even more dramatic example is the American Association of Retired Persons (AARP), which grew exponentially from 400,000 card-carrying members in 1960 to 33 million in 1993, becoming (after the Catholic Church) the largest private organization in the world. The national administrators of these organizations are among the most feared lobbyists in Washington, in large part because of their massive mailing lists of presumably loyal members.

These new mass-membership organizations are plainly of great political importance. From the point of view of social connectedness, however, they are sufficiently different from classic "secondary associations" that we need to invent a new label – perhaps "tertiary associations." For the vast majority of their members, the only act of membership consists in writing a check for dues or perhaps occasionally reading a newsletter. Few ever attend any meetings of such organizations, and most are unlikely ever (knowingly) to encounter any other member. The bond between any two members of the Sierra Club is less like the bond between any two members of a gardening club and more like the bond between any two Red Sox fans (or perhaps any two devoted Honda owners): they root for the same team and they share some of the same interests, but they are unaware of each other's existence. Their ties, in short, are to common symbols, common leaders, and perhaps common ideals, but not to one another. The theory of social capital argues that associational membership should, for example, increase social trust, but this prediction is much less straightforward with regard to membership in tertiary associations. From the point of view of social connectedness, the Environmental Defense Fund and a bowling league are just not in the same category.

If the growth of tertiary organizations represents one potential (but probably not real) counterexample to my thesis, a second countertrend is represented by the growing prominence of nonprofit organizations, especially nonprofit service agencies. This so-called third sector includes everything from Oxfam and the Metropolitan Museum of Art to the Ford Foundation and the Mayo Clinic. In other words, although most secondary associations are nonprofits, most nonprofit agencies are not secondary associations. To identify trends in the size of the nonprofit sector with trends in social connectedness would be another fundamental conceptual mistake.

A third potential countertrend is much more relevant to an assessment of social capital and civic engagement. Some able researchers have argued that the last few decades have witnessed a rapid expansion in "support groups" of various sorts. Robert Wuthnow reports that fully 40 percent of all Americans claim to be "currently involved in [a] small group that meets regularly and provides support or caring for those who participate in it." Many of these groups are religiously affiliated, but many others are not. For example, nearly 5 percent of Wuthnow's national sample claim to participate regularly in a "self-help" group, such as Alcoholics Anonymous, and nearly as many say they belong to book-discussion groups and hobby clubs.

The groups described by Wuthnow's respondents unquestionably represent an important form of social capital, and they need to be accounted for in any serious reckoning of trends in social connectedness. On the other hand, they do not typically play the same role as traditional civic associations. As Wuthnow emphasizes,

> Small groups may not be fostering community as effectively as many of their proponents would like. Some small groups merely provide occasions for individuals to focus on themselves in the presence of others. The social contract binding members together asserts only the weakest of obligations. Come if you have time. Talk if you feel like it. Respect everyone's opinion. Never criticize. Leave quietly if you become dissatisfied . . . We can imagine that [these small groups] really substitute for families, neighborhoods, and broader community attachments that may demand lifelong commitments, when, in fact, they do not.

All three of these potential countertrends – tertiary organizations, nonprofit organizations, and support groups – need somehow to be weighed against the erosion of conventional civic organizations. One way of doing so is to consult the General Social Survey.

Within all educational categories, total associational membership declined significantly between 1967 and 1993. Among the college-educated, the average number of group memberships per person fell from 2.8 to 2.0 (a 26 percent decline); among high-school graduates, the number fell from 1.8 to 1.2 (32 percent); and among those with fewer than 12 years of education, the number fell from 1.4 to 1.1 (25 percent). In other words, at all educational (and hence social) levels of American society, and counting all sorts of group memberships, the average number of associational memberships has fallen by about a fourth over the last quarter-century. Without controls for educational levels, the trend is not nearly so clear, but the central point is this: more Americans than ever before are in social circumstances that foster associational involvement (higher education, middle age, and so on), but nevertheless aggregate associational membership appears to be stagnant or declining.

Broken down by type of group, the downward trend is most marked for church-related groups, for labor unions, for fraternal and veterans' organizations, and for school-service groups. Conversely, membership in professional associations has risen over these years, although less than might have been predicted, given sharply rising educational and occupational levels. Essentially the same trends are evident for both men and women in the sample. In short, the available survey evidence confirms our earlier conclusion: American social capital in the form of civic associations has significantly eroded over the last generation.

GOOD NEIGHBORLINESS AND SOCIAL TRUST

I noted earlier that most readily available quantitative evidence on trends in social connectedness involves formal settings, such as the voting booth, the union hall, or the PTA. One glaring exception is so widely discussed as to require little comment here: the most fundamental form of social capital is the family, and the massive evidence of the loosening of bonds within the family (both extended and nuclear) is well known. This trend, of course, is quite consistent with – and may help to explain – our theme of social decapitalization.

A second aspect of informal social capital on which we happen to have reasonably reliable time-series data involves neighborliness. In each General Social Survey since 1974 respondents have been asked, "How often do you spend a social evening with a neighbor?" The proportion of Americans who socialize with their neighbors more than once a year has slowly but steadily declined over the last two decades, from 72 percent in 1974 to 61 percent in 1993. (On the other hand, socializing with "friends who do not live in your neighborhood" appears to be on the increase, a trend that may reflect the growth of workplace-based social connections.)

Americans are also less trusting. The proportion of Americans saying that most people can be trusted fell by more than a third between 1960, when 58 percent chose that alternative, and 1993, when only 37 percent did. The same trend is apparent in all educational groups; indeed, because social trust is also correlated with education and because educational levels have risen sharply, the overall decrease in social trust is even more apparent if we control for education.

Our discussion of trends in social connectedness and civic engagement has tacitly assumed that all the forms of social capital that we have discussed are themselves coherently correlated across individuals. This is in fact true. Members of associations are much more likely than nonmembers to participate in politics, to spend time with neighbors, to express social trust, and so on.

The close correlation between social trust and associational membership is true not only across time and across individuals, but also across countries. Evidence from the 1991 World Values Survey demonstrates the following: across the 35 countries in this survey, social trust and civic engagement are strongly correlated; the greater the density of associational membership in a society, the more trusting its citizens. Trust and engagement are two facets of the same underlying factor – social capital.

America still ranks relatively high by cross-national standards on both these dimensions of social capital. Even in the 1990s, after several decades' erosion, Americans are more trusting and more engaged than people in most other countries of the world. The trends of the past quarter-century, however, have apparently moved the United States significantly lower in the international rankings of social capital. The recent deterioration in American social capital has been sufficiently great that (if no other country changed its position in the meantime) another quarter-century of change at the same rate would bring the United States, roughly speaking, to the midpoint among all these countries, roughly equivalent to South Korea, Belgium, or Estonia today. Two generations' decline at the same rate would leave the United States at the level of today's Chile, Portugal, and Slovenia.

WHY IS US SOCIAL CAPITAL ERODING?

As we have seen, something has happened in America in the last two or three decades to diminish civic engagement and social connectedness. What could that "something" be? Here are several possible explanations, along with some initial evidence on each.

The movement of women into the labor force. Over these same two or three decades, many millions of American women have moved out of the home into paid employment. This is the primary, though not the sole, reason why the weekly working hours of the average American have increased significantly during these years. It seems highly plausible that this social revolution should have reduced the time and energy available for building social capital. For certain organizations, such as the PTA, the League of Women Voters, the Federation of Women's Clubs, and the Red Cross, this is almost certainly an important part of the story. The sharpest decline in women's civic participation seems to have come in the 1970s; membership in such "women's" organizations as these has been virtually halved since the late 1960s. By contrast, most of the decline in participation in men's organizations occurred about 10 years later; the total decline to date has been approximately 25 percent for the typical organization. On the other hand, the survey data imply that the aggregate declines for men are virtually as great as those for women. It is logically possible, of course, that the male declines might represent the knock-on effect of women's liberation, as dishwashing crowded out the lodge, but time-budget studies suggest that most husbands of working wives have assumed only a minor part of the housework. In short, something besides the women's revolution seems to lie behind the erosion of social capital.

Mobility: The "re-potting" hypothesis

Numerous studies of organizational involvement have shown that residential stability and such related phenomena as homeownership are clearly associated with greater civic engagement. Mobility, like frequent re-potting of plants, tends to disrupt root systems, and it takes time for an uprooted individual to put down new roots. It seems plausible that the automobile, suburbanization, and the movement to the Sun Belt have reduced the social rootedness of the average American, but one fundamental difficulty with this hypothesis is apparent: the best evidence shows that residential stability and homeownership in America have risen modestly since 1965, and are surely higher now than during the 1950s, when civic engagement and social connectedness by our measures was definitely higher.

Other demographic transformations

A range of additional changes have transformed the American family since the 1960s – fewer marriages, more divorces, fewer children, lower real wages, and so on. Each of these changes might account for some of the slackening of civic engagement, since married, middle-class parents are generally more socially involved than other people. Moreover, the changes in scale that have swept over the American economy in these years – illustrated by the replacement of the corner grocery by the supermarket and now perhaps of the supermarket by electronic shopping at home, or the replacement of community-based enterprises by outposts of distant multinational firms – may perhaps have undermined the material and even physical basis for civic engagement.

The technological transformation of leisure

There is reason to believe that deep-seated technological trends are radically "privatizing" or "individualizing" our use of leisure time and thus disrupting many opportunities for social-capital formation. The most obvious and probably the most powerful instrument of this revolution is television. Time-budget studies in the 1960s showed that the growth in time spent watching television dwarfed all other changes in the way Americans passed their days and nights. Television has made our communities (or, rather, what we experience as our communities) wider and shallower. In the language of economics, electronic technology enables individual tastes to be satisfied more fully, but at the cost of the positive social externalities associated with more primitive forms of entertainment. The same logic applies to the replacement of vaudeville by the movies and now of movies by the VCR. The new "virtual reality" helmets that we will soon don to be entertained in total isolation are merely the latest extension of this trend. Is technology thus driving a wedge between our individual interests and our collective interests? It is a question that seems worth exploring more systematically.

WHAT IS TO BE DONE?

The last refuge of a social-scientific scoundrel is to call for more research. Nevertheless, I cannot forbear from suggesting some further lines of inquiry.

We must sort out the dimensions of social capital, which clearly is not a unidimensional concept, despite language (even in this essay) that implies the contrary. What types of organizations and networks most effectively embody – or generate – social capital, in the sense of mutual reciprocity, the resolution of dilemmas of collective action, and the broadening of social identities? In this essay I have emphasized the density of associational life. In earlier work I stressed the structure of networks, arguing that "horizontal" ties represented more productive social capital than vertical ties.

Another set of important issues involves macrosociological crosscurrents that might intersect with the trends described here. What will be the impact, for example, of electronic networks on social capital? My hunch is that meeting in an electronic forum is not the equivalent of meeting in a bowling alley – or even in a saloon – but hard empirical research is needed. What about the development of social capital in the workplace? Is it growing in counterpoint to the decline of civic engagement, reflecting some social analogue of the first law of thermodynamics – social capital is neither created nor destroyed, merely redistributed? Or do the trends described in this essay represent a deadweight loss?

A rounded assessment of changes in American social capital over the last quarter-century needs to count the costs as well as the benefits of community engagement. We must not romanticize small-town, middle-class civic life in the America of the 1950s. In addition to the deleterious trends emphasized in this essay, recent decades have witnessed a substantial decline in intolerance and probably also in overt discrimination, and those beneficent trends may be related in complex ways to the erosion of traditional social capital. Moreover, a balanced accounting of the social-capital books would need to reconcile the insights of this approach with the undoubted insights offered by Mancur Olson and others who stress that closely knit social, economic, and political organizations are prone to inefficient cartelization and to what political economists term "rent seeking" and ordinary men and women call corruption.

Finally, and perhaps most urgently, we need to explore creatively how public policy impinges on (or might impinge on) social-capital formation. In some well-known instances, public policy has destroyed highly effective social networks and norms. American slum-clearance policy of the 1950s and 1960s, for example, renovated physical capital, but at a very high cost to existing social capital. The consolidation of country post offices and small school districts has promised administrative and financial efficiencies, but full-cost accounting for the effects of these policies on social capital might produce a more negative verdict. On the other hand, such past initiatives as the county agricultural-agent system, community colleges, and tax deductions for charitable contributions illustrate that government can encourage social-capital formation. Even a recent proposal in San Luis Obispo, California, to require that all new houses have front porches illustrates the power of government to influence where and how networks are formed.

The concept of "civil society" has played a central role in the recent global debate about the preconditions for democracy and democratization. In the newer democracies this phrase has properly focused attention on the need to foster a vibrant civic life in soils traditionally inhospitable to self-government. In the established democracies, ironically, growing numbers of citizens are questioning the effectiveness of their public institutions at the very moment when liberal democracy has swept the battlefield, both ideologically and geopolitically. In America, at least, there is reason to suspect that this democratic disarray may be linked to a broad and continuing erosion of civic engagement that began a quarter-century ago. High on our scholarly agenda should be the question of whether a comparable erosion of social capital may be under way in other advanced democracies, perhaps in different institutional and behavioral guises. High on America's agenda should be the question of how to reverse these adverse trends in social connectedness, thus restoring civic engagement and civic trust.

"The Creative Class"

from *The Rise of the Creative Class: And How It's Transforming Work, Leisure, Community and Everyday Life* (2002)

Richard Florida

Editors' Introduction

In *The Condition of the Working Class in 1844* (p. 46), and in subsequent collaborations with his colleague Karl Marx, Friedrich Engels announced the emergence of a new social class – the proletariat or industrial working class – that was destined to have a world-historical impact on the shape and content of human society at the time of the Industrial Revolution and the rise of the industrial city. In *The Rise of the Creative Class*, Richard Florida describes the emergence of a new socio-economic class, one that creates ideas and innovations rather than products and is the driving force of post-industrialism rather than industrialism. Florida asks us to ask ourselves: will the new "creative class" have as important and revolutionary an impact on the twenty-first-century information-based economy and society as the working class had in the nineteenth and twentieth centuries?

According to Florida, there are two layers to the creative class. First, there is a "Super-Creative Core" consisting of "scientists and engineers, university professors, poets and novelists, artists, entertainers, actors, designers and architects, as well as the thought leadership of modern society: nonfiction writers, editors, cultural figures, think-tank researchers, analysts and other opinion-makers." Second, there are "creative professionals" – those who "work in a wide range of knowledge-intensive industries such as high-tech sectors, financial services, the legal and health care professions, and business management" – as well as many technicians and paraprofessionals who now add "creative value" to an enterprise by having to think for themselves. All these, taken together, constitute a true economic class that "both underpins and informs its members' social, cultural and lifestyle choices."

Florida is quick to note that he is not using the term class to denote "the ownership of property, capital or the means of production." On the contrary, he argues, if we use those old Marxist categories, we are still talking about old-style, bourgeoisie-and-proletarian capitalism. In the new postmodern, post-industrial economic order, the "members of the Creative Class do not own and control any significant property in the physical sense. Their property – which stems from their creative capacity – is an intangible because it is literally in their heads."

Many cities have embraced Florida's thesis about the creative class. Eager to attract "creative class" residents, some cities have sponsored special arts districts and diversity festivals as a part of their redevelopment policies in an attempt to jump-start lagging economies. In some cases, such as Denver's LoDo neighborhood, arts-friendly policies – along with new light-rail transit and a downtown baseball stadium – succeeded in revivifying what had been a decaying warehouse district. In other cases, such as San Francisco's adoption of planning regulations supporting the building of "live-work" lofts for artists and other creatives, was arguably just another ploy on the part of housing developers with connections at City Hall.

Inevitably, a critical opposition to "creative class" theory has developed. Some have called Florida "elitist," and Steven Malanga, a senior fellow at the conservative Manhattan Institute, writing in the *Wall Street Journal*, has called Florida the peddler of "economic snake oil" and the developer of "trendy, New Age theories" that are just "plain wrong." It is lower taxes and public safety, not arts festivals and lively gay neighborhoods, according to Malanga, that attract the industries that bring high employment and robust tax revenues for municipalities.

Richard Florida is the director of the Martin Prosperity Institute at the Rotman School of Management at the University of Toronto. He has taught at MIT, Harvard, and Carnegie Mellon and is the founder of the Creative Group and Catalytix, two business-communications-and-strategy consulting firms. *The Rise of the Creative Class* received the Washington Monthly's Political Book Award for 2002 and was praised by the Harvard Business Review as one of the most important "breakthrough ideas" of recent socio-economic analysis. Florida has also published *The Flight of the Creative Class: The New Global Competition for Talent* (New York: HarperCollins, 2005), *Cities and the Creative Class* (New York: Routledge, 2005), *Who's Your City?* (New York: Basic Books, 2009), and *The Great Reset: How New Ways of Living and Working Drive Post-Crash Prosperity* (New York: HarperCollins, 2010).

For intellectual sources of Florida's thesis about the creative class, see Fritz Machlup, *The Production and Distribution of Knowledge in the United States* (Princeton, NJ: Princeton University Press, 1962), Daniel Bell, *The Coming of Post-Industrial Society* (New York: Basic Books, 1973), and Peter Drucker, *Post-Capitalist Society* (New York: Harper Business, 1995). For critiques of the Florida thesis, see Allen J. Scott, "Creative Cities: Conceptual Issues and Policy Questions" (*Journal of Urban Affairs*, 28, 2006) and Michele Heyman and Christopher Farticy, "It Takes a Village: A Test of the Creative Class, Social Capital and Human Capital Theories" (*Journal of Urban Affairs*, 44, 2009).

The rise of the Creative Economy has had a profound effect on the sorting of people into social groups or classes. Others have speculated over the years on the rise of new classes in the advanced industrial economies. During the 1960s, Peter Drucker and Fritz Machlup described the growing role and importance of the new group of workers they dubbed "knowledge workers." Writing in the 1970s, Daniel Bell pointed to a new, more meritocratic class structure of scientists, engineers, managers and administrators brought on by the shift from a manufacturing to a "postindustrial" economy. The sociologist Erik Olin Wright has written for decades about the rise of what he called a new "professional-managerial" class. Robert Reich more recently advanced the term "symbolic analysts" to describe the members of the workforce who manipulate ideas and symbols. All of these observers caught economic aspects of the emerging class structure that I describe here.

Others have examined emerging social norms and value systems. Paul Fussell presciently captured many that I now attribute to the Creative Class in his theory of the "X Class." Near the end of his 1983 book *Class* – after a witty romp through status markers that delineate, say, the upper middle class from "high proles" – Fussell noted the presence of a growing "X" group that seemed to defy existing categories:

> [Y]ou are not born an X person . . . you earn X-personhood by a strenuous effort of discovery in which curiosity and originality are indispensable. . . . The young flocking to the cities to devote themselves to "art," "writing," "creative work" – anything, virtually, that liberates them from the presence of a boss or superior – are aspirant X people. . . . If, as [C. Wright] Mills has said, the middle-class person is "always somebody's man," the X person is nobody's. . . . X people are independent-minded. . . . They adore the work they do, and they do it until they are finally carried out, "retirement" being a concept meaningful only to hired personnel or wage slaves who despise their work.

Writing in 2000, David Brooks outlined the blending of bohemian and bourgeois values in a new social grouping he dubbed the Bobos. My take on Brooks's synthesis . . . is rather different, stressing the very transcendence of these two categories in a new creative ethos.

The main point I want to make here is that the basis of the Creative Class is economic. I define it as an economic class and argue that its economic function both underpins and informs its members' social, cultural and lifestyle choices. The Creative Class consists of people who add economic value through their creativity. It thus includes a great many knowledge workers, symbolic analysts and professional and technical workers, but emphasizes their true role in the economy. My definition of class emphasizes the way people organize themselves into social groupings and common identities based principally on their economic function. Their social and cultural preferences, consumption and buying habits, and their social identities all flow from this.

I am not talking here about economic class in terms of the ownership of property, capital or the means of production. If we use class in this traditional Marxian sense, we are still talking about a basic structure of capitalists who own and control the means of production, and workers under their employ. But little analytical utility remains in these broad categories of bourgeoisie and proletarian, capitalist and worker. Most members of the Creative Class do not own and control any significant property in the physical sense. Their property – which stems from their creative capacity – is an intangible because it is literally in their heads. And it is increasingly clear from my field research and interviews that while the members of the Creative Class do not yet see themselves as a unique social grouping, they actually share many similar tastes, desires and preferences. This new class may not be as distinct in this regard as the industrial Working Class in its heyday, but it has an emerging coherence.

THE NEW CLASS STRUCTURE

The distinguishing characteristic of the Creative Class is that its members engage in work whose function is to "create meaningful new forms." I define the Creative Class as consisting of two components. The Super-Creative Core of this new class includes scientists and engineers, university professors, poets and novelists, artists, entertainers, actors, designers and architects, as well as the thought leadership of modern society: nonfiction writers, editors, cultural figures, think-tank researchers, analysts and other opinion-makers. Whether they are software programmers or engineers, architects or filmmakers, they fully engage in the creative process. I define the highest order of creative work as producing new forms or designs that are readily transferable and widely useful – such as designing a product that can be widely made, sold and used; coming up with a theorem or strategy that can be applied in many cases; or composing music that can be performed again and again. People at the core of the Creative Class engage in this kind of work regularly; it's what they are paid to do. Along with problem solving, their work may entail problem finding: not just building a better mousetrap, but noticing first that a better mousetrap would be a handy thing to have.

Beyond this core group, the Creative Class also includes "creative professionals" who work in a wide range of knowledge-intensive industries such as high-tech sectors, financial services, the legal and health care professions, and business management. These people engage in creative problem solving, drawing on complex bodies of knowledge to solve specific problems. Doing so typically requires a high degree of formal education and thus a high level of human capital. People who do this kind of work may sometimes come up with methods or products that turn out to be widely useful, but it's not part of the basic job description. What they *are* required to do regularly is think on their own. They apply or combine standard approaches in unique ways to fit the situation, exercise a great deal of judgment, perhaps try something radically new from time to time. Creative Class people such as physicians, lawyers and managers do this kind of work in dealing with the many varied cases they encounter. In the course of their work, they may also be involved in testing and designing new techniques, new treatment protocols, or new management methods and even develop such things themselves. As a person continues to do more of this latter work, perhaps through a career shift or promotion, that person moves up to the Super-Creative Core: producing transferable, widely usable new forms is now their primary function.

[. . .]

As the creative content of other lines of work increases – as the relevant body of knowledge becomes more complex, and people are more valued for their ingenuity in applying it – some now in the Working Class or Service Class may move into the Creative Class and even the Super-Creative Core. Alongside the growth in essentially creative occupations, then, we are also seeing growth in creative content across other occupations. A prime example is the secretary in today's pared-down offices. In many cases this person not only takes on a host of tasks once performed by a large secretarial staff, but becomes a true office manager – channeling flows of information, devising and setting up new systems, often making key decisions on the fly. This person contributes more than "intelligence" or computer skills. She or he adds creative value. Everywhere we look, creativity is increasingly valued. Firms and organizations value it for the results that it can produce and individuals value it as a route to self-expression and job satisfaction. Bottom line: As creativity becomes more valued, the Creative Class grows.

Not all workers are on track to join, however. For instance in many lower-end service jobs we find the trend running the opposite way; the jobs continue to be "de-skilled" or "de-creatified." For a counter worker at a fast-food chain, literally every word and move is dictated by a corporate template: "Welcome to Food Fix, sir, may I take your order? Would you like nachos with that?" This job has been thoroughly taylorized – the worker is given far less latitude for exercising creativity than the waitress at the old, independent neighborhood diner enjoyed. Worse yet, there are many people who do not have jobs, and who are being left behind because they do not have the background and training to be part of this new system.

[. . .]

COUNTING THE CREATIVE CLASS

It is one thing to provide a compelling description of the changing class composition of society, as writers like Bell, Fussell or Reich have done. But I believe it is also important to calibrate and quantify the magnitude of the change at hand. . . . Let's take a look at the key trends.

- The *Creative Class* now includes some *38.3* million Americans, roughly 30 percent of the entire U.S. workforce. It has grown from roughly *3* million workers in 1900, an increase of more than tenfold. At the turn of the twentieth century, the Creative Class made up just 10 percent of the workforce, where it hovered until 1950 when it began a slow rise; it held steady around 20 percent in the 1970s and *1980s*. Since that time, this new class has virtually exploded, increasing from less than *20* million to its current total, reaching *25* percent of the working population in 1991 before climbing to *30* percent by *1999*.
- At the heart of the Creative Class is the *Super-Creative Core*, comprising 15 million workers, or 12 percent of the workforce. It is made up of people who work in science and engineering, computers and mathematics, education, and the arts, design and entertainment, people who work in directly creative activity, as we have seen. Over the past century, this segment rose from less than 1 million workers in 1900 to *2.3* million in 1950 before crossing 10 million in 1991. In doing so, it increased its share of the workforce from *2.5* percent in 1900 to *5* percent in 1960, 8 percent in 1980 and 9 percent in 1990, before reaching 12 percent by 1999.
- The traditional *Working Class* has today *33* million workers, or a quarter of the U.S. workforce. It consists of people in production operations, transportation and materials moving, and repair and maintenance and construction work. The percentage of the workforce in working-class occupations peaked at 40 percent in 1920, where it hovered until 1950, before slipping to *36* percent in 1970, and then declining sharply over the past two decades.
- The *Service Class* includes 55.2 million workers or 43 percent of the U.S. workforce, making it the largest group of all. It includes workers in lower-wage, lower-autonomy service occupations such as health care, food preparation, personal care, clerical work and other lower-end office work. Alongside the decline of the Working Class, the past century has seen a tremendous rise in the Service Class, from 5 million workers in 1900 to its current total of more than ten times that amount.

It's also useful to look at the changing composite picture of the U.S. class structure over the twentieth century. In 1900, there were some 10 million people in the Working Class, compared to 2.9 million in the Creative Class and 4.8 million in the Service Class. The Working Class was thus larger than the two other classes combined. Yet the largest class at that time was agricultural workers, who composed nearly 40 percent of the workforce but whose numbers rapidly declined to just a very small percentage today. In 1920, the Working Class accounted for 40 percent of the workforce, compared to slightly more than 12 percent for the Creative Class and 21 percent for the Service Class.

In 1950, the class structure remained remarkably similar. The Working Class was still in the majority, with *25* million workers, some *40* percent of the workforce, compared to 10 million in the Creative Class (16.5 percent) and 18 million in the Service Class (30 percent). In relative terms, the Working Class was as large as it was in 1920 and bigger than it was in 1900. Though the Creative Class had grown slightly in percentage terms, the Service Class had grown considerably, taking up much of the slack coming from the steep decline in agriculture.

The tectonic shift in the U.S. class structure has taken place over the past two decades. In 1970, the Service Class pulled ahead of the Working Class, and by 1980 it was much larger (46 versus 32 percent), marking the first time in the twentieth century that the Working Class was not the dominant class. By 1999, both the Creative Class and the Service Class had pulled ahead of the Working Class. The Service Class, with 55 million workers (43.4 percent), was bigger in relative terms than the Working Class had been at any time in the past century.

These changes in American class structure reflect a deeper, more general process of economic and social change. The decline of the old Working Class is part and parcel of the decline of the industrial economy on which it was based, and of the social and demographic patterns upon which that old society was premised. The Working Class no longer has the hand it once did in setting the tone or establishing the values of American life – for that matter neither does the 1950s managerial class. Why, then, have the social functions of the Working Class not been taken over by the new largest class, the Service Class? As we have seen, the Service Class has little clout and its rise in numbers can be understood only alongside the rise of the Creative Class. The Creative Class – and the modern Creative Economy writ large – depends on this ever-larger Service Class to "outsource" functions that were previously provided within the family. The Service Class exists mainly as a supporting infrastructure for the Creative Class and the Creative Economy. The Creative Class also has considerably more economic power. Members earn substantially more than those in other classes. In 1999, the average salary for a member of the Creative Class was nearly $50,000 ($48,752) compared to roughly $28,000 for a Working Class member and $22,000 for a Service Class worker. . . .

I see these trends vividly played out in my own life. I have a nice house with a nice kitchen but it's often mostly a fantasy kitchen – I eat out a lot, with "servants" preparing my food and waiting on me. My house is clean, but I don't clean it, a housekeeper does. I also have a gardener and a pool service; and (when I take a taxi) a chauffeur. I have, in short, just about all the servants of an English lord except that they're not mine full-time and they don't live below stairs; they are part-time and distributed in the local area. Not all of these "servants" are lowly serfs. The person who cuts my hair is a very creative stylist much in demand, and drives a new BMW. The woman who cleans my house is a gem: I trust her not only to clean but to rearrange and suggest ideas for redecorating; she takes on these things in an entrepreneurial manner. Her husband drives a Porsche. To some degree, these members of the Service Class have adopted many of the functions along with the tastes and values of the Creative Class, with which they see themselves sharing much in common. Both my hairdresser and my housekeeper have taken up their lines of work to get away from the regimentation of large organizations; both of them relish creative pursuits. Service Class people such as these are close to the mainstream of the Creative Economy and prime candidates for reclassification.

CREATIVE CLASS VALUES

The rise of the Creative Class is reflected in powerful and significant shifts in values, norms and attitudes. Although these changes are still in process and certainly not fully played out, a number of key trends have been discerned by researchers who study values, and I have seen them displayed in my field research across the United States. Not all of these attitudes break with the past: Some represent a melding of traditional values and newer ones. They are also values that have long been associated with more highly educated and creative people. On the basis of my own interviews and focus groups, along with a close reading of statistical surveys conducted by others, I cluster these values along three basic lines.

Individuality. The members of the Creative Class exhibit a strong preference for individuality and self-statement. They do not want to conform to organizational or institutional directives and resist traditional group-oriented norms. This has always been the case among creative people from "quirky" artists to "eccentric" scientists. But it has now become far more pervasive. In this sense, the increasing nonconformity to organizational norms may represent a new mainstream value. Members of the Creative Class endeavor to create individualistic identities that reflect their creativity. This can entail a mixing of multiple creative identities.

Meritocracy. Merit is very strongly valued by the Creative Class, a quality shared with Whyte's class of organization men. The Creative Class favors hard work, challenge and stimulation. Its members have a propensity for goal-setting and achievement. They want to get ahead because they are good at what they do.

Creative Class people no longer define themselves mainly by the amount of money they make or their position in a financially delineated status order. While money may be looked upon as a marker of achievement, it is not the whole story. In interviews and focus groups, I consistently come across people valiantly trying to defy an economic class into which they were born. This is particularly true of the young descendants of the truly wealthy – the capitalist class – who frequently describe themselves as just "ordinary" creative people working on music, film or intellectual endeavors of one sort or another. Having absorbed the Creative Class value of merit, they no longer find true status in their wealth and thus try to downplay it.

There are many reasons for the emphasis on merit. Creative Class people are ambitious and want to move up based on their abilities and effort. Creative people have always been motivated by the respect of their peers. The companies that employ them are often under tremendous competitive pressure and thus cannot afford much dead wood on staff: everyone has to contribute. The pressure is more intense than ever to hire the best people regardless of race, creed, sexual preference or other factors.

But meritocracy also has its dark side. Qualities that confer merit, such as technical knowledge and mental discipline, are socially acquired and cultivated. Yet those who have these qualities may easily start thinking they were born with them, or acquired them all on their own, or that others just "don't have it." By papering over the causes of cultural and educational advantage, meritocracy may subtly perpetuate the very prejudices it claims to renounce. On the bright side, of course, meritocracy ties into a host of values and beliefs we'd all agree are positive – from faith that virtue will be rewarded, to valuing self-determination and mistrusting rigid caste systems. Researchers have found such values to be on the rise, not only among the Creative Class in the United States, but throughout our society and other societies.

Diversity and Openness. Diversity has become a politically charged buzzword. To some it is an ideal and rallying cry, to others a Trojan-horse concept that has brought us affirmative action and other liberal abominations. The Creative Class people I study use the word a lot, but not to press any political hot buttons. Diversity is simply something they value in all its manifestations. This is spoken of so often, and so matter-of-factly, that I take it to be a fundamental marker of Creative Class values. As my focus groups and interviews reveal, members of this class strongly favor organizations and environments in which they feel that anyone can fit in and can get ahead.

Diversity of peoples is favored first of all out of self-interest. Diversity can be a signal of meritocratic norms at work. Talented people defy classification based on race, ethnicity, gender, sexual preference or appearance. One indicator of this preference for diversity is reflected in the fact that Creative Class people tell me that at job interviews they like to ask if the company offers same-sex partner benefits, even when they are not themselves gay. What they're seeking is an environment open to differences. Many highly creative people, regardless of ethnic background or sexual orientation, grew up feeling like outsiders, different in some way from most of their schoolmates. They may have odd personal habits or extreme styles of dress. Also, Creative Class people are mobile and tend to move around to different parts of the country; they may not be "natives" of the place they live even if they are American-born. When they are sizing up a new company and community, acceptance of diversity and of gays in particular is a sign that reads "nonstandard people welcome here." It also registers itself in changed behaviors and organizational policies. For example, in some Creative Class centers like Silicon Valley and Austin, the traditional office Christmas party is giving way to more secular, inclusive celebrations. The big event at many firms is now the Halloween party: Just about anyone can relate to a holiday that involves dressing up in costume.

While the Creative Class favors openness and diversity, to some degree it is a diversity of elites, limited to highly educated, creative people. Even though the rise of the Creative Class has opened up new avenues of advancement for women and members of ethnic minorities, its existence has certainly failed to put an end to long-standing divisions of race and gender. Within high-tech industries in particular these divisions still seem to hold. The world of high-tech creativity doesn't include many African-Americans. Several of my interviewees noted that a typical high-tech company "looks like the United Nations minus the black faces." This is unfortunate but not surprising. For several reasons, U.S. blacks are under-represented in many professions, and this may be compounded today by the so-called digital divide – black families in the United States tend to be poorer than average, and thus their children are less likely to have access to computers. My own research shows a negative statistical correlation between concentrations of high-tech firms in a region and nonwhites as a percentage of the population, which is particularly disturbing in light of my other findings on the positive relationship between high-tech and other kinds of diversity – from foreign-born people to gays.

There are intriguing challenges to the kind of diversity that the members of the Creative Class are drawn to. Speaking of a small software company that had the usual assortment of Indian, Chinese, Arabic and other employees, an Indian technology professional said: "That's not diversity! They're all software engineers." Yet despite the holes in the picture, distinctive value changes are indeed afoot. . . .

"Visions of a New Reality: The City and the Emergence of Modern Visual Culture"

(1999)

Frederic Stout

A fundamental precept of Marxist cultural analysis is that superstructures of thought and artistic expression rest upon and derive from a material base rooted in social and economic reality. Thus, each historical era creates characteristic forms of expression and explanatory discourse that reflect, indeed construct, the social reality of the period. Lewis Mumford spoke to this process when, in *The City in History* (1961), he wrote that in "the Book of Job, one beholds Jerusalem; in Plato, Sophocles, and Euripides, Athens; in Shakespeare and Marlowe . . . Elizabethan London."

For the cities of the Industrial Revolution, a number of forms of expression and modes of critical analysis arose to make sense of the dramatic and rapidly changing social reality. In literature, Balzac and Dickens pioneered a tradition of social realism particularly focused on the struggles of the urban poor which later attracted such followers as Emile Zola, Theodore Dreiser, Mrs. Gaskell, Upton Sinclair . . . and thousands more. In the realm of social and political analysis, the use of statistical evidence based on survey data compiled by government commissions and philanthropic organizations, the construction of great explanatory theories (such as those of Marx and Weber), and the development of social science methodologies helped observers of the urban milieu make sense of the revolutionary changes that were taking place around them. And in the realm of art, a whole new kind of visual culture emerged rooted in the observation of the new urban reality, both social and physical. It was a culture that began with nineteenth-century popular illustrated journalism and evolved to include mainstream traditions in the twentieth-century history of photography and cinema.

In the history of the fine arts, urban social realism plays a brief but highly visible role. Although much artistic social activism was largely confined to the satirical printmakers (except in those countries where state-sponsored socialist realism became the prescribed orthodoxy), works like William Hogarth's depictions of eighteenth-century London and Honoré Daumier's glimpses of nineteenth-century Paris are as influential as they are familiar. In painting, however, the realism of Courbet and Millet quickly give way to the impressionist celebration of light and the post-impressionist analysis of pure form. In America, the urban imagery of The Eight (dubbed "The Ashcan School" by a hostile critic) was revolutionary and shocking at their group show in 1907 but strictly old hat by the time modern abstraction was introduced at the Armory Show in 1912. One explanation for this is that The Eight stood not at the beginning but at the end of a fifty-year social realist tradition in the visual arts, a tradition that took place not in the salons of the easel painters but in the pages of the illustrated newspapers and magazines that flourished as a ubiquitous element of urban popular culture during the last half of the nineteenth century. Several of the Ashcan School painters – among them John Sloan, William Glackens, Everett Shinn, and Edward Hopper – had previously worked as journalistic illustrators, and it is to the pages of the popular newspapers that one must turn to see the earliest representations of the modern industrial city.

Illustrated journalism began in England as early as the 1820s, and in America in the 1840s, in response to a growing demand for information and entertainment on the part of a marginally literate but fully enfranchised working class and petit bourgeoisie. The first images of cities were almost always long-range or bird's-eye views. These static images were imitative of a received landscape tradition that had been commonly applied to rural and wild nature scenes, but they conveyed relatively little information about urban subject matter and were soon replaced by more kinetic images and vignettes. "Contrast pictures" – both dual images such as "Music in the Street/Music in the Parlor" (Plate 10) from an 1868 issue of *The Illustrated News* and single images such as "The Hearth-stone of the Poor" (Plate 11) from an 1876 issue of *Harper's Weekly* by the exceptionally talented Sol Eytinge, Jr. – were particularly useful for comparing the lives of the urban rich and poor. Soon, a new kind of composite image emerged – "City Sketches" by the illustrator C.A. Barry from an 1855 issue of *Ballou's Pictorial Drawing-Room Companion* (Plate 12) is an interesting early example – that more accurately reflected the diverse, jumbled-together chaos that was the common experience of urban street life in the nineteenth century. Eventually, such composite pictures became a staple of popular illustrated journalism, and the step-by-step depiction of industrial processes became a favored topic, as in "Bicycles and Tricycles – How They Are Made" (Plate 13) from an 1887 issue of *Frank Leslie's Illustrated Newspaper*.

Handdrawn popular illustration died out around the turn of the century when a new technology of visual representation, photography, took over the journalistic duties of the newspaper illustrators. As a genre, popular illustration had accomplished much. It had created a visual culture embedded in the social reality of urban life and had urged visual art generally away from landscape toward cityscape, from stasis to kinesis. The contribution of photography would be even greater. Here was a technology of visual representation perfectly suited to its age. In the hands of journalists, it created a new and powerfully accurate kind of documentary record. In the hands of social activists, it created images of passion and outrage that could not be ignored. In the hands of the popular masses, it created snapshots and memories, both social and personal. And in the hands of artists – and who with a camera in his hands was not an artist? – it created art.

Photography is arguably the single most characteristic medium of expression and representation of the modern period, and a complete history of photography would include consideration of its use in science, medicine, criminology, education, and commerce as well as art. But as a medium of artistic expression, photography perfectly exemplified the spirit of the modern age, addressing every known genre and inventing new ones of its own. Nature and landscape, still life and the nude, staged heroic set pieces and candid genre scenes – all found new vitality in the hands of the masters of the new technology of visual representation. But it was urban subjects more than any others – the inescapable social and physical reality of the modern city – that captured photography's unique potential as an expressive medium and gave photography its historic and artistic *raison d'être*.

Although photography did not completely supplant handdrawn illustrations in journalism until the 1890s and early 1900s, the very first news photo may well have been "Burning Mills, Oswego, New York" (Plate 14), a daguerreotype by George N. Barnard that was printed in the *Oswego Daily Times* in 1853. While pictures of factory fires had been a staple of popular illustration, photography brought a new realism and intensity to the images presented to the public. Even today, dramatic pictures of big fires, even if they are from cities far away, have a special fascination and are often prominently placed in newspapers and the evening television news broadcasts.

The spectacle of the burning factory was a special case in a larger melodrama: the ongoing, day-to-day reality of modern urban life. Many of the very earliest photographs by Daguerre, Nadar, and Fox Talbot were simple street scenes made by the photographer simply turning his camera out the studio window. Like the "city sketches" type of popular illustration, these photographic street scenes conveyed a special sense of the vitality of urban life. Edward Anthony's "A Rainy Day on Broadway" (Plate 15) shows New York in 1859. The motion of those on the street and on the sidewalks is frozen in time – clearly Anthony's shutter speed was fast enough to eliminate most of the blurring that was so common in early photographs of moving subjects – but the image itself does not seem frozen. Rather, the image speaks directly to the viewer's own experience of street life, and the interaction of the viewed image and the viewer's imagination almost magically captures the hustle and bustle of New York street life.

Not all streets in the modern city were so pleasant. In "Bandits' Roost, 39½ Mulberry Street," of 1889 (Plate 16), Jacob Riis invites the viewer to look into one of New York's most notorious and dangerous back alleys. The author of *How the Other Half Lives* (1890), Riis was a crusading social reformer who had observed the dark side of the city as a police reporter for the local newspapers and he used his skills as a photographer to supplement his activism. Many middle-class viewers were undoubtedly repelled by images such as this one. It was the kind of view that one might catch fleetingly as one passed along a main thoroughfare, something to glance at quickly and then turn away. But Riis the reformer would have us not look away, and photography rivets the gaze on an unpleasant but inescapable social reality. In this, Riis's photographic work is directly linked to Friedrich Engels's descriptions of the Manchester slums in *The Condition of the Working Class in England in 1844*.

Jacob Riis was but one in a long and distinguished line of photographers who turned their art and their talents to the ends of political activism and social protest. One of the greatest of these visual documentors of the modern city was Lewis Hine. Hine's picture of a young girl tending a massive cloth-weaving machine in a North Carolina mill has become a nearly universal icon of industrial child labor, and his photographs of high-steel construction workers casually toiling hundreds of feet above the city are gut-wrenching images of working-class heroism. For a time, Hine worked at Ellis Island, the New York port of entry for immigrants to America. His portraits of the immigrants (Plate 17) are masterpieces of photographic humanism. Portraiture holds an important place in the history of photography. Whereas oil-painting portraiture had served the interests of the aristocracy and the haute bourgeoisie almost exclusively, the new medium allowed members of the broad urban middle class, and the new industrial working class as well, to record their images for posterity. And whereas the images of immigrants in popular illustration had tended toward racial and ethnic caricatures, Hine's portraits reveal the individuality and humanism behind the stereotypes. Another photograph depicting the immigrant condition – all the more powerful for its juxtaposition of social commentary and pure artistic composition – was Alfred Stieglitz's "The Steerage" of 1907 (Plate 18). Here, the interplay between the immigrants, mostly enveloped in shadows, and the physical structures of gangway and ladder combine simultaneously to fascinate the eye and touch the soul.

The Great Depression of the 1930s strengthened the documentary and social activist tendencies in photography that had been pioneered by Riis, Hine, and the photo-journalists, especially in work that had been commissioned by the Works Progress Administration and other New Deal agencies. The photographs of Southern sharecroppers by Walker Evans and the images of Harlem streetlife by Gordon Parkes are notable examples of the intersection of art and social protest. Dorothea Lange was another activist photographer who, like Evans, was drawn to the documentation of rural poverty. Her pictures of dispossessed dust bowl emigres are among the most famous and most powerful images of the Depression era. And in her picture of a roadside sign (Plate 19), published in a 1939 Farm Security Administration report on rural migration, Lange captured both the bleakness of the Depression and the kinds of fears and insecurities that underlay Frank Lloyd Wright's "Broadacre City" proposal – a homestead of at least one acre per person – and other back-to-the-land schemes.

The people of the city and their problems were the dominant, but not the sole, subject of urban documentary photography. The very physical presence of the modern city was often evoked through images of architecture and infrastructure. Stieglitz's moody impression of the Flatiron Building cloaked in mist is one well-known example, and Margaret Bourke-White's formalist, monumental depiction of Hoover Dam that appeared on the cover of the first *Life* magazine is another. Charles Sheeler's "Ford Plant, Detroit" of 1927 (Plate 20) demonstrates the way in which pure architectural form – an assemblage of strong verticals and diagonals revealed through light and shadow – could be read both as pure cubist composition and powerful, if somewhat distanced, social documentation.

If the architecture of the city could present images of foreboding power or lyrical freedom, it was nonetheless the lives of the people of the city themselves that became the constantly recurring theme of photojournalists and artists alike. The work of Weegee (Arthur Fellig) is a case in point. Although most of his pictures were made in the 1940s and 1950s and featured startling images of auto wrecks, transvestite bars, and gangland killings, the individual faces communicate directly to the viewer with a sense of direct, immediate reality that is the transcendent essence of urban life. "The Critic" (Plate 21), for example, was taken in 1943, but the people and the streetlife situation – so like the contrast pictures in popular illustration – are instantly recognizable as big-city icons of the painful urban conflation of luxury and poverty, indifference and despair.

The project of comprehending the modern city visually played, and continues to play, a central role in the history of art and consciousness. The great themes of the city – its kinetic activity, its juxtapositions and ironies, its massive forms and tiny details – provided the artist with a subject matter that could not be ignored and pioneered modes of visual perception and communication that were to fundamentally transform the nature of social life. Whereas popular illustrated journalism liberated visual images from the control of privileged elites, photography completed the democratization of the visual by placing it in the hands of the masses.

Eventually, the new modes of visual perception would be connected to narrative – particularly to the "urban narratives" of young rural innocents encountering the experience of the city for the first time, a plot line as old as Gilgamesh and given renewed meaning as a result of the millions of immigration stories that fueled the urbanization process during the nineteenth and twentieth centuries. And out of the interconnection of urban narrative and visual representation would come cinema, the ultimate realization of the kinetic imagery that urban life characterized and photography mirrored. In King Vidor's *The Crowd* of 1926 (Plate 22), we see the once-optimistic individual lost in the enveloping urban mass. In Fritz Lang's *Metropolis* of 1929 (Plate 23), we see the modern citadels of faceless power looming over the dehumanizing structures of class segregation and oppression. In both cases, the viewer confronts and subconsciously confirms the artist's perception of the reality of modern urban life. The confrontation may be discomfiting, but the ubiquity of modern visual culture makes the confrontations inescapable and memorable, perhaps more memorable even than actual encounters on the streets of the city. Writing about the historic city, Lewis Mumford found that they were characterized by an "urban drama" and that the dramatic dialogue was "one of the ultimate expressions of life in the city." In the modern city, it is the image – sometimes celebratory, sometimes haunting, always definitive in its explanatory value – that is paramount both as spectacle and revelation.

Plate 10 **"Music in the Street, Music in the Parlor"** (Unknown artist, 1868, *The Illustrated News*)

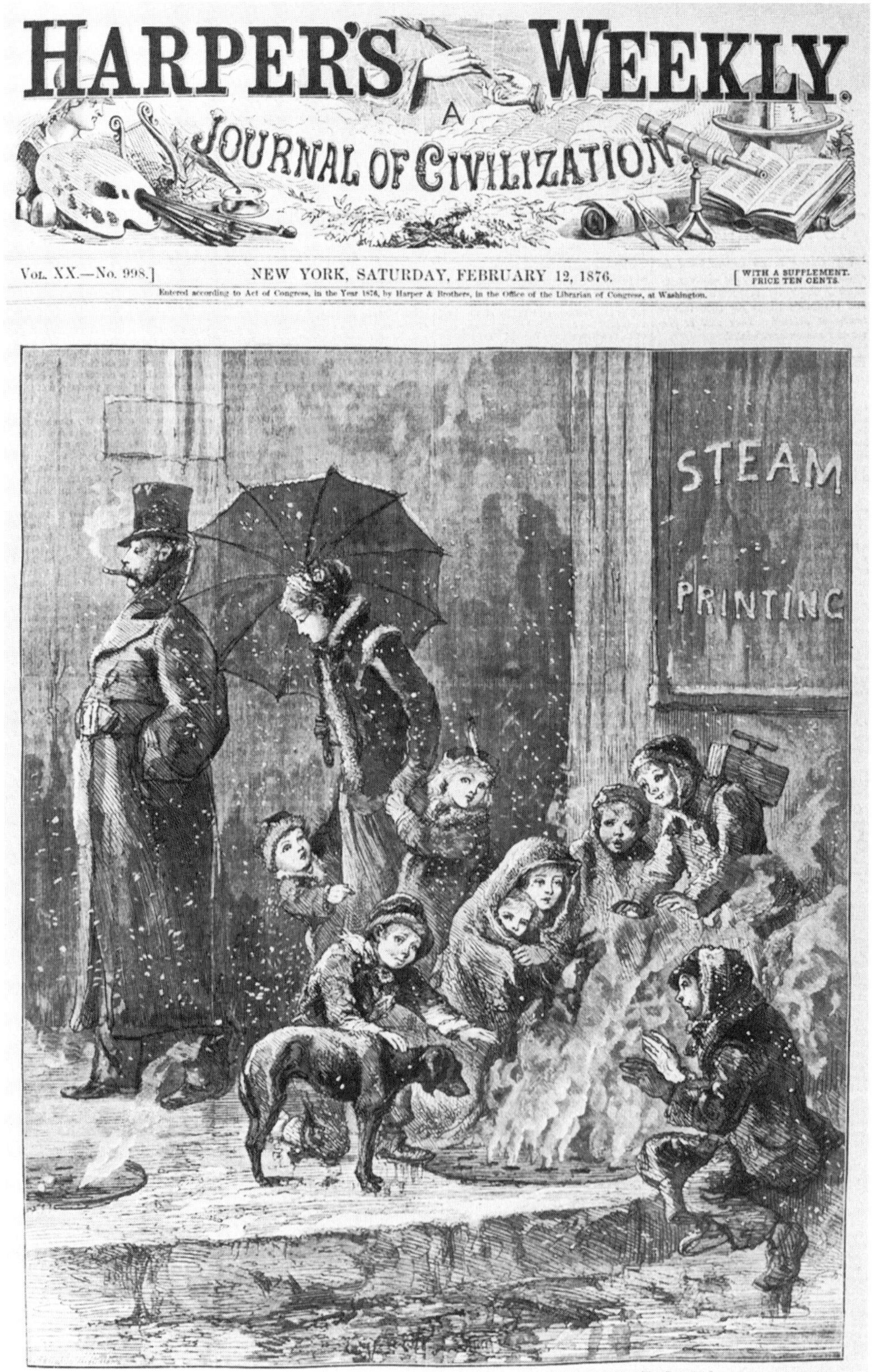

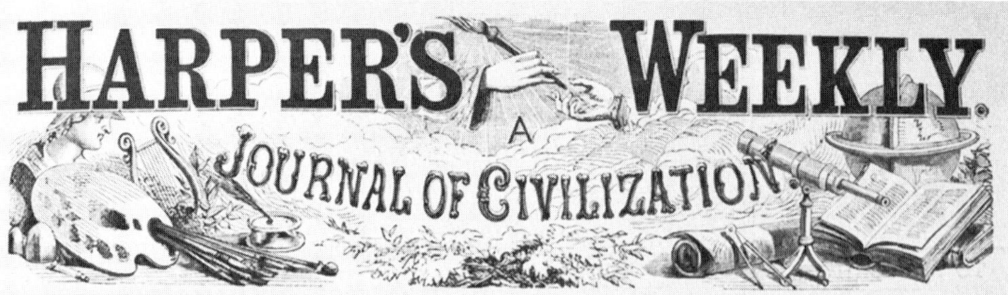

HARPER'S WEEKLY
A JOURNAL OF CIVILIZATION.
VOL. XX.—No. 998.]
NEW YORK, SATURDAY, FEBRUARY 12, 1876.
[WITH A SUPPLEMENT. PRICE TEN CENTS.
Entered according to Act of Congress, in the Year 1876, by Harper & Brothers, in the Office of the Librarian of Congress, at Washington.
STEAM
PRINTING
THE HEARTH-STONE OF THE POOR—WASTE STEAM NOT WASTED.—DRAWN BY SOL EYTINGE, JUN.—[SEE PAGE 131.]

Plate 11 **"The Hearth-Stone of the Poor"** (Sol Eytinge, Jr., 1876, *Harpers Weekly*)

208 BALLOU'S PICTORIAL DRAWING-ROOM COMPANION.

LIFE IN TOWN.

CITY SKETCHES

THE DOCK LOAFER

THE CHILDREN OF THE POOR.

THE CHARCOAL PEDLER.

EMIGRANTS.

SKATERS.

THE DUSTMAN.

THE WATCHMAN.

BOSTON STREET CHARACTERS.

[For description, see page 2[illegible]]

Plate 12 **"City Sketches"** (C. A. Barry, 1855, *Ballou's Pictorial Drawing-Room Companion*)

Plate 13 **"Bicycles and Tricycles – How They Are Made"** (Unknown artist, 1887, *Frank Leslie's Illustrated Newspaper*)

Plate 14 **"Burning Mills, Oswego, New York"** (George N. Barnard, 1853, *Oswego Daily Times*) (Courtesy George Eastman House)

Plate 15 **“A Rainy Day on Broadway” (Edward Anthony, 1859)** (Courtesy George Eastman House)

Plate 16 **"Bandits' Roost, 39½ Mulberry Street" (Jacob Riis, New York, 1889)** (The Jacob A. Riis collection # 101. Copyright © Museum of the City of New York)

WELSH COAL MINER

SLAVIC STEEL-WORKER

Photos by Hine

ITALIAN LABORER

IRISH IRON WORKER

Plate 17 **Ellis Island Immigrant Portraits (Lewis Hine, ca. 1910)**. In Grove S. Dow, *Social Problems of Today* (New York: Thomas Y. Crowell, 1925). Public domain

Plate 18 **"The Steerage" (Alfred Stieglitz, 1907)** (Courtesy of the Museum of Modern Art, New York)

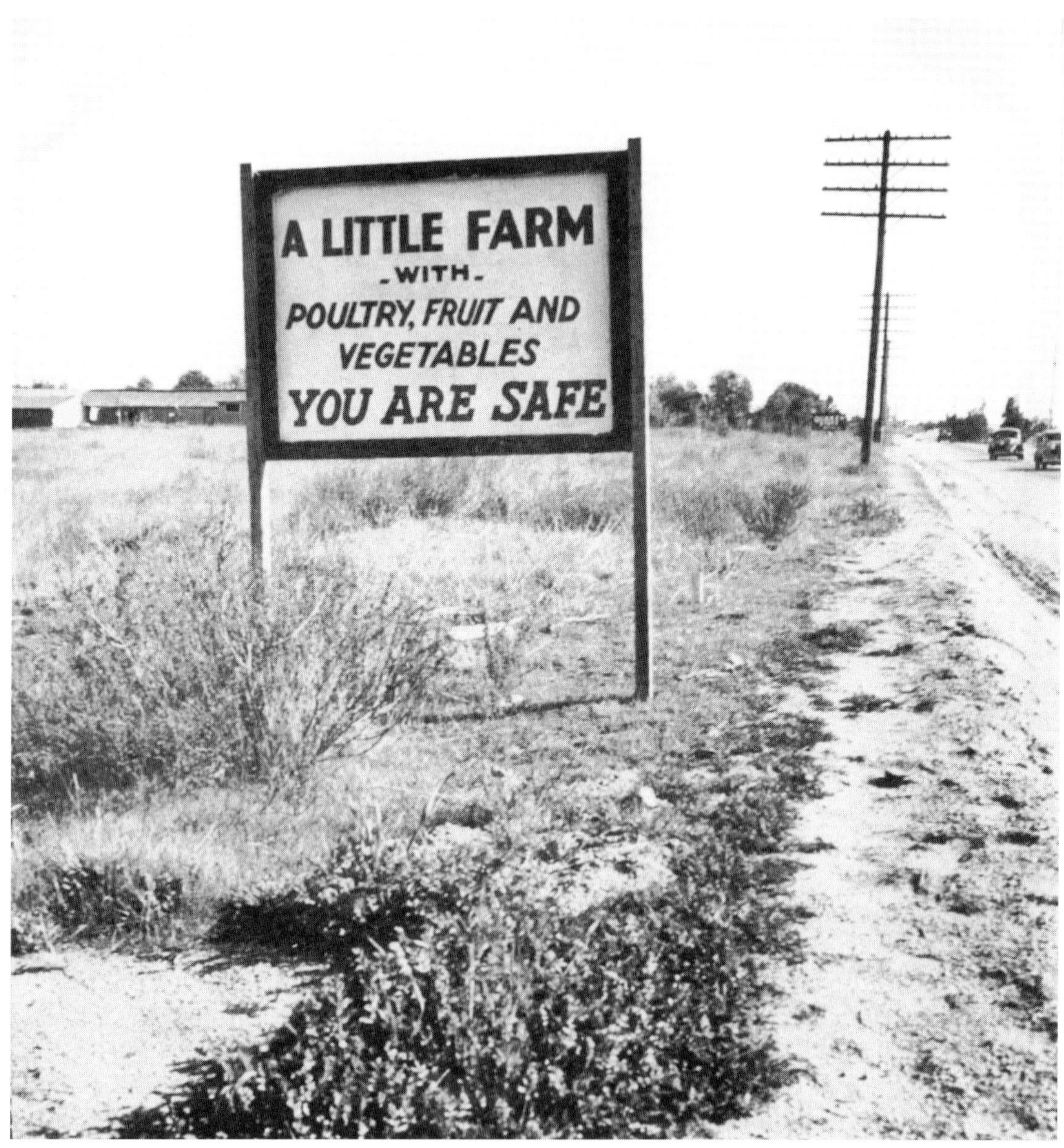

Plate 19 **Untitled (roadside sign) (Dorothea Lange, 1939)**. Farm Security Administration, US Department of Agriculture. Printed in C.E. Lively and Conrad Tauber, *Rural Migration in the United States* (Washington, DC: Works Progress Administration, 1939)

Plate 20 **"Ford Plant, Detroit" (Charles Sheeler, 1927)** (Courtesy of the Museum of Modern Art, New York)

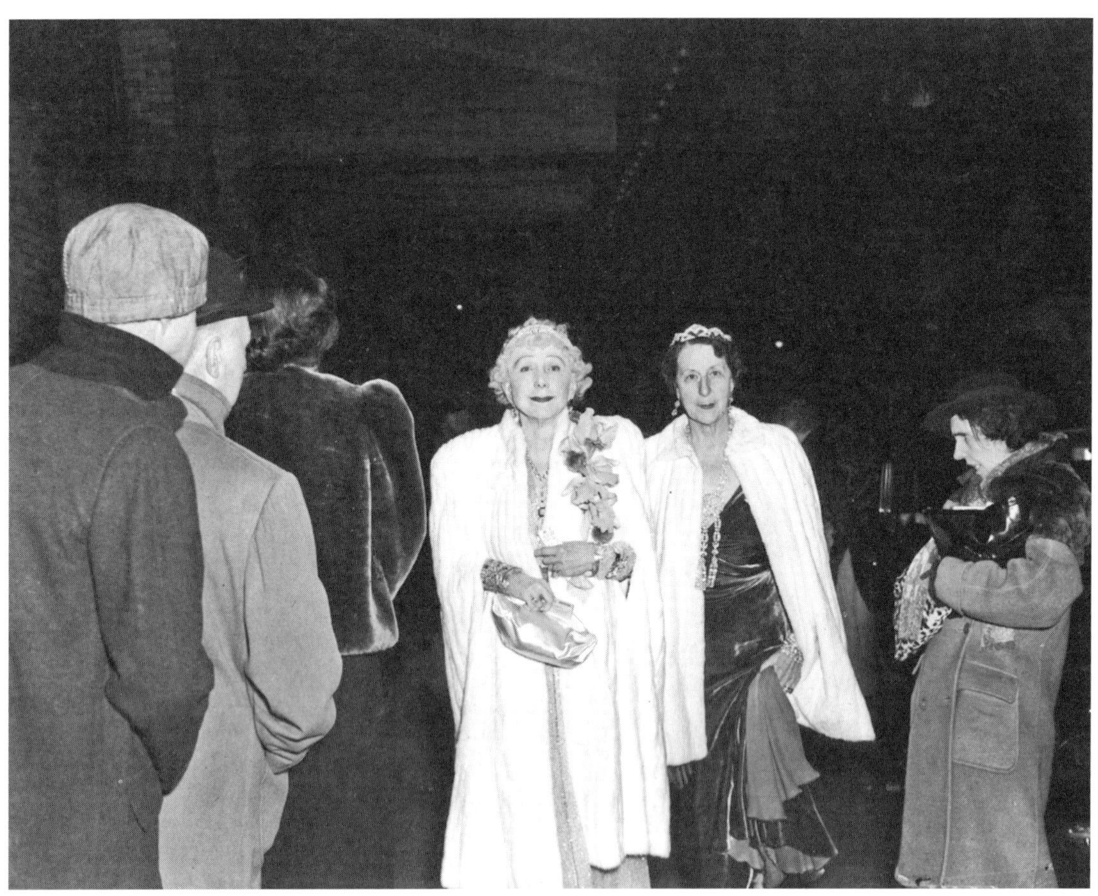

Plate 21 **"The Critic" (Weegee (Arthur Fellig), 1943)** (Copyright © 1994, International Center of Photography, New York. Bequest of Wilma Wilcox)

Plate 22 **Still from *The Crowd* (King Vidor, 1926)** (Courtesy of the Museum of Modern Art, New York, Film Stills Archive)

Plate 23 **Still from *Metropolis* (Fritz Lang, 1929)** (Courtesy of the Museum of Modern Art, New York, Film Stills Archive)

Urban space

MADISON AV.
CATHEDRAL
COLUMBUS
GRAMMAR
SCHOOL
JACKSON
BAPTIST
CHURCH

INTRODUCTION TO PART THREE

The cultural and social characteristics of cities described in Part Two display an astonishing variety over time from the complex Black society W.E.B. Dubois documented in late-nineteenth-century Philadelphia's Seventh Ward (p. 110), to the impersonal world of early-twentieth-century immigrant culture that formed the basis for Louis Wirth's theory of urbanism as a way of life (p. 96), to the yuppie creative class enclaves that Richard Florida feels are important to advanced urban economies (p. 143). The size, density, spatial distribution of functions, and physical form of cities described in Part One on the evolution of cities ranged from Ur's mud-brick walls and ziggurat (p. 31) to the marble agora of the Greek polis (p. 40); from the polluted slums of nineteenth-century Manchester (p. 46) to the high-tech research campuses, malls, and residential areas of present-day technoburbs (p. 75).

Part Three of *The City Reader* deals with both the physical and social aspects of urban space. It contains classic and contemporary writings on the spatial aspects of cities. Of all the social science disciplines and applied professional fields that deal with cities, geography is most centrally concerned with urban space. Part Three introduces material on urban geography that is much more fully developed in another reader in this series: Nicholas Fyfe and Judith Kenny (eds), *The Urban Geography Reader* (London: Routledge, 2005) in the Routledge Urban Reader Series contains selections dealing with topics as varied as race, class, gender, and sexual orientation; globalization; economic restructuring; politics, governance, and inequality; social movements and social conflict; urban form and symbolism; and technology.

Parts Five, Six, and Seven of this reader on Urban Planning History and Visions, Urban Planning Theory and Practice, and Perspectives on Urban Design return to material on urban space, but extend the discussion to interventions that shape the physical form and design of cities and regions.

For millennia physical geographers have studied and mapped physical and political features of the earth, and cultural geographers have studied spatial aspects of cultural and social phenomena. But it was not until the 1950s that urban geography emerged as an important subfield of geography. Nowadays urban geography has been reinvigorated by a new generation of geographers who define the discipline broadly and, along with other natural and social scientists and planners, use geographical information systems (GIS) software to do computerized spatial analysis with a power and precision unimaginable in the past. The selection by law professor Myron Orfield (p. 296) in part four describes how he used GIS to analyze metropolitan spatial inequality and develop proposals for greater spatial fiscal equity.

Scholars from many disciplines have contributed to our understanding of urban space. This part of *The City Reader* includes writings from the disciplines of sociology (Ernest Burgess) and history (Robert Bruegmann), and the professional fields of urban planning (Daphne Spain), landscape architecture (J.B. Jackson), and urban design (Ali Madanipour). Writers from these and other disciplines throughout this book inform our understanding of urban space.

Many cities have grown organically with little or no explicit overall plan or centrally controlled regulation. They contain vernacular architecture created by ordinary people without specialized training in the design professions, such as J.B. Jackson describes (p. 202). Other cities display a mix of organic growth and development planned at different periods in their history. The old crooked streets and irregular lots that constitute most of central Boston, Massachusetts, contrast with the regularity of Boston's Back Bay. There a swampy area was developed during the latter part of the nineteenth century following a plan by landscape architect Frederick Law Olmsted. Disorderly districts of London that have evolved since the Middle Ages contrast with the area rebuilt after the great London fire of 1666 which destroyed much of the city. Only a few entire cities, like Australia's capital city Canberra and Brazil's capital city Brasilia, have been built according to a consistent master plan – and even they and their suburbs have vernacular buildings that do not fit the plan.

It was sociologists, not geographers, who pioneered the modern field of urban geography. In academia "schools" of thought are sometimes identified with the work of a group of individuals in the same general location, who work together to some extent on a common project. One of the most celebrated examples is the Chicago school of sociology formed around the pioneering work of a group of American sociologists based at the University of Chicago in the 1920s and 1930s. A slim volume of ten essays titled *The City: Suggestions for Investigation of Human Behavior in the Urban* (Chicago, IL: University of Chicago Press, 1929) expresses core ideas of the Chicago school. In addition to path-breaking work on the social structure of cities by Louis Wirth (p. 96), Robert Park, Ernest W. Burgess and others, members of the Chicago school of urban sociology developed the first systematic theory about the physical form of cities. The first selection in Part Three, Ernest W. Burgess's "The Growth of the City" (p. 161), advances many provocative hypotheses.

Burgess was interested in understanding the internal structure of a single city, not the way in which multiple cities were related to each other. Burgess argued that there was an underlying social and economic logic to the physical form of cities. In his view they were organized in a series of concentric rings moving out from a central business district (CBD). Each ring had a distinct set of residents and functions. Physical form and human life were intimately linked. According to Burgess, cities were not static. Chicago school sociologists were strongly influenced by theory about the way in which plant and animal communities evolved and changed. Drawing on the social ecology perspective, Burgess envisioned cities as dynamic organisms with a constant flow of new residents coming into the inner rings and a flow from these rings outward over time. According to Burgess and other social ecologists, processes of invasion and succession, similar to what occurred in plant and animal communities occurred in cities as different ethnic groups, races, classes, and other social groups competed for space. Each part of an urban system had distinct characteristics and played a unique role in the total system. Each was related to the others and the whole system was in a state of constant flux.

Geographers, sociologists, land economists and others, inspired by Burgess, have studied city form during eighty years since Burgess's classic essay on the growth of the city was published. It is often juxtaposed with theories developed by real estate economist Homer Hoyt in the 1930s postulating an organization of cities in sectors radiating out along transportation corridors from a CBD, with an essay by University of Chicago geographers Chauncy Harris and Edward Ullman written in 1945 concluding that most cities have multiple nuclei rather than either a concentric zone or sectoral organization, and most recently with the competing paradigm developed by Michael Dear (p. 170) and other members of the Los Angeles school of urbanism.

In the late 1980s a group of loosely associated scholars, professionals, and advocates based in Southern California became convinced that what was happening in the region was symptomatic of a broader sociogeographic transformation taking place within the United States as a whole. Geographers Michael Dear, Steven Flusty, academics at the University of California, Los Angeles, and the University of Southern California, and other like-minded scholars in Southern California developed a robust and controversial school of thought referred to as the Los Angeles school of urbanism.

The LA school substitutes a postmodern view of urban process in place of the Chicago school's modernist perspectives on the city. Dear and Flusty use the term "Keno capitalism" – derived from a game in which outcomes are determined largely by chance – to describe new postmodern urban processes. They see urban evolution as a nonlinear, chaotic process, not the rational, deterministic process the Chicago school describes. Capital touches down as if by chance on a parcel of land. Land values soar in the area near the favored parcel. Other similar neighborhoods that did not get selected for new development decline. New developments are noncontiguous. City centers are grafted onto the landscape as an afterthought. Suburbanization bears no relationship to a core-related decentralization. Urban peripheries are organizing what remains of the center rather than the other way around. The global economy largely determines urban economic functions.

Burgess and Dear wrote about the way in which a single city or metropolitan region is organized. A related topic in the study of urban form concerns the way in which cities relate to one another in systems of cities. The seminal work related to systems of cities was written by a German economic geographer named Walter Christaller in 1933. Based on meticulous empirical research on telephone communications in Southern Germany, Christaller developed central place theory. In his view there is a hierarchy of economic functions from the most frequent, such as mom-and-pop grocery stores that are found even in the smallest human settlements, to very specialized functions such as a stock exchange that is likely to be found only in the largest city of a country or an entire region of the world region.

Part Eight on Cities in a Global Society picks up the debate on world city systems with descriptions by sociologist Saskia Sassen (p. 554), geographers Jonathan Beaverstock, Richard Smith, and Peter Taylor (p. 563) and interdisciplinary urbanists Neil Brenner and Roger Keil (p. 599) about the present-day global city network.

The debates about the internal structure of the city initiated by Burgess and his Chicago school colleagues are still raging now thanks to Michael Dear and the LA school, and the debate about world systems of cities initiated by Walter Christaller continues in the work of writers like Jonathan Beaverstock, Richard Smith, and Peter Taylor (p. 563), Saskia Sassen (p. 554), and Neil Brenner and Roger Keil (p. 599), which deal with economic and social aspects of spatial process at work on cities and regions. But how do these issues affect different social groups? University of Virginia urban planning professor Daphne Spain asks and answers the provocative question "What Happened to Gender Relations on the Way from Chicago to Los Angeles?" (p. 176).

Spain takes both Burgess and Dear and their colleagues in the Chicago and LA schools to task for neglecting gender as a factor in shaping the internal structure of cities. She notes that the role that women played in 1925 when Burgess's celebrated article was written – mostly as homemakers who were not employed outside of the home – had a profound impact on urban structure. Single-wage-earner households had less disposable income to afford single family homes than two-wage-earner households, contributing to the denser form of cities at that time. At that time women cooked, looked after children, and took care of extended families, reducing the need for fast food restaurants, professional childcare, and nursing homes for elderly people. At home much of the day, women reduced the need for formal security systems and gated communities. With a majority of women in the workforce today, the impact on metropolitan form is enormous. Higher income in two-wage-earner families contributes to urban sprawl. Women drive to work, school, and other destinations, contributing to the kind of spread-out urban form Dear describes. New institutions such as fast food restaurants and professional childcare centers perform roles individual women performed in former times.

The selection by landscape theorist J.B. Jackson (p. 202) also notes an underlying logic to the physical form, social structure, and function of cities. According to Jackson, small cities that evolved without city planners or even architects display an underlying regularity and logic. For Jackson the built environment of every place expresses the belief system and values of the humans who created it. Despite the fact that until recently most cities have grown with little or no formal planning or professional design, little scholarly attention was devoted to studying vernacular urban form until J.B. Jackson pioneered the study of vernacular landscapes. Jackson was fascinated with the physical form of barns, fences, billboards, and grain silos; pioneer settlements with stumps in the field and muddy roads; dying former railroad towns where the train no longer stops; humble mobile homes in rural New Mexico; and gas stations and frozen custard stands along the principal artery across the American Southwest, Highway 66. Jackson was one of the first to argue that a study of vernacular architecture provides important understanding of the culture and values of the people who have built it. According to Jackson, a close look at Highway 66 in the 1950s reveals a great deal about the worldview of the highway builders and residents along the highway. Nowadays, largely as a result of

Jackson's influence, there is a large literature describing and analyzing the function and cultural meaning of Las Vegas casinos, White Tower hamburger stands, suburban tract homes, billboards, and other vernacular architecture.

In his essay in this part (p. 186) University of Newcastle, England, urban design professor Ali Madanipour analyzes the way in which Europeans knowingly or unwittingly exclude people from other cultures from the full benefits of their societies. Madanipour distinguishes between economic discrimination, in which members of a group are excluded from access to employment, political discrimination, in which they are excluded from political power, and cultural exclusion in which the group members are marginalized from the symbols, meanings, rituals, and discourses of the dominant culture. Since exclusion often has a spatial dimension, Madanipour suggests a number of strategies to break down spatial exclusion and increase inclusion. Subsidized housing, for example, may permit low-income foreign immigrants to live in parts of a city they could not otherwise afford, giving them access to job opportunities and better education for their children.

In "Fortress L.A." (p. 195), social critic Mike Davis, another member of the LA school of urbanism, addresses the dark side of the postmodern metropolis. Like Michael Dear (p. 170), Davis also uses Los Angeles as his exemplar city. Davis describes a built environment complete with surveillance cameras, barrel-shaped park benches designed to keep people from sleeping on them, overhead sprinklers to douse the homeless, windowless concrete hotel walls facing streets, entrances to public buildings reminiscent of the fortifications in front of medieval European castles, and gated communities where the rich can be (or at least feel) safe from outsiders. All these artifacts provide a disturbing glimpse of Los Angeles and, by implication, postmodern cities emerging around the world.

"The Growth of the City: An Introduction to a Research Project"

from Robert E. Park, Ernest W. Burgess, and Roderick D. McKenzie, *The City* (1925)

Ernest W. Burgess

Editors' Introduction

Ernest W. Burgess (1886–1966) was a member of the famed sociology department at the University of Chicago in the 1920s and 1930s that set out to reinvent modern sociology by taking academic research to the streets and by using the city of Chicago itself as a "living laboratory" for the study of urban problems and social dynamics.

Throughout a long and productive career, Burgess addressed a whole series of issues that connected the social dynamics of the city as a whole with the lives of its citizens. He wrote extensively on issues related to marriage and the family, the relation of personality to social groups, and, in the final decades of his life, problems of elderly people. His most famous contribution to the study of the city was the 1925 essay reprinted here: "The Growth of the City."

Subtitled "An Introduction to a Research Project," Burgess's seminal analysis of the interrelation of the social growth and the physical expansion of modern cities helped foster the subfield of urban geography as well as urban sociology. Burgess focused on patterns within a single city (what is referred to as the internal structure of the city), rather than relationships among cities (systems of cities). Seeking to describe what he called the pulse of the community, Burgess devised a theory that was thoroughly organic, dynamic, and developmental.

In the expansion of the city, Burgess wrote, a process of distribution takes place, which sifts and sorts and relocates individuals and groups by residence and occupation. It was this dynamic process – process was one of Burgess's favorite words – that gives form and character to the city.

Chicago, at the time this selection was written, was a dynamic, rapidly growing city of recent immigrants. Chicago's wealth was built on its location as the receiving center for natural resources from the developing frontier. Rail lines from the parts of the United States that developed in the latter half of the nineteenth and first part of the twentieth centuries converged in Chicago and then continued east. Grain from the west, lumber from the north, and cattle from the southwest all came into Chicago to be sorted, processed, and shipped east. Miles of grain elevators, enormous lumberyards, and huge slaughterhouses were able to employ hundreds of thousands of unskilled immigrant workers. Traders, factory owners, and entrepreneurs grew rich. First- and second-generation immigrants who had prospered in the booming economy formed a middle class.

Central to Burgess's analysis of urban growth was his famous model based on a series of concentric circles that divided the city into five zones. The concentric zone model looks like a static map of Chicago's economic demography (a cross-sectional analysis), but the model is really a theoretical diagram of a dynamic process illustrating how urban social and economic structure changes over time (a longitudinal analysis). Burgess called the process of neighborhood change "succession," a term he borrowed from the science of plant ecology to describe urban metabolism and mobility. By borrowing terminology from the natural sciences, and by drawing analogies between the urban and the natural worlds, Burgess helped establish the study of social ecology as a distinct approach to understanding the underlying patterns of urban growth and development.

For all the problems and pathologies of urban life Burgess saw cities as progressing. Within a generation Chicago had morphed from a frontier town to a booming world metropolis. His model is logical and rational. He was convinced that there was an underlying logic to the social and economic structure of cities that could be understood scientifically. He saw the city itself as the driving force behind the region of which it was a part. While Chicago was a world city at the time, he believed Chicago's growth patterns were primarily endogenous, governed by local, rather than global, forces.

Following the publication of Burgess's essay, a number of urban theorists offered modifications and even refutations of the simple elegance of the concentric zone model. In 1939, real estate economist Homer Hoyt proposed a sectoral model for modern capitalist cities based on wedges of activity extending outward from the city center along transportation corridors. In 1945, geographers Chauncy Harris and Edward Ullman suggested a multiple nuclei model, arguing that cities developed around several, not just one, centers of economic activity. These authors' descriptions of their alternative models are described in *The Urban Geography Reader* (Fyfe and Kenny, 2005).

The Burgess model remains essential reading in urban geography and urban sociology courses and is still recognized as a brilliant and provocative piece of theoretical writing. Both Burgess's applied field work methods and his concentric zone model continue to inspire modern scholars and influence important works on urban space. Prior to joining the Harvard University faculty, William Julius Wilson was a member of the University of Chicago sociology department and his study of the urban underclass (p. 117) is in the tradition of using Chicago as a laboratory to study important issues related to class and race established by Burgess and others. Elijah Anderson (p. 127) was a student of Wilson's and is in the Chicago school of sociology tradition. Ali Madanipour's analysis of "Social Exclusion and Space" (p. 186) owes a profound debt to Burgess. J.B. Jackson (p. 202) shared Burgess's belief that there is an underlying logic to urban form that occurs even in the absence of formal planning. But Jackson emphasizes the influence of history and culture on the urban landscape.

The Burgess model has also been widely criticized. Michael Dear (p. 170) and many other modern writers either reject the Burgess model altogether or argue that if it accurately described urban processes in twentieth-century cities, it no longer does nowadays. Burgess's model is in the modernist tradition. It assumed a logical, rational set of processes with the central business district at the center emanating outward in concentric rings to the edge of the city and the suburbs beyond. But does this model adequately describe the process of urban change today? A competing, postmodernist school of thought, the Los Angeles school of urbanism thinks not. Michael Dear (p. 170) argues that present-day growth in Los Angeles is largely determined by development in the periphery of the region, not what happens in the CBD as Burgess argued. Global structural forces, Dear and other LA school theorists argue, determine metropolitan spatial structure, not decisions made by individuals acting as free agents as Burgess believed.

In criticizing Burgess for his lack of attention to gender issues, Daphne Spain (p. 176) and other writers question the premises of social ecology in which some social groups (Blacks, immigrants, women) are relegated to inferior status by deterministic forces beyond their control.

Computer technology, including statistical packages and geographic information systems, software, now makes it possible for present-day geographers and sociologists to summarize vast amounts of data and map the internal structure of cities in ever more sophisticated ways. Thus, understanding of the relationship between social groups and urban form pioneered by Burgess continues to advance by leaps and bounds.

Ernest W. Burgess was a professor of sociology at the University of Chicago and one of the most influential members of the Chicago school of sociology that included such luminaries as his office mate Louis Wirth, author of "Urbanism as a Way of Life" (p. 96), and Robert E. Park, who developed important sociological theories about immigration, assimilation, and social ecology. Burgess served as chair of the University of Chicago sociology department and as president of the American Sociological Society (1934), Sociological Research Association (1942), and the Social Science Research Council (1945–1946). He was managing editor of the *American Sociological Society* from 1921 to 1930, and editor of the *American Journal of Sociology* from 1936 to 1940.

This selection is taken from Robert E. Park, Ernest W. Burgess, and Roderick D. McKenzie, *The City* (Chicago, IL: University of Chicago Press, 1984, originally published in 1925).

The University of Chicago maintains Burgess's papers in its Special Collections Research Center. Burgess's most important writings are in Ernest W. Burgess, *Basic Writings of Ernest W. Burgess* (Chicago, IL: Community and Family Study Center, University of Chicago, 1974).

Nicholas Fyfe and Elizabeth Kenny (eds), *Urban Geography Reader* (London: Routledge, 2005) in the Routledge Urban Reader Series contains writings by Homer Hoyt on his sector model of the internal structure of the city and Chauncy Harris and Edward Ullman on their multiple nuclei theory of the internal structure of cities.

The outstanding fact of modern society is the growth of great cities. Nowhere else have the enormous changes which the machine industry has made in our social life registered themselves with such obviousness as in the cities. In the United States the transition from a rural to an urban civilization, though beginning later than in Europe, has taken place, if not more rapidly and completely, at any rate more logically in its most characteristic forms.

All the manifestations of modern life which are peculiarly urban – the skyscraper, the subway, the department store, the daily newspaper, and social work – are characteristically American. The more subtle changes in our social life, which in their cruder manifestations are termed "social problems," problems that alarm and bewilder us, such as divorce, delinquency, and social unrest, are to be found in their most acute forms in our largest American cities. The profound and "subversive" forces which have wrought these changes are measured in the physical growth and expansion of cities. That is the significance of the comparative statistics of Weber, Bucher, and other students.

These statistical studies, although dealing mainly with the effects of urban growth, brought out into clear relief certain distinctive characteristics of urban as compared with rural populations. The larger proportion of women to men in the cities than in the open country, the greater percentage of youth and middle-aged, the higher ratio of the foreign-born, the increased heterogeneity of occupation increase with the growth of the city and profoundly alter its social structure. These variations in the composition of population are indicative of all the changes going on in the social organization of the community. In fact, these changes are a part of the growth of the city and suggest the nature of the processes of growth.

The only aspect of growth adequately described by Bucher and Weber was the rather obvious process of the aggregation of urban population. Almost as overt a process, that of expansion, has been investigated from a different and very practical point of view by groups interested in city planning, zoning, and regional surveys. Even more significant than the increasing density of urban population is its correlative tendency to overflow, and so to extend over wider areas, and to incorporate these areas into a larger communal life. This paper, therefore, will treat first of the expansion of the city, and then of the less-known processes of urban metabolism and mobility which are closely related to expansion.

EXPANSION AS PHYSICAL GROWTH

The expansion of the city from the standpoint of the city plan, zoning, and regional surveys is thought of almost wholly in terms of its physical growth. Traction studies have dealt with the development of transportation in its relation to the distribution of population throughout the city. The surveys made by the Bell Telephone Company and other public utilities have attempted to forecast the direction and the rate of growth of the city in order to anticipate the future demands for the extension of their services. In the city plan the location of parks and boulevards, the widening of traffic streets, the provision for a civic center, are all in the interest of the future control of the physical development of the city.

This expansion in area of our largest cities is now being brought forcibly to our attention by the Plan for the Study of New York and Its Environs, and by the formation of the Chicago Regional Planning Association, which extends the metropolitan district of the city to a radius of 50 miles, embracing 4,000 square miles of territory. Both are attempting to measure expansion in order to deal with the changes that accompany city growth. In England, where more than one-half of the inhabitants live in cities having a population of 100,000 and over, the lively appreciation of the bearing of urban expansion on social organization is thus expressed by C.B. Fawcett:

> One of the most important and striking developments in the growth of the urban populations of the more advanced peoples of the world during the last few decades has been the appearance of a number of vast urban aggregates, or conurbations, far larger and more numerous than the great cities of any preceding age. These have usually been formed by the simultaneous expansion of a number of neighboring towns, which have grown out toward each other until they have reached a practical coalescence in one continuous urban area. Each such conurbation still has within it many nuclei of denser town growth, most of which represent the central areas of the various towns from which it has grown, and these nuclear patches are connected by the less densely urbanized areas which began as suburbs of these towns. The latter are still usually rather less continuously occupied by buildings, and often have many open spaces.
>
> These great aggregates of town dwellers are a new feature in the distribution of man over the earth. At the present day there are from thirty to forty of them, each containing more than a million people, whereas only a hundred years ago there were, outside the great centers of population on the waterways of China, not more than two or three. Such aggregations of people are phenomena of great geographical and social importance; they give rise to new problems in the organization of the life and well-being of their inhabitants and in their varied activities. Few of them have yet developed a social consciousness at all proportionate to their magnitude, or fully realized themselves as definite groupings of people with many common interests, emotions and thoughts.

In Europe and America the tendency of the great city to expand has been recognized in the term "the metropolitan area of the city," which far overruns its political limits, and, in the case of New York and Chicago, even state lines. The metropolitan area may be taken to include urban territory that is physically contiguous, but it is coming to be defined by that facility of transportation that enables a business man to live in a suburb of Chicago and to work in the loop, and his wife to shop at Marshall Field's and attend grand opera in the Auditorium.

EXPANSION AS A PROCESS

No study of expansion as a process has yet been made, although the materials for such a study and intimations of different aspects of the process are contained in city planning, zoning, and regional surveys. The typical processes of the expansion of the city can best be illustrated, perhaps, by a series of concentric circles, which may be numbered to designate both the successive zones of urban extension and the types of areas differentiated in the process of expansion [Figure 1].

[Figure 1] represents an ideal construction of the tendencies of any town or city to expand radially from its central business district – on the map "the Loop" (I). Encircling the downtown area there is normally an area in transition, which is being invaded by business and light manufacture (II). A third area (III) is inhabited by the workers in industries who have escaped from the area of deterioration (II) but who desire to live within easy access of their work. Beyond this zone is the "residential area" (IV) of high-class apartment buildings or of exclusive "restricted" districts of single family dwellings. Still farther, out beyond the city limits, is the commuters' zone: suburban areas, or satellite cities, within a thirty- to sixty-minute ride of the central business district.

This [figure] brings out clearly the main fact of expansion, namely, the tendency of each inner zone to extend its area by the invasion of the next outer zone. This aspect of expansion may be called *succession*, a process which has been studied in detail in plant ecology. If this [figure] is applied to Chicago, all four of these zones were in its early history included in the circumference of the inner zone, the present business district. The present boundaries of the area of deterioration were not many years ago those of the zone now inhabited by independent wage-earners, and within the memories of thousands of Chicagoans contained the residences of the "best families." It hardly needs to be added that neither Chicago nor any other city fits perfectly into this ideal scheme. Complications are introduced by the lake front, the Chicago River, railroad lines, historical factors in the location of industry, the relative degree of the resistance of communities to invasion, etc.

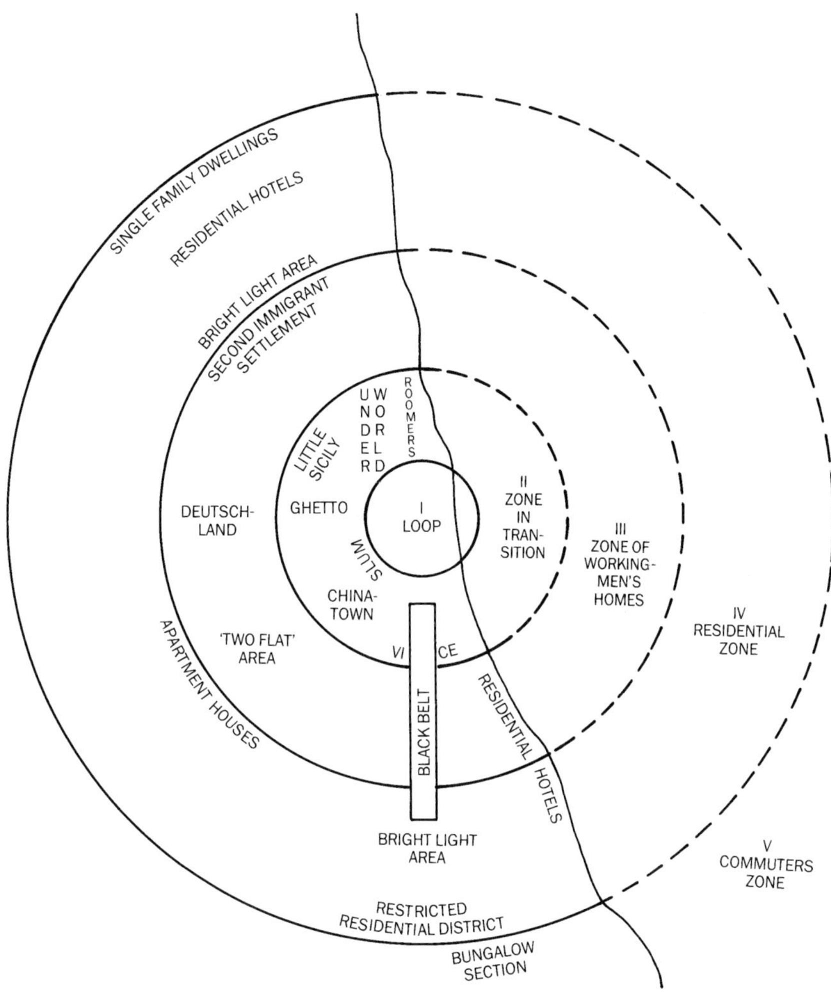

Figure 1

Besides extension and succession, the general process of expansion in urban growth involves the antagonistic and yet complementary processes of concentration and decentralization. In all cities there is the natural tendency for local and outside transportation to converge in the central business district. In the downtown section of every large city we expect to find the department stores, the skyscraper office buildings, the railroad stations, the great hotels, the theaters, the art museum, and the city hall. Quite naturally, almost inevitably, the economic, cultural, and political life centers here. The relation of centralization to the other processes of city life may be roughly gauged by the fact that over half a million people daily enter and leave Chicago's "loop." More recently sub-business centers have grown up in outlying zones. These "satellite loops" do not, it seems, represent the "hoped for" revival of the neighborhood, but rather a telescoping of several local communities into a larger economic unity. The Chicago of yesterday, an agglomeration of country towns and immigrant colonies, is under-going a process of reorganization into a centralized decentralized system of local communities coalescing into sub-business areas visibly or invisibly dominated by the central business district. The actual processes of what may be called centralized decentralization are now being studied in the development of the chain store, which is only one illustration of the change in the basis of the urban organization.

Expansion, as we have seen, deals with the physical growth of the city, and with the extension of the technical services that have made city life not only livable, but comfortable, even luxurious. Certain of these basic necessities of urban life are possible only through tremendous development of communal existence. Three millions of people in Chicago are dependent upon one unified water system, one giant gas company, and one huge electric light plant. Yet, like most of the other aspects of our communal urban life, this economic co-operation is an example of co-operation without a shred of what the "spirit of co-operation" is commonly thought to signify. The great public utilities are a part of the mechanization of life in great cities, and have little or no other meaning for social organization.

Yet the processes of expansion, and especially the rate of expansion, may be studied not only in the physical growth and business development, but also in the consequent changes in the social organization and in personality types. How far is the growth of the city, in its physical and technical aspects, matched by a natural but adequate readjustment in the social organization? What, for a city, is a normal rate of expansion, a rate of expansion with which controlled changes in the social organization might successfully keep pace?

SOCIAL ORGANIZATION AND DISORGANIZATION AS PROCESSES OF METABOLISM

These questions may best be answered, perhaps, by thinking of urban growth as a resultant of organization and disorganization analogous to the anabolic and katabolic processes of metabolism in the body. In what way are individuals incorporated into the life of a city? By what process does a person become an organic part of his society? The natural process of acquiring culture is by birth. A person is born into a family already adjusted to a social environment – in this case the modern city. The natural rate of increase of population most favorable for assimilation may then be taken as the excess of the birth-rate over the death-rate, but is this the normal rate of city growth? Certainly, modern cities have increased and are increasing in population at a far higher rate. However, the natural rate of growth may be used to measure the disturbances of metabolism caused by any excessive increase, as those which followed the great influx of southern Negroes into northern cities since the war. In a similar way all cities show deviations in composition by age and sex from a standard population such as that of Sweden, unaffected in recent years by any great emigration or immigration. Here again, marked variations, as any great excess of males over females, or of females over males, or in the proportion of children, or of grown men or women, are symptomatic of abnormalities in social metabolism.

Normally the processes of disorganization and organization may be thought of as in reciprocal relationship to each other, and as co-operating in a moving equilibrium of social order toward an end vaguely or definitely regarded as progressive. So far as disorganization points to reorganization and makes for more efficient adjustment, disorganization must be conceived not as pathological, but as normal. Disorganization as preliminary to reorganization of attitudes and conduct is almost invariably the lot of the newcomer to the city, and the discarding of the habitual, and often of what has been to him the moral, is not infrequently accompanied by sharp mental conflict and sense of personal loss. Oftener, perhaps, the change gives sooner or later a feeling of emancipation and an urge toward new goals.

In the expansion of the city a process of distribution takes place which sifts and sorts and relocates individuals and groups by residence and occupation. The resulting differentiation of the cosmopolitan American city into areas is typically all from one pattern, with only interesting minor modifications. Within the central business district or on an adjoining street is the "main stem" of "hobohemia," the teeming Rialto of the homeless migratory man of the Middle West. In the zone of deterioration encircling the central business section are always to be found the so-called "slums" and "bad lands," with their submerged regions of poverty, degradation, and disease, and their underworlds of crime and vice. Within a deteriorating area are rooming-house districts, the purgatory of "lost souls." Nearby is the Latin Quarter, where creative and rebellious spirits resort. The slums are also crowded to over flowing with immigrant colonies – the Ghetto, Little Sicily, Greek town, Chinatown – fascinatingly combining old world heritages and American adaptations. Wedging out from here is the Black Belt with its free and disorderly life. The area of deterioration, while essentially one of decay, of stationary or declining population, is also one of regeneration, as witness the mission, the settlement, the artists' colony, radical centers – all obsessed with the vision of a new and better world.

The next zone is also inhabited predominantly by factory and shop workers, but skilled and thrifty. This is an area of second immigrant settlement, generally of the second generation. It is the region of escape from the slum, the *Deutschland* of the aspiring Ghetto family. For *Deutschland* (literally "Germany") is the name given, half in envy, half in derision, to that region beyond the Ghetto where successful neighbors appear to be imitating German Jewish standards of living. But the inhabitant of this area in turn looks to the "Promised Land" beyond, to its residential hotels, its apartment-house region, its "satellite loops," and its "bright light" areas.

This differentiation into natural economic and cultural groupings gives form and character to the city. For segregation offers the group, and thereby the individuals who compose the group, a place and a role in the total organization of city life. Segregation limits development in certain directions, but releases it in others. These areas tend to accentuate certain traits, to attract and develop their kind of individuals, and so to become further differentiated.

The division of labor in the city likewise illustrates disorganization, reorganization and increasing differentiation. The immigrant from rural communities in Europe and America seldom brings with him economic skill of any great value in our industrial, commercial, or professional life. Yet interesting occupational selection has taken place by nationality, explainable more by racial temperament or circumstance than by old-world economic background as Irish policemen, Greek ice-cream parlors, Chinese laundries, Negro porters, Belgian janitors, etc.

The facts that in Chicago one million (996,589) individuals gainfully employed reported 509 occupations, and that over 1,000 men and women in *Who's Who* gave 116 different vocations give some notion of how in the city the minute differentiation of occupation "analyzes and sifts the population, separating and classifying the diverse elements." These figures also afford some intimation of the complexity and complication of the modern industrial mechanism and the intricate segregation and isolation of divergent economic groups. Interrelated with this economic division of labor is a corresponding division into social classes and into cultural and recreational groups. From this multiplicity of groups, with their different patterns of life, the person finds his congenial social world and – what is not feasible in the narrow confines of a village – may move and live in widely separated, and perchance conflicting, worlds. Personal disorganization may be but the failure to harmonize the canons of conduct of two divergent groups.

If the phenomena of expansion and metabolism indicate that a moderate degree of disorganization may and does facilitate social organization, they indicate as well that rapid urban expansion is accompanied by excessive increases in disease, crime, disorder, vice, insanity and suicide, rough indexes of social disorganization. But what are the indexes of the causes, rather than of the effects, of the disordered social metabolism of the city? The excess of the actual over the natural increase of population has already been suggested as a criterion. The significance of this increase consists in the immigration into a metropolitan city like New York and Chicago of tens of thousands of persons annually. Their invasion of the city has the effect of a tidal wave inundating first the immigrant colonies, the ports of first entry, dislodging thousands of inhabitants who overflow into the next zone, and so on and on until the momentum of the wave has spent its force on the last urban zone. The whole effect is to speed up expansion, to speed up industry, to speed up the "junking" process in the area of deterioration (II). These internal movements of the population become the more significant for study. What movement is going on in the city, and how may this movement be measured? It is easier, of course, to classify movement within the city than to measure it. There is the movement from residence to residence, change of occupation, labor turnover, movement to and from work, movement for recreation and adventure. This leads to the question: what is the significant aspect of movement for the study of the changes in city life? The answer to this question leads directly to the important distinction between movement and mobility.

MOBILITY AS THE PULSE OF THE COMMUNITY

Movement, per se, is not an evidence of change or of growth. In fact, movement may be a fixed and unchanging order of motion, designed to control a constant situation, as in routine movement. Movement that is significant for growth implies a change of movement in response to a new stimulus or situation. Change of movement of this type is called *mobility*. Movement of the nature of routine finds its typical expression in work. Change of movement, or mobility, is characteristically expressed in adventure. The great city, with its "bright lights," its emporiums of novelties and bargains, its palaces of amusement, its underworld of vice and crime, its risks of life and property from accident, robbery, and homicide, has become the region of the most intense degree of adventure and danger, excitement and thrill.

Mobility, it is evident, involves change, new experience, stimulation. Stimulation induces a response of the person to those objects in his environment which afford expression for his wishes. For the person, as for the physical organism, stimulation is essential to growth. Response to stimulation is wholesome so long as it is a correlated integral reaction of the entire personality. When the reaction is segmental, that is, detached from, and uncontrolled by, the organization of personality, it tends to become disorganizing or pathological. That is why stimulation for the sake of stimulation, as in the restless pursuit of pleasure, partakes of the nature of vice.

The mobility of city life, with its increase in the number and intensity of stimulations, tends inevitably to confuse and to demoralize the person. For an essential element in the mores and in personal morality is consistency, consistency of the type that is natural in the social control of the primary group. Where mobility is the greatest, and where in consequence primary controls break down completely, as in the zone of deterioration in the modern city, there develop areas of demoralization, of promiscuity, and of vice.

In our studies of the city it is found that areas of mobility are also the regions in which are found juvenile delinquency, boys' gangs, crime, poverty, wife desertion, divorce, abandoned infants, vice.

These concrete situations show why mobility is perhaps the best index of the state of metabolism of the city. Mobility may be thought of, in more than a fanciful sense, as the "pulse of the community." Like the pulse of the human body, it is a process which reflects and is indicative of all the changes that are taking place in the community, and which is susceptible of analysis into elements which may be stated numerically.

The elements entering into mobility may be classified under two main heads: (1) the state of mutability of the person, and (2) the number and kind of contacts or stimulations in his environment. The mutability of city populations varies with sex and age composition, and the degree of detachment of the person from the family and from other groups. All these factors may be expressed numerically. The new stimulations to which a population responds can be measured in terms of change of movement or of increasing contacts. Statistics on the movement of urban population may only measure routine, but an increase at a higher ratio than the increase of population measures mobility. In 1860 the horse-car lines of New York City carried about 50,000,000 passengers; in 1890 the trolley cars (and a few surviving horse-cars) transported about 500,000,000; in 1921, the elevated, subway, surface, and electric and steam suburban lines carried a total of more than 2,500,000,000 passengers. In Chicago the total annual rides per capita on the surface and elevated lines were 164 in 1890; 215 in 1900; 320 in 1910; and 338 in 1921. In addition, the rides per capita on steam and electric suburban lines almost doubled between 1916 (23) and 1921 (41), and the increasing use of the automobile must not be overlooked. For example, the number of automobiles in Illinois increased from 131,140 in 1915 to 833,920 in 1923.

Mobility may be measured not only by these changes of movement, but also by increase of contacts. While the increase of population of Chicago in 1912–22 was less than 25 percent (23.6 percent), the increase of letters delivered to Chicagoans was double that (49.6 percent) – from 693,048,196 to 1,038,007,854. In 1912 New York had 8.8 telephones; in 1922, 16.9 per 100 inhabitants. Boston had, in 1912, 10.1 telephones; ten years later, 19.5 telephones per 100 inhabitants. In the same decade the figures for Chicago increased from 12.3 to 21.6 per 100 population. But increase of the use of the telephone is probably more significant than increase in the number of telephones. The number of telephone calls in Chicago increased from 606,131,928 in 1914 to 944,010,586 in 1922, an increase of 55.7 percent, while the population increased only 13.4 percent.

Land values, since they reflect movement, afford one of the most sensitive indexes of mobility. The highest land values in Chicago are at the point of greatest mobility in the city, at the corner of State and Madison streets, in the Loop. A traffic count showed that at the rush period 31,000 people an hour, or 210,000 men and women in sixteen and one-half hours, passed the southwest corner. For over ten years land values in the Loop have been stationary but in the same time they have doubled, quadrupled and even sextupled in the strategic corners of the "satellite loops," an accurate index of the changes which have occurred. Our investigations so far seem to indicate that variations in land values, especially where correlated with differences in rents, offer perhaps the best single measure of mobility, and so of all the changes taking place in the expansion and growth of the city.

In general outline, I have attempted to present the point of view and methods of investigation which the department of sociology is employing in its studies in the growth of the city, namely, to describe urban expansion in terms of extension, succession, and concentration; to determine how expansion disturbs metabolism when disorganization is in excess of organization; and, finally, to define mobility and to propose it as a measure both of expansion and metabolism, susceptible to precise quantitative formulation, so that it may be regarded almost literally as the pulse of the community. In a way, this statement might serve as an introduction to any one of five or six research projects under way in the department. The project, however, in which I am directly engaged is an attempt to apply these methods of investigation to a cross-section of the city – to put this area, as it were, under the microscope, and so to study in more detail and with greater control and precision the processes which have been described here in the large. For this purpose the West Side Jewish community has been selected. This community includes the so-called "Ghetto," or area of first settlement, and Lawndale, the so-called "Deutschland," or area of second settlement. This area has certain obvious advantages for this study, from the standpoint of expansion, metabolism, and mobility. It exemplifies the tendency to expansion radially from the business center of the city. It is now relatively a homogeneous cultural group. Lawndale is itself an area in flux, with the tide of migrants still flowing in from the Ghetto and a constant egress to more desirable regions of the residential zone. In this area, too, it is also possible to study how the expected outcome of this high rate of mobility in social and personal disorganization is counteracted in large measure by the efficient communal organization of the Jewish community.

"The Los Angeles School of Urbanism: An Intellectual History"

Michael Dear

Editors' Introduction

In this provocative essay, University of California, Berkeley, professor of city and regional planning Michael Dear summarizes a radically different model of the logic behind the spatial structure of regions than Ernest Burgess's classical model. Dear's model reflects postmodernist thinking, particularly ideas of an important group of Southern California intellectuals identified as the Los Angeles school of urbanism. One good way to think about the internal structure of city-regions is by contrasting the Burgess model (representative of the Chicago school of sociology) and the LA school model (as described by Dear).

"Schools" of thought are sometimes defined when a group of individuals at some place and time develop a reasonably consistent body of ideas that is different enough from other ideas at the time that it qualifies as something special. Members of the "Dutch school" of painters in the Netherlands in the fifteenth and sixteenth centuries, for example, worked in a distinct style that set them apart from other artists of the time. Anyone comparing a Dutch landscape painting from that school would immediately see that it is dramatically different from, for example, a landscape painted by a painter in the nineteenth-century French Impressionist school of art. Within the Dutch school, Rembrandt's style was different from Johannes Vermeer's and while they were both impressionists, Claude Monet's painting style is different from Pierre-August Renoir's style.

Similarly, the Los Angeles school of urbanism consists of the work of a group of intellectual mavericks with different approaches, but who share enough in common that they self-identify and are identified by others as a distinct school of urbanist thought. Dear considers their point of view radically different from earlier modernist points of view.

The LA school of urbanism includes neo-Marxist geographers, left-wing urban sociologists, postmodernist architectural critics, labor historians and other Southern California intellectuals. Mike Davis, the author of "Fortress L.A." (p. 195), is a member of the LA school and coined the term "LA school".

LA school members consider themselves postmodernists. They distinguish their approach to cities from modernists like Le Corbusier (p. 336). In the selection that follows, Dear characterizes the Burgess concentric zone model of the internal structure of the city as in the modernist tradition.

According to Dear, key differences between modernist and postmodernist views as represented by the Burgess and LA school models are that:

- Burgess and other modernists view city-regions as coherent regional systems in which the central business district (CBD) organizes the rest of city space and the metropolitan hinterland beyond the formal city limits. In contrast postmodernist members of the LA school view city-regions as fragmented, with different areas influenced largely by global, rather than purely local, forces. LA school theorists argue that CBDs no longer act as centers defining the city-region. Rather, they argue that urban peripheries are organizing what remains of the center.

- According to Dear, Burgess believed that the personal choices of individuals shape overall urban conditions. In contrast Dear and LA theorists believe that great global structural forces determine metropolitan spatial structure. They believe that global corporate-dominated connectivity, is balancing or even offsetting individual-centered agency in urban processes.
- Burgess and his Chicago school colleagues held an essentially teleological view of urban evolution. They believed cities were evolving to ever more advanced and modern levels. The LA school questions that assumption. They see the evolution of cities as a nonlinear, chaotic process. They see many pathological aspects of postmodern LA that make it arguably a much less advanced city than many earlier cities. LA school members generally share Mike Davis's dystopianism (p. 195).
- Postmodernist concepts include the World City (a few urban centers controlling the world economy), the Dual City (increasingly polarized by race, class, income, and gender), the Hybrid City (characterized by new hybrid communities), and Cybercity (in which digital connectivity shapes all aspects of urban life).

The most compelling metaphor Dear and his colleague Steven Flusty use to describe the LA school paradigm is "Keno capitalism". Keno is a game in which a square in a rectangular grid is selected by chance. That event triggers activity in squares closest to the selected square. Different random squares on the board may be in play at any time. Squares which are furthest from the selected square(s) have little or activity. Players with "winning" squares, selected by chance win; players with squares furthest from the selected squares lose the game. Substituting a real-world land parcel for the Keno square, Dear and Flusty argue that in Los Angeles and other of the world's most dynamic metropolitan regions global development consortia choose land parcels for development nearly at random and inject a huge amount of capital to develop mega projects there. Often they might as well choose one parcel as another. Land values on and right around the chosen parcels skyrocket and the selection of the parcels touches off a whole string of development activities nearby. Other parcels in the metropolitan region – often in the periphery rather than the city itself – are also being selected, and frantic development activity occurs close to them as well. In between the Keno-like winning parcels little development activity occurs, or neighborhoods decline. A metropolitan region experiencing this form of development becomes, in Dear's words, "a noncontiguous collage of parcelized, consumption-oriented landscapes linked only by the (dis) information superhighway". An aerial photograph of such a region would look like a Keno board with apparently random spots of intense development here and there and little activity elsewhere. Dear notes that existing (modernist) forms are not completely irrelevant. Their past influence is discernible in postmodern landscapes and they continue to modestly influence the emerging spaces of postmodernity.

Postmodern urbanism as described by the LA school is characterized by edge cities, "privatopias" of homeowners' associations, "minoritization" (where the majority of the population is the non-white "other"), theme park environments, fortification, "containment centers" (prisons), and "technopoles" (geographical loci of high-tech production). Neo-Marxists members of the LA school agree with Engels (p. 46) that urban development is driven by capitalistic self-interest. Now that the scale of urbanization causes worldwide problems such as the loss of sustainability and global climate change described in Part Six on Urban Planning Theory and Practice, the consequences of Keno capitalism are potentially catastrophic.

Contrast Dear's postmodern perspective to Le Corbusier's modernist views (p. 336). Compare the first section of Nigel Taylor's article on planning theory (p. 386), which describes modernist planning theories such as the rational planning model, with the latter part of the selection which describes postmodern planning theories, such as Leonie Sandercock's concept of Cosmopolis. The rational planning model would arguably work much better in early-twentieth-century Los Angeles where development was occurring from the city center outward in predicable ways than in early-twenty-first-century Los Angeles, as described by Dear, Mike Davis (p. 195), and other LA school theorists. An approach like Sandercock's that assumes development is diverse, unpredictable, and largely influenced by exogenous world forces might be more appropriate in contemporary Los Angeles.

Michael Dear joined the faculty of the University of California, Berkeley, Department of City and Regional Planning in 2009. He is an honorary professor at the Bartlett School of Planning, University College, London. Prior to his current appointment he was a professor in the departments of geography and policy, planning, and development at the University of Southern California. Dear has degrees in geography, urban planning, and regional science. His research interests include homelessness, postmodern theory, Los Angeles, comparative urbanism, and the cultural geography of the US–Mexico borderlands. He was a founding editor of the journal *Society and Space*.

Dear's books include *From Chicago to L.A.: Making Sense of Urban Theory* (Thousand Oaks, CA: Sage, 2001), *The Postmodern Urban Condition* (Oxford: Blackwell, 2001), *Rethinking Los Angeles* (Thousand Oaks, CA: Sage, 1996), and *The Spaces of Postmodernity* (London: Wiley-Blackwell, 2000).

Other writings on postmodernist urban theory include Charles Jencks, *Heteropolis: Los Angeles, the Riots, and the Strange Beauty of Heteroarchitecture* (New York: St. Martin's, 1993), Allen J. Scott, *Technopolis: High-Technology Development and Regional Development in Southern California* (Berkeley, CA: University of California Press, 1993), Allen J. Scott and Edward Soja (eds), *The City: Los Angeles and Urban Theory at the End of the Twentieth Century* (Berkeley, CA: University of California Press, 1996), and Edward W. Soja, *Postmodern Geographies: The Reassertion of Space in Critical Social Theory* (New York: Verso, 1989).

The basic primer of the Chicago School was *The City*. Originally published in 1925, the book retains a tremendous vitality far beyond its interest as a historical document. I regard the book as emblematic of a modernist analytical paradigm that remained popular for most of the 20th century. Its assumptions included:

- a modernist view of the city as a unified whole, i.e., a coherent regional system in which the center organizes its hinterland;
- an individual-centered understanding of the urban condition; urban process in *The City* is typically grounded in the individual subjectivities of urbanites, their personal choices ultimately explaining the overall urban condition, including spatial structure, crime, poverty, and racism; and
- a linear evolutionist paradigm, in which processes lead from tradition to modernity, from primitive to advanced, from community to society, and so on.

There may be other important assumptions of the Chicago School, as represented in *The City*, that are not listed here. Finding them and identifying what is right or wrong about them is one of the tasks at hand, rather than excoriating the book's contributors for not accurately foreseeing some distant future.

The most enduring of the Chicago School models was the zonal or concentric ring theory, an account of the evolution of differentiated urban social areas by E. W. Burgess. Based on assumptions that included a uniform land surface, universal access to a single-centered city, free competition for space, and the notion that development would take place outward from a central core, Burgess concluded that the city would tend to form a series of concentric zones. The main ecological metaphors invoked to describe this dynamic were invasion, succession, and segregation, by which populations gradually filtered outward from the center as their status and level of assimilation progressed. The model was predicated on continuing high levels of immigration to the city.

At the core of Burgess' schema was the Central Business District (CBD), which was surrounded by a transitional zone, where older private houses were being converted to offices and light industry, or subdivided to form smaller dwelling units. This was the principal area to which new immigrants were attracted and it included areas of vice and unstable or mobile social groups. The transitional zone was succeed by a zone of working-men's homes, which included some of the city's oldest residential buildings inhabited by stable social groups. Beyond this, newer and larger dwellings were to be found, occupied by the middle classes. Finally, the commuters' zone was separate from the continuous built-up area of the city, where much of the zone's population was employed. Burgess' model was a broad generalization, and not intended to be taken too literally. He anticipated, for instance, that his schema would apply only in the absence of "opposing factors" such as local topography (in the case of Chicago, Lake Michigan). He also anticipated considerable internal variation within the different zones.

Other urbanists subsequently noted the tendency for cities to grow in star-shaped rather than concentric form, along highways that radiate from a center with contrasting land uses in the interstices. This observation gave rise to a sector theory of urban structure, an idea advanced in the late 1930s by Homer Hoyt (1933, 1939), who observed that once variations arose in land uses near the city center, they tended to persist as the city expanded. Distinctive sectors thus grew out from the CBD, often organized along major highways. Hoyt emphasized that "nonrational" factors could alter urban form, as when skillful promotion influenced the direction of speculative development. He also understood that older buildings could still reflect a concentric ring structure, and that sectors may not be internally homogeneous at one point in time.

The complexities of real-world urbanism were further taken up in the multiple nuclei theory of Chauncey Harris and Edward Ullman (1945). They proposed that cities have a cellular structure in which land uses develop around multiple growth-nuclei within the metropolis as a consequence of accessibility-induced variations in the land-rent surface and agglomeration (dis)economics. Harris and Ullman also allowed that real-world urban structure is determined by broader social and economic forces, the influence of history, and international influences. But whatever the precise reasons for their origin, once nuclei have been established, general growth forces reinforce their preexisting patterns.

Much of the urban research agenda of the 20th century has been predicated on the precepts of the concentric zone, sector, and multiple nuclei theories of urban structure. Their influences can be seen directly in factorial ecologies of intra-urban structure, land-rent models, studies of urban economies and diseconomies of scale, and designs for ideal cities and neighborhoods. The specific and persistent popularity of the Chicago concentric ring model is harder to explain, however, given the proliferation of evidence in support of alternative theories. The most likely reasons for its endurance are related to its beguiling simplicity and the enormous volume of publications produced by adherents of the Chicago School . . . .

In the final chapter of *The City*, Louis Wirth (1925) provided a magisterial review of the field of urban sociology, titled (with deceptive simplicity, and astonishing self-effacement) "A Bibliography of the Urban Community." But what Wirth did in this chapter, in a remarkably prescient way, was to summarize the fundamental premises of the Chicago School, and to isolate two fundamental features of the urban condition that were to rise to prominence at the beginning of the 21st century. Specifically, Wirth established that the city lies at the center of, and provides the organizational logic for, a complex regional hinterland based on trade:

> Far from being an arbitrary clustering of people and buildings, the city is the nucleus of a wider zone of activity from which it draws its resources and over which it exerts its influence. The city and its hinterland represent two phases of the same mechanism which may be analyzed from various points of view . . . .

He also noted that the development of satellite cities is characteristic of the latest phases of city growth, and that the location of such satellites can exert a determining influence upon the direction of growth:

> One of the latest phases of city growth is the development of satellite cities. These are generally industrial units growing up outside of the boundaries of the administrative city, which, however, are dependent upon the city proper for their existence. Often they become incorporated into the city proper after the city has inundated them, and thus lose their identity. The location of such satellites may exert a determining influence upon the direction of the city's growth. These satellites become culturally a part of the city long before they are actually incorporated into it . . . .

Wirth further observed that modern communications have transformed the world into a single mechanism, where the global and the local intersect decisively and continuously:

> With the advent of modern methods of communication the whole world has been transformed into a single mechanism of which a country or a city is merely an integral part. The specialization of function, which has been a concomitant of city growth, has created a state of interdependence of world-wide proportions. Fluctuations in the price of wheat on the Chicago Grain Exchange reverberate to the remotest part of the globe, and a new invention anywhere will soon have to be reckoned with at points far from its origin. The city has become a highly sensitive unit in this complex mechanism, and in turn acts as a transmitter of such stimulation as it receives to a local area. This is a true of economic and political as it is of social and intellectual life. . . .

And there, in a sense, you have it. From a few, relatively humble first steps, we gaze out over the abyss – the yawning gap of an intellectual fault line separating Chicago from Los Angeles. In a few short paragraphs, Wirth anticipated the pivotal moments that characterize Chicago-style urbanism – those primitives that eventually will separate it from an LA-style urbanism. He effectively foreshadowed avant la lettre the shift from what I term a "modern" to a "postmodern" city, and, in so doing, the necessity of the transition from the Chicago to the LA School. For it is no longer the center that organizes the urban hinterlands, but the hinterlands that determine what remains of the center. The imperatives of fragmentation have become the principal dynamic in contemporary cities; the 21st century's emerging world cities (including LA) are ground-zero loci in a communications-driven globalizing political economy.

The shift toward an LA School may be regarded as a move away from modernist perspectives on the city (à la Chicago School) to a postmodern view of urban process. We are all by now aware that the tenets of modernist thought have been undermined, discredited; in their place, a multiplicity of new ways of knowing have been substituted. Analogously, in postmodern cities, the logics of previous urbanisms have evaporated; and, in the absence of a single new imperative, multiple (ir)rationalities clamor to fill the vacuum. The LA School is distinguishable from the Chicago precepts (as noted above) by the following counterpropositions:

- Traditional concepts of urban form imagine the city organized around a central core; in a revised theory, the urban peripheries are organizing what remains of the center.
- A global, corporate-dominated connectivity is balancing, even offsetting, individual-centered agency in urban processes.
- A linear evolutionist urban paradigm has been usurped by a nonlinear, chaotic process that includes pathological forms such as common-interest developments (CIDs), and life-threatening environmental degradation (e.g. global warming).

In empirical terms, the urban dynamics driving these tendencies are by now well known. They include: *World City*: the emergence of a relatively few centers of command and control in a globalizing economy; *Dual City*: an increasing social polarization, i.e., the increasing gap between rich and poor, between nations, between the powerful and the powerless, between different ethnic, racial, and religious groupings, and between genders; *Hybrid City*: the ubiquity of fragmentation both in material and cognitive life, including the collapse of conventional communities, and the rise of new cultural categories and spaces, including especially cultural hybrids; and *Cybercity*: the challenges of the information age, especially the seemingly ubiquitous capacity of connectivity to supplant the constraints of place.

"Keno capitalism" is the synoptic term that Steven Flusty and I have adopted to describe the spatial manifestations that are consequent upon the (postmodern) urban condition implied by these assumptions. Urbanization is occurring on a quasi-random field of opportunities in which each space is (in principle) equally available through its connection with the information superhighway. . . . Capital touches down as if by chance on a parcel of land, ignoring the opportunities on intervening lots, thus sparking the development process. The relationship between development of one parcel and nondevelopment of another is a disjointed, seemingly unrelated affair. While not truly a random process, it is evident that the traditional, center-driven agglomeration economies that have guided urban development in the past no longer generally apply. Conventional city form, Chicago-style, is sacrificed in favor of a noncontiguous collage of parcelized, consumption-oriented landscapes devoid of conventional centers yet wired into electronic propinquity and nominally unified by the mythologies of the (dis)information superhighway. In such landscapes, "city centers" become almost an externality of fragmented urbanism; they are frequently grafted onto the landscape as a (much later) afterthought by developers and politicians concerned with identity and tradition. Conventions of "suburbanization" are also redundant in an urban process that bears no relationship to a core-related decentralization.

I am insisting on the term "postmodern" as a vehicle for examining LA urbanism for a number of reasons, even though many protagonists in the debates surrounding the LA School have explicitly distanced themselves from the precepts of postmodernism. I have long understood postmodernism as a concept that embraces three principal referents:

- A series of distinctive cultural and stylistic practices that are in and of themselves intrinsically interesting;
- The totality of such practices, viewed as a cultural ensemble characteristic of the contemporary epoch

of capitalism (often referred to as postmodernity); and

- A set philosophical and methodological discourses antagonistic to the precepts of Enlightenment thought, most particularly the hegemony of any single intellectual persuasion.

Implicit in each of these approaches is the notion of a "radical break," i.e., a discontinuity between past and present political, sociocultural and economic trends. My working hypothesis is that there is sufficient evidence to support the notion that we are witnessing a radical break in each of these three categories. This is the fundamental promise of the revolution prefigured by the LA School; this is why it is so revolutionary in its recapitulation of urban theory.

The localization (sometimes literally the concretization) of these diverse dynamics is creating the emerging time-space fabric of a postmodern society. This is not to suggest that existing (modernist) rationalities have been obliterated from the urban landscape or from our mind-sets; on the contrary, they persist as palimpsests of earlier logics, and continue to influence the emerging spaces of postmodernity. For instance, they are presently serving to consolidate the power of existing place-based centers of communication technologies, even as such technologies are supposed to liberate development from the constraints of place. However, newer urban places, such as LA, are being created by different intentionalities, just as older places such as Chicago are being overlain by the altered intentionalities of postmodernity. Nor am I suggesting that earlier theoretical logics have been (or should be) entirely usurped. For instance, in his revision of the Chicago School, Andrew Abbott . . . claimed that the "variables paradigm" of quantitative sociology has been exhausted, and that the "cornerstone of the Chicago vision was location" – points of departure that I regard as totally consistent with the time-space obsessions of the LA School of postmodern urbanism. Another example of overlap between modern and postmodern in current urban sociology is Michael Peter Smith's evocation of a transnational urbanism . . . .

REFERENCES

Abbott, A. (1999) *Department and Discipline: Chicago Sociology at One Hundred*. Chicago, IL: University of Chicago Press.

Burgess, E.W. (1925) The Growth of the City. In R.E. Park, E.W. Burgess, and R.McKenzie (eds), *The City: Suggestions of Investigation of Human Behavior in the Urban Environment*. Chicago, IL: University of Chicago Press, pp. 47–62.

Dear, M. and Flusty, S. (1998) Postmodern Urbanism. *Annals of the Association of American Geographers*, 88(1): 50–72.

Harris, C.D. and Ullman, E.L. (1945) The Nature of Cities. *Annals of the American Academy of Political and Social Science*, 242: 7–17.

Hoyt, H. (1933) *One Hundred Years of Land Values in Chicago*. Chicago, IL: University of Chicago Press.

Hoyt, H. (1939) *The Structure and Growth of Residential Neighborhoods in American Cities*. Washington, DC: United States Federal Housing Administration.

Smith, M.P. (2001) *Transnational Urbanism: Locating Globalization*. Oxford: Blackwell.

Wirth, L. (1925) A Bibliography of the Urban Community. In R.E. Park, E.W. Burgess, and R. McKenzie (eds), *The City: Suggestions of Investigation of Human Behavior in the Urban Environment*. Chicago, IL: University of Chicago Press, pp. 161–228.

"What Happened to Gender Relations on the Way from Chicago to Los Angeles?"

City and Community (2002)

Daphne Spain

Editors' Introduction

The important and controversial theories by Burgess (p. 161) and other members of the University of Chicago school of sociology and Dear (p. 170) and others in the LA school of urbanism propose many different reasons why cities are organized as they are. In the first selection in this part – written in 1925 – University of Chicago sociology professor Ernest W. Burgess proposes his "concentric zone" model for the logic underlying the internal structure of cities based on his observations of Chicago (p. 161). In the second selection, University of California, Berkeley, urban planning professor Michael Dear proposes a radically different explanation of the logic underlying the spatial organization of metropolitan regions today based on his observations of the Los Angeles region (p. 170). In the following essay University of Virginia urban planning professor Daphne Spain takes both schools to task for neglecting the role of women in their respective models. Spain argues that a gender perspective can make a valuable contribution to this debate (and by extension to all of urban theory). She proposes adding changing gender relations to the list of reasons behind the transformation of urban space from the modernist monocentric city to the postmodern polycentric metropolis.

The Chicago school faculty was almost entirely male and nearly all core members of the LA school are men. Not only were women not included in the two schools, but also Spain notes that both schools ignored women who were working on the same issues in the same city at the same time. For example, Ernest Burgess dismissed the work of his Chicago contemporary Jane Addams, the director of Hull House in one of Chicago's poorest communities. Addams knew first hand the needs of poor immigrant women (as well as men) from years providing social services to them. Jane Addams published articles in scholarly sociological journals and is now recognized as a seminal thinker. But Burgess and his colleagues dismissed her work as merely practical, rather than theoretical. Spain notes a similar failure of Los Angeles school intellectuals to reference the important work of their female colleagues.

When Park and Burgess were writing about Chicago at the beginning of the twentieth century, middle-class women were expected to stay home while their husbands or fathers went to work. As late as 1940 less than one-quarter of all women were in the labor force. But when Professor Spain wrote her article in 2001, 60 percent of all women were in the labor force, and it is the rare woman who does not work at some point in her life.

Spain's gender perspective provides important insights into the way in which women's entry into the labor market affects the spatial structure of cities and urban processes. The huge increase in female employment contributes to sprawl and the formation of edge cities by increasing the demand for vehicles. The transfer of domestic services from the home to the public sphere has exacerbated suburban sprawl as the demand for housing has increased and most new construction occurs at the urban periphery. More dual-wage-earner families are able to purchase low-density single-family homes on large lots. The majority of women drive to work alone. In addition to their commutes to work, women are more likely to drive to take children to school, take care of elderly parents, and shop. As women's time is increasingly occupied in paid employment, important services once performed by women at home are now performed by fast food restaurants, childcare centers, and assisted living institutions for elderly people. People have to eat several times a day. "Family" restaurants and fast food franchises have proliferated since the 1990s as a substitute for the individual family kitchen and dining room. Every strip development leading into every American city of any size has its own assortment of food outlets staffed by immigrants, teenagers, or retirees. While some private market-oriented theorists like Robert Bruegmann (p. 211) accept or even favor sprawl development, most theorists deplore this pattern.

Spain's gender perspective provides interesting contrasts to Mike Davis's theories (p. 195). As Spain points out, it helps explain the increase in gated communities ("privatopias"), which has occurred since the 1980s. Davis blames gated communities on racial and class prejudice and fear of crime. Spain notes that the increase correlates with the increase in women's labor force participation. Few middle-class families sought gated living when women were home all day to provide informal security, but as more than 60 percent of women now work outside the home, more and more households seek the security of a gated community with paid security personnel. Mike Davis focuses on mean streets and dangerous communities as the primary cause of fortress LA and what he calls, with characteristic bombast, "the carceral city." Davis identifies one public housing as a fenced-off war zone requiring identification for entry. Viewing this issue through her gender perspective, Spain reminds us that public housing is occupied predominantly by women. Davis identifies prisons as "containment centers" in the urban landscape, the masculine counterpart to public housing. In Spain's view, a gendered view of containment centers might also include daycare centers and retirement homes as places that hold economically marginal populations under supervision.

Female householders have created homes from which the man is (sometimes, Spain notes, only technically) absent. The growth in female householders has changed the metropolitan landscape primarily by creating a demand for more and different housing units. Women seeking to form their own households, for example, often need help getting established. Looking at the needs of residents in a city, an urban planner sensitive to Spain's gender perspective might be aware of the need for temporary shelters for female victims of domestic violence that a planner looking at data and formulating plans without thinking about gender would simply not be aware of. A planner for a redevelopment agency, public housing authority, or nonprofit housing development corporation using a gender perspective might see and advocate for buildings or units within buildings designed specifically to accommodate single women or female-headed households. Dolores Hayden, a feminist architect and planner who has written extensively about alternative building designs and plans that better accommodate women, has documented a variety of innovative designs of this type. Looking at low-rise residential development through a gender perspective, Claire Cooper Marcus and Wendy Sarkissian describe a variety of common-sense design principles that women residents themselves feel make housing fit their needs. Arranging kitchen areas where women can watch young children while they prepare meals, for example, is a very helpful design principle easily overlooked by architects and planners who do not consider gender issues.

In summary, Spain proposes that theorizing about gender relations and urban structure are similar. At the time Burgess was writing women's natural place was seen as at the center of the home. Burgess perceived the city as a centered organism around which various functions were rationally organized. Both those images are outdated. Women now fill a variety of roles both inside and outside the home, and the metropolitan area has become the site of scattered activities. What happened to gender relations on the way from Chicago to Los Angeles? According to Spain, the same thing that happened to urban form. They became less predictably centered and more diverse.

Daphne Spain is James M. Page Professor and chair of the Department of Urban and Environmental Planning in the School of Architecture at the University of Virginia. She is interested in the relationship between the built environment and social structure, with an emphasis on gender. One of her long-term research interests is the way in which groups of women change the urban environment. She is a member of the governing board of the Society for American Regional and Planning History, and a member of the editorial boards of the *Journal of the American Planning Association* and the *Journal of Urban Affairs*. She has received research grants from the Russell Sage Foundation and the Graham Foundation for Advanced Studies in the Fine Arts. Spain is currently working on a project titled "In the Spirit of Jane Addams: Moral Crusades and the American City," in which she argues that social movements shaped by moral values have an impact on urban form separate from the political economy.

Spain's books include *How Women Saved the City* (Minneapolis, MN: University of Minnesota Press, 2001), *Gendered Spaces* (Chapel Hill, NC: University of North Carolina Press, 1992), and *Balancing Act: Motherhood, Marriage and Employment among American Women*, with Suzanne Bianchi (New York: Russell Sage Foundation, 1996).

Other books on women in cities and city planning include Susan Fainstein and Lisa J. Servon (eds), *Gender and Planning: A Reader* (New Brunswick, NJ: Rutgers University Press, 2005), Dolores Hayden, *Redesigning the American Dream: Gender, Housing, and Family Life* (New York: Norton, 2002) and *The Grand Domestic Revolution: A History of Feminist Designs for American Homes, Neighborhoods and Cities* (Cambridge, MA: MIT Press, 1982), and Catherine Stimpson, Elsa Dixler, Martha Nelson, and Kathryn Yatrakis (eds), *Women and the American City* (Chicago, IL: University of Chicago Press, 1981).

■

. . . [T]he Chicago human ecologists described the monocentric city as an organism driven by population invasion and succession, while the Los Angeles postmodernists interpret globalization and economic restructuring as forces shaping the contemporary metropolis. In the intervening years numerous theories focused on transportation and communication technology, cultural practices, the political economy, growth coalitions, and public-private regimes as the key processes driving urban development. Curiously missing from this list of explanations, however, is the role of gender relations. The purpose of this essay is to bring the issue of gender into the debate about urban theory.

Neither the Chicago School at the beginning of the 20th century nor the Los Angeles School at its end adequately incorporated gender relations into theories of urban structure. Yet women's options in 1900 centered around the home, while their options in 2000 incorporated the workplace as well. The "walking city" of urban nostalgia still existed after home and work were separated *for men*. Only when women began to leave the home as well (in conjunction with the advent of the automobile) did the real spatial revolution begin.

World War II marked a turning point in the transformation of the monocentric industrial city into the polycentric informational metropolis. Central cities typically experienced growth before the War and declined thereafter. World War II also signaled the beginning of the "third period of crisis-generated urban restructuring". Soon thereafter, women's ability to achieve economic independence increased dramatically.

The subsequent restructuring of power within the home was surely as powerful an agent of urban change as the global economy. Indeed, the social movement for women's equality in industrialized nations has been called "the most important revolution because it goes to the roots of society and to the heart of who we are". Such a movement cannot change society without changing its cities as well.

A GENDER PERSPECTIVE

Gender relations are determined by women's status, which often responds to demographic changes. "Gender relations" refer to the beliefs, expectations, and behavior that characterize interactions between women and men. Traditional gender relations in the U.S. made women economically dependent on men because men engaged in paid labor while women performed unpaid work and bore primary responsibility for childcare. When Park and Burgess were writing about Chicago at the beginning of the 20th century, for example, middle-class women were expected to stay home while their husbands or fathers went to the office (Park, Burgess and McKenzie 1925). That is what they did; until 1940 less than one-quarter of all women were in the labor force. Many poor and minority women were employed in factories or as domestics, of course, but the ideal of separate spheres prevailed. Now, with 60 percent of all women in the labor force, it is the rare woman who does *not* work outside the home. For all races and ethnicities, the change in women's ability to earn a living affected gender relations by granting women greater economic power within, and outside, families.

Feminist scholars have long recognized the spatial consequences of gender relations for cities. . . . The separation between gender issues and urban theory is nothing new. Its seeds were sown nearly one hundred years ago in Chicago.

FROM CHICAGO TO LOS ANGELES

Gender relations at the beginning of the 20th century idealized separate spheres in which wives maintained the home and family while men earned a living. Domestic architecture reinforced these stereotypes by designating separate rooms for feminine and masculine activities. A woman's status was determined largely by whom she married. Relatively ineffective contraception made for large families, high maternal mortality, and short life expectancy. Few women attended college or earned professional degrees, and the one-fifth of women who were in the labor force in 1900 were typically unmarried, low-paid immigrants and African Americans. As a group, then, women's potential for economic independence was relatively low. Their options centered primarily around the home.

Some women were exceptions to this profile. They lived in cities, away from their families while they worked for wages, and they publicly demonstrated for the vote. A small minority of college-educated women created their own profession of settlement work, a combination of social work and progressive urban reform. The most notable settlement worker of all, Jane Addams, lived in Chicago's Hull House at the same time Robert Park and Ernest Burgess were developing their urban theories [p. 161]. Addams and Julia Lathrop documented deplorable conditions among immigrants in *Hull-House Maps and Papers* (1895). Yet Burgess considered their work only "the second stage in the trend of neighborhood work toward a scientific basis." Jane Addams published in sociological journals and her contemporaries in the University of Chicago's School of Social Service Administration wrote extensively about housing reform. The department of sociology dismissed their work, however, defining it as practical rather than theoretical.

Subsequent research has revealed a Chicago terrain invisible to Park and Burgess. Hull House and other settlements established public baths, playgrounds, kitchens, libraries, and kindergartens in the midst of Burgess's zone of transition. Boarding homes for "women adrift", YWCA-sponsored residences and vocational schools, and Catholic shelters for women and girls occupied the same landscape. But with the exception of the taxi-dance hall, where male patrons bought tickets to dance with women, members of the Chicago School virtually ignored gendered aspects of the city.

They could have learned something from Jane Addams. Her memoirs, published in 1910 as *Twenty Years at Hull House*, included astute observations about the impact of immigration on the city (Addams 1910). Addams and her colleagues provided care for children whose mothers worked in factories, organized women to demand better garbage disposal and street cleaning, taught adults how to speak English, and sponsored festivals celebrating ethnic heritage. Hull House met so many needs that it grew from an individual residence to an entire city block. Eventually within its walls were a gymnasium, nursery, music school, coffee house, theater, and rooms for working girls (Spain 2001). In the midst of the mundane, Addams recognized the sociological importance of her endeavor:

> The Settlement . . . is an experimental effort to aid in the solution of the social and industrial problems which are engendered by the modern conditions of life in a great city. It insists that these problems are not confined to any one portion of a city. It is an attempt to relieve, at the same time, the over-accumulation at one end of society and the destitution at the other.

This sounds like theory combined with practice, or praxis, in Marxist terms. In fact, Jane Addams and other settlement workers were decidedly leftist politically, which may be one reason their ideas failed to gain currency with members of the Chicago School.

[. . .]

Edward Soja (2000) and Michael Dear (2000), among others . . . propose that the absence of a central urban core is indicative of a fractured postmodern society. As society has become more fragmented by racial, ethnic, and gender diversity, the metropolis has assumed the form of a crazy quilt lacking a central focus. Los Angeles has eclipsed Chicago as the prototypical American city. Postmodern urban theory discards the human ecological models of the Chicago School, along with its positivist methodology, in favor of a philosophical, subjective interpretation of cities. Where Chicago sociologists saw the cooperation and benign competition characteristic of the industrial assembly line, the L.A. School sees the conflict and chaos associated with mobile capital and labor. According to Steven Flusty and Michael Dear, post-modern urbanism is characterized by edge cities, "privatopias" of homeowners' associations, "minoritization" (where the majority of the population is the non-white "other"), theme park environments, fortification, and "technopoles" (geographical loci of high-tech production). "Containment centers" (prisons) promote the image of the carceral city. Flusty and Dear invoke a gaming board metaphor they call "Keno capitalism" to describe a seemingly random pattern of development (Flusty and Dear 1999). They conclude that "conventional city form, Chicago style, is sacrificed in favor of a noncontiguous collage of parcelized, consumption-oriented landscapes devoid of conventional centers. . .". The processes accounting for all these changes include economic restructuring, globalization, and environmental politics.

Like Park and Burgess, Flusty and Dear could have learned something from women working in the same city at the same time they were developing their post-modern perspective. Architectural historian Dolores Hayden and urban planner Jacqueline Leavitt, both then teaching at UCLA, recognized the implications of the contemporary Women's Movement for gender and the city (Hayden 1980, 1981, 1984; Leavitt 1980). They wrote about space and gender, and they also engaged in the life of Los Angeles, as Jane Addams had in Chicago. Hayden was active in creating the Los Angeles Woman's Building in 1973, the same year in which David Harvey published *Social Justice and the City* (Harvey 1973). The Woman's Building was founded to provide "a social and physical place in the public world in which women can re-evaluate and re-create their gender identity, crossing boundaries of age, race, class, or ethnic origin".

While involved with the Woman's Building and other local projects, Hayden was publishing as well. In a 1980 essay titled "What would a non-sexist city be like?" (Hayden 1980), she advocated a Homemakers Organization for a More Egalitarian Society (HOMES). HOMES was a program through which existing suburban blocks of single-family houses could be modified to create accessory apartments, laundries, day care centers, and collective open space. Her later work dealt specifically with the mismatch between suburban housing built after World War II and women's changing status. Hayden's most recent book, *The Power of Place* (Hayden 1995), documents how she and others restored the history of women and minorities to Los Angeles's urban landscape.

UCLA professor Jacqueline Leavitt was a pioneer in the field of planning and gender. She challenged the gender bias in urban planning in the early 1980s, citing the small number of female planning professionals (Leavitt 1980). The lack of affordable housing for low-income women was also one of her priorities. For a national competition, Leavitt worked with architect Troy West to design cooperative housing for the elderly and single mothers. Some of Leavitt's most important research documented how women public housing residents in Los Angeles acquired the skills to make their spaces safer.

What did Chicago in 1900 have in common with Los Angeles in 2000 besides a disconnect between men and women studying the same city? Demographically quite a lot. Both cities were magnets for the major international immigration streams of their era. Immigrants moved through successive zones in Chicago, whereas they form a "heteropolis" in Los Angeles. Both cities attracted significant numbers of African Americans. "Race riots" in Chicago's Black Belt became "civil disturbances" in L.A.'s Watts. Both cities are stages on which the important issues of minority ethnic and racial status have been dramatized. In respect to women's status, though, Chicago and Los Angeles are a century apart. Women were still fighting for the franchise in 1900; by 2000 they could control their own fertility as well as vote. This crucial difference has implications for urban form.

Chicago in 1900 and Los Angeles in 2000 differed on at least four spatial dimensions: the presence of one center versus two or more; the location of activities; the level of density; and the direction of development. The industrial city a century ago had one Central Business District, mixed land uses that juxtaposed slaughterhouses and tenements, high population density, and the vertical profile of smokestacks and skyscrapers. In contrast, the contemporary informational metropolis consists of multiple centers, single-use zoning, low density, and a strong horizontal axis (see Table 1 and Figure 1). Most women's lives now include the home and workplace, which are separated by low-density, single-use zoning that contributes to suburban sprawl. As women have become more economically independent, their activities have both shaped and reflected the contemporary metropolis.

Figure 1 is an oversimplification to which there are obvious exceptions. Yet it serves well enough to

	Circa 1900	Circa 2000
Prototype	Industrial city	Informational metropolis
Number of centers	One	Two or more
Location of activities	Mixed	Separated
Density of population	High	Low
Direction of development	Vertical	Horizontal

Table 1 Spatial characteristics of urban form

Figure 1 Alternative models of urban form

summarize basic spatial differences before and after World War II. The War had an impact on more than urban form, however. It created a shortage of men, and thus had implications for gender relations. An imbalance in the sex ratio has certain predictable consequences for women's status. The absence of men during World War II opened new jobs for women, allowing them to receive the training and wages that eventually fostered independence. That independence was temporarily sidetracked by the economic and political necessity to employ thousands of returning veterans. During the 1950s women's labor force participation declined, the birth rate rose, and far more men than women attended college. But by the 1970s women's status began to change. Birth rates dropped, educational attainment rose, full-time labor force attachment increased, and more women headed their own households. These changes were facilitated by several federal policies.

POST-WORLD WAR II CHANGES IN WOMEN'S STATUS

As economic restructuring began to alter urban form, federal intervention involving reproductive rights and equal opportunity legislation started to enhance women's status. Highly effective oral contraception was introduced during the 1960s, and abortion was legalized with the Supreme Court decision of *Roe v. Wade* in 1973. Women became capable of making their own decisions about childbearing for the first time in history. *This* was a watershed. Demographers called it a "contraceptive revolution". The ability to control their fertility was only women's first step toward independence. The second step involved access to educational and financial resources.

Congress passed four significant pieces of equal opportunity legislation during the 1960s and 1970s. The Equal Pay Act of 1963 made it illegal to pay women and men different wages for the same job. Title IX of the Educational Amendments Act of 1972 prohibited sex discrimination in all public and private colleges receiving federal funds. The Equal Credit Opportunity Act of 1974 barred sex and marital-status discrimination in the credit process, and Section 303(b) of the Housing and Community Development Act of 1974 was amended to eliminate sex discrimination in housing and housing finance.

Combined with reproductive rights reform, equal opportunity laws provided women with powerful avenues for change. The first was *rising educational attainment.* In 1960 only six percent of adult women had a college degree; now nearly one-quarter of American women have graduated from college. As more women graduated from college, more joined the labor force. Gradually the schools and workplaces women shared with men became less spatially segregated. The history of education and employment in the U.S., in fact, has been characterized by declining spatial gender segregation and rising status for women.

The second change was *women's entry into the labor force.* So many women, including mothers, joined the labor force so rapidly that it soon became the norm for women to be employed outside the home. Between 1950 and the end of the century, the proportion of women in the labor force nearly doubled. Among married mothers with preschoolers, the proportion in the labor force rose from 12 to 64 percent between 1950 and 1997. The third trend to emerge was the *growth of female householders.* Prior to World War II,

	Circa 1900	Circa 2000
Prototype	Wife/mother	Employed mother
Fertility control	Ineffective	Effective
Percent with college degree	< 5%	25%
Percent in labor force	20%	60%
Percent of households maintained by women	13%	28%
Potential for independence	Low	High

Table 2 Indicators of women's status
Source: Solomon 1985, 64; U.S. Bureau of the Census 1975, 42 & 128; 1998, 61 & 167.

women maintained less than 15 percent of all households. By the end of the 20th century it was nearly 30 percent. Delayed marriage, longer life expectancy, high divorce rates, and rising rates of out-of-wedlock births all contributed to the increase in the number of female householders (see Table 2).

The modern Women's Movement that fueled these changes in women's status qualifies as one of those social movements that arise occasionally to "challenge the meaning of spatial structure and therefore attempt new functions and new forms". Manuel Castells defines an urban social movement as "collective actions consciously aimed at the transformation of the social interests and values embedded in the forms and functions of a historically given city" (Castells 1983). The Women's Movement met these criteria. It challenged the adage that a woman's place is in the home. The Women's Movement seems to have been overlooked as an agent of *urban* change, however. But why are gender relations any less powerful than economic restructuring or globalization as agents of spatial transformation?

NEW GENDER RELATIONS CREATE NEW URBAN SPACES

Having taken the L.A. School to task for ignoring gender, the next step is to incorporate gender into postmodern urban theory. Consider the concept of "privatopia", or gated communities administered by homeowners' associations. Dear estimates there are currently 150,000 homeowners' associations, and common-interest developments (CIDs) account for nearly ten percent of the American housing stock (Dear 2000). The U.S. currently has at least 20,000 gated communities, the vast majority of which have been built since the 1980s. Their increase correlates fairly strongly with the history of women's labor force involvement. Few middle-class families sought gated living when women were home all day to provide informal security. Whereas husbands once earned the income and wives had time to supervise children's play, most wives now have traded time at home for money. One of the costs has been the absence of neighborhood surveillance. Furthermore, services provided by homeowners' associations in gated communities are reminiscent of the work women volunteers performed one hundred years ago: landscaping of common grounds, garbage pickup, street cleaning, and maintaining parks and playgrounds were all part of the municipal housekeeping agenda that encouraged women to apply their domestic skills to the public sphere.

Mike Davis's (1990) concept of the city as a fortress presents another opportunity to incorporate gender. He focuses on mean streets and dangerous communities, identifying public housing as part of the carceral city. The Imperial Courts Housing Project in Los Angeles, for example, is a fenced-off war zone requiring identification for entry. He neglects to mention, however, that public housing is occupied predominantly by women and children. Thus danger is distributed disproportionately by both geography and gender. Leavitt's work with Los Angeles public housing residents recognized this and illustrated how resourceful women have been in creating a sense of safety and community. Davis also identifies prisons as "containment centers" in the urban landscape, the masculine counterpart to public housing. A gendered view of containment centers, however, might also include daycare centers and retirement homes as places that hold economically marginal populations under supervision.

Edge cities, a primary component of the postmodern metropolis, have evolved from the confluence of three conditions: (1) the dominance of automobiles and the need for parking; (2) the communications revolution; and (3) the entry of women in large numbers into the labor market. How, exactly, does women's market labor contribute to the formation of edge cities? One way, of course, is by increasing the demand for vehicles. The majority of Americans drive alone to work, and women are no exception. Most employed women also face two other issues: how to care for children or elderly parents and how to feed a family. Individual women's efforts to balance their family and work lives have collectively shaped the metropolitan area in significant ways. Important services once performed by women in the privacy (or seclusion) of the home have moved into the public arena: care of dependents and meal preparation. *Childcare facilities, assisted care institutions for the elderly, and eating establishments* all are providing services that were once a private responsibility.

When the proportion of married mothers with preschoolers nearly tripled in two decades, childcare became a public issue. The majority of working mothers in the 1960s depended on in-home babysitting provided by a relative or someone else; group care centers were rare. Over the decades, however, the childcare industry expanded to meet growing demand. By 1998 the U.S. had nearly 100,000 licensed childcare centers where 31 percent of preschoolers spent some part of their day.

Employed women with responsibility for elderly parents face similar concerns about care for dependents. Increased life expectancy means parents are living longer just as their daughters are committing more fully to the labor force. Employed women have less time (although theoretically more money) than their grandmothers had, making it possible to pay others to adopt tasks they once were expected to perform. In the last twenty-five years alone, the number of skilled nursing facilities has tripled. The "old-age home" of the last century has been replaced by nursing homes, retirement homes, and "assisted living" facilities, each label becoming more euphemistic as people live longer. Although few of the elderly currently live in one of these institutions, their numbers will surely grow as the population ages.

Housework can usually wait, and most studies of the division of household labor suggest that it does. But people have to eat several times a day. "Family" restaurants and fast food franchises have proliferated over the last twenty years as a substitute for the kitchen and dining room. Married couples with children now spend more than one-third of their food budget on meals outside the home. Every strip development leading into every American city of any size has its own assortment of food outlets staffed by immigrants, teenagers, or retirees – those who are marginal to the mainstream economy, just as women were when they prepared meals at home.

The transfer of domestic services from the home to the public sphere has exacerbated suburban sprawl since new construction occurs at the urban periphery. Zoning regulations that separate residential neighborhoods from commercial activities also have an impact on metropolitan form. Women typically need a car to get to work, deliver kids to daycare or soccer practice, and run household errands. The result has been a significant increase in the number of vehicles on the road. Since 1969, the rate of increase in household vehicles has been more than six times the rate of population growth.

One other type of urban space has been created by women's greater independence. Female householders have created homes from which the man is (sometimes only technically) absent. The growth in female householders has changed the metropolitan landscape primarily by creating a demand for more and different housing units. Young women who once moved straight from their parents' home into marriage now live independently for some years. Unless an unwed mother stays with her parents, she also must find a place to live. Every divorce splits one household into two. Women's longer life expectancy and lower remarriage rates mean they live alone longer after widowhood than men. Each of these new household types demands new housing.

Women seeking to form their own households often need help getting established. For example, temporary shelters for victims of domestic violence are a new addition to the urban landscape, although, to insure residents' safety, they are seldom identified as such. Boston and Minneapolis established transitional housing developments during the 1980s to bridge the gap between emergency shelter and permanent affordable housing for low-income women. Women and their children can live there for six months to two years while receiving childcare and job counseling. Women in Toronto developed and managed housing cooperatives to meet the needs of single mothers and elderly women.

New gender relations have transformed urban spaces in both the public realm and the private domain of the home. One hundred years ago, women were less visible in colleges and workplaces than they are today, while men were more visible in the typical home. Now women have moved into public spaces, and men have moved out of many private homes.

INTEGRATING GENDER INTO URBAN THEORY

Many factors have contributed to the transformation of urban space from the modernist monocentric city to the postmodern polycentric metropolis. According to Michael Dear, economic restructuring, globalization, and environmental politics are among the most important reasons. This essay proposes that changing gender relations should be added to the list. Ample opportunities existed in Chicago at the beginning of the 20th century, and in Los Angeles at the end of the century, to incorporate gender into urban theory. Yet work on gender and urban space has remained largely isolated in a parallel world of feminist scholarship.

Changing gender relations have shaped the metropolis in several ways. Women's ability to control fertility and achieve economic independence following World War II eventually had spatial implications. Care of dependents and meal preparation have moved out of the home and into the metropolis as women's labor force activity has increased. Childcare centers, assisted living facilities for the elderly, and franchise food chains have all contributed to suburban sprawl and the proliferation of edge cities. Although nurseries, old-age homes, and restaurants all existed at the beginning of the 20th century, only at its end did they become ubiquitous. The labor performed in these facilities is underpaid and relies on marginal workers – just what women were before World War II.

A gender perspective applied to current urban theory would count day care centers and retirement homes among the "containment centers" identified by Davis as part of the postmodern metropolis. It would also interpret the growth of gated communities (privatopias) as a consequence of women's entry into the labor force. Americans are not seeking a fortress to separate themselves from others as much as they are trying to replicate an era when mothers were home all day.

Some tenets of postmodern urban theory have direct corollaries with gender relations. Take one aspect of economic restructuring, that employees experience less job security than they once did. During the 1970s, when divorce rates were high, many wives also discovered that they received less job security than they had bargained for. These displaced homemakers were the rehearsal for downsizing and job lay-offs in the paid economy. The broken marriage contract that released women from the securities and responsibilities of marriage was a precursor to broken corporate loyalties. Or take the "dual city" metaphor of the underclass and overclass. A gender analysis would point out that there has always been a dual city consisting of women's free labor and men's paid labor. It was invisible because it existed under the same roof. The rise of the service sector has merely taken *unpaid* work out of the home and turned it into *underpaid* occupations throughout the metropolis.

Theorists of the Los Angeles School like to distance themselves from their Chicago ancestors, but they share one inescapable similarity. They both ignored the women who were working on the same issues in the same city at the same time. Just as Robert Park and Ernest Burgess barely acknowledged Jane Addams and Julia Lathrop, Michael Dear and Edward Soja have integrated little of Dolores Hayden's or Jacqueline Leavitt's perspectives into their own. After nearly a century, gender remains largely marginalized in urban theory.

In closing, I would like to propose that the way we think about gender relations and the way we theorize urban structure are similar. When we thought women's natural place was at the center of the home, we perceived the city as a centered organism around which various functions were rationally organized. Both those images are outdated. Women now fill a variety of roles both inside and outside the home, and the metropolitan area has become the site of scattered activities. What happened to gender relations on the way from Chicago to Los Angeles? The same thing that happened to urban form. They became less predictably centered and more diverse.

REFERENCES

Addams, Jane. (1910). *Twenty Years at Hull-House.* New York: Macmillan.

Addams, Jane, and Lathrop, Julia. (1895). *Hull-House Maps and Papers.* New York: Thomas Y. Crowell.

Castells, Manuel. (1983). *The City and the Grassroots: A Cross-Cultural Theory of Urban Social Movements.* Berkeley, CA: University of California Press.

Davis, Mike. (1990). *City of Quartz: Excavating the Future in Los Angeles.* New York: Vintage.

Dear, Michael. (2000). *The Postmodern Urban Condition.* Oxford: Blackwell.

Flusty, Steven, and Dear, Michael. (1999). "Invitation to a postmodern urbanism", in Robert Beauregard and Sophie Body-Gendrot (eds.) *The Urban Moment: Cosmopolitan Essays on the Late-20th Century City.* Thousand Oaks, CA: Sage, pp. 25–50.

Harvey, David. (1973). *Social Justice and the City.* Baltimore, MD: Johns Hopkins University Press.

Hayden, Dolores. (1980). "What would a non-sexist city be like? Speculations on housing, urban design, and human work", in Catharine Stimpson et al. (eds.) *Women and the American City.* Chicago, IL: University of Chicago Press, pp. 167–184.

Hayden, Dolores. (1981). *The Grand Domestic Revolution.* Cambridge, MA: MIT Press.

Hayden, Dolores. (1984). *Redesigning the American Dream.* New York: W.W. Norton.

Hayden, Dolores. (1995). *The Power of Place: Urban Landscapes as Public History.* Cambridge, MA: MIT Press.

Leavitt, Jacqueline. (1980). "The history, status, and concerns of women planners", in Catharine Stimpson et al. (eds.) *Women and the American City.* Chicago, IL: University of Chicago Press, pp. 223–227.

Park, Robert, Burgess, Ernest, and McKenzie, Roderick. (1925/1967). *The City.* Chicago, IL: University of Chicago Press.

Soja, Edward. (2000). *Postmetropolis: Critical Studies of Cities and Regions.* Oxford: Blackwell.

Solomon, Barbara. (1985). *In the Company of Educated Women.* New Haven, CT: Yale University Press.

Spain, Daphne. (2001). *How Women Saved the City.* Minneapolis, MN: University of Minnesota Press.

US Bureau of the Census. (1975). *Statistical Abstract of the United States.* Washington, DC: US Government Printing Office.

US Bureau of the Census. (1998). *Statistical Abstract of the United States.* Washington, DC: US Government Printing Office.

"Social Exclusion and Space"

from Ali Madanipour, Goran Cars, and Judith Allen (eds), *Social Exclusion in European Cities: Processes, Experiences, and Responses* (1998)

Ali Madanipour

Editors' Introduction

Exclusion of groups of city residents from access to all that the city has to offer on the basis of race, class, religion, income, gender, national origin, disability status, sexual orientation or some other characteristic has been and continues to be a pressing problem in cities throughout the world. University of Newcastle urban design professor Ali Madanipour's observations on spatial aspects of social exclusion in contemporary European cities is relevant to understanding social exclusion in cities everywhere in the world both nowadays and in the past.

Throughout history many of the most dynamic urban societies have welcomed foreigners and included them in the life of the city. H.D.F. Kitto notes that twenty-five centuries ago foreigners (*metics*) participated in most aspects of the life of the Greek polis (p. 40). They lived throughout the polis rather than in geographically segregated foreigners' neighborhoods, worked as merchants and trades people on an equal footing with Athenian citizens, and contributed significantly to the philosophical, scientific, literary, and artistic achievements of Athens' golden age. But they were not Athenian citizens and were excluded from participation in Athens' otherwise extraordinarily inclusive and democratic political institutions.

In his magisterial study titled *Cities and Civilization*, British planning professor Sir Peter Hall argues that the presence of a diverse group of foreigners or outsiders from the dominant culture has been a crucial ingredient in short periods of great cultural and technical efflorescence that characterize cities' golden ages. Hall describes, for example, how Jewish entrepreneurs who had previously worked in New York City's garment industry, were largely responsible for creating the motion picture industry. They were able to transfer understanding of how to respond quickly to the changing tastes of the United States' large lower income urban immigrant population they had learned in New York City's garment industry and quickly turn advances in technology to good advantage. Migrating to Hollywood, they created a new industry providing silent movies to a mass audience willing to spend a hard-earned nickel for Saturday night entertainment. Another of Hall's examples involves Blacks from the impoverished Mississippi River Delta. As they migrated up the Mississippi River to Chicago during the twentieth century, Blacks from the Delta brought blues music with them. Little blues clubs in Chicago's Black belt helped them cope with discrimination and the unsettling conditions of urban life. Blues music morphed into rock and roll and made a huge contribution to popular culture worldwide. Nowadays, Indian programmers in Silicon Valley, Chinese scientists in London, and Latin American novelists in New York City continue to enrich their host cultures and the entire world.

In many cities law and/or cultural norms have excluded some social groups at some time in history, including the present day. Racial discrimination was, and remains, an acute problem in many cities throughout the world. Black sociologist W.E.B. Du Bois describes in painful detail how Blacks in late-nineteenth-century Philadelphia were spatially isolated in just a few wards of the city and systematically barred from white schools, most public facilities, and well-paying jobs for which they were well qualified (p. 110). Friedrich Engels describes the brutal effect of class discrimination on working-class people in Manchester, England in 1844 (p. 46). Members of the Chicago school of sociology like Louis Wirth (p. 96) described the psychological damage, spatial separation and social exclusion immigrants from central and southern Europe experienced in early-twentieth-century Chicago. Mike Davis describes discrimination against poor people, minorities, and immigrants in contemporary Los Angeles (p. 195). Discrimination on the basis of ethnicity, religion, gender, and national origin continues in Europe and North America against Algerian, Pakistani, Turkish, East European, Mexican, and other groups.

As globalization continues to bring people from throughout the world into closer contact, and as the pace of immigration increases, the issue of exclusion becomes ever more pressing. In what different ways are some people excluded from participation in the life of the cities where they live? How is exclusion expressed in urban space? What can be done about it? These are questions Madanipour addresses.

Madanipour distinguishes between economic discrimination, in which members of a group are excluded from employment, political discrimination, in which they are excluded from political power by being denied voting rights or full political representation, and cultural exclusion in which the group members are marginalized from the symbols, meanings, rituals, and discourses of the dominant culture. Just as Sherry Arnstein (p. 238) sees citizen participation in decision-making as a "ladder" with rungs ascending from degrees of non-participation to full citizen power, Madanipour sees social exclusion as a continuum from complete lack of integration at one end of the spectrum to full integration into society at the other.

While some societal rules about exclusion are benign – the right of strangers to enter a person's home at will is unacceptable in almost all cultures – Madanipour argues that exclusion of groups from the opportunities and advantages that cities possess is both painful to members of the group and damaging to the society at large, which fails to take full advantage of talent available to it and wastes resources on conflict and social control.

Exclusion frequently has a spatial dimension. Members of a group are sometimes excluded from areas of a city by law as when medieval Venetian Jews were restricted to the city's ghetto and Chinese were prohibited from entering some parks in the European areas of nineteenth-century Shanghai. Even when people are legally free to enter areas of the city, as Mike Davis (p. 195) points out, subtle and not-so-subtle signs and cues may signal that members of a particular group are not welcome.

Madanipour suggests two potentially promising theoretical approaches to promote greater inclusion of marginalized groups into urban space: decommodifying space so that the private real estate market plays a less decisive role in where different groups are located within the city and deliberate city planning to despatialize social exclusion. Building inclusionary housing units for low- and moderate-income households in neighborhoods they could otherwise not afford is an example of the first strategy. In new inclusionary condominium developments in the United States, for example, sometimes some percentage of the units are reserved for sale to low and moderate-income households at below market cost. Mixed-use zoning to promote social diversity is an example of the second strategy Madanipour suggests. Some cities, for example, encourage a mix of market and below market rate housing units in the same area.

Madanipour concludes his analysis by advocating inclusionary practices, to assure that outsiders are more fully included in urban societies. He wants to break the trap of socio-spatial exclusion and provide more possibilities for inclusion.

Ali Madanipour teaches architecture, urban design and planning at the School of Architecture, Planning, and Landscape at the University of Newcastle in England. He is a founding member of Global Urban Research Unit at the University of Newcastle. Madanipour was born in Iran and practiced architecture before his academic career. His interests include design, development and management of cities, the social and psychological significance of urban space, processes that shape urban space, agencies of urban change, and implications of change for disadvantaged social groups and the environment. Madanipour's writings have been translated into German, Mandarin, Japanese, and Persian.

This selection is from *Social Exclusion in European Cities: Processes, Experiences, and Responses* (London: Jessica Kingsley, 1996), which Madanipour coedited with Goran Cars and Judith Allen. Madanipour's other books include *Whose Public Space?* (London: Routledge, 2010), *Tehran: The Making of a Metropolis* (New

York: Wiley, 1998), *Design of Urban Space: An Inquiry into a Socio-spatial Process* (New York: Wiley, 1996), *Public and Private Spaces of the City* (London: Routledge, 2003), and two coedited anthologies: *The Governance of Place*, coedited with Angela Hull and Patsy Healey (Aldershot, UK: Ashgate, 1996), and *Urban Governance, Institutional Capacity, and Social Milieux*, coedited with Goran Cars and Patsy Healey (Aldershot, UK: Ashgate, 2002).

For historical background on social exclusion in the United States, see Elizabeth Cobbs-Hoffman and Jon Gjerde, *Major Problems in American Immigration and Ethnic History*, 2nd edn (Boston, MA: Houghton Mifflin, 2006). Studies of contemporary race, class, and gender issues in the United States include Margaret L. Andersen and Patricia Hill Collins, *Race, Class, and Gender: An Anthology*, 7th edn (New York: Wadsworth, 2008), Roberta Fiske-Rusciano, *Experiencing Race, Class, and Gender in the United States* (New York: Wadsworth, 2008), Conrad Kottak and Kathryn Kozaitis, *On Being Different: Diversity and Multiculturalism in the North American Mainstream* (New York: McGraw-Hill, 2008), and Paula S. Rothenberg, *Race, Class, and Gender in the United States: An Integrated Study*, 7th edn (New York: Worth, 2006). A classic study of European immigration to the east coast is Oscar Lewis, *The Uprooted* (New York: Atlantic Monthly Press, 1951). Ronald Takaki, *Strangers from a Different Shore* (Boston, MA: Back Bay Books, 2003) is an excellent study of the Asian American immigration experience. Takaki, *A Different Mirror: A History of Multicultural America*, revised edn (Boston, MA: Back Bay Books, 2008) expands and elaborates on his earlier work.

This chapter concentrates on the relationship between social exclusion and space, exploring some of the frameworks which institute barriers to spatial practices. Its particular emphasis is on the way these barriers to movement are intertwined with social exclusionary processes. This shows that exclusion should be regarded as a socio-spatial phenomenon.

[. . .]

DIMENSIONS OF SOCIAL EXCLUSION

There is little disagreement on some of the major problems facing European cities. Challenges of competition from a global economy marked by a multiplicity of competitors and the European response in the form of moving into an integrative partnership are both aspects of globalization which have reshaped the social and spatial geography of cities. The restructuring of cities and societies, however, has been a costly exercise, as it has been parallel with a growing social divide, long-term unemployment and joblessness, especially for men, and casualization of work, undermining the quality of life for large groups of the population. These symptoms have led to concerns for the fragmentation of the social world, where some members of society are excluded in the 'mainstream' and where this exclusion is painful for the excluded and harmful for society as a whole.

Yet the concept of social exclusion still appears to be in need of clarification due to the variety of the cultural and political contexts in which it has been used. For some it is the question of poverty which should remain the focus of attention, while for others social exclusion makes sense in the broader perspective of citizenship and integration into the social context. Social exclusion, therefore, is not necessarily equated with economic exclusion, although this form of exclusion is often the cause of a wider suffering and deprivation.

As a concept, social exclusion still suffers from a lack of clarity, as it is interpreted and analysed differently. We come across a degree of ambiguity especially between poverty and social exclusion. Some researchers, who have concentrated on the problems of poverty, find social exclusion a vague concept which, for whatever reason, takes attention away from poverty and deprivation. Furthermore, it is argued that the concept of social exclusion is rooted in a certain intellectual and cultural tradition (Catholic, solidarity) and a particular welfare regime (corporatist) and as such is not shared by other (especially liberal) cultures and welfare regimes. On the other hand, those who find social exclusion a useful concept criticize an emphasis on poverty as too narrow. They seek to open the discussion to accommodate the general issues of social integration and citizenship. To confront this ambiguity and contradiction, we need to clarify the concept of social exclusion first.

The overall constitution of the social world is such that different forms of exclusion are fundamental to any social relationship. For example, the division of social life into public and private spheres means drawing boundaries round some spatial and temporal domains and excluding others from these domains. In this way, exclusion becomes an operating mechanism, an institutionalized form of controlling access: to places, to activities, to resources and information. Individual actions as well as legal, political and cultural structures rely heavily upon this operating mechanism and reproduce it constantly. Institutionally organized or individually improvised, it appears that we are all engaged in exclusionary processes that are essential for our social life.

Yet we know that, whatever their importance, these exclusionary processes work in close relationship with inclusionary activities to maintain a social fabric. Maintaining the continuity of the social world is only possible through a combination of and a fine balance between these two processes. At the individual level, seeking privacy without seeking social interaction would lead to isolation. At the social level, exclusion without inclusion would lead to a collapse of social structures. What is a negative state of affairs, therefore, is not exclusion in all its forms but an absence of inclusionary processes, a lack of a balance between exclusion and inclusion.

But what are the dimensions of the social world in which inclusion and exclusion take place? It is often mentioned that social exclusion is multidimensional. To be able to identify and analyse these dimensions, we should look at the dimensions of the social world in which exclusion and inclusion take place. We can identify economic, political and cultural arenas as the three broad spheres of social life in which social inclusion and exclusion are manifested and, therefore, can be analysed and understood.

In the economic arena, the main form of inclusion is access to resources, which is normally secured through employment. The main form of exclusion, therefore, is a lack of access to employment. Marginalization and long-term exclusion from the labour market lead to an absence of opportunity for production and consumption, which can in turn lead to acute forms of social exclusion.

Exclusion from the economic arena is often considered to be a crucial and painful form of exclusion. Poverty and unemployment are therefore frequently at the heart of most discussions of social exclusion, to the extent that poverty and economic exclusion are equated with social exclusion. There is a tendency in the literature to use these terms interchangeably. It is true that long-term economic exclusion can break down the political and cultural ties of the affected individuals and social groups. It is important, however, to note that there are other forms of social exclusion in political and cultural spheres.

In the political arena, the main form of inclusion is to have a stake in power, to participate in decision making. In European liberal democracies, inclusion is often ensured through voting and other processes associated with it. The most obvious form of social exclusion, therefore, is lack of political representation. This may take various forms: from the under-representation of women in parliaments and governments, to the complete exclusion of immigrant groups from political decision making; from the argument by smaller political parties for a new system of representation which would allow them a fairer share of power, to a withdrawal from political participation by those excluded in the economic and cultural arenas.

In the cultural arena, the main form of inclusion is to share a set of symbols and meanings. The most powerful of these have historically been language, religion and nationality. Some of the new sets of symbolic relationships include the way individual and group identities are formed through association with patterns of consumption, from necessities of daily life to cultural products. For example, in what has been termed a visual culture, aesthetics of social behaviour has become an essential part of social life. The main form of exclusion in the cultural arena, therefore, becomes a marginalization from these symbols, meanings, rituals and discourses. The forms of cultural exclusion vary widely, as experienced by minorities whose language, race, religion and lifestyle are different from those of the larger society.

Different social groups may experience varying degrees of these different but highly interrelated forms of social exclusion. The most acute forms of social exclusion, however, are those that simultaneously include elements of economic, political and cultural exclusion. The other end of the spectrum is occupied by citizens who are fully integrated in the mainstream of society through these three dimensions. Between these two extremes, there is a wide range of variations in which individuals and groups are included in some areas but excluded in others. A major trend is that more and more people suffer from anxiety and uncertainty, as there are ever larger numbers in transition from inclusion to exclusion.

SPATIALITY OF SOCIAL EXCLUSION

Social exclusion, therefore, should be understood in its political, economic and cultural dimensions. Exclusion from the political arena, i.e. the denial of participation in decision making, can alienate individuals and social groups. In the cultural arena, exclusion from common channels of cultural communication and integration can have similar effects. The exclusion from work and its impacts are widely known as undermining the ability of individuals and households to participate actively in social processes. When combined, these forms of exclusion can create an acute form of social exclusion which keeps the excluded at the very margins of the society, a phenomenon all too often marked by a clear spatial manifestation in deprived inner city or peripheral areas . . .

[. . .]

In the past, this spatiality of social exclusion had led to attempts to dismantle such pockets of deprivation without necessarily dismantling the causes of deprivation or the forces bringing them together in particular enclaves. The dismantling of spatial concentrations of deprivation has been a continuous trend: from Baron Haussmann's wide boulevards in the middle of poor neighbourhoods in the nineteenth century, to the slum clearance programmes and more subtle forms of housing management in the twentieth century. These have been attempts to despatialize social exclusion, which is evidence of its inherent and re-emerging spatiality. The latest form of despatialization and re-spatialization of social exclusion is homelessness, a process in which some groups are cut off from their previous socio-spatial contexts and are apparently without a home base. They, however, have clustered in particular parts of cities, spatializing again what was thought to be despatialized.

SPATIALITY AND DIFFERENCE

The absence of homogeneity is most apparent in cities, as they are sites of difference. Large cities have often grown by attracting people from around the country in which they are located or even from around the world. Cities have always been known as the meeting places of different people. As Aristotle noted: 'A city is composed of different kinds of men; similar people cannot bring a city into existence.' The unprecedented growth of cities since the nineteenth century has permanently brought forward the issue of difference in the city as a feature of urban life. Wirth, in his celebrated theory of urbanism, saw heterogeneity as a determining feature of the city, along with population size and density. For him, the city was a 'melting-pot of races, peoples, and cultures, and a most favourable breeding-ground of new biological and cultural hybrids'. In the city, individual differences have 'not only [been] tolerated but rewarded'. Such emphasis on the heterogeneity of cities has led to conceiving it as a world of strangers.

Two sets of reactions to the diversity in the city can be identified: there are those who have tried to impose an order onto it so that it becomes understandable and manageable and those who promote a celebration of diversity. However, both these reactions, which indeed represent modernist and postmodernist thinking, have been unable to deal with the issue of social marginalization and exclusion. Concentrations of disadvantage have remained in cities, despite the large-scale redevelopment schemes of the rationalist tendency and the more sensitive spatial transformations which followed. On the one hand, emphasis on the eradication of difference and seeing the city as a melting pot has led to undermining sensitivities and to disruption of lives. On the other hand, the emphasis on difference has led to social fragmentation and tribalism. Both have failed to cure the wounds of those living on the edge of the society.

BARRIERS TO SPATIAL PRACTICES

But how do we analyse space? There are many gaps and dilemmas associated with understanding space. From the centuries-old philosophical divide between absolute and relational space, to the gap between mental and real space, between physical and social space, between abstract and differential space, to the relationship between space and mass, space and time, and the variety of perspectives from which space can be studied, all bear the possibility of confusion and collision. It is possible to show, however, that to avoid the gaps and dilemmas associated with understanding space, we need to concentrate on the processes which produce the built environment. By analysing the intersection between space production and everyday life practices, we will be able to arrive at a dynamic understanding of space. We will then be able to understand and explain material space and its social and psychological contexts and attributes.

The question of social exclusion and integration, it can be argued, largely revolves around access. It is access to decision making, access to resources, and access to common narratives, which enable social integration. Many of these forms of access have clear spatial manifestations, as space is the site in which these different forms of access are made possible or denied. There is a direct relationship between our general sense of freedom and well-being with the choices open to us in our spatial practices. The more restricted our social options, the more restricted will be our spatial options, and the more excluded we feel or become. On the other hand, if we have a wide range of social options, we would have a wide range of places to go to, places for living, working and entertainment. Two extreme cases of the existence or absence of spatial freedom may be jetsetting executives versus prisoners. Whereas for one, the world may be shrinking to seem like a global village open for communication and interaction, for the other the world outside is large and out of reach. For most of us, however, our spaces are a continuum from accessible to non-accessible places. The space around us is a collection of open, closed or controlled places.

But how is the urban space organized and how are spatial practices controlled and regulated? We all have an understanding of the places where we can or cannot go, as over the years through our spatial practices, we have accumulated a knowledge about places and their patterns of accessibility. The physical organization of space, using elements from the natural or the built environment, has been socially and symbolically employed to put visible and strict limits on our spatial practices. For example, topography has always been used to institute difference and segregation, from ancient times when the hilltops were the place of gods for Greeks and Mesopotamians, to our own time when they are the living places of the rich and powerful. There is also a mental space, our perceptions of space. This may be regulated through codes and signs, preventing us from entering some spaces through outright warning or more subtle deterrents. Mental space may also be controlled through our fears and perceptions of activities in places. For example, we may be hesitant to enter an expensive-looking shopping centre if we do not have access to the resources needed for the activities there, even though there may not be any physical barriers which would prevent us from going there. A third form of barrier to our spatial behaviour is social control, which can range from legal prohibitions on entering places to constructing formal barriers along publicly recognized borders. National borders and public–private boundaries are examples of this form. A combination of formalized rules and regulations, informal codes and signs, and fears and desires control our spatial behaviour and alert us to the limitations on our access. Through these, we have come to know whether we can enter a place, are welcomed in another and excluded from others. More restrictions on our access to our surroundings would bring about the feeling of being trapped, alienated and excluded from our social space.

Space has, therefore, a major role in the integration or segregation of urban society. It is a manifestation of social relationships while affecting and shaping the geometries of these relationships. This leads us to the argument that social exclusion cannot be studied without also looking at spatial segregation and exclusion. Social cohesion or exclusion, therefore, are indeed socio-spatial phenomena . . . We know that all human societies have their own forms of social and spatial exclusion. So exclusionary processes per se are not the source of social fragmentation and disintegration. It is the absence of social integration which causes social exclusion, as individuals do not find the possibility and channels of participating in the mainstream society.

GLOBAL AND NATIONAL SPACE

National borders are the largest means of socio-spatial exclusion. The modern nation state exerts an exclusionary process along its boundaries, from lines on maps to barbed wires on the landscape. Those who are left outside need to go through special checks and controls to be allowed in. The same applies to those who are in and want to go out. The control of cross-border movement by the nation state, or by blocks of nation states as in the European Union, is a form of exclusion legitimated openly through political processes. A national territory, therefore, is a spatial manifestation of an institutionalized exclusionary process.

Other administrative boundaries, although potentially exclusionary, do not have such a forceful character, nor are they associated with such a degree of public awareness, such historical significance, or guarded by military might. No other form of exclusion has been associated with such high costs in human life, sacrifice and misery. Attempts to change or to protect national borders have inflicted the highest cost in human lives in the twentieth century, as experienced by two world wars and many regional conflicts. The birth of a nation state, when the multi-ethnic empires and states break up, can be a bloody process in which every means is used to exclude others. The surgical subdivision of national space, whether through external forces as in postwar Germany or by exploding internal forces as in the former Yugoslavia, has been equally difficult for those excluded from what they have regarded as their home.

In the national space enclosed within these boundaries, narratives of nationalism have been employed to legitimize the exclusion of others beyond these boundaries. Indeed, exclusionary narratives, which determine how 'we' are different from others, are often essential in binding individuals together as a group. The most dangerous of these narratives has been the rhetoric of hatred against other nations, races and groups. But there are many such exclusionary narratives which do not necessarily promote violence and hatred and still have a binding power. With these narratives, which often rely on a common historical experience, large groups of people have been associated with each other. The focal point of this association has been the nation state, which holds the power of controlling the national borders.

The narratives of nationalism attempt to create homogeneity out of an enormous diversity. As individuals have come together to create a democratic civil society, such narratives have helped the organization of modern democratic states . . .

[. . .]

NEIGHBOURHOODS, MARKETS AND REGULATION

At the local level, by following two processes, land and property development on the one hand and spatial planning on the other hand, we can see how a socio-spatial geometry of difference and segregation, which is the foundation of exclusion, emerges. We come across the term neighbourbood in a variety of distinct but interrelated usages. In one sense, the term is used loosely to address a locality. This daily usage is based on the images and understandings by individuals and groups of their surroundings. This is a view from below and, as such, can lead us to see a city as a collection of overlapping neighbourhoods. Research on people's perception of neighbourhood shows major differences according to age, gender, class and ethnicity. At the other end of the scale, there is a concept of neighbourhood from above, from the viewpoint of such experts as managers, planners and designers. Here neighbourhood refers to a particular part of a town and is used to understand urban structure and change in urban society. It is also used as a tool for management. From this viewpoint, the city is seen as a collection of segregated neighbourhoods.

Neighbourhoods as constituent parts of cities have long been the focus of attention by urban designers and planners. Drawing upon historic precedents and for practical reasons, neighbourhood has provided them with an intimate scale of the urban whole to understand and to deal with. Historically, neighbourhoods have been the sites and physical manifestations of close social relationships and so have been praised by town planners, especially those who have looked nostalgically to the feudal bonds of the medieval towns and the communal bonds of working-class neighbourhoods in the industrial city. A dichotomy emerged as a result of the unprecedented growth of the cities: between *gesellschaft* and *gemeinschaft*, between the alienation of the big city and the romanticized, small communities of towns and villages. To recreate the social cohesion of these small communities, it was thought, cities should be broken into smaller parts, into neighbourhoods. On the other hand, it was thought that the communitarianism of small neighbourhoods could overcome the individualism of the suburbs, those bourgeois utopias.

It is this association of neighbourhood as a physical entity with neighbourhood as a cohesive social unit that led to a series of reformist ideas throughout the twentieth century. From the widely used, and discredited, concept of neighbourhood unit, to Lynch's districts, which are still promoted to make cities legible, and today's urban villages and new urbanist neighbourhoods, there has been a long line of managerial attempts to promote social cohesion by spatial organization.

Along with this promotion of spatial subdivision by town planning, there has been a promotion of socio-spatial segregation by market forces through the ways in which space is produced, exchanged and used. The producers of space, such as volume housebuilders, tend to build in large-scale housing estates, creating an urban fabric which is a collection of different subdivisions. The land and property markets have operated so as to ensure the segregation of income groups and social classes. Commodification of space has led to different patterns of access to space and hence a differential spatial organization and townscape. Wherever there has been a tendency to decommodify space, as in the postwar social housing schemes, town planners and designers have ensured that a degree of spatial subdivision still prevailed.

We can therefore identify two processes: a land and property market which sees space as a commodity and tends to create socio-spatial segregation through differential access to this commodity, and a town planning and design tendency to regulate and rationalize space production by the imposition of some form of order. When we look at these two processes together, the picture which emerges is a collectivization of difference, of exclusion, which can lead to enclaves for the rich and the creation of new ghettos for the poor.

[. . .]

PUBLIC AND PRIVATE SPACE

Another form of socio-spatial exclusion, which is enforced with a rigour somewhat similar to the protection of national borders, is the separation between public and private territories. We guard our private spheres from intruders by whatever means, in some countries even legitimately by firearms. Privacy, private property and private space are intertwined, demarcated through a variety of objects and signs: from subtle variations of colour and texture to fences and high walls. Those who are in are entitled to be, excluding those who are not. This is an exclusionary process legitimized through public discourse, through custom or law. Violation of this exclusionary process is regarded as, at best, inconvenience and, at worst, crime. Public space, which is one of the manifestations of society's public sphere, is maintained by public agencies in the public interest and is accessible to the public. Access to public space, however, is subject to exclusionary processes. Public space is guarded from intrusion by private interests, a process which is regarded as essential for the health of the society. Some of the main currents in social and political thought that offer concepts of public space appear to stress the need to keep the public and private spheres distinctive and apart, despite the criticism that this idealizes the distinction.

[. . .]

The changing nature of development companies and the entry of the finance industry into built environment production and management has partly led to what is widely known as the privatization of space. Large-scale developers and financiers expect their commodities to be safe for investment and maintenance, hence their inclination to reduce as much as possible all the levels of uncertainty which could threaten their interests. This trend is parallel with the increasing fear of crime, rising competition from similar developments, and the rising expectations of the consumers, all encouraging the development of totally managed environments. What has emerged is an urban space where increasingly large sections are managed by private companies, as distinctive from those controlled by public authorities. Examples of these fragmented and privatized spaces are gated neighbourhoods, shopping malls and city centre walkways, under heavy private surveillance and separated from the public realm by controlled access and clear boundaries. This total management of parts of the city is in part an attempt to control crime. Crime acts as a counter-claim to space and as such is itself an exclusionary force, keeping many groups vulnerable and marginalized.

CONCLUSION: SOCIAL INTEGRATION AND SPATIAL FREEDOM

Social exclusion combines lack of access to resources, to decision making, and to common narratives. The multidimensional phenomenon of social exclusion finds spatial manifestation, in its acute forms, in deprived inner or peripheral urban areas. This spatiality of social exclusion is constructed through the physical organization of space as well as through the social control of space, as ensured by informal codes and signs and formal rules and regulations. These formal channels act at all scales of space. Global space is fragmented by national spaces, which have a tendency to deny difference and homogenize social groups. At the scale of local space, spatialization of social exclusion takes place through land and property markets. These markets tend to fragment, differentiate and commodify space through town planning mechanisms which tend to fragment, rationalize and manage space, and also through the legal and customary distinctions between the public and private spheres, with a constant tension between the two and a tendency for the privatization of space.

To break the trap of socio-spatial exclusion, one strategy could be to challenge these deep-seated forms of differentiation. We know, however, that wholesale challenges can be problematic themselves, as exemplified by attempts to redefine the public–private relationship in Eastern Europe. Furthermore, we know that any human society is likely to have some form of exclusionary process in its constitution. Nevertheless, it is true that the form of these exclusionary processes changes over time. A reflexive revisiting of the processes of differentiation is therefore a constantly necessary task. At the same time what is necessary and urgent is to institute and promote inclusionary processes, to strike a balance between exclusion and integration, to provide the possibility of integration and to break the trap of socio-spatial exclusion. We have seen that space is a major component part of social exclusion. Revisiting spatial barriers and promoting accessibility and more spatial freedom can therefore be the way spatial planning can contribute to promoting social integration.

"Fortress L.A."

from *City of Quartz: Excavating the Future in Los Angeles* (1990)

Mike Davis

Editors' Introduction

In this "visionary rant" acerbic Southern California social critic Mike Davis presents a dark vision of racial, ethnic, and class divisions and social conflict that characterize urban space in the prototypical metropolis of the future. Davis makes his working-class sympathies and anti-establishment bias perfectly clear on every page.

Mike Davis (b. 1946) is to contemporary Los Angeles what Friedrich Engels (p. 46) was to mid-nineteenth-century Manchester, England. Engels was a kind of explorer, reporting to an educated, middle-class audience about the horrors of Manchester and the miserable lives of the new industrial proletarian class of early modern capitalism. Similarly, 146 years later in 1990, neo-Marxist Davis explores the dark side of Los Angeles and reports on the hopelessness and despair of the post-industrial underclass, now largely defined by race, ethnicity, and gender to an audience largely comprised of young, disaffected intellectuals and academics. The parallels are striking. Davis is the heir to Engels because the contemporary metropolis – characterized by wealth and homelessness and divided against itself along the fault lines that separate gated communities and suburban enclaves from inner-city slums – is the heir to the geographical and social class divisions of the cities of the Industrial Revolution.

In *The Condition of the Working Class in England in 1844* (p. 46), Engels noted the boulevards that intersected Manchester and how the façades of those broad thoroughfares served to mask and disguise the hovels of the poor that lay beyond the view of middle-class commuters. Davis makes a similar point about contemporary Los Angeles. LA's freeways allow middle-class suburbanites to navigate the city without encountering the lives of the residents of the inner-city neighborhoods. Davis agrees with Ali Madanipour (p. 186) that the city itself has become a vast and continuous system of exclusionary signs that residents read and obey mostly on a subconscious level. Today's upscale, pseudo-public spaces, Davis writes, are full of invisible signs warning off the underclass "other." Although architectural critics are usually oblivious to how the built environment contributes to segregation, pariah groups – whether poor Latino families, young Black men, or elderly homeless white females – read the meaning immediately.

As both Engels and Davis described their respective paradigm cities, both reached the limits of language's ability to describe the physical and psychological conditions reported. Both writers were bitterly critical of the destruction of the physical environment they witnessed. Engels's description of the coal-blackened air and bubbling green miasma of industrial waste in Manchester's River Irk presage Davis's descriptions of ecological destruction and human-made disasters in contemporary Los Angeles. Engels compiled a mountain of personal observations, journalistic reports, and official survey data to create a catalog of social horror. Davis relies on a more emotional strategy. His overheated rhetorical excesses often overwhelm rational discourse.

Both men turned out to be prophets. During the twentieth century the oppressed proletariat in Russia, China, and much of the world led revolutions against the capitalist ruling class. The ghettos and barrios of South Central Los Angeles erupted into open rebellion in the Rodney King riots two years after the publication of *City of Quartz* after a video showed LA police beating King, a Black resident of the South Central neighborhood. Whereas

Engels saw massive social dislocations and systematically set about to fashion a theory of revolutionary socialism in response to the observed reality, Davis offers no similarly optimistic solution. Neither writer used modern quantitative research methods.

Compare Davis's insights regarding the underclass "other" with analysts of ghettoization W.E.B. Du Bois (p. 110), William Julius Wilson (p. 117), and Elijah Andersen (p. 127). Revisit Daphne Spain's criticism of Davis for neglecting women's issues (p. 176) and her comments on what a gender perspective would add. Consider liberal solutions to the urban problems Davis discusses proposed by William Julius Wilson (p. 117), Ali Madanipour (p. 186), Robert Putnam (p. 134), Harvey Molotch (p. 251), and Paul Davidoff (p. 435) and the conservative approach Michael Porter (p. 282) describes. Each of these writers proposes ways in which government officials, urban designers, and city planners might reduce social divisions.

Davis is in the tradition of writers who employ an eclectic culture studies approach to understanding cities. Other chapters in *City of Quartz* deal with aspects of Los Angeles as varied as the role of the Catholic Church and noir LA mystery novels. The culture studies approach to cities was pioneered by Scottish biologist Patrick Geddes before World War I. Geddes believe that every city's evolution should be understood in relation to the city's unique history and culture and that plans should be based on them. Geddes' disciple, Lewis Mumford (p. 91), popularized this approach in *The Culture of Cities* (New York: Harcourt Brace, 1938) and his best-selling *The City in History* (New York: Harcourt Brace, 1961). Sir Peter Hall's *Cities in Civilization* (New York: Pantheon, 1998) continues the culture studies tradition.

Mike Davis is a Los Angeles-based writer, social critic, and educator. He is a member of the "Los Angeles school of urbanism" – a like-thinking group of left-wing Los Angeles academics and activists that Michael Dear describes (p. 170). Davis coined the term "Los Angeles school" at the founding meeting of the group in 1987. Davis dropped out of Reed College in the mid-1960s to work as a truck driver, meat cutter, and anti-war activist. He obtained BA and MA degrees in history from the University of California, Los Angeles, in the 1970s. In 1992, two years after *City of Quartz* was published, Davis became a celebrity and his book a best-seller as Americans sought an explanation for race-related rioting in Los Angeles after a jury acquitted four white police officers who had been videotaped beating a black motorist named Rodney King. Six days of rioting left fifty-three Los Angeles residents dead and more than a billion dollars in property damage. Since that time Davis has continued to write and lecture. He is a distinguished professor of creative writing at the University of California, Riverside, and has taught at the University of California, Los Angeles, University of California, Irvine, and the Southern California Institute for Architecture. He is an editor of *The New Left Review* and contributes to the British monthly *Socialist Review*, the journal of the United Kingdom's Socialist Workers Party. Davis's many awards include appointment as a Getty Scholar at the Getty Research Institute (1996–1997), the Lannan Literary Award for Nonfiction (2007), and a prestigious MacArthur Fellowship (1998). MacArthur Fellowships – popularly referred to as "genius awards" – provide a select group of recipients unrestricted $500,000 grants over the course of five years to pursue creative work of their choosing.

This selection is from Mike Davis, *City of Quartz: Excavating the Future in Los Angeles* (London: Verso, 1990). *City of Quartz* was named the best book in urban politics for 1990 by the American Political Science Association, won the Isaac Deutscher Award from the London School of Economics, and was selected by a *San Francisco Examiner* poll as one of the ten best non-fiction books on the US West published in the twentieth century. It has been translated into eight languages.

Davis has written more than twenty books and one hundred articles about topics as varied as car bombs, avian flu, Las Vegas casinos, and, in two books of fiction for young adults, pirates, bats, mammoths, and dragons. Other of Davis's books about cities include *Governments of the Poor: Politics and Survival in the Global Slum*, with Forrest Hylton (London: Verso, 2007), *Planet of Slums* (London: Verso, 2006), *Dead Cities: A Natural History* (New York: New Press, 2002), *Las Vegas: The Grit Beneath the Glitter: Tales from the Real Las Vegas*, coedited with Hal Rothman (Berkeley, CA: University of California Press, 2002), and *The Ecology of Fear: Los Angeles and the Imagination of Disaster* (New York: Metropolitan Books, 1998). His book *Magical Urbanism: Latinos Reinvent the U.S. Big City* (London: Verso, 2001) contains some uncharacteristically positive comments about how immigrants are improving large US cities in addition to his characteristic slash and burn prose excoriating the oppression of Hispanic immigrants.

For more on the history and culture of Los Angeles and Southern California, consult Carey McWilliams' classic book, *Southern California: An Island on the Land*, 9th edn (Los Angeles, CA: Gibbs Smith, 1980), William Fulton, *The Reluctant Metropolis: The Politics of Urban Growth in Los Angeles* (Baltimore, MD: Johns Hopkins University Press, 2001), Robert M. Fogelson, *The Fragmented Metropolis: Los Angeles, 1850–1930* (Berkeley, CA: University of California Press, 1993), Robert Gottlieb, *Reinventing Los Angeles: Nature and Community in the Global City* (Cambridge, MA: MIT Press, 2007), Michael Dear, Jennifer Wolch, Manuel Pastor, and Peter Dreier (eds), *Up Against the Sprawl: Public Policy and the Making of Southern California* (Minneapolis, MN: University of Minnesota Press, 2004), and Rayner Banham, *Los Angeles: The Architecture of Four Ecologies* (London: Penguin, 1971).

For analyses of Los Angeles' troubled race relations, see Janet Abu-Lughod, *Race, Space, and Riots in Chicago, New York, and Los Angeles* (Oxford: Oxford University Press, 2007), Laura Pulido, *Black, Brown, Yellow, and Left: Radical Activism in Los Angeles* (Berkeley, CA: University of California Press, 2006), Min Song, *Strange Future: Pessimism and the 1992 Los Angeles Riots* (Durham, NC: Duke University Press, 2005), and Mark Baldassare (ed.), *The Los Angeles Riots: Lessons for the Urban Future* (Boulder, CO: Westview Press, 1994).

The carefully manicured lawns of Los Angeles' Westside sprout forests of ominous little signs warning: "Armed Response!" Even richer neighborhoods in the canyons and hillsides isolate themselves behind walls guarded by gun-toting private police and state-of-the-art electronic surveillance. Downtown, a publicly subsidized "urban renaissance" has raised the nation's largest corporate citadel, segregated from the poor neighborhoods around it by a monumental architectural glacis. In Hollywood, celebrity architect Frank Gehry, renowned for his "humanism," apotheosizes the siege look in a library designed to resemble a foreign-legion fort. In the Westlake district and the San Fernando Valley the Los Angeles Police barricade streets and seal off poor neighborhoods as part of their "war on drugs." In Watts, developer Alexander Haagen demonstrates his strategy for recolonizing inner-city retail markets: a panopticon shopping mall surrounded by staked metal fences and a substation of the LAPD in a central surveillance tower. Finally, on the horizon of the next millennium, an ex-chief of police crusades for an anti-crime "giant eye" – a geo-synchronous law enforcement satellite – while other cops discreetly tend versions of "Garden Plot," a hoary but still viable 1960s plan for a law-and-order armageddon.

Welcome to post-liberal Los Angeles, where the defense of luxury lifestyles is translated into a proliferation of new repressions in space and movement, undergirded by the ubiquitous "armed response." This obsession with physical security systems, and, collaterally, with the architectural policing of social boundaries, has become a zeitgeist of urban restructuring, a master narrative in the emerging built environment of the 1990s. Yet contemporary urban theory, whether debating the role of electronic technologies in precipitating "postmodern space," or discussing the dispersion of urban functions across poly-centered metropolitan "galaxies," has been strangely silent about the militarization of city life so grimly visible at the street level. Hollywood's pop apocalypses and pulp science fiction have been more realistic, and politically perceptive, in representing the programmed hardening of the urban surface in the wake of the social polarizations of the Reagan era. Images of carceral inner cities (*Escape from New York, Running Man*), high-tech police death squads (*Blade Runner*), sentient buildings (*Die Hard*), urban bantustans (*They Live!*), Vietnam-like street wars (*Colors*), and so on, only extrapolate from actually existing trends.

Such dystopian visions grasp the extent to which today's pharaonic scales of residential and commercial security supplant residual hopes for urban reform and social integration. The dire predictions of Richard Nixon's 1969 National Commission on the Causes and Prevention of Violence have been tragically fulfilled: we live in "fortress cities" brutally divided between "fortified cells" of affluent society and "places of terror" where the police battle the criminalized poor. The "Second Civil War" that began in the long hot summers of the 1960s has been institutionalized into the very structure of urban space. The old liberal paradigm of social control, attempting to balance repression with reform, has long been superseded by a rhetoric of social warfare that calculates the interests of the urban poor and the middle classes as a zero-sum game. In cities like Los Angeles, on the bad edge of postmodernity, one observes an unprecedented tendency to merge urban design, architecture and the police apparatus into a single, comprehensive security effort.

This epochal coalescence has far-reaching consequences for the social relations of the built environment. In the first place, the market provision of "security" generates its own paranoid demand. "Security" becomes a positional good defined by income access to private "protective services" and membership in some hardened residential enclave or restricted suburb. As a prestige symbol – and sometimes as the decisive borderline between the merely well-off and the "truly rich" – "security" has less to do with personal safety than with the degree of personal insulation, in residential, work, consumption and travel environments, from "unsavory" groups and individuals, even crowds in general.

Secondly, as William Whyte has observed of social intercourse in New York, "fear proves itself." The social perception of threat becomes a function of the security mobilization itself, not crime rates. Where there is an actual rising arc of street violence, as in Southcentral Los Angeles or Downtown Washington D.C., most of the carnage is self-contained within ethnic or class boundaries. Yet white middle-class imagination, absent from any firsthand knowledge of inner-city conditions, magnifies the perceived threat through a demonological lens. Surveys show that Milwaukee suburbanites are just as worried about violent crime as inner-city Washingtonians, despite a twentyfold difference in relative levels of mayhem. The media, whose function in this arena is to bury and obscure the daily economic violence of the city, ceaselessly throw up spectres of criminal underclasses and psychotic stalkers. Sensationalized accounts of killer youth gangs high on crack and shrilly racist evocations of marauding Willie Hortons foment the moral panics that reinforce and justify urban apartheid.

Moreover, the neo-military syntax of contemporary architecture insinuates violence and conjures imaginary dangers. In many instances the semiotics of so-called "defensible space" are just about as subtle as a swaggering white cop. Today's upscale, pseudo-public spaces – sumptuary malls, office centers, culture acropolises, and so on – are full of invisible signs warning off the underclass "Other." Although architectural critics are usually oblivious to how the built environment contributes to segregation, pariah groups – whether poor Latino families, young Black men, or elderly homeless white females – read the meaning immediately.

THE DESTRUCTION OF PUBLIC SPACE

The universal and ineluctable consequence of this crusade to secure the city is the destruction of accessible public space. The contemporary opprobrium attached to the term "street person" is in itself a harrowing index of the devaluation of public spaces. To reduce contact with untouchables, urban redevelopment has converted once vital pedestrian streets into traffic sewers and transformed public parks into temporary receptacles for the homeless and wretched. The American city, as many critics have recognized, is being systematically turned inside out – or, rather, outside in. The valorized spaces of the new megastructures and super-malls are concentrated in the center, street frontage is denuded, public activity is sorted into strictly functional compartments, and circulation is internalized in corridors under the gaze of private police.

The privatization of the architectural public realm, moreover, is shadowed by parallel restructurings of electronic space, as heavily policed, pay-access "information orders," elite databases and subscription cable services appropriate parts of the invisible agora. Both processes, of course, mirror the deregulation of the economy and the recession of non-market entitlements. The decline of urban liberalism has been accompanied by the death of what might be called the "Olmstedian vision" of public space. Frederick Law Olmsted, it will be recalled, was North America's Haussmann, as well as the Father of Central Park. In the wake of Manhattan's "Commune" of 1863, the great Draft Riot, he conceived public landscapes and parks as social safety-valves, mixing classes and ethnicities in common (bourgeois) recreations and enjoyments. As Manfredo Tafuri has shown in his well-known study of Rockefeller Center, the same principle animated the construction of the canonical urban spaces of the La Guardia–Roosevelt era.

This reformist vision of public space – as the emollient of class struggle, if not the bedrock of the American *polis* – is now as obsolete as Keynesian nostrums of full employment. In regard to the "mixing" of classes, contemporary urban America is more like Victorian England than Walt Whitman's or La

Guardia's New York. In Los Angeles, once-upon-a-time a demi-paradise of free beaches, luxurious parks, and "cruising strips," genuinely democratic space is all but extinct. The Oz-like archipelago of Westside pleasure domes – a continuum of tony malls, arts centers and gourmet strips – is reciprocally dependent upon the social imprisonment of the third-world service proletariat who live in increasingly repressive ghettoes and barrios. In a city of several million yearning immigrants, public amenities are radically shrinking, parks are becoming derelict and beaches more segregated, libraries and playgrounds are closing, youth congregations of ordinary kinds are banned, and the streets are becoming more desolate and dangerous.

Unsurprisingly, as in other American cities, municipal policy has taken its lead from the security offensive and the middle-class demand for increased spatial and social insulation. De facto disinvestment in traditional public space and recreation has supported the shift of fiscal resources to corporate-defined redevelopment priorities. A pliant city government – in this case ironically professing to represent a bi-racial coalition of liberal whites and Blacks – has collaborated in the massive privatization of public space and the subsidization of new, racist enclaves (benignly described as "urban villages"). Yet most current, giddy discussions of the "postmodern" scene in Los Angeles neglect entirely these overbearing aspects of counter-urbanization and counter-insurgency. A triumphal gloss – "urban renaissance," "city of the future," and so on – is laid over the brutalization of inner-city neighborhoods and the increasing South Africanization of its spatial relations. Even as the walls have come down in Eastern Europe, they are being erected all over Los Angeles.

The observations that follow take as their thesis the existence of this new class war (sometimes a continuation of the race war of the 1960s) at the level of the built environment. Although this is not a comprehensive account, which would require a thorough analysis of economic and political dynamics, these images and instances are meant to convince the reader that urban form is indeed following a repressive function in the political furrows of the Reagan–Bush era. Los Angeles, in its usual prefigurative mode, offers an especially disquieting catalogue of the emergent liaisons between architecture and the American police state.

THE FORBIDDEN CITY

The first militarist of space in Los Angeles was General Otis of the *Times*. Declaring himself at war with labor, he infused his surroundings with an unrelentingly bellicose air:

> He called his home in Los Angeles the Bivouac. Another house was known as the Outpost. The *Times* was known as the Fortress. The staff of the paper was the Phalanx. The *Times* building itself was more fortress than newspaper plant, there were turrets, battlements, sentry boxes. Inside he stored fifty rifles.

A great, menacing bronze eagle was the *Times*'s crown; a small, functional cannon was installed on the hood of Otis's touring car to intimidate onlookers. Not surprisingly, this overwrought display of aggression produced a response in kind. On 1 October 1910 the heavily fortified *Times* headquarters – citadel of the open shop on the West Coast – was destroyed in a catastrophic explosion blamed on union saboteurs.

Eighty years later, the spirit of General Otis has returned to subtly pervade Los Angeles' new "post-modern" Downtown: the emerging Pacific Rim financial complex which cascades, in rows of skyscrapers, from Bunker Hill southward along the Figueroa corridor. Redeveloped with public tax increments under the aegis of the powerful and largely unaccountable Community Redevelopment Agency (CRA), the Downtown project is one of the largest postwar urban designs in North America. Site assemblage and clearing on a vast scale, with little mobilized opposition, have resurrected land values, upon which big developers and off-shore capital (increasingly Japanese) have planted a series of billion-dollar, block-square megastructures: Crocker Center, the Bonaventure Hotel and Shopping Mall, the World Trade Center, the Broadway Plaza, Arco Center, CitiCorp Plaza, California Plaza, and so on. With historical landscapes erased, with megastructures and superblocks as primary components, and with an increasingly dense and self-contained circulation system, the new financial district is best conceived as a single, demonically self-referential hyperstructure, a Miesian skyscape raised to dementia.

Like similar megalomaniac complexes, tethered to fragmented and desolated Downtowns (for instance, the Renaissance Center in Detroit, the Peachtree and Omni Centers in Atlanta, and so on), Bunker Hill and the Figueroa corridor have provoked a storm of liberal objections against their abuse of scale and composition, their denigration of street landscape, and their confiscation of so much of the vital life activity of the center, now sequestered within subterranean concourses or privatized malls. Sam Hall Kaplan, the crusty urban critic of the *Times*, has been indefatigable in denouncing the anti-pedestrian bias of the new corporate citadel, with its fascist obliteration of street frontage. In his view the superimposition of "hermetically sealed fortresses" and air-dropped "pieces of suburbia" has "dammed the rivers of life" Downtown.

Yet Kaplan's vigorous defense of pedestrian democracy remains grounded in hackneyed liberal complaints about "bland design" and "elitist planning practices." Like most architectural critics, he rails against the oversights of urban design without recognizing the dimension of foresight, of explicit repressive intention, which has its roots in Los Angeles' ancient history of class and race warfare. Indeed, when Downtown's new "Gold Coast" is viewed en bloc from the standpoint of its interactions with other social areas and landscapes in the central city, the "fortress effect" emerges, not as an inadvertent failure of design, but as deliberate socio-spatial strategy.

The goals of this strategy may be summarized as a double repression: to raze all association with Downtown's past and to prevent any articulation with the non-Anglo urbanity of its future. Everywhere on the perimeter of redevelopment this strategy takes the form of a brutal architectural edge or glacis that defines the new Downtown as a citadel vis-à-vis the rest of the central city. Los Angeles is unusual amongst major urban renewal centers in preserving, however negligently, most of its circa 1900–30 Beaux Arts commercial core. At immense public cost, the corporate headquarters and financial district was shifted from the old Broadway–Spring corridor six blocks west to the greenfield site created by destroying the Bunker Hill residential neighborhood. To emphasize the "security" of the new Downtown, virtually all the traditional pedestrian links to the old center, including the famous Angels' Flight funicular railroad, were removed.

The logic of this entire operation is revealing. In other cities developers might have attempted to articulate the new skyscape and the old, exploiting the latter's extraordinary inventory of theaters and historic buildings to create a gentrified history – a gaslight district, Faneuil Market or Ghirardelli Square – as a support to middle-class residential colonization. But Los Angeles' redevelopers viewed property values in the old Broadway core as irreversibly eroded by the area's very centrality to public transport, and especially by its heavy use by Black and Mexican poor. In the wake of the Watts rebellion, and the perceived Black threat to crucial nodes of white power (spelled out in lurid detail in the McCone Commission Report), resegregated spatial security became the paramount concern. The Los Angeles Police Department abetted the flight of business from Broadway to the fortified redoubts of Bunker Hill by spreading scare literature typifying Black teenagers as dangerous gang members.

As a result, redevelopment massively reproduced spatial apartheid. The moat of the Harbor Freeway and the regraded palisades of Bunker Hill cut off the new financial core from the poor immigrant neighborhoods that surround it on every side. Along the base of California Plaza, Hill Street became a local Berlin Wall separating the publicly subsidized luxury of Bunker Hill from the lifeworld of Broadway, now reclaimed by Latino immigrants as their primary shopping and entertainment street. Because politically connected speculators are now redeveloping the northern end of the Broadway corridor (sometimes known as "Bunker Hill East"), the CRA is promising to restore pedestrian linkages to the Hill in the 1990s, including the Angels' Flight incline railroad. This, of course, only dramatizes the current bias against accessibility – that is to say, against any spatial interaction between old and new, poor and rich, except in the framework of gentrification or recolonization. Although a few white-collars venture into the Grand Central Market – a popular emporium of tropical produce and fresh foods – Latino shoppers or Saturday strollers never circulate in the Gucci precincts above Hill Street. The occasional appearance of a destitute street nomad in Broadway Plaza or in front of the Museum of Contemporary Art sets off a quiet panic; video cameras turn on their mounts and security guards adjust their belts.

Photographs of the old Downtown in its prime show mixed crowds of Anglo, Black and Latino pedestrians of different ages and classes. The contemporary Downtown "renaissance" is designed to make such heterogeneity virtually impossible. It is intended not just to "kill the street" as Kaplan fears, but to "kill the crowd," to eliminate that democratic admixture on the pavements and in the parks that Olmsted believed was America's antidote to European class polarizations. The Downtown hyperstructure – like some Buckminster Fuller post-Holocaust fantasy – is programmed to ensure a seamless continuum of middle-class work, consumption and recreation, without unwonted exposure to Downtown's working-class street environments. Indeed the totalitarian semiotics of ramparts and battlements, reflective glass and elevated pedways, rebukes any affinity or sympathy between different architectural or human orders. As in Otis's fortress *Times* building, this is the archisemiotics of class war.

"The Almost Perfect Town"

Landscape (1952)

John Brinckerhoff (J. B.) Jackson

Editors' Introduction

John Brinckerhoff Jackson (1909–1997) was a historian of landscapes as physical, cultural, and conceptual artifacts. To Jackson, the landscape was the totality of the natural and human-made environments that simultaneously surround and infuse all forms of human activity. He was particularly fascinated with vernacular built environments – every type of structure created by ordinary human beings.

In "The Almost Perfect Town," written in the early 1950s, Jackson describes a mythical Optimo City located in the American Southwest, but which could just as easily have been found almost anywhere in North America, Europe, Australia, and even parts of Asia and Africa beyond the margins of the great metropolitan regions. Optimo is a small place surrounded by rural land. Jackson observes that the world of Optimo City is still complete, precisely because the ties between country and town have not yet been broken. It is also important that Optimo has a history, however slender, that can be read in its architecture, in the layout of its streets, and in its traditional rivalry with (also mythical) Apache Center twenty miles away.

Jackson's Optimo City provides a revealing snapshot of small-town America in the 1950s before freeway construction, mass auto-ownership, and the proliferation of highway-based shopping centers. It also represents a timeless way of thinking about urban space. Savoring the peculiarities of this prototypical city representative of a specific time and place is characteristic of Jackson's humanistic approach to the study of landscapes. This kind of close observation builds empathy with people and places that enriches understanding and can inform good urban planning and sound public policy.

Jackson uses his loving, elegiac description of Optimo City as a way of criticizing developments in city planning after World War II that threatened to destroy local communities in the name of economic progress. The same forces are at work in China, India, the United Arab Emirates, and other rapidly changing societies as the present-day world urbanizes.

Jackson compares Optimo's unexceptional Courthouse Square to the Spanish plazas and the great European public squares and plazas that Camillo Sitte (p. 476) described as socially unifying communal centers. He identifies many vernacular examples of attractive "space between buildings" that Jan Gehl (p. 530) would like. Jackson noted with dismay that some of Optimo's business leaders in 1950 wanted to tear down the courthouse to build a parking lot and to replace it with the typical bureaucratic modernism of so many contemporary civic centers. Contrary to Jackson's view, many small towns have destroyed much of their charm in the sixty-plus years since this selection was written. Architects and planners like Peter Calthorpe and William Fulton (p. 360) and others associated with the Congress of the New Urbanism (p. 356) seek to recapture the human scale and charm of small towns such as Optimo city through neo-tradition plans and architecture.

Jackson may be regarded as an heir to the nineteenth-century French and English landscape gardening traditions and to the pioneering park-building of the United States' pioneer landscape architect, Frederick Law Olmsted (p. 321). What sets Jackson clearly apart from those traditional roots, however, is that, for Jackson, the urban landscape is not only a city's parks, public gardens, official buildings, and tree-lined boulevards, but also its highways, its shopping malls, diners, run-down warehouse districts, two-bedroom tract houses, and its slums. Jackson sees these many elements of the human-made landscape not only as physical objects, but also as social constructs full of meaning and moral implication. It is hardly surprising, then, that J.B. Jackson's approximate version of urban utopia – his almost perfect place – is the commonplace, vernacular, Main Street environment of a typical American small town.

John Brinkerhoff Jackson was born in 1909 in France to independently wealthy American parents. His childhood was divided between Washington, DC, his uncle's farm in rural Wagon Mound, New Mexico, elite Choate and Deerfield private preparatory schools in small, historic New England towns, and boarding schools in France and Switzerland where he became fluent in French, German, and Spanish. Jackson studied art and was proficient in sketching landscapes. After completing a BA degree in history and literature at Harvard in 1932 Jackson attended the University of Wisconsin's experimental college, where lectures by Lewis Mumford (p. 91) influenced his approach to architecture and urbanism. Later Jackson dabbled in architecture at the Massachusetts Institute of Technology, worked as a newspaper reporter, studied commercial drawing in Vienna, traveled through central Europe by motorcycle, and worked as a cowboy in Cimarron, New Mexico. As a military intelligence officer during World War II, Jackson's command of French and German and his European experience propelled him quickly to the rank of major. Traveling with allied forces in France, Jackson studied the works of French geographers and combed the libraries of resident landowners for geographical information to advise about troop movements.

After other travels and false starts, finally, in the spring of 1951, nineteen years after he graduated from college, Jackson launched *Landscape* magazine. *Landscape*'s articles were written without footnotes and aimed at general readers. For the first two years Jackson wrote almost all of the articles under the name J.B. Jackson and various pseudonyms. He also published his translations of French, German, Spanish, and Italian landscape journal articles and excerpts from books on the cultural landscape, previously unavailable to English language readers. *Landscape* contained the odd mix of observation of ordinary landscapes and philosophical musings on the meaning of vernacular architecture that Jackson had constructed from his eclectic background and which characterized all of his later writings. It attracted few readers and consistently lost money. But Lewis Mumford (p. 91) and other influential intellectuals applauded Jackson's innovative approach to the cultural landscape. In 1967 Jackson began teaching large lecture courses on the history of American and European landscapes at the University of California, Berkeley's landscape architecture and geography departments and in 1969 similar courses at Harvard in visual and environmental studies and landscape architecture. He relinquished his role as editor of *Landscape* magazine in 1968. Jackson retired from Berkeley in 1977 and Harvard in 1979. He continued writing short, unpretentious, brilliant essays on cultural landscapes until his death in 1996 at the age of 87. This selection is from *Landscape* (the journal that Jackson founded) and is reprinted in *A Sense of Place: A Sense of Time* (New Haven, CT: Yale University Press, 1996).

Other of Jackson's books and collections of essays include *Landscapes* (Amherst, MA: University of Massachusetts Press, 1970), *American Space* (New York: Norton, 1972), *Discovering the Vernacular Landscape* (New Haven, CT: Yale University Press, 1984), *The Necessity for Ruins, and Other Topics* (Amherst, MA: University of Massachusetts Press, 1980). The journal *Landscape* that Jackson founded and edited for many years contains additional writings by J.B. Jackson himself (under his own name and pseudonyms) and articles exploring the meaning of landscape by the many academic and applied landscape architects, urban planners, landscape designers, historians, and others that Jackson inspired.

A complete bibliography of Jackson's writing by Helen Lefkowitz Horowitz is in J.B. Jackson and Helen Lefkowitz Horowitz, *Landscape in Sight: Looking at America* (New Haven, CT: Yale University Press, 2000). Jackson's papers are housed in the Center for Southwest Studies, University Libraries, University of New Mexico, Albuquerque.

Books based on Jackson's approach by his former student and successor at Harvard, John R. Stilgoe (the Robert and Lois Orchard Professor of the History of Landscape at Harvard) include *Common Landscape of America, 1580 to 1845* (New Haven, CT: Yale University Press, 1983), *Outside Lies Magic: Regaining History and Awareness in Everyday Places* (New York: Walker, 1998), *Train Time: Railroads and the Imminent Reshaping of the United States Landscape* (Charlottesville, VA: University of Virgina Press), *Borderland: Origins of the*

American Suburb, 1820–1939 (New Haven, CT: Yale University Press, 1990), and *Alongshore* (New Haven, CT: Yale University Press).

Books based on Jackson's approach by his former student and successor at University of California, Berkeley, architecture professor Paul Groth include *Living Downtown: The History of Residential Hotels in the United States* (Berkeley, CA: University of California Press, 1994) and anthologies titled *Everyday America: Cultural Landscape Studies after J. B. Jackson*, coedited with Chris Wilson (Berkeley, CA: University of California Press, 2003) and *Understanding Ordinary Landscapes*, coedited with Todd Bressi (New Haven, CT: Yale University Press, 1997).

Other books on vernacular architecture include Robert Venturi, Denise Scott Brown, and Steven Izenour, *Learning from Las Vegas: The Forgotten Symbolism of Architectural Form* (Cambridge, MA: MIT Press, 1972), Jim Heimann, *California Crazy: Roadside Vernacular Architecture* (San Francisco, CA: Chronicle Books, 1981), Dell Upton and John M. Vlach, *Common Places: Readings in American Vernacular* (Atlanta, GA: University of Georgia Press, 1986), Bernard Rudofsky, *Architecture without Architects: A Short Introduction to Non-pedigreed Architecture* (Albuquerque, NM: University of New Mexico Press, 1987), Ronald W. Haase, *Classic Cracker: Florida's Wood-frame Vernacular Architecture* (Sarasota, FL: Pineapple Press, 1992), *Front Yard America: The Evolution and Meanings of a Vernacular Domestic Landscape* (Bowling Green, OH: Bowling Green State University, 1993), Thomas Carter (ed.), *Images of an American Land: Vernacular Architecture in the Western United States* (Albuquerque, NM: University of New Mexico Press, 1997), Colleen Josephine Sheehy, Fred E.H. Schroeder, and John Chase, *Glitter Stucco and Dumpster Diving: Reflections on Building Production in the Vernacular City* (London: Verso, 1998), Henri Glassie, *Vernacular Architecture* (Bloomington, IN: Indiana University Press, 2000), Paul Oliver, *Built to Meet Needs* (Oxford: Architectural Press, 2006), Herbert Gottfried and Jan Jennings, *American Vernacular Architecture and Regional Design: Cultural Process and Environmental Response* (New York: Norton, 2009), William Kingston Heath, *Vernacular Buildings and Interiors: 1870–1960* (New York: Architectural Press, 2009), and Paul Oliver (ed.), *Encyclopedia of Vernacular Architecture of the World* (Cambridge: Cambridge University Press, 1998).

> Optimo City (pop. 10,783, alt. 2,100 ft.), situated on a small rise overlooking the N. branch of the Apache River, is a farm and ranch center served by a spur of the S.P. County seat of Sheridan Co. Optimo City (originally established in 1843 as Ft. Gaffney). It was the scene of a bloody encounter with a party of marauding Indians in 1857. (See marker on courthouse lawn.) It is the location of a state Insane Asylum, of a sorghum processing plant and an overall factory. Annual County Fair and Cowboy Roundup Sept. 4. The highway now passes through a rolling countryside devoted to grain crops and cattle raising.

Thus would the state guide dispose of Optimo City and hasten on to a more spirited topic if Optimo City as such existed. Optimo City, however, is not one town, it is a hundred or more towns, all very much alike, scattered across the United States from the Alleghenies to the Pacific, most numerous west of the Mississippi and south of the Platte. When, for instance, you travel through Texas and Oklahoma and New Mexico and even parts of Kansas and Missouri, Optimo City is the blur of filling stations and motels you occasionally pass; the solitary traffic light, the glimpse up a side street of an elephantine courthouse surrounded by elms and sycamores, the brief congestion of mud-spattered pickup trucks that slows you down before you hit the open road once more. And fifty miles farther on Optimo City's identical twin appears on the horizon, and a half dozen more Optimos beyond that until at last, with some relief, you reach the metropolis with its new housing developments and factories and the cluster of downtown skyscrapers.

Optimo City, then, is actually a very familiar feature of the American landscape. But since you have never stopped there except to buy gas, it might be well to know it a little better. What is there to see? Not a great deal, yet more than you would at first suspect.

Optimo, being after all an imaginary average small town, has to have had an average small-town history, or at least a Western version of that average. The original Fort Gaffney (named after some inconspicuous worthy in the U.S. Army) was really little more than a stockade on a bluff overlooking a ford in the river; a few roads or trails in the old days straggled out into the plain (or desert as they called it then), lost heart and disappeared two or three miles from town. Occasionally even today someone digs up a fragment of the palisade or a bit of rust-eaten hardware in the backyards of the houses near the center of town, and the historical society possesses what it claims is the key to the principal gate. But on the whole, Optimo City is not much interested in its martial past. The fort as a military installation ceased to exist during the Civil War, and the last of the pioneers died a half century ago before anyone had the historical sense to take down his story. And when the county seat was located in the town the name was changed from Fort Gaffney with its frontier connotation to Optimo, which means (so the townspeople will tell you) "I hope for the best" in Latin.

What Optimo is really proud of even now is its identity as county seat. Sheridan County (and you will do well to remember that it was NOT named after the notorious Union general but after Horace Sheridan, an early member of the territorial legislature; Optimo still feels strongly about what it calls the War between the States) was organized in the 1870s and there ensued a brief but lively competition for the possession of the courthouse between Optimo and the next largest settlement, Apache Center, twenty miles away. Optimo City won, and Apache Center, a cowtown with one paved street, is not allowed to forget the fact. The football and basketball games between the Optimo Cougars and the Apache Braves are still characterized by a very special sort of rivalry. No matter how badly beaten Optimo City often is, it consoles itself by remembering that it is still the county seat, and that Apache Center, in spite of the brute cunning of its team, has still only one street paved. We shall presently come back to the meaning of that boast.

To get on with the history of Optimo.

THE INFLEXIBLE GRIDIRON

Aided by the state and Army engineers, the city fathers, back in the 1870s, surveyed and laid out the new metropolis. As a matter of course they located a square or public place in the center of the town and eventually they built their courthouse in the middle of the square; such having been the layout of every county seat these Western Americans had ever seen. Streets led from the center of each side of the square, being named Main Street North and South, and Sheridan Street East and West. Eventually these four streets and the square were surrounded by a gridiron pattern of streets and avenues – all numbered or lettered, and all of them totally oblivious of the topography of the town. Some streets charge up impossibly steep slopes, straight as an arrow; others lead off into the tangle of alders and cottonwoods near the river and get lost.

Strangely enough, this inflexibility in the plan has had some very pleasant results. South Main Street, which leads from the square down to the river, was too steep in the old days for heavily laden wagons to climb in wet weather, so at the foot of it on the flats near the river those merchants who dealt in farm produce and farm equipment built their stores and warehouses. The blacksmith and welder, the hay and grain supply, and finally the auction ring and the farmers' market found South Main the best location in town for their purpose – which purpose being primarily dealing with out-of-town farmers and ranchers. And when, after considerable pressure on the legislature and much resistance from Apache Center (which already had a railroad), the Southern Pacific built a spur to Optimo, the depot was naturally built at the foot of South Main. And of course the grain elevator and the stockyards were built near the railroad. The railroad spur was intended to make Optimo into a manufacturing city, and never did; all that ever came was a small overall factory and a plant for processing sorghum with a combined payroll of about 150. Most of the workers in the two establishments are Mexicans from south of the border – locally referred to in polite circles as "Latinos" or "Hispanos." They have built for themselves flimsy little houses under the cottonwoods and next to the river. "If ever we have an epidemic in Optimo," the men at the courthouse remark, "it will break out first of all in those Latino shacks." But they have done nothing as yet about providing them with better houses, and probably never will.

DOWNTOWN AND UPTOWN

Depot, market, factories, warehouses, slum – these features, combined with the fascination of the river bank and stockyards and the assorted public of railroaders and Latinos and occasional ranch hands – have all given South Main a very definite character: easy-going, loud, colorful, and perhaps during fair week or at shipping time a little disreputable. Boys on the Cougar football squad have specific orders to stay away from South Main, but they don't. Actually the whole of Optimo looks on the section with indulgence and pride; it makes the townspeople feel that they understand metropolitan problems when they can compare South Main with the New York waterfront.

North Main, up on the heights beyond the Courthouse Square and past the two or three blocks of retail stores, is (on the other hand) the very finest part of Optimo. The northwestern section of town, with its tree-shaded streets, its view over the river and the prairie, its summer breezes, has always been identified with wealth and fashion as Optimo understands them. Colonel Ephraim Powell (Confederate Army, Ret., owner of some of the best ranch country in the region) built his bride a handsome limestone house with a slate roof and a tower, and Walter Slymaker, proprietor of Slymaker's Mercantile and of the grain elevator, not to be outdone, built an even larger house farther up Main; so did Hooperson, first president of the bank. There are a dozen such houses in all, stone or Milwaukee brick with piazzas (or galleries, as the old timers still call them) and large, untidy gardens around them. It is worth noting, by the way, that the brightest claim to aristocratic heritage is this: grandfather came out West for his health. New England may have its "Mayflower" and "Arabella," east Texas its Three Hundred Founding Families, New Mexico its Conquistadores; but Optimo is loyal to the image of the delicate young college graduate who arrived by train with his law books, his set of Dickens, his taste for wine, and the custom of dressing for dinner. This legendary figure has about seen his day in the small talk of Optimo society, and the younger generation frankly doubts his having ever existed; but he (or his ghost) had a definite effect on local manners and ways of living. At all events, because of this memory Optimo looks down on those Western mining towns where Sarah Bernhardt and de Reszke and Oscar Wilde seem to have played so many one-night stands in now-vanished opera houses.

A WORLD IN ITSELF

Wickedness – or the suggestion of wickedness – at one end of Main, affluence and respectability at the other. How about Sheridan Street running East and West? That is where you'll find most of the stores; in the first four or five blocks on either side of the Courthouse Square. They form a rampart: narrow brick houses, most of them two stories high with elaborate cornices and long narrow windows; all of them devoid of modern commercial graces of chromium and black glass and artful window display, all of them ugly but all of them pretty uniform; and so you have on Sheridan Street something rarely seen in urban America: a harmonious and restful and dignified business section. Only eight or ten blocks of it in all, to be sure; turn any corner and you are at once in a residential area.

Here there is block after block of one-story frame houses with trees in front and picket fences or hedges; no sidewalk after the first block or so; a hideous church (without a cemetery of course); a small-time auto repair shop in someone's back yard; dirt roadway; and if you follow the road a few blocks more – say to 10th Street (after that there are no more signs) – you are likely to see a tractor turn into someone's drive with wisps of freshly cut alfalfa clinging to the vertical sickle bar. The countryside is that close to the heart of Optimo City, farmers are that much part of the town. And the glimpse of the tractor (like the glimpse of a deer or a fox driven out of the hills by a heavy winter) restores for a moment a feeling for an old kinship that seemed to have been destroyed forever. But this is what makes Optimo, the hundreds of Optimos throughout America, so valuable; the ties between country and town have not yet been broken. Limited though it may well be in many ways, the world of Optimo City is still complete.

The center of this world is Courthouse Square, with the courthouse, ponderous, barbaric, and imposing, in the center of that. The building and its woebegone little park not only interrupts the vistas of Main and Sheridan – it was intended to do this – it also interrupts the flow of traffic in all four directions. A sluggish eddy of vehicles and pedestrians is the result, Optimo's animate existence slowed and intensified. The houses on the four sides of the square are of the same vintage (and the same general architecture) as the monument in their midst: mid-nineteenth-century brick or stone; cornices like the brims of hats, fancy dripstones over the arched windows like eyebrows; painted blood-red or mustard-yellow or white; identical except for the six-story Gaffney Hotel and the classicism of the First National Bank.

Every house has a tin roof porch extending over the sidewalk, a sort of permanent awning which protects passersby and incidentally conceals the motley of store windows and signs. To walk around the square and down Sheridan Street under a succession of these galleries or metal awnings, crossing the strips of bright sunlight between the roofs of different height, is one of the delights of Optimo – one of its amenities in the English use of that word. You begin to understand why the Courthouse Square is such a popular part of town.

SATURDAY NIGHTS – BRIGHT LIGHTS

Saturday, of course, is the best day for seeing the full tide of human existence in Sheridan County. The rows of parked pickups are like cattle in a feed lot; the sidewalks in front of Slymaker's Mercantile, the Ranch Cafe, Sears, the drugstore, resound to the mincing steps of cowboy boots; farmers and ranchers, thumbs in their pants pockets, gather in groups to lament the drought (there is always a drought) and those men in Washington, while their wives go from store to movie house to store. Radios, jukeboxes, the bell in the courthouse tower; the teenagers doing "shave-and-a-haircut; bay rum" on the horns of their parents' cars as they drive round and round the square. The smell of hot coffee, beer, popcorn, exhaust, alfalfa, cow manure. A man is trying to sell a truckload of grapefruit so that he can buy a truckload of cinderblocks to sell somewhere else. Dogs; 10-year-old cowboys firing cap pistols at each other. The air is full of pigeons, floating candy wrappers, the flat strong accent erroneously called Texan.

All these people are here in the center of Optimo for many reasons – for sociability first of all, for news, for the spending and making of money; for relaxation. "Jim Guthrie and wife were in town last week, visiting friends and transacting business," is the way the *Sheridan Sentinel* describes it; and almost all of Jim Guthrie's business takes place in the square. That is one of the peculiarities of Optimo and one of the reasons why the square as an institution is so important. For it is around the square that the oldest and most essential urban (or county) services are established. Here are the firms under local control and ownership, those devoted almost exclusively to the interest of the surrounding countryside. Upstairs are the lawyers, doctors, dentists, insurance firms, the public stenographer, the Farm Bureau. Downstairs are the bank, the prescription drugstore, the newspaper office, and of course Slymaker's Mercantile and the Ranch Cafe.

INFLUENCE OF THE COURTHOUSE

Why have the chain stores not invaded this part of town in greater force? Some have already got a foothold, but most of them are at the far end of Sheridan or even out on the Federal Highway. The presence of the courthouse is partly responsible. The traditional services want to be as near the courthouse as they can, and real-estate values are high. The courthouse itself attracts so many out-of-town visitors that the problem of parking is acute. The only solution that occurs to the enlightened minds of the Chamber of Commerce is to tear the courthouse down, use the place for parking, and build a new one somewhere else. They have already had an architect draw a sketch of a new courthouse to go at the far end of Main Street: a chaste concrete cube with vertical motifs between the windows – a fine specimen of bureaucrat modernism. But the trouble is, where to get the money for a new courthouse when the old one is still quite evidently adequate and in constant use?

If you enter the courthouse you will be amazed by two things: the horrifying architecture of the place, and the variety of functions it fills. Courthouse means of course courtrooms, and there are two of those. Then there is the office of the County Treasurer, the Road Commissioner, the School Board, the Agricultural Agent, the Extension Agent, Sanitary Inspector, and usually a group of Federal agencies as well – PMA, Soil Conservation, FHA and so on. Finally the Red Cross, the Boy Scouts, and the District Nurse. No doubt many of these offices are tiresome examples of government interference in private matters; just the same, they are for better or worse part of almost every farmer's and rancher's business, and the courthouse, in spite of all the talk about county consolidation, is a more important place than ever.

As it is, the ugly old building has conferred upon Optimo a blessing which many larger and richer American towns can envy: a center for civic activity and a symbol for civic pride – something as different from the modern "civic center" as day is from night. Contrast the array of classic edifices, lost in the midst of waste space, the meaningless pomp of flagpoles and war memorials and dribbling fountains of any American city from San Francisco to Washington with the animation and harmony and the almost domestic intimacy of Optimo Courthouse Square, and you have a pretty good measure of what is wrong with much American city planning: civic consciousness has been divorced from everyday life, put in a special zone all by itself. Optimo City has its zones; but they are organically related to one another.

Doubtless the time will never come that the square is studied as a work of art. Why should it be? The craftsmanship in the details, the architecture of the building, the notions of urbanism in the layout of the square itself are all on a very countrified level. Still, such a square as this has dignity and even charm. The charm is perhaps antiquarian – a bit of rural America of seventy-five years ago; the dignity is something else again. It derives from the function of the courthouse and the square, and from its peculiarly national character.

COMMUNAL CENTER

The practice of erecting a public building in the center of an open place is in fact pretty well confined to America – more specifically to nineteenth-century America. The vast open areas favored by eighteenth-century European planners were usually kept free of construction, and public buildings – churches and palaces and law-courts – were located to face these squares; to command them, as it were. But they were not allowed to interfere with the original open effect. Even the plans of eighteenth-century American cities, such as Philadelphia and Reading and Savannah and Washington, always left the square or public place intact. Spanish America, of course, provides the best illustrations of all; the plaza, nine times out of ten, is surrounded by public buildings, but it is left free. Yet almost every American town laid out after (say) 1820 deliberately planted a public building in the center of its square. Sometimes it was a school, sometimes a city hall, more often a courthouse, and it was always approachable from all four sides and always as conspicuous as possible.

Why? Why did these pioneer city fathers go counter to the taste of the past in this matter? One guess is as good as another. Perhaps they were so proud of their representative institutions that they wanted to give their public buildings the best location available. Perhaps frontier America was following an aesthetic movement, already at that date strong in Europe, that held that an open space was improved when it contained some prominent free-standing object – an obelisk or a statue or a triumphal arch. However that may have been, the pioneer Americans went Europe one better, and put the largest building in town right in the center of the square.

Thus the square ceased to be thought of in nineteenth-century America as a vacant space; it became a container or (if you prefer) a frame. A frame, so it happened, not merely for the courthouse, but for all activity of a communal sort. Few aesthetic experiments have ever produced such brilliantly practical results. A society which had long since ceased to rally around the individual leader and his residence and which was rapidly tiring of rallying around the meetinghouse or church all at once found a new symbol: local representative government, or the courthouse. A good deal of flagwaving resulted – as European travelers have always told us – and a good deal of very poor "representational" architecture; but Optimo acquired something to be proud of, something to moderate that American tendency to think of every town as existing entirely for money-making purposes.

SYMBOL OF INDEPENDENCE

At this juncture the protesting voice of the Chamber of Commerce is heard. "One moment. Before you finish with our courthouse you had better hear the other side of the question. If the courthouse were torn down we would not only have more parking space – sorely needed in Optimo – we would also get funds for widening Main Street into a four-lane highway. If Main Street were widened Optimo could attract many new businesses catering to tourists and other transients – restaurants and motels and garages and all sorts of drive-in establishments. In the last ten years" (continues the Chamber of Commerce) "Optimo has grown by twelve hundred. Twelve hundred! At that rate we'll still be a small town of less than twenty thousand in 1999. But if we had new businesses we'd grow fast and have better schools and a new hospital, and the young people wouldn't move to the cities. Or do you expect Optimo to go on depending on a few hundred tight-fisted farmers and ranchers for its livelihood?" The voice, now shaking with emotion, adds something about "eliminating" South Main by means of an embankment and a clover leaf and picnic grounds for tourists under the cottonwoods where the Latinos still reside.

These suggestions are very sensible ones on the whole. Translate them into more general terms and what they amount to is this: if we want to get ahead, the best thing to do is break with our own past, become as independent as possible of our immediate environment and at the same time become almost completely dependent for our well-being on some remote outside resource. Whatever you may think of such a program, you cannot very well deny that it has been successful for a large number of American towns. Think of the hayseed communities which have suddenly found themselves next to an oil field or a large factory or an Army installation, and which have cashed in on their good fortune by transforming themselves overnight, turning their backs on their former sources of income, and tripling their population in a few years! It is true that these towns put all their eggs in one basket, that they are totally at the mercy of some large enterprise quite beyond their control. But think of the freedom from local environment; think of the excitement and the money! Given the same circumstances – and the Southwest is full of surprises still – why should Optimo not do the same?

A COMMON DESTINY

Because there are many different kinds of towns just as there are many different kinds of men, a development which is good for one kind can be death on another. Apache Center (to use that abject community as an example), with its stockyards and its one paved street and its very limited responsibility to the county, as a community might well become a boom-town and no one would be worse off. Optimo seems to have a different destiny. For almost a hundred years – a long time in this part of the world – it has been identified with the surrounding landscape and been an essential part of it. Whatever wealth it possesses has come from the farms and ranches, not from the overall factory or from tourists. The bankers and merchants will tell you, of course, that without their ceaseless efforts and their vision the countryside could never have existed; the farmers and ranchers consider Optimo's prosperity and importance entirely their own creation. Both parties are right to the extent that the town is part of the landscape – one might even say part of every farm, since much farm business takes place in the town itself.

Now if Optimo suddenly became a year-round tourist resort, or the overall capital of the Southwest; what would happen to that relationship, do you suppose? It would vanish. The farmers and ranchers would soon find themselves crowded out, and would go elsewhere for those services and benefits which they now enjoy in Optimo. And as for Optimo itself, it would soon achieve the flow of traffic, the new store fronts, the housing developments, the payrolls and bank accounts it cannot help dreaming about; and in the same process achieve a total social and physical dislocation, and a loss of a sense of its own identity. County Seat of Sheridan County? Yes; but much more important: Southwestern branch of the "American Cloak and Garment Corporation"; or the LITTLE TOWN WITH THE BIG WELCOME – 300 tourist beds which, when empty for one night out of three, threaten bankruptcy to half the town.

As of the present, Optimo remains pretty much as it has been for the last generation. The Federal Highway still bypasses the center (what a roadblock, symbolical as well as actual, that courthouse is!); so if you want to see Optimo, you had better turn off at the top of the hill near the watertower of the lunatic asylum – now called Fairview State Rest Home, and with the hideous high fence around it torn down. The dirt road eventually becomes North Main. The old Slymaker place is still intact. The Powell mansion, galleries and all, belongs to the American Legion, and a funeral home has taken over the Hooperson house. Then comes downtown Optimo; and then the courthouse, huge and graceless, in detail and proportion more like a monstrous birdhouse than a monument. Stop here. You'll find nothing of interest in the stores, and no architectural gems down a side street. Even if there were, no one would be able to point them out. The historical society, largely in the hands of ladies, thinks of antiquity in terms of antiques, and art as anything that looks pretty on the mantelpiece.

The weather is likely to be scorching hot and dry, with a wild ineffectual breeze in the elms and sycamores. You'll find no restaurant in town with atmosphere – no chandeliers made out of wagon wheels, no wall decorations of famous brands, no bar disguised as the Hitching Rail or the Old Corral. Under a high ceiling with a two-bladed fan in the middle, you'll eat ham hock and beans, hot bread, iced tea without lemon, and like it or go without. But as compensation of sorts at the next table there will be two ranchers eating with their hats on, and discussing the affairs, public and private, of Optimo City. To hear them talk, you'd think they owned the town.

That's about all. There's the market at the foot

of South Main, the Latino shacks around the overall factory, a grove of cottonwoods, and the Apache River (North Branch) trickling down a bed ten times too big; and then the open country. You may be glad to have left Optimo behind.

Or you may have liked it, and found it pleasantly old-fashioned. Perhaps it is; but it is in no danger of dying out quite yet. As we said to begin with, there is another Optimo City fifty miles farther on. The country is covered with them. Indeed they are so numerous that it sometimes seems as if Optimo and rural America were one and indivisible.

"The Causes of Sprawl"

from *Sprawl: A Compact History* (2005)

Robert Bruegmann

Editors' Introduction

No spatial policy issue has preoccupied urbanists more than urban sprawl. Spread out, low density, suburban development patterns are the norm in virtually every American metropolitan area. Each decennial U.S. census since World War II shows that new residential development is occurring at average densities lower than the average density of already built up areas of metropolitan regions. As Robert Fishman describes (p. 75) the eastern seaboard of the United States now has stretches of low-density technoburbia stretching from north of Boston to the tip of Florida. While European cities are generally more compact than their North American counterparts, the same pattern is discernible in Europe. In China, India, and fast-growing megacities around the world, the scale is different, but the pattern of relatively low-density development around existing urban cores is also common. Shanghai and Mumbai, for example, have extensive lower density residential development around their very dense centers.

Why do we have sprawl? Is sprawl good or bad? Should we do something to control it? You may well be surprised by Robert Bruegmann's opinions on these questions.

Most people hate sprawl, at least when asked their opinion of sprawl in the abstract rather than how it might affect their own ability to own a single family home with a large yard and one car for each adult member of their family. They deplore the loss of open space and farmland, long drives, traffic congestion, and the boring uniformity of suburban tract developments. Most academic social scientists, landscape architects, architects, and urban planners also condemn sprawl for using up prime farm land, threatening plant and animal communities, contributing to air pollution and global climate change, requiring expensive new infrastructure, increasing commute times, fostering racial and class segregation, isolating women, and contributing to a host of other ills.

Bruegmann does not agree. He argues that we have sprawl because that is what people want: a natural market response to the desires of millions of individuals. In this selection Bruegmann describes and then attacks most of the liberal explanations for why sprawl occurs and in the process advances a defense of sprawl.

Do Americans have sprawl because of their frontier roots and anti-urban bias? That's one common explanation. Bruegmann disagrees. He notes that the amount of space per capita in European cities and American cities is converging, despite the fact that Europeans never experienced a frontier of unlimited land. For all their pride in lovely urban places, Bruegmann says, the French and Italians today are exhibiting the same kind of anti-urban bias as Americans.

How about racism? Do Americans have sprawl because middle- and upper-income whites have fled central cities to get away from poor Blacks and foreign immigrants? Maybe in some cases. But Bruegmann makes three counterarguments: first, that relatively homogenous cities like Minneapolis – where most residents come from Scandanavian stock and there are few Blacks – are sprawling about as much as other U.S. cities, second, middle- and upper-income Blacks have been just as eager as their white counterparts to move out to suburbs, and third, spatial ethnic and income segregation is prevalent worldwide.

At the core of the disagreement between Bruegmann and critics of sprawl is differing degrees of acceptance of private market forces. One of the arguments in favor of public intervention in cities that economist Wilbur Thompson advances (p. 274) is in cases of "market failure" where, left unfettered, private markets produce results that are damaging to society. According to this line of argument, the private market fails because self-serving individuals maximize their individual well-being at the expense of others. Everyone seeks to own a single-family house in a low-density suburb regardless of how his or her cumulative decisions will affect the region. Greedy developers force people to live in suburbs in order to maximize their own profits. Bruegmann questions all these common views. He notes that at the turn of the century, housing advocates attacked greedy developers for crowding people into city neighborhoods like New York's Lower East Side. At that time reformers like Clarence Perry (p. 486) advocated housing at much lower densities on the fringes of the New York Metropolitan region or in satellite communities. In other words early reformers favored sprawl to meet people's needs. Bruegmann rejects the notion that residents of suburbs have been forced or duped into living in low-density suburban developments rather than choosing to live there. In his view developers build suburbs because that is what people want.

Another set of explanations blames government for sprawl. Proponents of this view argue that spending large amounts of federal money on highways (rather than subways, light rail lines, or other public transportation) made auto-dependent sprawl development inevitable. Low interest rate government-backed government loans for single-family homes were a carrot for lower-middle income people to settle in low-density suburbs rather than cities. Allowing homeowners to write off their mortgage interest payments induced marginal homebuyers to buy suburban homes. Government's failure to regulate redlining (when private banks refused to make mortgage loans in risky inner city neighborhoods) and even engaging in redlining themselves, provided sticks forcing people to abandon viable inner city neighborhoods.

But are these arguments true? Bruegmann notes that the mortgage interest write-off applies to inner city housing equally to housing in suburbs. He argues that banks are happy to invest wherever they can make money. As evidence that highway construction as not an anti-urban conspiracy Bruegmann notes that many cities had highway construction plans long before the federal aid highway program of the 1950s and most cities welcomed highway construction.

Finally Bruegmann rejects the argument that technology, and specifically the invention of the automobile, is responsible for sprawl. He notes that Los Angeles had already assumed its low-density, polycentric form by 1920 before mass auto ownership. Los Angeles' electric streetcar system makes this possible.

If all of these liberal explanations – a frontier ethic, anti-urban bias, racism, greedy individuals and greedy developers, bad government policies, fiscal and regulatory carrots and sticks that punished people who sought to live in cities and rewarded people who did not, and the automobile – are not convincing explanations, what then is the reason we have sprawl?

Bruegmann gives two fundamental explanations: affluence and democratic institutions. In his view people want to live in low-density suburbs. As incomes rise more people can afford to do so. Democratic institutions allow people to choose for themselves, and people choose to live in low density, sprawling developments.

Robert Bruegmann is a professor of architecture and art history in the College of Architecture and Planning at the University of Illinois, Chicago. He also holds an appointment in the University of Illinois at Chicago's Program in Urban Planning and Policy. His fields of research and teaching are architectural, urban, landscape, and planning history and historic preservation. Professor Bruegmann has taught at the University of Pennsylvania, Philadelphia College of the Arts, the Massachusetts Institute of Technology, and Columbia University. He has worked for the Historic American Buildings Survey and Historic American Engineering Record of the National Park Service.

This selection is from Robert Bruegmann, *Sprawl: A Compact History* (Chicago, IL: University of Chicago Press, 2005). Other of Bruegmann's books include *Benicia: Portrait of an Early California Town: An Architectural History 1846 to the Present* (San Francisco, CA: 101 Productions, 1980), *Holabird & Roche/Holabird & Root: An Illustrated Catalog of Works 1910–1940*, three volumes (New York: Garland with Chicago Historical Society, 1991), and an edited volume titled *Modernism at Mid-Century: The Architecture of the United States Air Force Academy* (Chicago, IL: University of Chicago Press, 1994).

University of Southern California planning professors Peter Gordon and Harry W. Richardson agree with Bruegmann that sprawl is a rational market response and essentially desirable. Their views are summarized in "The

Debate on Sprawl and Compact Cities: Thoughts Based on the Congress of New Urbanism Charter," in H.S. Geyer (ed.), *Handbook of Urban Policy* (Cheltenham, UK: Edward Elgar, 2010).

Most urbanists and planners disagree with Bruegmann. Dolores Hayden and Jim Wark, *A Field Guide to Sprawl* (New York: Norton, 2006) is a readable overview with photographs that illustrate the extent and variety of sprawl.

A good statement of the position that sprawl is economically inefficient and produces severe negative externalities for commuting, the environment, and society is Robert Burchell, Anthony Downs, Sahan Mukherji, and Barbara McCann, *Sprawl Costs: Economic Impacts of Unchecked Development* (Washington, DC: Island Press, 2005). See also Anthony Flint, *This Land: The Battle over Sprawl and the Future of America* (Baltimore, MD: Johns Hopkins University Press, 2006). A description of state growth management programs to reduce sprawl is Jerry Weitz, *Sprawl Busting: State Programs to Guide Growth* (Chicago, IL: Planners Press, 1999).

Histories of American suburbanization (and the resultant sprawl) include Kenneth T. Jackson, *Crabgrass Frontier: The Suburbanization of the United States* (Oxford: Oxford University Press, 1987), Andrés Duany, Elizabeth Plater-Zyberk, and Jeff Speck, *Suburban Nation: The Rise of Sprawl and the Decline of the American Dream* (Boston, MA: North Point Press, 2001), and Dolores Hayden, *Building Suburbia: Green Fields and Urban Growth, 1820–2000* (New York: Vintage, 2004).

. . . What causes sprawl? The answers to this question have been remarkably varied and contradictory. Let's consider briefly several of these, starting first with those that assume that sprawl is peculiarly American and attempt to explain why the United States is different from other places and then moving to more general explanations.

ANTI-URBAN ATTITUDES AND RACISM AS A CAUSE

A number of observers, usually highbrow Europeans or Americans who live and work in the central city, account for the massive amount of sprawl in the United States by claiming that it is the result of national character traits. American cities are so different from European cities, they say, because Americans are at heart anti-urban, attached to unfettered individualism, low-density living, and automobile usage. But . . . the history of urban decentralization seems to suggest that many of the supposed differences in American and European cities and suburbs are less the result of inherent differences in these societies than a matter of timing. Cities on both continents are, if anything, converging when it comes to space used per capita, automobile ownership, or other similar measures. All of this casts considerable doubt on the theory that Americans are uniquely anti-urban.

In fact, it is probably only possible to call Americans anti-urban if one accepts a specific set of assumptions about urbanity made by members of a small cultural elite. This group likes to think of urbanity as the kind of life lived by people in apartments in dense city centers that contain major highbrow cultural institutions. In these dense centers, they believe, citizens are more tolerant and cosmopolitan because of their constant interaction with other citizens unlike themselves. It is this definition of urbanity – in many cases based on an idealized vision of the European city of the late eighteenth and nineteenth centuries – that many Americans and, increasingly, citizens throughout the world reject or, more often, simply ignore. If they thought about it at all, they wouldn't agree that highbrow culture is necessarily better than their own middle-class culture, and they would probably have little patience with the argument that they would be more tolerant if they lived in apartment buildings on densely built city streets or were forced to interact with people they would rather avoid. Most Americans do not like the dirt and disorder that characterized historic nineteenth-century industrial cities, and they may be indifferent if not hostile to the clubby culture of the downtown elite cultural groups, but there is little evidence that suburbanites are opposed to urbanity. They only want to rearrange the physical elements to make life more convenient and pleasant for themselves and to avoid the things that made nineteenth-century industrial cities so unpleasant for people who did not have a great deal of money.

It is true that some suburbanites see their environment as the opposite of the old central city, peripheral to their everyday lives and just another exit on the freeway. However, it is likely that the majority considers these two places as good for different things. For them, suburbia is a good place to live, work, and raise children, while downtown is a place to see ballgames, go to a nightclub, visit a museum, or do some special Christmas shopping. As the old downtowns remake themselves as tourist destinations and places of entertainment, it appears that they have, if anything, become a more valued part of the larger urban world.

Another common explanation of the growth of American suburbs and the rise of sprawl is that it was caused by white flight fueled by racism. Although no one would deny that race has played a key role in many aspects of American life, it is significant that urban areas with small minority populations like Minneapolis have sprawled in much the same way as urban areas with large minority populations like Chicago. It is also the case that when they have become affluent enough to do so, African Americans have been just as willing as their white counterparts to move out to the suburbs. The suburbs they choose are often ones with largely African American population. This suggests that there is no simple relationship between race and sprawl.

Nor is it plausible to suggest that the segregating out by income level, race, and ethnicity is peculiarly American. These kinds of segregation have been visible not just in American suburbs but in cities and suburbs all over the world, particularly when large disparities in income is a major factor. It was certainly the case in all nineteenth-century industrial cities and today, whether in the old public housing of suburban Stockholm or Paris or the favelas and shantytowns of São Paulo or Istanbul, segregation of immigrants and poorer residents by skin color, religion, and income level is a pervasive feature of contemporary urban life.

ECONOMIC FACTORS AND THE CAPITALIST SYSTEM AS A CAUSE

Probably the single most common explanation of sprawl is that it has been a direct by-product of an insufficiently regulated capitalist system. This argument rests in great part on two dubious propositions. The first is that economic forces are the prime factor in human interactions, the driving force in most aspects of life, and everything else is secondary. In fact, although economic conditions have always had a strong relationship to urban forms, the history we have reviewed suggests that this influence is much less direct and obvious than many people believe. Similar urban forms can evolve in very different economic circumstances; different urban forms can accompany similar economic circumstances. Further, the history we have reviewed suggests that urban form is not just an effect but also a cause of economic conditions. Every decision by every individual about where he or she lives or works or plays will have repercussions throughout the system.

The second dubious notion is that there are many circumstances in which the capitalist system inherently doesn't work well, leading to "market failure" and unhappy results on the ground. Many individuals have claimed that sprawl is a logical result of capitalism because this kind of economic system induces buyers and sellers to act in ways to further their own good even at the expense of their neighbors or the common good. So, for example, many families, each acting to secure for themselves a location at the very edge of the urban area so they can enjoy proximity to nature, could produce a situation where only a handful will be able to enjoy the view, and even they will soon be outflanked. Or, it has often been claimed, developers, left to themselves, will maximize their profits by building at low densities no matter what their customers might actually want because building detached single-family houses is more profitable than building apartment buildings. Some observers claim that this fixation with the bottom line will inexorably produce settlement patterns that are inefficient, socially and environmentally harmful, ugly, or all of these. Therefore, government must intervene to produce better results. The kind of behavior that puts personal advantage over common good is hardly limited to matters economic, however. The same homebuyers who might try to maximize their personal advantage in buying a suburban house are the voters who elect government officials and who push for land-use regulations that will benefit them, often at the price of other parts of the population. Is it logical to think that landowners would suddenly act in a completely different fashion when they engage in political rather than economic transactions? Nor is the kind of behavior that puts personal interest above community welfare peculiar to low-density settlements. The resident of a central city who tries to block the badly needed expansion of a hospital next door to his apartment building because it would block his view is acting in a similar fashion. So it seems illogical to make any close link between the capitalist system and sprawl.

The notion that sprawl is the inevitable unhappy result of laissez-faire capitalism, moreover, turns on its head the analysis of reformers in the nineteenth and early twentieth centuries who were convinced that unregulated private forces would lead inexorably to excessively high densities. Housing advocate Benjamin Marsh, for example, bitterly attacked developers in 1910 for crowding as many people as possible into a single acre in order to maximize their profits. He was particularly indignant over the claim of some developers who argued that high density helped create community. He considered this to be no more than a cynical justification for greed and stated that the best solution for people of modest incomes was to move out of dense cities into detached houses surrounded by their own gardens.

Another problem with the private-market-as-cause-of-sprawl argument is that places like London were already sprawling in the seventeenth century, long before there was a fully developed consumer market for land. Or, looked at from a different vantage point, there is the fact that the development patterns in many cities and villages at the end of the nineteenth century, all widely admired by anti-sprawl activists today, were achieved primarily by private builders with relatively little governmental intervention while during the last several decades, during a period when the amount of intervention by government agencies in the land development process has increased dramatically, there has been a rising chorus of complaint. This might suggest that although there may indeed be market failures they are not necessarily more harmful than the "government failures" that have been caused by attempts to regulate the market.

Despite some basic problems with the argument, the theme of capitalism causing sprawl has led to the creation of a major edifice of historical argument. One recurrent theme of anti-sprawl reformers in the United States is that Americans never really chose to live in the suburbs. In the extreme form of this argument, Americans were forced to settle there by some powerful cabal of big business with the complicity of government. A remarkable case of the willingness of anti-sprawl critics to believe this despite all evidence to the contrary can be found in the persistence of the urban myth of the General Motors conspiracy.

This theory, popularized by a man named Bradford Snell in the 1970s, was an attempt to prove that American cities lost their streetcar systems because General Motors deliberately bought up the lines in order to close them down. As many authors have demonstrated, this theory was never plausible. General Motors may indeed have bought streetcar lines in a few cities, and some individuals at General Motors may conceivably have wanted to destroy a given streetcar system, but in the larger picture, the role of the automobile company was almost certainly insignificant. The streetcar has yielded to the less expensive and more flexible bus in virtually every city in the developed world, and most affluent cities, European and American, abandoned their streetcar systems with or without any intervention by General Motors.

The persistence of this story as the explanation of the demise of public transportation, as reinterpreted, for example, in the movie *Who Framed Roger Rabbit*, is explained by how conveniently it seems to encapsulate an entire worldview. From this point of view, the needs of ordinary city dwellers have been systematically denied in favor of the interests of greedy private entrepreneurs in league with corrupt public officials. Now greedy entrepreneurs and corrupt public officials there certainly are, and at times they undoubtedly have run roughshod over the needs of ordinary citizens. However, blaming greedy entrepreneurs, particularly real estate developers, for sprawl is highly problematic. Developers, if they possess anything like the guile attributed to them by the anti-sprawl crusaders, would be perfectly able to make money in the city as well as in the suburbs.

They would certainly be able to make money building at high density, as Benjamin Marsh believed, and as the condominium-conversion boom of the 1970s seems to prove. In fact, developers have often been the group most vocally opposed to large-lot zoning; they know that raising densities on a given piece of land can result in more units and higher profits.

A recent version of the attempt to explain urban form by the inherent nature of the capitalist system is the widespread idea that sprawl has some relation to the increasing globalization of markets. Of course it is true that changes in market conditions will have repercussions on the land, but attempts to describe the built environment of a particular city or part of a city as the result of globalization have, to date, rarely been very useful. In the end, whether a bank is owned locally or by a multinational corporation headquartered in a distant country, the dynamics of local real estate markets seem to play out in similar ways.

GOVERNMENT AS A CAUSE

Another group of observers, particularly in the United States, has tended to look at the other side of the equation and blame government failure, meaning bad policies at the local, state, and national level, for fostering sprawl. The federal government, they say, fueled sprawl through homeowner subsidies, highway programs, infrastructure subsidies, and federal income tax deductions. Some anti-sprawl reformers go so far as to say that it was federal policies, not the private market, that all but forced tens of millions of Americans to live in the suburbs in single-family houses. According to this line of reasoning, if the federal government had not built superhighways, subsidized suburban infrastructure, fostered long-term self-amortized mortgages, initiated federal mortgage insurance, allowed "redlining" of neighborhoods, and provided massive tax breaks for suburban homeowners, many city dwellers would have preferred to remain in large multistory apartment buildings in the dense central city rather than move to a single-family house in the suburbs.

None of these arguments is very convincing. First of all, the notion that the federal government, through the Interstate Highway Act, was responsible for advocating and planning these roads is misleading. Most cities and urban areas had extensive plans for superhighways in place already in the 1930s; many of them had allocated large sums of county and state money to begin construction of these roads long before the federal interstate highway program of the mid-1950s. These roads were heavily supported by central city interests because they were considered an important way to rejuvenate the city. Given the strong rebound of many of these cities in recent years, it is altogether possible that, at some point in the near future, most people will conclude that they were actually largely beneficial for central cities.

Another common assertion is that federal agencies starting in the 1930s specifically discriminated against city neighborhoods by introducing new low-interest self-amortizing mortgages that were made available to suburbanites but denied to many city dwellers. In fact, while agencies of the federal government helped to bolster the popularity of the long-term self-amortizing mortgage, this was not a new government invention of the 1930s or one that was specifically aimed at suburbanites. The self-amortizing mortgage had been used by private savings and loan associations in the early twentieth century and was common by the end of the 1920s. It is a policy that could have benefited any homeowner, whether in the central city or the suburbs. Nor was governmental "redlining" as important as anti-sprawl historians have claimed. The term "redlining" refers to a practice a line was drawn around certain neighborhoods, particularly poor and racially changing neighborhoods, that were considered too risky to lend in. The reason banks started this kind of policy was quite logical: to prevent financial losses in places where houses were likely to lose value. Federal agencies undoubtedly played a role in continuing and systematizing redlining. But neither the government nor the banks were doing anything either new or necessarily prejudicial to urban neighborhoods. Most banks, like most businesses, were perfectly happy to invest money in any part of the city or suburbs where they could make a profit. Their conclusion that older and racially changing neighborhoods, whether in the city or the suburbs, would inevitably see a drop in real estate values may have been too sweeping, and there probably was prejudice involved in rating the neighborhoods, but there was, in fact, a great deal of evidence over many years indicating that property values did tend to drop as neighborhoods got older and experienced ethnic or racial turnover. No amount of regulatory control would have altered this fact of life or made this kind of loan less risky.

In fact, for a great many relatively poor buyers – white or black – throughout urban America, redlining wasn't an issue at all because the option of a bank loan for them was never a serious possibility. Instead buyers, whether Polish immigrant workers in the Back of the Yards neighborhood of Chicago or the African Americans living in central Saint Louis, were forced to rely on help from their extended families or from institutions like churches or they turned to "contract buying," a practice where the seller provides the financing. The terms imposed on contract buyers were often onerous and unfair to the purchasers, but for many buyers, in neighborhoods with or without formal redlining, it was often the only way they could own property. It did allow many poor families to buy their own houses and apparently was an important mechanism in achieving an unprecedented rate of nearly 50 percent homeownership in the United States before World War I.

The final, and most important, federal policy blamed for sprawl has been homeowner deductions in the federal income tax. These deductions have undoubtedly had a major impact on all aspects of

American life. However, the United States is far from the only country with such provisions. Many other nations have instituted similar incentives. Furthermore, these tax incentives were clearly not part of any plot to entice city dwellers to the suburbs. They were part of the federal tax code from its earliest days. The intention of the deduction for mortgage interest and for local property taxes, for example, was to avoid the taxing of money that was arguably not part of income because it either was already going for taxes or would go to other parties who would pay tax on it. Other observers have argued that homeowners reap another windfall in the tax code because they don't have to pay taxes on the "rent" that they would have to pay to a landlord if they didn't own the property. According to this line of thought, for tax purposes homeowners should be treated both as investors and occupants. For their investment in the property to be treated the same as any other investment, they would have to pay taxes on their investment income, in this case the net income that would remain after they deduct all expenses from income, which would be primarily the rent they pay themselves.

There is little doubt that homeowner tax deductions have fueled a great deal of suburban residential construction, but this does not mean that it inherently favors the suburbs or larger lots in the suburbs. That the American tax code favors wealthy homeowners over poorer homeowners and all homeowners over renters is quite true, and perhaps should be amended or repealed on those grounds, but the advantages of the deduction are not tied to any geographical location. The deduction could have been used for any house, whether in city or suburbs. It is conspicuous, for example, that these tax deductions only became important for many people when incomes and tax rates increased dramatically after World War II. By the 1960s, when these deductions had became a really significant feature for many Americans, legislation was already in place to allow the deductions on any kind of single-family unit, whether a house in the suburbs or a condominium in a high-rise downtown. Most large American cities in the 1960s had a considerable supply of vacant land or land with relatively inexpensive buildings that could have been redeveloped at higher densities. In just these years, moreover, there was a boom in conversions of rental housing to condominiums.

Thus, if the demand had existed, construction in American cities could have outpaced construction in the suburbs, and mortgage interest deductions taken by city dwellers could have dwarfed those taken by residents of suburban areas. The reason that they did not is probably because most middle-class Americans in the late twentieth century had little interest in staying in the city if they could buy a larger and less expensive home in the suburbs. It is quite likely that the homeowner deductions have fueled some of the growth in house sizes since World War II by making them relatively more affordable. It might be logical to assume that the deductions would also have led to similar, automatic increases in lot sizes, but the fact that the total value of the deductions has risen dramatically while the average size of suburban lots has declined over the last fifty years suggests that the link between homeowner deductions and sprawl is weak.

In short, none of these governmental policies connected with home ownership explain sprawl. For this reason, it is not surprising that already by the end of the twenties, well before any of the federal policies that supposedly favored homeowners in the suburbs took hold, the move to the suburbs was in full swing, and close to half of all American families were able to own their own home. Even more striking is that, even with the mushrooming value of these incentives for homeowners, the rate of homeownership increased only from 50 to 67 percent during the second half of the twentieth century despite two of the most important boom periods in American history and the massive growth of low-density suburbs.

Another favorite explanation of the federal influence on sprawl is that it was caused by the government spending more federal dollars in the suburbs than in the central cities. In fact, it might be true that more money in recent decades has been spent on infrastructure projects in suburban areas than ones that are located in the central city. This is not surprising, however, since this is where the vast majority of metropolitan residents now live and where the vast majority of growth is taking place. To prove that this is inequitable would require a much more elaborate accounting than the typical studies to date in which a piece of freeway that was constructed in a suburban area gets entered into the one column and any road built in the city shows up in the other. For one thing, most transportation networks still converge on the central city and serve it. For another, this accounting would fail to consider the value of total federal expenditures over history. In any such accounting, the spending by the federal government since the eighteenth century for ports and railroads, bridges and highways, universities and hospitals located primarily in the central cities would have to be factored in. Looking beyond infrastructure, if all spending by the federal government is taken into account, federal spending today goes more heavily per capita to central cities than to suburbs, primarily because of the enormous price tag of social security payments, which go primarily to an older population that remains disproportionately in the central cities.

Other observers lay the blame for sprawl more with the states and local governments. The states, they say, have mostly refused to invoke the authority over land use reserved to them by the Constitution to compel local governments to plan rationally. In the case of local governments, it has been argued, building codes, zoning regulations, subdivision ordinances, and municipal rivalries fuel sprawl. In the most cynical interpretation of the evidence, some observers suggest that sprawl is all but inevitable in the current system because developers merely buy local politicians who will vote for sprawl. However, even if one believed that developers were this powerful, this conclusion would only be plausible if sprawl were inherently more profitable than building at higher densities, which is far from self-evident.

More moderate critics point especially to the use of zoning provisions that segregate land uses, restrict mixed-use developments, and impose minimum lot-size requirements. It is true that if all land-use restrictions were abolished, American cities might redevelop at somewhat higher densities and with more mixed use. Knowing exactly how this would play out, however, is virtually impossible because the cause-and-effect factors here are so difficult to disentangle. For example, it is clear that zoning itself cannot be blamed for most of the sprawl that has occurred because sprawl was well underway long before zoning became common in American cities, which only started to happen in the 1920s. Most early zoning ordinances, moreover, did not try to foster any new pattern of development. Instead, they extrapolated from historical patterns. This included the kind of sorting out of land uses in neighborhoods that had been underway for at least a century, as those who could afford to do so increasingly left crowded neighborhoods with incompatible land uses at the center to settle in neighborhoods at the edge where residential land was better protected by deed restrictions and other private covenants against industrial pollution and noxious land uses. What most zoning did was to take these private tools, make them public, rationalize them, and extend them across the entire city.

For this reason, most parts of Houston, which has historically been hostile to zoning, look and function very much like corresponding parts of other cities developed at the same time. In Houston, rather than zoning, it has been subdivision regulations and building codes that have mandated many of the features commonly found in suburban developments. But, like zoning, these provisions were mostly an extension and regularization of earlier private practices. Were these regulations the cause of urban form or were they the result of many years of experimentation with the kind of building patterns that city dwellers wanted? An important piece of testimony on this subject can be found in the history of private mechanisms to control the communities. In Houston, as elsewhere, the most important of these was the deed covenant, which could regulate everything from the size and shape of the building to the kind of people who could buy the property. Even when there was no zoning at all, wealthy individuals could and did protect their single-family neighborhoods by going to the courts at the first sign of what they considered an undesirable land use. One of the chief functions of zoning was to give a much larger part of the population the same kinds of control over their environment that the wealthy had always enjoyed.

A final reason that zoning has not had the effect that many people have claimed for it is that, so often when there has been a conflict between market demand and the zoning code, it has been the zoning codes that have given way. Because these changes have happened incrementally, typically a few parcels at a time and over many years, it has been difficult to document the overall effect of these changes. Still, it is quite possible to make some educated guesses. For example, given the current situation of rising density and declining lot sizes at the suburban edge of many American cities, it is clear either that zoning was not what caused such large lot sizes in earlier decades or that zoning has changed as necessary to accommodate market realities. Ironically, one place where many people now agree that zoning has genuinely had an effect in increasing sprawl is precisely in those suburban and exurban jurisdictions where anti-sprawl advocates were successful in introducing large-lot zoning in an effort to try to stop sprawl by making subdivision more difficult. Large-lot zoning, particularly favored since the

1960s, almost certainly forced many homeowners to buy more land than they otherwise would have wanted, leading to lower densities than would have been the case without the regulations. In short, the role of zoning in sprawl is much more ambiguous than the existing anti-sprawl literature would imply.

Another charge has been that the fragmentation of governments in metropolitan areas into many municipal jurisdictions has led to a situation where these governments compete with each other for new development rather than working together to plan for a less sprawling future. However, the idea that a fragmentation of local governments causes sprawl is not at all clear in actual practice. It is true that Saint Louis, which has a relatively small central city and a large number of suburban districts, has become one of the most decentralized cities in the United States and has experienced widespread abandonment in the central city and massive sprawl. By way of contrast, Melbourne or Sydney, Australia, places with even smaller central cities, have been held up as models of anti-sprawl. At the opposite end of the spectrum, central cities that occupy most of their urban region are not necessarily more compact. Tucson, Indianapolis, and Jacksonville, for example, occupy large parts of their metropolitan area, but they are all very low in density and are dispersed.

TECHNOLOGY AS A CAUSE

Another favored explanation for sprawl is that it was caused by new communications and transportation technologies. One of the most common explanations of the changes in city form in the past two centuries is the notion that the railroad tended to concentrate growth then the automobile dispersed it. This is, we are told, the primary reason the dense city of the early twentieth century yielded to the highly dispersed postwar city in the same period as mass transportation yielded to the automobile. This argument, plausible as it sounds at first glance, actually leaves a great deal unexplained. In the first place, as we have seen, the automobile did not directly replace any sort of mass transportation; what it more directly replaced was the private carriage. In fact, it would be more accurate to say that private transportation and mass transportation have coexisted and developed together through the nineteenth and twentieth centuries as the automobile replaced the private carriage and as the bus replaced the streetcar, which in turn replaced the cable car and horse-drawn street railway.

It is true that the use of private means of transportation has soared while the use of mass transportation has remained steady or declined in the same period of time that the population dispersed in virtually every major metropolitan area in the twentieth century. But this does not prove any simple cause-and-effect relationship. There is no more reason to think that the automobile causes decentralization than to believe that rail transportation can only work to centralize cities. As we have seen, the outward dispersal of urban population started centuries before the advent of the automobile. Certainly by the early twentieth century, suburbanization was in full swing using rail transportation as a principal means of dispersal. The Los Angeles region had become one of the most decentralized, dispersed, multicentered urban places the world had seen already by the time of the First World War, well before the impact of the private automobile was felt in any really significant way. It was the steam railroad, the cable car, the streetcar, and the interurban rail system that had made this possible. Even more important, the Los Angeles region has become dramatically denser since the 1950s in an era when the vast majority of people have relied on the private automobile. The fast-increasing rate of automobile ownership at the very heart of some of the densest urban regions in the affluent world today offers proof that high automobile ownership does not automatically lead to low densities.

In a similar way, it is not really logical to blame postwar urban freeways for sprawl. These roads were heavily supported by central-city interests because these individuals believed that these roads, like the railroads before them, would reinforce the centrality of the downtown and make it easier for people from throughout the region to get to it. In fact, they did make getting downtown much quicker. Also like the railroads, they made leaving town simpler, but there is no particular reason to think that the decentralization caused by roads has been any different in kind than that caused by the railroads. In fact, both caused some dispersal and both caused some centralization. The amount of each depended on a great many other factors and millions of individual choices.

If one were willing to believe in simple cause-and-effect relationships in urban development, one could turn the entire transportation argument on its head. From this perspective, the individual desires of large numbers of families wishing to live at lower densities could be seen as the primary cause of the growth in the successive development of the carriage industry, the railroad, public transportation, and finally the automobile industry. Each of these means of transportation did, in fact, give families increased mobility. It is this enormous increase in mobility, and not any specific means of transport, that has been a key factor in the large population dispersal that we have chronicled. What we can conclude is that although this increase in mobility certainly made sprawl possible it did not necessarily cause it.

AFFLUENCE AND DEMOCRATIC INSTITUTIONS AS A CAUSE

Perhaps a better way of looking at the causes of sprawl is to leave aside for the moment the question of why cities sprawled and instead to ask what were the forces that worked against sprawl and kept cities from dispersing even more than they did before the mid-nineteenth century. After all, in many ways it is puzzling that so many people would have chosen to live uncomfortably on top of one another in walled cities for such a long time when there was attractive land all around the city. According to Thomas Sieverts, this would seem to be a very unnatural condition, quite opposed to the "natural" habitat for man that appears to be neither the completely open field nor the enclosed forest but the areas that lie at the border between the two. [In *Cities without Cities* German architect Thomas Sieverts] addresses the decentralization of the compact historical European city and examines the new form of urbanity which has spread across the world describable as the urbanized landscape or the landscaped city. Sieverts calls this the *Zwischenstadt*, or "in-between city". In like fashion, he suggests, humans seem to favor neither a high degree of compaction nor a high degree of diffusion but a moderate clustering. If so, the compact historical city, such as seen in Europe before the nineteenth century, may turn out to have been an aberration, a short "interlude" in urban history. What sustained the compact city even beyond the period when it was necessary for defense, Sieverts suggests, was the concerted effort of a small elite of individuals and institutions who erected a system of "priest kings and religious associations, temples and churches, walls and markets, feudalism and the guilds." It was perhaps the wane of these forces in the seventeenth and eighteenth centuries even more than the advent of the railroad, telecommunications, and other innovations of the nineteenth century that really made sprawl possible.

Although sprawl has developed differently at different times and in different places, the history of sprawl suggests that the two factors that seem to track most closely with sprawl have been increasing affluence and political democratization. In places where citizens have become more affluent and have enjoyed basic economic and political rights, more people have been able to gain for themselves the benefits once reserved for wealthier citizens. I believe that the most important of these can be defined as privacy, mobility, and choice.

By privacy, I mean the ability to control one's own surroundings. This might take the form of a co-op apartment on Fifth Avenue in New York with a doorman at the sidewalk and a chauffeured car at the ready, or it could take the form of a modest house on a small plot of ground in the suburb. One of the major reasons the suburban house has been so successful is that it has been a way to obtain many of the advantages of privacy enjoyed by the millionaire on Fifth Avenue at much less cost.

By mobility I mean both personal and social mobility. Where, in the nineteenth century, only the richest and most powerful urban dwellers could maintain their own carriages and get around urban areas on their own power at will, by the end of the 1920s private transportation was in reach of tens of millions of middle-class suburbanites particularly in the United States. The option of using an automobile has given city dwellers around the world an enormously increased mobility. City dwellers everywhere travel on average vastly more than they did at the beginning of twentieth century. This physical mobility has allowed a dramatic expansion of educational and employment opportunities. In turn, this has led to increased social and economic mobility.

Finally there is choice, perhaps the most important element of all and the most hotly disputed. Many members of cultural elites are not interested in hearing about the benefits of increased choice for the population at large because they believe that ordinary citizens, given a choice, will usually make the wrong one. Sprawl has certainly increased choices for ordinary citizens. At the turn of the century, it was primarily wealthy families who had multiple options in their living, working, and recreational settings. An affluent

New York banker and his family could live in many different communities in the city or its suburbs. They could summer in the Adirondacks or at Newport, winter in Florida or on the French Riviera. They had the luxury of ignoring their neighbors and choosing their friends elsewhere. Today, even the most humble American middle-class family enjoys many of these choices. And even if the alternatives aren't thrilling, the very fact of having choices at all makes virtually any situation more tolerable. The most convincing answer to the question of why sprawl has persisted over so many centuries seems to be that a growing number of people have believed it to be the surest way to obtain some of the privacy, mobility, and choice that once were available only to the wealthiest and most powerful members of society.

It would not be wise to conclude from this, however, that affluence causes sprawl. The fact that some of the wealthiest individuals in every large city continue to live at very high densities at the center suggests that affluence is compatible with many different settlement patterns. If everyone became wealthy enough, it is quite possible that a large number might want to live in places like Park Avenue in New York or an apartment in the sixteenth arrondissement in Paris and that new urban districts would be built to accommodate this demand. In the case of urban areas and sprawl, as in the case of virtually any vast and complicated human or natural system, there is very little simple cause and effect. Rather, there are innumerable forces, always acting on each other in complex and unpredictable ways.

PART FOUR

Urban politics, governance, and economics

INTRODUCTION TO PART FOUR

The material on urban society and culture in Part Two and urban space in Part Three raises important issues about how cities should be governed and the economy of cities. The conflicts related to gated communities and sprinklers that douse homeless people that Mike Davis discusses in Part Two (p. 195) pose political questions: what should government, particularly local government, do when different groups want to use urban space in different ways? Richard Florida (p. 143) argues that to compete economically, cities must attract creative individuals. The selections by William Julius Wilson (p. 117) in Part Two and Ali Madanipour (p. 186) in Part Three illustrate how sociological issues of race and class and issues of urban space are intimately related to economic questions and questions of urban politics and governance. If Mexican immigrants in Los Angeles cannot vote because of their questionable immigration status, their voices will not be heard and local policies will not be responsive to them. The economy of a Parisian suburb will suffer if Franco-Algerian residents are discriminated against in the local labor market. Part Four focuses on these questions of urban politics, governance, and economics. In Elizabeth Strom and John Mollenkopf (eds), *The Urban Politics Reader* (London: Routledge, 2006) in the Routledge Urban Reader Series, there are additional readings on urban politics.

A good starting point in an examination of urban politics, governance, and economics is to examine the way in which social differences and disagreements about the use of urban space play out in the real world. Geographer David Harvey's selection (p. 230) nicely bridges the material in Part Two on urban society and culture, Part Three on urban space, and the material in this part on urban politics, governance, and economics.

While Harvey is a geographer, the issues he discusses are essentially political ones. Harvey sees cities as centers of conflict based on ideology, race, gender, and individual and group interests. Civil rights groups, feminists, Republicans, environmentalists, business people, individual property owners, and punks have different interests and values. The land that business people want for a new office building may be the exact same land that environmentalists feel should be a park and that skateboarders feel is perfect for skateboarding. These differences in interests and values inevitably lead to conflict among different groups in cities. The processes used to resolve these kinds of conflicts are political processes. Harvey is sympathetic to the militant particularism of some groups who focus on their specific interest, locality, or the house or apartment where they live, but feels that narrow values are more effective if they are generalized. How disparate values and interests are generalized and expressed moves us into the domain of urban politics.

Many of the best writings about urban politics are by political scientists. Urban politics is a distinct subfield within the social science discipline of political science. But scholars from other academic disciplines are also interested in urban politics. For example, political sociology is a subfield within sociology and political economy is a subfield within economics.

There is a large literature about the role of political machines and bosses in nineteenth- and early-twentieth-century American city politics. First-hand observers like settlement house leader Jane Addams, former boss George Washington Plunkitt, and scholars, like Lord Bryce – the British Ambassador to the United States in the late nineteenth century and a formidable student of American culture – observed that local government in most American cities was dominated by coalitions of Irish and other immigrants. Uneducated, narrow-minded, and corrupt political "bosses" were able to marshal enough votes from members of their ethnic group to take control of city government. Lord Bryce coined the term political "machine" to refer to the organizations that these bosses ran. The bosses were extremely able and hard working. They provided jobs, patronage, and basic social services to their constituents. By the 1930s local boss rule was largely a thing of the past and most American local government political machines had ceased to exist. But the legacy of bosses and machines is still visible in the ethnic politics of present-day cities.

In the 1950s and 1960s, seminal work by sociologist Floyd Hunter and political scientist Robert Dahl began a major debate between two schools of thought about how representative local government in the United States really is. In his book *Community Power Structure* (1953), based on research in Atlanta, Georgia, Hunter concluded that in Atlanta, and by implication other cities, a small, interlocking elite consisting of key business people and members of established and socially prominent families made all the really important decisions about Atlanta, including the governmental decisions. They sat on the boards of each other's corporations, intermarried, chatted at the same social clubs, and ran things in their interest. Atlanta's elite did not see themselves as selfish or short sighted. They believed that the decisions they were making were in the best interest of everyone in Atlanta – that they served a common good. But Hunter questions these assumptions. He suggests that other groups in Atlanta had different ideas of what collective decisions would be best for the city. Because Hunter concluded that a small elite ruled Atlanta, his model of urban community power is referred to as the elitist model of community power. In *Who Governs?* (1961) Dahl reached almost diametrically opposite conclusions. He concluded that local political power in New Haven, Connecticut, and by implication other cities, was fragmented. Dahl used a research method called "decision analysis." He and his students analyzed who was involved in making important decisions. To do this they looked at the racial, ethnic, gender, and occupation characteristics of members of successive New Haven city councils and task forces involved in different decisions. They attended meetings to observe who was present and how influential they were, read local newspapers, and combed through records related to decisions about urban renewal, the location of a hospital, freeway construction, and other matters that were important to the city. Dahl and his students concluded that many different people from a variety of walks of life were involved in decision-making in New Haven and that many different people influenced the outcome of different decisions. As a result, Dahl advanced a pluralist model of urban community power. The competing models of Dahl and Hunter stimulated debate between elitists and pluralists and the further elaboration of theories of urban politics. Some theorists critical of pluralist interpretations of community power focused on structural features of global capitalism. Marxists in particular feel that the global capitalist system really determines what city governments can and cannot do. Structuralists argue that if bankers in New York and London decide to pull investment out of Reykjavik, Iceland, or Vilnius, Lithuania, the city governments there cannot stop them. And if billions of dollars flow out of Reykjavik and Vilnius, the city governments there cannot build roads or schools or carry out other parts of their political agendas.

Clarence Stone, a professor of public policy and political science at George Washington University in Washington, DC, and others have developed "regime theory" to explain the way in which coalitions of elected officials and others work together to carry out local political agendas.

The second selection in this part addresses the question of how to make pluralism actually work. During the 1960s Sherry Arnstein was the chief advisor on citizen participation in the US Department of Housing and Urban Development (HUD) Model Cities Program, a federal government program that provided billions of dollars to help lower-income city neighborhoods develop and implement physical and social programs. While the US Model Cities Program no longer exists, the issue of how to involve citizens in local decision-making that affects them is important everywhere in the world (though often ignored by authoritarian and insensitive governments). Based on her observations of how much US local governments actually included citizens in decision-making in the late 1960s and early 1970s, Arnstein developed a theoretical model using a ladder with different rungs as a metaphor for degrees of citizen participation in decision-making. Arnstein begins her selection on the ladder of citizen participation (p. 238) by noting that the idea of citizen participation is a little like eating spinach: no one is against it in principle because it is good for you. But whether or not city dwellers participate effectively in government programs that affect them and their neighborhoods varies greatly.

A little background will help clarify Arnstein's selection. Many urban renewal programs in the United States during the 1950s and 1960s were intended to, and did, tear down low-income, minority neighborhoods and remove the residents only to replace their homes and community institutions with office buildings, luxury housing, garages, and other developments totally unrelated to the former residents. The gritty, but communal, Italian neighborhood in Boston's West End that Jane Jacobs lovingly describes (p. 105) was leveled by Boston's West End urban renewal project. The street ballet she describes has been replaced by Starbucks coffee shops and smart boutiques. In the urban renewal program, invitations to neighborhood residents to help decide what the urban renewal project should be like were usually a sham. In sharp contrast the US War on Poverty in the late 1960s emphasized maximum feasible participation of the poor. Many decisions about how to use federal anti-poverty funds were made by residents themselves. Many local poverty programs lacked the capacity to manage programs or dissolved into internal feuding. Cutting local elected officials and established agencies out of the loop to design programs and manage anti-poverty funds created a major political backlash. The Model Cities Program, where Arnstein worked, sought to strike a balance between the top–down urban renewal approach in which there was no real citizen participation in decisions and the War on Poverty model in which neighborhood groups had real power, but often fought among themselves, wasted money, and accomplished little.

Arnstein asks readers to imagine a ladder with different rungs of citizen participation from lowest to highest. The rungs of the ladder range from non-participation (manipulation and therapy) at the bottom of the ladder to partnership and citizen control at the top. While Arnstein herself favors citizen control as the ultimate objective of urban programs, she considers all but the bottom two rungs of her ladder useful to varying degrees. In the decades since Arnstein's classic article was written, there has been a succession of urban development programs in the United States, Western European countries, and elsewhere in the world. Many urban revitalization programs call for some degree of citizen participation. Arnstein's ladder is useful in understanding how well-intentioned local government officials can foster meaningful citizen participation in urban programs if they choose to and how citizens can assert their views if they do not.

While Arnstein focuses on urban politics at the neighborhood level, urban politics typically involves decisions at the city level or decisions that involve multiple levels of government. To get things done at the level of a city, metropolitan area, region, or state requires interest groups working together to form coalitions. The third selection in this part, by Harvey Molotch (p. 251), a professor of social and cultural analysis in New York University's sociology department, argues that pro-growth coalitions of business leaders, civic boosters, property owners, investors in local-oriented financial institutions and their allies – what he calls "growth machines" – dominate city government decisions about public expenditures, land use, and urban social life. The political and economic essence of virtually any given locality is growth. While different members of growth machines may favor one type of development or another, they all want to foster a good business climate that retains and attracts business. Members of growth machines believe and work to convince others that growth is good for everyone in the community, but Molotch is not convinced that there is a direct link between economic growth and jobs. A final topic that Molotch addresses in this 1975 article is the emergence of a countercoalition. While US cities were overwhelmingly pro-growth at the time, Molotch already detected emerging anti-growth sentiment. Some citizens were noticing that growth tended to increase traffic congestion and pollution. It often overburdened local schools. Tax rates in cities that had successfully pursued pro-growth strategies were often the same or higher than in ones that had not. Coalitions in some progressive towns and cities – typically with a large university population – were already agitating for the kinds of sustainable urban development the Brundtland Commission report (p. 351) urged more than a decade later or green urbanist and carbon-neutral development that Timothy Beatley (p. 446) and Stephen Wheeler (p. 458) place at the top of planners' agendas today.

One of the most important, and certainly among the most sensitive, local government functions everywhere in the world is police work. In "Broken Windows" (p. 263), James Q. Wilson and George L. Kelling propose a theory about how crime comes to dominate declining neighborhoods and what to do about it. Wilson and Kelling emphasize the importance of citizen perceptions of crime. They argue that if residents think crime has increased, they will become more reclusive and less involved in the community whether or not crime has actually increased in the neighborhood. The withdrawal of citizens will in turn open the door to real crime. A single broken window may be a trivial problem, Wilson and Kelling argue, but if it is not fixed, the signal that no one cares enough to fix it will lead to more broken windows and then to drug dealing and violent crime. The remedy Wilson and Kelling suggest – which has been widely adopted since this article was written – is community policing. Community policing typically involves assigning police officers to a specific neighborhood for a long enough time for them to get to know it well. Rather than just driving by in police cars, community police regularly patrol the neighborhood on foot. By observing what is going on at street level and chatting with neighborhood residents, they will come to understand community norms and values. They will be able to distinguish between what Elijah Anderson (p. 127) describes as harmless "decent" neighborhood residents and potentially dangerous "street" people. Wilson and Kelling go on to advance the controversial view that these community police should informally enforce community norms of appropriate civil behavior as the neighborhood itself defines them, even if that calls for extralegal or perhaps illegal controls. Wilson and Kelling note that a transitional neighborhood may be inhabited both by neighborhood regulars and by strangers. The regulars in turn consist of what Wilson and Kelling term decent folk, and derelicts and drunks who are not so decent but know their place. So long as questionable street behavior stays within neighborhood-defined norms, Wilson and Kelling argue that community police should look the other way. They believe that the police should leave the well-known and harmless local drunk alone even if he is technically breaking the law. But if rowdy teenagers, prostitutes, dope dealers, or other strangers violate community norms – regardless of whether they actually violate the law – they argue that community police should intervene, perhaps by ordering the strangers to leave even if there is no legal basis for such an order. This will keep windows from being broken (actually and symbolically) and by implication will prevent the spread of serious crime.

Politics, economics, and public finance are intimately connected. The final three selections in this part introduce important ideas about the way in which urban economies work and how urban economics concepts can produce better local government decision-making, greater fiscal equality in metropolitan regions, and help central cities capitalize on the advantages they possess to compete effectively in a global economy.

Wilbur Thompson is largely responsible for creating the field of urban economics. His 1965 *Preface to Urban Economics* was the standard text dozens of economics departments used as they launched urban economics courses in the 1960s and 1970s. In the selection included in this part (p. 274), Thompson introduces concepts that were new at the time but have since been thoroughly incorporated into mainstream urban economics thinking. Thompson argues that the failure to consider the real costs of publicly provided goods and services often leads to inefficient, sometimes irrational, public policy. Thompson provides examples of how using pricing can produce better public policy. He classifies the kinds of goods local governments provide into several key types: public goods, like highways, that need to be provided on a collective, rather than an individual, basis; merit goods, like polio vaccinations, that society deems so important that they should be free and universally available at the public expense; and payments, like food stamps, made to redistribute income from one group such as well-off taxpayers to another group such as the indigent poor. Thompson considers these all justifiable bases for public expenditures and supports them if they are carefully thought out and consciously applied. But where local governments do not think through the rationale for a public expenditure clearly – based on these and related economic concepts – Thompson argues that public programs may be distorted and public funds wasted. Thompson was an early proponent of pricing public goods. Where the goods are scarce – like space on a congested highway – he advocates using price to ration the scarce goods. Thompson pioneered the idea of congestion pricing, for example charging higher highway and bridge tolls during commuting hours when there is very high demand for them and they are likely to be congested. A new idea at the time, congestion pricing is now widely used.

Harvard business professor Michael Porter (p. 283), approaches the issue of urban inequality and what to do about it from a private-sector perspective. Porter believes that the economic and social health of inner cities depends upon economic development by the private sector. To succeed, he argues, economic development must be based on the economic self-interest of private firms rather than phony businesses propped up by government subsidies and preference programs. The key to success, according to Porter, lies in capitalizing on the competitive advantage that inner-city neighborhoods possess: strategic location, local market demand, their capacity to integrate with regional business clusters, and human resources. Porter urges an economic, not a social, model for development. He counsels corporations to shift their philanthropic priorities away from providing social services, such as daycare or food for homeless people, to providing managerial expertise to develop neighborhood economies.

The final section in this part, by Minnesota law professor Myron Orfield (p. 296) turns from the basic economic issues Thompson discusses to the question of metropolitan fiscal equity. Metropolitan regions are fragmented into many small local government jurisdictions. The growth machines in these jurisdictions that Molotch describes (p. 251) compete with each other in a zero-sum game to attract the most economically desirable land uses – the ones that will produce the greatest net revenue. Since the playing field is far from level, some jurisdictions succeed in attracting most of the revenue-generating land uses. They can maintain low taxes and still provide superior local services. The jurisdictions that lose out in the competition have limited fiscal capacity. They must levy higher taxes, provide fewer services, or both. Orfield describes how he used statistical analysis of data on local government characteristics and geographical information systems to analyze metropolitan fiscal disparities and develop a typology of different kinds of communities, including different categories of "at risk" communities. Building on a theory of what he calls metropolitics, Orfield describes how communities with common interests can band together to change laws and policies to increase metropolitan fiscal equity. Orfield believes that stable, cooperative regions are essential for the well-being of society and advances specific policies to increase regional fiscal equity such as metropolitan tax base-sharing programs, coordinated infrastructure planning, and land use reform.

"Contested Cities: Social Process and Spatial Form"

from Nick Jewson and Susanne MacGregor (eds), *Transforming Cities* (1997)

David Harvey

Editors' Introduction

Urban politics is a distinct subfield within the discipline of political science and much good writing on urban politics and governance comes from political scientists. But questions of politics and governance also interest geographers like David Harvey, sociologists like Harvey Molotch, (p. 251), lawyers like Myron Orfield (p. 296), social workers like Sherry Arnstein (p. 238), business professors like Michael Porter (p. 282) and scholars from other academic disciplines and professional fields.

Harvey starts this selection by reminding us of the massive urbanization of the human population that Kingsley Davis describes (p. 230). Cities are critical to understanding the current human condition. Yet, Harvey notes, cities as a category are strikingly absent from many discussions of modernization, modernity, postmodernity, and capitalist and industrial society. He would like to see cities included in discussions of these topics.

Harvey emphasizes the importance of thinking about cities in terms of processes rather than just things. For him cities are sites of conflict based on race, ideology, gender, and other social categories. Harvey thinks dialectically. He argues that processes are both shaped by time and place and shape time and place. For example, when communist Chinese leader Mao Zedong decided to send 20 million urban Chinese intellectuals out of Shanghai, Beijing, Guangzhou, and other large cities into the countryside during the Great Proletarian Cultural Revolution, this was a social process shaped by a specific place (China) at a specific time (during the Great Proletarian Cultural Revolution). For a while, this social process produced a distinct kind of chaotic and austere revolutionary city. Historical sites and buildings with European architecture were destroyed; massive public works projects, bomb shelters, and monuments to revolutionary heroes were built. The Cultural Revolution effectively ended with Mao's death in 1976. Some evidence of what was built (and destroyed) during the Great Proletarian Cultural Revolution period is still evident in Chinese cities, but Chinese society and politics are remarkably fluid and much has changed, as witnessed by the modern skyline of Shanghai, plate 44 in the global cities plate section at the end of Part Eight.

Some processes lead to very durable outcomes. The decisions to build the Three Mile Island nuclear power plant, dam the Yangtze River, or rebuild the World Trade Center site produce physical forms and social structures that last for a very long time. What looks like a good solution to urban development to one generation, such as a massive highway construction program to solve transportation problems, building identical suburban tract housing to deal with housing needs, or creating a nuclear power plant to meet urban energy needs, may appear very problematic to the next generation. Uncritical support for the automobile culture that Kenneth Jackson describes (p. 65), the all-out pro-growth policies of urban growth machines that Harvey Molotch documented, (p. 251) and the resource-consuming development the Brundtland Commission condemned (p. 351) were little questioned a generation ago, but are currently the subject of much critical comment. This is why Harvey urges planners and policy-makers to design flexible, adjustable cities and to encourage fluid social processes that can change over time.

Two additional themes in Harvey's selection deal with community and the relationship between the natural and built environment. Louis Wirth (p. 96), Robert Putnam (p. 134), Ali Madanipour (p. 186), Jane Jacobs (p. 105), Peter Calthorpe and William Fulton (p. 360), Clarence Perry (p. 428), and many other writers represented in this book are concerned that cities alienate people from each other and break down community. All these writers favor agendas to build greater community in cities. Harvey raises some important questions about community building. He argues that many community-building projects are recipes for isolation. The positive identification of some groups is often achieved by first defining other groups as "the other" – devalued and semi-human. For example, people within a gated community in Los Angeles of the kind that Mike Davis describes (p. 195) may become a community that play golf together and share drinks at their country club, safe within the gates. They will not be alienated and isolated from each other. But this kind of community is based on walling themselves off from lower-income people and people of color outside the gates of their artificial community.

Harvey criticizes the belief that good design will solve social process problems. Rather, Harvey argues, the social processes underlying even the best designed, community-enhancing place need to be cultivated and sustained. Just building a remarkable physical community like Ebenezer Howard's garden city of Letchworth, England, or a New Urbanist city like Seaside, Florida, based on principles espoused by the Congress of the New Urbanism (p. 356) will not, Harvey argues, create community.

Harvey notes that community activism is often built around militant particularisms in which a group coheres around a value, like social justice or environmental conservation, related to a very narrow and very local concern. For example, groups may focus on getting higher wages for low-paid garment workers at the Ajax garment factory or saving an endangered salamander in the Bolinas lagoon. These militant particularisms may contribute to society, but Harvey argues that they will be much more useful if they spill over into wider, more universal concerns: for example, if militant environmentalists generalize their concern for one breed of salamander into a more universal (and perhaps more balanced) environmental movement; or if the militant garment workers generalize their concerns into a broad movement for all workers to be paid a living wage. Harvey critiques the artificial distinction some environmentalists draw between the natural and the built environments. He agrees with Stephen Wheeler (p. 458) and Timothy Beatley (p. 446) that the natural and built environments need to evolve together. The sustainable urban development, green urbanism, design with nature, and ecological design movements all follow this notion of harmony between the natural and built environments.

Contrast geographer Harvey's views on the limited impact that design can have with Jane Jacobs (p. 105) and with writers in the urban design section of this book like Jan Gehl (p. 530) and Clarence Perry (p. 428). While none of these writers are rigid environmental determinists who believe that the environment determines human behavior, all of them believe that good design can help communities function better socially. Clarence Perry and Jan Gehl both advocate street designs to limit through traffic, reduce speeds, and increase safety for pedestrians, particularly children. Oscar Newman, an architect who devoted his life to studying the relationship between architectural design, crime, and safety developed principles for designing "defensible space" to greatly reduce crime even in high crime, low rent public housing projects.

David Harvey (b. 1935) is Professor of Geography and Environmental Engineering at Johns Hopkins University; Senior Research Fellow, St Peter's College, University of Oxford; and Miliband Visiting Fellow, London School of Economics and Political Science. From 1987 to 1993 he was Halford McKinder Professor of Geography at Oxford. Harvey is currently working on questions of environmental justice, globalization, alternative modes of urbanization, and uneven geographical development within a globalizing world.

This selection is from Nick Jewson and Susanne MacGregor (eds), *Transforming Cities* (London: Routledge, 1997). Books by Harvey include *Spaces of Capital* (London: Routledge, 2002), *Spaces of Hope* (Berkeley, CA: University of California Press, 2000), *Justice, Nature, and the Geography of Difference* (Oxford: Blackwell, 1996), *The Urban Experience* (Malden, MA: Blackwell, 1990), *The Condition of Postmodernity: An Enquiry into the Origins of Cultural Change* (Oxford: Blackwell, 1989), *The Urbanization of Capital: Studies in the History and Theory of Capitalist Urbanization* (Baltimore, MD: Johns Hopkins University Press, 1985), *Consciousness and the Urban Experience: Studies in the History and Theory of Capitalist Urbanization* (Baltimore, MD: Johns

Hopkins University Press, 1985), *The Limits to Capital* (Oxford: Blackwell, 1982), *Social Justice and the City* (Oxford: Basil Blackwell, 1973, new edition 1988), and *Explanation in Geography* (London: Edward Arnold, 1969).

Classical and contemporary writings on urban culture that include postmodernist and Marxist writings are contained in Malcolm Miles and Tim Hall (with Ian Borden) (eds), *The City Cultures Reader*, 2nd edn (London: Routledge, 2004).

Other postmodernist writings on social processes and spatial form include Henri Lefebvre, *Production of Space* (Oxford: Blackwell, 1991) and *Writings on Cities* (Oxford: Blackwell, 1995), Edward Soja, *Postmodern Geographies* (London: Verso, 1997) and *Postmetropolis* (London: Blackwell, 2000), Mike Craig and Nigel Thrift (eds), *Thinking Space* (London: Routledge, 2000), and Fredric Jameson, *Postmodernism or the Cultural Logic of Capitalism* (Durham, NC: Duke University Press, 1992).

■

At the beginning of this century, there were little more than a dozen or so cities in the world with more than a million people. They were all in the advanced capitalist countries and London, by far the largest of them, had just under 7 million. At the beginning of this century too, no more than 7 per cent of the world's population could reasonably be classified as 'urban'. By the year 2000 there may be as many as 500 cities with more than a million inhabitants. The largest of them (like Tokyo, São Paulo, Bombay and possibly Shanghai) will boast populations of more than 20 million, trailed by a score of cities, mostly in the so-called developing countries, with upwards of 10 million. Sometime early next century, if present trends continue, more than half of the world's population will be classified as urban rather than rural.

The twentieth century has been the century of urbanization. There has been a massive reorganization of the world's population, of its political and its institutional structures and of the very ecology of the earth.

These observations immediately suggest some fundamental questions. First, given these transformations, why is it that the urban so frequently disappears from our discussions of broader political-economic processes and social trends? Most of the writing about our recent history has failed to take into account this massive reorganization and its consequences. The urban rarely appears as a salient category in our analyses. The crucial categories seem to be those of modernization, modernity, post-modernity, capitalist and industrial society. So what has happened to the category 'urban'? This question is important because the qualities of urban living in the next century will define the qualities of life for the mass of humanity. And all political-economic processes we observe are mediated through the filter of urban organization. Discussions of contemporary politics, for example, often proceed as if a concept like that of 'democracy' can remain unaffected by urban transformations when, plainly, there is a huge difference between democracy in ancient Athens and democracy in contemporary São Paulo.

If we think about the likely qualities of life in the next century by projecting forward current trends in our cities, most commentators would end up with a somewhat dystopian view. We are producing marginalization, disempowerment, alienation, pollution and degradation. It might be said that this is nothing new and that, in the nineteenth century, conditions were even worse. In the past, however, urbanization and the consequences of urbanization were taken rather more seriously than they are today. In the late nineteenth century, the bourgeoisie at least had some notion that cities were important places and, therefore, that urban reform was necessary. This generated a bourgeois reform movement – from Birmingham to Chicago – which included figures such as Jane Addams, Octavia Hill, Charles Booth, Patrick Geddes, Ebenezer Howard and many others. All of these had some vision for the future and a clear grasp of the need for reform. The nineteenth century faced the difficulties of the urban in a very positive and powerful way. It blended socialist sentiments, anarchist ideas, notions of bourgeois reformism and social responsibility into a programmatic attempt to clean up the cities. The 'gas and water socialism' of the late nineteenth and early twentieth centuries did a great deal to improve the conditions of urban life for the mass of the population. There are many contemporary analysts who, armed with the insights of Foucault, will assert that these innovations were merely about social control, which indeed in part they were. But having acknowledged this point, I think we have also to recognize that a significant proportion of the population found itself living in better circumstances as a result. Moreover, inherent in these interventions was a visionary notion of an alternative city – a city beautiful with facilities and services that would, indeed, pacify alienated populations.

Some of that concern would be helpful to have back in our cities right now. In the past, capital regarded cities as important places which had to be efficiently organized and where social controls needed to operate in some sort of meaningful way. We now find that capital is no longer concerned about cities. Capital needs fewer workers and much of it can move all over the world, deserting problematic places and populations at will. As a result, the coalition between big capital and bourgeois reformism has disappeared. Moreover, the bourgeoisie itself seems to have lost much of its guilty conscience about cities. It has, I think, concluded there is little to fear from socialist revolution, and so has attenuated its engagement with reformism. Increasingly the wealthy seal themselves off in those fanciful, gated communities – which are being built all over the United States – that enable the bourgeoisie to cut themselves off from what their representatives call by the hateful term 'the underclass'. 'The underclass' is left inside the ghetto, along with drugs, AIDS, epidemics of tuberculosis and much else. In this new politics, the poor no longer matter. The marginalization of the poor is accompanied by a blasé indifference on the part of the rich and powerful.

This blasé indifference is a matter of great concern. Accordingly I would like to highlight some fundamental questions and beliefs about the role of the city in political, economic, social and ecological life. In defining that role, we are also formulating a notion of the kind of cities we would like to construct into the next century.

I would like to begin with a fundamental methodological question: what is the relationship between process and form? . . . In my own work – from the standpoint of historical, geographical materialism and very strongly in the dialectical tradition – one of the rules of engagement which I have always tried to follow is to say that process takes precedence over things. We should focus on processes rather than things and we should think of things as products of processes. From this standpoint, we have to ask some fundamental questions about the nature of the categories we use to describe the world. Most of the categories we use tend to be 'thing' categories. If instead we examine dynamics and processes, we may try to do so by conceiving them as relationships between pre-existing things. But if things too are not pre-existing, but are actually constituted in some way by a process, then you have to have a rather different vision. This transformation in our way of thought seems to me absolutely essential if we are going to get to the heart of what the city is about.

Tony Leeds, an urban anthropologist, towards the end of his life wrote this:

> In earlier years I thought of society . . . as a structure of positions, roles, statuses, groups, institutions and so on, all given shape . . . by the cultures on which they draw. Process I saw as 'forces', movement, connection, pressures, taking place in and among these loci or nodes of organization, peopled by individuals. Although this still seems largely true to me, it has also come to seem a static view – more societal order than societal becoming . . . Since it does not seem inherent in nature . . . that these loci exist, it seems unacceptable simply to take them as axiomatic; rather we must search for ways to account for their appearances and forms. More and more, the problems of becoming . . . have led me to look at society as continuous process out of which structure or order precipitates in the forms of the loci listed above.

This, then, is a conceptualization in which process takes priority over things and which focuses on the way in which things get precipitated out of process.

Two terms or words deserve closer examination in our discussions. One is 'urbanization' – which we can convert into the 'urban process' or the 'urbanizing process' or the 'urbanization process'. The other is a 'thing-type' word – 'the city'. It is important to consider the relationship between the urbanizing process and this thing called the city. Now, from a dialectical standpoint, the relationship between process and thing becomes complicated because things, once constituted, have the habit of affecting the very processes which constituted them. The ways that particular 'thing-like structures' (such as political-administrative territories, built environments, fixed networks of social relations) precipitate out of fluid social processes and the fixed forms these things then assume have a powerful influence upon the way that social processes can operate. Moreover, different fixed forms have been precipitated out at different historical moments and assume qualities reflective of social processes at work in particular times and places. The result is an urban environment constituted as a palimpsest, a series of layers constituted and constructed at different historical moments all superimposed upon each other. The question then becomes how does the life process work in and around all of those things which have been constituted at different historical periods? How are new meanings given to them? How are new possibilities constructed? I suggest that attention to this relationship between process and form will help us understand why the urban has been neglected and, furthermore, will enable us to change completely the terms of the debate.

In this vein, I want to suggest that the reduction of the urban – or the portrayal of the city as a minor feature of social organization – can only occur when particular assumptions are made about the nature of space and time or space/time. There are three different ways of understanding spatiality or space/time that are worth noting here. The first way is the absolute notion of space/time – attributable to Newton, Descartes and Kant – in which space and time are mere containers of social action. They are passive, neutral containers. These passive, neutral containers simply allow us to locate the action which is occurring. I would like to suggest that there is a parallel here with thinking that conceives of cities as passive, neutral containers of processes and contests. These ways of thinking focus on contestations occurring within the city – the city happens to be the mere site of a process of contestation (over gender, race, class or whatever). A radically different approach is one which sees the city not so much as a site of contestation but as something to be constructed and in which the contestation is over the construction, or framing, of the city itself. What would that imply about notions of space and time?

There is a well-known alternative to the absolute view of space/time: that is, the relative view attributed mainly to Einstein and worked on by others since. In this view, space and time, although they are still containers, are not neutral with respect to the processes they contain. Metrics of space and time can and do vary depending upon the nature of the processes under consideration. In geography, this idea has been adapted to think of different ways of measuring and mapping distances. Physical distance is different from distance measured in terms of the cost or time taken to move between points, and in the last two cases the space described is not necessarily Euclidean. Different metrics yield different maps of the space/time co-ordinates within which social interaction occurs.

A third perspective on space/time that I have employed – indeed it was incorporated in *Social Justice in the City* more than twenty years ago – is a relational view . . . This view is that space and time do not exist outside of process: process defines space/time. Each particular kind of process will define its own distinctive spatio-temporality. Our studies should, therefore, aim to explain the way in which different processes define spatio-temporality, and then, having defined that spatio-temporality, find themselves bound by its rules in certain kinds of ways. Moreover, our cities are constituted not by one but by multiple spatio-temporalities, producing multiple frameworks within which conflictual social processes are worked out.

From this standpoint, we have to take very seriously the notion . . . that space and time are not simply constituted by but are also constitutive of social processes. This is also true for the urban. The urban and the city are not simply constituted by social processes, they are constitutive of them. We have to understand that dialectic in order to appreciate how urbanization is constructed and produces all of these thing-like configurations which we call cities – with political organization, social organization and physical structures. We have to appreciate better the centrality of that moment of urban construction, which is fundamental to how the social process operates. In exactly the same way, we have to take seriously the idea of that moment of construction of spatio-temporality, which then defines how the system itself will operate. From this standpoint, it is possible to reposition the urban as fundamental in contemporary debates. At the same time we transform our notion of urbanization. We would abandon the view of the urban as simply a site or a container of social action in favour of the idea that it is, in itself, a set of conflictual heterogeneous processes which are producing spatio-temporalities as well as producing things, structures and permanencies in ways which constrain the nature of the social process. Social processes, in giving rise to things, create the things which then enhance the nature of those particular social processes.

One outcome may be that we find ourselves stuck for a very long time with a particular kind of social process. An example would be nuclear power. Once nuclear power stations exist all sorts of things follow. If a nuclear power station goes on the blink, can you imagine calling a town meeting to discuss democratically what to do about it? The answer is no, you can't. In these circumstances, we are immediately driven back to the realms of expert knowledge and expert decision-making. So a thing has been created which for as long as it lasts – which is going to be a very long time – is by its very nature going to be basically undemocratic in terms of the sort of social process that supports it. Here is a social process that has defined a certain spatio-temporality for the next 10,000 years, which in turn implies perpetuation of a certain kind of social order if it is not to unravel in highly destructive ways.

We have to be thinking in these kinds of terms about the nature of cities. What kinds of cities we create, how we create them, how flexible they are, how adjustable they can be: these are the questions we need to ask in order to understand better the relationship between process and thing. Our aim and objective should be to liberate emancipatory processes of social change. In so doing, however, we must understand that liberatory impulses and politics are always going to be contained and constrained by the nature of things which have been produced in the past.

This, then, is my first major point. We have to reconceptualize the urban as the production of space and the production of spatio-temporality, understood as a dialectical relationship between process and thing.

The second major point I would like to make concerns the currently widespread invocation of the word 'community'. It too entails an exploration of process/thing relationships. One of the aspects of much contemporary debate about the urban which I find particularly striking is the tendency when faced with all sorts of difficulties again and again to reach into this bag called 'community', on the assumption that 'community' is going to save us all. Community, endowed with salving powers, is perceived as capable of redeeming the mess which we are creating in our cities. This mode of thinking goes all the way from Prince Charles and the construction of urban villages through to communitarian philosophies that, it is believed, will save us from crass individualistic materialism.

There is here too an issue about the relationship between the thing called community and the processes which constitute it. What kinds of processes constitute community? Is a community, once constituted, going to liberate or imprison further social processes? A lot of community construction projects are, in the end, a recipe for isolation. They isolate groups from the city as a whole. They move them towards a fragmented notion of what the urban process is about. Here I find myself in agreement with Iris Marion Young when she says:

> Racism, ethnic chauvinism and class devaluation, I suggest, grow partly from the desire for community . . . Practically speaking, such mutual understanding can be approximated only within a homogeneous group that defines itself by common attributes. Such common identification, however, entails reference also to those excluded. In the dynamics of racism and ethnic chauvinism in the United States today, the positive identification of some groups is often achieved by first defining other groups as the other, the devalued, semihuman.

What, then, are the implications of current notions of community? In answer, I would like to propose a dialectical view of relationships between process and community.

I think it is important to acknowledge that a lot of community activism is absolutely fundamental to many forms of social struggle. As a form of mobilization of power of people in place it can sometimes be extremely important and extremely useful. Community activism can simply be a way of containing discontent but it can also be a very important moment in more general mobilization. In this context, we have to think about the construction of community not as an end in itself but as a moment in a process. Here I refer to critiques of the nineteenth-century thinking which I described earlier. There were two flaws in that thinking. The first was the belief that, somehow or other, the proper design of things would solve all of the problems in the social process. It was assumed that if you could just build your urban village, like Ebenezer Howard, or your Radiant City, like Le Corbusier, then the thing would have the power to keep the process forever in harmonious state. The problem of these thinkers was not that they had a totalizing vision or subscribed to master narratives or indulged in master planning. Their problem was not that they had a conception of the city or the social process as a whole. Their problem was that they took this notion of thing and gave it power over the process. Their second flaw was that they did much the same with community. Much of the ideology that came out of Geddes and Ebenezer Howard was precisely about the construction of community. In particular, the construction of communities which were fixed and had certain qualities with respect to class and gender relations. Once again, the domination of things seems to me to be the fundamental flaw.

What then is the significance of community mobilization? The concept I wish to use here is the one that Raymond Williams tentatively suggested, and which he then shrank away from, but which I want to resurrect. It is what Williams calls 'militant particularism'. This idea suggests that almost all radical movements have their origin in some place, with a particular set of issues which people are pursuing and following. The key issue is whether that militant particularism simply remains localized or whether, at some point or other, it spills over into some more universal construction. Williams suggested that the whole history of socialism had to be read as a series of militant particularisms which generated what he described as the extraordinary claim that there is an alternative kind of society, called socialist, which would be a universal kind of condition to which we could all reasonably aspire. In other words, in this view foundational values and beliefs were discovered in particular struggles and then translated onto a broader terrain of conflict. It seems to me that the notion of community, viewed in this way, can be a positive moment within a political process. However, it is only a positive moment if it ceases to be an end in itself, ceases to be a thing which is going to solve all of our problems, and starts to be a moment in this process of broader construction of a more universal set of values which are going to be about how the city is going to be as a whole.

The third major point I am going to make is this: until very recently there was almost no mention of cities in the ecological literature. Cities were always regarded as the high point of the pollution and plundering of planet Earth. The environment was equated with nature; it was certainly not the built environment of cities. There is something curious about ecological rhetoric here (although I am probably misrepresenting some of the current thinking because it is getting a bit more sophisticated). Ecological rhetoric is committed to a totalizing perspective in the sense that, quite rightly, it perceives that everything relates to everything else. However, it has also failed to address the environment of cities and the 50 per cent of the world's population that are living in urban circumstances.

Why is it that we tend to think of the built environment of cities as somehow or other not being the environment? Where did that separation come from? Again it comes back to the notion that there is a thing called a city, which has various qualities and attributes, that is not part of a process. It seems to me that we have to think of environment and environmental modification as a fundamental process which we have always been engaged in and will always continue to be engaged in. The environmental modification process then has to be understood as producing certain kinds of structures and things, such as fields, forests and cities. That environmental modification process cannot be separated from the whole question of urban living. There is, it seems to me, nothing particularly anti-ecological about cities. Why should we think of them that way? When does the built, constructed environment end and 'the natural environment' begin? Where does society begin and nature end? Go and look in a field of wheat and say where nature begins and society ends. You can't do it.

Here too, then, there is a dichotomy which works its way through our thinking, which we have to challenge, in which the relationship between processes and things is fundamental. We have to pay serious attention to the nature of the ecological modification process, and understand it not as something which is simply resident in nature. For example, one of the major ecological variables at work in the world right now is money flow. Just think of what would happen to the ecosystems of the world if the money flow stopped. How many ecosystems of this world are actually supported by money flow? Vast areas of the world would undergo radical ecological change if the money supply or commodity exchange was suddenly cut off or stopped. Some radical ecologists appear to relish such an outcome, as a transformation back to some ecologically sustainable condition in which the alienation of self from nature can be overcome by human beings treading far more lightly on the surface of the earth. But I believe we must pursue much more positive ecological politics. Ecological transformations are an inevitable facet of how human beings live their lives and construct their historical geographies. Urbanization is an ecological process and we desperately need creative ways to think and act on that relation. Conversely, it is impossible to talk of ecological politics without concomitantly examining urban processes in all their complexity and fullness.

We have to move the urban, and the urbanizing process, into a more central position in our debates and discussions about ecological, social, political and economic change. From this standpoint, there are a number of myths that we have to confront and contest.

The first myth is the simple idea that when we have got the economy right then we can spend money to get our cities right. This sort of thinking takes the view that cities are relatively unimportant: when we have got enough money and we have organized ourselves right then we can spend a little time fixing them up. From my perspective that is entirely the wrong way round. Getting things economically right in our cities is the path towards economic change and economic development, even to economic growth. To treat the cities as the secondary feature of this whole dynamic is essentially wrong.

The same is true with respect to social relations. We should not wait upon some great political revolution to tell us how to reorganize our cities in a socialist or eco-feminist or some other way. No, what we have to do is to work on the nature of the social relations in the cities. If there is going to be a revolution, it is going to be a long revolution, located within the urban process. That long revolution of social relations is going to have to comprise a steady working out, over a long period of time, of transformations. Here, I think again, community mobilization and the transformation of militant particularism have a vital role to play, enabling us to find the universal concerns that exist within a realm of difference. There is a certain dialectic here of unity and difference, universal and particular, which has to be worked out. We should not retain the notion of community as particularity or difference. We have to transcend those particularities and look for a negotiation of universalities through which to talk about how the cities of the future should be.

The point is not to see cities as anti-ecological. Cities are fundamental ecological features in themselves and the processes that build cities are ecological processes. The world of ecology and that of cities are part and parcel of each other; what we have to do is link them together much more strongly, in a more programmatic way. It is only in those terms that we can really push towards a full understanding of . . . 'contested cities'. This issue is not simply about contestation inside cities but more importantly concerns contests over the construction and framing of cities – especially what they are going to be in the future.

"A Ladder of Citizen Participation"

Journal of the American Institute of Planners (1969)

Sherry Arnstein

Editors' Introduction

Local city government is important and plural actors at the city level can influence the outcome of policies and programs that affect peoples lives. Public, private, and nonprofit sector actors often work together to make decisions about local projects. In the United States and Western democracies, local governments establish policy, regulate land use, and influence development, but most land is privately owned and most development occurs with private funds. But how, exactly, should citizens participate in local government decision-making? Guidance as to how this might best be done comes from this classic article by Sherry Arnstein titled "A Ladder of Citizen Participation."

Arnstein uses the metaphor of a ladder to describe gradations of citizens' participation in urban programs that affect their lives. The ladder represents a hypothetical model. Arnstein makes clear her own personal commitment to a redistribution of power from haves to have-nots by empowering the poor and powerless.

At the lowest level of Arnstein's ladder are two forms of nonparticipation, which she terms manipulation and therapy. According to Arnstein, some governmental organizations contrive phony forms of participation, which are really aimed at getting citizens to accept a predetermined course of action. While gullible citizens may think they are participating in decision-making at these lowest levels of the ladder, Arnstein says they really are not. They are simply being used by decision-makers. Almost at the bottom of the ladder is another form of nonparticipation, which Arnstein identifies as therapy. Decision-makers get people together to allegedly participate in decision-making, but really in order to preach to them about their personal shortcomings. The intent is to cure participants of attitudes and behaviors that local government officials do not like under the guise of seeking their advice. Arnstein brands this form of nonparticipation both dishonest and arrogant. Like manipulation, participating in what are really therapy sessions under the guise of participation is worse than no participation at all.

Legitimate, but low, rungs of the ladder are informing and consultation. Arnstein considers informing citizens of the facts about a government program and their rights, responsibilities and options is a good first step, particularly if it is designed to go beyond a one-way flow of information. Consultation – getting citizens' opinions – is even better if the process is honest and citizens' opinions are really considered. Surveys, for example, may provide real input from citizens to decision-makers. But if a survey is the only form of participation, that would not go far in assuring that citizen views really carry weight. Higher up the ladder is placation, in which government gives in to some citizen demands. But having government merely throw complaining citizens some crumbs to placate them is not really a satisfactory form of participation.

The highest rungs on Arnstein's ladder are partnership, three rungs from the top, delegated power, one rung below the top, and citizen control at the very top of the ladder. During the US War on Poverty in the 1960s, local governments delegated power to run programs to some citizen groups, giving them full control over policy, funding, hiring, and other decisions. Delegated power and citizens' control have been rare since that time. Opponents of citizen control advance many of the arguments that Arnstein identifies, that citizen control arguably balkanizes public services, may be costly and inefficient, can reward opportunistic citizen hustlers, and may just be symbolic politics. Nowadays partnerships between public, private, and nonprofit organizations are popular. Arnstein places true partnerships relatively high on her eight-rung ladder. Partnerships represent a redistribution of power arrived at through negotiation along the lines John Forester (p. 421) describes. Where the odd bedfellows of local government, private corporations, and neighborhood nonprofit community-based organizations form joint planning and decision-making structures, citizen views can have real weight. Just like partnership among businesses or between countries, local partnerships like this may have stresses and strains and each party will have to give a little. But if they are maintained, everyone's interests are considered.

Both Sherry Arnstein and Paul Davidoff (p. 435) were engaged liberals who wrote their classic statements about citizen participation and advocacy planning in the late 1960s. Compare the approach of Davidoff, the lawyer who argues in favor of skilled professionals advocating on behalf of powerless clients, with the approach of Arnstein, the social work professional, who favors empowering individuals and communities by involving them directly in planning and decision-making.

Arnstein says that citizen participation is like eating spinach – everyone is in favor of it in principle. But are there limits? In *The Environmental Protection Hustle*, Massachusetts Institute of Technology professor Bernard Frieden provides good case study evidence that citizens often stall needed projects claiming concern for the environment, when they really want to protect their own property values and privileged status. The vociferous participation of NIMBYs (Not-In-My-Back-Yard) often torpedo needed projects, drag out approvals, and impose costs on public and private entities that are trying to get things done. Some political scientists point to the phenomenon of hyperpluralism in some American cities, where there are so many contending groups and so much attention to participation that it is difficult to get anything done.

Arnstein (1929–1997) was born in New York City and grew up in Los Angeles. She studied physical education at the University of California, Los Angeles and worked briefly as a social worker, did community relations work for a hospital, and worked as a magazine editor before joining the staff of the Kennedy Administration's Commission on Juvenile Delinquency in 1963, where she helped communities develop programs to improve job opportunities, housing, and schools. Arnstein became a special assistant to the assistant secretary of the Department of Health, Education and Welfare (HEW), where she planned a federal strategy to desegregate hospitals. When the Model Cities Program was created in 1966, Arnstein became the chief adviser at the United States Department of Housing and Urban Development (HUD) on citizen participation not only for the Model Cities Program but also for the entire agency. After her work at HEW and HUD, Arnstein worked with the consulting firm of Arthur D. Little as a public policy analyst and project manager in technology assessment, especially as it applied to health care. Arnstein later served for ten years as the Executive Director of the American Association of Colleges of Osteopathic Medicine (AACOM) until her death in 1976.

"A Ladder of Citizen Participation" was published in the *Journal of the American Planning Association* in 1968. It has been reprinted over eighty times and translated into a number of different languages. Arnstein coauthored *Perspectives on Technology Assessment* with Alexander N. Christakis (Jerusalem: Science and Technology Publishers, 1976).

Other books on citizen participation in urban planning and programs include Cliff Zukin, Scott Keeter, Molly Andolina, Krista Jenkins, and Michael X. Delli Carpini, *A New Engagement? Political Participation, Civic Life, and the Changing American Citizen* (New York: Oxford University Press, 2006), James L. Creighton, *The Public Participation Handbook: Making Better Decisions through Citizen Involvement* (San Francisco, CA: Jossey-Bass, 2005), Thomas Ehrlich, *Public Policymaking in a Democratic Society: A Guide to Civic Engagement* (Armonk, NY: M.E. Sharpe, 2002), Henry Sanoff, *Community Participation Methods in Design and Planning* (New York: Wiley, 1999), and John F. Forester, *The Deliberative Practitioner: Encouraging Participatory Planning Processes* (Cambridge, MA: MIT Press, 1999).

British planning professor Patsy Healey has developed an important body of theoretical literature on collaborative planning, which involves not only citizens, but also other stakeholders: *Collaborative Planning: Shaping Places in Fragmented Societies*, 2nd edn (New York: Palgrave Macmillan, 2006).

Peter Marris and Martin Rein's classic *Dilemmas of Social Reform*, 2nd edn (Chicago, IL: University of Chicago Press, 1982) describes community-based urban programs and articulates a philosophy of social change that influenced US urban policy in the 1960s. Two very different views on the US War on Poverty are Sar Levitan's sympathetic *The Great Society's Poor Law* (Baltimore, MD: Johns Hopkins University Press, 1969), and Daniel Patrick Moynihan's highly critical *Maximum Feasible Misunderstanding* (New York: Free Press, 1969). The US Model Cities program, its antecedents, and the initial phase of the successor Community Development Block Grant program are discussed in Bernard J. Frieden and Marshal Kaplan, *The Politics of Neglect: Urban Aid from Model Cities to Revenue Sharing* (Cambridge, MA: MIT Press, 1975).

Books on public participation in urban planning and programs in Europe include Thomas Zitel and Ditmar Fuchs, *Participatory Democracy and Political Participation: Can Democracy Reform Bring Citizens Back In?* (London: Routledge, 2006). James Barlow, *Public Participation in Urban Development: The European Experience* (Washington, DC: Brookings Institution, 1995), *Community Involvement in Planning and Development Processes* (London: HMSO, 1995), and Albert Mabileau, *Local Politics and Participation in Britain and France* (Cambridge: Cambridge University Press, 1990).

The idea of citizen participation is a little like eating spinach: no one is against it in principle because it is good for you. Participation of the governed in their government is, in theory, the cornerstone of democracy – a revered idea that is vigorously applauded by virtually everyone. The applause is reduced to polite handclaps, however, when this principle is advocated by the have-not blacks, Mexican Americans, Puerto Ricans, Indians, Eskimos, and whites. And when the have-nots define participation as redistribution of power, the American consensus on the fundamental principle explodes into many shades of outright racial, ethnic, ideological, and political opposition.

There have been many recent speeches, articles, and books which explore in detail *who* are the have-nots of our time. There has been much recent documentation of *why* the have-nots have become so offended and embittered by their powerlessness to deal with the profound inequities and injustices pervading their daily lives. But there has been very little analysis of the content of the current controversial slogan: "citizen participation" or "maximum feasible participation." In short: *What* is citizen participation and what is its relationship to the social imperatives of our time?

Citizen participation is citizen power

Because the question has been a bone of political contention, most of the answers have been purposely buried in innocuous euphemisms like "self-help" or "citizen involvement." Still others have been embellished with misleading rhetoric like "absolute control" which is something no one – including the President of the United States – has or can have. Between understated euphemisms and exacerbated rhetoric, even scholars have found it difficult to follow the controversy. To the headline reading public, it is simply bewildering.

My answer to the critical *what* question is simply that citizen participation is a categorical term for citizen power. It is the redistribution of power that enables the have-not citizens, presently excluded from the political and economic processes, to be deliberately included in the future. It is the strategy by which the have-nots join in determining how information is shared, goals and policies are set, tax resources are allocated, programs are operated, and benefits like contracts and patronage are parceled out. In short, it is the means by which they can induce significant social reform which enables them to share in the benefits of the affluent society.

EMPTY REFUSAL VERSUS BENEFIT

There is a critical difference between going through the empty ritual of participation and having the real power needed to affect the outcome of the process. This difference is brilliantly capsulized in a poster painted last spring [1968] by the French students to explain the student-worker rebellion. (See Figure 1.) The poster highlights the fundamental point that participation without redistribution of power is an empty and frustrating process for the powerless. It allows the powerholders to claim that all sides were considered, but makes it possible for only some of those sides to benefit. It maintains the status quo. Essentially, it is what has been happening in most of the 1,000

Figure 1 French student poster. In English, "I participate, you participate, he participates, we participate, you participate . . . they profit"

Community Action Programs, and what promises to be repeated in the vast majority of the 150 Model Cities programs.

Types of participation and "nonparticipation"

A typology of eight *levels* of participation may help in analysis of this confused issue. For illustrative purposes the eight types are arranged in a ladder pattern with each rung corresponding to the extent of citizens' power in determining the end product. (See Figure 2.)

The bottom rungs of the ladder are (1) *Manipulation* and (2) *Therapy*. These two rungs describe levels of "nonparticipation" that have been contrived by some to substitute for genuine participation. Their real objective is not to enable people to participate in planning or conducting programs, but to enable powerholders to "educate" or "cure" the participants. Rungs 3 and 4 progress to levels of "tokenism" that allow the have-nots to hear and to have a voice: (3) *Informing* and (4) *Consultation*. When they are proffered by power-holders as the total extent of participation, citizens may indeed hear and be heard. But under these conditions they lack the power to insure that their views will be *heeded* by the powerful. When participation is restricted to these levels, there is no follow-through, no "muscle," hence no assurance of changing the status quo. Rung (5) *Placation* is simply a higher level tokenism because the groundrules allow have-nots to advise, but retain for the powerholders the continued right to decide.

Further up the ladder are levels of citizen power with increasing degrees of decision-making clout. Citizens can enter into a (6) *Partnership* that enables them to negotiate and engage in trade-offs with traditional power holders. At the topmost rungs, (7) *Delegated Power* and (8) *Citizen Control*, have-not citizens obtain the majority of decision-making seats, or full managerial power.

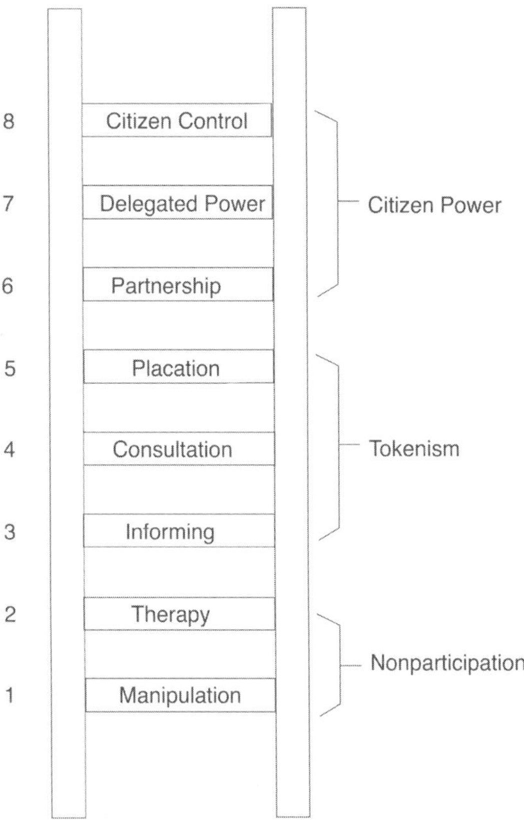

Figure 2 Eight rungs on the ladder of citizen participation

Obviously, the eight-rung ladder is a simplification, but it helps to illustrate the point that so many have missed – that there are significant gradations of citizen participation. Knowing these gradations makes it possible to cut through the hyperbole to understand the increasingly strident demands for participation from the have-nots as well as the gamut of confusing responses from the powerholders.

Though the typology uses examples from federal programs such as urban renewal, anti-poverty, and Model Cities, it could just as easily be illustrated in the church, currently facing demands for power from priests and laymen who seek to change its mission; colleges and universities which in some cases have become literal battlegrounds over the issue of student power; or public schools, city halls, and police departments (or big business which is likely to be next on the expanding list of targets). The underlying issues are essentially the same – "nobodies" in several arenas are trying to become "somebodies" with enough power to make the target institutions responsive to their views, aspirations, and needs.

LIMITATIONS OF THE TYPOLOGY

The ladder juxtaposes powerless citizens with the powerful in order to highlight the fundamental divisions between them. In actuality, neither the have-nots nor the powerholders are homogeneous blocs. Each group encompasses a host of divergent points of view, significant cleavages, competing vested interests, and splintered subgroups. The justification for using such simplistic abstractions is that in most cases the have-nots really do perceive the powerful as a monolithic "system," and powerholders actually do view the have-nots as a sea of "those people," with little comprehension of the class and caste differences among them.

It should be noted that the typology does not include an analysis of the most significant roadblocks to achieving genuine levels of participation. These roadblocks lie on both sides of the simplistic fence. On the powerholders' side, they include racism, paternalism, and resistance to power redistribution. On the have-nots' side, they include inadequacies of the poor community's political socioeconomic infrastructure and knowledge-base, plus difficulties of organizing a representative and accountable citizens' group in the face of futility, alienation, and distrust.

Another caution about the eight separate rungs on the ladder: In the real world of people and programs, there might be 150 rungs with less sharp and "pure" distinctions among them. Furthermore, some of the characteristics used to illustrate each of the eight types might be applicable to other rungs. For example, employment of the have-nots in a program or on a planning staff could occur at any of the eight rungs and could represent either a legitimate or illegitimate characteristic of citizen participation. Depending on their motives, powerholders can hire poor people to coopt them, to placate them, or to utilize the have-nots' special skills and insights. Some mayors, in private, actually boast of their strategy in hiring militant black leaders to muzzle them while destroying their credibility in the black community.

Characteristics and illustrations

It is in this context of power and powerlessness that the characteristics of the eight rungs are illustrated by examples from current federal social programs.

1. MANIPULATION

In the name of citizen participation, people are placed on rubberstamp advisory committees or advisory boards for the express purpose of "educating" them or engineering their support. Instead of genuine citizen participation, the bottom rung of the ladder signifies the distortion of participation into a public relations vehicle by powerholders.

This illusory form of "participation" initially came into vogue with urban renewal when the socially elite were invited by city housing officials to serve on Citizen Advisory Committees (CACs). Another target of manipulation were the CAC subcommittees on minority groups, which in theory were to protect the rights of Negroes in the renewal program. In practice, these subcommittees, like their parent CACs, functioned mostly as letterheads, trotted forward at appropriate times to promote urban renewal plans (in recent years known as Negro removal plans).

At meetings of the Citizen *Advisory* Committees, it was the officials who educated, persuaded, and advised the citizens, not the reverse. Federal guidelines for the renewal programs legitimized the manipulative agenda by emphasizing the terms "information-gathering," public relations," and "support" as the explicit functions of the committees.

This style of nonparticipation has since been applied to other programs encompassing the poor. Examples of this are seen in Community Action Agencies (CAAs) which have created structures called "neighborhood councils" or "neighborhood advisory groups." These bodies frequently have no legitimate function or power. The CAAs use them to "prove" that "grass-roots people" are involved in the program. But the program may not have been discussed with "the people." Or it may have been described at a meeting in the most general terms; "We need your signatures on this proposal for a multiservice center which will house, under one roof, doctors from the health department, workers from the welfare department, and specialists from the employment service."

The signatories are not informed that the $2 million-per-year center will only refer residents to the same old waiting lines at the same old agencies across town. No one is asked if such a referral center is really needed in his neighborhood. No one realizes that the contractor for the building is the mayor's brother-in-law, or that the new director of the center will be the same old community organization specialist from the urban renewal agency.

After signing their names, the proud grassrooters dutifully spread the word that they have "participated" in bringing a new and wonderful center to the neighborhood to provide people with drastically needed jobs and health and welfare services. Only after the ribbon-cutting ceremony do the members of the neighborhood council realize that they didn't ask the important questions, and that they had no technical advisors of their own to help them grasp the fine legal print. The new center, which is open 9 to 5 on weekdays only, actually adds to their problems. Now the old agencies across town won't talk with them unless they have a pink paper slip to prove that they have been referred by "their" shiny new neighborhood center.

Unfortunately, this chicanery is not a unique example. Instead it is almost typical of what has been perpetrated in the name of high-sounding rhetoric like "grassroots participation." This sham lies at the heart of the deep-seated exasperation and hostility of the have-nots toward the powerholders.

One hopeful note is that, having been so grossly affronted, some citizens have learned the Mickey Mouse game, and now they too know how to play. As a result of this knowledge, they are demanding genuine levels of participation to assure them that public programs are relevant to their needs and responsive to their priorities.

2. THERAPY

In some respects group therapy, masked as citizen participation, should be on the lowest rung of the ladder because it is both dishonest and arrogant. Its administrators – mental health experts from social workers to psychiatrists – assume that powerlessness is synonymous with mental illness. On this assumption, under a masquerade of involving citizens in planning, the experts subject the citizens to clinical group therapy. What makes this form of "participation" so invidious is that citizens are engaged in extensive activity, but the focus of it is on curing them of their "pathology" rather than changing the racism and victimization that create their "pathologies."

Consider an incident that occurred in Pennsylvania less than one year ago. When a father took his seriously ill baby to the emergency clinic of a local hospital, a young resident physician on duty instructed him to take the baby home and feed it sugar water. The baby died that afternoon of pneumonia and dehydration. The overwrought father complained to the board of the local Community Action Agency. Instead of launching an investigation of the hospital to determine what changes would prevent similar deaths or other forms of malpractice, the board invited the father to attend the CAA's (therapy) child-care sessions for parents, and promised him that someone would "telephone the hospital director to see that it never happens again."

Less dramatic, but more common examples of therapy, masquerading as citizen participation, may be seen in public housing programs where tenant groups are used as vehicles for promoting control-your-child or cleanup campaigns. The tenants are brought together to help them "adjust their values and attitudes to those of the larger society." Under these ground rules, they are diverted from dealing with such important matters as: arbitrary evictions; segregation of the housing project; or why there is a three-month time lapse to get a broken window replaced in winter.

The complexity of the concept of mental illness in our time can be seen in the experiences of student/civil rights workers facing guns, whips, and other forms of terror in the South. They needed the help of socially attuned psychiatrists to deal with their fears and to avoid paranoia.

3. INFORMING

Informing citizens of their rights, responsibilities, and options can be the most important first step toward legitimate citizen participation. However, too frequently the emphasis is placed on a one-way flow of information – from officials to citizens – with no channel provided for feedback and no power for negotiation. Under these conditions, particularly when information is provided at a late stage in planning, people have little opportunity to influence the program designed "for their benefit." The most frequent tools used for such one-way communication are the news media, pamphlets, posters, and responses to inquiries.

Meetings can also be turned into vehicles for one-way communication by the simple device of providing superficial information, discouraging questions, or giving irrelevant answers. At a recent Model Cities citizen planning meeting in Providence, Rhode Island, the topic was "tot-lots." A group of elected citizen representatives, almost all of whom were attending three to five meetings a week, devoted an hour to a discussion of the placement of six tot-lots. The neighborhood is half black, half white. Several of the black representatives noted that four tot-lots were proposed for the white district and only two for the black. The city official responded with a lengthy, highly technical explanation about costs per square foot and available property. It was clear that most of the residents did not understand his explanation. And it was clear to observers from the Office of Economic Opportunity that other options did exist which, considering available funds, would have brought about a more equitable distribution of facilities. Intimidated by futility, legalistic jargon, and prestige of the official, the citizens accepted the "information" and endorsed the agency's proposal to place four lots in the white neighborhood.

4. CONSULTATION

Inviting citizens' opinions, like informing them, can be a legitimate step toward their full participation. But if consulting them is not combined with other modes of participation, this rung of the ladder is still a sham since it offers no assurance that citizen concerns and ideas will be taken into account. The most frequent methods used for consulting people are attitude surveys, neighborhood meetings, and public hearings.

When powerholders restrict the input of citizens' ideas solely to this level, participation remains just a window-dressing ritual. People are primarily perceived as statistical abstractions, and participation is measured by how many come to meetings, take brochures home, or answer a questionnaire. What citizens achieve in all this activity is that they have "participated in participation." And what powerholders achieve is the evidence that they have gone through the required motions of involving "those people."

Attitude surveys have become a particular bone of contention in ghetto neighborhoods. Residents are increasingly unhappy about the number of times per week they are surveyed about their problems and hopes. As one woman put it: "Nothing ever happens with those damned questions, except the surveyor gets $3 an hour, and my washing doesn't get done that day." In some communities, residents are so annoyed that they are demanding a fee for research interviews.

Attitude surveys are not very valid indicators of community opinion when used without other input from citizens. Survey after survey (paid for out of anti-poverty funds) has "documented" that poor housewives most want tot-lots in their neighborhood where young children can play safely. But most of the women answered these questionnaires without knowing what their options were. They assumed that if they asked for something small, they might just get something useful in the neighborhood. Had the mothers known that a free prepaid health insurance plan was a possible option, they might not have put tot-lots so high on their wish lists.

A classic misuse of the consultation rung occurred at a New Haven, Connecticut, community meeting held to consult citizens on a proposed Model Cities grant. James V. Cunningham, in an unpublished report to the Ford Foundation, described the crowd as large and mostly hostile:

> Members of The Hill Parents Association demanded to know why residents had not participated in drawing up the proposal. CAA director Spitz explained that it was merely a proposal for seeking Federal planning funds – that once funds were obtained, residents would be deeply involved in the planning. An outside observer who sat in the audience described the meeting this way:

"Spitz and Mel Adams ran the meeting on their own. No representatives of a Hill group moderated or even sat on the stage. Spitz told the 300 residents that this huge meeting was an example of 'participation in planning.' To prove this, since there was a lot of dissatisfaction in the audience, he called for a 'vote' on each component of the proposal. The vote took this form: 'Can I see the hands of all those in favor of a health clinic? All those opposed?' It was a little like asking who favors motherhood."

It was a combination of the deep suspicion aroused at this meeting and a long history of similar forms of "window-dressing participation" that led New Haven residents to demand control of the program.

By way of contrast, it is useful to look at Denver where technicians learned that even the best intentioned among them are often unfamiliar with, and even insensitive to, the problems and aspirations of the poor. The technical director of the Model Cities program has described the way professional planners assumed that the residents, victimized by high-priced local storekeepers, "badly needed consumer education." The residents, on the other hand, pointed out that the local storekeepers performed a valuable function. Although they overcharged, they also gave credit, offered advice, and frequently were the only neighborhood place to cash welfare or salary checks. As a result of this consultation, technicians and residents agreed to substitute the creation of needed credit institutions in the neighborhood for a consumer education program.

5. PLACATION

It is at this level that citizens begin to have some degree of influence though tokenism is still apparent. An example of placation strategy is to place a few hand-picked "worthy" poor on boards of Community Action Agencies or on public bodies like the board of education, police commission, or housing authority. If they are not accountable to a constituency in the community and if the traditional power elite hold the majority of seats, the have-nots can be easily outvoted and outfoxed. Another example is the Model Cities advisory and planning committees. They allow citizens to advise or plan ad infinitum but retain for powerholders the right to judge the legitimacy or feasibility of the advice. The degree to which citizens are actually placated, of course, depends largely on two factors: the quality of technical assistance they have in articulating their priorities; and the extent to which the community has been organized to press for those priorities.

It is not surprising that the level of citizen participation in the vast majority of Model Cities programs is at the placation rung of the ladder or below. Policymakers at the Department of Housing and Urban Development (HUD) were determined to return the genie of citizen power to the bottle from which it had escaped (in a few cities) as a result of the provision stipulating "maximum feasible participation" in poverty programs. Therefore, HUD channeled its physical-social-economic rejuvenation approach for blighted neighborhoods through city hall. It drafted legislation requiring that all Model Cities' money flow to a local City Demonstration Agency (CDA) through the elected city council. As enacted by Congress, this gave local city councils final veto power over planning and programming and ruled out any direct funding relationship between community groups and HUD.

HUD required the CDAs to create coalition, policy-making boards that would include necessary local powerholders to create a comprehensive physical-social plan during the first year. The plan was to be carried out in a subsequent five-year action phase. HUD, unlike OEO, did not require that have-not citizens be included on the CDA decision-making boards. HUD's Performance Standards for Citizen Participation only demanded that "citizens have clear and direct access to the decision-making process."

Accordingly, the CDAs structured their policy-making boards to include some combination of elected officials; school representatives; housing, health, and welfare officials; employment and police department representatives; and various civic, labor, and business leaders. Some CDAs included citizens from the neighborhood. Many mayors correctly interpreted the HUD provision for "access to the decision-making process" as the escape hatch they sought to relegate citizens to the traditional advisory role.

Most CDAs created residents' advisory committees. An alarmingly significant number created citizens' policy boards and citizens' policy committees which are totally misnamed as they have either no policy-making function or only a very limited authority. Almost every CDA created about a dozen planning committees or task forces on functional lines: health, welfare, education, housing, and unemployment.

In most cases, have-not citizens were invited to serve on these committees along with technicians from relevant public agencies. Some CDAs, on the other hand, structured planning committees of technicians and parallel committees of citizens.

In most Model Cities programs, endless time has been spent fashioning complicated board, committee, and task force structures for the planning year. But the rights and responsibilities of the various elements of those structures are not defined and are ambiguous. Such ambiguity is likely to cause considerable conflict at the end of the one-year planning process. For at this point, citizens may realize that they have once again extensively "participated" but have not profited beyond the extent the powerholders decide to placate them.

Results of a staff study (conducted in the summer of 1968 before the second round of seventy-five planning grants were awarded) were released in a December 1968 HUD bulletin. Though this public document uses much more delicate and diplomatic language, it attests to the already cited criticisms of non-policy-making policy boards and ambiguous complicated structures, in addition to the following findings:

1. Most CDAs did not negotiate citizen participation requirements with residents.
2. Citizens, drawing on past negative experiences with local powerholders, were extremely suspicious of this new panacea program. They were legitimately distrustful of city hall's motives.
3. Most CDAs were not working with citizens' groups that were genuinely representative of model neighborhoods and accountable to neighborhood constituencies. As in so many of the poverty programs, those who were involved were more representative of the upwardly mobile working-class. Thus their acquiescence to plans prepared by city agencies was not likely to reflect the views of the unemployed, the young, the more militant residents, and the hard-core poor.
4. Residents who were participating in as many as three to five meetings per week were unaware of their minimum rights, responsibilities, and the options available to them under the program. For example, they did not realize that they were not required to accept technical help from city technicians they distrusted.
5. Most of the technical assistance provided by CDAs and city agencies was of third-rate quality, paternalistic, and condescending. Agency technicians did not suggest innovative options. They reacted bureaucratically when the residents pressed for innovative approaches. The vested interests of the old-line city agencies were a major – albeit hidden – agenda.
6. Most CDAs were not engaged in planning that was comprehensive enough to expose and deal with the roots of urban decay. They engaged in "meetingitis" and were supporting strategies that resulted in "projectitis," the outcome of which was a "laundry list" of traditional programs to be conducted by traditional agencies in the traditional manner under which slums emerged in the first place.
7. Residents were not getting enough information from CDAs to enable them to review CDA developed plans or to initiate plans of their own as required by HUD. At best, they were getting superficial information. At worst, they were not even getting copies of official HUD materials.
8. Most residents were unaware of their rights to be reimbursed for expenses incurred because of participation – babysitting, transportation costs, and so on. The training of residents, which would enable them to understand the labyrinth of the federal–state–city systems and networks of subsystems, was an item that most CDAs did not even consider.

These findings led to a new public interpretation of HUD's approach to citizen participation. Though the requirements for the seventy-five "second-round" Model City grantees were not changed, HUD's twenty-seven-page technical bulletin on citizen participation repeatedly advocated that cities share power with residents. It also urged CDAs to experiment with subcontracts under which the residents' groups could hire their own trusted technicians.

A more recent evaluation was circulated in February 1969 by OSTI, a private firm that entered into a contract with OEO to provide technical assistance and training to citizens involved in Model Cities programs in the north-east region of the country. OSTI's report to OEO corroborates the earlier study. In addition it states:

> In practically no Model Cities structure does citizen participation mean truly shared decision-making, such that citizens might view themselves as "the partners in this program . . ."

In general, citizens are finding it impossible to have a significant impact on the comprehensive planning which is going on. In most cases the staff planners of the CDA and the planners of existing agencies are carrying out the actual planning with citizens having a peripheral role of watchdog and, ultimately, the "rubber stamp" of the plan generated. In cases where citizens have the direct responsibility for generating program plans, the time period allowed and the independent technical resources being made available to them are not adequate to allow them to do anything more than generate very traditional approaches to the problems they are attempting to solve.

In general, little or no thought has been given to the means of insuring continued citizen participation during the stage of implementation. In most cases, traditional agencies are envisaged as the implementors of Model Cities programs and few mechanisms have been developed for encouraging organizational change or change in the method of program delivery within these agencies or for insuring that citizens will have some influence over these agencies as they implement Model Cities programs . . . By and large, people are once again being planned *for*. In most situations the major planning decisions are being made by CDA staff and approved in a formalistic way by policy boards.

6. PARTNERSHIP

At this rung of the ladder, power is in fact redistributed through negotiation between citizens and power-holders. They agree to share planning and decision-making responsibilities through such structures as joint policy boards, planning committees, and mechanisms for resolving impasses. After the groundrules have been established through some form of give-and-take, they are not subject to unilateral change.

Partnership can work most effectively when there is an organized power-base in the community to which the citizen leaders are accountable; when the citizens' group has the financial resources to pay its leaders reasonable honoraria for their time-consuming efforts; and when the group has the resources to hire (and fire) its own technicians, lawyers, and community organizers. With these ingredients, citizens have some genuine bargaining influence over the outcome of the plan (as long as both parties find it useful to maintain the partnership). One community leader described it "like coming to city hall with hat on head instead of in hand."

In the Model Cities program only about fifteen of the so-called first generation of seventy-five cities have reached some significant degree of power-sharing with residents. In all but one of those cities, it was angry citizen demands, rather than city initiative, that led to the negotiated sharing of power. The negotiations were triggered by citizens who had been enraged by previous forms of alleged participation. They were both angry and sophisticated enough to refuse to be "conned" again. They threatened to oppose the awarding of a planning grant to the city. They sent delegations to HUD in Washington. They used abrasive language. Negotiation took place under a cloud of suspicion and rancor.

In most cases where power has come to be shared it was *taken by the citizens*, not given by the city. There is nothing new about that process. Since those who have power normally want to hang onto it, historically it has had to be wrested by the powerless rather than proffered by the powerful.

Such a working partnership was negotiated by the residents in the Philadelphia model neighborhood. Like most applicants for a Model Cities grant, Philadelphia wrote its more than 400-page application and waved it at a hastily called meeting of community leaders. When those present were asked for an endorsement, they angrily protested the city's failure to consult them on preparation of the extensive application. A community spokesman threatened to mobilize a neighborhood protest *against* the application unless the city agreed to give the citizens a couple of weeks to review the application and recommend changes. The officials agreed.

At their next meeting, citizens handed the city officials a substitute citizen participation section that changed the groundrules from a weak citizens' advisory role to a strong shared power agreement. Philadelphia's application to HUD included the citizens' substitution word for word. (It also included a new citizen prepared introductory chapter that changed the city's description of the model neighborhood from a paternalistic description of problems to a realistic analysis of its strengths, weaknesses, and potentials.) Consequently, the proposed policy-making committee of the Philadelphia CDA was revamped to give five out of eleven seats to the residents' organization, which is called the Area Wide Council (AWC). The AWC obtained a subcontract from the CDA for more than $20,000 per month, which it used to maintain the neighborhood organization, to pay citizen leaders $7 per meeting for their planning services, and to pay the salaries of a staff of community organizers, planners, and other technicians. AWC has the power to initiate plans of its own, to engage in joint planning with CDA committees, and to review plans initiated by city agencies. It has a veto power in that no plans may be submitted by the CDA to the city council until they have been reviewed, and any differences of opinion have been successfully negotiated with the AWC. Representatives of the AWC (which is a federation of neighborhood organizations grouped into sixteen neighborhood "hubs") may attend all meetings of CDA task forces, planning committees, or sub-committees.

Though the city council has final veto power over the plan (by federal law), the AWC believes it has a neighborhood constituency that is strong enough to negotiate any eleventh-hour objections the city council might raise when it considers such AWC proposed innovations as an AWC Land Bank, an AWC Economic Development Corporation, and an experimental income maintenance program for 900 poor families.

7. DELEGATED POWER

Negotiations between citizens and public officials can also result in citizens achieving dominant decision-making authority over a particular plan or program. Model City policy boards or CAA delegate agencies on which citizens have a clear majority of seats and genuine specified powers are typical examples. At this level, the ladder has been scaled to the point where citizens hold the significant cards to assure accountability of the program to them. To resolve differences, powerholders need to start the bargaining process rather than respond to pressure from the other end.

Such a dominant decision-making role has been attained by residents in a handful of Model Cities including Cambridge, Massachusetts; Dayton and Columbus, Ohio; Minneapolis, Minnesota; St. Louis, Missouri; Hartford and New Haven, Connecticut; and Oakland, California.

In New Haven, residents of the Hill neighborhood have created a corporation that has been delegated the power to prepare the entire Model Cities plan. The city, which received a $117,000 planning grant from HUD, has subcontracted $110,000 of it to the neighborhood corporation to hire its own planning staff and consultants. The Hill Neighborhood Corporation has eleven representatives on the twenty-one-member CDA board which assures it a majority voice when its proposed plan is reviewed by the CDA.

Another model of delegated power is separate and parallel groups of citizens and powerholders, with provision for citizen veto if differences of opinion cannot be resolved through negotiation. This is a particularly interesting coexistence model for hostile citizen groups too embittered toward city hall – as a result of past "collaborative efforts" – to engage in joint planning.

Since all Model Cities programs require approval by the city council before HUD will fund them, city councils have final veto powers even when citizens have the majority of seats on the CDA Board. In Richmond, California, the city council agreed to a citizens' counter-veto, but the details of that agreement are ambiguous and have not been tested.

Various delegated power arrangements are also emerging in the Community Action Program as a result of demands from the neighborhoods and OEO's most recent instruction guidelines which urged CAAs "to exceed (the) basic requirements" for resident participation. In some cities, CAAs have issued subcontracts to resident dominated groups to plan and/or operate one or more decentralized neighborhood program components like a multipurpose service center or a Headstart program. These contracts usually include an agreed upon line-by-line budget and program specifications. They also usually include a specific statement of the significant powers that have been delegated, for example: policy-making; hiring and firing; issuing subcontracts for building, buying, or leasing. (Some of the subcontracts are so broad that they verge on models for citizen control.)

8. CITIZEN CONTROL

Demands for community controlled schools, black control, and neighborhood control are on the increase. Though no one in the nation has absolute control, it is very important that the rhetoric not be confused with intent. People are simply demanding that degree of power (or control) which guarantees that participants or residents can govern a program or an institution, be in full charge of policy and managerial aspects, and be able to negotiate the conditions under which "outsiders" may change them.

A neighborhood corporation with no intermediaries between it and the source of funds is the model most frequently advocated. A small number of such experimental corporations are already producing goods and/or social services. Several others are reportedly in the development stage, and new models for control will undoubtedly emerge as the have-nots continue to press for greater degrees of power over their lives.

Though the bitter struggle for community control of the Ocean Hill-Brownsville schools in New York City has aroused great fears in the headline reading public, less publicized experiments are demonstrating that the have-nots can indeed improve their lot by handling the entire job of planning, policy-making, and managing a program. Some are even demonstrating that they can do all this with just one arm because they are forced to use their other one to deal with a continuing barrage of local opposition triggered by the announcement that a federal grant has been given to a community group or an all black group.

Most of these experimental programs have been capitalized with research and demonstration funds from the Office of Economic Opportunity in cooperation with other federal agencies. Examples include:

1. A $1.8 million grant was awarded to the Hough Area Development Corporation in Cleveland to plan economic development programs in the ghetto and to develop a series of economic enterprises ranging from a novel combination shopping-center-public-housing project to a loan guarantee program for local building contractors. The membership and board of the nonprofit corporation is composed of leaders of major community organizations in the black neighborhood.
2. Approximately $1 million ($595,751 for the second year) was awarded to the Southwest Alabama Farmers' Cooperative Association (SWAFCA) in Selma, Alabama, for a ten-county marketing cooperative for food and livestock. Despite local attempts to intimidate the coop (which included the use of force to stop trucks on the way to market) first year membership grew to 1,150 farmers who earned $52,000 on the sale of their new crops. The elected coop board is composed of two poor black farmers from each of the ten economically depressed counties.
3. Approximately $600,000 ($300,000 in a supplemental grant) was granted to the Albina Corporation and the Albina Investment Trust to create a black-operated, black-owned manufacturing concern using inexperienced management and unskilled minority group personnel from the Albina district. The profitmaking wool and metal fabrication plant will be owned by its employees through a deferred compensation trust plan.
4. Approximately $800,000 ($400,000 for the second year) was awarded to the Harlem Commonwealth Council to demonstrate that a community-based development corporation can catalyze and implement an economic development program with broad community support and participation. After only eighteen months of program development and negotiation, the council will soon launch several large-scale ventures including operation of two supermarkets, an auto service and repair center (with built-in manpower training program), a finance company for families earning less than $4,000 per year, and a data processing company. The all black Harlem-based board is already managing a metal castings foundry.

Though several citizen groups (and their mayors) use the rhetoric of citizen control, no Model City can meet the criteria of citizen control since final approval power and accountability rest with the city council.

Daniel P. Moynihan argues that city councils are representative of the community, but Adam Walinsky illustrates the nonrepresentativeness of this kind of representation:

> Who . . . exercises "control" through the representative process? In the Bedford-Stuyvesant ghetto of New York there are 450,000 people – as many as in the entire city of Cincinnati, more than in the entire state of Vermont. Yet the area has only one high school, and 80 per cent of its teenagers are dropouts; the infant mortality rate is twice the national average; there are over 8000 buildings abandoned by everyone but the rats, yet the area received not one dollar of urban renewal funds during the entire first 15 years of that program's operation; the unemployment rate is known only to God.

F O U R

Members of the growth machine see the main objective of expenditure of public funds, regulation of land use, and other government action as promotion of growth. Thus, given a choice between spending local own-source property tax revenue on a women's center or a one-stop office where potential investors can learn about business opportunities in the city, members of the growth coalition would fund the one-stop office.

Members of a growth machine may have different priorities from each other. For example, executives for a large department store may want the growth machines to spend property tax revenue on a new municipal parking structure near their store, while the city's largest bank might like to see the city invest in a new park that will boost property values. A third member of the same growth coalition might want money spent to sweeten an urban renewal deal. But all members of the growth machine want to foster a good business climate that attracts or retains business.

Molotch notes that members of growth machines do not see their vision as selfish and self-serving. They believe that growth is good for everyone in the community. In their view growth will make it possible to reduce local property tax rates and will generate increased property tax revenue that can be used to improve schools, libraries, parks, and other amenities; retain jobs that might otherwise disappear and create new and better paying jobs so that city residents can find work and the city has lower unemployment and needs to spend less on social welfare; and that economic multiplier effects will help small businesses employ more staff and sell more goods and services. Molotch questions these assumptions. His research suggests that tax rates in cities where growth machines have been successful stayed about the same as before. He does not believe that there is a direct link between economic growth and jobs. New stadium projects touted as magnets to help local business, for example, often turn out to be money sinks.

Molotch's article was written in 1976, when the consensus that growth is important was more widely accepted than it is nowadays. While US cities were overwhelmingly pro-growth at the time, Molotch's prediction that anti-growth sentiment would grow has proven prescient. While there were only a few effective anti-growth coalitions at the time that Molotch wrote this selection (located mainly in university towns), growth management is now a dominant concern in many cities – particularly in affluent, liberal communities in environmentally sensitive areas. Countercoalitions that oppose growth are now powerful forces in many cities. In Europe green urbanism is an important force. And, as Timothy Beatley describes (p. 446), green urbanism – unknown to most of the world at the time the Brundtland Commission published its report in 1987 (p. 351) – is now widely incorporated into city planning. At least at a rhetorical level, as Stephen Wheeler describes (p. 458), many nations and many cities are now committed to plans and policies to reduce their carbon footprints, promote alternative energy sources, and make responsible stewardship of natural resources a priority over unquestioned growth.

Review Orfield's description of metropolitan fiscal inequities (p. 296). Wouldn't growth provide jobs and property tax revenue to the "at risk" communities that Orfield, a liberal, wants to help? Contrast Molotch's characterization of coalitions that support growth as similar to old-fashioned political machines that care little about the real interests of poor and working-class people with Michael Porter's views on the competitive advantage of the inner city (p. 282). Porter believes that helping indigenous entrepreneurs succeed is the best way to help the urban poor. Porter believes in many of the values Molotch describes as characteristic of growth machines, but argues that they can be used to help improve the lives of inner-city residents.

Harvey Molotch (b. 1940) holds a dual appointment at New York University as Professor of Sociology, and Professor of Metropolitan Studies within the Department of Social and Cultural Analysis. He received a PhD in Sociology from the University of Chicago in 1968. Molotch taught at the University of California, Santa Barbara, from 1968 to 2003. He has been a visiting professor at the State University of New York, Stonybrook, the University of Essex, the University of Lund, Sweden, University of Washington, and Northwestern University. In 1998–1999 he was Centennial Professor at the London School of Economics. He has also been a Fellow of the Center for Advanced Studies at Stanford University and the Russell Sage Foundation. In addition to urban development and political economy, Molotch's research interests have included studies of race (white flight), environmental degradation (an oil spill in Santa Barbara, California), media (journalist writings as the product of social structures), mechanisms of interactional inequalities, including human conversation (what gaps in conversation reveal about human interaction), and the sociology of architecture, design, and consumption (studying the way in which social structures produce material things), colloquially "where stuff comes from" as his latest book is titled. In 2003

Molotch received a lifetime career achievement in urban and community scholarship award from the Urban and Community Studies Section of the American Sociological Association.

This selection "The City as a Growth Machine: Towards a Political Economy of Place" was published in the *American Journal of Sociology*, 82 (September, 1976), 309–332.

Molotch developed his ideas on the city as a growth machine in *Urban Fortunes*, with John Logan (Berkeley, CA: University of California Press, 1987). *Urban Fortunes* won the Robert Ezra Park Award of the Urban and Community Sociology section of the American Sociological Association (ASA) as the best book of 1987 and the ASA's Distinguished Scholarly Contribution to Sociology Award in 1990.

Other books by Molotch include *Managed Integration: Dilemmas of Doing Good in the City* (Berkeley, CA: University of California Press, 1972), and *Where Stuff Comes from: How Toasters, Toilets, Cars, Computers and Many Other Things Come to Be as They Are* (New York: Routledge, 2003).

Other books on growth machines include John Mollenkopf, *The Contested City* (Princeton, NJ: Princeton University Press), David Wilson, *The Urban Growth Machine* (Albany, NY: State University of New York Press, 2007), Barbara Ferman, *Challenging the Growth Machine* (Lawrence, KS: University of Kansas Press, 1996), and Natalie McPherson, *Machines and Economic Growth: The Implications for Growth Theory of the History of the Industrial Revolution* (Santa Barbara, CA: Greenwood Press, 1994).

Conventional definitions of "city," "urban place," or "metropolis" have led to conventional analyses of urban systems and urban-based social problems. Usually traceable to Wirth's classic and highly plausible formulation of "numbers, density and heterogeneity," there has been a continuing tendency, even in more recent formulations, e.g. to conceive of place quite apart from a crucial dimension of social structure: power and social class hierarchy. Consequently, sociological research based on the traditional definitions of what an urban place is has had very little relevance to the actual, day-to-day activities of those at the top of local power structure whose priorities set the limits within which decisions affecting land use, the public budget, and urban social life come to be made. It has not been very apparent from the scholarship of urban social science that land, the basic stuff of place, is a market commodity providing wealth and power, and that some very important people consequently take a keen interest in it. Thus, although there are extensive literatures on community power as well as on how to define and conceptualize a city or urban place, there are few notions available to link the two issues coherently, focusing on the urban settlement as a political economy.

This paper aims toward filling this need. I speculate that the political and economic essence of virtually any given locality, in the present American context, is *growth*. I further argue that the desire for growth provides the key operative motivation toward consensus for members of politically mobilized local elites, however split they might be on other issues, and that a common interest in growth is the overriding commonality among important people in a given locale – at least insofar as they have any important local goals at all. Further, this growth imperative is the most important constraint upon available options for local initiative in social and economic reform.

It is thus that I argue that the very essence of a locality is its operation as a growth machine.

The clearest indication of success at growth is a constantly rising urban-area population – a symptom of a pattern ordinarily comprising an initial expansion of basic industries followed by an expanded labor force, a rising scale of retail and wholesale commerce, more far-flung and increasingly intensive land development, higher population density, and increased levels of financial activity. Although throughout this paper I index growth by the variable population growth, it is this entire syndrome of associated events that is meant by the general term "growth." I argue that the means of achieving this growth, of setting off this chain of phenomena, constitute the central issue for those serious people who care about their locality and who have the resources to make their caring felt as a political force. The city is, for those who count, a growth machine.

THE HUMAN ECOLOGY: MAPS AS INTEREST MOSAICS

I have argued elsewhere that any given parcel of land represents an interest and that any given locality is thus an aggregate of land-based interests. That is, each landowner (or person who otherwise has some interest in the prospective use of a given piece of land) has in mind a certain future for that parcel which is linked somehow with his or her own well-being. If there is a simple ownership, the relationship is straightforward: to the degree to which the land's profit potential is enhanced, one's own wealth is increased. In other cases, the relationship may be more subtle: one has interest in an adjacent parcel, and if a noxious use should appear, one's own parcel may be harmed. More subtle still is the emergence of concern for an aggregate of parcels: one sees that one's future is bound to the future of a larger area, that the future enjoyment of financial benefit flowing from a given parcel will derive from the general future of the proximate aggregate of parcels. When this occurs, there is that "we feeling" which bespeaks of community. We need to see each geographical map – whether of a small group of land parcels, a whole city, a region, or a nation – not merely as a demarcation of legal, political, or topographical features, but as a mosaic of competing land interests capable of strategic coalition and action.

Each unit of a community strives, at the expense of the others, to enhance the land-use potential of the parcels with which it is associated. Thus, for example, shopkeepers at both ends of a block may compete with one another to determine in front of which building the bus stop will be placed. Or, hotel owners on the north side of a city may compete with those on the south to get a convention center built nearby. Likewise, area units fight over highway routes, airport locations, campus developments, defense contracts, traffic lights, one-way street designations, and park developments. The intensity of group consciousness and activity waxes and wanes as opportunities for and challenges to the collective good rise and fall; but when these coalitions are of sufficiently enduring quality, they constitute identifiable, ongoing communities. Each member of a community is simultaneously the member of a number of others; hence, communities exist in a nested fashion (e.g., neighborhood within city, within region), with salience of community level varying both over time and circumstance. Because of this nested nature of communities, subunits which are competitive with one another at one level (e.g., in an interblock dispute over where the bus stop should go) will be in coalition at a higher level (e.g., in an intercity rivalry over where the new port should go). Obviously, the anticipation of potential coalition acts to constrain the intensity of conflict at more local loci of growth competition.

Hence, to the degree to which otherwise competing land-interest groups collude to achieve a common land-enhancement scheme, there is community – whether at the level of a residential block club, a neighborhood association, a city or metropolitan chamber of commerce, a state development agency, or a regional association. Such aggregates, whether constituted formally or informally, whether governmental political institutions or voluntary associations, typically operate in the following way: an attempt is made to use government to gain those resources which will enhance the growth potential of the area unit in question. Often, the governmental level where action is needed is at least one level higher than the community from which the activism springs. Thus, individual landowners aggregate to extract neighborhood gains from the city government; a cluster of cities may coalesce to have an effective impact on the state government, etc. Each locality, in striving to make these gains, is in competition with other localities because the degree of growth, at least at any given moment, is finite. The scarcity of developmental resources means that government becomes the arena in which land-use interest groups compete for public money and attempt to mold those decisions which will determine the land-use outcomes. Localities thus compete with one another to gain the *preconditions* of growth. Historically, U.S. cities were created and sustained largely through this process; it continues to be the significant dynamic of contemporary local political economy and is critical to the allocation of public resources and the ordering of local issue agendas.

Government decisions are not the only kinds of social activities which affect local growth chances; decisions made by private corporations also have major impact. When a national corporation decides to locate a branch plant in a given locale, it sets the conditions for the surrounding land-use pattern. But even here, government decisions are involved: plant-location decisions are made with reference to such issues as labor costs, tax rates, and the costs of obtaining raw materials and transporting goods to markets. It is government decisions (at whatever level) that help determine the cost of access to markets and raw materials. This is especially so in the present era of raw material subsidies (e.g., the mineral depletion allowance) and reliance on government approved or subsidized air transport, highways, railways, pipelines, and port developments. Government decisions influence the cost of overhead expenses (e.g., pollution abatement requirements, employee safety standards), and government decisions affect the costs of labor through indirect manipulation of unemployment rates, through the use of police to constrain or enhance union organizing, and through the legislation and administration of welfare laws.

Localities are generally mindful of these governmental powers and, in addition to creating the sorts of physical conditions which can best serve industrial growth, also attempt to maintain the kind of "business climate" that attracts industry: for example, favorable taxation, vocational training, law enforcement, and "good" labor relations. To promote growth, taxes should be "reasonable," the police force should be oriented toward protection of property, and overt social conflict should be minimized. Increased utility and government costs caused by new development should be borne (and they usually are) by the public at large, rather than by those responsible for the "excess" demand on the urban infrastructure. Virtually any issue of a major business magazine is replete with ads from localities of all types (including whole countries) trumpeting their virtues in just these terms to prospective industrial settlers. In addition, a key role of elected and appointed officials becomes that of "ambassador" to industry, to communicate, usually with appropriate ceremony, these advantages to potential investors.

I aim to make the extreme statement that this organized effort to affect the outcome of growth distribution is the essence of local government as a dynamic political force. It is not the only function of government, but it is the key one and, ironically, the one most ignored. Growth is not, in the present analysis, merely one among a number of equally important concerns of political process. Among contemporary social scientists, perhaps only Murray Edelman has provided appropriate conceptual preparation for viewing government in such terms. Edelman contrasts two kinds of politics. First there is the "symbolic" politics which comprises the "big issues" of public morality and the symbolic reforms featured in the headlines and editorials of the daily press. The other politics is the process through which goods and services actually come to be distributed in the society. Largely unseen, and relegated to negotiations within committees (when it occurs at all within a formal government body), this is the politics which determines who, in *material terms*, gets what, where, and how. This is the kind of politics we must talk about at the local level: it is the politics of distribution, and land is the crucial (but not the only) variable in this system.

The people who participate with their energies, and particularly their fortunes, in local affairs are the sort of persons who – at least in vast disproportion to their representation in the population – have the most to gain or lose in land-use decisions. Prominent in terms of numbers have long been the local businessmen, particularly property owners and investors in locally oriented financial institutions, who *need* local government in their daily money-making routines. Also prominent are lawyers, syndicators, and realtors who need to put themselves in situations where they can be most useful to those with the land and property resources. Finally, there are those who, although not directly involved in land use, have their futures tied to growth of the metropolis as a whole. At least, when the local market becomes saturated one of the few possible avenues for business expansion is sometimes the expansion of the surrounding community itself.

This is the general outline of the coalition that actively generates the community "we feeling" (or perhaps more aptly, the "our feeling") that comes to be an influence in the politics of a given locality. It becomes manifest through a wide variety of techniques. Government funds support "boosterism" of various sorts: the Chamber of Commerce, locality-promotion ads in business journals and travel publications, city-sponsored parade floats, and stadia and other forms of support for professional sports teams carrying the locality name. The athletic teams in particular are an extraordinary mechanism for instilling a spirit of civic jingoism regarding the "progress" of the locality. A stadium filled with thousands (joined by thousands more at home before the TV) screaming for Cleveland or Baltimore (or whatever) is a scene difficult to fashion otherwise. This enthusiasm can be drawn upon, with a glossy claim of creating a "greater Cleveland," "greater Baltimore," etc., in order to gain general acceptance for local growth-oriented programs. Similarly, public school curricula, children's essay contests, soapbox derbies, spelling contests, beauty pageants, etc., help build an ideological base for local boosterism and the acceptance of growth. My conception of the territorial bond among humans differs from those cast in terms of primordial instincts: instead, I see this bond as socially organized and sustained, at least in part, by those who have a use for it. I do not claim that there are no other sources of civic jingoism and growth enthusiasm in American communities, only that the growth-machine coalition mobilizes what is there, legitimizes and sustains it, and channels it as a political force into particular kinds of policy decisions.

The local institution which seems to take prime responsibility for the sustenance of these civic resources – the metropolitan newspaper – is also the most important example of a business which has its interest anchored in the aggregate growth of the locality. Increasingly, American cities are one-newspaper (metropolitan daily) towns (or one-newspaper-company towns), and the newspaper business seems to be one kind of enterprise for which expansion to other locales is especially difficult . . . A paper's financial status (and that of other media to a lesser extent) tends to be wed to the size of the locality. As the metropolis expands, a larger number of ad lines can be sold on the basis of the increasing circulation base. The local newspaper thus tends to occupy a rather unique position: like many other local businesses, it has an interest in growth, but unlike most, its critical interest is not in the specific geographical pattern of that growth. That is, the crucial matter to a newspaper is not whether the additional population comes to reside on the north side or south side, or whether the money is made through a new convention center or a new olive factory. The newspaper has no axe to grind, except the one axe which holds the community elite together: growth. It is for this reason that the newspaper tends to achieve a statesman-like attitude in the community and is deferred to as something other than a special interest by the special interests. Competing interests often regard the publisher or editor as a general community leader, as an ombudsman and arbiter of internal bickering and, at times, as an enlightened third party who can restrain the short-term profiteers in the interest of more stable, long-term, and properly planned growth. [The] paper becomes the reformist influence, the "voice of the community," restraining the competing subunits, especially the small-scale, arriviste "fast-buck artists" among them. The papers are variously successful in their continuous battle with the targeted special interests. The media attempt to attain these goals not only through the kind of coverage they develop and editorials they write but also through the kinds of candidates they support for local office. The present point is not that the papers control the politics of the city, but rather that one of the sources of their special influence is their commitment to growth per se, and growth is a goal around which all important groups can rally.

Thus it is that, although newspaper editorialists have typically been in the forefront expressing sentiment in favor of "the ecology," they tend nevertheless to support growth-inducing investments for their regions. The *New York Times* likes office towers and additional industrial installations in the city even more than it loves the environment. The *Los Angeles Times* editorializes against narrow-minded profiteering at the expense of the environment but has also favored the development of the supersonic transport because of the "jobs" it would lure to Southern California. The papers do tend to support "good planning principles" in some form because such good planning is a long-term force that makes for even more potential future growth. If the roads are not planned wide enough, their narrowness will eventually strangle the increasingly intense uses to which the land will be put. It just makes good sense to plan, and good planning for "sound growth" thus is the key "environmental policy" of the nation's local media and their statesmen allies. Such policies of "good planning" should not be confused with limited growth or conservation: they more typically represent the opposite sort of goal.

Often leaders of public or quasi-public agencies (e.g., universities, utilities) achieve a role similar to that of the newspaper publisher: they become growth "statesmen" rather than advocates for a certain type or intralocal distribution of growth. A university may require an increase in the local urban population pool to sustain its own expansion plans and, in addition, it may be induced to defer to others in the growth machine (bankers, newspapers) upon whom it depends for the favorable financial and public-opinion environment necessary for institutional enhancement.

There are certain persons, ordinarily conceived of as members of the elite, who have much less, if any, interest in local growth. Thus, for example, there are branch executives of corporations headquartered elsewhere who, although perhaps emotionally sympathetic with progrowth outlooks, work for corporations which have no vested interest in the growth of the locality in question. Their indirect interest is perhaps in the existence of the growth ideology rather than growth itself. It is that ideology which in fact helps make them revered people in the area (social worth is often defined in terms of number of people one employs) and which provides the rationale for the kind of local governmental policies most consistent with low business operating costs. Nonetheless, this interest is not nearly as strong as the direct growth interests of developers, mortgage bankers, etc., and thus . . . there is a tendency for such executives to play a lesser local role than the parochial, homegrown businessmen whom they often replace.

Thus, because the city is a growth machine, it draws a special sort of person into its politics. These people – whether acting on their own or on behalf of the constituency which financed their rise to power – tend to be businessmen and, among businessmen, the more parochial sort. Typically, they come to politics not to save or destroy the environment, not to repress or liberate the blacks, not to eliminate civil liberties or enhance them. They may end up doing any or all of these things once they have achieved access to authority, perhaps as an inadvertent consequence of making decisions in other realms. But these types of symbolic positions are derived from the fact of having power – they are typically not the dynamics which bring people to power in the first place. Thus, people often become "involved" in government, especially in the local party structure and fund raising, for reasons of land business and related processes of resource distribution. Some are "statesmen" who think in terms of the growth of the whole community rather than that of a more narrow geographical delimitation. But they are there to wheel and deal to affect resource distribution through local government. As a result of their position, and in part to develop the symbolic issues which will enable them (in lieu of one of their opponents or colleagues) to maintain that position of power, they get interested in such things as welfare cheating, busing, street crime, and the price of meat. This interest in the symbolic issues is thus substantially an after-effect of a need for power for other purposes. This is not to say that such people don't "feel strongly" about these matters – they do sometimes. It is also the case that certain moral zealots and "concerned citizens" go into politics to right symbolic wrongs; but the money and other supports which make them viable as politicians is usually non-symbolic money.

Those who come to the forefront of local government (and those to whom they are directly responsive), therefore, are not statistically representative of the local population as a whole, nor even representative of the social classes which produce them. The issues they introduce into public discourse are not representative either. As noted by Edelman (1964), the distributive issues, the matters which bring people to power, are more or less deliberately dropped from public discourse. The issues which are allowed to be discussed and the positions which the politicians take on them derive from the world views of those who come from certain sectors of the business and professional class and the need which they have to whip up public sentiment without allowing distributive issues to become part of public discussion. It follows that any political change which succeeded in replacing the land business as the key determinant of the local political dynamic would simultaneously weaken the power of one of the more reactionary political forces in the society, thereby affecting outcomes with respect to those other symbolic issues which manage to gain so much attention. Thus, should such a change occur, there would likely be more progressive positions taken on civil liberties, and less harassment of welfare recipients, social "deviants," and other defenseless victims.

LIABILITIES OF THE GROWTH MACHINE

Emerging trends are tending to enervate the locality growth machines. First is the increasing suspicion that in many areas, at many historical moments, growth benefits only a small proportion of local residents. Growth almost always brings with it the obvious problems of increased air and water pollution, traffic congestion, and overtaxing of natural amenities. These dysfunctions become increasingly important and visible as increased consumer income fulfills people's other needs and as the natural cleansing capacities of the environment are progressively overcome with deleterious material. While it is by no means certain that growth and increased density inevitably bring about social pathologies, growth does make such pathologies more difficult to deal with. For example, the larger the jurisdiction, the more difficult it becomes to achieve the goal of school integration without massive busing schemes. As increasing experience with busing makes clear, small towns can more easily have interracial schools, whether fortuitously through spatial proximity or through managed programs.

In addition, the weight of research evidence is that growth often costs existing residents more money. Evidently, at various population levels, points of diminishing returns are crossed such that additional increments lead to net revenue losses. A 1970 study of Palo Alto, California, indicated that it was substantially cheaper for that city to acquire at full market value its foothill open space than to allow it to become an "addition" to the tax base. A study of Santa Barbara, California, demonstrated that additional population growth would require higher property taxes, as well as higher utility costs. Similar results on the costs of growth have been obtained in studies of Boulder, Colorado, and Ann Arbor, Michigan. Systematic analyses of government costs as a function of city size and growth have been carried out under a number of methodologies, but the use of the units of analysis most appropriate for comparison (urban areas) yields the finding that the cost is directly related both to size of place and rate of growth, at least for middle-size cities. Especially significant are per capita police costs, which virtually all studies show to be positively related to both city size and rate of growth.

Although damage to the physical environment and costs of utilities and governmental services may rise with size of settlement, "optimal" size is obviously determined by the sorts of values which are to be maximized. It may indeed be necessary to sacrifice clean air to accumulate a population base large enough to support a major opera company. But the essential point remains that growth is certainly less of a financial advantage to the taxpayer than is conventionally depicted, and that most people's values are, according to the survey evidence more consistent with small places than large. Indeed, it is rather clear that some substantial portion of the migrations to the great metropolitan areas of the last decade has been more in spite of people's values than because of them. In the recent words of Sundquist (1975):

> The notion commonly expressed that Americans have "voted with their feet" in favor of the great cities is, on the basis of every available sampling, so much nonsense. . . . What is called "freedom of choice" is, in sum, freedom of employer choice or, more precisely, freedom of choice for that segment of the corporate world that operates mobile enterprises. The real question, then, is whether freedom of corporate choice should be automatically honored by government policy at the expense of freedom of individual choice where those conflict.

Taking all the evidence together, it is certainly a rather conservative statement to make that under many circumstances growth is a liability financially and in quality of life for the majority of local residents. Under such circumstances, local growth is a transfer of quality of life and wealth from the local general public to a certain segment of the local elite. To raise the question of wisdom of growth in regard to any specific locality is hence potentially to threaten such a wealth transfer and the interests of those who profit by it.

THE PROBLEMS OF JOBS

Perhaps the key ideological prop for the growth machine, especially in terms of sustaining support from the working-class majority, is the claim that growth "makes jobs." This claim is aggressively promulgated by developers, builders, and chambers of commerce; it becomes a part of the statesman talk of editorialists and political officials. Such people do not speak of growth as useful to profits – rather, they speak of it as necessary for making jobs. But local growth does not, of course, make jobs: it distributes jobs. The United States will see next year the construction of a certain number of new factories, office units, and highways – regardless of where they are put. Similarly, a given number of automobiles, missiles, and lampshades will be made, regardless of where they are manufactured. Thus, the number of jobs in this society, whether in the building trades or any other economic sector, will be determined by rates of investment return, federal decisions affecting the money supply, and other factors having very little to do with local decision making. All that a locality can do is to attempt to guarantee that a certain proportion of newly created jobs will be in the locality in question. Aggregate employment is thus unaffected by the outcome of this competition among localities to "make" jobs.

The labor force is essentially a single national pool; workers are mobile and generally capable of taking advantage of employment opportunities emerging at geographically distant points. As jobs develop in a fast-growing area, the unemployed will be attracted from other areas in sufficient numbers not only to fill those developing vacancies but also to form a work-force sector that is continuously unemployed. Thus, just as local growth does not affect aggregate employment, it likely has very little long-term impact upon the local rate of unemployment. Again, the systematic evidence fails to show any advantage to growth: there is no tendency for either larger places or more rapidly growing ones to have lower unemployment rates than other kinds of urban areas. In fact, the tendency is for rapid growth to be associated with higher rates of unemployment.

This pattern of findings is vividly illustrated through inspection of relevant data on the most extreme cases of urban growth: those SMSAs which experienced the most rapid rates of population increase over the last two intercensus decades[1] . . . In the case of both decade comparisons, half of the urban areas had unemployment rates above the national figure for all SMSAs.

Even the 25 slowest-growing (1960–70) SMSAs failed to experience particularly high rates of unemployment. [A]lthough all were places of net migration loss less than half of the SMSAs of this group had unemployment rates above the national mean at the decade's end.

Just as striking is the comparison of growth and unemployment rates for all SMSAs in California during the 1960–66 period – a time of general boom in the state . . . [A]mong all California metropolitan areas there is no significant relationship . . . between 1960–66 growth rates and the 1966 unemployment rate. . . . [W]hile there is a wide divergence in growth rates across metropolitan areas, there is no comparable variation in the unemployment rates, all of which cluster within the relatively narrow range of 4.3–6.5 per cent. Consistent with my previous argument, I take this as evidence that the mobility of labor tends to flatten out cross-SMSA unemployment rates, regardless of widely diverging rates of locality growth. Taken together, the data indicate that local population growth is no solution to the problem of local unemployment.

It remains possible that for some reason certain specific rates of growth may be peculiarly related to lower rates of unemployment and that the measures used in this and cited studies are insensitive to these patterns. Similarly, growth in certain types of industries may be more likely than growth in others to stimulate employment without attracting migrants. It may also be possible that certain population groups, by reason of cultural milieu, are less responsive to mobility options than others and thus provide bases for exceptions to the general argument I am advancing. The present analysis does not preclude such future findings but does assert, minimally, that the argument that growth makes jobs is contradicted by the weight of evidence that is available.

I conclude that for the average worker in a fast-growing region job security has much the same status as for a worker in a slower-growing region: there is a surplus of workers over jobs, generating continuous anxiety over unemployment and the effective depressant on wages which any lumpenproletariat of unemployed and marginally employed tends to exact. Indigenous workers likely receive little benefit from the growth machine in terms of jobs; their "native" status gives them little edge over the "foreign" migrants seeking the additional jobs which may develop. Instead, they are interchangeable parts of the labor pool, and the degree of their job insecurity is expressed in the local unemployment rate, just as is the case for the nonnative worker. Ironically, it is probably this very anxiety which often leads workers, or at least their union spokespeople, to support enthusiastically employers' preferred policies of growth. It is the case that an actual decline in local job opportunities, or economic growth not in proportion to natural increase, might induce the hardship of migration. But this price is not the same as, and is less severe than, the price of simple unemployment. It could also rather easily be compensated through a relocation subsidy for mobile workers, as is now commonly provided for high-salaried executives by private corporations and in a limited way generally by the federal tax deduction for job-related moving expenses.

Workers' anxiety and its ideological consequences emerge from the larger fact that the United States is a society of constant substantial joblessness, with unemployment rates conservatively estimated by the Department of Commerce at 4–8 per cent of that portion of the work force defined as ordinarily active. There is thus a game of musical chairs being played at all times, with workers circulating around the country, hoping to land in an empty chair at the moment the music stops. Increasing the stock of jobs in any one place neither causes the music to stop more frequently nor increases the number of chairs relative to the number of players. The only way effectively to ameliorate this circumstance is to create a full-employment economy, a comprehensive system of drastically increased unemployment insurance, or some other device which breaks the connection between a person's having a livelihood and the remote decisions of corporate executives. Without such a development, the fear of unemployment acts to make workers politically passive (if not downright supportive) with respect to land-use policies, taxation programs, and antipollution nonenforcement schemes which, in effect, represent income transfers from the general public to various sectors of the elite. Thus, for many reasons, workers and their leaders should organize their political might more consistently not as part of the growth coalitions of the localities in which they are situated, but rather as part of national movements which aim to provide full employment, income security, and programs for taxation, land use, and the environment which benefit the vast majority of the population. They tend not to be doing this at present.

THE PROBLEM OF NATURAL INCREASE

Localities grow in population not simply as a function of migration but also because of the fecundity of the existing population. Some means are obviously needed to provide jobs and housing to accommodate such growth – either in the immediate area or at some distant location. There are ways of handling this without compounding the environmental and budgetary problems of existing settlements. First, there are some localities which are, by many criteria, not overpopulated. Their atmospheres are clean, water supplies plentiful, and traffic congestion nonexistent. In fact, in certain places increased increments of population may spread the costs of existing road and sewer systems over a larger number of citizens or bring an increase in quality of public education by making rudimentary specialization possible. In the state of California, for example, the great bulk of the population lives on a narrow coastal belt in the southern two-thirds of the state. Thus the northern third of the state consists of a large unpopulated region rich in natural resources, including electric power and potable water. The option chosen in California, as evidenced by the state aqueduct, was to move the water from the uncrowded north to the dense, semiarid south, thus lowering the environmental qualities of both regions, and at a substantial long-term cost to the public budget. The opposite course of action was clearly an option. The point is that there are relatively underpopulated areas in this country which do not have "natural" problems of inaccessibility, ugliness, or lack of population-support resources. Indeed, the nation's most severely depopulated areas, the towns of Appalachia, are in locales of sufficient resources and are widely regarded as aesthetically appealing; population out-migration likely decreased the aesthetic resources of both the migrants to and residents of Chicago and Detroit, while resulting in the desertion of a housing stock and utility infrastructure designed to serve a larger population. Following from my more general perspective, I see lack of population in a given area as resulting from the political economic decisions made to populate other areas instead. If the process were rendered more rational, the same investments in roads, airports, defense plants, etc., could be made to effect a very different land-use outcome. Indeed, utilization of such deliberate planning strategies is the practice in some other societies and shows some evidence of success; perhaps it could be made to work in the United States as well.

As a long-term problem, natural increase may well be phased out. American birth rates have been steadily decreasing for the last several years, and we are on the verge of a rate providing for zero population growth. If a stable population actually is achieved, a continuation of the present interlocal competitive system will result in the proliferation of ghost towns and unused capital stocks as the price paid for the growth of the successful competing units. This will be an even more clearly preposterous situation than the current one, which is given to produce ghost towns only on occasion.

THE EMERGING COUNTERCOALITION

Although growth has been the dominant ideology in most localities in the United States, there has always been a subversive thread of resistance. Treated as romantic, or as somehow irrational, this minority long was ignored, even in the face of accumulating journalistic portrayals of the evils of bigness. But certainly it was an easy observation to make that increased size was related to high levels of pollution, traffic congestion, and other disadvantages. Similarly, it was easy enough to observe that tax rates in large places were not generally less than those in small places; although it received little attention, evidence that per capita government costs rise with population size was provided a generation ago. But few took note, though the very rich, somehow sensing these facts to be the case, managed to reserve for themselves small, exclusive meccas of low density by tightly imposing population ceilings (e.g., Beverly Hills, Sands Point, West Palm Beach, Lake Forest).

In recent years, however, the base of the antigrowth movement has become much broader and in some localities has reached sufficient strength to achieve at least toeholds of political power. The most prominent cases seem to be certain university cities (Palo Alto, Santa Barbara, Boulder, Ann Arbor), all of which have sponsored impact studies documenting the costs of additional growth. Other localities which have imposed growth controls tend also to be places of high amenity value (e.g., Ramapo, N.Y.; Petaluma, Calif.; Boca Raton, Fla.). The antigrowth sentiment has become an important part of the politics of a few large cities (e.g., San Diego) and has been the basis of important political careers at the state level (including the governorship) in Oregon, Colorado, and Vermont. Given the objective importance of the issue and the evidence on the general costs of growth, there is nothing to prevent antigrowth coalitions from similarly gaining power elsewhere – including those areas of the country which are generally considered to possess lower levels of amenity. Nor is there any reason, based on the facts of the matter, for these coalitions not to further broaden their base to include the great majority of the working class in the localities in which they appear.

But, like all political movements which attempt to rely upon volunteer labor to supplant political powers institutionalized through a system of vested economic interest, antigrowth movements are probably more likely to succeed in those places where volunteer reform movements have a realistic constituency – a leisured and sophisticated middle class with a tradition of broad-based activism, free from an entrenched machine. At least, this appears to be an accurate profile of those places in which the antigrowth coalitions have already matured.

Systematic studies of the social make up of the antigrowth activists are only now in progress, but it seems that the emerging countercoalition is rooted in the recent environmental movements and relies on a mixture of young activists (some are veterans of the peace and civil rights movements), middle-class professionals, and workers, all of whom see their own tax rates as well as life-styles in conflict with growth. Important in leadership roles are government employees and those who work for organizations not dependent on local expansion for profit, either directly or indirectly. In the Santa Barbara antigrowth movements, for example, much support is provided by professionals from research and electronics firms, as well as branch managers of small "high-technology" corporations. Cosmopolitan in outlook and pecuniary interest, they use the local community only as a setting for life and work, rather than as an exploitable resource. Related to this constituency are certain very wealthy people (particularly those whose wealth derives from the exploitation of nonlocal environments) who continue a tradition (with some modifications) of aristocratic conservation.

Should it occur, the changes which the death of the growth machine will bring seem clear enough with respect to land-use policy. Local governments will establish holding capacities for their regions and then legislate, directly or indirectly, to limit population to those levels. The direction of any future development will tend to be planned to minimize negative environmental impacts. The so-called natural process . . . of land development which has given American cities their present shape will end as the political and economic foundations of such processes are undermined. Perhaps most important, industrial and business land users and their representatives will lose, at least to some extent, the effectiveness of their threat to locate elsewhere should public policies endanger the profitability they desire. As the growth machine is destroyed in many places, increasingly it will be the business interests who will be forced to make do with local policies, rather than the local populations having to bow to business wishes. New options for taxation, creative land-use programs, and new forms of urban services may thus emerge as city government comes to resemble an agency which asks what it can do for its people rather than what it can do to attract more people. More specifically, a given industrial project will perhaps be evaluated in terms of its social utility – the usefulness of the product manufactured either to the locality or to the society at large. Production, merely for the sake of local expansion, will be less likely to occur. Hence, there will be some pressure to increase the use value of the country's production apparatus and for external costs of production to be borne internally.

When growth ceases to be an issue, some of the investments made in the political system to influence and enhance growth will no longer make sense, thus changing the basis upon which people get involved in government. We can expect that the local business elites – led by land developers and other growth-coalition forces – will tend to withdraw from local politics. This vacuum may then be filled by a more representative and, likely, less reactionary activist constituency. It is noteworthy that where antigrowth forces have established beachheads of power, their programs and policies have tended to be more progressive than their predecessors' – on all issues, not just on growth. In Colorado, for example, the environmentalist who led the successful fight against the Winter Olympics also successfully sponsored abortion reform and other important progressive causes. The environmentally based Santa Barbara "Citizens Coalition" (with city government majority control) represents a fusion of the city's traditional left and counterculture with other environmental activists. The result of the no-growth influence in localities may thus be a tendency for an increasing progressiveness in local politics. To whatever degree local politics is the bedrock upon which the national political structure rests (and there is much debate here), there may follow reforms at the national level as well. Perhaps it will then become possible to utilize national institutions to effect other policies which both solidify the death of the growth machine at the local level and create national priorities consistent with the new opportunities for urban civic life. These are speculations based upon the questionable thesis that a reform-oriented, issue-based citizens' politics can be sustained over a long period. The historical record is not consistent with this thesis; it is only emerging political trends in the most affected localities and the general irrationality of the present urban system that suggest the alternative possibility is an authentic future.

EDITORS' NOTE

1 SMSA stands for Standard Metropolitan Statistical Area – an outdated term used by the United States Census between 1959 and 1983 to define metropolitan areas in the United States. An SMSA consisted of one or more entire counties containing at least one city (or twin cities) having a population of 50,000 or more, plus adjacent metropolitan counties economically and socially integrated with the central city.

REFERENCES

Edelman, Murray. (1964). *The Symbolic Uses of Politics.* Urbana, IL: University of Illinois Press.

Sundquist, James. (1975). *Dispersing Population: What America Can Learn from Europe.* Washington, DC: Brookings Institution.

"Broken Windows"

Atlantic Monthly (1982)

James Q. Wilson and George L. Kelling

Editors' Introduction

Why is urban crime a problem in inner-city neighborhoods? What can and should government do about it? What is the proper role for police working in high crime areas and marginal areas at risk of increased crime? These are issues that James Q. Wilson and George L. Kelling address in this selection.

Sociologist Elijah Anderson found that violence wreaks havoc daily on the lives of residents in the poor inner-city Black Chicago and New York neighborhoods he studied and that it increasingly spills over into downtown and residential middle-class areas (p. 127). Anderson found the "code of the street" in these areas reflected a profound lack of faith in the police, who are generally viewed as representing the dominant white society and not caring to protect inner-city residents. Anderson concluded that the great majority of the residents in the poor Black neighborhoods he studied were "decent folk" – civilly disposed, socially conscious, and self-reliant men and women who want their children to value education, get decent jobs, and enjoy the future. In Anderson's view, the criminal element in these neighborhoods – among the most desperate and alienated people in the inner city – have a "street" orientation and view the police as unworthy of respect and deserving of little or no moral authority. How can the police distinguish between decent residents and criminals? What should they do to reduce crime and make people feel safer? Can they gain the trust or at least the respect of the Black urban underclass?

Wilson and Kelling studied police behavior in troubled neighborhoods. Kelling spent many months walking Newark, New Jersey, neighborhoods with local police officers, observing what was going on in the neighborhoods and how the police actually handled neighborhood problems. This is a good example of qualitative research. Wilson, then a Harvard professor, worked with Kelling to jointly develop this classic article. Wilson and Kelling were particularly interested in police discretion and how the police handled troublesome behavior at the borderline between acceptable and unacceptable, legal and illegal. Wilson and Kelling are interested as much in neighborhood residents' perceptions of crime as in crime itself. A particular focus of their research was to develop a model of policing that neighborhood residents (at least those residents the authors term decent folk) would support. Based on their research, they advanced a controversial theory of neighborhood transition and an influential model for community policing that has since been implemented in many communities.

The central metaphor of this selection is of a single broken window. Imagine a city neighborhood like Boston's West End as described by Jane Jacobs (p. 105) – a diverse, viable, exciting urban neighborhood, but with some crime. Most of the people in the neighborhood Jane Jacobs saw each morning as the street ballet begins are regulars, people who live or work in the neighborhood or are frequently in the area. Some are strangers. The behavior of some people in the neighborhood might startle or even scare middle- and upper-class Bostonians, but it is understood and tolerated by neighborhood regulars. Some of the people in the area, however, are dangerous and engage in criminal behavior: drug dealing, prostitution, theft, and assault. Civil order exists, but it is fragile. Imagine that someone breaks a single window in the neighborhood. According to Wilson and Kelling, how the police respond to that trivial but unacceptable act is fraught with consequences. One response is to do nothing. After all, police are busy working on serious crimes and municipal budgets are tight. But doing nothing in response to a broken window, Wilson and Kelling argue, will signal that no one cares about unacceptable conduct. It will be an invitation for people to break more windows. Neighborhood residents will start avoiding one another, stop participating in neighborhood block parties, and eventually cower in their own homes.

Wilson and Kelling argue that historically neighborhood residents judged the success of police activity by whether it succeeded in maintaining order. Governments often tolerated (or encouraged) police to keep order through informal means without much regard to legal niceties. A drunk might have a legal right to sit on a neighborhood park bench. But if his presence sufficiently disturbed decent neighborhood residents, the local cop was encouraged to make him move elsewhere. Gang members in Chicago's crime-ridden Robert Taylor Homes might have a legal right to loiter by a playground, but the authors argue that decent project residents would want the local police to get rid of them ("kick ass" is the term Wilson and Kelling use). Joe Dickens, one of the people Elijah Anderson profiles (p. 132), might have a legal right to let his kids "rip and run" unsupervised making a racket on the street with their big wheel bikes while he and his buddies drink beer and play loud rap music, but, under Wilson's theory, his decent neighbors would like the police to force Dickens to control his children and turn down the volume on the boombox. Wilson and Kelling take the controversial position that broad police discretion to enforce neighborhood standards is desirable.

James Q. Wilson (b. 1931) is the Ronald Reagan Professor of Public Policy at Pepperdine University in California. From 1961 to 1987, he was the Shattuck Professor of Government at Harvard and from 1986 until 1997 the James Collins Professor of Management at the University of California, Los Angeles (UCLA). He has written extensively on politics, economics, and criminology. He chaired President Johnson's White House Task Force on Crime (1966), was on President Nixon's National Advisory Commission on Drug Abuse Prevention (1972–1973), was a member of the Attorney General's Task Force on Violent Crime under President Reagan (1981), and on President George Bush's Foreign Intelligence Advisory Board (1985–1990). Wilson is the chairman of the Council of Academic Advisors of the conservative American Enterprise Institute. He is a former president of the American Political Science Association (APSA) and received the APSA's James Madison award for a career of distinguished scholarship, the John Gaus award for exemplary scholarship in the fields of political science and public administration and a lifetime achievement award in 2001. He was awarded the Presidential Medal of Freedom by President George W. Bush in 2003.

George L. Kelling (b. 1935) is a Senior Fellow at the Manhattan Institute, a professor in the School of Criminal Justice at Rutgers University, and a fellow in the Kennedy School of Government at Harvard University. Kelling is currently researching organizational change in policing and the development of comprehensive community crime prevention programs. Kelling has a PhD in social work from the University of Wisconsin, Madison and practiced social work as a childcare worker, probation officer, and administering residential care programs for aggressive and disturbed youths. Working at the Police Foundation in the early 1970s, Kelling conducted several large-scale experiments in policing, including the Newark Foot Patrol Experiment, from which much of the source material for "Broken Windows" is drawn.

"Broken Windows" was published in *The Atlantic Monthly*, 249(3) in March, 1982. The broken windows theory is further developed in George L. Kelling and Catherine M. Coles, *Fixing Broken Windows* (New York: Martin Kessler, 1996).

Other books by James Q. Wilson related to crime prevention include *Crime*, coedited with Joan Petersilia (San Francisco, CA: ICS Press, 1995), *Crime and Public Policy* (San Francisco, CA: ICS Press, 1983), *Drugs and Crime*, coedited with Michael Tonry (Chicago, IL: University of Chicago Press, 1990), *Families, Schools, and Delinquency Prevention*, coedited with Glenn C. Loury (New York: Springer-Verlag, 1987), *Understanding and Controlling Crime*, coedited with David P. Farrington and Lloyd E. Ohlin (New York: Springer-Verlag, 1986), *Thinking about Crime* (New York: Basic Books, 1983), *Varieties of Police Behavior* (Cambridge, MA: Harvard University Press, 1968), and *Crime and Human Nature* (New York: Free Press, 1998).

Other notable books by Wilson include *American Government*, 12th edn, with John J. Dilulio, Jr. (Boston, MA: Houghton Mifflin, 2011), *Bureaucracy: What Government Agencies Do and Why They Do It* (New York: Basic Books, 2000), and *The Moral Sense* (New York: Free Press, 1997).

Books on community policing include Peter Grabosky (ed.), *Community Policing and Peacekeeping.* (Boca Raton, FL: CRC, 2009), Dominique Wisler and Ihekwoaba D. Onwudiwe (eds), *Community Policing: International*

Patterns and Comparative Perspectives (Boca Raton, FL: CRC, 2009), Larry K. Gaines, *Community Policing: A Contemporary Perspective*, 5th edn (Indianapolis, IN: Anderson, 2008), Willard M. Oliver, *Community-Oriented Policing: A Systemic Approach to Policing*, 4th edn (New York: Prentice Hall, 2007), Linda S. Miller and Kären M. Hess, *Community Policing: Partnerships for Problem Solving* (New York: Wadsworth, 2007), Jeremy M. Wilson, *Community Policing in America* (New York: Routledge, 2006), Elizabeth M. Watson, Alfred R. Stone, and Stuart M. DeLuca, *Strategies for Community Policing* (Upper Saddle River, NJ: Prentice-Hall, 1998), and Wesley G. Skogan and Susan M. Hartnett, *Community Policing, Chicago Style* (New York: Oxford University Press, 1997).

In the mid-1970s, the state of New Jersey announced a "Safe and Clean Neighborhoods Program," designed to improve the quality of community life in twenty-eight cities. As part of that program, the state provided money to help cities take police officers out of their patrol cars and assign them to walking beats. The governor and other state officials were enthusiastic about using foot patrol as a way of cutting crime, but many police chiefs were skeptical. Foot patrol, in their eyes, had been pretty much discredited. It reduced the mobility of the police, who thus had difficulty responding to citizen calls for service, and it weakened headquarters control over patrol officers.

Many police officers also disliked foot patrol, but for different reasons: it was hard work, it kept them outside on cold, rainy nights, and it reduced their chances for making a "good pinch." In some departments, assigning officers to foot patrol had been used as a form of punishment. And academic experts on policing doubted that foot patrol would have any impact on crime rates; it was, in the opinion of most, little more than a sop to public opinion. But since the state was paying for it, the local authorities were willing to go along.

Five years after the program started, the Police Foundation, in Washington, DC, published an evaluation of the foot-patrol project. Based on its analysis of a carefully controlled experiment carried out chiefly in Newark, the foundation concluded, to the surprise of hardly anyone, that foot patrol had not reduced crime rates. But residents of the foot-patrolled neighborhoods seemed to feel more secure than persons in other areas, tended to believe that crime had been reduced, and seemed to take fewer steps to protect themselves from crime (staying at home with the doors locked, for example). Moreover, citizens in the foot-patrol areas had a more favorable opinion of the police than did those living elsewhere. And officers walking beats had higher morale, greater job satisfaction, and a more favorable attitude toward citizens in their neighborhoods than did officers assigned to patrol cars.

These findings may be taken as evidence that the skeptics were right – foot patrol has no effect on crime; it merely fools the citizens into thinking that they are safer. But in our view, and in the view of the authors of the Police Foundation study (of whom Kelling was one), the citizens of Newark were not fooled at all. They knew what the foot-patrol officers were doing, they knew it was different from what motorized officers do, and they knew that having officers walk beats did in fact make their neighborhoods safer.

But how can a neighborhood be "safer" when the crime rate has not gone down – in fact, may have gone up? Finding the answer requires first that we understand what most often frightens people in public places. Many citizens, of course, are primarily frightened by crime, especially crime involving a sudden, violent attack by a stranger. This risk is very real, in Newark as in many large cities. But we tend to overlook or forget another source of fear – the fear of being bothered by disorderly people. Not violent people, nor, necessarily, criminals, but disreputable or obstreperous or unpredictable people: panhandlers, drunks, addicts, rowdy teenagers, prostitutes, loiterers, the mentally disturbed.

What foot-patrol officers did was to elevate, to the extent they could, the level of public order in these neighborhoods. Though the neighborhoods were predominantly black and the foot patrolmen were mostly white, this "order-maintenance" function of the police was performed to the general satisfaction of both parties.

One of us (Kelling) spent many hours walking with Newark foot-patrol officers to see how they defined "order" and what they did to maintain it. One beat was typical: a busy but dilapidated area in the heart of Newark, with many abandoned buildings, marginal shops (several of which prominently displayed knives and straight-edged razors in their windows), one large department store, and, most important, a train station and several major bus-stops. Though the area was run-down, its streets were filled with people, because it was a major transportation center. The good order of this area was important not only to those who lived and worked there but also to many others, who had to move through it on their way home, to supermarkets, or to factories.

The people on the street were primarily black; the officer who walked the street was white. The people were made up of "regulars" and "strangers." Regulars included both "decent folk" and some drunks and derelicts who were always there but who "knew their place." Strangers were, well, strangers, and viewed suspiciously, sometimes apprehensively. The officer – call him Kelly – knew who the regulars were, and they knew him. As he saw his job, he was to keep an eye on strangers, and make certain that the disreputable regulars observed some informal but widely understood rules. Drunks and addicts could sit on the stoops, but could not lie down. People could drink on side streets, but not at the main intersection. Bottles had to be in paper bags. Talking to, bothering, or begging from people waiting at the bus stop was strictly forbidden. If a dispute erupted between a businessman and a customer, the businessman was assumed to be right, especially if the customer was a stranger. If a stranger loitered, Kelly would ask him if he had any means of support and what his business was; if he gave unsatisfactory answers, he was sent on his way. Persons who broke the informal rules, especially those who bothered people waiting at bus stops, were arrested for vagrancy. Noisy teenagers were told to keep quiet.

These rules were defined and enforced in collaboration with the "regulars" on the street. Another neighborhood might have different rules, but these, everybody understood, were the rules for this neighborhood. If someone violated them, the regulars not only turned to Kelly for help but also ridiculed the violator. Sometimes what Kelly did could be described as "enforcing the law," but just as often it involved taking informal or extralegal steps to help protect what the neighborhood had decided was the appropriate level of public order. Some of the things he did probably would not withstand a legal challenge.

A determined skeptic might acknowledge that a skilled foot-patrol officer can maintain order but still insist that this sort of "order" has little to do with the real sources of community fear – that is, with violent crime. To a degree, that is true. But two things must be borne in mind. First, outside observers should not assume that they know how much of the anxiety now endemic in many big-city neighborhoods stems from a fear of "real" crime and how much from a sense that the street is disorderly, a source of distasteful, worrisome encounters. The people of Newark, to judge from their behavior and their remarks to interviewers, apparently assign a high value to public order, and feel relieved and reassured when the police help them maintain that order.

Second, at the community level, disorder and crime are usually inextricably linked, in a kind of developmental sequence. Social psychologists and police officers tend to agree that if a window in a building is broken *and is left unrepaired*, all the rest of the windows will soon be broken. This is as true in nice neighborhoods as in run-down ones. Window-breaking does not necessarily occur on a large scale because some areas are inhabited by determined window-breakers whereas others are populated by window-lovers; rather, one unrepaired broken window is a signal that no one cares, and so breaking more windows costs nothing. (It has always been fun.)

Philip Zimbardo, a Stanford psychologist, reported in 1969 on some experiments testing the broken-window theory. He arranged to have an automobile without license plates parked with its hood up on a street in the Bronx and a comparable automobile on a street in Palo Alto, California. The car in the Bronx was attacked by "vandals" within ten minutes of its "abandonment." The first to arrive were a family – father, mother, and young son – who removed the radiator and battery. Within twenty-four hours, virtually everything of value had been removed. Then random destruction began – windows were smashed, parts torn off, upholstery ripped. Children began to use the car as a playground. Most of the adult "vandals" were well-dressed, apparently clean-cut whites. The car in Palo Alto sat untouched for more than a week. Then Zimbardo smashed part of it with a sledgehammer. Soon, passersby were joining in. Within a few hours, the car had been turned upside down and utterly destroyed. Again, the "vandals" appeared to be primarily respectable whites.

Untended property becomes fair game for people out for fun or plunder, and even for people who ordinarily would not dream of doing such things and who probably consider themselves law-abiding. Because of the nature of community life in the Bronx – its anonymity, the frequency with which cars are abandoned and things are stolen or broken, the past experience of "no one caring" – vandalism begins much more quickly than it does in staid Palo Alto, where people have come to believe that private possessions are cared for, and that mischievous behavior is costly. But vandalism can occur anywhere once communal barriers – the sense of mutual regard and the obligations of civility – are lowered by actions that seem to signal that "no one cares."

We suggest that "untended" behavior also leads to the breakdown of community controls. A stable neighborhood of families who care for their homes, mind each other's children, and confidently frown on unwanted intruders can change, in a few years or even a few months, to an inhospitable and frightening jungle. A piece of property is abandoned, weeds grow up, a window is smashed. Adults stop scolding rowdy children; the children, emboldened, become more rowdy. Families move out, unattached adults move in. Teenagers gather in front of the corner store. The merchant asks them to move; they refuse. Fights occur. Litter accumulates. People start drinking in front of the grocery; in time, an inebriate slumps to the sidewalk and is allowed to sleep it off. Pedestrians are approached by panhandlers.

At this point it is not inevitable that serious crime will flourish or violent attacks on strangers will occur. But many residents will think that crime, especially violent crime, is on the rise, and they will modify their behavior accordingly. They will use the streets less often, and when on the streets will stay apart from their fellows, moving with averted eyes, silent lips, and hurried steps. "Don't get involved." For some residents, this growing atomization will matter little, because the neighborhood is not their "home" but "the place where they live." Their interests are elsewhere; they are cosmopolitans. But it will matter greatly to other people, whose lives derive meaning and satisfaction from local attachments rather than worldly involvement; for them, the neighborhood will cease to exist except for a few reliable friends whom they arrange to meet.

Such an area is vulnerable to criminal invasion. Though it is not inevitable, it is more likely that here, rather than in places where people are confident they can regulate public behavior by informal controls, drugs will change hands, prostitutes will solicit, and cars will be stripped. That the drunks will be robbed by boys who do it as a lark, and the prostitutes' customers will be robbed by men who do it purposefully and perhaps violently. That muggings will occur.

Among those who often find it difficult to move away from this are the elderly. Surveys of citizens suggest that the elderly are much less likely to be the victims of crime than younger persons, and some have inferred from this that the well-known fear of crime voiced by the elderly is an exaggeration: perhaps we ought not to design special programs to protect older persons; perhaps we should even try to talk them out of their mistaken fears. This argument misses the point. The prospect of a confrontation with an obstreperous teenager or a drunken panhandler can be as fear-inducing for defenseless persons as the prospect of meeting an actual robber; indeed, to a defenseless person, the two kinds of confrontation are often indistinguishable. Moreover, the lower rate at which the elderly are victimized is a measure of the steps they have already taken – chiefly, staying behind locked doors – to minimize the risks they face. Young men are more frequently attacked than older women, not because they are easier or more lucrative targets but because they are on the streets more.

Nor is the connection between disorderliness and fear made only by the elderly. Susan Estrich, of the Harvard Law School, has recently gathered together a number of surveys on the sources of public fear. One, done in Portland, Oregon, indicated that three-fourths of the adults interviewed cross to the other side of a street when they see a gang of teenagers; another survey, in Baltimore, discovered that nearly half would cross the street to avoid even a single strange youth. When an interviewer asked people in a housing project where the most dangerous spot was, they mentioned a place where young persons gathered to drink and play music, despite the fact that not a single crime had occurred there. In Boston public housing projects, the greatest fear was expressed by persons living in the buildings where disorderliness and incivility, not crime, were the greatest. Knowing this helps one understand the significance of such otherwise harmless displays as subway graffiti. As Nathan Glazer has written, the proliferation of graffiti, even when not obscene, confronts the subway rider with the "inescapable knowledge that the environment he must endure for an hour or more a day is uncontrolled and uncontrollable, and that anyone can invade it to do whatever damage and mischief the mind suggests."

In response to fear, people avoid one another, weakening controls. Sometimes they call the police. Patrol cars arrive, an occasional arrest occurs, but crime continues and disorder is not abated. Citizens complain to the police chief, but he explains that his department is low on personnel and that the courts do not punish petty or first-time offenders. To the residents, the police who arrive in squad cars are either ineffective or uncaring; to the police, the residents are animals who deserve each other. The citizens may soon stop calling the police, because "they can't do anything."

The process we call urban decay has occurred for centuries in every city. But what is happening today is different in at least two important respects. First, in the period before, say, World War II, city dwellers – because of money costs, transportation difficulties, familial and church connections – could rarely move away from neighborhood problems. When movement did occur, it tended to be along public-transit routes. Now mobility has become exceptionally easy for all but the poorest or those who are blocked by racial prejudice. Earlier crime waves had a kind of built-in self-correcting mechanism: the determination of a neighborhood or community to reassert control over its turf. Areas in Chicago, New York, and Boston would experience crime and gang wars, and then normalcy would return, as the families for whom no alternative residences were possible reclaimed their authority over the streets.

Second, the police in this earlier period assisted in that reassertion of authority by acting, sometimes violently, on behalf of the community. Young toughs were roughed up, people were arrested "on suspicion" or for vagrancy, and prostitutes and petty thieves were routed. "Rights" were something enjoyed by decent folk, and perhaps also by the serious professional criminal, who avoided violence and could afford a lawyer.

This pattern of policing was not an aberration or the result of occasional excess. From the earliest days of the nation, the police function was seen primarily as that of a night watchman: to maintain order against the chief threats to order – fire, wild animals, and disreputable behavior. Solving crimes was viewed not as a police responsibility but as a private one. In the March, 1969, *Atlantic*, one of us (Wilson) wrote a brief account of how the police role had slowly changed from maintaining order to fighting crimes. The change began with the creation of private detectives (often ex-criminals), who worked on a contingency-fee basis for individuals who had suffered losses. In time, the detectives were absorbed into municipal police agencies and paid a regular salary; simultaneously, the responsibility for prosecuting thieves was shifted from the aggrieved private citizen to the professional prosecutor. This process was not complete in most places until the twentieth century.

In the 1960s, when urban riots were a major problem, social scientists began to explore carefully the order-maintenance function of the police, and to suggest ways of improving it – not to make streets safer (its original function) but to reduce the incidence of mass violence. Order-maintenance became, to a degree, coterminous with "community relations." But, as the crime wave that began in the early 1960s continued without abatement throughout the decade and into the 1970s, attention shifted to the role of the police as crime-fighters. Studies of police behavior ceased, by and large, to be accounts of the order-maintenance function and became, instead, efforts to propose and test ways whereby the police could solve more crimes, make more arrests, and gather better evidence. If these things could be done, social scientists assumed, citizens would be less fearful.

A great deal was accomplished during this transition, as both police chiefs and outside experts emphasized the crime-fighting function in their plans, in the allocation of resources, and in deployment of personnel. The police may well have become better crime-fighters as a result. And doubtless they remained aware of their responsibility for order. But the link between order-maintenance and crime-prevention, so obvious to earlier generations, was forgotten.

That link is similar to the process whereby one broken window becomes many. The citizen who fears the ill-smelling drunk, the rowdy teenager, or the importuning beggar is not merely expressing his distaste for unseemly behavior; he is also giving voice to a bit of folk wisdom that happens to be a correct generalization – namely, that serious street crime flourishes in areas in which disorderly behavior goes unchecked. The unchecked panhandler is, in effect, the first broken window. Muggers and robbers, whether opportunistic or professional, believe they reduce their chances of being caught or even identified if they operate on streets where potential victims are already intimidated by prevailing conditions. If the neighborhood cannot keep a bothersome panhandler from annoying passersby, the thief may reason, it is even less likely to call the police to identify a potential mugger or to interfere if the mugging actually takes place.

Some police administrators concede that this process occurs, but argue that motorized-patrol officers can deal with it as effectively as foot-patrol officers. We are not so sure. In theory, an officer in a squad car can observe as much as an officer on foot; in theory, the former can talk to as many people as the latter. But the reality of police–citizen encounters is powerfully altered by the automobile. An officer on foot cannot separate himself from the street people; if he is approached, only his uniform and his personality can help him manage whatever is about to happen. And he can never be certain what that will be – a request for directions, a plea for help, an angry denunciation, a teasing rcmark, a confused babble, a threatening gesture.

In a car, an officer is more likely to deal with street people by rolling down the window and looking at them. The door and the window exclude the approaching citizen; they are a barrier. Some officers take advantage of this barrier, perhaps unconsciously, by acting differently if in the car than they would on foot. We have seen this countless times. The police car pulls up to a corner where teenagers are gathered. The window is rolled down. The officer stares at the youths. They stare back. The officer says to one, "C'mere." He saunters over, conveying to his friends by his elaborately casual style the idea that he is not intimidated by authority "What's your name?" "Chuck." "Chuck who?" "Chuck Jones." "What'ya doing, Chuck?" "Nothin'." "Got a P.O. [parole officer]?" "Nah." "Sure?" "Yeah." "Stay out of trouble, Chuckie." Meanwhile, the other boys laugh and exchange comments among themselves, probably at the officer's expense. The officer stares harder. He cannot be certain what is being said, nor can he join in and, by displaying his own skill at street banter, prove that he cannot be "put down." In the process, the officer has learned almost nothing, and the boys have decided the officer is an alien force who can safely be disregarded, even mocked.

Our experience is that most citizens like to talk to a police officer. Such exchanges give them a sense of importance, provide them with the basis for gossip, and allow them to explain to the authorities what is worrying them (whereby they gain a modest but significant sense of having "done something" about the problem). You approach a person on foot more easily, and talk to him more readily, than you do a person in a car. Moreover, you can more easily retain some anonymity if you draw an officer aside for a private chat. Suppose you want to pass on a tip about who is stealing handbags, or who offered to sell you a stolen TV. In the inner city, the culprit, in all likelihood, lives nearby. To walk up to a marked patrol car and lean in the window is to convey a visible signal that you are a "fink."

The essence of the police role in maintaining order is to reinforce the informal control mechanisms of the community itself. The police cannot, without committing extraordinary resources, provide a substitute for that informal control. On the other hand, to reinforce those natural forces the police must accommodate them. And therein lies the problem.

Should policc activity on the street be shaped, in important ways, by the standards of the neighborhood rather than by the rules of the state? Over the past two decades, the shift of police from order-maintenance to law-enforcement has brought them increasingly under the influence of legal restrictions, provoked by media complaints and enforced by court decisions and departmental orders. As a consequence, the order-maintenance functions of the police are now governed by rules developed to control police relations with suspected criminals. This is, we think, an entirely new development. For centuries, the role of the police as watchmen was judged primarily not in terms of its compliance with appropriate procedures but rather in terms of its attaining a desired objective. The objective was order, an inherently ambiguous term but a condition that people in a given community recognized when they saw it. The means were the same as those the community itself would employ, if its members were sufficiently determined, courageous, and authoritative. Detecting and apprehending criminals, by contrast, was a means to an end, not an end in itself; a judicial determination of guilt or innocence was the hoped-for result of the law-enforcement mode. From the first, the police were expected to follow rules defining that process, though states differed in how stringent the rules should be. The criminal-apprehension process was always understood to involve individual rights, the violation of which was unacceptable because it meant that the violating officer would be acting as a judge and jury – and that was not his job. Guilt or innocence was to be determined by universal standards under special procedures.

Ordinarily, no judge or jury ever sees the persons caught up in a dispute over the appropriate level of neighborhood order. That is true not only because most cases are handled informally on the street but also because no universal standards are available to settle arguments over disorder, and thus a judge may not be any wiser or more effective than a police officer. Until quite recently in many states, and even today in some places, the police make arrests on such charges as "suspicious person" or "vagrancy" or "public drunkenness" – charges with scarcely any legal meaning. These charges exist not because society wants judges to punish vagrants or drunks but because it wants an officer to have the legal tools to remove undesirable persons from a neighborhood when informal efforts to preserve order in the streets have failed.

Once we begin to think of all aspects of police work as involving the application of universal rules under special procedures, we inevitably ask what constitutes an "undesirable person" and why we should "criminalize" vagrancy or drunkenness. A strong and commendable desire to see that people are treated fairly makes us worry about allowing the police to rout persons who are undesirable by some vague or parochial standard. A growing and not-so-commendable utilitarianism leads us to doubt that any behavior that does not "hurt" another person should be made illegal. And thus many of us who watch over the police are reluctant to allow them to perform, in the only way they can, a function that every neighborhood desperately wants them to perform.

This wish to "decriminalize" disreputable behavior that "harms no one" – and thus remove the ultimate sanction the police can employ to maintain neighborhood order – is, we think, a mistake. Arresting a single drunk or a single vagrant who has harmed no identifiable person seems unjust, and in a sense it is. But failing to do anything about a score of drunks or a hundred vagrants may destroy an entire community. A particular rule that seems to make sense in the individual case makes no sense when it is made a universal rule and applied to all cases. It makes no sense because it fails to take into account the connection between one broken window left untended and a thousand broken windows. Of course, agencies other than the police could attend to the problems posed by drunks or the mentally ill, but in most communities – especially where the "deinstitutionalization" movement has been strong – they do not.

The concern about equity is more serious. We might agree that certain behavior makes one person more undesirable than another, but how do we ensure that age or skin color or national origin or harmless mannerisms will not also become the basis for distinguishing the undesirable from the desirable? How do we ensure, in short, that the police do not become the agents of neighborhood bigotry?

We can offer no wholly satisfactory answer to this important question. We are not confident that there is a satisfactory answer, except to hope that by their selection, training, and supervision, the police will be inculcated with a clear sense of the outer limit of their discretionary authority. That limit, roughly, is this – the police exist to help regulate behavior, not to maintain the racial or ethnic purity of a neighborhood.

Consider the case of the Robert Taylor Homes in Chicago, one of the largest public-housing projects in the country. It is home for nearly 20,000 people, all black, and extends over ninety-two acres along South State Street. It was named after a distinguished black who had been, during the 1940s, chairman of the Chicago Housing Authority. Not long after it opened, in 1962, relations between project residents and the police deteriorated badly. The citizens felt that the police were insensitive or brutal; the police, in turn, complained of unprovoked attacks on them. Some Chicago officers tell of times when they were afraid to enter the Homes. Crime rates soared.

Today, the atmosphere has changed. Police–citizen relations have improved – apparently, both sides learned something from the earlier experience. Recently, a boy stole a purse and ran off. Several young persons who saw the theft voluntarily passed along to the police information on the identity and residence of the thief, and they did this publicly, with friends and neighbors looking on. But problems persist, chief among them the presence of youth gangs that terrorize residents and recruit members in the project. The people expect the police to "do something" about this, and the police are determined to do just that.

But do what? Though the police can obviously make arrests whenever a gang member breaks the law, a gang can form, recruit, and congregate without breaking the law. And only a tiny fraction of gang-related crimes can be solved by an arrest; thus, if an arrest is the only recourse for the police, the residents' fears will go unassuaged. The police will soon feel helpless, and the residents will again believe that the police "do nothing." What the police in fact do is to chase known gang members out of the project. In the words of one officer, "We kick ass."

Project residents both know and approve of this. The tacit police–citizen alliance in the project is reinforced by the police view that the cops and the gangs are the two rival sources of power in the area, and that the gangs are not going to win.

None of this is easily reconciled with any conception of due process or fair treatment. Since both residents and gang members are black, race is not a factor. But it could be. Suppose a white project confronted a black gang, or vice versa. We would be apprehensive about the police taking sides. But the substantive problem remains the same: how can the police strengthen the informal social-control mechanisms of natural communities in order to minimize fear in public places? Law enforcement, per se, is no answer. A gang can weaken or destroy a community by standing about in a menacing fashion and speaking rudely to passersby without breaking the law.

We have difficulty thinking about such matters, not simply because the ethical and legal issues are so complex but because we have become accustomed to thinking of the law in essentially individualistic terms. The law defines *my* rights, punishes *his* behavior, and is applied by *that* officer because of *this* harm. We assume, in thinking this way, that what is good for the individual will be good for the community, and what doesn't matter when it happens to one person won't matter if it happens to many. Ordinarily, those are plausible assumptions. But in cases where behavior that is tolerable to one person is intolerable to many others, the reactions of the others – fear, withdrawal, flight – may ultimately make matters worse for everyone, including the individual who first professed his indifference.

It may be their greater sensitivity to communal as opposed to individual needs that helps explain why the residents of small communities are more satisfied with their police than are the residents of similar neighborhoods in big cities. Elinor Ostrom and her co-workers at Indiana University compared the perception of police services in two poor, all-black Illinois towns – Phoenix and East Chicago Heights – with those of three comparable all-black neighborhoods in Chicago. The level of criminal victimization and the quality of police–community relations appeared to be about the same in the towns and the Chicago neighborhoods, but the citizens living in their own villages were much more likely than those living in the Chicago neighborhoods to say that they do not stay at home for fear of crime, to agree that the local police have "the right to take any action necessary" to deal with problems, and to agree that the police "look out for the needs of the average citizen." It is possible that the residents and the police of the small towns saw themselves as engaged in a collaborative effort to maintain a certain standard of communal life, whereas those of the big city felt themselves to be simply requesting and supplying particular services on an individual basis.

If this is true, how should a wise police chief deploy his meager forces? The first answer is that nobody knows for certain, and the most prudent course of action would be to try further variations on the Newark experiment, to see more precisely what works in what kinds of neighborhoods. The second answer is also a hedge – many aspects of order-maintenance in neighborhoods can probably best be handled in ways that involve the police minimally, if at all. A busy, bustling shopping center and a quiet, well-tended suburb may need almost no visible police presence. In both cases, the ratio of respectable to disreputable people is ordinarily so high as to make informal social control effective.

Even in areas that are in jeopardy from disorderly elements, citizen action without substantial police involvement may be sufficient. Meetings between teenagers who like to hang out on a particular corner and adults who want to use that corner might well lead to an amicable agreement on a set of rules about how many people can be allowed to congregate, where, and when.

Where no understanding is possible – or if possible, not observed – citizen patrols may be a sufficient response. There are two traditions of communal involvement in maintaining order. One, that of the "community watchmen," is as old as the first settlement of the New World. Until well into the nineteenth century, volunteer watchmen, not policemen, patrolled their communities to keep order. They did so, by and large, without taking the law into their own hands – without, that is, punishing persons or using force. Their presence deterred disorder or alerted the community to disorder that could not be deterred. There are hundreds of such efforts today in communities all across the nation. Perhaps the best known is that of the Guardian Angels, a group of unarmed young persons in distinctive berets and T-shirts, who first came to public attention when they began patrolling the New York City subways but who claim now to have chapters in more than thirty American cities. Unfortunately, we have little information about the effect of these groups on crime. It is possible, however, that whatever their effect on crime, citizens find their presence reassuring, and that they thus contribute to maintaining a sense of order and civility.

The second tradition is that of the "vigilante." Rarely a feature of the settled communities of the East, it was primarily to be found in those frontier towns that grew up in advance of the reach of government. More than 350 vigilante groups are known to have existed; their distinctive feature was that their members did take the law into their own hands, by acting as judge, jury, and often executioner as well as policeman. Today, the vigilante movement is conspicuous by its rarity, despite the great fear expressed by citizens that the older cities are becoming "urban frontiers." But some community-watchmen groups have skirted the line, and others may cross it in the future. An ambiguous case, reported in *The Wall Street Journal*, involved a citizens' patrol in the Silver Lake area of Belleville, New Jersey. A leader told the reporter, "We look for outsiders." If a few teenagers from outside the neighborhood enter it, "we ask them their business," he said. "If they say they're going down the street to see Mrs. Jones, fine, we let them pass. But then we follow them down the block to make sure they're really going to see Mrs. Jones."

Though citizens can do a great deal, the police are plainly the key to order-maintenance. For one thing, many communities, such as the Robert Taylor Homes, cannot do the job by themselves. For another, no citizen in a neighborhood, even an organized one, is likely to feel the sense of responsibility that wearing a badge confers. Psychologists have done many studies on why people fail to go to the aid of persons being attacked or seeking help, and they have learned that the cause is not "apathy" or "selfishness" but the absence of some plausible grounds for feeling that one must personally accept responsibility. Ironically, avoiding responsibility is easier when a lot of people are standing about. On streets and in public places, where order is so important, many people are likely to be "around," a fact that reduces the chance of any one person acting as the agent of the community. The police officer's uniform singles him out as a person who must accept responsibility if asked. In addition, officers, more easily than their fellow citizens, can be expected to distinguish between what is necessary to protect the safety of the street and what merely protects its ethnic purity.

But the police forces of America are losing, not gaining, members. Some cities have suffered substantial cuts in the number of officers available for duty. These cuts are not likely to be reversed in the near future. Therefore, each department must assign its existing officers with great care. Some neighborhoods are so demoralized and crime-ridden as to make foot patrol useless; the best the police can do with limited resources is respond to the enormous number of calls for service. Other neighborhoods are so stable and serene as to make foot patrol unnecessary. The key is to identify neighborhoods at the tipping point where the public order is deteriorating but not unreclaimable, where the streets are used frequently but by apprehensive people, where a window is likely to be broken at any time, and must quickly be fixed if all are not to be shattered.

Most police departments do not have ways of systematically identifying such areas and assigning officers to them. Officers are assigned on the basis of crime rates (meaning that marginally threatened areas are often stripped so that police can investigate crimes in areas where the situation is hopeless) or on the basis of calls for service (despite the fact that most citizens do not call the police when they are merely frightened or annoyed). To allocate patrols wisely, the department must look at the neighborhoods and decide, from first-hand evidence, where an additional officer will make the greatest difference in promoting a sense of safety.

One way to stretch limited police resources is being tried in some public-housing projects. Tenant organizations hire off-duty police officers for patrol work in their buildings. The costs are not high (at least not per resident), the officer likes the additional income, and the residents feel safer. Such arrangements are probably more successful than hiring private watchmen, and the Newark experiment helps us understand why. A private security guard may deter crime or misconduct by his presence, and he may go to the aid of persons needing help, but he may well not intervene – that is, control or drive away someone challenging community standards. Being a sworn officer – a "real cop" – seems to give one the confidence, the sense of duty, and the aura of authority necessary to perform this difficult task.

Patrol officers might be encouraged to go to and from duty stations on public transportation and, while on the bus or subway car, enforce rules about smoking, drinking, disorderly conduct, and the like. The enforcement need involve nothing more than ejecting the offender (the offense, after all, is not one with which a booking officer or a judge wishes to be bothered).

Perhaps the random but relentless maintenance of standards on buses would lead to conditions on buses that approximate the level of civility we now take for granted on airplanes.

But the most important requirement is to think that to maintain order in precarious situations is a vital job. The police know this is one of their functions, and they also believe, correctly, that it cannot be done to the exclusion of criminal investigation and responding to calls. We may have encouraged them to suppose, however, on the basis of our oft-repeated concerns about serious, violent crime, that they will be judged exclusively on their capacity as crime-fighters. To the extent that this is the case, police administrators will continue to concentrate police personnel in the highest-crime areas (though not necessarily in the areas most vulnerable to criminal invasion), emphasize their training in the law and criminal apprehension (and not their training in managing street life), and join too quickly in campaigns to decriminalize "harmless" behavior (though public drunkenness, street prostitution, and pornographic displays can destroy a community more quickly than any team of professional burglars).

Above all, we must return to our long-abandoned view that the police ought to protect communities as well as individuals. Our crime statistics and victimization surveys measure individual losses, but they do not measure communal losses. Just as physicians now recognize the importance of fostering health rather than simply treating illness, so the police – and the rest of us – ought to recognize the importance of maintaining, intact, communities without broken windows.

"The City as a Distorted Price System"

Psychology Today (1968)

Wilbur Thompson

Editors' Introduction

The subfield of urban economics is relatively new. Unlike urban geography, which has existed since ancient times, urban sociology, which originated in late-nineteenth-century Germany and flowered in the 1920s and 1930s with the Chicago school sociologists like Ernest W. Burgess (p. 161) and Louis Wirth (p. 96), or the field of urban planning in which university-level courses were first taught in 1909, economics departments began to teach courses in urban economics only after a remarkable book by Wilbur Thompson, *A Preface to Urban Economics*, was published in 1965.

Scholars from a specific discipline often feel people trained in the discipline will be the best decision-makers. Architect Frank Lloyd Wright had the fanciful vision that a county architect should be the key policy-maker at the local level (p. 345). Similarly, Thompson proposes the idea that cities should have a "city economist" to shape city policy. While cities do not now have a position like that, economists within government and the private and nonprofit sectors now bring the theory and practice of urban economics and public finance that Thompson and other economist have developed to bear on decision-making.

Three concepts Thompson introduces are fundamental in urban economics: the idea of collectively consumed public goods, merit goods designed to encourage desired behaviors, and payments to redistribute income.

Public goods are provided free by government for the use of everyone. Air pollution control, police, and city streets are examples. Thompson sees a place for this kind of good in private free-market economies. Public goods cost money. There is a price associated with providing them. But unlike the cost of goods in the private market, the price of these public goods is often implicit, rather than explicit. You don't think how much it will cost you to drive from home to school on a public street or think, "Oh, it's a hot muggy day, I'll be paying a lot for government to control air pollution today." Of course your tax dollars are paying for the street and the air pollution control, but in complicated ways, invisible to the ordinary consumer, and little related to market realities. The failure to think through intended policy and price public goods accordingly, Thompson argues, leads to muddled and sometimes irrational policy. An example Thompson gives of how inattention to price can lead to bad public policy involves fireproofing. A rational public policy might be to encourage homeowners to invest in fireproofing their homes in order to reduce the risk of fire. This costs each individual homeowner money, and a fireproofed home will be worth more than it was before the fireproofing. Since homeowners' property taxes are based on the value of their homes, property taxes will go up. Having their property tax go up penalizes prudent homeowners, whose investment reduces the cost of firefighters responding to a fire or the risk of a fire on their property spreading to adjacent properties. If other homeowners in the same neighborhood do nothing to fireproof their houses and let them deteriorate into firetraps, they will be rewarded by lower property taxes. An alternative approach would be to reward homeowners who fireproof with some form of tax abatement or tax credit and to penalize non-performing homeowners by increasing their taxes (and perhaps using the extra money to improve the fire department so it can handle the increased risk they are imposing on society).

Merit goods are goods that government provides for free because there has been a collective decision that they are so important that everyone should have them regardless of ability to pay. Free polio shots are an example. With rare exceptions, governments provide children free polio shots. Governments virtually everywhere want children immunized against polio regardless of their parents' ability to pay because any child who gets polio is likely to be permanently crippled for life, bringing suffering to the child and the family and huge medical bills that someone will have to pay. The crippled child will probably never be able to work and will be a permanent burden on his or her family and the public. This tragedy could have been averted for a few pennies in a single, simple, painless vaccination. Thompson calls merit goods a case of the majority playing God, and coercing the minority by the use of bribes to change their behavior. Some people oppose some merit goods (even, in rare instances, polio vaccinations) on these grounds, so Thompson argues that merit goods should be used sparingly in situations where they are very important and command near universal support. Thompson argues that paying for merit goods, such as polio immunizations, is a legitimate governmental activity that should not be subject to market pricing in rare cases where the public interest requires the majority to force everyone to cooperate.

Payments to redistribute income are a third kind of payment that is not governed by price. The classic example Thompson gives is welfare payments to the indigent, such as a monthly payment to a blind, elderly person with no assets or income. One group (taxpayers) pays; another group (indigent, elderly, blind people) receive the good. In communist countries and some countries with a culture that supports a large welfare state such as Sweden, government may pay for virtually all of the costs of health and education for everyone and provide a reasonably high minimum welfare payment to provide for indigent people. The United States, United Kingdom, and most other countries have a much more restricted policy regarding income redistribution. They dislike large, cumbersome bureaucracies and distrust the ability of government to manage redistributive policies. Accordingly, they rely primarily on private market solutions to meet almost all basic human needs.

Thompson feels that, if clearly identified and intended, each of these three types of goods – public goods, merit goods, and redistributive payments – can serve a useful role. Often, however, he feels that there is too little thought about the purpose of such payments and conceptual sloppiness about what is intended. If there were a city economist in a city, she might, for example, force a city council to think through just how much of a subsidy they care to give to a museum. While the city council may regard museum visits as a legitimate public good that should be supported by public tax dollars, a careful examination might lead them to decide, for example, that there should be a differential fee structure so that students and senior citizens can visit the museum free, but other museum visitors will have to pay an entrance charge. They might conclude that free museum admission is just not as important as polio shots, road maintenance, or aid to indigent, elderly blind individuals. People with resources, not government, should decide how much visiting a museum is worth to them. Alternatively, the city economist would convince the city council that a low museum entrance fee is necessary when the museum first opens in order to lure patrons, but once patronage is established, higher fees are in order. More extreme examples of public goods that Thompson argues should be carefully scrutinized, but often are not, include municipal golf courses and marinas. Does it make sense for taxpayers to subsidize golfers and yacht owners? Are these merit goods? Public goods? Goods worthy of redistributive policy?

Thompson was an early advocate of using price to ration scarce goods, particularly use of highways and parking spaces. Since this seminal article was written, congestion pricing has become common in many parts of the world. Tolls are set high during peak commute hours. People who value their mobility highly (affluent commuters rushing to get to work on time) will be willing to pay the high toll cost and will appreciate the lack of congestion. People for whom mobility at that time is less important (friends planning to get together to work out at the gym) may choose to exercise later and save the higher toll.

Wilbur Thompson (b. 1923) is an economist who defined the field of urban economics. This selection appeared in the popular US publication *Psychology Today* in August 1968. His seminal urban economics book is *A Preface to Urban Economics* (Baltimore, MD: Johns Hopkins University Press, 1965). Thompson also wrote *An Econometric Model of Postwar State Industrial Development* (Detroit, MI: Wayne State University Press, 1959).

The leading contemporary urban economics text is Arthur O. O'Sullivan, *Urban Economics*, 7th edn (Chicago, IL: McGraw-Hill, 2008). Other urban economics texts include John McDonald and Daniel MacMillan, *Urban Economics: Theory and Policy* (Oxford: Wiley-Blackwell, 2006), Brendan O'Flaherty, *City Economics* (Cambridge,

MA: Harvard University Press, 2005), and Robert W. Wassmer (ed.), *Readings in Urban Economics: Issues and Public Policy* (Oxford: Blackwell, 2000).

Jane Jacobs, *The Economy of Cities* (New York: Vintage, 1970) is a provocative book that sees cities as incubators of new economic ideas and argues in favor of experimentation.

The failure to use price – as an *explicit* system – in the public sector of the metropolis is at the root of many, if not most, of our urban problems. Price, serving its historic functions, might be used to ration the use of existing facilities, to signal the desired directions of new public investment, to guide the distribution of income, to enlarge the range of public choice and to change tastes and behavior. Price performs such functions in the private market place, but it has been virtually eliminated from the public sector. We say "virtually eliminated" because it does exist but in an implicit, subtle, distorted sense that is rarely seen or acknowledged by even close students of the city, much less by public managers. Not surprisingly, this implicit price system results in bad economics.

We think of the property tax as a source of public revenue, but it can be reinterpreted as a price. Most often, the property tax is rationalized on "ability-to-pay" grounds with real property serving as a proxy for income. When the correlation between income and real property is challenged, the apologist for the property tax shifts ground and rationalizes it as a "benefit" tax. The tax then becomes a "price" which the property owner pays for the benefits received – fire protection, for example. But this implicit "price" for fire services is hardly a model of either efficiency or equity. Put in a new furnace and fireproof your building (reduce the likelihood of having a fire) and your property tax (fire service premium) goes up, let your property deteriorate and become a firetrap and your fire protection premium goes down! One bright note is New York City's one-year tax abatement on new pollution-control equipment; a timid step but in the right direction.

Often "urban sprawl" is little more than a color word which reflects (betrays?) the speaker's bias in favor of high population density and heavy interpersonal interaction – his "urbanity." Still, typically, the price of using urban fringe space has been set too low – well below the full costs of running pipes, wires, police cars and fire engines farther than would be necessary if building lots were smaller. Residential developers are, moreover, seldom discouraged (penalized by price) from "leap frogging" over the contiguous, expensive vacant land to build on the remote, cheaper parcels. Ordinarily, a flat price is charged for extending water or sewers to a new household regardless of whether the house is placed near to or far from existing pumping stations.

Again, the motorist is subject to the same license fees and tolls, if any, for the extremely expensive system of streets, bridges, tunnels, and traffic controls he enjoys, regardless of whether he chooses to drive downtown at the rush hour and thereby pushes against peak capacity or at off-peak times when it costs little or nothing to serve him. To compound this distortion of prices, we usually set the toll at zero. And when we do charge tolls, we quite perversely cut the commuter (rush-hour) rate below the off-peak rate.

It is not enough to point out that the motorist supports roadbuilding through the gasoline tax. The social costs of noise, air pollution, traffic control and general loss of urban amenities are borne by the general taxpayer. In addition, drivers during off-peak hours overpay and subsidize rush-hour drivers. Four lanes of expressway or bridge capacity are needed in the morning and evening rush hours where two lanes would have served if movements had been random in time and direction: that is, near constant in average volume. The peak-hour motorists probably should share the cost of the first two lanes and bear the full cost of the other two that they alone require. It is best to begin by carefully distinguishing where market tests are possible and where they are not. Otherwise, the case for applying the principles of price is misunderstood; either the too-ardent advocate overstates his case or the potential convert projects too much. In either case, a "disenchantment" sets in that is hard to reverse.

Much of the economics of the city is "public economies," and the pricing of urban public services poses some very difficult and even insurmountable problems. Economists have, in fact, erected a very elegant rationalization of the public economy almost wholly on the nonmarketability of public goods and services. While economists have perhaps oversold the inapplicability of price in the public sector, let us begin with what we are not talking about.

The public economy supplies "collectively consumed" goods, those produced and consumed in one big indivisible lump. Everyone has to be counted in the system, there is no choice of *in* or *out*. We cannot identify individual benefits, therefore we cannot exact a *quid pro quo*. We cannot exclude those who would not pay voluntarily; therefore we must turn to compulsory payments: taxes. Justice and air-pollution control are good examples of collectively consumed public services.

A second function of the public economy is to supply "merit goods." Sometimes the majority of us become a little paternalistic and decide that we know what is best for all of us. We believe some goods are especially meritorious, like education, and we fear that others might not fully appreciate this truth. Therefore, we produce these merit goods, at considerable cost, but offer them at a zero price. Unlike the first case of collectively consumed goods, we could sell these merit goods. A schoolroom's doors can be closed to those who do not pay, *quite unlike justice*. But we choose to open the doors wide to ensure that no one will turn away from the service because of its cost, and then we finance the service with compulsory payments. Merit goods are a case of the majority playing God, and "coercing" the minority by the use of bribes to change their behavior.

A third classic function of government is the redistribution of income. Here we wish to perform a service for one group and charge another group the cost of that service. Welfare payments are a clear case. Again, any kind of a private market or pricing mechanism is totally inappropriate; we obviously do not expect welfare recipients to return their payments. Again, we turn to compulsory payments: taxes. In sum, the private market may not be able to process certain goods and services (pure "public goods"), or it may give the "wrong" prices ("merit goods"), or we simply do not want the consumer to pay (income-redistributive services).

But the virtual elimination of price from the public sector is an extreme and highly simplistic response to the special requirements of the public sector. Merit goods may be subsidized without going all the way to zero prices. Few would argue for full-cost admission prices to museums, but a good case can be made for moderate prices that cover, say, their daily operating costs, (e.g., salaries of guards and janitors, heat and light).

Unfortunately, as we have given local government more to do, we have almost unthinkingly extended the tradition of "free" public services to every new undertaking, despite the clear trend in local government toward the assumption of more and more functions that do not fit the neat schema above. The provision of free public facilities for automobile movement in the crowded cores of our urban areas can hardly be defended on the grounds that: (a) motorists could not be excluded from the expressways if they refused to pay the toll, or (b) the privately operated motor vehicle is an especially meritorious way to move through densely populated areas, or (c) the motorists cannot afford to pay their own way and that the general (property) taxpayers should subsidize them. And all this applies with a vengeance to municipal marinas and golf courses.

PRICES TO RATION THE USE OF EXISTING FACILITIES

We need to understand better the rationing function of price as it manifests itself in the urban public sector: how the demand for a temporarily (or permanently) fixed stock of a public good or service can be adjusted to the supply. At any given time the supply of street, bridge, and parking space is fixed; "congestion" on the streets and a "shortage" of parking space express demand greater than supply at a zero price, a not too surprising phenomenon. Applying the market solution, the shortage of street space at peak hours ("congestion") could have been temporarily relieved (rationalized) by introducing a short-run rationing price to divert some motorists to other hours of movement, some to other modes of transportation, and some to other activities.

Public goods last a long time and therefore current additions to the stock are too small to relieve shortages quickly and easily. *The rationing function of price is probably more important in the public sector where it is customarily ignored than in the private sector where it is faithfully expressed.*

Rationing need not always be achieved with money, as when a motorist circles the block over and over looking for a place to park. The motorist who is not willing to "spend time" waiting and drives away forfeits the scarce space to one who will spend time (luck averaging out). The parking "problem" may be reinterpreted as an implicit decision to keep the money price artificially low (zero or a nickel an hour in a meter) and supplement it with a waiting cost or time price. The problem is that we did not clearly understand and explicitly agree to do just that.

The central role of price is to allocate – across the board – scarce resources among competing ends to the point where the value of another unit of any good or service is equal to the incremental cost of producing that unit. Expressed loosely, in the long run we turn from using prices to dampen demand to fit a fixed supply to adjusting the supply to fit the quantity demanded, at a price which reflect the production costs.

Prices which ration also serve to signal desired new directions in which to reallocate resources. If the rationing price exceeds those costs of production which the user is expected to bear directly, more resources should ordinarily be allocated to that activity. And symmetrically a rationing price below the relevant costs indicates an uneconomic provision of that service in the current amounts. Rationing prices reveal the intensity of the users' demands. How much is it really worth to drive into the heart of town at rush hour or launch a boat? In the long run, motorists and boaters should be free to choose, in rough measure, the amount of street and dock space they want and for which they are willing to pay. But, as in the private sector of our economy, free choice would carry with it full (financial) responsibility for that choice.

We need also to extend our price strategy to "factor prices"; we need a sophisticated wage policy for local public employees. Perhaps the key decision in urban development pertains to the recruiting and assignment of elementary- and secondary-school teachers. The more able and experienced teachers have the greater range of choice in post and quite naturally they choose the newer schools in the better neighborhoods, after serving the required apprenticeship in the older schools in the poorer neighborhoods. Such a pattern of migration certainly cannot implement a policy of equality of opportunity.

This author argued six years ago that egalitarianism in the public school system has been overdone; even the army recognizes the role of price when it awards extra "jump pay" to paratroopers, only a slightly more hazardous occupation than teaching behind the lines. Besides, it is male teachers whom we need to attract to slum schools, both to serve as father figures where there are few males at home and to serve quite literally as disciplinarians. It is bad economics to insist on equal pay for teachers everywhere throughout the urban area when males have a higher productivity in some areas and when males have better employment opportunities outside teaching – higher "opportunity costs" that raise their supply price. It is downright silly to argue that "equal pay for equal work" is achieved by paying the same money wage in the slums as in the suburbs.

About a year ago, on being offered premium salaries for service in ghetto schools, the teachers rejected, by name and with obvious distaste, any form of "jump pay." One facile argument offered was that they must protect the slum child from the stigma of being harder to teach, a nicety surely lost on the parents and outside observers. One suspects that the real reason for avoiding salary differentials between the "slums and suburbs" is that the teachers seek to escape the hard choice between the higher pay and the better working conditions. *But that is precisely what the price system is supposed to do: equalize sacrifice.*

PRICES TO GUIDE THE DISTRIBUTION OF INCOME

A much wider application of tolls, fees, fines, and other "prices" would also confer greater control over the distribution of income for two distinct reasons. First, the taxes currently used to finance a given public service create *implicit* and *unplanned* redistribution of income. Second, this drain on our limited supply of tax money prevents local government from undertaking other programs with more *explicit* and *planned* redistributional effects.

More specifically, if upper-middle- and upper-income motorists, golfers, and boaters use subsidized public streets, golf links, and marinas more than in proportion to their share of local tax payments from which the subsidy is paid, then these public activities redistribute income toward greater inequality. Even if these activities were found to be neutral with respect to the distribution of income, public provision of these discretionary services comes at the expense of a roughly equivalent expenditure on the more classic public services: protection, education, public health, and welfare.

Self-supporting public golf courses are so common and marinas are such an easy extension of the same principle that it is much more instructive to test the faith by considering the much harder case of the public museum: "culture." Again, we must recall that it is the middle- and upper-income classes who typically visit museums, so that free admission becomes, in effect, redistribution toward greater inequality, to the extent that the lower-income nonusers pay local taxes (e.g., property taxes directly or indirectly through rent, local sales taxes). The low prices contemplated are not, moreover, likely to discourage attendance significantly and the resolution of special cases (e.g., student passes) seems well within our competence.

Unfortunately, it is not obvious that "free" public marinas as tennis courts pose foregone alternatives – "opportunity costs." If we had to discharge a teacher or policeman every time we built another boat dock or tennis court, we would see the real cost of these public services. But in a growing economy, we need only not hire another teacher or policeman and that is not so obvious. In general, then, given a binding local budget constraint – scarce tax money – to undertake a local public service that is unequalizing or even neutral in income redistribution is to deny funds to programs that have the desired distributional effect, and is to lose control over equity.

Typically, in oral presentations at question time, it is necessary to reinforce this point by rejoining, "No, I would not put turnstiles in the playgrounds in poor neighborhoods, rather it is only because we do put turnstiles at the entrance to the playgrounds for the middle- and upper-income groups that we will be able to 'afford' playgrounds for the poor."

PRICES TO ENLARGE THE RANGE OF CHOICE

But there is more at stake in the contemporary chaos of hidden and unplanned prices than "merely" efficiency and equity. *There is no urban goal on which consensus is more easily gained than the pursuit of great variety and choice – "pluralism."* The great rural to urban migration was prompted as much by the search for variety as by the decline of agriculture and rise of manufacturing. Wide choice is seen as the saving grace of bigness by even the sharpest critics of the metropolis. Why, then, do we tolerate far less variety in our big cities than we could have? We have lapsed into a state of tyranny by the majority, in matters of both taste and choice.

In urban transportation the issue is not, in the final analysis, whether users of core-area street space at peak hours should or should not be required to pay their own way in full. The problem is, rather, that by not forcing a direct *quid pro quo* in money, we implicitly substitute a new means of payment – time in the transportation services "market." The peak-hour motorist does pay in full, through congestion and time delay. But *implicit choices* blur issues and confuse decision making.

Say we were carefully to establish how many more dollars would have to be paid in for the additional capacity needed to save a given number of hours spent commuting. The *majority* of urban motorists perhaps would still choose the present combination of "underinvestment" in highway, bridge and parking facilities, with a compensatory heavy investment of time in slow movement over these crowded facilities. Even so, a substantial minority of motorists do prefer a different combination of money and time cost. A more affluent, long-distance commuter could well see the current level of traffic congestion as a real problem and much prefer to spend more money to save time. If economies of scale are so substantial that only one motorway to town can be supported, or if some naturally scarce factor (e.g., bridge or tunnel sites) prevents parallel transportation facilities of different quality and price, then the preferences of the minority must be sacrificed to the majority interest and we do have a real "problem." But, ordinarily, in large urban areas there are a number of near-parallel routes to town, and an unsatisfied minority group large enough to justify significant differentiation of one or more of these streets and its diversion to their use. Greater choice through greater scale is, in fact, what bigness is all about.

The simple act of imposing a toll, at peak hours, on one of these routes would reduce its use, assuming that nearby routes are still available without user charges, thereby speeding movement of the motorists who remain and pay. The toll could be raised only to the point where some combination of moderately rapid movement and high physical output were jointly optimized. Otherwise the outcry might be raised that the public transportation authority was so elitist as to gratify the desire of a few very wealthy motorists for very rapid movement, heavily overloading the "free" routes. It is, moreover, quite possible, even probable, that the newly converted, rapid-flow, toll-route would handle as many vehicles as it did previously as a congested street and not therefore spin off any extra load on the free routes.

Our cities cater, at best, to the taste patterns of the middle-income class, as well they should, *but not so exclusively*. This group has chosen, indirectly through clumsy and insensitive tax-and-expenditure decisions and ambiguous political processes, to move about town flexibly and cheaply, but slowly, in private vehicles. Often, and almost invariably in the larger urban areas, we would not have to encroach much on this choice to accommodate also those who would prefer to spend more money and less time, in urban movement. In general, we should permit urban residents to pay in their most readily available "currency" – time or money.

Majority rule by the middle class in urban transportation has not only disenfranchised the affluent commuter, but more seriously it has debilitated the low-fare, mass transit system on which the poor depend. The effect of widespread automobile ownership and use on the mass transportation system is an oft-told tale: falling bus and rail patronage leads to less frequent service and higher overhead costs per trip and often higher fares which further reduce demand and service schedules. Perhaps two-thirds or more of the urban residents will tolerate and may even prefer slow, cheap automobile movement. But the poor are left without access to many places of work – the suburbanizing factories in particular – and they face much reduced opportunities for comparative shopping, and highly constrained participation in the community life in general. A truly wide range of choice in urban transportation would allow the rich to pay for fast movement with money, the middle-income class to pay for the privacy and convenience of the automobile with time, and the poor to economize by giving up (paying with) privacy.

A more sophisticated price policy would expand choice in other directions. Opinions differ as to the gravity of the water-pollution problem near large urban areas. The minimum level of dissolved oxygen in the water that is needed to meet the standards of different users differs greatly, as does the incremental cost that must be incurred to bring the dissolved oxygen levels up to successively higher standards. The boater accepts a relatively low level of "cleanliness" acquired at relatively little cost. Swimmers have higher standards attained only at much higher cost. Fish and fisherman can thrive only with very high levels of dissolved oxygen acquired only at the highest cost. Finally, one can imagine an elderly convalescent or an impoverished slum dweller or a confirmed landlubber who is not at all interested in the nearby river. What, then, constitutes "clean"?

A majority rule decision, whether borne by the citizen directly in higher taxes or levied on the industrial polluters and then shifted on to the consumer in higher produce prices, is sure to create a "problem." If the pollution program is a compromise – a halfway measure – the fisherman will be disappointed because the river is still not clean enough for his purposes and the landlubbers will be disgruntled because the program is for "special interests" and he can think of better uses for his limited income. Surely, we can assemble the managerial skills in the local public sector needed to devise and administer a structure of user charges that would extend choice in outdoor recreation, consistent with financial responsibility, with lower charges for boat licenses and higher charges for fishing licenses.

Perhaps the most fundamental error we have committed in the development of our large cities is that we have too often subjected the more affluent residents to petty irritations which serve no great social purpose, then turned right around and permitted this same group to avoid responsibilities which have the most critical and pervasive social ramifications. It is a travesty and a social tragedy that we have prevented the rich from buying their way out of annoying traffic congestion – or at least not helped those who are long on money and short on time arrange such an accommodation. Rather, we have permitted them, through political fragmentation and flight to tax havens, to evade their financial and leadership responsibilities for the poor of the central cities. That easily struck goal, "pluralism and choice," will require much more managerial sophistication in the local public sector than we have shown to date.

PRICING TO CHANGE TASTES AND BEHAVIOR

Urban managerial economies will probably also come to deal especially with "developmental pricing" analogous to "promotional pricing" in business. Prices below cost may be used for a limited period to create a market for a presumed "merit good." The hope would be that the artificially low price would stimulate consumption and that an altered *expenditure pattern* (practice) would lead in time to an altered *taste pattern* (preference), as experience with the new service led to a fuller appreciation of it. Ultimately, the subsidy would be withdrawn, whether or not tastes changed sufficiently to make the new service self-supporting – provided, of course, that no permanent redistribution of income was intended.

For example, our national parks had to be subsidized in the beginning and this subsidy could be continued indefinitely on the grounds that these are "merit goods" that serve a broad social interest. But long experience with outdoor recreation has so shifted tastes that a large part of the costs of these parks could now be paid for by a much higher set of park fees.

It is difficult, moreover, to argue that poor people show up at the gates of Yellowstone Park, or even the much nearer metropolitan area regional parks, in significant number, so that a subsidy is needed to continue provision of this service for the poor. A careful study of the users and the incidence of the taxes raised to finance our parks may even show a slight redistribution of income toward greater inequality.

Clearly, this is not the place for an economist to pontificate on the psychology of prices but a number of very interesting phenomena that seem to fall under this general heading deserve brief mention. A few simple examples of how charging a price changes behavior are offered, but left for others to classify.

In a recent study of depressed areas, the case was cited of a community-industrial-development commission that extended its fund-raising efforts from large business contributors to the general public in a supplementary "nickel and dime" campaign. They hoped to enlist the active support of the community at large, more for reasons of public policy than for finance. But even a trivial financial stake was seen as a means to create broad and strong public identification with the local industrial development programs and to gain their political support.

Again, social-work agencies have found that even a nominal charge for what was previously a free service enhances both the self-respect of the recipient and his respect for the usefulness of the service. Paradoxically, we might experiment with higher public assistance payments coupled to *nominal* prices for selected public health and family services, personal counseling, and surplus foods.

To bring a lot of this together now in a programmatic way, we can imagine a very sophisticated urban public management beginning with below-cost prices on, say, the new rapid mass transit facility during the promotional period of luring motorists from their automobiles and of "educating" them on the advantages of a carefree journey to work. Later, if and when the new facility becomes crowded during rush hours and after a taste of this new transportation mode has become well established, the "city economist"' might devise a-three-price structure of fares: the lowest fare for regular off-peak use, the middle fare for regular peak use (tickets for commuters), and the highest fare for the occasional peak-time user. Such a schedule would reflect each class's contribution to the cost of having to carry standby capacity.

If the venture more than covered its costs of operation, the construction of additional facilities would begin. Added social benefits in the form of a cleaner, quieter city or reduced social costs of traffic control and accidents could be included in the cost accounting ("cost-benefit analysis") underlying the fare structure. But below-cost fares, taking care to count social as well as private costs, would not be continued indefinitely except for merit goods or when a clear income-redistribution end is in mind. And, even then, not without careful comparison of the relative efficiency of using the subsidy money in alternative redistributive programs. We need, it would seem, not only a knowledge of the economy of the city, but some very knowledgeable city economists as well.

"The Competitive Advantage of the Inner City"

Harvard Business Review (1995)

Michael Porter

Editors' Introduction

Harvard Business School professor Michael Porter declares that the economic distress of inner-city neighborhoods may be the most pressing issue facing the United States. Like his Harvard colleague William Julius Wilson (p. 117), Porter sees lack of jobs as the root cause of crime, drug abuse, dysfunctional families, and other social problems. And like Wilson, Porter feels that the government response to the problem of inner-city decline and lack of jobs has been ineffective in the United States and other countries. But where Wilson advocates government support of full employment programs, Porter believes the best solution to inner city problems lies in identifying and taking advantage of their particular strengths.

Porter characterizes the current approach to inner-city problems as based on a social model aimed at the individual. As economist Wilbur Thompson describes (p. 274), market-oriented countries like the United States rely primarily on private business to meet social needs. They reserve subsidized housing, food stamps, free medical assistance, and other redistributive programs only for "the truly disadvantaged". Similarly corporate philanthropy has generally sought to meet the individual needs of the poorest members of society. Scandinavian countries mix free-market approaches and a much more extensive welfare state. Communist countries, agreeing with Engels' critique of capitalism (p. 46), have abolished markets (at least nominally) and allocate resources through more centralized political systems with varying degrees of success.

Porter argues that government programs to create jobs in inner-city neighborhoods and train local residents for them have been fragmented and inefficient. He believes that programs consisting of subsidies, preference programs, and expensive efforts to stimulate economic activity have not worked well. In Porter's view, governments have dumped money on small, marginal inner-city businesses that could not turn a profit without government help. These businesses in turn have often hired or required private sector contractors to hire incompetent workers. Or they have spent large sums to get hopelessly blighted redevelopment areas and badly contaminated brownfield areas back into usable condition. Often the subsidized firms fail, the workers hired through preference programs are fired when the subsidies run out, and the brownfields and redevelopment project areas sit vacant or are developed as showcase projects that burn up money better spent elsewhere.

Porter argues for a new economic model of inner-city revitalization. He favors private, for-profit initiatives based on economic self-interest, rather than artificial inducements, charity, and government mandates. Such an approach will work, Porter argues, only if it takes advantage of the true competitive advantages of the inner city.

Porter considers the view that inner-city real estate or labor costs are sufficiently lower to make a compelling reason for firms to locate in inner-city neighborhoods rather than in suburban or exurban locations to be a myth. Rather, he argues, there are four true advantages to inner-city locations: first, their strategic location, second, local market demand the areas themselves possess, third, possibilities of integration with regional job clusters, and fourth, an industrious labor force that is eager to work. Firms that can best exploit these inner-city advantages,

Porter believes, can turn a profit without government assistance and may find inner-city locations the best place to do business.

Unlike Wilson, who emphasizes the loss of jobs that people with limited education can perform, Porter argues that there still are (or could be) plenty of firms in the inner city that can use unskilled labor for warehouse and production-line workers, truck drivers, retail, and other unskilled jobs.

Porter places a large part of the blame for the high costs and difficulties of doing business in inner cities on government regulation and anti-business attitudes. He argues that local governments can and should improve the economic climate and make inner-city neighborhoods more attractive for private investment by reducing regulation. These are essentially the strategies that Harvey Molotch found that growth machines favored (p. 251). Reducing regulation to encourage business reinvestment in distressed inner city neighborhoods has been a theme in British Enterprise Zone and US Empowerment Zone and Enterprise Community programs. In these programs a blighted inner city area is designated. Within the boundaries of the zone, government provides a mixture of incentives to encourage businesses to locate within the zone. Some of the incentives are monetary, such as subsidies or tax relief. Others involve regulatory relief, such as allowing industries to avoid costly environmental regulations such as technology to reduce greenhouse gas emissions or to comply with strict and costly occupational health and safety requirements. The idea is that industries will locate within the zones and employ local residents. With jobs and income residents will be able to afford better housing, food, clothing, and health care. They will not be dependent on local government for welfare assistance. And with pride in earning a living, many social ills will disappear.

While their diagnosis of the problem is similar, what would William Julius Wilson think of Porter's characterization of the social model and preference programs? Given historic racism, bad schools, and the lack of skills many inner-city Black ghetto residents possess, aren't preference programs necessary? Would Porter's prescriptions reach people with a "street culture" who chose to live the "gangsta lifestyle" that Elijah Anderson describes (p. 127)?

Michael Porter (b. 1947) is the Bishop William Lawrence University Professor of Business Administration at the Harvard Business School and the director of Harvard's Institute for Strategy and Competitiveness. Since 2000 he has been a University Professor – the highest professional recognition that Harvard awards a faculty member. As well as his MBA from Harvard Business School and a PhD in Business Economics from Harvard, Porter holds fourteen honorary doctorates. Porter's graduate course on competitiveness is taught in partnership with more than eighty other universities, many of them in developing nations. Porter also leads Harvard Business School's New CEO Workshop, a program for newly appointed CEOs of the world's largest and most complex organizations. In the early 1990s Porter turned his attention to inner-city problems. In 1994, he founded and remains chair of the Initiative for a Competitive Inner City (ICIC), a nonprofit private-sector organization that assists inner-city business development across the United States. Recently Porter has extended his thinking about competitiveness and strategic planning to the fields of health care and environmental management. In addition to advising private corporations, and national governments, Professor Porter serves as senior strategy advisor to the Boston Red Sox, a major league baseball team.

Professor Porter extended the ideas advanced in this selection in "New Strategies for Inner-City Economic Development," *Economic Development Quarterly*, 11(1) (February 1997).

Porter's most influential book, *Competitive Strategy: Techniques for Analyzing Industries and Competitors*, originally published in Cambridge, MA in 1998 by the Harvard Business School Press, is in its sixty-third printing and has been translated into nineteen languages. *Competitive Advantage: Creating and Sustaining Superior Performance*, originally published in New York in 1985 by the Free Press, is in its thirty-eighth printing. Other of Porter's eighteen books include *Michael E. Porter on Competition*, updated and expanded edition (Boston, MA: Harvard Business School Press, 2008), *Competitive Strategy: Techniques for Analyzing Industries and Competitors* (New York: Free Press, 1998), *On Competition* (Boston, MA: Harvard Business School Press, 1998), and *The Competitive Advantage of Nations* (New York: Free Press, 1990).

Books on community-based economic development – generally less oriented to private sector approaches than Porter – include Edward J. Blakeley and Nancy Green Leigh, *Planning Local Economic Development: Theory and Practice*, 4th edn (Thousand Oaks, CA: Sage, 2009), Paul Ong and Anastasia Loukaitou-Sideris (eds), *Jobs and Economic Development in Minority Communities* (Philadelphia, PA: Temple University Press, 2006), Sammis White, Richard D. Bingham, and Edward W. Hill (eds), *Financing Economic Development in the 21st Century* (Armonk, NY: M.E. Sharpe, 2003), and Joan Fitzgerald and Nancey Green Leigh, *Economic Revitalization: Cases and Strategies for City and Suburb* (Thousand Oaks, CA: Sage, 2002). See also the large literature on social entrepreneurship such as David Bornstein, *How to Change the World: Social Entrepreneurs and the Power of New Ideas*, updated edition (Oxford: Oxford University Press, 2008), Alex Nicholls (ed.), *Social Entrepreneurship: New Models of Sustainable Social Change* (Oxford: Oxford University Press, 2008) and Peter C. Brinckerhoff, *Social Entrepreneurship: The Art of Mission-Based Venture Development* (Hoboken, NJ: Wiley, 2000).

The economic distress of America's inner cities may be the most pressing issue facing the nation. The lack of businesses and jobs in disadvantaged urban areas fuels not only a crushing cycle of poverty but also crippling social problems, such as drug abuse and crime. And, as the inner cities continue to deteriorate, the debate on how to aid them grows increasingly divisive.

The sad reality is that the efforts of the past few decades to revitalize the inner cities have failed. The establishment of a sustainable economic base – and with it employment opportunities, wealth creation, role models, and improved local infrastructure – still eludes us despite the investment of substantial resources.

Past efforts have been guided by a social model built around meeting the needs of individuals. Aid to inner cities, then, has largely taken the form of relief programs such as income assistance, housing subsidies, and food stamps, all of which address highly visible – and real – social needs.

Programs aimed more directly at economic development have been fragmented and ineffective. These piecemeal approaches have usually taken the form of subsidies, preference programs, or expensive efforts to stimulate economic activity in tangential fields such as housing, real estate, and neighborhood development. Lacking an overall strategy, such programs have treated the inner city as an island isolated from the surrounding economy and subject to its own unique laws of competition. They have encouraged and supported small, subscale businesses designed to serve the local community but ill equipped to attract the community's own spending power, much less export outside it. In short, the social model has inadvertently undermined the creation of economically viable companies. Without such companies and the jobs they create, the social problems will only worsen.

The time has come to recognize that revitalizing the inner city will require a radically different approach. While social programs will continue to play a critical role in meeting human needs and improving education, they must support – and not undermine – a coherent economic strategy. The question we should be asking is how inner-city-based businesses and nearby employment opportunities for inner-city residents can proliferate and grow. A sustainable economic base can be created in the inner city, but only as it has been created elsewhere: through private, for-profit initiatives and investment based on economic self-interest and genuine competitive advantage – not through artificial inducements, charity, or government mandates.

We must stop trying to cure the inner city's problems by perpetually increasing social investment and hoping for economic activity to follow. Instead, an economic model must begin with the premise that inner-city businesses should be profitable and should be positioned to compete on a regional, national, and even international scale. These businesses should be capable not only of serving the local community but also of exporting goods and services to the surrounding economy. The cornerstone of such a model is

 (Portions of the original article which have been omitted from this edited selection are indicated by ellipses in the form of three dots for material within a single paragraph, or three dots enclosed in brackets for one or more entire paragraphs.)

to identify and exploit the competitive advantages of inner cities that will translate into truly profitable businesses.

Our policies and programs have fallen into the trap of redistributing wealth. The real need – and the real opportunity – is to create wealth.

TOWARDS A NEW MODEL: LOCATION AND BUSINESS DEVELOPMENT

Economic activity in and around inner cities will take root if it enjoys a competitive advantage and occupies a niche that is hard to replicate elsewhere. If companies are to prosper, they must find a compelling competitive reason for locating in the inner city. A coherent strategy for development starts with that fundamental economic principle, as the contrasting experiences of the following companies illustrate.

Alpha Electronics (the company's name has been disguised), a 28-person company that designed and manufactured multimedia computer peripherals, was initially based in lower Manhattan. In 1987, the New York City Office of Economic Development set out to orchestrate an economic "renaissance" in the South Bronx by inducing companies to relocate there. Alpha, a small but growing company, was sincerely interested in contributing to the community and eager to take advantage of the city's willingness to subsidize its operations. The city, in turn, was happy that a high-tech company would begin to stabilize a distressed neighborhood and create jobs. In exchange for relocating, the city provided Alpha with numerous incentives that would lower costs and boost profits. It appeared to be an ideal strategy.

By 1994, however, the relocation effort had proved a failure for all concerned. Despite the rapid growth of its industry, Alpha was left with only 8 of its original 28 employees. Unable to attract high-quality employees to the South Bronx or to train local residents, the company was forced to outsource its manufacturing and some of its design work. Potential suppliers and customers refused to visit Alpha's offices. Without the city's attention to security, the company was plagued by theft.

What went wrong? Good intentions notwithstanding, the arrangement failed the test of business logic. Before undertaking the move, Alpha and the city would have been wise to ask themselves why none of the South Bronx's thriving businesses was in electronics. The South Bronx as a location offered no specific advantages to support Alpha's business, and it had several disadvantages that would prove fatal. Isolated from the lower Manhattan hub of computer-design and software companies, Alpha was cut off from vital connections with customers, suppliers, and electronic designers.

In contrast, Matrix Exhibits, a $2.2 million supplier of trade-show exhibits that has 30 employees, is thriving in Atlanta's inner city. When Tennessee-based Matrix decided to enter the Atlanta market in 1985, it could have chosen a variety of locations. All the other companies that create and rent trade show exhibits are based in Atlanta's suburbs. But the Atlanta World Congress Center, the city's major exhibition space, is just a six-minute drive from the inner city, and Matrix chose the location because it provided a real competitive advantage. Today Matrix offers customers superior response time, delivering trade-show exhibits faster than its suburban competitors. Matrix benefits from low rental rates for warehouse space – about half the rate its competitors pay for similar space in the suburbs – and draws half its employees from the local community. The commitment of local police has helped the company avoid any serious security problems. Today Matrix is one of the top five exhibition houses in Georgia.

Alpha and Matrix demonstrate how location can be critical to the success or failure of a business. Every location – whether it be a nation, a region, or a city – has a set of unique local conditions that underpin the ability of companies based there to compete in a particular field. The competitive advantage of a location does not usually arise in isolated companies but in clusters of companies – in other words, in companies that are in the same industry or otherwise linked together through customer, supplier, or similar relationships. Clusters represent critical masses of skill, information, relationships, and infrastructure in a given field. Unusual or sophisticated local demand gives companies insight into customers' needs. Take Massachusetts's highly competitive cluster of information-technology industries: it includes companies specializing in semiconductors, workstations, supercomputers, software, networking equipment, databases, market research, and computer magazines.

Clusters arise in a particular location for specific historical or geographic reasons – reasons that may cease to matter over time as the cluster itself becomes powerful and competitively self-sustaining. In successful clusters such as Hollywood, Silicon Valley, Wall Street, and Detroit, several competitors often push one another to improve products and processes. The presence of a group of competing companies contributes to the formation of new suppliers, the growth of companies in related fields, the formation of specialized training programs, and the emergence of technological centers of excellence in colleges and universities. The clusters also provide newcomers with access to expertise, connections, and infrastructure that they in turn can learn and exploit to their own economic advantage.

If locations (and the events of history) give rise to clusters, it is clusters that drive economic development. They create new capabilities, new companies, and new industries. I initially described this theory of location in *The Competitive Advantage of Nations* (Free Press, 1990), applying it to the relatively large geographic areas of nations and states. But it is just as relevant to smaller areas such as the inner city. To bring the theory to bear on the inner city, we must first identify the inner city's competitive advantages and the ways inner-city businesses can forge connections with the surrounding urban and regional economies.

THE TRUE ADVANTAGES OF THE INNER CITY

The first step toward developing an economic model is identifying the inner city's true competitive advantages. There is a common misperception that the inner city enjoys two main advantages: low-cost real estate and labor. These so-called advantages are more illusory than real. Real estate and labor costs are often higher in the inner city than in suburban and rural areas. And even if inner cities were able to offer lower-cost labor and real estate compared with other locations in the United States, basic input costs can no longer give companies from relatively prosperous nations a competitive edge in the global economy. Inner cities would inevitably lose jobs to countries like Mexico or China, where labor and real estate are far cheaper.

Only attributes that are unique to inner cities will support viable businesses. My ongoing research of urban areas across the United States identifies four main advantages of the inner city: strategic location, local market demand, integration with regional clusters, and human resources. Various companies and programs have identified and exploited each of those advantages from time to time. To date, however, no systematic effort has been mounted to harness them.

Strategic location

Inner cities are located in what *should* be economically valuable areas. They sit near congested high-rent areas, major business centers, and transportation and communications nodes. As a result, inner cities can offer a competitive edge to companies that benefit from proximity to downtown business districts, logistical infrastructure, entertainment or tourist centers, and concentrations of companies.

Local market demand

The inner-city market itself represents the most immediate opportunity for inner-city-based entrepreneurs and businesses. At a time when most other markets are saturated, inner-city markets remain poorly served – especially in retailing, financial services, and personal services. In Los Angeles, for example, retail penetration per resident in the inner city compared with the rest of the city is 35% in supermarkets, 40% in department stores, and 50% in hobby, toy, and game stores.

The first notable quality of the inner-city market is its size. Even though average inner-city incomes are relatively low, high population density translates into an immense market with substantial purchasing power. Boston's inner city, for example, has an estimated total family income of $3.4 billion.

Spending power per acre is comparable with the rest of the city despite a 21% lower average household income level than in the rest of Boston, and, more significantly, higher than in the surrounding suburbs. In addition, the market is young and growing rapidly, owing in part to immigration and relatively high birth rates.

Integration with regional clusters

The most exciting prospects for the future of inner-city economic development lie in capitalizing on nearby regional clusters: those unique-to-a-region collections of related companies that are competitive nationally and even globally. For example, Boston's inner city is next door to world-class financial-services and health-care clusters. South Central Los Angeles is close to an enormous entertainment cluster and a large logistical-services and wholesaling complex.

The ability to access competitive clusters is a very different attribute – and one much more far reaching in economic implication – than the more generic advantage of proximity to a large downtown area with concentrated activity. Competitive clusters create two types of potential advantages. The first is for business formation. Companies providing supplies, components, and support services could be created to take advantage of the inner city's proximity to multiple nearby customers in the cluster . . .

The second advantage of these clusters is the potential they offer inner-city companies to compete in downstream products and services. For example, an inner-city company could draw on Boston's strength in financial services to provide services tailored to inner-city needs – such as secured credit cards, factoring, and mutual funds – both within and outside the inner city in Boston and elsewhere in the country.

[. . .]

Human resources

The inner city's fourth advantage takes on a number of deeply entrenched myths about the nature of its residents. The first myth is that inner-city residents do not want to work and opt for welfare over gainful employment. Although there is a pressing need to deal with inner-city residents who are unprepared for work, most inner-city residents are industrious and eager to work. For moderate-wage jobs ($6 to $10 per hour) that require little formal education (for instance, warehouse workers, production-line workers, and truck drivers), employers report that they find hard-working, dedicated employees in the inner city. For example, a company in Boston's inner-city neighborhood of Dorchester bakes and decorates cakes sold to supermarkets throughout the region. It attracts and retains area residents at $7 to $8 per hour (plus contributions to pensions and health insurance) and has almost 100 local employees. The loyalty of its labor pool is one of the factors that has allowed the bakery to thrive.

Admittedly, many of the jobs currently available to inner-city residents provide limited opportunities for advancement. But the fact is that they are jobs; and the inner city and its residents need many more of them close to home. Proposals that workers commute to jobs in distant suburbs – or move to be near those jobs – underestimate the barriers that travel time and relative skill level represent for inner-city residents. Moreover, in deciding what types of businesses are appropriate to locate in the inner city, it is critical to be realistic about the pool of potential employees. Attracting high-tech companies might make for better press, but it is of little benefit to inner-city residents. Recall the contrasting experiences of Alpha Electronics and Matrix Exhibits. In the case of Alpha, there was a complete mismatch between the company's need for highly skilled professionals and the available labor pool in the local community. In contrast, Matrix carefully considered the available workforce when it established its Atlanta office. Unlike the Tennessee headquarters, which custom-designs and creates exhibits for each client, the Atlanta office specializes in rentals made from prefabricated components – work requiring less-skilled labor, which can be drawn from the inner- city. Given the workforce, low-skill jobs are realistic and economically viable: they represent the first rung on the economic ladder for many individuals who otherwise would be unemployed. Over time, successful job creation will trigger a self-reinforcing process that raises skill and wage levels.

The second myth is that the inner city's only entrepreneurs are drug dealers. In fact, there is a real capacity for legitimate entrepreneurship among inner-city residents, most of which has been channeled into the provision of social services. For instance, Boston's inner city has numerous social service providers as well as social, fraternal, and religious organizations. Behind the creation and building of those organizations is a whole cadre of local entrepreneurs who have responded to intense local demand for social services and to funding opportunities provided by government, foundations, and private sector sponsors. The challenge is to redirect some of that talent and energy toward building for-profit businesses and creating wealth.

The third myth is that skilled minorities, many of whom grew up in or near inner cities, have abandoned their roots. Today's large and growing pool of talented minority managers represents a new generation of

New Model	Old Model
Economic: create wealth	*Social: redistribute wealth*
Private sector	Government and social service organizations
Profitable businesses	Subsidized businesses
Integration with the regional economy	Isolation from the larger economy
Companies that are export oriented	Companies that serve the local community
Skilled and experienced minorities engaged in building businesses	Skilled and experienced minorities engaged in the social service sector
Mainstream, private sector institutions enlisted	Special institutions created
Inner-city disadvantages addressed directly	Inner-city disadvantages counterbalanced
Government focuses on improving the environment for business	Government directly involved with providing services or funding

Figure 1 Inner-city economic development

potential inner-city entrepreneurs. Many have been trained at the nation's leading business schools and have gained experience in the nation's leading companies. Approximately 2,800 African Americans and 1,400 Hispanics graduate from M.B.A. programs every year compared with only a handful 20 years ago. Thousands of highly trained minorities are working at leading companies such as Morgan Stanley, Citibank, Ford, HewlettPackard, and McKinsey & Company. Many of these managers have developed the skills, net-work, capital base, and confidence to begin thinking about joining or starting entrepreneurial companies in the inner city . . .

THE REAL DISADVANTAGES OF THE INNER CITY

The second step toward creating a coherent economic strategy is addressing the very real disadvantages of locating businesses in the inner city. The inescapable fact is that businesses operating in the inner city face greater obstacles than those based elsewhere. Many of those obstacles are needlessly inflicted by government. Unless the disadvantages are addressed directly, instead of indirectly through subsidies or mandates, the inner city's competitive advantages will continue to erode.

Land

Although vacant property is abundant in inner cities, much of it is not economically usable. Assembling small parcels into meaningful sites can be prohibitively expensive and is further complicated by the fact that a number of city, state, and federal agencies each control land and fight over turf . . .

Building costs

The cost of building in the inner city is significantly higher than in the suburbs because of the costs and delays associated with logistics, negotiations with *community* groups, and strict urban regulations: restrictive zoning, architectural codes, permits, inspections, and government-required union contracts and minority setasides. Ironically, despite the desperate need for new projects, construction in inner cities is far more regulated than it is in the suburbs – a legacy of big city politics and entrenched bureaucracies.

[. . .]

Other costs

Compared with the suburbs, inner cities have high costs for water, other utilities, workers' compensation, health care, insurance, permitting and other fees, real estate and other taxes, OSHA compliance, and neighborhood hiring requirements. For example, Russer Foods, a manufacturing company located in Boston's inner city, operates a comparable plant in upstate New York. The Boston plant's expenses are 55% higher for workers' compensation, 50% higher for family medical insurance, 166% higher for unemployment insurance, 340% higher for water, and 67% higher for electricity. High costs like these drive away companies and hold down wages. Some costs, such as those for workers' compensation, apply to the state or region as a whole. Others, such as real estate taxes, apply citywide. Still others, such as property insurance, are specific to the inner city. All are devastating to maintaining fragile inner-city companies and to attracting new businesses.

It is an unfortunate reality that many cities – because they have a greater proportion of residents dependent on welfare, Medicaid, and other social programs – require higher government spending and, as a result, higher corporate taxes. The resulting tax burden feeds a vicious cycle – driving out more companies while requiring even higher taxes from those that remain. Cities have been reluctant to challenge entrenched bureaucracies and unions, as well as inefficient and outdated government departments, all of which unduly raise city costs.

Finally, excessive regulation not only drives up building and other costs but also hampers almost all facets of business life in the inner city, from putting up an awning over a shop window to operating a pushcart to making site improvements. Regulation also stunts inner-city entrepreneurship, serving as a formidable barrier to small and start-up companies. Restrictive licensing and permitting, high licensing fees, and archaic safety and health regulations create barriers to entry into the very types of businesses that are logical and appropriate for creating jobs and wealth in the inner city.

Security

Both the reality and the perception of crime represent profound impediments to urban economic development. First, crime against property raises costs. For example, the Shops at Church Square, an inner-city strip shopping center in Cleveland, Ohio, spends $2 per square foot more than a comparable suburban center for a full-time security guard, increased lighting, and continuous cleaning – raising overall costs by more than 20%. Second, crime against employees and customers creates an unwillingness to work in and patronize inner-city establishments and restricts companies' hours of operation. Fear of crime ranks among the most important reasons why companies opening new facilities failed to consider inner-city locations and why companies already located in the inner city left. Currently, police devote most of their resources to the security of residential areas, largely overlooking commercial and industrial sites.

Infrastructure

Transportation infrastructure planning, which today focuses primarily on the mobility of residents for shopping and commuting, should consider equally the mobility of goods and the ease of commercial transactions. The most critical aspects of the new economic model – the importance of the location of the inner city, the connections between inner-city businesses and regional clusters, and the development of export-oriented businesses – require the presence of strong logistical links between inner-city business sites and the surrounding economy. Unfortunately, the business infrastructure of the inner city has fallen into disrepair. The capacity of roads, the frequency and location of highway on-ramps and off-ramps, the links to downtown, and the access to railways, airports, and regional logistical networks are inadequate.

Employee skills

Because their average education levels are low, many inner-city residents lack the skills to work in any but the most unskilled occupations. To make matters worse, employment opportunities for less educated workers have fallen markedly. In Boston between 1970 and 1990, for example, the percentage of jobs held by people without high school diplomas dropped from 29% to 7%, while those held by college graduates climbed from 18% to 44%. And the unemployment rate for African–American men aged 16 to 64 with less than a high school education in major northeastern cities rose from 19% in 1970 to 57% in 1990.

Management skills

The managers of most inner-city companies lack formal business training. That problem, however, is not unique to the inner city; it is a characteristic of small businesses in general. Many individuals with extensive work histories but little or no formal managerial training start businesses. Inner-city companies without well trained managers experience a series of predictable problems that are similar to those that affect many small businesses: weaknesses in strategy development, market segmentation, customer-needs evaluation, introduction of information technology, process design, cost control, securing or restructuring financing, interaction with lenders and government regulatory agencies, crafting business plans, and employee training. Local community colleges often offer management courses, but their quality is uneven, and entrepreneurs are hard-pressed for time to attend them.

Capital

Access to debt and equity capital represents a formidable barrier to entrepreneurship and company growth in inner-city areas.

First, most inner-city businesses still suffer from poor access to debt funding because of the limited attention that mainstream banks paid them historically. Even in the best of circumstances, small business lending is only marginally profitable to banks because transaction costs are high relative to loan amounts. Many banks remain in small-business lending only to attract deposits and to help sell other more profitable products.

The federal government has made several efforts to address the inner city's problem of debt capital. As a result of legislation like the Community Reinvestment Act, passed in order to overcome bias in lending, banks have begun to pay much more attention to inner-city areas. In Boston, for example, leading banks are competing fiercely to lend in the inner city – and some claim to be doing so profitably. Direct financing efforts by government, however, have proved ineffective. The proliferation of government loan pools and quasi-public lending organizations has produced fragmentation, market confusion, and duplication of overhead. Business loans that would provide scale to private sector lenders are siphoned off by these organizations, many of which are high-cost, bureaucratic, and risk-averse. In the end, the development of high-quality private sector expertise in inner-city business financing has been undermined.

Second, equity capital has been all but absent. Inner-city entrepreneurs often lack personal or family savings and networks of individuals to draw on for capital. Institutional sources of equity capital are scarce for minority-owned companies and have virtually ignored inner-city business opportunities.

Attitudes

A final obstacle to companies in the inner city is antibusiness attitudes. Some workers perceive businesses as exploitative, a view that guarantees poor relations between labor and management. Equally debilitating are the antibusiness attitudes held by community leaders and social activists. These attitudes are the legacy of a regrettable history of poor treatment of workers, departures of companies, and damage to the environment. But holding on to these views today is counterproductive. Too often, community leaders mistakenly view businesses as a means of directly meeting social needs; as a result, they have unrealistic expectations for corporate involvement in the community . . .

Demanding linkage payments and contributions and stirring up antibusiness sentiment are political tools that brought questionable results in the past when owners had less discretion about where they chose to locate their companies. In today's increasingly competitive business environment, such tactics will serve only to stunt economic growth.

Overcoming the business disadvantages of the inner city as well as building on its inherent advantages will require the commitment and involvement of business, government, and the nonprofit sector. Each will have to abandon deeply held beliefs and past approaches. Each must be willing to accept a new model for the inner city based on an economic rather than social perspective. The private sector, nongovernment or social service organizations, must be the focus of the new model.

The new role of the private sector

The economic model challenges the private sector to assume the leading role. First, however, it must adopt new attitudes toward the inner city. Most private sector initiatives today are driven by preference programs or charity. Such activities would never stand on their own merits in the marketplace. It is inevitable, then, that they contribute to growing cynicism. The private sector will be most effective if it focuses on what it does best: creating and supporting economically viable businesses built on true competitive advantage. It should pursue four immediate opportunities as it assumes its new role.

1. *Create and expand business activity in the inner city.* The most important contribution companies can make to inner cities is simply to do business there. Inner cities hold untapped potential for profitable businesses. Companies and entrepreneurs must seek out and seize those opportunities that build on the true advantages of the inner city. In particular, retailers, franchisers, and financial services companies have immediate opportunities. Franchises represent an especially attractive model for inner-city entrepreneurship because they provide not only a business concept but also training and support.

Businesses can learn from the mistakes that many outside companies have made in the inner city. One error is the failure of retail and service businesses to tailor their goods and services to the local market . . .

Another common mistake is the failure to build relationships within the community and to hire locally. Hiring local residents builds loyalty from neighborhood customers, and local employees of retail and service businesses can help stores customize their products. Evidence suggests that companies that were perceived to be in touch with the community had far fewer security problems, whether or not the owners lived in the community.

[. . .]

2. *Establish business relationships with inner city companies.* By entering into joint ventures or customer–supplier relationships, outside companies will help inner-city companies by encouraging them to export and by forcing them to be competitive. In the long run, both sides will benefit . . . Such relationships, based not on charity but on mutual self-interest, are sustainable ones; every major company should develop them.

3. *Redirect corporate philanthropy from social services to business-to-business efforts.* Countless companies give many millions of dollars each year to worthy inner-city social-service agencies. But philanthropic efforts will be more effective if they also focus on building business-to-business relationships that, in the long run, will reduce the need for social services.

First, corporations could have a tremendous impact on training. The existing system for job training in the United States is ineffective. Training programs are fragmented, overhead intensive, and disconnected from the needs of industry. Many programs train people for nonexistent jobs in industries with no projected growth. Although reforming training will require the help of government, the private sector must determine how and where resources should be allocated to ensure that the specific employment needs of local and regional businesses are met. Ultimately, employers, not government, should certify all training programs based on relevant criteria and likely job availability.

Training programs led by the private sector could be built around industry clusters located in both the inner city (for example, restaurants, food service, and food processing in Boston) and the nearby regional economy (for example, financial services and health care in Boston). Industry associations and trade groups, supported by government incentives, could sponsor their own training programs in collaboration with local training institutions.

[. . .]

Second, the private sector could make an equally substantial impact by providing management assistance to inner-city companies. As with training, current programs financed or operated by the government are inadequate. Outside companies have much to offer companies in the inner city: talent, know-how, and contacts. One approach to upgrading management skills is to emphasize networking with companies in the regional economy that either are part of the same cluster (customers, suppliers, and related businesses) or have expertise in needed areas. An inner-city company could team up with a partner in the region who provides management assistance; or a consortium of companies with a required expertise, such as information technology, could provide assistance to inner-city businesses in need of upgrading their systems.

[. . .]

4. *Adopt the right model for equity capital investments.* The investment community – especially venture capitalists – must be convinced of the viability of investing in the inner city. There is a small but growing number of minority-oriented equity providers (although none specifically focus on inner cities). A successful model for inner-city investing will probably not look like the familiar venture-capital model created primarily for technology companies. Instead, it may resemble the equity funds operating in the emerging economies of Russia or Hungary – investing in such mundane but potentially profitable projects as supermarkets and laundries. Ultimately, inner-city-based businesses that follow the principles of competitive advantage will generate appropriate returns to investors – particularly if aided by appropriate incentives, such as tax exclusions for capital gains and dividends for qualifying inner-city businesses.

The new role of government

To date, government has assumed primary responsibility for bringing about the economic revitalization of the inner city. Existing programs at the federal, state, and local levels designed to create jobs and attract businesses have been piecemeal and fragmented at best. Still worse, these programs have been based on subsidies and mandates rather than on marketplace realities. Unless we find new approaches, the inner city will continue to drain our rapidly shrinking public coffers.

Undeniably, inner cities suffer from a long history of discrimination. However, the way for government to move forward is not by looking behind. Government can assume a more effective role by supporting the private sector in new economic initiatives. It must shift its focus from direct involvement and intervention to creating a favorable environment for business. This is not to say that public funds will not be necessary. But subsidies must be spent in ways that do not distort business incentives, focusing instead on providing the infrastructure to support genuinely profitable businesses. Government at all levels should focus on four goals as it takes on its new role.

1. *Direct resources to the areas of greatest economic need.* The crisis in our inner cities demands that they be first in line for government assistance. This may seem an obvious assertion. But the fact is that many programs in areas such as infrastructure, crime prevention, environmental cleanup, land development, and purchasing preference spread funds across constituencies for political reasons. For example, most transportation infrastructure spending goes to creating still more attractive suburban areas. In addition, a majority of preference-program assistance does not go to companies located in low-income neighborhoods.

[. . .]

Unfortunately, the qualifying criteria for current government assistance programs are not properly designed to channel resources where they are most needed. Preference programs support business based on the race, ethnicity, or gender of their owners rather than on economic need. In addition to directing resources away from the inner city, such race-based or gender-based distinctions reinforce inappropriate stereotypes and attitudes, breed resentment, and increase the risk that programs will be manipulated to serve unintended populations. Location in an economically distressed area and employment of a significant percentage of its residents should be the qualification for government assistance and preference programs. Shifting the focus to economic distress in this way will help enlist all segments of the private sector in the solutions to the inner city's problems.

2. *Increase the economic value of the inner city as a business location.* In order to stimulate economic development, government must recognize that it is a part of the problem. Today its priorities often run counter to business needs. Artificial and outdated government-induced costs must be stripped away in the effort to make the inner city a profitable location for business. Doing so will require rethinking policies and programs in a wide range of areas . . .

Indeed, there are numerous possibilities for reform. Imagine, for example, policy aimed at eliminating the substantial land and building cost penalties that businesses face in the inner city. Ongoing rent subsidies run the risk of attracting companies for which an inner-city location offers no other economic value. Instead, the goal should be to provide building-ready sites at market prices. A single government entity could be charged with assembling parcels of land and with subsidizing demolition, environmental cleanup, and other costs. The same entity could also streamline all aspects of building – including zoning, permitting, inspections, and other approvals.

That kind of policy would require further progress on the environmental front. A growing number of cities – including Detroit, Chicago, Indianapolis, Minneapolis, and Wichita, Kansas – have successfully developed so-called brownfield urban areas by making environmental cleanup standards more flexible depending on land use, indemnifying land owners against additional costs if contamination is found on a site after a cleanup, and using tax-increment financing to help fund cleanup and redevelopment costs.

Government entities could also develop a more strategic approach to developing transportation and communications infrastructures, which would facilitate the fluid movement of goods, employees, customers, and suppliers within and beyond the inner city. Two projects in Boston are prime examples: first, a new exit ramp connecting the inner city to the nearby Massachusetts Turnpike, which in turn connects to the surrounding region and beyond; and a direct access road to the harbor tunnel, which connects to Logan International Airport. Though inexpensive, both projects are stalled because the city does not have a clear vision of their economic importance.

3. *Deliver economic development programs and services through mainstream, private sector institutions.* There has been a tendency to rely on small community-based nonprofits, quasi-governmental organizations, and special-purpose entities, such as community development banks and specialized small-business investment corporations, to provide capital and business-related services. Social service institutions have a role, but it is not this. With few exceptions nonprofit and government organizations cannot provide the quality of training, advice, and support to substantial companies that mainstream, private sector organizations can. Compared with private sector entities such as commercial banks and venture capital companies, special-purpose institutions and non-profits are plagued by high overhead costs; they have difficulty attracting and retaining high-quality personnel, providing competitive compensation, or offering a breadth of experience in dealing with companies of scale.

[. . .]

The most important way to bring debt and equity investment to the inner city is by engaging the private sector. Resources currently going to government or quasi-public financing would be better channeled through other private financial institutions or directed at recapitalizing minority-owned banks focusing on the inner city, provided that there were matching private sector investors. Minority-owned banks that have superior knowledge of the inner city market could gain a competitive advantage by developing business-lending expertise in inner-city areas.

As in lending, the best approach to increase the supply of equity capital to the inner cities is to provide private sector incentives consistent with building economically sustainable businesses. One approach would be for both federal and state governments to eliminate the tax on capital gains and dividends from long-term equity investments in inner-city-based businesses or subsidiaries that employ a minimum percentage of inner-city residents. Such tax incentives, which are based on the premise of profit, can play a vital role in speeding up private sector investment. Private sector sources of equity will be attracted to inner-city investment only when the creation of genuinely profitable businesses is encouraged.

4. *Align incentives built into government programs with true economic performance.* Aligning incentives with business principles should be the goal of every government program. Most programs today would fail such a test. For example, preference programs in effect guarantee companies a market. Like other forms of protectionism, they dull motivation and retard cost and quality improvement. A 1988 General Accounting Office report found that within six months of graduating from the Small Business Association's purchasing preference program, 30% of the companies had gone out of business. An additional 58% of the remaining companies claimed that the withdrawal of the SBA's support had had a devastating impact on business. To align incentives with economic performance, preference programs should be rewritten to require an increasing amount of non-set-aside business over time.

Direct subsidies to businesses do not work. Instead, government funds should be used for site assembly, extra security, environmental clean-up, and other investments designed to improve the business environment. Companies then will be left to make decisions based on true profit.

The new role of community-based organizations

Recently, there has been renewed activity among community-based organizations (CBOs) to become directly involved in business development. CBOs can, and must, play an important supporting role in the process. But choosing the proper strategy is critical, and many CBOs will have to change fundamentally the way they operate. While it is difficult to make a general set of recommendations to such a diverse group of organizations, four principles should guide community-based organizations in developing their new role.

1. *Identify and build on strengths.* Like every other player, CBOs must identify their unique competitive advantages and participate in economic development based on a realistic assessment of their capabilities, resources, and limitations. Community-based organizations have played a much-needed role in developing low-income housing, social programs, and civic infrastructure. However, while there have been a few notable successes, the vast majority of businesses owned or managed by CBOs have been failures. Most CBOs lack the skills, attitudes, and incentives to advise, lend to, or operate substantial businesses. They were able to master low-income housing development, in which there were major public subsidies and a vacuum of institutional capabilities. But, when it comes to financing and assisting for-profit business development, CBOs simply can't compete with existing private sector institutions.

Moreover, CBOs naturally tend to focus on community entrepreneurship: small retail and service businesses that are often owned by neighborhood residents. The relatively limited resources of CBOs, as well as their focus on relatively small neighborhoods, is not well-suited to developing the more substantial companies that are necessary for economic vitality.

Finally, the competitive imperatives of for-profit business activity will raise inevitable conflicts for CBOs whose mission rests with the community. Turning down local residents in favor of better qualified outside entrepreneurs, supporting necessary layoffs or the dismissal of poorly performing workers, assigning prime sites for business instead of social uses, and approving large salaries to successful entrepreneurs and managers are only a handful of the necessary choices. Given these organizations' roots in meeting the social needs of neighborhoods, it will be difficult for them to put profit ahead of their traditional mission.

2. *Work to change workforce and community attitudes.* Community-based organizations have a unique advantage in their intimate knowledge of and influence within inner-city communities, and they can use that advantage to help promote business development. CBOs can help create a hospitable environment for business by working to change community and workforce attitudes and acting as a liaison with residents to quell unfounded opposition to new businesses . . .

3. *Create work-readiness and job-referral systems.* Community-based organizations can play an active role in preparing, screening, and referring employees to local businesses. A pressing need among many inner city residents is work-readiness training, which includes communication, self-development, and workplace practices. CBOs, with their intimate knowledge of the local community, are well equipped to provide this service in close collaboration with industry . . .

CBOs can also help inner-city residents by actively developing screening and referral systems. Admittedly, some inner-city-based businesses do not hire many local residents. The reasons are varied and complex but seem to revolve around a few bad experiences that owners have had with individual employees and their work attitudes, absenteeism, false injury claims, or drug use . . .

4. *Facilitate commercial site improvement and development.* Community-based organizations (especially community development corporations) can also leverage their expertise in real estate and act as a catalyst to facilitate environmental cleanup and the development of commercial and industrial property . . .

OVERCOMING IMPEDIMENTS TO PROGRESS

This economic model provides a new and comprehensive approach to reviving our nation's distressed urban communities. However, agreeing on and implementing it will not be without its challenges. The private sector, government, inner-city residents, and the public at large all hold entrenched attitudes and prejudices about the inner city and its problems. These will be slow to change. Rethinking the inner city in economic rather than social terms will be uncomfortable for many who have devoted years to social causes and who view profit and business in general with suspicion. Activists accustomed to lobbying for more government resources will find it difficult to embrace a strategy for fostering wealth creation. Elected officials used to framing urban problems in social terms will be resistant to changing legislation, redirecting resources, and taking on recalcitrant bureaucracies. Government entities may find it hard to cede power and control accumulated through past programs. Local leaders who have built social service organizations and merchants who have run mom-and-pop stores could feel threatened by the creation of new initiatives and centers of power. Local politicians schooled in old-style community organizing and confrontational politics will have to tread unfamiliar ground in facilitating co-operation between business and residents.

These changes will be difficult ones for both individuals and institutions. Nonetheless, they must be made. The private sector, government, and community-based organizations all have vital new parts to play in revitalizing the economy of the inner city. Businesspeople, entrepreneurs, and investors must assume a lead role; and community activists, social service providers, and government bureaucrats must support them. The time has come to embrace a rational economic strategy and to stem the intolerable costs of outdated approaches.

"Metropolitics"

Myron Orfield

Editors' Introduction

In this selection, newly written for this edition of *The City Reader*, University of Minnesota law professor Myron Orfield begins by describing the dramatically different characteristics of municipalities that are lumped together under the catchall term "suburb." He describes how widely their needs and ability to raise money differ. Orfield makes the case for equalizing revenue and services across jurisdictions, and suggests reforms that will promote greater fiscal equity in metropolitan regions. Underlying Orfield's analysis are creative use of statistical analysis and spatial analysis using geographical information systems (GIS) software. His recommendations are also based on an approach to thinking about the politics of metropolitan regions that he calls metropolitics. While Orfield's data and discussion are based on metropolitan regions in the United States, similar disparities among local governments exist in many other countries and his approach has nearly universal relevance.

Local government in metropolitan areas of the United States is fragmented into dozens or hundreds of separate cities and counties. Fragmentation is often extreme. For example, the Pittsburgh, Pennsylvania, metropolitan region is governed by 418 separate local governments. Usually there is one large core city in the region with many smaller suburban governments.

Orfield applied a type of multivariate statistical analysis called cluster analysis to group similar municipalities into categories based on empirical measures of their demographics, fiscal characteristics, and other factors that affect the cost of providing local services. Orfield did spatial analysis of this metropolitan data using GIS software to display similarities and differences among the different types of suburbs in map form.

Orfield distinguishes among six types of suburbs, three of which he considers to be "at risk." He named the three types of at-risk suburbs: at-risk segregated communities, at-risk older communities, and at-risk low-density communities. He named the three other types: bedroom-developing communities, affluent job centers, and very affluent job centers. Each type of suburb faces some challenges, but the nature of the challenges and their severity are quite different.

Each layer of government in the US federal system is assigned governmental functions and given access to some revenue sources. In the United States, the federal government receives most of its revenue from the federal income tax and has responsibility for national defense, diplomacy, the postal service, and other concerns of national interest. States raise revenue from sources such as state sales taxes and spend it on state-level projects like state highways. At the local level cities and counties raise most of their revenue from the local property tax – an annual excise tax based on the assessed value of land and buildings in the community. They use their local property tax receipts and other own-source revenue to provide police and fire services, streets, sanitation, parks, libraries, and other local services. Many school districts are highly dependent on property tax revenue to pay the cost of K–12 education.

Access to revenue at the local level is not equal. Nor do all citizens in a region pay proportionally or receive comparable local services. Some jurisdictions have much greater tax capacity than others. For example, some jurisdictions have a great deal of high value property that can yield a large amount of property tax revenue This means that they can provide more and better services than jurisdictions without a comparable revenue source, tax citizens less, or both. Moreover, it is often the wealthiest jurisdictions with the least need that are best off financially. A segregated at-risk community like East Palo Alto, California, with a large low-income minority population and little business or industry to tax, needs a great deal of revenue for social welfare expenditures, but has little own-source revenue to meet these needs. A neighboring very affluent job center like Mountain View, California, with many wealthy residents and the corporate headquarters of Google and dozens of high-tech companies, collects a large amount of property tax revenue, but has little need for social welfare expenditures (though, as Orfield points out, rapid growth in communities like Mountain View creates needs for expenditures to deal with growth). Moreover, there is a vicious cycle in which communities like East Palo Alto that lose out in the competition for desirable revenue sources early on are unable to compete for desirable development thereafter. Rich jurisdictions grow richer; poor ones are often locked into a cycle of decline. Fiscal zoning and tax-base competition encourage concentrations of poor families in communities that are the least able to generate the revenues they need.

While cluster analysis provides a solid, scientific basis for classifying jurisdictions, it is difficult for non-experts to grasp the statistical output. But the implications of such analysis displayed in maps are easy to interpret. The revolution in GIS and related spatial information technologies introduced in Part Three on Urban Space make it possible to produce easily understandable maps that powerfully demonstrate metropolitan disparities. Orfield's GIS maps of the percentage of elementary students eligible for free lunches by school in the Denver region and the tax capacity per household of municipalities in the central area of the New York region illustrate the power of GIS for this purpose.

Local governments have the legal authority to regulate land use within their boundaries and, as Orfield points out, often use fiscal zoning to encourage land uses they consider desirable and discourage others. For example, a jurisdiction that relies primarily on the local property tax for its revenue may encourage only high-value local land uses such as commercial, industrial, and high-end residential uses. They want new development to pay its own way by generating as much property tax revenue as the development costs in services or ideally to decrease current residents' financial burden. They want neighboring jurisdictions to house low- and moderate- income households, particularly if they have school-aged children, because these households typically require more in expenditures than they generate in revenue. They might encourage housing for affluent childless senior citizens that pays more in property taxes than it consumes in municipal expenditures, but discourage housing for families with young children who will add to the school district's expenses.

The final section of Orfield's selection states his own normative point of view, describes an innovative theory about metropolitan politics, and makes policy recommendations. Orfield considers competition for revenue sources among local governments wasteful and shortsighted. He favors reducing metropolitan fiscal inequality. He argues that a healthy society needs stable, cooperative regions and that fiscal equity will benefit society as a whole. His main concern is to reduce the burden on at-risk segregated and at-risk older communities and increase services to their residents. He feels the current fragmented and competitive governance structure of metropolitan regions is unfair and inefficient.

Orfield summarizes a theory of metropolitics that he has developed. He argues that once common interests of a group of municipalities have been determined using statistical analysis and spatial analysis using GIS, different clusters of communities should work together to change the system in ways that will benefit them.

Orfield concludes the selection by suggesting policies to promote regional equity, such as state revenue sharing programs that distribute state revenues to local governments based significantly on need and metropolitan tax base sharing programs that share tax resources within a single region. Other policies that Orfield discusses include regional review and coordination of local planning, land use reform, coordinated infrastructure planning, regional housing planning, and metropolitan governance reform.

Myron Orfield is a law professor at the University of Minnesota Law School, associate director of the University of Minnesota's Institute on Race and Poverty, and a non-resident senior fellow at the Brookings Institution in Washington, DC. He is an authority on civil rights, state and local government, state and local finance, land use, regional governance, and legislative process. Orfield served five terms in the Minnesota House of Representatives and one term in the Minnesota Senate. During his legislative career, Orfield was the architect of a series of important changes in land use, fair housing, and school and local government aid programs in Minnesota.

Orfield's theories are more fully developed in *American Metropolitics: The New Suburban Reality* (Washington, DC: Brookings Institution Press, 2002) and *Metropolitics: A Regional Agenda for Community and Stability* (Washington, DC: Brookings Institution Press, 1999).

Other books exploring metropolitan issues are Peter Dreier, John Mollenkopf, and Todd Swanstrom, *Place Matters: Metropolitics for the Twenty-First Century*, 2nd revised edn (Kansas City, KS: University Press of Kansas Press, 2005), Robert Lang, *Edgeless Cities: Exploring the Elusive Metropolis* (Washington, DC: Brookings Institution Press, 2002), and Peter Calthorpe, *The Next American Metropolis: Ecology, Community, and the American Dream*, 3rd edn (New York: Princeton Architectural Press, 1997).

■

In the months leading up to national elections, the eyes of political pollsters and pundits are typically on the United States' suburbs. The notion that the suburbs are where elections are won or lost has become an unassailable *idée fixe* in contemporary politics. While there is certainly some truth to this premise, it obscures the more complex reality that "the suburbs" are in fact a remarkably diverse collection of communities with a broad range of differing strengths and weaknesses.

A close look at the United States' twenty-five largest metropolitan areas shows that far from being a monolith, the suburbs actually comprise several distinct types. Some inner-ring suburban communities suffer from the same urban ills that afflict inner cities, such as poverty and racial segregation. Many developing suburbs on the fringes of metropolitan areas are experiencing explosive population growth but have limited resources to pay for the schools, sewers, and roads that this growth requires. Still others enjoy the tax benefits of large concentrations of office space and high-end housing, but are plagued by traffic congestion and degradation of the open space that made them attractive places to live in the first place.

The prevailing catch-as-catch-can pattern of metropolitan development, which encourages wasteful intra-regional competition and environmentally damaging land use, hurts all types of suburbs. Socioeconomic segregation, fiscal inequality, and sprawl plague virtually every metropolitan area, and appear to be growing worse in most of them. At least 40 percent of the metropolitan population resides in suburbs with social or fiscal challenges severe enough to be considered "at risk" in our classification. Another 25 percent lives in rapidly developing communities that are struggling to keep up with their explosive growth with limited financial resources.

Regional government reform is needed to stem this tide. Though the obstacles are formidable, there is reason for optimism. Every type of metropolitan community – from central cities wrestling with poverty and other social ills to the affluent outer-ring suburbs beset by traffic congestion and runaway development – stands to benefit from these reforms. Political parties and leaders who can persuade metropolitan voters to act in their long-term self-interest on these issues will be rewarded with far greater gains than those chasing the vagaries of shifting polls

THE NEW SUBURBAN REALITY

In the inner-ring Chicago suburb of Cicero, where a visit by Martin Luther King once precipitated a violent protest against housing integration, nonwhite students are now in the majority. In the mid-1990s in Cherokee County, an Atlanta suburb comprised largely of bedroom-developing communities, students often attended schools set up in trailers as their communities had neither the tax base nor other resources to build new schools for a growing population. At the same time, schools were closing for lack of students in the region's core. Lopatcong Township, New Jersey, an area at the fringes of the New York region making the transition from rural to suburban, is defending its 2003 ordinance to limit multifamily dwellings to two bedrooms, effectively zoning out families with children in order to keep school enrollment (and costs) down. The proliferation of large-lot housing developments in suburban Macomb County, Michigan, has contaminated a nearby lake due to a rash of failed septic systems, which will cost between $2 billion and $4 billion to convert to sewer.

These examples reflect the fragmentation that lies at the heart of the United States' new suburban reality. If the suburbs were ever a homogeneous bastion of untroubled prosperity, they certainly are no longer. Evidence for this goes well beyond the anecdotal. An analysis of the twenty-five largest metropolitan areas demonstrates that varying social and economic pressures have led to the emergence of distinct types of suburban communities that differ from one another in identifiable ways.

A method known as cluster analysis was used to group suburban areas according to several measures of their fiscal characteristics (specifically, their ability to raise tax revenue and the change over time in that ability) as well as key factors that directly or indirectly affect the cost of providing local services (including poverty levels, population density and growth, age of housing, and racial composition). The cluster analysis identified six types of communities, three of which face economic or social challenges severe enough to be considered "at risk."

The health of any community is largely a function of whether it has adequate resources to meet its particular needs. Two important factors used in the cluster analysis are school populations, which affect the "needs" side of the ledger, and tax capacity, on the "resources" side. Schools are a powerful indicator of a community's current health and of its future well-being. As the number of poor children in a community's schools grows, middle-class families' demand for housing in the community softens, and housing prices reflect this decline. Families with school-age children are likely to leave first because changes in the schools affect them most. Some non-poor families may choose to stay in the community but put their children in private schools, though few households can afford the additional expense for long. A community with schools in transition may also draw "empty-nesters" and other non-poor households without school-age children. Poverty rates among school-age children therefore tend to rise more quickly than the overall poverty rate.

Although poverty and its consequences underlie economic segregation, it is difficult to separate poverty from race and ethnicity, particularly for African Americans and Latinos, who are strongly discriminated against in the housing market. Sadly, an analysis of racial data for elementary school students in the twenty-five largest metropolitan areas shows that once the minority share in a community's schools increases to a threshold level (10–20 percent), racial transition accelerates until minority percentages reach very high levels (greater than 80 percent).

While trends in a community's school population indicate critical local needs, local tax capacity is a good measure of the ability to raise revenues to meet those needs. Communities with copious tax resources have low tax rates and great services. Resource-poor communities have just the opposite. Why is this? Think of it this way: if a community's tax wealth per household is $100, a 10 percent tax rate raises $10 per household for services; if tax wealth is $1,000 per household, the same rate raises $100. No matter how

BOX 1 THE TRUTH ABOUT WHITE FLIGHT

The close relationship between racially segregated communities and areas of concentrated poverty has been used to support flawed conclusions about African Americans and Latinos. Some people, associating an influx of minorities into a community with social and economic decline, conclude that minority residents somehow contribute less than whites to a community's health and stability. Nowhere was this tragic misconception better illustrated than in a segment from the television news magazine NBC Dateline about the white-collar Chicago suburb of Matteson, Illinois, 20 miles south of the Loop. In the early to mid-1990s, black middle-class families began to move to Matteson, a community of large, attractive suburban homes, open space, and good schools. These blacks were, by most important demographic measures, at least the socioeconomic equals of Matteson's white residents. Some were, in fact, better off than Matteson's whites. But as soon as black households became a significant percentage of the population, there was a sudden sell-off of homes by white residents. Asked why they were moving, the white sellers replied, "Because crime is increasing." On the evidence, neither claim was true. School test scores and the crime rate remained unchanged. However, once the white residents left, demand for middle-class housing in Matteson cooled, because the black middle class was not large enough to sustain market demand. Not only did the schools become more segregated, but also they became much poorer. This is why "white flight" invariably means poverty: this tragic sequence of events has played itself out in countless suburbs across the United States.

smart administrators are, and no matter how much reorganization they do, they cannot avoid this basic truth.

One of the three at-risk suburban types identified by the cluster analysis is comprised of aging communities that have very low tax capacity, high municipal costs, and – most distinctively – high concentrations of minority children in the public schools. As a group, these *at-risk segregated communities* had per-household tax capacities that were less than two-thirds of the metropolitan area average, and the slowest growth in tax capacity of all the suburban types. On the cost side, this group had very high poverty rates (nearly twice the regional average), lower-than-average population growth, aging housing stock, a population density almost four times the regional average, and a higher percentage of minority children in the public schools than even the central cities.

The at-risk segregated communities are some of metropolitan America's worst places to live. Poor and segregated, they have a fraction of the resources of the central cities they surround. In 1994, the taxes on a $100,000 house in the at-risk segregated suburb of Maywood, Illinois, were $4,672. This level of taxation would support local school spending of $3,350 per pupil. In Kenilworth, an affluent suburb to the north, the taxes would be $2,688, yet this lower rate, applied to the whole tax base, would support almost three times the level of spending per pupil. Similarly, business taxes on a 100,000-square-foot office building in booming DuPage County were $212,639, compared with $468,000 in south suburban Cook County.

A second category of at-risk communities – made up mostly of inner-ring suburbs and outlying cities that have been swallowed up by metropolitan growth – has older housing stock than any of the other suburban groups. Like the at-risk segregated communities, these *at-risk older communities* have relatively low tax capacity and tax-capacity growth, and even higher density, but they also have relatively low levels of poverty and of minority children in public schools.

These places often stand cheek by jowl with the at-risk segregated suburbs, and there is often a strongly defended racial line between them. In fact, though, the at-risk segregated and older communities have many common concerns. Both groups have slow (or even negative) population growth, relatively meager local resources, and struggling commercial districts. Their main street corridors and commercial districts cannot attract new, big businesses that could easily build on greenfield sites. Despite these commonalities, segregated and older at-risk suburbs have not formed a cohesive political whole, probably because they are often divided on the issue of race (see Table 1).

Many communities included in the third at-risk group are exurbs on the fringes of the metropolitan areas that are making the transition from rural to suburban. These *at-risk low-density communities* share the characteristics of low tax capacity and low-tax-capacity growth with the other at-risk suburbs, but they differ in other important ways. Many are just beginning the transition from rural or farm land to suburban development patterns. Their relatively low fiscal resources are thus stretched thin by demands for new infrastructure and the other accoutrements of growth. Compared to most other suburban areas, they must also cope with significantly higher-than-average poverty.

The fourth suburban type represents what many would regard as the quintessential suburb. *Bedroom-developing communities* have rapidly growing populations that tend to be white and relatively affluent. Density is low, housing is new, and tax capacity is just below average but growing at an average rate. Although this group contained about a quarter of the population of the metropolitan areas studied, it had nearly 60 percent of the population growth in those areas. Though not experiencing the social stress of some of the at-risk communities, bedroom-developing suburbs must manage the costs of a high rate of population growth with only average (or below average) local resources (see Tables 2 and 3).

Both the at-risk low-density and the bedroom-developing communities share fiscal pressures arising from school and infrastructure finance. In all the large metropolitan areas, the student-to-household ratio in these two types of communities is much higher than the regional average. Because of this ratio and their (at best) average tax base, these suburbs often have the lowest per-pupil spending in metropolitan United States. Developmental infrastructure such as roads and sewers can also present large challenges for the at-risk and bedroom-developing suburbs.

The last two classifications include many of the so-called "edge cities": suburban communities with vast amounts of open space and more jobs than bedrooms. *Affluent job centers* (and the even more prosperous *very affluent job centers*) reap the benefits of extraordinary tax bases – capacities of more than two and five times the regional averages respectively – that are growing

Municipality type	Number of municipalities	Tax capacity	Change in tax capacity	Free lunch eligible	Density	Population growth	Age of housing	Minority percentage
At-risk, segregated	348	66%	93%	175%	369%	97%	108%	209%
At-risk, older communities	391	74	96	59	735	98	110	35
At-risk, low-density	1,104	66	96	103	104	102	97	65
Bedroom-developing	2,152	90	100	32	83	106	85	16
Affluent job centers	625	212	105	27	97	105	88	26
Very affluent job centers	91	525	102	39	46	101	91	38
Central cities	30	101	97	193	452	94	125	207
All suburban	4,711	106	99	61	164	104	92	45

Table 1 Characteristics of the community types

Sources: National Center for Education Statistics, US Bureau of the Census, and various state and local government agencies (fiscal data).

All variables except "Number of municipalities" are expressed as percentages of metropolitan area averages. "Population growth" and "Change in tax capacity" were calculated as the ratio of 1998 levels to 1993 levels.

	Dissimilarity indexes		
	1992	1997	% Change
Atlanta	50	52	4
Boston	n.a.	55	n.a.
Chicago	94	95	1
Cincinnati	59	57	−3
Cleveland	62	64	3
Dallas/Ft Worth	51	51	0
DC/Baltimore	53	51	−4
Denver	48	55	15
Detroit	60	60	0
Houston	39	39	0
Kansas City	54	53	−2
Los Angeles	54	57	6
Miami	49	50	2
Milwaukee	66	63	−5
Mpls/St Paul	42	48	14
NY/Newark	n.a.	66	n.a.
Philadelphia	n.a.	51	n.a.
Phoenix	n.a.	n.a.	n.a.
Pittsburgh	43	39	−9
Portland	36	50	39
St Louis	46	60	30
San Diego	51	51	0
SF/Oakland	48	53	10
Seattle	34	38	12
Tampa	32	36	13
25 metropolitan area average	51	54	6

Table 2 Segregation by income in elementary schools dissimilarity indexes for 1992 and 1997

Source: National Center for Education Statistics.

at rates outstripping regional averages. Collectively, they have more than four times the office space per household of any other group of suburbs, more even than central cities. At the same time, cost factors such as poverty and age of housing are well below regional averages. As might be expected, the political and business leaders in these communities work hard to maintain their quality of life, and, of types of suburbs, they are the ones that have revolted most successfully against growth and sprawl.

These places might seem to have it all: affluent residents, a high tax base, an average number of children, and very low poverty. However, the mass of jobs in these two types of communities also has its downside. First, because many workers cannot afford the local housing, these beehives of local activity generally have intense traffic congestion. Second, because land becomes so valuable, it is often difficult to maintain open space.

Well over half (over 56 percent) of the suburban population of the twenty-five largest metropolitan areas lived in at-risk communities. Yet they controlled only 38 percent of local tax capacity in the suburbs. Conversely, the two clusters of affluent job centers accounted for less than 10 percent of the suburban population, but had 22 percent of the local tax capacity. Poverty levels and other cost factors diverge in equally dramatic fashion. These disparities point to a widening gulf between "have" and "have not" suburbs.

In fact, quantitative analyses show that both economic and racial segregation in US schools rose during the 1990s. Dissimilarity indexes (general measures of the degree of segregation) show that metropolitan areas with increased economic and racial segregation in elementary schools between 1992 and 1997 outnumbered metro areas with reduced segregation during those same years.

Tax-base inequality also increased during the 1990s. A general measure of inequality in tax bases known as the Gini coefficient indicated an average increase of about 8 percent in the twenty-five largest metropolitan areas between 1993 and 1998, with eighteen of the metro areas showing increases in inequality.

Comparing the Gini coefficients for the twenty-five largest US metropolitan areas in 1998 to the economic and racial dissimilarity indexes for the same cities in 1997 shows just how closely tax-base inequality in a metropolitan area correlates with income and racial segregation. Seven of the ten metropolitan areas with the most unequal tax-base distributions are also among the ten areas with the greatest degree of income segregation in schools and nine of the ten are among the ten areas showing the greatest degrees of racial segregation in schools (see Table 4).

Urban sprawl indicators also correlate strongly with measures of segregation and inequality. Regions where population density in the urbanized areas declined the most tend to show the greatest degrees of racial segregation and tax-base inequality. Comparing the sprawl data with tax capacity data for different types of communities shows that sprawl affects the fiscal health of sprawling communities. The average tax capacity

	Elementary schools 1992	Elementary schools 1997	Elementary schools % change	Metropolitan population 1990	Metropolitan population 2000	Metropolitan population % change
Atlanta	66	67	2	69	66	–4
Boston	67	66	–1	71	66	–7
Chicago	76	75	–1	85	81	–5
Cincinnati	76	77	1	77	75	–3
Cleveland	76	76	0	83	77	–7
Dallas/Ft Worth	58	58	0	64	59	–8
DC/Baltimore	65	65	0	66	57	–14
Denver	53	55	4	65	62	–5
Detroit	81	82	1	88	85	–3
Houston	46	45	–2	68	68	0
Kansas City	67	70	4	73	69	–5
Los Angeles	56	57	2	74	68	–8
Miami	60	60	0	73	74	1
Milwaukee	65	69	6	83	82	–1
Mpls/St Paul	54	53	–2	64	58	–9
NY/Newark	72	71	–1	82	82	0
Philadelphia	66	67	2	77	72	–6
Phoenix	53	56	6	52	44	–15
Pittsburgh	70	69	–1	71	67	–6
Portland	42	40	–5	66	48	–27
St Louis	66	69	5	79	74	–6
San Diego	44	46	5	59	54	–8
SF/Oakland	45	48	7	65	61	–6
Seattle	40	39	–3	58	50	–14
Tampa	37	35	–5	71	64	–10
25 metropolitan area average	60	61	1	71	67	–7

Table 3 Racial segregation in metropolitan populations and elementary schools dissimilarity indexes in selected years

Source: National Center for Education Statistics and 2000 Census of Population data compiled by the Mumford Center, State University of New York at Albany.

for at-risk, low-density suburbs in the twelve metropolitan areas with the greatest degrees of sprawl is 60 percent of the regional average; in the thirteen metro areas with the least sprawl, the average capacity is 78 percent of the regional average. Likewise, the capacities for bedroom-developing suburbs are 82 percent of the regional average in sprawling metro areas, and 101 percent of the average in more contained areas. Clearly, the suburban areas most directly affected by sprawl are fiscally stronger relative to the rest of their metropolitan areas in regions where growth is managed more effectively (see Maps 1, 2 and 3).

Metropolitan area	1993 Gini coefficient	1998 Gini coefficient	1993–1998 Change in Gini coefficient
Atlanta	0.16	0.17	2 %
Boston	0.21	0.25	16
Chicago	0.26	0.27	2
Cincinnati	0.31	0.36	15
Cleveland	0.21	0.24	14
Dallas/Ft Worth	0.17	0.19	10
DC	0.25	0.22	–12
Denver	0.20	0.21	8
Detroit	0.23	0.21	–5
Houston	0.13	0.15	15
Kansas City	0.32	0.25	–22
Los Angeles	0.20	0.22	9
Miami	0.19	0.21	10
Milwaukee	0.25	0.27	6
Mpls St Paul	0.18	0.17	–1
New York	0.24	0.23	–5
Philadelphia	0.28	0.33	20
Phoenix	0.11	0.15	38
Pittsburgh	0.26	0.26	2
Portland	0.11	0.15	30
St Louis	0.32	0.37	15
San Diego	0.10	0.11	1
San Francisco	0.15	0.17	15
Seattle	0.11	0.21	99
Tampa	0.13	0.13	2
25 metropolitan area average	0.20	0.22	8

Table 4 1993 and 1998 Gini coefficients, tax capacity per household

Sources: Various state and local government agencies

THE ROAD TO REFORM

The many challenges facing the United States' metropolitan areas can be attacked effectively only through a coordinated, regional approach. Concentrated poverty and community disinvestments, among the most important of the countless factors feeding metropolitan sprawl, are related to incentives built into public policies for metropolitan development. These incentives include tax policies that promote wasteful competition among local governments, transportation and infrastructure investment patterns that subsidize sprawling development, and fragmented governance that makes thoughtful and efficient land-use planning more difficult.

Fortunately, the foundations for positive change are, to a large extent, already in place. Regional tax reform, which involves a more equitable fiscal relationship among the cities in a metropolitan area, has its roots in the state school-aid systems that exist in virtually every state in the country. Land-use reform to combat sprawl is a growing issue in the nation, and sixteen

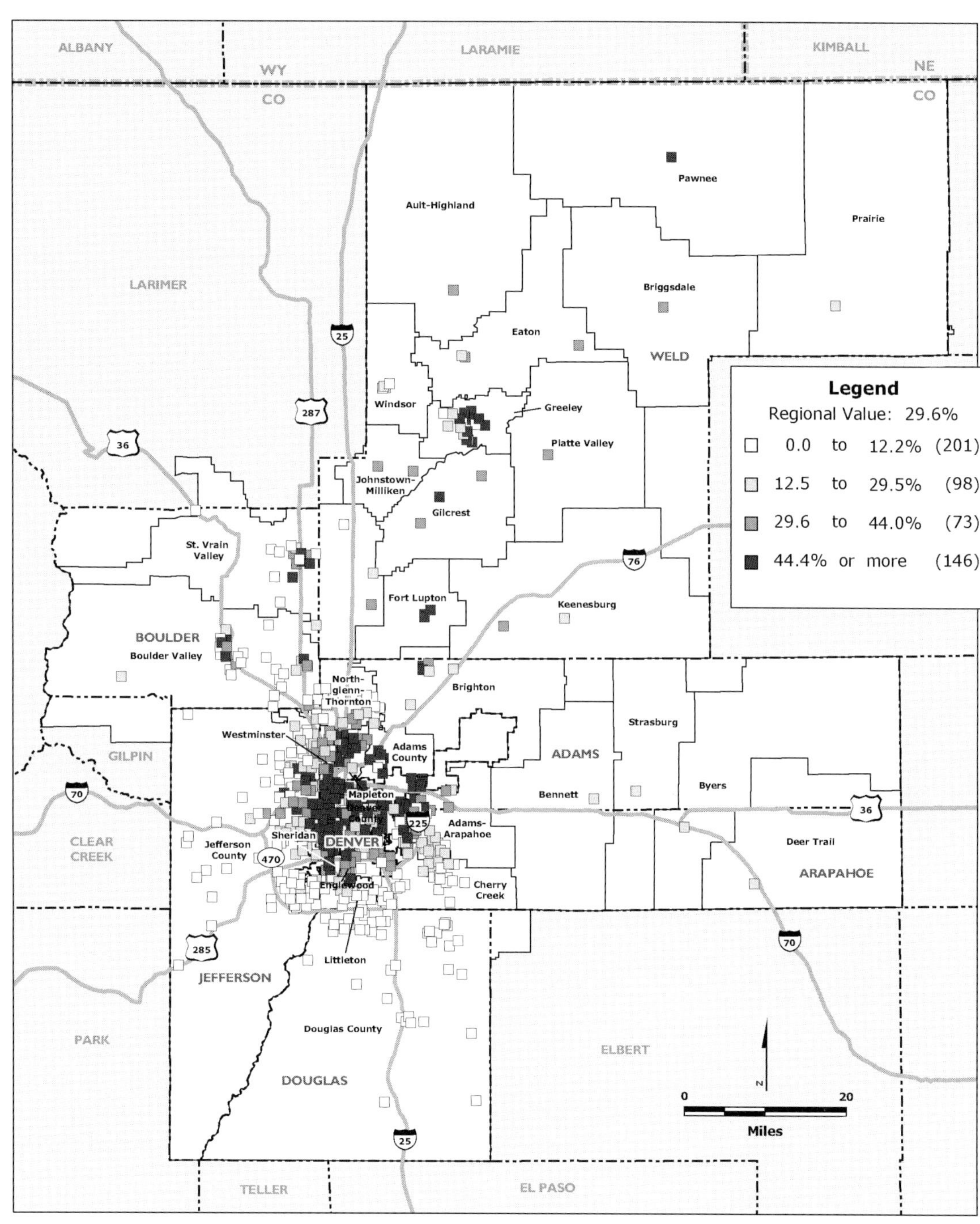

Map 1 Denver region: Percentage of students eligible for free or reduced lunch by elementary school, 2001

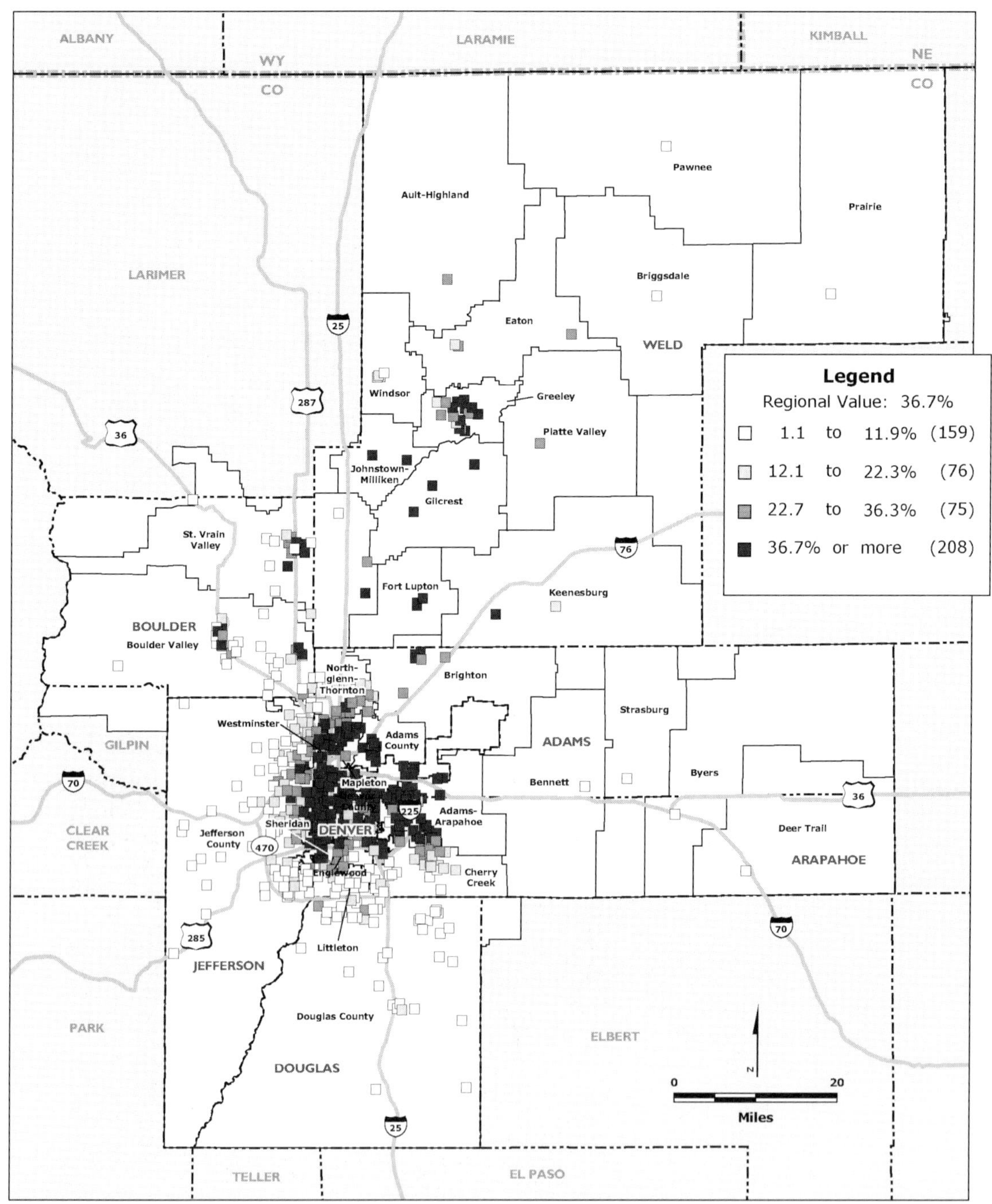

Map 2 Denver region: Percentage of non-Asian minority elementary students by school, 2001

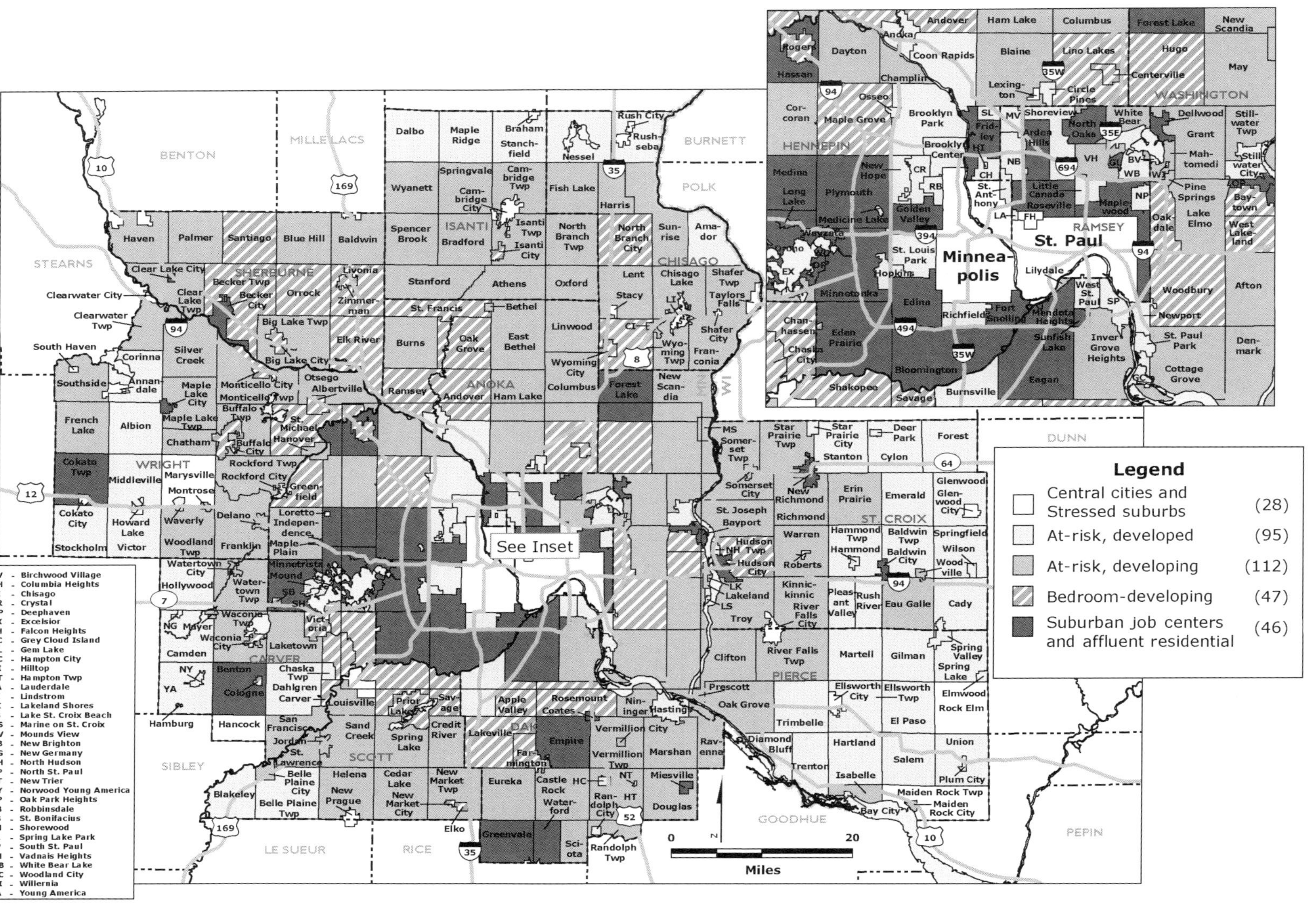

Map 3 Twin cities region: community classification

states have already adopted comprehensive growth management acts. Federal law has required that regional governments coordinate hundreds of millions of transportation dollars in every region in the country the challenge now is to make these existing regional governments more effective and more accountable to the people they serve.

Tax reform

Under the fiscal system that currently holds sway in most regions of the country, local governments have strong incentives to adopt policies and regulations designed to serve their own short-term economic interest at the expense of their own long-term health and the well-being of the region as a whole.

One way that local governments do this is through "fiscal zoning," a deliberate attempt by a government to reap fiscal dividends from new development by limiting the types of land uses within its jurisdiction. Because property taxes are the most significant form of revenue for most local governments, they have a direct incentive to tailor their land-use regulations to encourage development of high-value commercial, industrial, and residential properties that generate relatively little in public costs, and to discourage development of lower-value properties such as affordable housing that create a need for higher public expenditures.

When played out over an entire metropolitan area, this fiscal zoning process can significantly influence where people can afford to live, the types and quality of public services they receive from their local government, and the presence or absence of employment opportunities near their homes.

Another aspect of local governments' short-sighted pursuit of positive fiscal dividends is the wasteful and biased competition for desirable commercial and industrial properties. It is wasteful because one community's gain is likely to be another community's loss. The resources expended in such competition typically do not enhance the overall regional economy, but only shuffle activity from one place to another. It is biased because it creates the potential for a vicious, self-reinforcing cycle of decline in places that "lose" early in the game. As a locality loses activities that generate positive fiscal dividends, it must either raise taxes on its remaining tax base to maintain services at existing levels or reduce services at existing tax rates. Either choice further reduces the locality's ability to compete for additions to its tax base or to keep its existing base.

Fiscal zoning and tax-base competition tend to concentrate families and individuals with the greatest need for public services in communities that are the least able to generate the revenue to provide those services. Conversely, those who can afford to live where they choose (and therefore are less in need of public services) are increasingly concentrated in communities that have managed to successfully attract the development of large, expensive homes and other revenue-generating land uses. The result is a widening gap between communities with low tax capacities and high costs, on the one hand, and those with high tax capacities and low costs on the other.

The arguments for tax reform are primarily efficiency arguments. Attenuating the link between growth in particular types of local land uses and the tax base available to produce local services reduces wasteful competition. Providing financial incentives for particular types of development that provide regional benefits but do not generate local fiscal dividends can improve the functioning of regional housing and labor markets

An essential part of creating a stable, cooperative region is to gradually equalize the resources of local governments with land-use planning powers. In addition to improving equity, which will allow central cities, at-risk suburbs, and many bedroom-developing suburbs to lower taxes and improve services, it will reduce the competition between places, give communities real fiscal incentives to cooperate, and make regional land-use planning easier to achieve.

Many states attempt to reduce fiscal inequity among jurisdictions through revenue-sharing programs that distribute a portion of the revenue from one or more state taxes to local governments through a variety of formulas. Although most revenue-sharing programs began with a simple per-capita approach, they now generally place greater emphasis on the communities' needs, typically determined by characteristics such as tax base, revenues, spending, or some combination of the three. Equity measures improve for all but two of the twenty-five largest metropolitan areas when aid is added to local tax capacity. However, the effects of aid vary considerably, ranging from a 63 percent change for the better in the inequality measure to an 11 percent change for the worse.

Tax-base sharing, an alternative way to reduce tax-base inequities, has several advantages over the patchwork quilt of aid programs common to most states. Unlike separate programs that distribute state revenues to counties, cities, townships, and special districts, tax-base sharing simply redistributes the common base from which each local jurisdiction derives its revenues. It also helps to equalize the resources available to local governments without removing local control over tax rates. Further, by requiring local governments to relinquish some of their fiscal dividend from new commercial/industrial development, tax-base sharing reduces the incentive to waste taxpayer dollars by stealing it away from other communities. Similarly, including residential property in tax-base sharing dilutes local governments' incentives to use fiscal zoning or its substitutes to restrict residential development to "profitable" types of housing, making cooperative, efficient land-use planning easier.

With tax-base sharing, a portion of each locality's tax base (or growth in tax base) is contributed to a regional pool and redistributed according to criteria such as tax capacity, service cost or need indicators, or land-use decisions. In Minneapolis-St Paul, the only metropolitan area for which tax-base sharing legislation has actually been enacted, local tax-base disparities were reduced by roughly 20 percent by the program in the year 2000. Simulations for other metropolitan areas show that tax-base sharing is a much more cost-effective means of reducing tax-base equity than existing aid programs. Tax-base sharing reduces disparities by two percentage points for each percentage point of shared revenues, while current aid programs reduce disparities by just half of a percentage point for each percentage point of aid.

Reforms in these policy areas need not be radical. All states provide at least some financial support to local governments. A reform agenda can begin with incremental improvements in the way current aid is allocated. Tax-base-sharing programs can be designed to capture a portion of tax-base growth, as occurred in the Twin Cities, rather than part of existing tax bases, allowing regions to reap the efficiency benefits immediately while the redistributive impacts grow more slowly.

Land-use reform

Individual communities can do little to deal with the underlying regional forces contributing to sprawling development patterns. While local development moratoriums, slowdowns, or other local restrictions may seem like a good strategy for reducing the negative impacts of increased development, ultimately they only throw development farther out to surrounding communities eager to attract additional development to add to their tax base and help them keep up with the costs of their residential growth. In many cases, these surrounding communities are at-risk low-density and bedroom-developing communities trying to keep up with their growing costs.

A number of states have tried to tackle the difficulties associated with purely local land-use planning through some form of statewide planning. At present, sixteen states have a land-use planning system in place; ten of these states actually require comprehensive local planning, while the other six encourage it. Oregon led the way with the passage of its Land Use Act in 1973. This landmark legislation requires each of the state's cities and counties to adopt a long-range, comprehensive plan for development consistent with the state's specified planning goals.

Another popular strategy employed by states to combat sprawling development has been to authorize and encourage the use of various "smart growth" tools. Common growth-management tools include the urban growth boundary, which prevents or limits development outside a designated area; the urban

BOX 2 HOW NOT TO CURB GROWTH

Efforts by individual communities to unilaterally curb development within their boundaries often end up contributing to sprawl instead of reducing it. In 1972, the San Francisco region city of Petaluma decided to slow growth by limiting the number of building permits issued annually. This caused a dramatic increase in housing demand in farther-out Santa Rosa. According to US Census figures, the population of the Santa Rosa area nearly doubled between 1970 and 1980. In the end, Santa Rosa had to build new roads and sewers, and residents of Petaluma were forced to deal with the increased traffic through their community.

service area, which limits provision of public services such as sewerage and water to a designated area; designated areas where growth will be focused; and concurrency, which requires adequate public infrastructure to be in place before or at the same time as development occurs. These can be effective tools. Misused or used in isolation without complementary policies in the non-developing portions of regions, however, they can contribute to low-density, dispersed development instead of preventing it.

Smart-growth planning also attempts to protect agricultural lands and open space from development, maintaining the amenity value of such areas and preserving them for future generations. To this end, many states and regions create agricultural district programs, purchase agriculture conservation easements or development rights through state land trust funds, and allow the transfer of development rights from a rural to an urban location. These land-preservation tools, though well intentioned, are extremely costly and cannot on their own truly change the nature of US development patterns.

Effective regional land-use reform hinges on three elements: coordinated infrastructure planning, a regional housing plan, and regional review and coordination of local planning.

Coordinated infrastucture planning

Piecemeal provision of the basic infrastructure that guides regional investment and development patterns is a major contributor to inefficient, sprawling development, congested roadways, and environmental strains. Regionalizing infrastructure provision and planning helps guide development in more efficient and equitable ways. It can, for instance, help reduce per capita costs throughout the region by creating an orderly pattern of development. Transportation investments are an especially important part of regional infrastructure that should be coordinated with other investments, and giving a regional agency authority over transportation investments is one way to help achieve this goal.

Regional housing plan

A regional strategy to reduce zoning, financial, and other barriers to the development of affordable housing is the logical first step toward the goal of mixed-income housing in every community within a region. The housing industry has long argued that regulatory barriers such as large lot sizes, prohibitions on multi-family housing, and assorted fees hurt the natural marketplace for affordable housing. Removing such barriers is a step that the building community can accept, and is a way to develop a relationship with an important private-sector actor in land development. Fair-share requirements ensure that all places contribute to the regionwide supply of affordable housing. These programs allocate to each city a part of the region's affordable housing, on the basis of the jurisdiction's population, previous efforts to create affordable housing, and job availability. An effective fair-share housing program seeks a sustainable balance of lower-cost and more expensive housing in all areas of the region, whether they are greenfield suburban sites or gentrifying neighborhoods

Regional review and coordination of local planning

Because much land-use and infrastructure planning is best provided at the local level, regional land-use reform requires a coordinated framework in which local governments develop comprehensive land-use plans that are consistent with state or regional planning goals. Ideally, these goals are clearly laid out and applicable to all communities within the region, and any local plans and policies inconsistent with these goals may be challenged in court or in special forums created for such adjudication. There should be strong penalties for noncompliance, such as financial sanctions or the loss of authority to make land-use decisions and to grant building permits.

Metropolitan governance reform

The fragmentation of metropolitan areas into many local governments is not only a barrier to effective growth management, but also a leading cause of racial and economic segregation, sprawl, and fiscal disparities within those areas. In regions without a shared tax base or dominant central city, competing jurisdictions often duplicate infrastructure and services that could be provided more cost effectively in older suburbs and central cities. Duplication of services and infrastructure in turn contributes to fiscal, social, and environmental stresses in the at-risk communities at the core of metropolitan regions as well as in those at the edge. Zoning incentives to attract high-value residential and commercial development result in exclusive neighborhoods, segregated by race and income. Meanwhile, the new office and commercial centers in suburban edge cities siphon customers and resources from established business districts and allow the commuter zone to expand, further inducing sprawl.

Recognizing fragmentation's negative effects, a number of regions have acted to bring a greater regional focus to local governance. Metropolitan planning organizations (MPOs) are the most widespread form of regional governance in the United States today. MPOs were created by Congress in the 1970s to address the growing transportation challenges in metropolitan regions. Given broad powers to guide regional growth through long-range transportation planning and the allocation of federal transportation funds to individual jurisdictions, the MPOs in the United States' twenty-five largest metropolitan areas are, in a very real sense, special-purpose regional governing bodies.

However, MPOs are not directly accountable to voters and do not always make their transportation investments with social separation, sprawl, and fiscal inequities in mind. Without broader authority and a mandate to address these assorted issues comprehensively, MPOs are limited in what they can accomplish on regional concerns.

Several regional councils and associations designated as MPOs have, either by state mandate or through their own initiative, taken on myriad other functions, attempting to fill the void in regional governance created by political fragmentation. Some of the most common duties taken on by MPO staff include air quality conformity planning, local and regional economic development initiatives, land-use plan review and coordination, ride-share services, and regional demographic and economic forecasting.

A strong, accountable regional governing body is an essential part of a comprehensive regional reform plan. The following strategies will help to ensure the long-term viability of any regional governing body, whether an MPO with expanded authority or some other regional body.

- **Strategy 1:** Apportion voting membership by population. Decisions on how and where to spend taxpayer dollars for regional investments should be made in a fair and equitable manner, giving equal representation to all types of communities and residents in a region.
- **Strategy 2:** Hold direct elections for voting members. Direct elections of members of regional governing bodies would make regional decision-making more open and participatory. Even without expanding the current scope of MPO powers, direct election of MPO boards would create a legitimate forum for the discussion of regional issues. Any increase in MPOs' powers would make direct election even more important.
- **Strategy 3:** Broaden and deepen public awareness of how transportation investments contribute to or alleviate social separation and sprawl.

BOX 3 REGIONALISM AT WORK

Two regions – Portland, Oregon, and Minneapolis-St Paul – have vested significant and comprehensive planning powers in a single regional government body. Portland Metro controls development patterns through its administration of the state-mandated regional urban growth boundary. The Twin Cities Metropolitan Council regulates the expansion of its Metropolitan Urban Service Area through its authority to plan for and permit extensions to the regional sewer system.

These formal powers, complemented by council members' accountability to the governor in Minnesota and directly to the voters in Portland, give these regional governments political leverage that other metropolitan planning organizations and regional councils lack. Unlike most MPOs, members of the Portland and Twin Cities councils are unaffiliated with local governments and state agencies. This detachment from parochial interests gives Metro and the Met Council unique freedom to focus exclusively on regional needs and concerns.

Regional bodies should be required to evaluate their transportation decisions to determine whether they worsen or alleviate social separation and sprawling development patterns in the region.

Strategy 4: Broaden the scope of land-use planning. MPOs or another regional body should develop an advisory land-use plan for the region that embodies a vision for efficiently coordinating all major forms of developmental infrastructure. These advisory land-use plans might offer cities incentives to submit for review comprehensive plans covering such issues as sustainable development, affordable housing, and public transit (see Table 5).

Metropolitan area	Tax capacity Gini coefficient	Gini coefficient after tax base sharing	% change	Gini coefficient after aid	% change
Atlanta	0.17	0.13	−21	0.17	3
Boston	0.25	0.20	−20	0.19	−22
Chicago	0.27	0.22	−20	0.17	−36
Cincinnati	0.36	0.29	−20	0.35	−2
Cleveland	0.24	0.20	−19	0.22	−9
Dallas/Ft Worth	0.19	0.15	−21	n.a	n.a
DC	0.22	0.18	−21	0.17	−24
Denver	0.21	0.17	−19	0.20	−7
Detroit	0.21	0.17	−21	0.24	11
Houston	0.15	0.12	−22	n.a	n.a
Kansas City	0.25	0.20	−21	0.22	−11
Los Angeles	0.22	0.18	−19	0.15	−33
Miami	0.21	0.17	−18	0.17	−18
Milwaukee	0.27	0.22	−18	0.10	−63
Mpls St Paul	0.17	n.a	n.a	0.17	−3
New York	0.23	0.18	−22	0.18	−22
Philadelphia	0.33	0.28	−16	0.26	−21
Phoenix	0.15	0.12	−21	0.09	−41
Pittsburgh	0.26	0.21	−19	0.25	−4
Portland	0.15	0.12	−18	0.13	−12
St Louis	0.37	0.29	−20	0.24	−36
San Diego	0.11	0.08	−20	0.08	−20
San Francisco	0.17	0.14	−20	0.13	−27
Seattle	0.21	0.17	−21	0.20	−7
Tampa	0.13	0.11	−19	0.12	−14
25 Metropolitan Area Average	0.22	0.178	−20	0.182	−17

Table 5 Revenue capacity equity before and after aid from state governments and tax base sharing

Making the case for regional reform

Economists and others have made the important point that regional cooperation helps every community, but the parochial costs and benefits of regional reforms vary by community type within metropolitan areas. Therefore, making the case for regionalism requires an understanding of the nature of the different suburban community types and the ways they may benefit from the various reforms.

The at-risk developed suburbs

The case for regional reform to present to the at-risk segregated and at-risk older suburbs is simple. Regional equity gives them lower taxes and better services. In the at-risk developed suburbs, taxes are comparatively high for the mix of services provided. In states and regions without substantial state-supported school equity, these taxes can be the highest in metropolitan United States. Simulations of property tax-sharing throughout the country show the older suburbs as the largest net gainers of resources of any of the subregions. New equity resources could help older suburbs shore up and improve aging infrastructure, clean up brownfield sites, reconfigure abandoned malls or industrial facilities, invest in housing in declining neighborhoods, and give underfunded schools a boost. If the equity is sufficiently comprehensive, such measures could be taken even as the local tax rates were being reduced.

The residential resources of at-risk developed suburbs are often deteriorating or threatened by rapid change on their borders. A strong, well-implemented housing plan that requires newer suburbs to take more responsibility for affordable housing is the only way to avoid this downward transition. Such a plan takes pressure off the older suburbs and prevents the concentration of poverty and decline in these places. Once older declining suburbs understand that they already have more than their fair share of affordable housing, they can use a good regional housing plan as a powerful defensive strategy to maintain their communities' stability.

Without regional solutions, the future of these at-risk places is bleak. With their low fiscal capacity and lack of amenities, they have little hope of improving their position in a competitive regional economy. If they cut taxes, they cannot generate the revenues needed to deal with their old infrastructure or poverty problems in their schools. If they raise taxes to deal with these challenges, they cannot attract businesses or homeowners. In the end, these places have no haven outside regional cooperation.

The developing suburbs

At-risk low-density and bedroom-developing suburbs have three compelling reasons to support regional co-operation. First, it will reduce their taxes and increase their services, most notably in terms of schools. Second, it will help them get the infrastructure they need for safe and orderly development. Third, it will provide a better alternative to local unilateral growth moratoriums or slow-growth action to respond to the increasingly negative reaction within these communities to the development status quo.

While bedroom-developing communities are places of comparatively low poverty and diversity, their children-per-household ratio is very high. Throughout the country, at-risk low-density suburbs spend less per pupil than districts in other types of metropolitan communities. Through school equity and almost any form of tax sharing, both of these types of developing communities can be among the largest recipients of per-student aid. And as with the older suburbs, regional fiscal equity can also allow these places to have lower tax rates.

In chasing after development to make up for the lack of a local tax base, developing communities tend to neglect the provision of infrastructure that will eventually be needed but will be more costly to provide retroactively once development is in place. Regionalism provides assistance for infrastructure in developing communities through equity, which can give them money to build infrastructure as well as to relieve cash-flow crises that force them to seek development at any price, and through sharing regional infrastructure costs. By pooling regional resources, and creating regional funds and bonding authorities, regionalism can get infrastructure to these communities in a cost-effective way.

Sprawl is another problem of particular concern to residents of bedroom-developing suburbs. Most of the local initiatives to curb growth have been in these places. But a single community can have little effect on the growth of a region. Acting alone, a community not only is unlikely to solve its own growth-related problems but also is likely to impose higher costs on the region when it tries. In the end, regional or statewide planning to protect open space and create a regional growth boundary has been more effective than unilateral action. Regionally funded transit commuting alternatives are among the most promising ways to respond to growing congestion. A cooperative regional approach that encourages affordable housing close to affluent job centers is also likely to be more helpful than local NIMBY (Not In My Back Yard) approaches.

Affluent job centers

Despite their low poverty rates and high fiscal capacities, affluent job centers are not immune from problems caused by the prevailing pattern of regional development. Because they are intense centers of job growth, these communities are often troubled by higher rates of congestion than other suburban areas, particularly in the United States' fast-growth regions. Open space is harder to preserve in these communities, because land becomes very valuable. In the most extreme cases, suburban "edge cities" can become as densely urban and congested as city business districts.

Some of the most celebrated and extreme fights against status-quo development patterns have occurred in this small group of suburbs. Here, too, regionalism presents the only possible response to these concerns, the only real way to maintain a suburban/rural edge, and the only plausible plan for dealing with traffic congestion. It is the only way to have an effect on a neighboring community's poor decisions.

Today's metropolitan politics are based on an inaccurate model of poor cities and rich suburbs. It does not acknowledge that almost half of the US population lives in places that have finished developing and have increasing urban problems. Nor does it come to terms with the fiscal pressure of growth and the public's increasing discontent with sprawl and loss of open space. A new metropolitics must understand the diversity of US suburbs and build a broad bipartisan movement for greater regional cooperation. If metropolitics does not succeed, our metropolitan regions will continue to become more unequal, and more energy will be spent growing against ourselves.

Plate 24 **New Urbanist neo-traditionalism: street in Seaside, Florida.** Designed by the firm of Andrés Duany and Elizabeth Plater-Zyberk, a postmodernist streetscape in Seaside, Florida, recaptures the sense and scale of the traditional American small town. People on their front porches are close enough to the sidewalks to converse with pedestrians.

Plate 25 **Iconic architecture: Guggenheim Museum, Bilbao, Spain.** Beginning with the citadels and towers of the earliest cities, iconic architectural forms have always been a part of the urban environment. Spectacular buildings like Frank Gehry's Guggenheim Museum in Bilbao, Spain, provide what have been called identity-signatures for cities in the age of globalization.

Plate 26 **The mall has it all: Mall of America, Minneapolis.** The suburban shopping center, sometimes called the "new downtown," is both an extravaganza of commercialism and a new kind of social and entertainment center. This one, the Mall of America in Minneapolis, Minnesota, has shops, restaurants, and even a roller-coaster.

Plate 27 **The persistence of tradition.** Many Islamic cities are gleamingly modern, but others retain a traditional, almost medieval flavor. In Fez, Morocco, a crowd fills a walled courtyard to watch a street entertainer. (Photograph by Paul V. Turner)

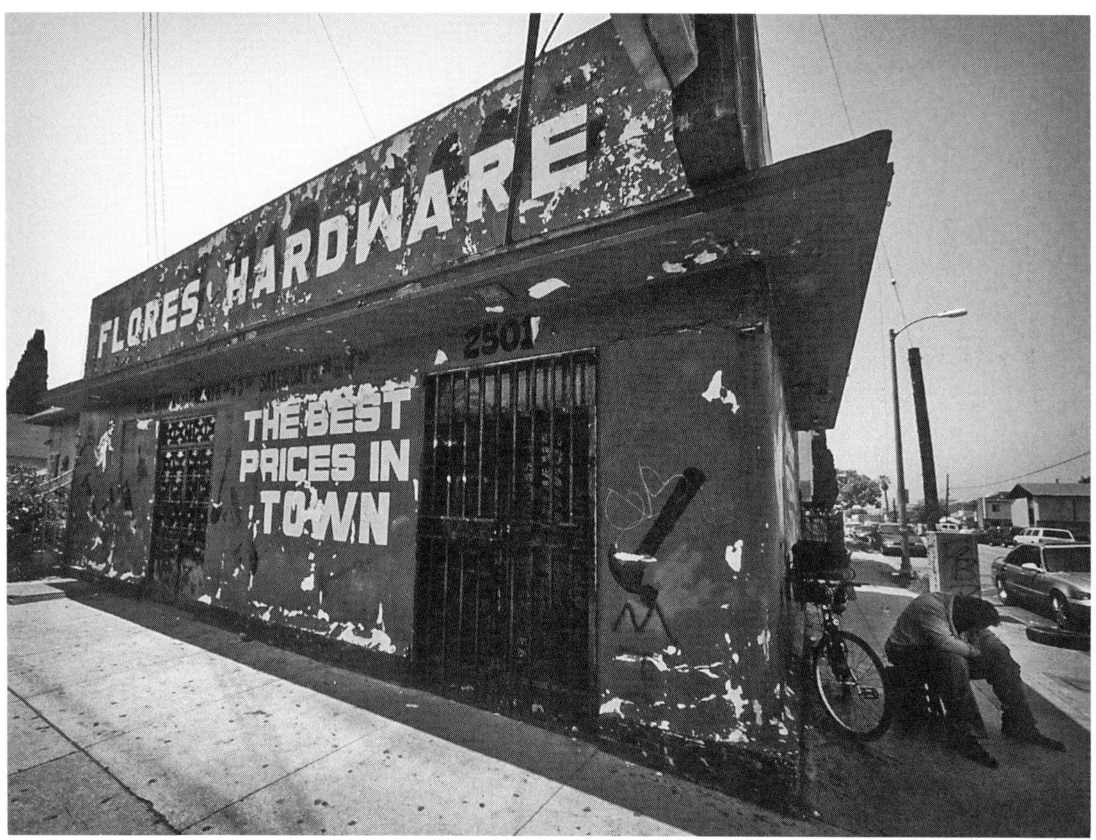

Plate 28 **The persistence of poverty and decay.** Amid the splendor and wealth of new global power centers and the comfort of suburban residential communities, inner-city decay persists, even dominates. Graffiti, barred windows, and litter mark this barrio corner in Los Angeles. (Photograph by Ken Alexander)

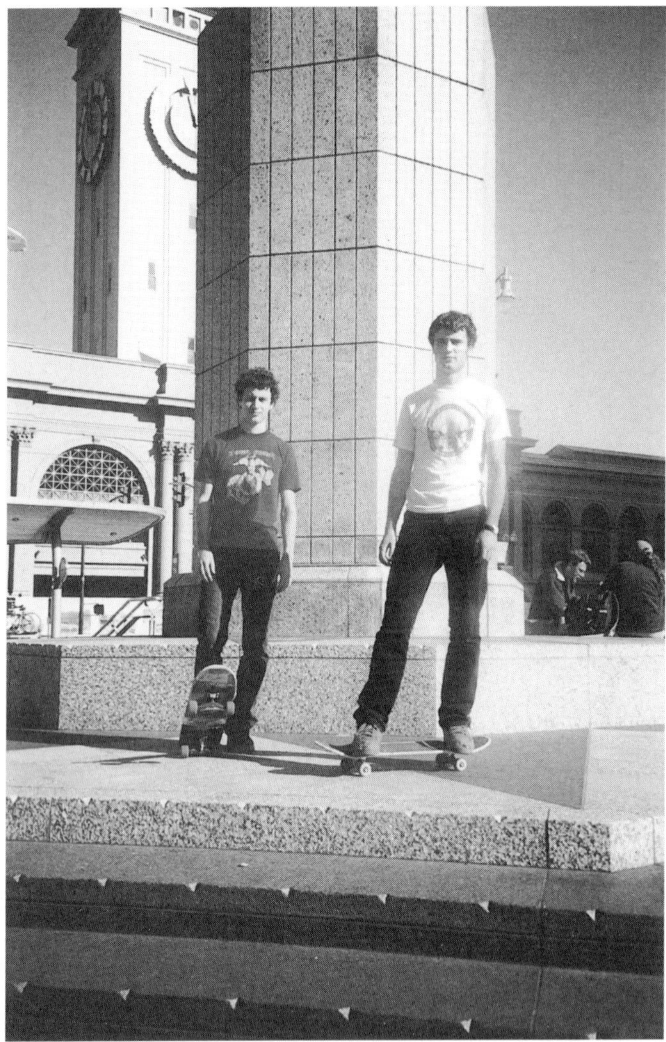

Plate 29 **The streets belong to the kids.** Skateboarders may seem like a relatively recent phenomenon, but youth has always found ways to exploit and even expropriate adult public space. These young masters of their domain are on San Francisco's waterfront Embarcadero. (Photograph by Frederic Stout)

Plate 30 **The streets belong to the people.** Street protests and demonstrations – whether for civil rights, the economic interests of workers, or anti-war coalitions – have made public space a frequent location for insurgent politics that can sometimes lead to revolution. This woodblock print by the Belgian artist Frans Masereel, from his graphic novel *Die Stadt* (*The City*, 1925), captures the power of this ancient and honorable tradition.

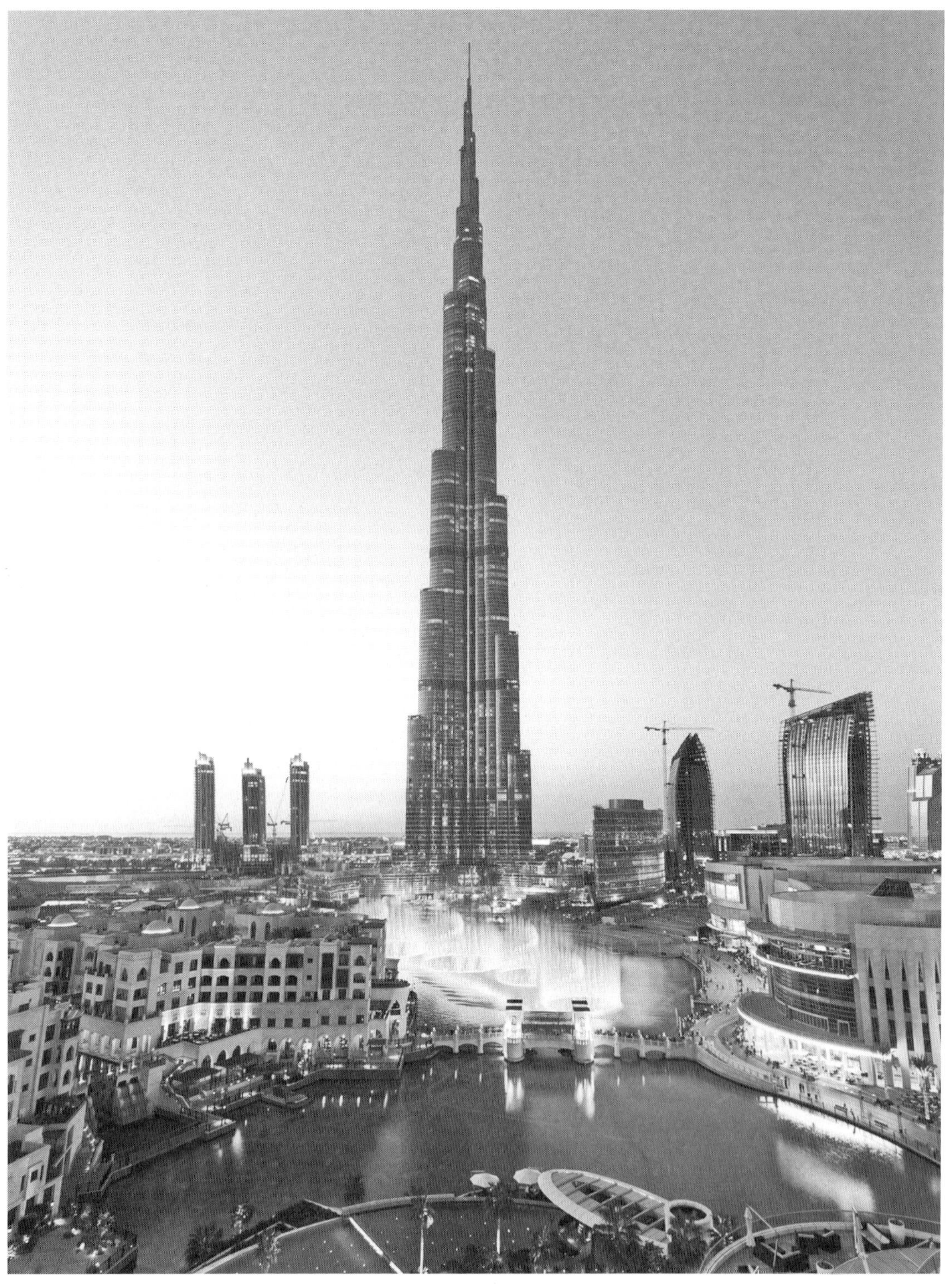

Plate 31 **Urban pride: architecture as symbolic power.** The Burj Khalifa, designed by Adrian Smith (then of the firm Skidmore, Owings, and Merrill in Chicago), is the world's tallest building and proclaims Dubai to be one of the new power centers of the global economy.

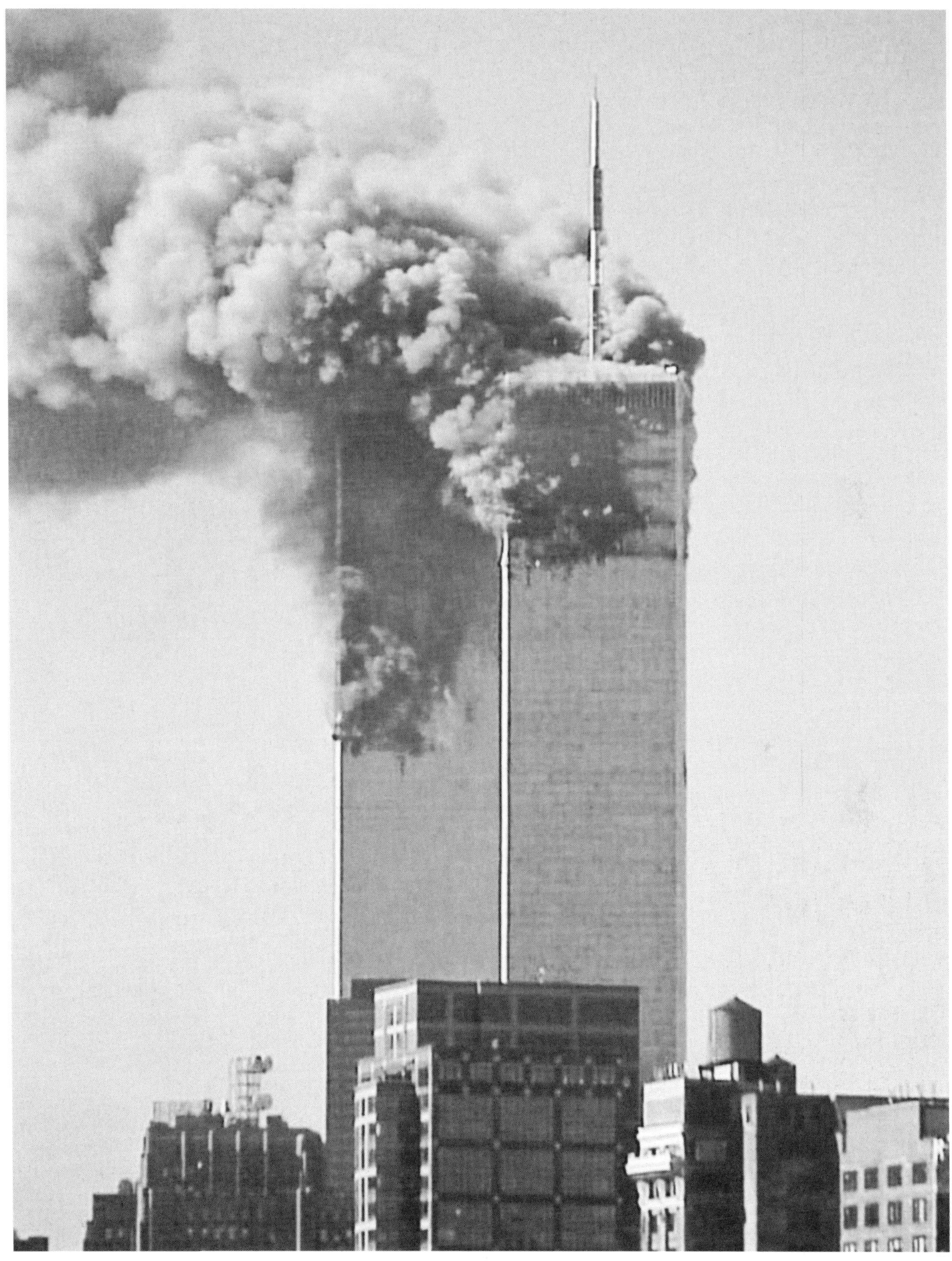

Plate 32 **Urban Terror: World Trade Center, New York, September 11, 2001.** The twin towers of the World Trade Center in New York were at one time the tallest buildings in the world but then became the target for terrorists committed to bringing down the symbols of global economic power and Western cultural dominance. (Photograph by Wanda McCormick, Readio.com)

Urban planning history and visions

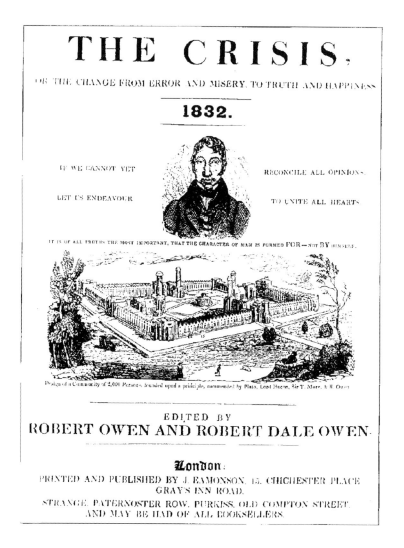

THE CRISIS,

OR THE CHANGE FROM ERROR AND MISERY, TO TRUTH AND HAPPINESS

1832.

IF WE CANNOT YET RECONCILE ALL OPINIONS,

LET US ENDEAVOUR TO UNITE ALL HEARTS.

IT IS OF ALL TRUTHS THE MOST IMPORTANT, THAT THE CHARACTER OF MAN IS FORMED FOR—NOT BY HIMSELF.

Design of a Community of 2,000 Persons, founded upon a principle, commended by Plato, Lord Bacon, Sir T. More, & R. Owen

EDITED BY
ROBERT OWEN AND ROBERT DALE OWEN.

London:
PRINTED AND PUBLISHED BY J. EAMONSON, 15, CHICHESTER PLACE
GRAY'S INN ROAD.
STRANGE, PATERNOSTER ROW, PURKISS, OLD COMPTON STREET,
AND MAY BE HAD OF ALL BOOKSELLERS.

INTRODUCTION TO PART FIVE

The effects of urban planning are perhaps the greatest – and, at the same time, the most invisible – influences on human life and culture. In the words of Percival Goodman and Paul Goodman, the co-authors of *Communitas: Means of Livelihood and Ways of Life* (1947), we hardly realize as we go about the daily round of our lives "that somebody once drew some lines on a piece of paper who might have drawn otherwise" and that "now, as engineer and architect once drew, people have to walk and live."

When the Sumerian kings built the walls of Eridu and Uruk, they engaged in acts of urban planning and thus determined how their people would "walk and live." The walls provided safety and protection for the people of the city and also defined the new political unity of the city-state. The associated roads, bridges, irrigation systems, and centers for market and ceremonial functions – all served a dual function in that they met the practical social needs of the urban population in general and fulfilled the power aspirations of the god-king and priestly elites in particular. The ancient citadels were centers of religious meaning, as well as economic and political power, and thus a third component of urban planning – an idealized, often spiritual vision of what constitutes the best possible state of human existence – was present at the very beginning of city building.

The origins of modern urban planning are complex. On one level, modern planning is a direct extension of the ancient and pre-modern models: imposing order on nature for the health, safety, and amenity of the urban masses, for the political benefit of the urban elites, and for a way of expressing each culture's highest spiritual ideals. On another level, however, modern planning is far more complex than anything that had ever gone on before. Modern planning operates, by and large, in a politically and economically pluralistic environment, making every alteration of the physical arrangements of the city a complex negotiation between competing interests. And the practice of modern urban planning also takes place at a stage of human development when the planner's defining goal is no longer merely to impose human order on nature, but to continuously impose order on the city itself.

All the goals and functions of planning – both the ancient holdovers and the modern elaborations – are present in the first planning responses to the urban conditions associated with the Industrial Revolution. As Friedrich Engels (p. 46) and other contemporary observers described, the cities of the new industrialism were characterized by horrendous overcrowding, ubiquitous misery, and despair. There were daily threats to the public health and safety, not only for the impoverished working class, but also for the capitalist middle class. These conditions gave rise to movements for housing reform, to great advances in the technologies of water supply and sewage disposal, and to the emergence of middle-class suburbs. They also led to the construction of model "company towns" by various industrial firms in both Europe and America, and eventually to the development of a modern urban planning profession.

Reviewing the history of urban planning in the nineteenth century, Richard LeGates and Frederic Stout (the co-editors of this volume) have written that "the classic texts of early urban planning history often seem surprisingly modern." An example of the surprising modernity of early urban planning is the nineteenth-century parks movement, especially the work of Frederick Law Olmsted (p. 321), which gave rise to something very like comprehensive urban planning practice. Projects like Central Park in New York represented a transplantation and democratization of European landscape gardening traditions, to be sure, but Olmsted's goal was not merely to bring nature into the city. Rather, Olmsted repeatedly appealed to the political and economic leadership of American cities to create parks that would achieve a whole range of public benefits: they would contribute to the public health by serving as the "lungs" of the city; they would be practical and necessary additions to the physical infrastructure of the metropolis, providing a general recreation ground; their ponds and reservoirs would serve as adjuncts to municipal water-supply systems; and they would soften and tame human nature, by providing wholesome alternatives to the vulgar street amusements, bars, and brothels that daily tempted poor and working-class youth.

Olmsted was a visionary and a reformer, but he was also a successful businessman and a canny political operative capable of offering his clients useful strategic advice on how to fund and build constituencies in favor of large municipal projects. Somewhat less practical, but even more visionary, were a group of architects, planners, and activists who may be termed, collectively, the utopian modernists. Three of these – Ebenezer Howard, Le Corbusier, and Frank Lloyd Wright – define the mainstream of that utopian tradition. Not one of them had his utopian vision realized in its entirety, but each had an enormous influence on the way contemporary cities, and city life, developed in the twentieth century. A fourth, the Spanish engineer and planner Arturo Soria y Mata, is influential for his vision of the relationship between transportation systems and land use. Soria's vision of a "linear city" imagined urban developments built along a central spine containing an electric streetcar line and other utilities.

Ebenezer Howard (p. 328) prided himself on being "the inventor of the Garden City idea," and his tireless devotion to the project of decongesting the modern metropolis by building small, self-contained, green-belted cities in the rural countryside is one of the marvels of modern urban planning history. The quaintly Victorian illustrations and diagrams in *Garden Cities of Tomorrow* illustrates Howard's vision of "a group of slumless, smokeless cities." Howard originally wanted his Garden Cities to be cooperatively owned. He wanted the surrounding greenbelt to be much larger than the built-up part of the city itself. And he wanted his cities to be economically independent, not commuter suburbs. In the process of actually building Letchworth and Welwyn – the two Garden Cities constructed before his death in 1924 – Howard had to compromise many of his original goals. Building lots and businesses were privately owned; the greenbelt became more of a park than an extensive rural buffer zone; and neither of the original Garden Cities ever became a fully independent economic entity. Nonetheless, as the plan for Welwyn illustrates, these were fully planned communities which embodied many of Howard's ideals. The Garden City experiment gave rise to a larger movement of town planning, and disciples of Howard spread his ideas and his example worldwide.

Charles-Edouard Jeanneret, better known as Le Corbusier (p. 336), was another utopian visionary who never saw his ideal plans fully developed but who was enormously influential nonetheless. Le Corbusier wanted his "Contemporary City of Three Million" to be a series of exquisite towers, geometrically arranged in a surrounding park, and he spent years looking for governmental and industrial sponsors for his plan. Many "Corbusian" high-rise urban developments have been built throughout the world. Indeed, the "International Style" of modern architecture and the principles of the International Congress of Modern Architecture (CIAM), which Le Corbusier pioneered, have become global standards of urban development. But in almost every case, the surrounding park has been compromised away in the process of realization. In case after case, the tower in the park has become the tower without the park or, even worse, the tower in the parking lot!

While Le Corbusier was issuing his manifestos and shocking the architectural and planning establishment with his modernist plans, American planners like Clarence Perry (p. 486), Clarence Stein, and Henry Wright were also wrestling with the problem of how to adapt urban form to the automobile. Perry, an architect and educator, published his seminal work on "The Neighborhood Unit", reprinted in Part Seven on Perspectives on Urban Design. Perry envisioned compact, school-centered, neighborhoods for nuclear families with cars and worked out land use and street designs to accommodate enough households to support a primary school surrounded by streets engineered for slow-moving traffic almost exclusively from residents of the neighborhood itself. In their influential plan for Radburn, New Jersey, Stein and Wright invented and implemented a series of planning concepts including superblocks, residential culs-de-sac, and the separation of pedestrian and vehicular traffic. Frank Lloyd Wright (p. 345), the originator of the visionary "Broadacre City" plan of 1935, was also responding to the automobile. Wright called for a city composed of family homesteads – one full acre per person – and the withering away of dense and crowded traditional cities. The private automobile, Wright thought, would virtually abolish distance and allow for a new kind of community based on individualism, families and self-reliance. Some argue that what actually became of Wright's Broadacre was sprawl suburbia. One acre per person became one-eighth acre per family or less; the core cities refused to wither away; the transportation monoculture of the automobile became a new form of dependency, rather than a technology of liberation; and the family-oriented suburban community became problematic at best, an object of ridicule at worst, producing the kind of "drive-in culture" described by Kenneth T. Jackson (p. 65). More recently, defenders of suburbia like Robert Bruegmann (p. 211) have argued that what many disdain as "sprawl" is a natural, popular, and almost inevitable response to the challenge of housing ever-larger urban populations.

The utopian visionaries were more than just planners, if they can be said to be planners at all. Even Ebenezer Howard, the most moderate of the group, was a dreamer. Together, the utopian modernists concerned themselves with great philosophical issues such as the connection between humankind and nature, the relationship of city plan to moral reform, and the role of urban design and technologies to the evolutionary transformation of society. It would be left to more practical men and women – the actual members of the urban planning profession as it developed in the twentieth century – to address the real-world problems of ever-changing cities and metropolitan regions. If the utopian modernists established the lofty goals, the professional planners – whose work is described in Part Six on Urban Planning Theory and Practice – attended to the details. Still, the role of visionary projections of better lives through better urban planning persists as an important motivating force in contemporary urban planning. Establishing a good planning vision and sticking to it can have profound positive impacts. Redevelopment projects like the famous "Paseo del Rio" of San Antonio, Texas, took blighted areas – in this case a river-turned-drainage-ditch – that disfigured the downtown and turned them into community attractions. Like many other cities, San Antonio developed a vision of turning a problem area into a magnificent location of social amenities and recreational activities. Nowadays, the Paseo provides a pleasant place to sit, stroll, paddle, and shop. Boston, Massachusetts worked the same kind of urban planning magic by collaborating with developer James Rouse to turn a seedy and obsolete marketplace around Quincy Market into a magnificent center for strolling, shopping, dining, and cultural events. Similarly, San Francisco's famous Ghirardelli Square was once an abandoned chocolate factory.

In the 1970s and 1980s, a new idea took hold in planning circles, the idea of sustainable development. Closely allied with environmentalism and "green" politics, "sustainability" became an ubiquitous catchword in planning discussions. The idea was defined and raised to a level of worldwide prominence by the publication in 1987 of the Report of the World Commission on Environment and Development, commonly known as the Brundtland Report (p. 351) because the WCED chairperson was Gro Harlem Brundtland, the prime minister of Norway.

According to the Brundtland Report, sustainable development is "development that meets the needs of the present without compromising the ability of future generations to meet their own needs." Concerned about "the biosphere's ability to absorb the effect of human activities," the WCED defined humanity's basic needs as "food, clothing, shelter, and jobs" and called on the nations of the world to pay special attention to "the largely unmet needs of the world's poor, which should be given overriding priority." In the wake of the Brundtland Report, the world's nations have been asked to cut back on industrial production and the emission of greenhouse gases that affect the climate, and cities around the world have been asked to adopt "green policies" relating to transportation, energy use, resource management, and sprawl. As Timothy Beatley describes (p. 446), many European cities are now pursuing policies suggested by the Brundtland Report.

The tradition of visionary urban planning – especially the idea of sustainability – is very much alive and well at the present time. In addition to visions of sustainable urban development discussed throughout this volume – most particularly the work of Timothy Beatley and Stephen Wheeler (p. 458) – the "New Urbanism" is both a visionary planning and design movement and a profit-making, business-oriented process of real-estate development that began by emphasizing a "new traditionalism" based on small-town scale for new communities from Florida to California and that has now spread worldwide.

The principles of the New Urbanist movement are laid out in detail in the Charter of the New Urbanism (p. 356). Originally promulgated in 1993 – but reflecting ideas that hark back to the work of Ebenezer Howard, Clarence Perry, Jane Jacobs, and many others – the Charter details the goals and strategies of an urban planning and design movement that began in the United States and has now become influential worldwide. The New Urbanism, as practiced by Peter Calthorpe (p. 360), Andrés Duany, Elizabeth Plater-Zyberk, and dozens of other architects and planners, fervently embraces the natural environment, metropolitanism, and the ideals of social justice and participatory democracy. It just as ardently opposes the wastefulness of suburban sprawl, inner-city decay, and agricultural deterioration. There are many, diverse approaches to New Urbanist planning, but the Charter calls for adherence to a set of broad guidelines at the metropolitan scale, at the neighborhood or district scale, and at the intimate scale of houses and streets. The metropolis as a whole should encourage density and infill, support a wide range of transit options, and view every element of the urban planning process as parts of "one interrelated community-building challenge." Neighborhoods should be compact, mixed use, and "pedestrian friendly," and the architecture of homes should respect the local climate and community history in a way that creates environments that are safe, accessible, and open.

Following along on the principles of sustainable development, William Fulton, an urban planner, and Peter Calthorpe, an architect, planner, and co-founder of the Congress for the New Urbanism, advocate a vision of a new "Regional City" in which land use and transportation systems are designed together in harmony with the natural environment to eliminate the blight of suburban sprawl and produce livable communities. One of Calthorpe's earliest visions was for an urban development model based on "Pedestrian Pockets," towns and neighborhoods small enough to allow every resident to walk to the business center. Over time, these evolved into the kind of transit-oriented development (TODs) that are currently being built not as a further extension of traditional suburbia but in and for what he was variously called "The Next American Metropolis" and "The Post-Suburban Metropolis." Calthorpe's Pedestrian Pockets, transit-oriented developments, and Regional City conceptions look backward to the Garden Cities of Ebenezer Howard – especially in their use of green-belting and light-rail mass-transit options – but they look forward to an entirely new relationship between city and region, between individual and community. In short, the New Urbanist vision is a response to a future already in the process of becoming. Characterized by "a dramatic shift in the nature and location of our workplace and a fundamental change in the character of our increasingly diverse households," it is the latest, and one of the most successful, examples of how originally visionary ideas in planning and design have a way of becoming contemporary urban reality.

"Public Parks and the Enlargement of Towns"

American Social Science Association (1870)

Frederick Law Olmsted

Editors' Introduction

Frederick Law Olmsted (1822–1903) has been called "America's great pioneer landscape architect," and, during his lifetime, he was widely recognized as one of the most influential public figures in the nation. Along with his business partner, the English-born architect Calvert Vaux, Olmsted originated and dominated the urban parks movement, pioneered the development of planned suburbs, and laid out scores of public and private institutions. Central Park in New York, illustrated as it looked in 1863 in Plate 33, remains his best known masterpiece. The designs for Riverside, Illinois (outside Chicago), the Boston park system, the Capitol grounds in Washington, DC, the 1893 World's Fair, and the campus of Stanford University in California are equally impressive contributions to the built environment.

Olmsted began his career practicing and writing about farming, then turned his talents to journalism and, in the 1850s, published a series of books describing the society and economy of the slave states of the American South (collected into one volume as *The Cotton Kingdom* in 1861). With this background, it is hardly surprising that Olmsted thoroughly imbued his art of landscape architecture with a wide variety of social and political, as well as cultural, concerns.

"Public Parks and the Enlargement of Towns" was originally presented as an address to the American Social Science Association, meeting at the Lowell Institute, Boston, in 1870. In it, Olmsted provides a number of specific guidelines for parks and parkways and suggests ways to overcome political resistance to public funding for parks and planned urban growth. Most importantly, however, he lays out the political and philosophical case for public parks in terms of three great moral imperatives: first, the need to improve public health by sanitation measures and the use of trees to combat air and water pollution; second, the need to combat urban vice and social degeneration, particularly among the children of the urban poor; and third, the need to advance the cause of civilization by the provision of urban amenities that would be democratically available to all.

Both as a practitioner and as a theorist, Olmsted anticipated many of the principal concerns of urban planning, both infrastructural and social, down to the present day. Indeed, behind the somewhat convoluted Victorianisms of his prose lies a strikingly modern mind. In the design of the Garden City Ebenezer Howard (p. 328) borrowed directly from Olmsted, and even plans so fundamentally different as those of Frank Lloyd Wright (p. 345) and Le Corbusier (p. 336) owe a debt to Olmsted insofar as they recognize and address the central problem of the relationship between nature and the built urban environment. As one of the founders of modern landscape architecture and integrated urban design, Olmsted's work and thought invite comparison with all those who came after him in the profession, either as practitioners or critics. Although the contemporary New Urbanists like Peter Calthorpe and William Fulton (p. 360) and the advocates of sustainable planning like the World Commission on Environment and Development (p. 351), Timothy Beatley (p. 446), and Stephen Wheeler (p. 458) go well beyond the parks movement in their comprehensive vision of the city–nature relationship, Olmsted and the nineteenth-century park builders can still be regarded as pioneers of a new way of looking at the urban built environment.

A selection of Olmsted's most important writings may be found in S.B. Sutton (ed.), *Civilizing American Cities: Writings on City Landscapes by Frederick Law Olmsted* (Cambridge, MA: MIT Press, 1971). Johns Hopkins University has published most of Olmsted's work in the multivolume *Collected Papers of Frederick Law Olmsted* (Baltimore, MD: Johns Hopkins University Press, 1977–1992). Biographies of Olmsted and commentary on his work include Laura Wood Roper, *FLO: A Biography of Frederick Law Olmsted* (Baltimore, MD: Johns Hopkins University Press, 1973), Elizabeth Stevenson, *Park Maker: A Life of Frederick Law Olmsted* (New York: Macmillan, 1977), Charles E. Beveridge, Paul Rocheleau, and David Larkin, *Frederick Law Olmsted: Designing the American Landscape* (New York: Rizzoli, 1995), and Witold Rybezynski, *A Clearing in the Distance: Frederick Law Olmsted and America in the Nineteenth Century* (New York: Scribner, 1999). Also of interest are the pictures of Olmsted parks by noted photographer Lee Friedlander collected in *Photographs: Frederick Law Olmsted Landscapes* (New York: Distributed Art Publishers, 2008).

Galen Cranz, *The Politics of Park Design: A History of Urban Parks in America* (Cambridge, MA: MIT Press, 1982) is a superb overview that places Olmsted's planning and landscape design achievements in the context of a larger movement for urban social reform. See also Cynthia Zaitzevsky, *Frederick Law Olmsted and the Boston Park System* (Cambridge, MA: Belknap Press, 1992), and Susan L. Klaus, *Modern Arcadia: Frederick Law Olmsted, Jr. and the Plan for Forest Hill Gardens* (Cambridge, MA: MIT Press, 2002). For information on La Villette in Paris and other great European parks, see Topos, *Parks: Green Spaces in European Cities* (New York: Princeton Architectural Press, 2002). Also of interest are Peter Harnik, *Inside City Parks* (Washington, DC: Urban Land Institute, 2000), and Terence Young, *Building San Francisco's Parks, 1850–1930* (Baltimore, MD: Johns Hopkins University Press, 2004).

We have reason to believe, then, that towns which of late have been increasing rapidly on account of their commercial advantages, are likely to be still more attractive to population in the future; that there will in consequence soon be larger towns than any the world has yet known, and that the further progress of civilization is to depend mainly upon the influences by which men's minds and characters will be affected while living in large towns.

Now, knowing that the average length of the life of mankind in towns has been much less than in the country, and that the average amount of disease and misery and of vice and crime has been much greater in towns, this would be a very dark prospect for civilization, if it were not that modern Science has beyond all question determined many of the causes of the special evils by which men are afflicted in towns, and placed means in our hands for guarding against them. It has shown, for example, that under ordinary circumstances, in the interior parts of large and closely built towns, a given quantity of air contains considerably less of the elements which we require to receive through the lungs than the air of the country or even of the outer and more open parts of a town, and that instead of them it carries into the lungs highly corrupt and irritating matters, the action of which tends strongly to vitiate all our sources of vigor – how strongly may perhaps be indicated in the shortest way by the statement that even metallic plates and statues corrode and wear away under the atmosphere influences which prevail in the midst of large towns, more rapidly than in the country.

The irritation and waste of the physical powers which result from the same cause, doubtless indirectly affect and very seriously affect the mind and the moral strength; but there is a general impression that a class of men are bred in towns whose peculiarities are not perhaps adequately accounted for in this way. We may understand these better if we consider that whenever we walk through the denser part of a town, to merely avoid collision with those we meet and pass upon the sidewalks, we have constantly to watch, to foresee, and to guard against their movements. This involves a consideration of their intentions, a calculation of their strength and weakness, which is not so much for their benefit as our own. Our minds are thus brought into close dealings with other minds without any friendly flowing toward them, but rather a drawing from them. Much of the intercourse between men when engaged in the pursuits of commerce has the same tendency – a tendency to regard others in a hard if not always hardening way. Each detail of observation and of the process of thought required in this kind of intercourse or contact of minds is so slight and so common in the experience of towns-people that they are seldom conscious of it. It certainly involves some expenditure nevertheless. People from the country are even conscious of the effect on their nerves and minds of the street contact – often complaining that they feel confused by it; and if we had no relief from it at all during our waking hours, we should all be conscious of suffering from it. It is upon our opportunities of relief from it, therefore, that not only our comfort in town life, but our ability to maintain a temperate, good-natured, and healthy state of mind, depends. This is one of many ways in which it happens that men who have been brought up, as the saying is, in the streets, who have been most directly and completely affected by town influences, so generally show, along with a remarkable quickness of apprehension, a peculiarly hard sort of selfishness. Every day of their lives they have seen thousands of their fellow-men, have met them face to face, have brushed against them, and yet have had no experience of anything in common with them.

[. . .]

It is practically certain that the Boston of today is the mere nucleus of the Boston that is to be. It is practically certain that it is to extend over many miles of country now thoroughly rural in character, in parts of which farmers are now laying out roads with a view to shortening the teaming distance between their wood-lots and a railway station, being governed in their courses by old property lines, which were first run simply with reference to the equitable division of heritages, and in other parts of which, perhaps, some wild speculators are having streets staked off from plans which they have formed with a rule and pencil in a broker's office, with a view, chiefly, to the impressions they would make when seen by other speculators on a lithographed map. And by this manner of planning, unless views of duty or of interest prevail that are not yet common, if Boston continues to grow at its present rate even for but a few generations longer, and then simply holds its own until it shall be as old as the Boston in Lincolnshire now is, more men, women, and children are to be seriously affected in health and morals than are now living on this Continent.

Is this a small matter – a mere matter of taste; a sentimental speculation?

It must be within the observation of most of us that where, in the city, wheel-ways originally twenty-feet wide were with great difficulty and cost enlarged to thirty, the present width is already less nearly adequate to the present business than the former was to the former business; obstructions are more frequent, movements are slower and oftener arrested, and the liability to collision is greater. The same is true of sidewalks. Trees thus have been cut down, porches, bow-windows, and other encroachments removed, but every year the walk is less sufficient for the comfortable passing of those who wish to use it.

It is certain that as the distance from the interior to the circumference of towns shall increase with the enlargement of their population, the less sufficient relatively to the service to be performed will be any given space between buildings.

In like manner every evil to which men are specially liable when living in towns, is likely to be aggravated in the future, unless means are devised and adapted in advance to prevent it.

Let us proceed, then, to the question of means, and with a seriousness in some degree befitting a question, upon our dealing with which we know the misery or happiness of many millions of our fellow-beings will depend.

We will for the present set before our minds the two sources of wear and corruption which we have seen to be remediable and therefore preventable. We may admit that commerce requires that in some parts of a town there shall be an arrangement of buildings, and a character of streets and of traffic in them which will establish conditions of corruption and of irritation, physical and mental. But commerce does not require the same conditions to be maintained in all parts of a town.

Air is disinfected by sunlight and foliage. Foliage also acts mechanically to purify the air by screening it. Opportunity and inducement to escape at frequent intervals from the confined and vitiated air of the commercial quarter, and to supply the lungs with air screened and purified by trees, and recently acted upon by sunlight, together with opportunity and inducement to escape from conditions requiring vigilance, wariness, and activity toward other men, – if these could be supplied economically, our problem would be solved.

In the old days of walled towns all tradesmen lived under the roof of their shops, and their children and apprentices and servants sat together with them in the evening about the kitchen fire. But now that the dwelling is built by itself and there is greater room, the inmates have a parlor to spend their evening in; they spread carpets on the floor to gain in quiet, and hang drapery in their windows and papers on their walls to gain in seclusion and beauty. Now that our towns are built without walls, and we can have all the room that we like, is there any good reason why we should not make some similar difference between parts which are likely to be dwelt in, and those which will be required exclusively for commerce?

Would trees, for seclusion and shade and beauty, be out of place, for instance, by the side of certain of our streets? It will, perhaps, appear to you that it is hardly necessary to ask such a question, as throughout the United States trees are commonly planted at the sides of streets. Unfortunately they are seldom so planted as to have fairly settled the question of the desirableness of systematically maintaining trees under these circumstances. In the first place, the streets are planned, wherever they are, essentially alike. Trees are planted in the space assigned for sidewalks, where at first, while they are saplings and the vicinity is rural or suburban, they are not much in the way, but where, as they grow larger, and the vicinity becomes urban, they take up more and more space, while space is more and more required for passage. That is not all. Thousands and tens of thousands are planted every year in a manner and under conditions as nearly certain as possible either to kill them outright, or to so lessen their vitality as to prevent their natural and beautiful development, and to cause premature decrepitude. Often, too, as their lower limbs are found inconvenient, no space having been provided for trees in laying out the street, they are deformed by butcherly amputations. If by rare good fortune they are suffered to become beautiful, they still stand subject to be condemned to death at any time, as obstructions in the highway.

What I would ask is, whether we might not with economy make special provision in some of our streets – in a twentieth or a fiftieth part, if you please, of all – for trees to remain as a permanent furniture of the city? I mean, to make a place for them in which they would have room to grow naturally and gracefully. Even if the distance between the houses should have to be made half as much again as it is required to be in our commercial streets, could not the space be afforded? Out of town space is not costly when measures to secure it are taken early. The assessments for benefit where such streets were provided for, would, in nearly all cases, defray the cost of the land required. The strips of ground required for the trees, six, twelve, twenty feet wide, would cost nothing for paving or flagging.

The change both of scene and of air which would be obtained by people engaged for the most part in the necessarily confined interior commercial parts of the town, on passing into a street of this character after the trees have become stately and graceful, would be worth a good deal. If such streets were made still broader in some parts, with spacious malls, the advantage would be increased. If each of them were given the proper capacity, and laid out with laterals and connections in suitable directions to serve as a convenient trunk line of communication between two large districts of the town or the business centre and the suburbs, a very great number of people might thus be placed every day under influences counteracting those with which we desire to contend.

These, however, would be merely very simple improvements upon arrangements which are in common use in every considerable town. Their advantages would be incidental to the general uses of streets as they are. But people are willing very often to seek recreations as well as receive it by the way. Provisions may indeed be made expressly for public recreations, with certainty that if convenient they will be resorted to.

We come then to the question: what accommodations for recreation can we provide which shall be so agreeable and so accessible as to be efficiently attractive to the great body of citizens, and which, while giving decided gratification, shall also cause those who resort to them for pleasure to subject themselves, for the time being, to conditions strongly counteractive to the special, enervating conditions of the town?

In the study of this question all forms of recreation may, in the first place, be conveniently arranged under two general heads. One will include all of which the predominating influence is to stimulate exertion of any part or parts needing it; the other, all which cause us to receive pleasure without conscious exertion. Games chiefly of mental skill, as chess, or athletic sports, as baseball, are examples of means of recreation of the first class, which may be termed that of *exertive* recreation; music and the fine arts generally of the second or *receptive* division.

Considering the first by itself, much consideration will be needed in determining what classes of exercises may be advantageously provided for. In the Bois de Boulogne there is a race-course; in the Bois de Vincennes a ground for artillery target-practice. Military parades are held in Hyde Park. A few cricket clubs are accommodated in most of the London parks, and swimming is permitted in the lakes at certain hours. In the New York Park, on the other hand, none of these exercises are provided for or permitted, except that the boys of the public schools are given the use on holidays of certain large spaces for ball playing. It is considered that the advantage to individuals which would be gained in providing for them would not compensate for the general inconvenience and expense they would cause.

I do not propose to discuss this part of the subject at present, as it is only necessary to my immediate purpose to point out that if recreations requiring large spaces to be given up to the use of a comparatively small number, are not considered essential, numerous small grounds so distributed through a large town that some one of them could be easily reached by a short walk from every house, would be more desirable than a single area of great extent, however rich in landscape attractions it might be. Especially would this be the case if the numerous local grounds were connected and supplemented by a series of trunk-roads or boulevards such as has already been suggested.

Proceeding to the consideration of receptive recreations, it is necessary to ask you to adopt and bear in mind a further subdivision, under two heads, according to the degree in which the average enjoyment is greater when a large congregation assembles for a purpose of receptive recreation, or when the number coming together is small and the circumstances are favorable to the exercise of personal friendliness.

The first I shall term *gregarious*; the second, *neighborly*. Remembering that the immediate matter in hand is a study of fitting accommodations, you will, I trust, see the practical necessity of this classification.

Purely gregarious recreation seems to be generally looked upon in New England society as childish and savage, because, I suppose, there is so little of what we call intellectual gratification in it. We are inclined to engage in it indirectly, furtively, and with complication. Yet there are certain forms of recreation, a large share of the attraction of which must, I think, lie in the gratification of the gregarious inclination, and which, with those who can afford to indulge in them, are so popular as to establish the importance of the requirement.

If I ask myself where I have experienced the most complete gratification of this instinct in public and out of doors, among trees, I find that it has been in the promenade of the Champs-Élysées. As closely following it I should name other promenades of Europe, and our own upon the New York parks. I have studiously watched the latter for several years. I have several times seen fifty thousand people participating in them; and the more I have seen of them, the more highly have I been led to estimate their value as means of counteracting the evils of town life.

Consider that the New York Park and the Brooklyn Park are the only places in those associated cities where, in this eighteen hundred and seventieth year after Christ, you will find a body of Christians coming together, and with an evident glee in the prospect of coming together, all classes largely represented, with a common purpose, not at all intellectual, competitive with none, disposing to jealousy and spiritual or intellectual pride toward none, each individual adding by his mere presence to the pleasure of all others, all helping to the greater happiness of each. You may thus often see vast numbers of persons brought closely together, poor and rich, young and old, Jew and Gentile. I have seen a hundred thousand thus congregated, and I assure you that though there have been not a few that seemed a little dazed, as if they did not quite understand it, and were, perhaps, a little ashamed of it, I have looked studiously but vainly among them for a single face completely unsympathetic with the prevailing expression of good nature and light-heartedness.

Is it doubtful that it does men good to come together in this way in pure air and under the light of heaven, or that it must have an influence directly counteractive to that of the ordinary hard, hustling working hours of town life?

You will agree with me, I am sure, that it is not, and that opportunity, convenient, attractive opportunity, for such congregation, is a very good thing to provide for, in planning the extension of a town.

[. . .]

Think that the ordinary state of things to many is at this beginning of the town. The public is reading just now a little book in which some of your streets of which you are not proud are described. Go into one of those red cross streets any fine evening next summer, and ask how it is with their residents. Oftentimes you will see half a dozen sitting together on the door-steps or, all in a row, on the curb-stones, with their feet in the gutter; driven out of doors by the closeness within; mothers among them anxiously regarding their children who are dodging about at their play, among the noisy wheels on the pavement.

Again, consider how often you see young men in knots of perhaps half a dozen in lounging attitudes rudely obstructing the sidewalks, chiefly led in their little conversation by the suggestions given to their minds by what or whom they may see passing in the street, men, women, or children, whom they do not know and for whom they have no respect or sympathy. There is nothing among them or about them which is adapted to bring into play a spark of admiration, of delicacy, manliness, or tenderness. You see them presently descend in search of physical comfort to a brilliantly lighted basement, where they find others of their sort, see, hear, smell, drink, and eat all manner of vile things.

Whether on the curb-stones or in the dram-shops, these young men are all under the influence of the same impulse which some satisfy about the tea-table with neighbors and wives and mothers and children, and all things clean and wholesome, softening, and refining.

If the great city to arise here is to be laid out little by little, and chiefly to suit the views of land-owners, acting only individually, and thinking only of how what they do is to affect the value in the next week or the next year of the few lots that each may hold at the time, the opportunities of so obeying this inclination as at the same time to give the lungs a bath of pure sunny air, to give the mind a suggestion of rest from the devouring eagerness and intellectual strife of town life, will always be few to any, to many will amount to nothing.

But is it possible to make public provision for recreation of this class, essentially domestic and secluded as it is?

It is a question which can, of course, be conclusively answered only from experience. And from experience in some slight degree I shall answer it. There is one large American town, in which it may happen that a man of any class shall say to his wife, when he is going out in the morning: "My dear, when the children come home from school, put some bread and butter and salad in a basket, and go to the spring under the chestnut-tree where we found the Johnsons last week. I will join you there as soon as I can get away from the office. We will walk to the dairy-man's cottage and get some tea, and some fresh milk for the children, and take our supper by the brook-side"; and this shall be no joke, but the most refreshing earnest.

There will be room enough in the Brooklyn Park, when it is finished, for several thousand little family and neighborly parties to bivouac at frequent intervals through the summer, without discommoding one another, or interfering with any other purpose, to say nothing of those who can be drawn out to make a day of it, as many thousand were last year. And although the arrangements for the purpose were yet very incomplete, and but little ground was at all prepared for such use, besides these small parties, consisting of one or two families, there came also, in companies of from thirty to a hundred and fifty, somewhere near twenty thousand children with their parents, Sunday-school teachers, or other guides and friends, who spent the best part of a day under the trees and on the turf, in recreations of which the predominating element was of this neighborly receptive class. Often they would bring a fiddle, flute, and harp, or other music. Tables, seats, shade, turf, swings, cool spring-water, and a pleasing rural prospect, stretching off half a mile or more each way, unbroken by a carriage road or the slightest evidence of the vicinity of the town, were supplied them without charge and bread and milk and ice-cream at moderate fixed charges. In all my life I have never seen such joyous collections of people. I have, in fact, more than once observed tears of gratitude in the eyes of poor women, as they watched their children thus enjoying themselves.

The whole cost of such neighborly festivals, even when they include excursions by rail from the distant parts of the town, does not exceed for each person, on an average, a quarter of a dollar; and when the arrangements are complete, I see no reason why thousands should not come every day where hundreds come now to use them; and if so, who can measure the value, generation after generation, of such provisions for recreation to the over-wrought, much-confined people of the great town that is to be?

For this purpose neither of the forms of ground we have heretofore considered are at all suitable. We want a ground to which people may easily go after their day's work is done, and where they may stroll for an hour, seeing, hearing, and feeling nothing of the bustle and jar of the streets, where they shall, in effect, find the city put far away from them. We want the greatest possible contrast with the streets and the shops and the rooms of the town which will be consistent with convenience and the preservation of good order and neatness. We want, especially, the greatest possible contrast with the restraining and confining conditions of the town, those conditions which compel us to walk circumspectly, watchfully, jealously, which compel us to look closely upon others without sympathy. Practically, what we most want is a simple, broad, open space of clean greensward, with sufficient play of surface and a sufficient number of trees about it to supply a variety of light and shade. This we want as a central feature. We want depth of wood enough about it not only for comfort in hot weather, but to completely shut out the city from our landscapes.

The word *park*, in town nomenclature, should, I think, be reserved for grounds of the character and purpose thus described.

[. . .]

A park fairly well managed near a large town, will surely become a new center of that town. With the determination of location, size, and boundaries should therefore be associated the duty of arranging new trunk routes of communication between it and the distant parts of the town existing and forecasted.

These may be either narrow informal elongations of the park, varying say from two to five hundred feet in width, and radiating irregularly from it, or if, unfortunately, the town is already laid out in the unhappy way that New York and Brooklyn, San Francisco and Chicago, are, and, I am glad to say, Boston is not, on a plan made long years ago by a man who never saw a spring-carriage, and who had a conscientious dread of the Graces, then we must probably adopt formal Park-ways. They should be so planned and constructed as never to be noisy and seldom crowded, and so also that the straightforward movement of pleasure-car carriages need never be obstructed, unless at absolutely necessary crossings, by slow-going heavy vehicles used for commercial purposes. If possible, also, they should be branched or reticulated with other ways of a similar class, so that no part of the town should finally be many minutes' walk from some one of them; and they should be made interesting by a process of planting and decoration, so that in necessarily passing through them, whether in going to or from the park, or to and from business, some substantial recreative advantage may be incidentally gained. It is a common error to regard a park as something to be produced complete in itself, as a picture to be painted on canvas. It should rather be planned as one to be done in fresco, with constant consideration of exterior objects, some of them quite at a distance and even existing as yet only in the imagination of the painter.

I have thus barely indicated a few of the points from which we may perceive our duty to apply the means in our hands to ends far distant, with reference to this problem of public recreations. Large operations of construction may not soon be desirable, but I hope you will agree with me that there is little room for question, that reserves of ground for the purposes I have referred to should be fixed upon as soon as possible, before the difficulty of arranging them, which arises from private building, shall be greatly more formidable than now.

"Author's Introduction" and "The Town–Country Magnet"

from *Garden Cities of To-morrow* (1898/1902)

Ebenezer Howard

Editors' Introduction

A stenographer by trade, Ebenezer Howard (1850–1928) was a quiet, modest, self-effacing man – "a man without credentials or connections," as one biographer put it – who nevertheless managed to change the world. Born in London, Howard early experienced the pollution, congestion, and social dislocations of the modern industrial metropolis. After a year in the United States (as a homesteader in Nebraska), he returned to England in 1876 and became involved in political movements and discussion groups addressing what was then termed "the Social Question." Howard was influenced by a number of radical theorists and visionaries including the social reformer Robert Owen, the utopian novelist Edward Bellamy, and the single tax advocate Henry George. He published *To-morrow: A Peaceful Path to Real Reform* in 1898 (now better known under its 1902 title, *Garden Cities of To-morrow*) and methodically set about convincing people of the beauty and utility of "the Garden City idea."

Although Howard's plan may seem quaintly Victorian to the modern reader, the ideas he put forth were revolutionary at the time. Indeed, Howard's ideas of urban decentralization, zoning for different uses, the integration of nature into cities, green-belting, and the development of self-contained "New Town" communities outside crowded central cities illustrated in Plate 35 laid the groundwork for the entire tradition of modern city planning.

Unlike many other utopian dreamers, Howard lived to see his plans actually put into action, if in a somewhat compromised form. In his own lifetime, the Garden Cities of Letchworth and Welwyn were built in England. Later, the Garden City idea spread to continental Europe, to the United States by way of the New Deal in the 1930s, and to much of the rest of the world.

Howard's argument begins with a protest against urban overcrowding, the one issue upon which, he writes, "men of all parties" are "well-nigh universally agreed." He then explains why "the people continue to stream into the already overcrowded cities" by reference to "the town magnet," that combination of jobs and amenities that characterizes the modern metropolis. Arrayed against this urban magnetic force is "the country magnet," the appealing features of the more natural, but increasingly desolate, rural districts. Finally, Howard describes his own plan, a new kind of human community based on "the town–country magnet," which is the best of both worlds.

As detailed in his famous concentric-ring diagram (which, he is careful to warn, is "a diagram only," not an actual site plan), the centre of a Garden City is to be a central park containing important public buildings and surrounded by a "Crystal Palace" ring of retail stores. The entire city of approximately 1,000 acres, serving a population of 32,000, would be encircled by a permanent agricultural greenbelt of some 5,000 acres, and the new cities would be connected with central "Social Cities" and each other by a system of railroad lines, forming a metropolitan region.

Howard's ideas about the evils of overcrowding are similar to those of Friedrich Engels (p. 46), and his solution to the problem invites comparison with the very different solutions proposed by Le Corbusier (p. 336) and Frank Lloyd Wright (p. 345). Direct followers of Howard include Patrick Geddes and Lewis Mumford (p. 91), who helped to spread the Garden City idea throughout Europe and America. More recently, Peter Calthorpe (p. 360) has effectively reinvented the Garden City idea in California as the Regional City in the form of green-belted, suburban "Pedestrian Pockets" and transit-oriented developments (TODs) linked to central cities (and each other) by a network of light-rail transportation systems.

Garden Cities of To-morrow remains a readable and relevant book. It is available as the second volume of Richard T. LeGates and Frederic Stout (eds), *Early Urban Planning* (nine volumes, London: Routledge/Thoemmes, 1998) and in earlier editions by Attic Books (1985), Eastbourne (1985), MIT Press (1965), Faber & Faber (1960, 1951, and 1946). The original edition appeared under the title *To-morrow: A Peaceful Path to Real Reform* (Sonnenschein, 1898), and an elegant new edition is now available, under the original title, edited by Peter Hall and Colin Ward (London: Routledge, 2003).

Biographies of Ebenezer Howard include Robert Beevers, *The Garden City Utopia: A Critical Biography of Ebenezer Howard* (New York: St. Martin's, 1988), and Dugald Macfadyen, *Sir Ebenezer Howard and the Town Planning Movement* (Manchester: Manchester University Press, 1933; reprinted Cambridge, MA: MIT Press, 1970). Excellent accounts of Howard and the Garden City movement may be found in Robert Fishman, *Urban Utopias in the Twentieth Century* (New York: Basic Books, 1977) and Peter Hall, *Cities of Tomorrow* (Oxford: Blackwell, 1988).

Additional books about Ebenezer Howard and the Garden City movement include Standish Meacham, *Regaining Paradise: Englishness and the Early Garden City Movement* (New Haven, CT: Yale University Press, 1999), Peter Geoffrey Hall and Colin Ward, *Sociable Cities: The Legacy of Ebenezer Howard* (New York: Wiley, 1998), Stephen V. Ward (ed.), *The Garden City: Past, Present and Future* (London: E & FN Spon, 1992), Stanley Buder, *Visionaries and Planners: The Garden City Movement and the Modern Community* (Oxford: Oxford University Press, 1990), and Kermit Parsons and David Schuyler (eds), *From the Garden City to Green Cities: The Legacy of Ebenezer Howard* (Baltimore, MD: Johns Hopkins University Press, 2002).

AUTHOR'S INTRODUCTION

In these days of strong party feeling and of keenly contested social and religious issues, it might perhaps be thought difficult to find a single question having a vital bearing upon national life and well-being on which all persons, no matter of what political party, or of what shade of sociological opinion, would be found to be fully and entirely agreed . . .

[. . .]

There is, however, a question in regard to which one can scarcely find any difference of opinion . . . It is wellnigh universally agreed by men of all parties, not only in England, but all over Europe and America and our colonies, that it is deeply to be deplored that the people should continue to stream into the already over-crowded cities, and should thus further deplete the country districts.

All . . . are agreed on the pressing nature of this problem, all are bent on its solution, and though it would doubtless be quite Utopian to expect a similar agreement as to the value of any remedy that may be proposed, it is at least of immense importance that, on a subject thus universally regarded as of supreme importance, we have such a consensus of opinion at the outset. This will be the more remarkable and the more hopeful sign when it is shown, as I believe will be conclusively shown in this work, that the answer to this, one of the most pressing questions of the day, makes of comparatively easy solution many other problems which have hitherto taxed the ingenuity of the greatest thinkers and reformers of our time. Yes, the key to the problem how to restore the people to the land – that beautiful land of ours, with its canopy of sky, the air that blows upon it, the sun that warms it, the rain and dew that moisten it – the very embodiment of Divine love for man – is indeed a *Master Key*, for it is the key to a portal through which, even when scarce ajar, will be seen to pour a flood of light on the problems of intemperance, of excessive toil, of restless anxiety, of grinding poverty – the true limits of Governmental interference, ay, and even the relations of man to the Supreme Power.

It may perhaps be thought that the first step to be taken towards the solution of this question – how to restore the people to the land – would involve a careful consideration of the very numerous causes which have hitherto led to their aggregation in large cities. Were this the case, a very prolonged enquiry would be necessary at the outset. Fortunately, alike for writer and for reader, such an analysis is not, however, here requisite, and for a very simple reason, which may be stated thus: Whatever may have been the causes which have operated in the past, and are operating now, to draw the people into the cities, those causes may all be summed up as "attractions"; and it is obvious, therefore, that no remedy can possibly be effective which will not present to the people, or at least to considerable portions of them, greater "attractions" than our cities now possess, so that the force of the old "attractions" shall be overcome by the force of new "attractions" which are to be created. Each city may be regarded as a magnet, each person as a needle; and, so viewed, it is at once seen that nothing short of the discovery of a method for constructing magnets of yet greater power than our cities possess can be effective for redistributing the population in a spontaneous and healthy manner.

So presented, the problem may appear at first sight to be difficult, if not impossible, of solution. "What", some may be disposed to ask, "can possibly be done to make the country more attractive to a workaday people than the town – to make wages, or at least the standard of physical comfort, higher in the country than in the town; to secure in the country equal possibilities of social intercourse, and to make the prospects of advancement for the average man or woman equal, not to say superior, to those enjoyed in our large cities?" The issue one constantly finds presented in a form very similar to that. The subject is treated continually in the public press, and in all forms of discussion, as though men, or at least working men, had not now, and never could have, any choice or alternative, but either, on the one hand, to stifle their love for human society – at least in wider relations than can be found in a straggling village – or, on the other hand, to forgo almost entirely all the keen and pure delights of the country. The question is universally considered as though it were now, and for ever must remain, quite impossible for working people to live in the country and yet be engaged in pursuits other than agricultural; as though crowded, unhealthy cities were the last word of economic science; and as if our present form of industry, in which sharp lines divide agricultural from industrial pursuits, were necessarily an enduring one. This fallacy is the very common one of ignoring altogether the possibility of alternatives other than those presented to the mind. There are in reality not only, as is so constantly assumed, two alternatives – town life and country life – but a third alternative, in which all the advantages of the most energetic and active town life, with all the beauty and delight of the country, may be secured in perfect combination; and the certainty of being able to live this life will be the magnet which will produce the effect for which we are all striving – the spontaneous movement of the people from our crowded cities to the bosom of our kindly mother earth, at once the source of life, of happiness, of wealth, and of power. The town and the country may, therefore, be regarded as two magnets, each striving to draw the people to itself – a rivalry which a new form of life, partaking of the nature of both, comes to take part in. This may be illustrated by a diagram (Figure 1) of "The Three Magnets", in which the chief advantages of the Town and of the Country are set forth with their corresponding drawbacks, while the advantages of the Town–Country are seen to be free from the disadvantages of either.

The Town magnet, it will be seen, offers, as compared with the Country magnet, the advantages of high wages, opportunities for employment, tempting prospects of advancement, but these are largely counterbalanced by high rents and prices. Its social opportunities and its places of amusement are very alluring, but excessive hours of toil, distance from work, and the "isolation of crowds" tend greatly to reduce the value of these good things. The well-lit streets are a great attraction, especially in winter, but the sunlight is being more and more shut out, while the air is so vitiated that the fine public buildings, like the sparrows, rapidly become covered with soot, and the very statues are in despair. Palatial edifices and fearful slums are the strange, complementary features of modern cities.

The Country magnet declares herself to be the source of all beauty and wealth; but the Town magnet mockingly reminds her that she is very dull for lack of society, and very sparing of her gifts for lack of capital. There are in the country beautiful vistas, lordly parks, violet-scented woods, fresh air, sounds of rippling water; but too often one sees those threatening words, "Trespassers will be prosecuted". Rents, if estimated by the acre, are certainly low, but such low rents are the natural fruit of low wages rather than a cause of substantial comfort; while long hours and lack of amusements forbid the bright sunshine and the pure air to gladden the hearts of the people. The one industry, agriculture, suffers frequently from excessive rainfalls; but this wondrous harvest of the clouds is seldom properly in-gathered, so that, in times of drought,

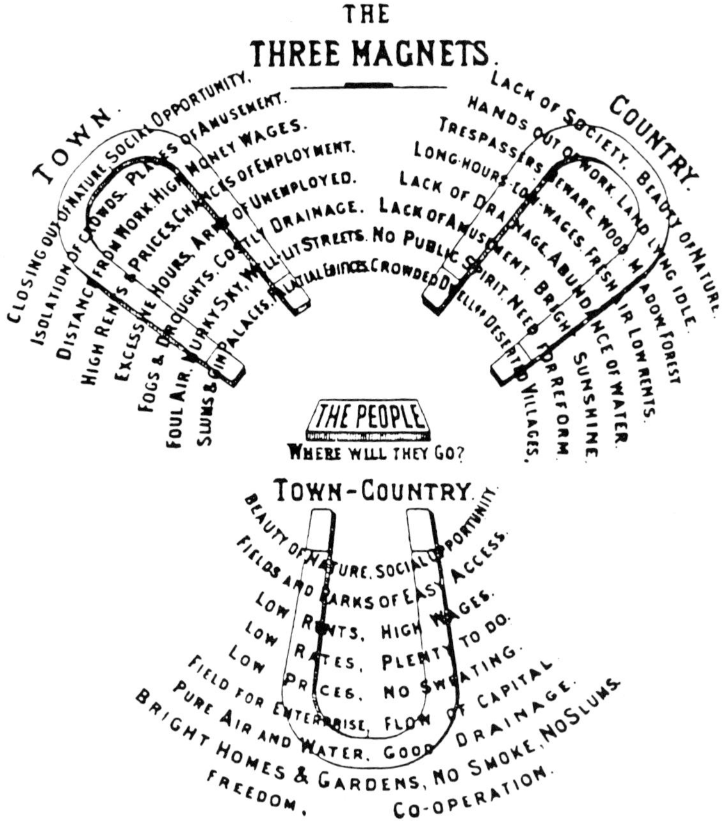

Figure 1

there is frequently, even for drinking purposes, a most insufficient supply. Even the natural healthfulness of the country is largely lost for lack of proper drainage and other sanitary conditions, while, in parts almost deserted by the people, the few who remain are yet frequently huddled together as if in rivalry with the slums of our cities.

But neither the Town magnet nor the Country magnet represents the full plan and purpose of nature. Human society and the beauty of nature are meant to be enjoyed together. The two magnets must be made one. As man and woman by their varied gifts and faculties supplement each other, so should town and country. The town is the symbol of society – of mutual help and friendly co-operation, of fatherhood, motherhood, brotherhood, sisterhood, of wide relations between man and man – of broad, expanding sympathies – of science, art, culture, religion. And the country! The country is the symbol of God's love and care for man. All that we are and all that we have comes from it. Our bodies are formed of it; to it they return. We are fed by it, clothed by it, and by it are we warmed and sheltered. On its bosom we rest. Its beauty is the inspiration of art, of music, of poetry. Its forces propel all the wheels of industry. It is the source of all health, all wealth, all knowledge. But its fullness of joy and wisdom has not revealed itself to man. Nor can it ever, so long as this unholy, unnatural separation of society and nature endures. Town and country *must be married*, and out of this joyous union will spring a new hope, a new life, a new civilization. It is the purpose of this work to show how a first step can be taken in this direction by the construction of a Town–Country magnet; and I hope to convince the reader that this is practicable, here and now, and that on principles which are the very soundest, whether viewed from the ethical or the economic standpoint.

I will undertake, then, to show how in "Town–Country" equal, nay better, opportunities of social intercourse may be enjoyed than are enjoyed in any crowded city, while yet the beauties of nature may encompass and enfold each dweller therein; how higher wages are compatible with reduced rents and rates; how abundant opportunities for employment and bright prospects of advancement may be secured for all; how capital may be attracted and wealth created; how the most admirable sanitary conditions may be ensured; how beautiful homes and gardens may be seen on every hand; how the bounds of freedom may be widened, and yet all the best results of concert and co-operation gathered in by a happy people.

The construction of such a magnet, could it be effected, followed, as it would be, by the construction of many more, would certainly afford a solution of the burning question set before us by Sir John Gorst, "how to back the tide of migration of the people into the towns, and to get them back upon the land".

[. . .]

THE TOWN–COUNTRY MAGNET

The reader is asked to imagine an estate embracing an area of 6,000 acres, which is at present purely agricultural, and has been obtained by purchase in the open market at a cost of £40 an acre, or £240,000. The purchase money is supposed to have been raised on mortgage debentures, bearing interest at an average rate not exceeding 4 per cent. The estate is legally vested in the names of four gentlemen of responsible position and of undoubted probity and honour, who hold it in trust, first, as a security for the debenture-holders, and, secondly, in trust for the people of Garden City, the Town–Country magnet, which it is intended to build thereon. One essential feature of the plan is that all ground rents, which are to be based upon the annual value of the land, shall be paid to the trustees, who, after providing for interest and sinking fund, will hand the balance to the Central Council of the new municipality, to be employed by such Council in the creation and maintenance of all necessary public works – roads, schools, parks, etc. The objects of this land purchase may be stated in various ways, but it is sufficient here to say that some of the chief objects are these: To find for our industrial population work at wages of *higher purchasing power*, and to secure healthier surroundings and more regular employment. To enterprising manufacturers, co-operative societies, architects, engineers, builders, and mechanicians of all kinds, as well as to many engaged in various professions, it is intended to offer a means of securing new and better employment for their capital and talents, while to the agriculturists at present on the estate as well as to those who may migrate thither, it is designed to open a new market for their produce close to their doors. Its object is, in short, to raise the standard of health and comfort of all true workers of whatever grade – the means by which these objects are to be achieved being a healthy, natural, and economic combination of town and country life, and this on land owned by the municipality.

Garden City, which is to be built near the centre of the 6,000 acres, covers an area of 1,000 acres, or a sixth part of the 6,000 acres, and might be of circular form, 1,240 yards (or nearly three-quarters of a mile) from centre to circumference. (Figure 2 is a ground plan of the whole municipal area, showing the town in the centre; and Figure 3, which represents one section or ward of the town, will be useful in following the description of the town itself – *a description which is, however, merely suggestive, and will probably be much departed from . . .*)

Six magnificent boulevards – each 120 feet wide – traverse the city from centre to circumference, dividing it into six equal parts or wards. In the centre is a circular space containing about five and a half acres, laid out as a beautiful and well-watered garden; and, surrounding this garden, each standing in its own ample grounds, are the larger public buildings – town hall, principal concert and lecture hall, theatre, library, museum, picture-gallery, and hospital.

The rest of the large space encircled by the "Crystal Palace" is a public park, containing 145 acres, which includes ample recreation grounds within very easy access of all the people.

Running all round the Central Park (except where it is intersected by the boulevards) is a wide glass arcade called the "Crystal Palace", opening on to the park. This building is in wet weather one of the favourite resorts of the people, whilst the knowledge that its bright shelter is ever close at hand tempts people into Central Park, even in the most doubtful of weathers. Here manufactured goods are exposed for sale, and here most of that class of shopping which requires the joy of deliberation and selection is done. The space enclosed by the Crystal Palace is, however, a good deal larger than is required for these purposes, and a considerable part of it is used as a Winter Garden – the whole forming a permanent exhibition of a most

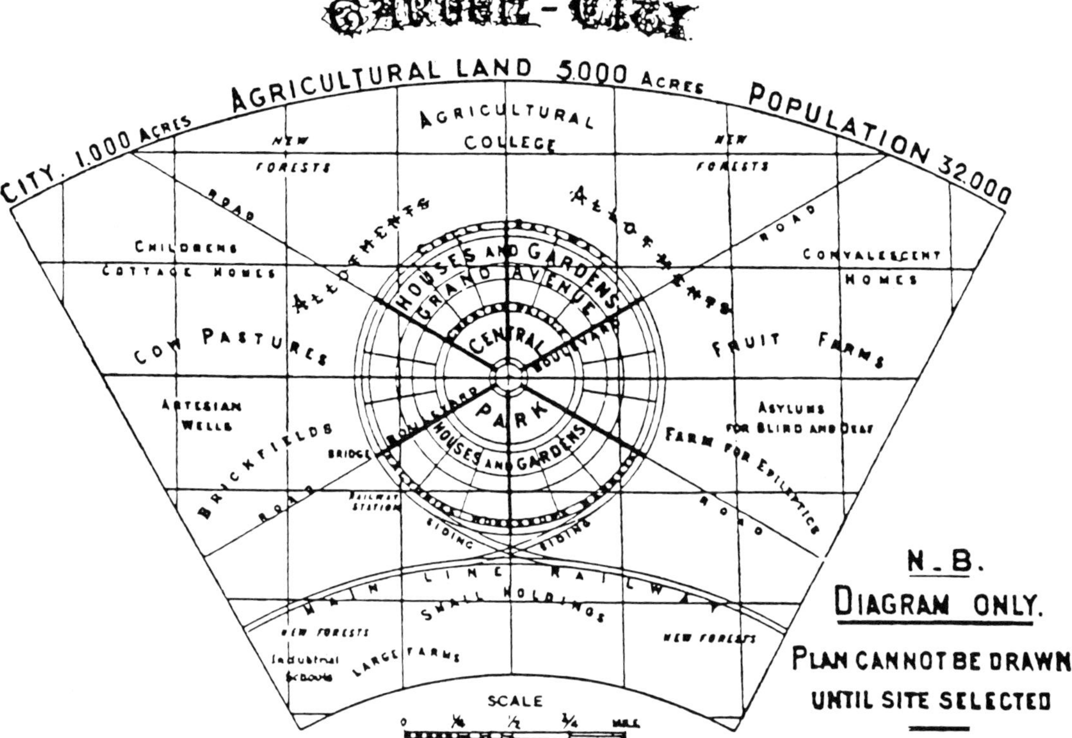

Figure 2

attractive character, whilst its circular form brings it near to every dweller in the town – the furthest removed inhabitant being within 600 yards.

Passing out of the Crystal Palace on our way to the outer ring of the town, we cross Fifth Avenue – lined, as are all the roads of the town, with trees – fronting which, and looking on to the Crystal Palace, we find a ring of very excellently built houses, each standing in its own ample grounds; and, as we continue our walk, we observe that the houses are for the most part built either in concentric rings, facing the various avenues (as the circular roads are termed), or fronting the boulevards and roads which all converge to the centre of the town. Asking the friend who accompanies us on our journey what the population of this little city may be, we are told about 30,000 in the city itself, and about 2,000 in the agricultural estate, and that there are in the town 5,500 building lots of an *average* size of 20 feet × 130 feet – the minimum space allotted for the purpose being 20 × 100. Noticing the very varied architecture and design which the houses and groups of houses display – some having common gardens and co-operative kitchens – we learn that general observance of street line or harmonious departure from it are the chief points as to house building, over which the municipal authorities exercise control, for, though proper sanitary arrangements are strictly enforced, the fullest measure of individual taste and preference is encouraged.

Walking still toward the outskirts of the town, we come upon "Grand Avenue". This avenue is fully entitled to the name it bears, for it is 420 feet wide, and, forming a belt of green upwards of three miles long, divides that part of the town which lies outside Central Park into two belts. It really constitutes an additional park of 115 acres – a park which is within 240 yards of the furthest removed inhabitant. In this splendid avenue six sites, each of four acres, are occupied by public schools and their surrounding playgrounds and gardens, while other sites are reserved for churches, of such denominations as the religious beliefs of the people may determine, to be erected and maintained out of the funds of the worshippers and their friends. We observe that the houses fronting on Grand

Avenue have departed (at least in one of the wards – that of which Figure 3 is a representation) – from the general plan of concentric rings, and, in order to ensure a longer line of frontage on Grand Avenue, are arranged in crescents – thus also to the eye yet further enlarging the already splendid width of Grand Avenue.

On the outer ring of the town are factories, warehouses, dairies, markets, coal yards, timber yards, etc., all fronting on the circle railway, which encompasses the whole town, and which has sidings connecting it with a main line of railway which passes through the estate. This arrangement enables goods to be loaded direct into trucks from the warehouses and work shops, and so sent by railway to distant markets, or to be taken direct from the trucks into the warehouses or factories; thus not only effecting a very great saving in regard to packing and cartage, and reducing to a minimum loss from breakage, but also, by reducing the traffic on the roads of the town, lessening to a very marked extent the cost of their maintenance. The smoke fiend is kept well within bounds in Garden City; for all machinery is driven by electric energy, with the result that the cost of electricity for lighting and other purposes is greatly reduced.

The refuse of the town is utilized on the agricultural portions of the estate, which are held by various individuals in large farms, small holdings, allotments, cow pastures, etc.; the natural competition of these various methods of agriculture, tested by the willingness of occupiers to offer the highest rent to the municipality, tending to bring about the best system of husbandry, or, what is more probable, the best systems adapted for various purposes. Thus it is easily conceivable that it may prove advantageous to grow wheat in very large fields, involving united action under a capitalist farmer, or by a body of co-operators; while the cultivation of vegetables, fruits, and flowers, which requires closer and more personal care, and more of the artistic and inventive faculty, may possibly be best dealt with by individuals, or by small groups of individuals having a common belief in the efficacy and value of certain dressings, methods of culture, or artificial and natural surroundings.

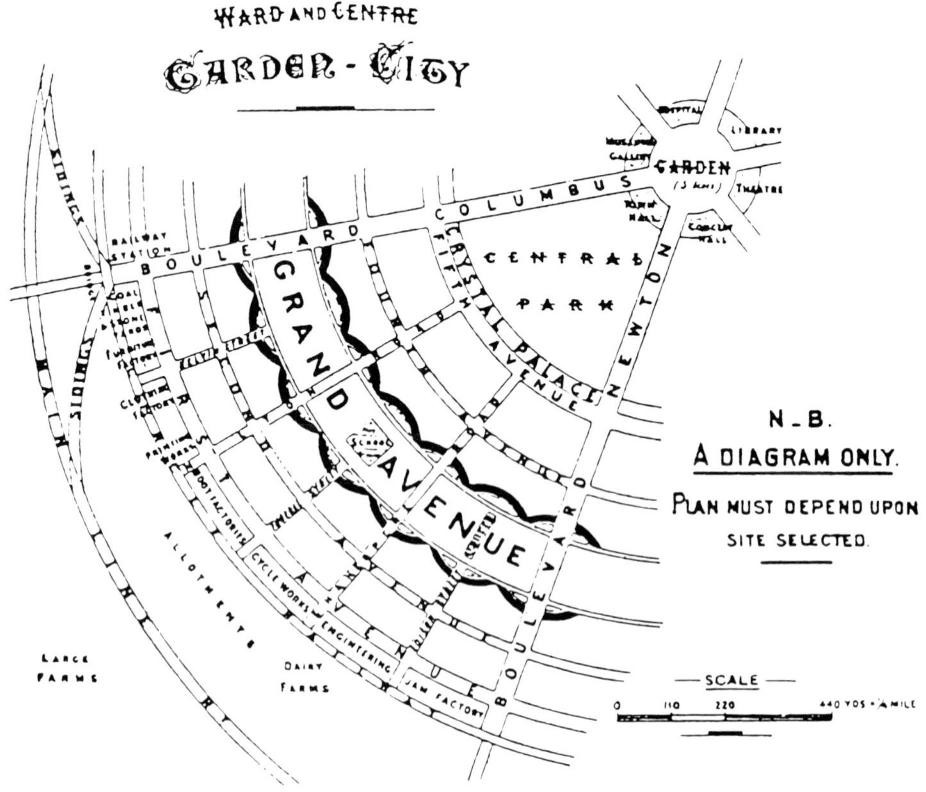

Figure 3

This plan, or, if the reader be pleased to so term it, this absence of plan, avoids the dangers of stagnation or dead level, and, though encouraging individual initiative, permits of the fullest co-operation, while the increased rents which follow from this form of competition are common or municipal property, and by far the larger part of them are expended in permanent improvements.

While the town proper, with its population engaged in various trades, callings, and professions, and with a store or depot in each ward, offers the most natural market to the people engaged on the agricultural estate, inasmuch as to the extent to which the towns-people demand their produce they escape altogether any railway rates and charges; yet the farmers and others are not by any means limited to the town as their only market, but have the fullest right to dispose of their produce to whomsoever they please. Here, as in every feature of the experiment, it will be seen that it is not the area of rights which is contracted, but the area of choice which is enlarged.

This principle of freedom holds good with regard to manufacturers and others who have established themselves in the town. These manage their affairs in their own way, subject, of course, to the general law of the land, and subject to the provision of sufficient space for workmen and reasonable sanitary conditions. Even in regard to such matters as water, lighting, and telephonic communication – which a municipality, if efficient and honest, is certainly the best and most natural body to supply – no rigid or absolute monopoly is sought; and if any private corporation or any body of individuals proved itself capable of supplying on more advantageous terms, either the whole town or a section of it, with these or any commodities the supply of which was taken up by the corporation, this would be allowed. No really sound system of *action* is in more need of artificial support than is any sound system of *thought*. The area of municipal and corporate action is probably destined to become greatly enlarged; but, if it is to be so, it will be because the people possess faith in such action, and that faith can be best shown by a wide extension of the area of freedom.

Dotted about the estate are seen various charitable and philanthropic institutions. These are not under the control of the municipality, but are supported and managed by various public-spirited people who have been invited by the municipality to establish these institutions in an open healthy district, and on land let to them at a pepper-corn rent, it occurring to the authorities that they can the better afford to be thus generous, as the spending power of these institutions greatly benefits the whole community. Besides, as those persons who migrate to the town are among its most energetic and resourceful members, it is but just and right that their more helpless brethren should be able to enjoy the benefits of an experiment which is designed for humanity at large.

"A Contemporary City"

from *The City of Tomorrow and its Planning* (1929)

Le Corbusier

Editors' Introduction

Le Corbusier (1887–1965) was one of the founding fathers of the Modernist movement and of what has come to be known as the International Style in architecture. Painter, architect, city planner, philosopher, author of revolutionary cultural manifestos – Le Corbusier exemplified the energy and efficiency of the Machine Age. His was the bold, nearly mystical rationality of a generation that was eager to accept the scientific spirit of the twentieth century on its own terms and to throw off all pre-existing ties – political, cultural, conceptual – with what it considered an exhausted, outmoded past.

Born Charles-Edouard Jeanneret-Gris, Le Corbusier grew up in the Swiss town of La Chaux-de-Fonds, noted for its watch-making industry. He took his famous pseudonym after he moved to Paris to pursue a career in art and architecture. From the first, his designs for modern houses – he called them "machines for living" – were strikingly original, and many people were shocked by the spare cubist minimalism of his designs. The real shock, however, came in 1922 when Le Corbusier presented the public with his plan for "A Contemporary City of Three Million People." Laid out in a rigidly symmetrical grid pattern, the city consisted of neatly spaced rows of identical, strictly geometrical skyscrapers, as illustrated in Plate 37. This was not the city of the future, Le Corbusier insisted, but the city of today. It was to be built on the Right Bank, after demolishing several hundred acres of the existing urban fabric of Paris!

The "Contemporary City" proposal certainly caught the attention of the public, but it did not win Le Corbusier many actual urban planning commissions. Throughout the 1920s, 1930s, and 1940s, he sought out potential patrons wherever he could find them – the industrial capitalists of the Voisin automobile company, the communist rulers of the Soviet Union, and the fascist Vichy government of occupied France – mostly without success. Le Corbusier's real impact came not from cities he designed and built himself but from cities that were built by others incorporating the planning principles that he pioneered. Most notable among these was the notion of "the skyscraper in the park," an idea that is today ubiquitous. Whether in relatively complete examples like Brasilia and Chandigar, India (where new cities were built from scratch), or in partial examples such as the skyscraper parks and the high-rise housing blocks that have been built in cities worldwide, the Le Corbusier vision has truly transformed the global urban environment.

Le Corbusier's "Contemporary City" plan has often been contrasted to Frank Lloyd Wright's "Broadacre" (p. 345), and the comparison of a thoroughly centralized versus a thoroughly decentralized plan is indeed striking. Le Corbusier's boldness invites comparison with the original optimism of the post-World War II reconstruction and redevelopment efforts and even with the work of such visionary megastructuralists as Paolo Soleri. Some, however, have seen in the hyper-rationality of the pure Corbusian ideal an elitism and rigid class structure that runs counter to the democratic tradition. Lewis Mumford (p. 91), Jane Jacobs (p. 105), and Peter Hall (p. 373) may be counted as three of the severest critics. Allan Jacobs and Donald Appleyard's "Urban Design Manifesto" (p. 518) deliberately takes the form of a Le Corbusier pronouncement but rejects his program, opting instead for lively streets, participatory planning, and the integration of old buildings into the new urban fabric. Beneath all the sparkling clarity of Le Corbusier's urban designs are questions that must forever remain conjectural: how would democratic politics be practiced in a Corbusian city? What would social relationships be like amid the gleaming towers? Many of the "mega-urban regions" of Asia (p. 590) seem to rely on Corbusian principles, but would the "creative class" that Richard Florida (p. 143) writes about be comfortable in a Corbusian city? How would the space-of-flows/space-of-place relationships envisioned by Manuel Castells (p. 572) work, or not work, within a Corbusian environment? And what is it about the Corbusian skyscraper as a characteristic cultural form of modern Western urbanism that made the twin towers of the World Trade Center a target for the Al-Qaeda Islamist attacks on September 11, 2001?

Le Corbusier's writings include *The City of Tomorrow and its Planning* (New York: Dover, 1987, translated by Frederich Etchells from *Urbanisme*, 1929), *Concerning Town Planning*, (New Haven, CT: Yale University Press, 1948, translated by Clive Entwistle from *Propos d'Urbanisme*, 1946), and *L'Urbanisme des Trois Etablissements Humaines* (Paris: Editions de Minuit, 1959).

Excellent accounts of Le Corbusier's ideas may be found in Robert Fishman, *Urban Utopias in the Twentieth Century* (New York: Basic Books, 1977), Peter Hall, *Cities of Tomorrow* (Oxford: Blackwell, 1988), Kenneth Frampton, *Le Corbusier: Architect of the Twentieth Century* (New York: Abrams, 2002, with photographs by Roberto Schezen), and Nicholas Fox Weber, *Le Corbusier: A Life* (New York: Knopf, 2008). For a closer look at some of Le Corbusier's most important urban planning projects, consult Vikramaditya Prakash, *Chandigarh's Le Corbusier: The Struggle for Modernity in Postcolonial India* (Seattle, WA: University of Washington Press, 2002) and Klaus-Peter Gast and Arthur Ruegg, *Le Corbusier: Paris-Chandigarh* (Boston, MA: Birkhäuser, 2000). Jean-Louis Cohen, *Le Corbusier, 1887–1965* (Los Angeles, CA: Taschen, 2005) contains excellent photographs of Le Corbusier's architectural work from all periods as does *Le Corbusier Le Grand* by the editors of Phaidon Press (London: 2008). For background on modernism as a movement, consult Richard Weston, *Modernism* (London: Phaidon, 2001), Christopher Wilk, *Modernism: Designing a New World* (London: Victoria & Albert Museum, 2006), and Peter Gay, *Modernism: The Lure of Heresy* (New York: Norton, 2007). For the broader cultural and literary background of the movement, consult Pericles Lewis, *The Cambridge Introduction to Modernism* (Cambridge: Cambridge University Press, 2007).

The existing congestion in the centre must be eliminated.

The use of technical analysis and architectural synthesis enabled me to draw up my scheme for a contemporary city of three million inhabitants. The result of my work was shown in November 1922 at the Salon d'Automne in Paris. It was greeted with a sort of stupor; the shock of surprise caused rage in some quarters and enthusiasm in others. The solution I put forward was a rough one and completely uncompromising. There were no notes to accompany the plans, and, alas! not everybody can read a plan. I should have had to be constantly on the spot in order to reply to the fundamental questions which spring from the very depths of human feelings. Such questions are of profound interest and cannot remain unanswered. When at a later date it became necessary that this book should be written, a book in which I could formulate the new principles of Town Planning, I resolutely decided *first of all* to find answers to these fundamental questions. I have used two kinds of argument: first, those essentially human ones which start from the mind or the heart or the physiology of our sensations as a basis; secondly, historical and statistical arguments. Thus I could keep in touch with what is fundamental and at the same time be master of the environment in which all this takes place.

In this way I hope I shall have been able to help my reader to take a number of steps by means of which he can reach a sure and certain position. So that when I unroll my plans I can have the happy assurance that his astonishment will no longer be stupefaction nor his fears mere panic.

[. . .]

A CONTEMPORARY CITY OF THREE MILLION INHABITANTS

Proceeding in the manner of the investigator in his laboratory, I have avoided all special cases, and all that may be accidental, and I have assumed an ideal site to begin with. My object was not to overcome the existing state of things, but *by constructing a theoretically water-tight formula to arrive at the fundamental principles of modern town planning*. Such fundamental principles, if they are genuine, can serve as the skeleton of any system of modern town planning; being as it were the rules according to which development will take place. We shall then be in a position to take a special case, no matter what: whether it be Paris, London, Berlin, New York or some small town. Then, as a result of what we have learnt, we can take control and decide in what direction the forthcoming battle is to be waged. For the desire to rebuild any great city in a modern way is to engage in a formidable battle. Can you imagine people engaging in a battle without knowing their objectives? Yet that is exactly what is happening. The authorities are compelled to do something, so they give the police white sleeves or set them on horseback, they invent sound signals and light signals, they propose to put bridges over streets or moving pavements under the streets; more garden cities are suggested, or it is decided to suppress the tramways, and so on. And these decisions are reached in a sort of frantic haste in order, as it were, to hold a wild beast at bay. That beast is the great city. It is infinitely more powerful than all these devices. And it is just beginning to wake. What will to-morrow bring forth to cope with it?

We must have some rule of conduct.

We must have fundamental principles for modern town planning.

Site

A level site is the ideal site [for the contemporary city (Figure 1)]. In all those places where traffic becomes over-intensified the level site gives a chance of a normal solution to the problem. Where there is less traffic, differences in level matter less.

The river flows far away from the city. The river is a kind of liquid railway, a goods station and a sorting house. In a decent house the servants' stairs do not go through the drawing room – even if the maid is charming (or if the little boats delight the loiterer leaning on a bridge).

Population

This consists of the citizens proper; of suburban dwellers; and of those of a mixed kind.

(a) Citizens are of the city: those who work and live in it.
(b) Suburban dwellers are those who work in the outer industrial zone and who do not come into the city: they live in garden cities.
(c) The mixed sort are those who work in the business parts of the city but bring up their families in garden cities.

To classify these divisions (and so make possible the transmutation of these recognized types) is to attack the most important problem in town planning, for such a classification would define the areas to be allotted to these three sections and the delimitation of their boundaries. This would enable us to formulate and resolve the following problems:

1 The *City*, as a business and residential centre.
2 The *Industrial City* in relation to the *Garden Cities* (i.e. the question of transport).
3 The *Garden Cities* and the *daily transport* of the workers.

Our first requirement will be an organ that is compact, rapid, lively and concentrated: this is the City with its well organized centre. Our second requirement will be another organ, supple, extensive and elastic; this is the *Garden City* on the periphery. Lying between these two organs, we must *require the legal establishment* of that absolute necessity, a protective zone which allows of extension, *a reserved zone* of woods and fields, a fresh-air reserve.

Density of population

The more dense the population of a city is the less are the distances that have to be covered. The moral, therefore, is that we must *increase the density of the centres of our cities, where business affairs are carried on.*

Lungs

Work in our modern world becomes more intensified day by day, and its demands affect our nervous system in a way that grows more and more dangerous. Modern toil demands quiet and fresh air, not stale air.

The towns of to-day can only increase in density at the expense of the open spaces which are the lungs of a city.

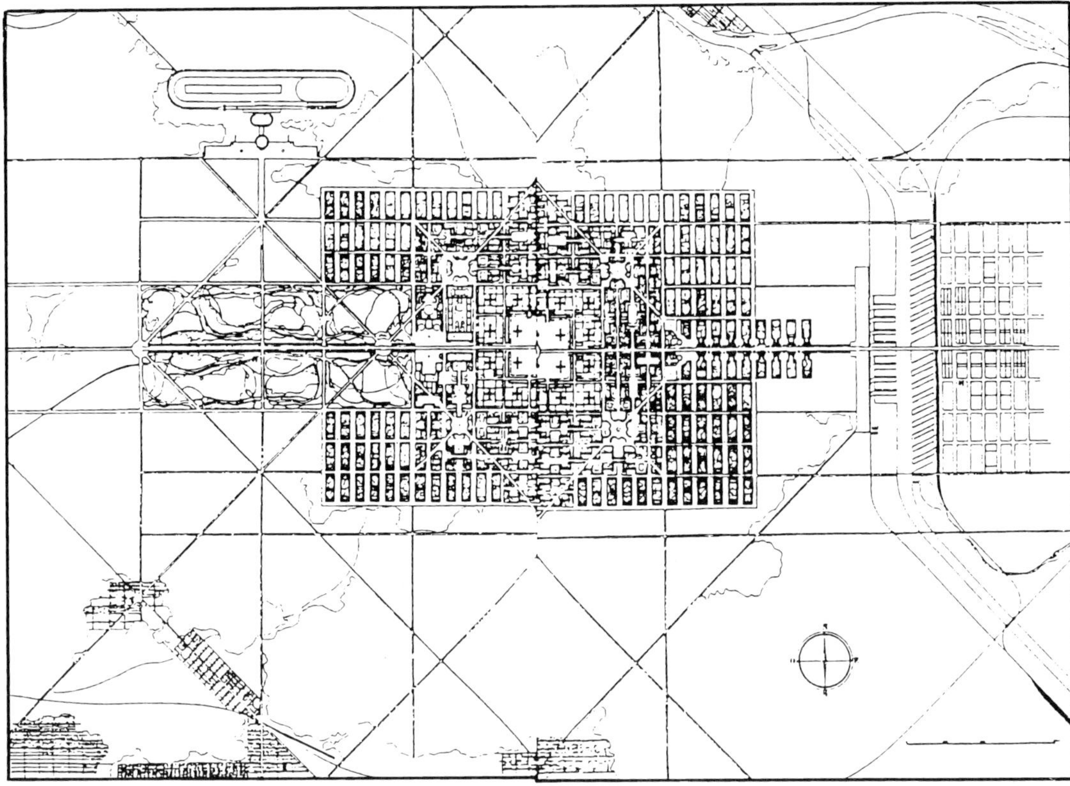

Figure 1

We must *increase the open spaces and diminish the distances to be covered.* Therefore the centre of the city must be constructed *vertically.*

The city's residential quarters must no longer be built along "corridor-streets", full of noise and dust and deprived of light.

It is a simple matter to build urban dwellings away from the streets, without small internal courtyards and with the windows looking on to large parks; and this whether our housing schemes are of the type with "set-backs" or built on the "cellular" principle.

The street

The street of to-day is still the old bare ground which has been paved over, and under which a few tube railways have been run.

The modern street in the true sense of the word is a new type of organism, a sort of stretched-out workshop, a home for many complicated and delicate organs, such as gas, water and electric mains. It is contrary to all economy, to all security, and to all sense to bury these important service mains. They ought to be accessible throughout their length. The various storeys of this stretched-out workshop will each have their own particular functions. If this type of street, which I have called a "workshop", is to be realized, it becomes as much a matter of construction as are the houses with which it is customary to flank it, and the bridges which carry it over valleys and across rivers.

The modern street should be a masterpiece of civil engineering and no longer a job for navvies.

The "corridor-street" should be tolerated no longer, for it poisons the houses that border it and leads to the construction of small internal courts or "wells".

Traffic

Traffic can be classified more easily than other things.

To-day traffic is not classified – it is like dynamite flung at hazard into the street, killing pedestrians. Even so, *traffic does not fulfil its function*. This sacrifice of the pedestrian leads nowhere.

If we classify traffic we get:

(a) Heavy goods traffic.
(b) Lighter goods traffic, i.e. vans, etc., which make short journeys in all directions.
(c) Fast traffic, which covers a large section of the town.

Three kinds of roads are needed, and in superimposed storeys:

(a) Below-ground there would be the street for heavy traffic. This storey of the houses would consist merely of concrete piles, and between them large open spaces which would form a sort of clearing-house where heavy goods traffic could load and unload.
(b) At the ground floor level of the buildings there would be the complicated and delicate network of the ordinary streets taking traffic in every desired direction.
(c) Running north and south, and east and west, and forming the two great axes of the city, there would be great *arterial roads for fast one-way traffic* built on immense reinforced concrete bridges 120 to 180 yards in width and approached every half-mile or so by subsidiary roads from ground level. These arterial roads could therefore be joined at any given point, so that even at the highest speeds the town can be traversed and the suburbs reached without having to negotiate any cross-roads.

The number of existing streets should be diminished by two-thirds. The number of crossings depends directly on the number of streets; and *cross-roads are an enemy to traffic*. The number of existing streets was fixed at a remote epoch in history. The perpetuation of the boundaries of properties has, almost without exception, preserved even the faintest tracks and footpaths of the old village and made streets of them, and sometimes even an avenue . . . The result is that we have cross-roads every fifty yards, even every twenty yards or ten yards. And this leads to the ridiculous traffic congestion we all know so well.

The distance between two bus stops or two tube stations gives us the necessary unit for the distance between streets, though this unit is conditional on the speed of vehicles and the walking capacity of pedestrians. So an average measure of about 400 yards would give the normal separation between streets, and make a standard for urban distances. My city is conceived on the gridiron system with streets every 400 yards, though occasionally these distances are subdivided to give streets every 200 yards.

This triple system of superimposed levels answers every need of motor traffic (lorries, private cars, taxis, buses) because it provides for rapid and *mobile* transit.

Traffic running on fixed rails is only justified if it is in the form of a convoy carrying an immense load; it then becomes a sort of extension of the underground system or of trains dealing with suburban traffic. *The tramway has no right to exist in the heart of the modern city.*

If the city thus consists of plots about 400 yards square, this will give us sections of about 40 acres in area, and the density of population will vary from 50,000 down to 6,000, according as the "lots" are developed for business or for residential purposes. The natural thing, therefore, would be to continue to apply our unit of distance as it exists in the Paris tubes to-day (namely, 400 yards) and to put a station in the middle of each plot.

Following the two great axes of the city, two "storeys" below the arterial roads for fast traffic, would run the tubes leading to the four furthest points of the garden city suburbs, and linking up with the metropolitan network . . . At a still lower level, and again following these two main axes, would run the one-way loop systems for suburban traffic, and below these again the four great main lines serving the provinces and running north, south, east and west. These main lines would end at the Central Station, or better still might be connected up by a loop system.

The station

There is only one station. The only place for the station is in the centre of the city. It is the natural place for it, and there is no reason for putting it anywhere else. The railway station is the hub of the wheel.

The station would be an essentially subterranean building. Its roof, which would be two storeys above the natural ground level of the city, would form the aerodrome for aero-taxis. This aerodrome (linked up with the main aerodrome in the protected zone) must be in close contact with the tubes, the suburban lines, the main lines, the main arteries and the administrative services connected with all these . . .

The plan of the city

The basic principles we must follow are these:

1 We must de-congest the centres of our cities.
2 We must augment their density.
3 We must increase the means for getting about.
4 We must increase parks and open spaces.

At the very centre we have the *station* with its landing stage for aero-taxis.

Running north and south, and east and west, we have the *main arteries* for fast traffic, forming elevated roadways 120 feet wide.

At the base of the sky-scrapers and all round them we have a great open space 2,400 yards by 1,500 yards, giving an area of 3,600,000 square yards, and occupied by gardens, parks and avenues. In these parks, at the foot of and round the sky-scrapers, would be the restaurants and cafes, the luxury shops, housed in buildings with receding terraces: here too would be the theatres, halls and so on; and here the parking places or garage shelters.

The sky-scrapers are designed purely for business purposes.

On the left we have the great public buildings, the museums, the municipal and administrative offices. Still further on the left we have the "Park" (which is available for further logical development of the heart of the city).

On the right, and traversed by one of the arms of the main arterial roads, we have the warehouses, and the industrial quarters with their goods stations.

All around the city is the *protected zone* of woods and green fields.

Further beyond are the *garden cities*, forming a wide encircling band.

Then, right in the midst of all these, we have the *Central Station*, made up of the following elements:

(a) The landing-platform; forming an aerodrome of 200,000 square yards in area.
(b) The entresol or mezzanine; at this level are the raised tracks for fast motor traffic: the only crossing being gyratory.
(c) The ground floor where are the entrance halls and booking offices for the tubes, suburban lines, main line and air traffic.
(d) The "basement": here are the tubes which serve the city and the main arteries.
(e) The "sub-basement": here are the suburban lines running on a one-way loop.
(f) The "sub-sub-basement": here are the main lines (going north, south, east and west).

The city

Here we have twenty-four sky-scrapers capable each of housing 10,000 to 50,000 employees; this is the business and hotel section, etc., and accounts for 400,000 to 600,000 inhabitants.

The residential blocks, of the two main types already mentioned, account for a further 600,000 inhabitants.

The garden cities give us a further 2,000,000 inhabitants, or more.

In the great central open space are the cafes, restaurants, luxury shops, halls of various kinds, a magnificent forum descending by stages down to the immense parks surrounding it, the whole arrangement providing a spectacle of order and vitality.

Density of population

(a) The sky-scraper: 1,200 inhabitants to the acre.
(b) The residential blocks with set-backs: 120 inhabitants to the acre. These are the luxury dwellings.
(c) The residential blocks on the "cellular" system, with a similar number of inhabitants.

This great density gives us our necessary shortening of distances and ensures rapid inter-communication.

Note. The average density to the acre of Paris in the heart of the town is 146, and of London 63; and of the over-crowded quarters of Paris 213, and of London 169.

Open spaces

Of the area (a), 95 per cent of the ground is open (squares, restaurants, theatres).

Of the area (b), 85 per cent of the ground is open (gardens, sports grounds).

Of the area (c), 48 per cent of the ground is open (gardens, sports grounds).

Educational and civic centres, universities, museums of art and industry, public services, county hall

The "Jardin anglais". (The city can extend here, if necessary.)

Sports grounds: Motor racing track, Racecourse, Stadium, Swimming baths, etc.

The protected zone (which will be the property of the city), with its aerodrome

A zone in which all building would be prohibited; reserved for the growth of the city as laid down by the municipality: it would consist of woods, fields, and sports grounds. The forming of a "protected zone" by continual purchase of small properties in the immediate vicinity of the city is one of the most essential and urgent tasks which a municipality can pursue. It would eventually represent a tenfold return on the capital invested.

Industrial quarters: types of buildings employed

For business: sky-scrapers sixty storeys high with no internal wells or courtyards . . .

Residential buildings with "set-backs", of six double storeys; again with no internal wells: the flats looking on either side on to immense parks.

Residential buildings on the "cellular" principle, with "hanging gardens", looking on to immense parks; again no internal wells. These are "service-flats" of the most modern kind.

Garden cities: their aesthetic, economy, perfection and modern outlook

A simple phrase suffices to express the necessities of tomorrow: WE MUST BUILD IN THE OPEN.

The lay-out must be of a purely geometrical kind, with all its many and delicate implications.

[. . .]

The city of to-day is a dying thing because it is not geometrical. To build in the open would be to replace our present haphazard arrangements, *which are all we have to-day*, by a *uniform* lay-out. Unless we do this *there is no salvation*.

The result of a true geometrical lay-out is *repetition*. The result of repetition is a *standard*, the perfect form (i.e. the creation of standard types). A geometrical lay-out means that mathematics play their part.

There is no first-rate human production but has geometry at its base. It is of the very essence of Architecture. To introduce uniformity into the building of the city we must *industrialize building*. Building is the one economic activity which has so far resisted industrialization. It has thus escaped the march of progress, with the result that the cost of building is still abnormally high.

The architect, from a professional point of view, has become a twisted sort of creature. He has grown to love irregular sites, claiming that they inspire him with original ideas for getting round them. Of course he is wrong. For nowadays the only building that can be undertaken must be either for the rich or built at a loss (as, for instance, in the case of municipal housing schemes), or else by jerry-building and so robbing the inhabitant of all amenities. A motor-car which is achieved by mass production is a masterpiece of comfort, precision, balance and good taste. A house built to order (on an "interesting" site) is a masterpiece of incongruity – a monstrous thing.

If the builder's yard were reorganized on the lines of standardization and mass production we might have gangs of workmen as keen and intelligent as mechanics.

The mechanic dates back only twenty years, yet already he forms the highest caste of the working world.

The mason dates . . . from time immemorial! He bangs away with feet and hammer. He smashes up everything round him, and the plant entrusted to him falls to pieces in a few months. The spirit of the mason must be disciplined by making him part of the severe and exact machinery of the industrialized builder's yard.

The cost of building would fall in the proportion of 10 to 2.

The wages of the labourers would fall into definite categories; to each according to his merits and service rendered.

The "interesting" or erratic site absorbs every creative faculty of the architect and wears him out. What results is equally erratic: lopsided abortions; a specialist's solution which can only please other specialists.

We must build *in the open*: both within the city and around it.

Then having worked through every necessary technical stage and using absolute ECONOMY, we shall be in a position to experience the intense joys of a creative art which is based on geometry.

THE CITY AND ITS AESTHETIC

(The plan of a city which is here presented is a direct consequence of purely geometric considerations.)

A new unit *on a large scale* (400 yards) inspires everything. Though the gridiron arrangement of the streets every 400 yards (sometimes only 200) is uniform (with a consequent ease in finding one's way about), no two streets are in any way alike. This is where, in a magnificent contrapuntal symphony, the forces of geometry come into play.

Suppose we are entering the city by way of the Great Park. Our fast car takes the special elevated motor track between the majestic sky-scrapers: as we approach nearer there is seen the repetition against the sky of the twenty-four sky-scrapers; to our left and right on the outskirts of each particular area are the municipal and administrative buildings; and enclosing the space are the museums and university buildings.

Then suddenly we find ourselves at the feet of the first sky-scrapers. But here we have, not the meagre shaft of sunlight which so faintly illumines the dismal streets of New York, but an immensity of space. The whole city is a Park. The terraces stretch out over lawns and into groves. Low buildings of a horizontal kind lead the eye on to the foliage of the trees. Where are now the trivial *Procuracies*? Here is the *city* with its crowds living in peace and pure air, where noise is smothered under the foliage of green trees. The chaos of New York is overcome. Here, bathed in light, stands the modern city [Figure 2].

Our car has left the elevated track and has dropped its speed of sixty miles an hour to run gently through the

Figure 2 A Contemporary city

residential quarters. The "set-backs" permit of vast architectural perspectives. There are gardens, games and sports grounds. And sky everywhere, as far as the eye can see. The square silhouettes of the terraced roofs stand clear against the sky, bordered with the verdure of the hanging gardens. The uniformity of the units that compose the picture throw into relief the firm lines on which the far-flung masses are constructed. Their outlines softened by distance, the sky-scrapers raise immense geometrical facades all of glass, and in them is reflected the blue glory of the sky. An overwhelming sensation. Immense but radiant prisms.

And in every direction we have a varying spectacle: our "gridiron" is based on a unit of 400 yards, but it is strangely modified by architectural devices! (The "set-backs" are in counterpoint, on a unit of 600 × 400.)

The traveller in his airplane, arriving from Constantinople or Pekin it may be, suddenly sees appearing through the wavering lines of rivers and patches of forests that clear imprint which marks a city which has grown in accordance with the spirit of man: the mark of the human brain at work.

As twilight falls the glass sky-scrapers seem to flame.

This is no dangerous futurism, a sort of literary dynamite flung violently at the spectator. It is a spectacle organized by an Architecture which uses plastic resources for the modulation of forms seen in light.

A city made for speed is made for success.

"Broadacre City: A New Community Plan"

Architectural Record (1935)

Frank Lloyd Wright

Editors' Introduction

For more than half a century, the question "Who is the greatest American architect?" could have only one answer: Frank Lloyd Wright (1867–1959). First with his revolutionary "prairie houses" that seemed to grow directly out of the Midwest landscape with their long, low cantilevered rooflines, and later with such masterpieces as the Imperial Hotel in Tokyo, the Guggenheim Museum in New York, and the breathtaking "Fallingwater" house in Western Pennsylvania, Wright became the spokesman for "organic architecture" and a style of building that expressed "the nature of the materials."

To many, Wright's architecture and "the architecture of American democracy" were synonymous. As an unabashed egotist and a pioneer in the field of media celebrity, Wright encouraged the popular identification of himself with the American spirit. He cultivated an imperious image of plain-speaking, anti-collectivist democracy and sought personally to embody the notion of radical individualism. As an artistic genius, Wright despised the popular philistinism of his day and attributed the observable decline of American popular culture to "the mobocracy" and to the unprincipled bankers and politicians who served its interests. By the 1920s and 1930s, Wright had become a social revolutionary but not, characteristically, of the socialist Left. Rather, Wright called for a radical transformation of American society to restore earlier Emersonian and Jeffersonian virtues. The physical embodiment of that utopian vision was Broadacre City. Wright unveiled his model of Broadacre City, illustrated in Plate 39, at Rockefeller Center, New York, in 1935. The article reprinted here represents his first and clearest statement of the revolutionary proposal whereby every citizen of the United States would be given a minimum of one acre of land per person, with the family homestead being the basis of civilization, and with government reduced to nothing more than a county architect who would be in charge of directing land allotments and the construction of basic community facilities. Many at the time thought the idea was totally outlandish, but Broadacre (and the small, efficient "Usonian" house) proved to be prophetic as sprawling suburban regions transformed the American landscape during the second half of the twentieth century.

Wright believed that two inventions – the telephone and the automobile – made the old cities "no longer modern," and he fervently looked forward to the day when dense, crowded agglomerations like New York and Chicago would wither and decay. In their place, Americans would reinhabit the rural landscape (and reacquire the rural virtues of individual freedom and self-reliance) with a "city" of independent homesteads in which people would be isolated enough from one another to insure family stability but connected enough, through modern telecommunications and transportation, to achieve a real sense of community. Borrowing an idea from the anarchist philosopher Kropotkin, Wright believed that the citizens of Broadacre should pursue a combination of manual and intellectual work every day, thus achieving a human wholeness that modern society and the modern city had destroyed. He also believed that a system of personal freedom and dignity through land ownership was the way to guarantee social harmony and avoid class struggle. Broadacre City invites immediate comparison with the very different models of Ebenezer Howard's Garden City (p. 328) and with Le Corbusier's cities based on towers in a park (p. 336). Intriguingly, the overall population density of Broadacre, on the one hand, and the Garden City and Corbusian visions, on the other, were not all that different, depending on the actual acreage of the surrounding parkland or greenbelt. Both Wright's and Le Corbusier's plans are wedded to the automobile, one vision seeing a centralizing, the other a decentralizing, effect. But the most revealing comparisons are with Robert Fishman's description of the post-suburban "technoburbs" (p. 75), Melvin Webber's prediction of a "post-urban age" (p. 549), and Manuel Castells's concept of "the space of flows" (p. 572). Considering the nature of the global cities described in Part Eight of this volume, one cannot help but wonder whether what Wright envisioned in 1935 may actually be realized, with the help of computer-based telecommunications and the possibility of "telecommuting" to work over the Internet in the twenty-first century.

This selection is from *Architectural Record*, 77 (April, 1935). For more on Broadacre City see Robert Fishman, *Urban Utopias of the Twentieth Century* (New York: Basic Books, 1977). John Sergeant, *Frank Lloyd Wright's Usonian Houses: The Case for Organic Architecture* (New York: Whitney Library of Design, 1984) is also useful, and William Allin Storer, *A Frank Lloyd Wright Companion* (Chicago, IL: University of Chicago Press, 1994) is an impressive and definitive reference book.

Three excellent biographies of Wright are Meryle Secrest, *Frank Lloyd Wright: A Biography* (Chicago, IL: University of Chicago Press, 1998), Brendan Gill, *Many Masks: A Life of Frank Lloyd Wright* (New York: Da Capo Press, 1998), and Ada Louise Huxtable, *Frank Lloyd Wright: A Life* (New York: Viking, 2004). For good overviews of Wright's work, see David Larkin and Bruce Brooks Pfeiffer (eds), *Frank Lloyd Wright: The Masterworks* (New York: Rizzoli, 1993), Neil Levine, *The Architecture of Frank Lloyd Wright* (Princeton, NJ: Princeton University Press, 1996), and Roger Friedland and Harold Zellman, *The Fellowship: The Untold Story of Frank Lloyd Wright and the Taliesin Fellowship* (New York: Regan, 2006). The very best sources on Wright are Wright himself, although his writing style is often quirky and hyperbolic. Of particular interest are *When Democracy Builds* (Chicago, IL: University of Chicago Press, 1945), *Genius and the Mobocracy* (New York: Duell, Sloan & Pearce, 1949), and *The Living City* (New York: Horizon, 1958).

Given the simple exercise of several inherently just rights of man, the freedom to decentralize, to redistribute and to correlate the properties of the life of man on earth to his birthright – the ground itself – and Broadacre City becomes reality.

As I see Architecture, the best architect is he who will devise forms nearest organic as features of human growth by way of changes natural to that growth. Civilization is itself inevitably a form but not, if democracy is sanity, is it necessarily the fixation called "academic." All regimentation is a form of death which may sometimes serve life but more often imposes upon it. In Broadacres all is symmetrical but it is seldom obviously and never academically so.

Whatever forms issue are capable of normal growth without destruction of such pattern as they may have. Nor is there much obvious repetition in the new city. Where regiment and row serve the general harmony of arrangement both are present, but generally, both are absent except where planting and cultivation are naturally a process or walls afford a desired seclusion. Rhythm is the substitute for such repetitions everywhere. Wherever repetition (standardization) enters, it has been modified by inner rhythms either by art or by nature as it must, to be of any lasting human value.

The three major inventions already at work building Broadacres, whether the powers that over-built the old cities otherwise like it or not, are:

1 The motor car: general mobilization of the human being.
2 Radio, telephone and telegraph: electrical intercommunication becoming complete.
3 Standardized machine-shop production: machine invention plus scientific discovery.

The price of the major three to America has been the exploitation we see everywhere around us in waste and in ugly scaffolding that may now be thrown away. The price has not been so great if by way of popular government we are able to exercise the use of three inherent rights of any man:

1 His social right to a direct medium of exchange in place of gold as a commodity: some form of social credit.
2 His social right to his place on the ground as he has had it in the sun and air: land to be held only by use and improvements.
3 His social right to the ideas by which and for which he lives: public ownership of invention and scientific discoveries that concern the life of the people.

The only assumption made by Broadacres as ideal is that these three rights will be the citizen's so soon as the folly of endeavoring to cheat him of their democratic values becomes apparent to those who hold (feudal survivors or survivals), as it is becoming apparent to the thinking people who are held blindly abject or subject against their will.

The landlord is no happier than the tenant. The speculator can no longer win much at a game about played out. The present success-ideal, placing, as it does, premiums upon the wolf, the fox and the rat in human affairs and above all, upon the parasite, is growing more evident every day as a falsity just as injurious to the "successful" as to the victims of such success. Well – sociologically, Broadacres is release from all that fatal "success" which is, after all, only excess. So I have called it a new freedom for living in America. It has thrown the scaffolding aside. It sets up a new ideal of success.

In Broadacres, by elimination of cities and towns the present curse of petty and minor officialdom, government, has been reduced to one minor government for each county. The waste motion, the back and forth haul, that today makes so much idle business is gone. Distribution becomes automatic and direct, taking place mostly in the region of origin. Methods of distribution of everything are simple and direct. From the maker to the consumer by the most direct route.

Coal (one-third the tonnage of the haul of our railways) is eliminated by burning it at the mines and transferring that power, making it easier to take over the great railroad rights of way; to take off the cumbersome rolling stock and put the right of way into general service as the great arterial on which truck traffic is concentrated on lower side lanes, many lanes of speed traffic above and monorail speed trains at the center, continuously running. Because traffic may take off or take on at any given point, these arterials are traffic not dated but fluescent. And the great arterial as well as all the highways become great architecture, automatically affording within their structure all necessary storage facilities of raw materials, the elimination of all unsightly piles of raw material.

In the hands of the state, but by way of the county, is all redistribution of land – a minimum of one acre going to the childless family and more to the larger family as effected by the state. The agent of the state in all matters of land allotment or improvement, or in matters affecting the harmony of the whole, is the architect. All building is subject to his sense of the whole as organic architecture. Here architecture is landscape and landscape takes on the character of architecture by way of the simple process of cultivation.

All public utilities are concentrated in the hands of the state and county government as are matters of administration, patrol, fire, post, banking, license and record, making politics a vital matter to everyone in the new city instead of the old case where hopeless indifference makes "politics" a grafter's profession.

In the buildings for Broadacres no distinction exists between much and little, more and less. Quality is in all, for all, alike. The thought entering into the first or last estate is of the best. What differs is only individuality and extent. There is nothing poor or mean in Broadacres.

Nor does Broadacres issue any dictum or see any finality in the matter either of pattern or style.

Organic character is style. Such style has myriad forms inherently good. Growth is possible to Broadacres as a fundamental form, not as mere accident of change but as integral pattern unfolding from within.

Here now may be seen the elemental units of our social structure [Figure 1]: the correlated farm, the factory – its smoke and gases eliminated by burning coal at places of origin, the decentralized school, the various conditions of residence, the home offices, safe traffic, simplified government. All common interests take place in a simple coordination wherein all are employed: *little* farms, *little* homes for industry, *little* factories, *little* schools, a *little* university going to the people mostly by way of their interest in the ground, *little* laboratories on their own ground for professional men. And the farm itself, notwithstanding its animals, becomes the most attractive unit of the city. The

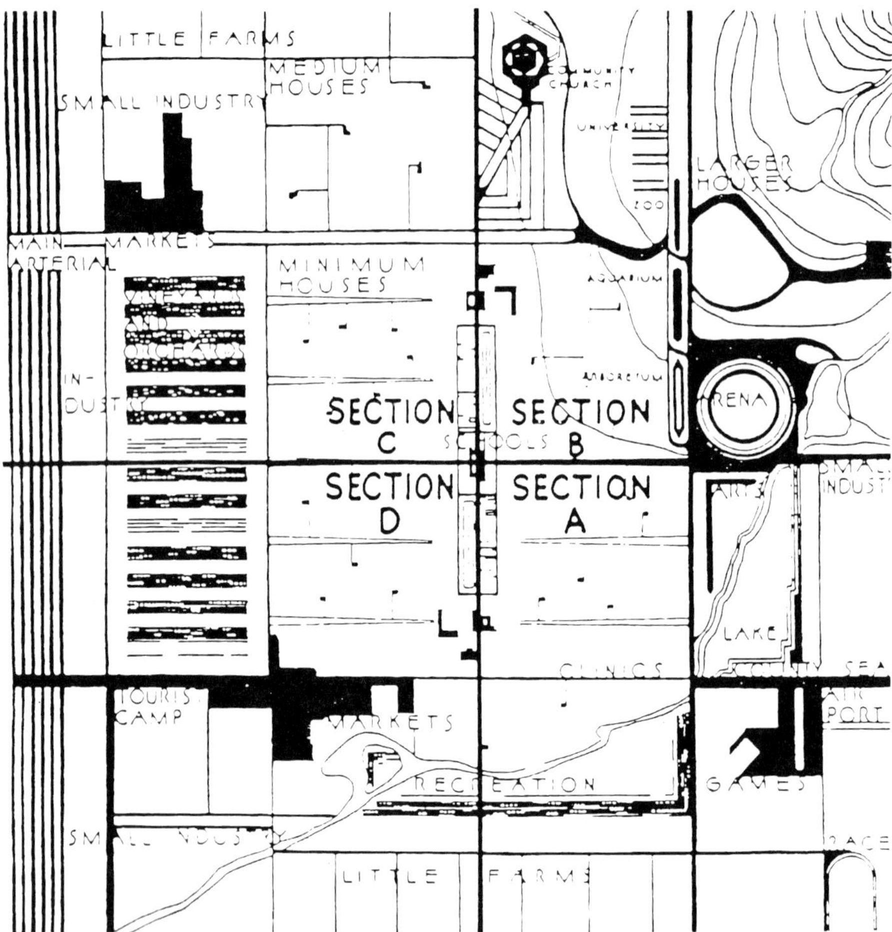

Figure 1

husbandry of animals at last is in decent association with them and with all else as well. True farm relief.

To build Broadacres as conceived would automatically end unemployment and all its evils forever. There would never be labor enough nor could underconsumption ever ensue. Whatever a man did would be done – obviously and directly – mostly by himself in his own interest under the most valuable inspiration and direction: under training, certainly, if necessary. Economic independence would be near, a subsistence certain; life varied and interesting.

Every kind of builder would be likely to have a jealous eye to the harmony of the whole within broad limits fixed by the county architect, an architect chosen by the county itself. Each county would thus naturally develop an individuality of its own. Architecture – in the broad sense – would thrive.

In an organic architecture the ground itself predetermines all features; the climate modifies them; available means limit them; function shapes them.

Form and function are one in Broadacres. But Broadacres is no finality! The model shows four square miles of a typical countryside developed on the acre as unit according to conditions in the temperate zone and accommodating some 1,400 families. It would swing north or swing south in type as conditions, climate and topography of the region changed.

In the model the emphasis has been placed upon diversity in unity, recognizing the necessity of cultivation as a need for formality in most of the planting. By a simple government subsidy certain specific acres or groups of acre units are, in every generation, planted to useful trees, meantime beautiful, giving privacy and various rural divisions. There are no rows of trees alongside the roads to shut out the view. Rows where they occur are perpendicular to the road or the trees are planted in groups. Useful trees like white pine, walnut, birch, beech, fir, would come to maturity as well as fruit and nut trees and they would come as a profitable crop meantime giving character, privacy and comfort to the whole city. The general park is a flowered meadow beside the stream and is bordered with ranks of trees, tiers gradually rising in height above the flowers at the ground level. A music-garden is sequestered from noise at one end. Much is made of general sports and festivals by way of the stadium, zoo, aquarium, arboretum and the arts.

The traffic problem has been given special attention, as the more mobilization is made a comfort and a facility the sooner will Broadacres arrive. Every Broadacre citizen has his own car. Multiple-lane highways make travel safe and enjoyable. There are no grade crossings nor left turns on grade. The road system and construction is such that no signals nor any lamp-posts need be seen. No ditches are alongside the roads. No curbs either. An inlaid purfling over which the car cannot come without damage to itself takes its place to protect the pedestrian.

In the affair of air transport Broadacres rejects the present airplane and substitutes the self-contained mechanical unit that is sure to come: an aerator capable of rising straight up and by reversible rotors able to travel in any given direction under radio control at a maximum speed of, say, 200 miles an hour, and able to descend safely into the hexacomb from which it arose or anywhere else. By a doorstep if desired.

The only fixed transport trains kept on the arterial are the long-distance monorail cars traveling at a speed (already established in Germany) of 220 miles per hour. All other traffic is by motor car on the twelve lane levels or the triple truck lanes on the lower levels which have on both sides the advantage of delivery direct to warehousing or from warehouses to consumer. Local trucks may get to warehouse-storage on lower levels under the main arterial itself. A local truck road parallels the swifter lanes.

Houses in the new city are varied: make much of fireproof synthetic materials, factory-fabricated units adapted to free assembly and varied arrangement, but do not neglect the older nature-materials wherever they are desired and available. House-holders' utilities are nearly all planned in prefabricated utility stacks or units, simplifying construction and reducing building costs to a certainty. There is the professional's house with its laboratory, the minimum house with its workshop, the medium house ditto, the larger house and the house of machine-age luxury. We might speak of them as a one-car house, a two-car house, a three-car house and a five-car house. Glass is extensively used as are roofless rooms. The roof is used often as a trellis or a garden. But where glass is extensively used it is usually for domestic purposes in the shadow of protecting overhangs.

Copper for roofs is indicated generally on the model as a permanent cover capable of being worked in many appropriate ways and giving a general harmonious color effect to the whole.

Electricity, oil and gas are the only popular fuels. Each land allotment has a pit near the public lighting fixture where access to the three and to water and sewer may be had without tearing up the pavements.

The school problem is solved by segregating a group of low buildings in the interior spaces of the city where the children can go without crossing traffic. The school building group includes galleries for loan collections from the museum, a concert and lecture hall, small gardens for the children in small groups and well-lighted cubicles for individual outdoor study: there is a small zoo, large pools and green playgrounds.

This group is at the very center of the model and contains at its center the higher school adapted to the segregation of the students into small groups.

This tract of four miles square, by way of such liberal general allotment determined by acreage and type of ground, including apartment buildings and hotel facilities, provides for about 1,400 families at, say, an average of five or more persons to the family.

To reiterate: the basis of the whole is general decentralization as an applied principle and architectural reintegration of all units into one fabric; free use of the ground held only by use and improvements; public utilities and government itself owned by the people of Broadacre City; privacy on one's own ground for all and a fair means of subsistence for all by way of their own work on their own ground or in their own laboratory or in common offices serving the life of the whole.

There are too many details involved in the model of Broadacres to permit complete explanation. Study of the model itself is necessary study. Most details are explained by way of collateral models of the various types of construction shown: highway construction, left turns, crossovers, underpasses and various houses and public buildings.

Anyone studying the model should bear in mind the thesis upon which the design has been built by the Taliesin Fellowship, built carefully not as a finality in any sense but as an interpretation of the changes inevitable to our growth as a people and a nation.

Individuality established on such terms must thrive. Unwholesome life would get no encouragement and the ghastly heritage left by over-crowding in overdone ultra-capitalistic centers would be likely to disappear in three or four generations. The old success ideals having no chance at all, new ones more natural to the best in man would be given a fresh opportunity to develop naturally.

"Towards Sustainable Development"

from *Our Common Future* (1987)

World Commission on Environment and Development
(The Brundtland Commission)

Editors' Introduction

The physical city is a human-made construct, but the relationship between the city and the surrounding natural environment has always helped to define the character and quality of urban life. If the very first cities were expressions of "hydraulic civilizations" based on the control of water for irrigation, and if the cities of the Industrial Revolution ushered in a new age of unprecedented environmental degradation, it was unpolluted nature – or at least the *idea* of unpolluted nature – that offered a continuing source of intellectual regeneration and moral comparison. The utopian visionaries of the nineteenth and twentieth centuries – Frederick Law Olmsted (p. 321), Ebenezer Howard (p. 328), Le Corbusier (p. 336), and Frank Lloyd Wright (p. 345) – created city plans that balanced the man-made urban areas with parks, public gardens, or surrounding rural zones. And in recent years, the forces of nature ecology, green urbanism, and sustainability have grown in importance and become dominant forces in urban policy and planning.

In 1968, Stanford University biologist Paul Ehrlich published *The Population Bomb*, predicting global overcrowding and persistent starvation in the underdeveloped nations by the 1980s. In 1970, the first Earth Day was celebrated in San Francisco. And in 1972, the Club of Rome published its influential report on *The Limits to Growth*, arguing that the world's governments and economic powers needed to begin cutting back on overproduction and overconsumption in the face of uneven development, exploding population growth, and declining resources. All these developments suggested a growing shift toward environmentalism as the new reform paradigm at the very time that capitalism was globalizing and the ideological certainties of the 1960s Left were beginning to lose favor. Then, in 1987, came *Our Common Future*, the report of the United Nations-sponsored World Commission on Environment and Development (WCED) and the clarion call for "sustainable development."

The WCED report is commonly called the Brundtland Report, so named after the chairperson of the Commission, Gro Harlem Brundtland of Norway, the only prime minister of a major country to have previously served as environment minister. In its "Call to Action," the Brundtland Report argued that during the twentieth century "the relationship between the human world and the planet that sustains it has undergone a profound change" and that the increased "rate of change is outstripping the ability of . . . our current capabilities." As a result, the world needs to embrace the concept of sustainable development, defined as "development that meets the needs of the present without compromising the ability of future generations to meet their own needs." Such sustainable development, the report continues, "contains within it . . . the concept of 'needs,' in particular the needs of the world's poor, to which overriding priority should be given."

Some skeptics dismissed the Brundtland Report as just one more attack on free-market capitalism in favor of massive government regulation of the economy. And some radical environmentalists even claimed that "sustainable development" was an oxymoron favored by big business interests to portray capitalism as ecologically benign. But the momentum behind the environmental vision of sustainability sparked by the Brundtland Report continued to grow with the Rio Declaration on Environment and Development of 1992, the Kyoto Protocol on global climate change of 1997–1999, and the World Summit on Sustainable Development in Johannesburg in 2002. Nowadays, concern about climate change and carbon dioxide emissions caused by fossil fuels has caused governments to look for new, cleaner sources of energy. And in urban planning, the idea of "sustainability" has been broadly adopted as one of the key concepts behind urban development projects worldwide, but especially in Europe and North America and most particularly in the work of New Urbanists like Andrés Duany, Elizabeth Plater-Zyberk, Peter Calthorpe, and William Fulton (p. 360).

For descriptions in this volume of sustainable development applied to urban contexts, consult Timothy Beatley's "Planning for Sustainability in European Cities" (p. 446) and Stephen Wheeler's "Urban Planning and Global Climate Change" (p. 458). For a more expanded range of views on urban sustainability, consult Stephen Wheeler and Timothy Beatley (eds), *The Sustainable Urban Development Reader* (London: Routledge, 2004). The literature on environmentalism in general is huge and ubiquitous. For the best dissenting view, consult Bjorn Lumborg, *The Skeptical Environmentalist: Measuring the Real State of the World* (Cambridge: Cambridge University Press, 2001).

A CALL FOR ACTION

Over the course of this century, the relationship between the human world and the planet that sustains it has undergone a profound change.

When the century began, neither human numbers nor technology had the power radically to alter planetary systems. As the century closes, not only do vastly increased human numbers and their activities have that power, but major, unintended changes are occurring in the atmosphere, in soils, in waters, among plants and animals, and in the relationships among all of these. The rate of change is outstripping the ability of scientific disciplines and our current capabilities to assess and advise. It is frustrating the attempts of political and economic institutions, which evolved in a different, more fragmented world, to adapt and cope. It deeply worries many people who are seeking ways to place those concerns on the political agendas.

The onus lies with no one group of nations. Developing countries face the obvious life-threatening challenges of desertification, deforestation, and pollution, and endure most of the poverty associated with environmental degradation. The entire human family of nations would suffer from the disappearance of rain forests in the tropics, the loss of plant and animal species, and changes in rainfall patterns. Industrial nations face the life-threatening challenges of toxic chemicals, toxic wastes, and acidification. All nations may suffer from the releases by industrialized countries of carbon dioxide and of gases that react with the ozone layer, and from any future war fought with the nuclear arsenals controlled by those nations. All nations will have a role to play in changing trends, and in righting an international economic system that increases rather than decreases inequality, that increases rather than decreases numbers of poor and hungry.

The next few decades are crucial. The time has come to break out of past patterns. Attempts to maintain social and ecological stability through old approaches to development and environmental protection will increase instability. Security must be sought through change. The Commission has noted a number of actions that must be taken to reduce risks to survival and to put future development on paths that are sustainable. Yet we are aware that such a reorientation on a continuing basis is simply beyond the reach of present decision-making structures and institutional arrangements, both national and international.

This Commission has been careful to base our recommendations on the realities of present institutions, on what can and must be accomplished today. But to keep options open for future generations, the present generation must begin now, and begin together.

To achieve the needed changes, we believe that an active follow-up of this report is imperative. It is with this in mind that we call for the UN General Assembly, upon due consideration, to transform this report into a UN Programme on Sustainable Development. Special follow-up conferences could be initiated at the regional level. Within an appropriate period after the presentation of this report to the General Assembly, an international conference could be convened to review progress made, and to promote follow-up arrangements that will be needed to set benchmarks and to maintain human progress.

First and foremost, this Commission has been concerned with people – of all countries and all walks of life. And it is to people that we address our report. The changes in human attitudes that we call for depend on a vast campaign of education, debate, and public participation. This campaign must start now if sustainable human progress is to be achieved.

The members of the World Commission on Environment and Development came from 21 very different nations. In our discussions, we disagreed often on details and priorities. But despite our widely differing backgrounds and varying national and international responsibilities, we were able to agree to the lines along which change must be drawn.

We are unanimous in our conviction that security, wellbeing, and very survival of the planet depend on such changes, now.

A THREATENED FUTURE

The Earth is one but the world is not. We depend on one biosphere for sustaining our lives. Yet each community, each country, strives for survival and prosperity with little regard for its impact on others. Some consume the Earth's resources at a rate that would leave little for future generations. Others, many more in number, consume far too little and live with the prospect of hunger, squalor, disease, and early death.

Yet progress has been made. Throughout much of the world, children born today can expect to live longer and be better educated than their parents. In many parts, the newborn can also expect to attain a higher standard of living in a wider sense. Such progress provides hope as we contemplate the improvements still needed, and also as we face our failures to make this Earth a safer and sounder home for us and for those who are to come.

The failures that we need to correct arise both from poverty and from the short-sighted way in which we have often pursued prosperity. Many parts of the world are caught in a vicious downwards spiral: Poor people are forced to overuse environmental resources to survive from day to day, and their impoverishment of their environment further impoverishes them, making their survival ever more difficult and uncertain. The prosperity attained in some parts of the world is often precarious, as it has been secured through farming, forestry, and industrial practices that bring profit and progress only over the short term.

Societies have faced such pressures in the past and, as many desolate ruins remind us, sometimes succumbed to them. But generally these pressures were local. Today the scale of our interventions in nature is increasing and the physical effects of our decisions spill across national frontiers. The growth in economic interaction between nations amplifies the wider consequences of national decisions. Economics and ecology bind us in ever-tightening networks. Today, many regions face risks of irreversible damage to the human environment that threaten the basis for human progress.

These deepening interconnections are the central justification for the establishment of this Commission. We traveled the world for nearly three years, listening. At special public hearings organized by the Commission, we heard from government leaders, scientists, and experts, from citizens' groups concerned about a wide range of environment and development issues, and from thousands of individuals – farmers, shanty-town residents, young people, industrialists, and indigenous and tribal peoples.

We found everywhere deep public concern for the environment, concern that has led not just to protests but often to changed behaviour. The challenge is to ensure that these new values are more adequately reflected in the principles and operations of political and economic structures.

We also found grounds for hope: that people can cooperate to build a future that is more prosperous, more just, and more secure; that a new era of economic growth can be attained, one based on policies that sustain and expand the Earth's resource base; and that the progress that some have known over the last century can be experienced by all in the years ahead. But for this to happen, we must understand better the symptoms of stress that confront us, we must identify the causes, and we must design new approaches to managing environmental resources and to sustaining human development.

SYMPTOMS AND CAUSES

Environmental stress has often been seen as the result of the growing demand on scarce resources and the pollution generated by the rising living standards of the relatively affluent. But poverty itself pollutes the environment, creating environmental stress in a different way. Those who are poor and hungry will often destroy their immediate environment in order to survive: They will cut down forests, their livestock will overgraze grasslands; they will overuse marginal land; and in growing numbers they will crowd into congested cities. The cumulative effect of these changes is so far-reaching as to make poverty itself a major global scourge.

On the other hand, where economic growth has led to improvements in living standards, it has sometimes been achieved in ways that are globally damaging in the longer term. Much of the improvement in the past has been based on the use of increasing amounts of raw materials, energy, chemicals, and synthetics and on the creation of pollution that is not adequately accounted for in figuring the costs of production processes. These trends have had unforeseen effects on the environment. Thus today's environmental challenges arise both from the lack of development and from the unintended consequences of some forms of economic growth.

[. . .]

Sustainable development is development that meets the needs of the present without compromising the ability of future generations to meet their own needs. It contains within it two key concepts:

- the concept of 'needs', in particular the essential needs of the world's poor, to which overriding priority should be given; and
- the idea of limitations imposed by the state of technology and social organization on the environment's ability to meet present and future needs.

Thus the goals of economic and social development must be defined in terms of sustainability in all countries – developed or developing, market-oriented or centrally planned. Interpretations will vary, but must share certain general features and must flow from a consensus on the basic concept of sustainable development and on a broad strategic framework for achieving it. Development involves a progressive transformation of economy and society. A development path that is sustainable in a physical sense could theoretically be pursued even in a rigid social and political setting. But physical sustainability cannot be secured unless development policies pay attention to such considerations as changes in access to resources and in the distribution of costs and benefits. Even the narrow notion of physical sustainability implies a concern for social equity between generations, a concern that must logically be extended to equity within each generation.

THE CONCEPT OF SUSTAINABLE DEVELOPMENT

The satisfaction of human needs and aspirations is the major objective of development. The essential needs of vast numbers of people in developing countries – for food, clothing, shelter, jobs – are not being met, and beyond their basic needs these people have legitimate aspirations for an improved quality of life. A world in which poverty and inequity are endemic will always be prone to ecological and other crises. Sustainable development requires meeting the basic needs of all and extending to all the opportunity to satisfy their aspirations for a better life.

Living standards that go beyond the basic minimum are sustainable only if consumption standards everywhere have regard for long-term sustainability. Yet many of us live beyond the world's ecological means, for instance in our patterns of energy use. Perceived needs are socially and culturally determined, and sustainable development requires the promotion of values that encourage consumption standards that are within the bounds of the ecologically possible and to which all can reasonably aspire.

Meeting essential needs depends in part on achieving full growth potential, and sustainable development clearly requires economic growth in places where such needs are not being met. Elsewhere, it can be consistent with economic growth, provided the content of growth reflects the broad principles of sustainability and non-exploitation of others. But growth by itself is not enough. High levels of productive activity and widespread poverty can coexist, and can endanger the environment. Hence sustainable development requires that societies meet human needs both by increasing productive potential and by ensuring equitable opportunities for all. An expansion in numbers can increase the pressure on resources and slow the rise in living standards in areas where deprivation is widespread. Though the issue is not merely one of population size but of the distribution of resources, sustainable development can only be pursued if demographic developments are in harmony with the changing productive potential of the ecosystem.

A society may in many ways compromise its ability to meet the essential needs of its people in the future – by overexploiting resources, for example. The direction of technological developments may solve some immediate problems but lead to even greater ones. Large sections of the population may be marginalized by ill-considered development.

Settled agriculture, the diversion of watercourses, the extraction of minerals, the emission of heat and noxious gases into the atmosphere, commercial forests, and genetic manipulation are all examples of human intervention in natural systems during the course of development. Until recently, such interventions were small in scale and their impact limited. Today's interventions are more drastic in scale and impact, and more threatening to life-support systems both locally and globally. This need not happen. At a minimum, sustainable development must not endanger the natural systems that support life on Earth: the atmosphere, the waters, the soils, and the living beings.

Growth has no set limits in terms of population or resource use beyond which lies ecological disaster. Different limits hold for the use of energy, materials, water, and land. Many of these will manifest themselves in the form of rising costs and diminishing returns, rather than in the form of any sudden loss of a resource base. The accumulation of knowledge and the development of technology can enhance the carrying capacity of the resource base. But ultimate limits there are, and sustainability requires that long before these are reached, the world must ensure equitable access to the constrained resource and reorient technological efforts to relieve the pressure.

Economic growth and development obviously involve changes in the physical ecosystem. Every ecosystem everywhere cannot be preserved intact. A forest may be depleted in one part of a water-shed and extended elsewhere, which is not a bad thing if the exploitation has been planned and the effects on soil erosion rates, water regimes, and genetic losses have been taken into account. In general, renewable resources like forests and fish stocks need not be depleted provided the rate of use is within the limits of regeneration and natural growth. But most renewable resources are part of a complex and interlinked ecosystem, and maximum sustainable yield must be defined after taking into account system-wide effects of exploitation.

As for nonrenewable resources, like fossil fuels and minerals, their use reduces the stock available for future generations. But this does not mean that such resources should not be used. In general the rate of depletion should take into account the criticality of that resource, the availability of technologies for minimizing depletion, and the likelihood of substitutes being available. Thus land should not be degraded beyond reasonable recovery. With minerals and fossil fuels, the rate of depletion and the emphasis on recycling and economy of use should be calibrated to ensure that the resource does not run out before acceptable substitutes are available. Sustainable development requires that the rate of depletion of nonrenewable resources should foreclose as few future options as possible.

Development tends to simplify ecosystems and to reduce their diversity of species. And species, once extinct, are not renewable. The loss of plant and animal species can greatly limit the options of future generations; so sustainable development requires the conservation of plant and animal species.

So-called free goods like air and water are also resources. The raw materials and energy of production processes are only partly converted to useful products. The rest comes out as wastes. Sustainable development requires that the adverse impacts on the quality of air, water, and other natural elements are minimized so as to sustain the ecosystem's overall integrity.

In essence, sustainable development is a process of change in which the exploitation of resources, the direction of investments, the orientation of technological development, and institutional change are all in harmony and enhance both current and future potential to meet human needs and aspirations.

"Charter of the New Urbanism"

Congress for the New Urbanism (1993)

Editors' Introduction

The sustainability principles espoused by the Brundtland Commission (p. 351) helped to give weight and authority to urban environmentalist organizations worldwide, none more so than an innovative planning and design movement called the New Urbanism. The Chicago-based Congress for a New Urbanism was officially established in 1993, but the immediate origin of the movement was a meeting at the Awahnee Hotel in Yosemite Valley, California, in 1991. There, an extraordinary collective of visionary architects and designers – among them, Peter Calthorpe, Andrés Duany, Elizabeth Plater-Zyberk, Michael Corbett, Stafanos Polyzoides, Daniel Solomon, and Elizabeth Moule – met with a number of California policy makers to promulgate the Awahnee Principles for future urban development along ecologically sound lines. Many of those Principles became elements of the Charter of the New Urbanism. The movement made swift gains throughout the 1990s – becoming a favored model of the US Department of Housing and Urban Development during the Clinton Administration – and New Urbanist projects were built throughout the United States and Canada. In 2003, an allied Council for European Urbanism was established in the UK, actively encouraged by HRH the Prince of Wales, and the New Urbanism spread worldwide with projects in France, Portugal, Sweden, Italy, Belgium, the Netherlands, Australia, New Zealand, South Africa, and China.

The heart of the Charter of the New Urbanism consists of twenty-seven principles – nine in each of three broad categories – preceded by a kind of preamble that establishes the visionary, almost utopian, goals of the movement. The preamble begins by asserting that all of the present-day urban ills – inner-city decay, suburban sprawl, the deterioration of agricultural and wilderness lands, even race- and class-based segregation – are parts of "one interrelated community-building challenge." It goes on to call for the "restoration of existing urban centers" and the transformation of "sprawling suburbs into communities of real neighborhoods and diverse districts." New Urbanism recognizes "that physical solutions by themselves will not solve social and economic problems," but it insists that "a coherent and supportive physical framework" is a necessary, if not sufficient, precondition for urban progress and that such progress must be achieved "through citizen-based participatory planning and design."

The twenty-seven principles of the Charter address contemporary urban planning issues at a much finer level of detail, beginning with an examination of cities and towns at the metropolitan scale. "The metropolitan region," it states, is defined by natural topography and represents "a fundamental economic unit of the contemporary world." These metropolitan-scale principles go on to call for urban growth boundaries that do not "blur or eradicate the edges of the metropolis," intensive "infill development" within existing cities, region-wide revenue sharing, and a wide range of transportation options that "maximize access and mobility . . . while reducing dependence upon the automobile." The next set of principles examines the needs of "the neighborhood, the district, and the corridor." The Charter calls for neighborhoods that are "compact, pedestrian-friendly, and mixed-use" so that "many activities of daily living" can be within walking distance. In addition, neighborhoods should contain local shopping districts (not distant malls), parks, and community schools. Finally, another nine principles look at "the block, the street, and the building," calling for an architecture that "transcends style," that grows from "local climate, topography, history, and building practice," and that creates environments characterized by safety, accessibility, and openness.

The New Urbanism is not without its critics. Some have dismissed it as a "New Suburbanism" that addresses the issues of the young middle-class – double-income, no kids families called DINKs – but that has no relevance for inner-city neighborhoods. Others feel that the "new traditionalism" tendencies of many New Urbanist developments feel artificial and too carefully, too strictly planned, and one critic claimed that the emphasis on openness and accessibility leads to "crime-friendly neighborhoods." But for all this, the New Urbanism has proven to be a long-lived and ever-evolving movement. Unlike most of the twentieth-century planning movements, the New Urbanism is not tied to the ambitions and pretensions of a single individual. Rather like the Garden City movement that Ebenezer Howard pioneered but did not monopolize, the New Urbanism has attracted a large number of practitioners, and the movement has various wings and branches that continually question and inform the movement's mainstream. For example, although the Charter of the New Urbanism calls for a reasonable mix of transit options, including the private automobile, one somewhat alarmist video documentary is titled *The End of Suburbia: Oil Depletion and the Collapse of the American Dream* (2004). Yet another video, *New Urban Cowboy: Toward a New Pedestrianism* (2008), appears to be a publicity vehicle for the producer's campaign for the Florida governorship! For a more sober critique, see David Harvey, "The New Urbanism and the Communitarian Trap," *Harvard Design Magazine*, 1 (1997), pp. 68–69.

The literature on the New Urbanism is as rich and varied as the movement itself. Peter Katz, *The New Urbanism: Towards an Architecture of Community* (New York: McGraw-Hill, 1994), and Doug Kelbaugh, *Common Place: Toward Neighborhood and Regional Design* (Seattle, WA: University of Washington Press, 1997) provide overviews of designs by Calthorpe and other New Urbanists. Kenneth B. Hall and Gerald A. Porterfield, *Community by Design: New Urbanism for Suburbs and Small Communities* (New York: McGraw-Hill Professional, 2001), and E. Talen, *New Urbanism and American Planning: The Conflict of Cultures* (New York: Routledge, 2005) offer detailed analyses of the New Urbanist movement. The 69-page *New Urbanism: Peter Calthorpe versus Lars Lerup: Michigan Debates on Urbanism* (Ann Arbor, MI: University of Michigan Press, 2005) is a lively and scintillating exchange of views with an afterword by editor Robert Fishman. See also James Howard Kunstler, *The Geography of Nowhere* (New York: Simon & Schuster, 1993) and, *Home from Nowhere: Remaking Our Everyday World for the Twenty-first Century* (New York: Simon & Schuster, 1998) for a popular account of Calthorpe and other New Urbanists' work. Also of interest are John Dutton, *New American Urbanism: Re-forming the Suburban Metropolis* (Milan: Skira, 2001), Todd W. Bressi (ed.), *The Seaside Debates: A Critique of the New Urbanism* (New York: Rizzoli, 2002), and Gabriele Tagliaventi, *New Urbanism* (Florence: Alinea, 2002). An excellent place to begin any research on the New Urbanist movement is Michael Leccese and Kathleen McCormick (eds) *Charter of the New Urbanism* (New York: McGraw-Hill, 1999).

THE CONGRESS FOR THE NEW URBANISM views disinvestment in central cities, the spread of placeless sprawl, increasing separation by race and income, environmental deterioration, loss of agricultural lands and wilderness, and the erosion of society's built heritage as one interrelated community-building challenge.

WE STAND for the restoration of existing urban centers and towns within coherent metropolitan regions, the reconfiguration of sprawling suburbs into communities of real neighborhood and diverse districts, the conservation of natural environments, and the preservation of our built legacy.

WE RECOGNIZE that physical solutions by themselves will not solve social and economic problems, but neither can economic vitality, community stability, and environmental health be sustained without a coherent and supportive physical framework.

WE ADVOCATE the restructuring of public policy and development practices to support the following principles: neighborhoods should be diverse in use and population; communities should be designed for the pedestrian and transit as well as the car; cities and towns should be shaped by physically defined and universally accessible public spaces and community institutions; urban places should be framed by architecture and landscape design that celebrate local history, climate, ecology, and building practice.

WE REPRESENT a broad-based citizenry, composed of public and private sector leaders, community activists, and multidisciplinary professionals. We are committed to reestablishing the relationship between the art of building and the making of community, through citizen-based participatory planning and design.

WE DEDICATE ourselves to reclaiming our homes, blocks, streets, parks, neighborhoods, districts, towns, cities, regions, and environment.

We assert the following principles to *guide public policy, development practice, urban planning, and design:*

The region: Metropolis, city, and town

1. Metropolitan regions are finite places with geographic boundaries derived from topography, watersheds, coastlines, farmlands, regional parks, and river basins. The metropolis is made of multiple centers that are cities, towns, and villages, each with its own identifiable center and edges.
2. The metropolitan region is a fundamental economic unit of the contemporary world. Governmental cooperation, public policy, physical planning, and economic strategies must reflect this new reality.
3. The metropolis has a necessary and fragile relationship to its agrarian hinterland and natural landscapes. The relationship is environmental, economic, and cultural. Farmland and nature are as important to the metropolis as the garden is to the house.
4. Development patterns should not blur or eradicate the edges of the metropolis. Infill development within existing urban areas conserves environmental resources, economic investment, and social fabric, while reclaiming marginal and abandoned areas. Metropolitan regions should develop strategies to encourage such infill development over peripheral expansion.
5. Where appropriate, new development contiguous to urban boundaries should be organized as neighborhoods and districts, and be integrated with the existing urban pattern. Noncontiguous development should be organized as towns and villages with their own urban edges, and planned for a jobs/housing balance, not as bedroom suburbs.
6. The development and redevelopment of towns and cities should respect historical patterns, precedents and boundaries.
7. Cities and towns should bring into proximity a broad spectrum of public and private uses to support a regional economy that benefits people of all incomes. Affordable housing should be distributed throughout the region to match job opportunities and to avoid concentrations of poverty.
8. The physical organization of the region should be supported by a framework of transportation alternatives. Transit, pedestrian, and bicycle systems should maximize access and mobility throughout the region while reducing dependence upon the automobile.
9. Revenues and resources can be shared more cooperatively among the municipalities and centers within regions to avoid destructive competition for tax base and to promote rational coordination of transportation, recreation, public services, housing, and community institutions.

The neighborhood, the district, and the corridor

1. The neighborhood, the district, and the corridor are the essential elements of development and redevelopment in the metropolis. They form identifiable areas that encourage citizens to take responsibility for their maintenance and evolution.
2. Neighborhoods should be compact, pedestrian-friendly, and mixed-use. Districts generally emphasize a special single use, and should follow the principles of neighborhood design when possible. Corridors are regional connectors of neighborhoods and districts; they range from boulevards and rail lines to rivers and parkways.
3. Many activities of daily living should occur within walking distance, allowing independence to those who do not drive, especially the elderly and the young. Interconnected networks of streets should be designed to encourage walking, reduce the number and length of automobile trips, and conserve energy.
4. Within neighborhoods, a broad range of housing types and price levels can bring people of diverse ages, races, and incomes into daily interaction, strengthening the personal and civic bonds essential to an authentic community.
5. Transit corridors, when properly planned and coordinated, can help organize metropolitan structure and revitalize urban centers. In contrast, highway corridors should not displace investment from existing centers.
6. Appropriate building densities and land uses should be within walking distance of transit stops, permitting public transit to become a viable alternative to the automobile.

7. Concentrations of civic, institutional, and commercial activity should be embedded in neighborhoods and districts, not isolated in remote, single-use complexes. Schools should be sized and located to enable children to walk or bicycle to them.
8. The economic health and harmonious evolution of neighborhoods, districts, and corridors can be improved through graphic urban design codes that serve as predictable guides for change.
9. A range of parks, from tot-lots and village greens to ball fields and community gardens, should be distributed within neighborhoods. Conservation areas and open lands should be used to define and connect different neighborhoods and districts.

The block, the street, and the building

1. A primary task of all urban architecture and landscape design is the physical definition of streets and public spaces as places of shared use.
2. Individual architectural projects should be seamlessly linked to their surroundings. This issue transcends style.
3. The revitalization of urban places depends on safety and security. The design of streets and buildings should reinforce safe environments, but not at the expense of accessibility and openness.
4. In the contemporary metropolis, development must adequately accommodate automobiles. It should do so in ways that respect the pedestrian and the form of public space.
5. Streets and squares should be safe, comfortable, and interesting to the pedestrian. Properly configured, they encourage walking and enable neighbors to know each other and protect their communities.
6. Architecture and landscape design should grow from local climate, topography, history, and building practice.
7. Civic buildings and public gathering places require important sites to reinforce community identity and the culture of democracy. They deserve distinctive form, because their role is different from that of other buildings and places that constitute the fabric of the city.
8. All buildings should provide their inhabitants with a clear sense of location, weather and time. Natural methods of heating and cooling can be more resource-efficient than mechanical systems.
9. Preservation and renewal of historic buildings, districts, and landscapes affirm the continuity and evolution of urban society.

"Designing the Region" and "Designing the Region is Designing the Neighborhood"

from *The Regional City: Planning for the End of Sprawl* (2001)

Peter Calthorpe and William Fulton

Editors' Introduction

What is the most appropriate scale for urban planning? The home? The street? The neighborhood? The city? Or the entire metropolitan region? In *The Regional City: Planning for the End of Sprawl*, Peter Calthorpe and William Fulton argue that every element of the planned environment interacts with every other and that planners embracing the goal of sustainability and New Urbanist concepts should follow a holistic, ecological approach that harmonizes the intimate scale of the neighborhood with the metropolitan scale of the region.

Peter Calthorpe (b. 1949), founder of the architectural firm of Calthorpe Associates based in Berkeley, California, is an urban futurist rooted in both the realities of the present and the traditions of the past. An author, practicing architect, and former academic who taught at the University of California, Berkeley, and elsewhere, Calthorpe is one of the founders of the Congress for the New Urbanism. William Fulton is president of the widely respected Solimar Research Group of Ventura, California, a firm that provides governments, foundations, and planning agencies with empirical research on land use, metropolitan growth, and related policy concerns. Founded in 1999, Solimar specializes in urban growth boundaries and what it calls "urban containment policy" and what others call "Smart Growth." In *The Regional City*, Calthorpe and Fulton present a subtly nuanced exploration of the New Urbanism at its best, a unified vision that combines earlier ideas about neighborhoods as "Pedestrian Pockets" and "transit-oriented developments" in a sweeping regional conception of contemporary urban form.

As a leading proponent of ecology and environmentalism as applied to urban design, Calthorpe has long been a prophet of a new kind of twenty-first-century community: the intimate, human-scale model that is descended directly from Ebenezer Howard's Garden Cities (p. 328) and that quickly became an ubiquitous element of New Urbanist planning and design, first in the United States and later in Europe and elsewhere. Calthorpe began his career by writing and lecturing extensively on the necessity of ecologically sensitive design and energy-efficient building based on the application of passive solar techniques. For a time, he was a partner of Sim Van der Ryn, the California state architect in the administration of progressive – some say "New Age" – governor Jerry Brown. Together, they published *Sustainable Communities* (1986), an influential volume that helped spread the ideas of solar energy, recycling, appropriate technology, and environmentalist approaches to urban planning and design.

In *The Pedestrian Pocket Book* (1989), edited by Doug Kelbaugh, Calthorpe examined the "profound mismatch between the old suburban patterns . . . and the post-industrial culture in which we now find ourselves." The similarities between his solution and the one proposed by Howard in 1898 were striking. Both the Garden Cities and the Pedestrian Pockets were to be surrounded by greenbelts of permanent agricultural land. Both were relatively dense developments, allowing residents to walk to the urban center in a short period of time. Both combined residential, commercial, and workplace elements, and where the Garden City was served by a railroad connection, the Pedestrian Pocket avoided the typical suburban monoculture of the automobile by a system of light-rail transit connectors.

Although an important source of the New Urbanism style of community development is the Garden City, the real application of the Calthorpe plan is in the blossoming new ring of suburban development currently springing up around the old metropolitan cores, the area that journalist Joel Garreau has dubbed "Edge City" and which Robert Fishman calls the "technoburbs" (p. 75). In *The Next American Metropolis*, Calthorpe refined and matured the Pedestrian Pocket idea to fit the emerging realities of Edge City technoburbia with what he calls "transit-oriented developments" (TODs). And in *The Regional City*, he and Fulton present nothing less than "a new paradigm of community . . . that leads from the Edge City to the Regional City." This, they argue, offers a conception that presents a fuller, richer social, economic, and civic life than is possible in the cultural sterilities of Edge City as it has developed in the final decades of the twentieth century. Signaling a clear break with traditional modernism as exemplified by the theories of Le Corbusier (p. 336), Calthorpe and Fulton argue that the Regional City expresses a "set of principles rooted more in ecology than in mechanics," focusing on issues of human scale, diversity, and conservation rather than rationality and mechanical efficiency.

Calthorpe is the most practical of urban visionaries because his visions represent "a response to a transformation that has already expressed itself: the transformation from the industrial forms of segregation and centralization to the decentralized and integrated forms of the post industrial era." This, he writes, is the result of "a culture adjusting itself" to new realities. Moreover, Calthorpe and Fulton are comfortable recognizing that the creation of a new urban paradigm will require the clear and careful application of state power operating through local and regional planning agencies. They write that "the idea that a region can be 'designed' is central to the Regional City" and call for an active role on the part of central governments, especially in the provision of transit systems, open-space, and development financing.

Andrés Duany and Elizabeth Plater-Zyberk, Jaime Correa, Steven Peterson and Barbara Littenberg, Mark Schimmenti, Daniel Solomon, and a number of other architects and planners are designing human-scale communities with design aspects similar to the ones Calthorpe and Fulton espouse. Because their architecture draws on traditional small town elements, these New Urbanist architects are sometimes referred to as neotraditionalists, and some of their most famous designs – notably the town of Seaside and the Disney-sponsored community of Celebration in Florida – have been roundly criticized as artificial enclaves for the well-to-do that care more for nostalgia and middle-class amenity than the needs of working people and depressed inner-city neighborhoods.

The selection here reprinted is from Peter Calthorpe and William Fulton, *The Regional City: Planning for the End of Sprawl* (Washington, DC: Island Press, 2001). Other books by Peter Calthorpe include *Sustainable Communities*, with Sym Van der Ryn (San Francisco, CA: Sierra Club Books, 1986), and *The Next American Metropolis* (New York: Princeton Architectural Press, 1993). Other books by William Fulton include *The Reluctant Metropolis* (Baltimore, MD: Johns Hopkins University Press, 2001) and *Guide to California Planning*, 2nd edn (Point Arena, CA: Solano Press, 2005).

For background readings on the New Urbanism, consult the Editors' Introduction to the Charter of the New Urbanism (p. 356). For readings pertaining to the work of Andrés Duany and Elizabeth Plater-Zyberk – the most prominent New Urbanist architects and developers other than Calthorpe – consult Duany, Plater-Zyberk and Jeff Speck, *Suburban Nation: The Rise of Sprawl and the Decline of the American Dream* (New York: North Point Press, 2001). An earlier book describing their theories is Duany and Plater-Zyberk, *Towns and Townmaking Principles* (New York: Rizzoli, 1991). Duany also coauthored *The Smart Growth Manual*, with Jeff Speck (New York: McGraw-Hill, 2002) and edited *The New Civic Art: Elements of Town Planning* (New York: Rizzoli, 2002). One of the earliest Duany developments was Seaside, Florida. For a discussion of how that paradign New Urbanist town was designed and built, see David Mohney and Keller Easterling (eds), *Seaside: Making a Town in America* (New York: Princeton Architectural Press, 1991). Critiques of Seaside – both positive and negative – are contained in Tod Bressi (ed.), *The Seaside Debates: A Critique of the New Urbanism* (New York: Rizzoli, 2002).

DESIGNING THE REGION

At the heart of creating concrete visions for the Regional City is the notion that they can be "designed." We use the term "design" not in the typical sense of artistically configuring a physical form but to imply a process that synthesizes many disciplines. Regional design is an act that integrates multiple facets at once: the demands of the region's ecology, its economy, its history, its politics, its regulations, its culture, and its social structure. And its results are specific physical forms as well as abstract goals and policies – regional maps and neighborhood urban design standards as well as implementation strategies, governmental policies, and financing mechanisms.

Too often we plan and engineer rather than design. Engineering tends to optimize isolated elements without regard for the larger system, whereas planning tends to be ambiguous, leaving the critical details of place making to chance. If we merely plan and engineer, we forfeit the possibility of developing a "whole systems" approach or a "design" that recognizes the trade-offs between isolated efficiencies and integrated parts.

The engineering mentality often reduces complex, multifaceted problems to one measurable dimension. For example, traffic engineers optimize road size for auto capacity without considering the trade-off of neighborhood scale, walkability, or beauty. Civil engineers efficiently channelize our streams without considering recreational, ecological, or esthetic values. Commercial developers optimize the delivery of goods without balancing the social need of neighborhoods for local identity and meeting places. Again and again we sacrifice the synergy of the whole for the efficiency of the parts.

The idea that a region or even a neighborhood could or should be "designed" is central to the Regional City. We need to acknowledge that we can direct our growth and that such action can include complex trade-offs as well as unexpected synergies. The common impression is that our neighborhoods, towns, or regions evolve organically (and somewhat mysteriously). They are the product of invisible market forces or the summation of technical imperatives. There also is the illusion that these forces cannot and should not be tampered with. Planning failed in the past; therefore it will fail in the future.

The real illusion, of course, is that we cannot control the form of our communities. Historically, design played a large role in shaping our forms of settlement. The template that underlies much of our suburban growth was designed in the thirties by Frank Lloyd Wright with his Broadacre Cities plans and Clarence Stein's Greenbelt towns. These were then bastardized and codified by the HUD minimum property standards of the 1950s. The template for urban redevelopment was developed about the same time by Le Corbusier and a European group of architects called CIAM (Congrès Internationaux d'Architecture Moderne). Their vision of superblocks and high-rise development became the basis of our urban renewal programs of the 1960s.

The problem is not that our suburbs and cities are lacking design but that they are designed according to failed principles with flawed implementation. They are designed in accord with modernist principles and implemented by specialists. The modernist principles of specialization, standardization, and mass production in emulating our industrial economy had a severe effect on the character of our neighborhoods and regions.

At the neighborhood scale, specialization meant that each land use – residential, retail, commercial, or civic – was isolated and developed by "experts" who optimized their particular zones without any responsibility for the whole. Regional specialization meant that each area within the region could play an independent role: suburbs for the middle class and new businesses, cities for the poor and declining industries, and countryside for nature and agriculture.

As a complement to specialization, standardization led to the homogenization of our communities, a blindness to history and the demise of unique ecological systems. A "one size fits all" mentality of efficiency overrode the special qualities of place and community.

Mass production (in housing, transportation, offices, and so forth) upends the delicate balance between local enterprise, regional systems, and global networks. The logic of mass production moves relentlessly toward ever-increasing scales, which in turn reinforces the specialization and standardization of everyday life.

Against this modern alliance of specialization, standardization, and mass production stands a set of principles rooted more in ecology than in mechanics. They are the principles of diversity, conservation, and human scale. Diversity at each scale calls for more complex, differentiated communities shaped from the unique qualities of place and history. Conservation implies care for existing resources whether natural, social, or institutional. And the principle of human scale brings the individual back into a picture increasingly fashioned around remote and mechanistic concerns.

These alternative principles apply equally to the social, economic, and physical dimensions of communities. For example, the social implications of human scale may mean more police officers walking a beat rather than hovering overhead in a helicopter; the economic implications of human scale may mean economic policies that support small local business rather than major industries and corporations; and the physical implications of human scale may be realized in the form and detail of buildings as they relate to the street. Unlike the standard governmental categories of economic development, housing, education, and health services, each of these design principles incorporates physical design, social programs, and economic strategies. These principles, then, are the ones that we believe should form the foundation of a new regional and neighborhood design ethic.

Human scale

For several generations, the design of buildings, the planning of communities, and the growth of our institutions have exemplified the view that "bigger is better." Efficiency was correlated with large, centralized organizations and processes. Now the idea of decentralized networks of smaller working groups and more personalized institutions is gaining currency in both government and business. Efficiency is correlated with nimble, small working groups, not large hierarchical institutions.

Certainly, the reality of our time is a complex mix of both of these trends. For example, we have ever-larger retail outlets at the same time that Main Streets are making a comeback. Some businesses are growing larger and more centralized while the "new economy" is bursting with small-scale start-ups and intimate working groups. Housing production is diversifying home types at the same time that it consolidates into larger financing packages. Both directions are evolving at the same time, and the shape of our communities will have to accommodate this complex reality.

Yet people are reacting to an imbalance between these two forces. The building blocks of our communities – schools, local shopping areas, housing subdivisions, apartment complexes, and office parks – have all grown into forms that defy human scale. And we are witnessing a reaction to this lack of scale in many ways. People uniformly long for an architecture that puts detail and identity back into what have too often become generic, if functional, buildings. They desire the character and scale of a walkable street, complete with shade trees and buildings that orient windows and entries their way. They idealize Main Street shopping areas and historic urban districts.

Human scale is a design principle that responds simultaneously to simple human desires and the emerging ethos of the new decentralized economics. The focus on human scale represents a shift away from top-down social programs, from characterless housing projects, and from more and more remote institutions. In its most concrete expression, human scale is the stoop of a townhouse or the front porch of a home rather than the stairwell of a high-rise or the garage door of a tract home. Human scale in economics means supporting individual entrepreneurs and local businesses. Human scale in community means a strong neighborhood focus and an environment that encourages everyday interaction.

Diversity

Diversity has multiple meanings and profound implications. It has the most challenging implications for the social, environmental, and economic dimensions of community planning. Perhaps its most obvious outcome is the creation of communities that are diverse in use and in population. As a planning axiom, it calls for a return to mixed-use neighborhoods that contain a broad range of uses as well as a broad range of housing types and people.

The four fundamental elements of community – civic places, commercial uses, housing opportunities, and natural systems – define the physical elements of diversity at any scale. As a physical principle, diversity in neighborhoods ensures that destinations are close at hand and that the shared institutions of community are integrated. It also implies an architecture rich in character and streetscapes that vary with place and use.

As a social principle, diversity is controversial and perhaps the most challenging of all. It implies creating neighborhoods that provide for a large range in age group, household type, income, and race. As already stated, neighborhoods have always (to a greater or lesser degree) been defined by commonalties even if energized by differences. But today we have reached an extreme: age, income, family size, and race are all divided into discrete market segments and locations that are built independently. Complete housing integration may be a distant goal, but inclusive neighborhoods that broaden the economic range, expand the mix of age and household types, and open the door to racial integration are feasible and desirable.

Diversity is a principle with significant economic implications. Gone are the days when economic-revitalization efforts focused on a single industry or a major governmental program. A more ecological understanding of industry clusters has emerged. This sensibility validates the notion that a range of complementary but differing enterprises (large and small, local, regional, and global) are important to maintaining a robust economy, and that now more than ever, the quality of life and the urbanism of a place, as well as the more traditional economic factors, play a significant role in the emerging economy.

Finally, diversity is a fundamental principle that can help to guide the preservation of local and regional ecologies. Clearly, understanding the complex nature of the existing or stressed habitats and watershed systems mandates a different approach to open-space planning. Active recreation, agriculture, and habitat preservation are often at odds. A broad range of open-space types, from the most active to the most passive, must be integrated in neighborhood and regional designs. Diversity in use, diversity in population, diversity in enterprise, and diversity in natural systems are fundamental to the Regional City.

Conservation

Conservation implies many things in community design beyond husbanding resources and protecting natural systems; it implies preserving and restoring the cultural, historic, and architectural assets of a place as well. Conservation calls for designing communities and buildings that require fewer resources – less energy, less land, less waste, and fewer materials – but it also implies caring for what we have and developing an ethic of reuse and repair – in both our physical and our social realms.

The principle of conservation and its complements, restoration and preservation, should be applied to the built environment as well as to the natural environment – not only to our historic building stock and neighborhood institutions, but also to human resources and human history. Communities should strive to conserve their cultural identity, physical history, and unique natural systems. Restoration and conservation are more than environmental themes; they are an approach to the way that we think about community at the regional and local levels.

Conserving resources has many obvious implications in community planning. Foremost are the quantities of farmlands and natural systems displaced by sprawling development and the quantity of auto travel required to support it. Even within more compact, walkable communities, conservation of resources can lead to new design strategies. The preservation of waterways and on-site water-treatment systems can add identity and natural amenities at the same time that they conserve water quality. Energy-conservation strategies in buildings often lead to environments that are climate responsive and unique to place.

Conserving the historic buildings and institutions of a neighborhood can preserve the icons of community identity. Restoring and enhancing the vernacular architecture of a place can simultaneously reduce energy costs, reestablish local history, and create jobs. Although the preservation movement has made great strides with landmark buildings, it is correct now in extending its agenda beyond building facades to the social fabric of neighborhoods and to the ecology of the communities that are the lifeblood of historic districts.

Conserving human resources is another implication of this fundamental principle. In too many of our communities, poverty, lack of education, and declining job opportunities lead to a tragic waste of human potential. As we have begun to see, communities are not viable when concentrations of poverty turn them into a wasteland of despair and crime. In this context, "conservation" takes on a larger meaning – the restoration and rehabilitation of human potential wherever it is being squandered and overlooked. There should be no natural or cultural environments that are disposable or marginalized. Conservation and restoration are practical undertakings that can be economically strengthening and socially enriching.

DESIGNING THE REGION IS DESIGNING THE NEIGHBORHOOD

What happens to regions or neighborhoods if they are "designed" according to these principles? An interesting set of parallel design strategies emerges at both the regional and the neighborhood levels. First and foremost, the region and its elements – the city, suburbs, and their natural environment – should be conceived as a unit, just as the neighborhood and its elements – housing, shops, open space, civic institutions, and businesses – should be designed as a unit. Treating each element separately is endemic to many of the problems that we now face. Just as a neighborhood needs to be developed as a whole system, the region must be treated as a human ecosystem, not a mechanical assembly.

Seen as this integrated whole, the region can be designed in much the same way as we would design a neighborhood. That the whole, the region, would be similar to its most basic pieces, its neighborhoods, is an important analogy. Both need protected natural systems, vibrant centers, human-scale circulation systems, a common civic realm, and integrated diversity. Developing such an architecture for the region creates the context for healthy neighborhoods, districts, and city centers. Developing such an architecture for the neighborhood creates the context for regions that are sustainable, integrated, and coherent. The two scales have parallel features that reinforce one another.

Major open-space corridors within the region, such as rivers, ridge lands, wetlands, or forests, can be seen as a "village green" at a megascale – the commons of the region. These natural commons establish an ecological identity as the basis of a region's character. Similarly, the natural systems and shared open spaces at the neighborhood scale are fundamental to its identity and character. A neighborhood's natural systems, like the region's, are as much a part of its commons as its civic institutions or commercial center.

Just as a neighborhood needs a vital center to serve as the crossroads of a local community, the region needs a vital central city to serve as its cultural heart and as a link to the global economy. In the Edge City metropolis, both types of centers are failing. In the suburbs, what were village centers of human proportions are overcome by remote discount centers and relentless commercial strips. In the central cities, poverty and disinvestment erode historic neighborhood communities. Both fall prey to specialized enterprises oriented to mass distribution rather than the local community. Like the commons, healthy centers, both urban and suburban, are fundamental to local and regional coherence.

Regional and neighborhood design has other parallels. Pedestrian scale within the neighborhood – walkable streets and nearby destinations – has a partner in transit systems at the regional scale. Transit can organize the region in much the same way as a street network orders a neighborhood. Transit lines focus growth and redevelopment in the region just as main streets can focus a neighborhood. Crossing local and metropolitan scales, transit supports the life of the pedestrian within each neighborhood and district by providing access to regional destinations. In a complementary fashion, pedestrian-friendly neighborhoods support transit by providing easy access for riders, not cars. The two scales, if designed as parallel strategies, reinforce each other.

As we have pointed out, diversity is a fundamental design principle for both the neighborhood and the region. A diverse population and job base within a region supports a resilient economy and a rich culture in much the same way that diverse uses and housing in a neighborhood support a complex and active community. The suburban trend to segregate development by age and income translates at the regional level into an increasing spatial and economic polarization . . . Both trends can be countered by policies that support inclusionary housing and mixed-use environments.

These parallels across scales are not merely coincidence. The fundamental nature of a culture and economy expresses itself at many scales simultaneously. Sprawl and our lack of regional structure is a manifestation of an older and quite different paradigm. Since World War II, our economy and culture have accelerated their movement toward the industrial qualities of mass production, standardization, and specialization. The massive suburbanization that marks this period is the direct expression of these qualities. As a counterpoint, the principles and concurrences just outlined define a new paradigm of community and growth, one that leads from the Edge City to the Regional City.

PART SIX

Urban planning theory and practice

INTRODUCTION TO PART SIX

Contemporary urban planning has come a long way from its origins in the Olmsted-inspired parks movement, the visionary plans of Ebenezer Howard and the Garden Cities movement, Daniel Burnham's monumental City Beautiful projects, the prescient regional plans of eccentric Scottish biologist Patrick Geddes, Le Corbusier and his modernist followers, Frank Lloyd Wright's brilliant Broadacre City vision, and a host of other imagined and implemented plans discussed in Part Five on Urban Planning History and Visions. Nowadays urban and regional planning (or town and country planning as it is called in the UK) has matured into an important profession with its own body of theory and set of professional practices. Part Six focuses on the theory and practice of urban planning today.

If the city is the stage on which the human drama is played out, urban planners are the stagehands. Large local governments may employ dozens or even hundreds of professionally trained planners. Even most small and medium-sized cities and towns now have city planners and some sort of explicit plans for their future development. In addition to professionals whose formal education is in urban and regional planning, planning staffs are likely to include architects and urban designers, geographers, economists, civil engineers, transportation experts, environmental professionals, computer experts, and staff trained in negotiation and other communicative planning skills. Urban planners work at the regional, state, and national levels of government and in the private and nonprofit sectors as well as at the local level.

Urban plans are grounded in analysis of local conditions and contain a vision of an urban future the citizens, local elected officials, planning staff, and consultants consider desirable. The best city plans reflect the culture and history of their city region and respect vernacular planning and architecture.

Urban plans vary greatly in approach, content, sophistication, comprehensiveness, time frame, and format. Plans developed by the Los Angeles City Planning Commission or planners in Curitiva, Brazil, run to many volumes built on mountains of data and sophisticated analysis. The town plan for a small town, on the other hand, may contain a common-sense description of the town's situation and some practical suggestions worked out by the residents and local elected officials under the direction of a part-time planning consultant.

The degree to which planners involve citizens in the urban planning process varies, depending on the planning culture of the city. Most urban planning now involves significant citizen participation at the middle and sometimes upper rungs of Sherry Arnstein's ladder of citizen participation (p. 238)

Urban planning draws on social science and design as well as specialized knowledge related to land use, transportation, environmental and other planning specialties. Planning methods rely heavily on quantitative social science methods, but can also involve qualitative methods. In addition to analytic skills planners need verbal, written, and visual communication skills and the ability to work with people.

The substance of city plans varies as widely as the planning culture of the cities that produce them. The plan for a small city like J.B. Jackson's hypothetical Optimo City (p. 202) might advance a narrow, business-as-usual vision for its future that envisions tearing down the courthouse and building a parking garage to attract more off-highway business. By contrast, green city plans in some of the most environmentally sensitive European cities that Timothy Beatley describes (p. 446) are filled with imaginative ideas for use of public bicycle depots, wind and solar power generation, community gardens, co-generation, gray water systems, and recycling. The city and regional plans developed by planners who subscribe to the ideas of

Peter Calthorpe (p. 360) and the Charter of the New Urbanism (p. 356) reflect a distinct set of New Urbanist values.

Studying cities at any scale from observing a single neighborhood through mastering complex urban modeling is interesting work. The opportunity to make plans as humble as a one-street traffic calming plan for a small town to planning an entire new city for 3 million inhabitants in China is an exciting enterprise.

The first two selections in this part describe planning theory. Sir Peter Hall's "The City of Theory" (p. 373) discusses the evolution and current status of twentieth-century urban planning theory. Nigel Taylor's "Anglo-American Town Planning Theory Since 1945: Three Significant Developments But No Paradigm Shifts" (p. 383) recaps and supplements the information in Hall's selection and provides Taylor's own interpretation of the post-World War II evolution of planning theory.

In the first half of the twentieth century, urban planning was mostly an elitist, ivory-tower exercise that paid little attention to plan implementation. According to Peter Hall (p. 373), planning theory at that time was preoccupied with how to create stable cities geared to a static world. During this "golden age," the planner was free from political interference and serenely sure of his technical capacities. He (male) produced new town plans strongly influenced by the Garden Cities and City Beautiful movements. Theoretically perfect physical plans were drawn in excruciating detail only to gather dust. While an advance over earlier static architectural city planning, this ivory-tower general planning approach was never very practical and was no longer defensible after World War II as the pace of urban development and urban change accelerated.

As long as half a century ago, computers promised to revolutionize urban planning practice. During the "systems revolution" of the 1960s, urban planners input mountains of data into mainframe computers and wrote computer programs to model traffic flows, land conversion, and the relationship between different "systems" that make up a city. The systems-planning theorists believed that empirical data and computer logic could provide the optimum solution to all planning problems. Nowadays, most planners see urban planning as a normative enterprise in which many different alternatives are possible and there is no single "best" solution to a given planning problem. As a result many present-day planning theorists acknowledge the value of accurate empirical information and computer analysis, but focus on planning processes that will facilitate plans to accommodate a variety of different interests. While computers are used extensively in urban planning today, the choices of what urban futures should be must fall to humans. Planners use spreadsheets, computerized statistical packages, and geographical information systems (GIS) and computer assisted design (CAD) software extensively in their work. But they must ultimately decide what plans to suggest based on community values and local politics as well as the results of analysis using these tools. Planning, like politics, is the art of the possible.

As Hall (p. 373) and Taylor (p. 383) describe, since the 1950s, urban planning theory has been buffeted by a series of conflicting approaches proposed by Marxists, advocacy planners, equity planners, pluralists, disjointed incrementalists, probabilistic planners, systems planners, green urbanists, ecological designers, feminist planners, and communicative action theorists. Perhaps as a result of such a variety of approaches, a humbler, pluralistic, more realistic and flexible approaches to urban planning theory have now emerged.

University of North Carolina planning professors Edward Kaiser and David Godschalk (p. 399) provide a good introduction to urban planning practice. They trace the evolution of twentieth-century land use planning and describe the current status of mainstream land use planning using the metaphor of a tree with a sturdy trunk and many branches. By the 1950s, according to Kaiser and Godschalk, there was a general consensus that urban general plans should be long-term, visionary documents, charting the desired physical form of the city. Physical land use planning was their primary purpose, rather than social or economic planning. In practice, Kaiser and Godschalk conclude, the general plan trunk of the urban land use planning tree is still at the core of most urban general plans but has now branched into management and policy plans as well as physical design.

Planning no longer takes place in ivory towers. As planning has become more relevant it has also become more conflictual. John Forester (p. 421) and Paul Davidoff (p. 435) describe conflicting values in urban planning and suggest approaches that recognize the pluralism and conflict. Citizens and decision-makers did not care much about unrealistic static physical designs or utopian general plans developed in the first half of the twentieth century because these were mostly academic exercises that did not directly affect people. But they care a great deal about plans that propose locating a hazardous waste disposal site near them, restricting the way in which they can use land they own (reducing its value), razing a historic church, developing open space, polluting a stream, or building a highway in their neighborhood. Even if they recognize the importance of a locally unwanted land use (LULU) being built somewhere, they do not want it built near them. There is a word for this: Not In My Back Yard. NIMBYs (noun) oppose development near them that would negatively affect them.

Cornell planning professor John Forester describes how planners actually interact with neighborhood residents, local elected officials, interest groups, and private developers (p. 421). Forester provides perhaps the best view we have of what current urban planning is really like as perceived by planners themselves. Based on interviews with dozens of practicing planners, Forester describes planners' day-to-day activities, the nature of and limitations on their power, and the strategies they actually employ to get things done. Planners negotiate, mediate, resolve conflict, and serve as diplomats shuttling back and forth between competing factions. They can bring a gender perspective to their work. Forester is interested in equity planning that will distribute resources such as housing, open space, and transportation options fairly and applauds the efforts of many planners to redirect market forces and to empower people and communities poorly served by the private market. Planning practitioners who study Forester's theoretical writings on planning in the face of conflict have much to learn that can help them be more effective.

Planner and lawyer Paul Davidoff proposes an approach to urban planning that recognizes that different groups compete in the planning process (p. 435). Unlike systems planners who believe mathematical modeling of data can produce a "best" solution to a planning problem, Davidoff sees planning as essentially a normative political process in which competing values contend. Davidoff envisioned a kind of planning practice in which city planners would act as advocates, particularly for poor people and disenfranchised groups. He invented a name for this kind of planning: advocacy planning. Davidoff and many planners whom he inspired saw advocacy planning as one way to bring about non-violent social change. Davidoff's concern with social justice is an enduring one. From the early efforts of nineteenth-century reformers advocating on behalf of slum residents in the new industrial cities, through the New Deal, New Frontier, and Great Society programs of the twentieth century, to present-day planners inspired by social justice ideals, progressives have always made a connection between urban planning and social justice. A recent incarnation of this tradition is equity planning, a subfield developed by Cleveland State University professor and former Cleveland city planning director Norman Krumholz.

Another enduring value in urban planning has been a concern to harmonize the built environment with the natural environment. By the middle of the nineteenth century, park planners like Joseph Paxton in England and Frederick Law Olmsted in the United States were working hard to bring nature into crowded cities. In the early twentieth century, Scottish biologist and planner Patrick Geddes had worked out an elaborate scheme for regional planning that reflected different ecosystems. Similar thinking informed Ian McHarg's theories in his classic book *Design with Nature* (1969). Now sustainable urban planning as proposed by the Brundtland Commission (p. 351), green urbanism as described by Timothy Beatley (p. 446), ecological design, and planning for carbon-neutral cities as described by Steven Wheeler (p. 458) continue to advance theory about planning cities in harmony with nature.

The selection in Part Six by Timothy Beatley describes green urbanism in Europe and what planners worldwide can learn from it. University of Virginia planning professor Timothy Beatley describes what environmentally conscious cities in Europe have actually done to retain their compact form, promote public transit, reduce auto dependency, substitute renewable energy sources for non-renewable energy sources, and support pedestrians (p. 446). Beatley identifies core aspects of the European sustainable urban development agenda, provides specific examples of exemplary green practices that some European cities have implemented, and eloquently argues that green urbanism is possible and can produce livable cities that respect the natural environment of the Earth. Beatley's selection combines theory and practice. It is both realistic about urban environmental challenges and optimistic about what can be done.

The final selection in Part Six by University of California, Davis, professor of landscape architecture and urban design, Stephen Wheeler (p. 458) addresses perhaps the most fundamental planning problems confronting the world today – how to plan cities in such a way that global climate change will be slowed or stopped and the effects of climate change that has already occurred can be mitigated. After a long period of denial and inaction, there is a virtually unanimous scientific consensus that human activity is warming the earth at an alarming pace. Scientific measurement of melting polar ice caps and glaciers, data on the rise in sea level, and changes in ecosystems in many parts of the world show that many areas of the world are becoming warmer. Ice is melting, oceans are rising, and conditions are becoming dryer. But the changes are far from uniform. Some regions of the earth are becoming wetter, colder, or more susceptible to extreme weather conditions. Solutions to global climate change will require radical changes in global energy uses and a level of international cooperation unthinkable in the recent past. Millions of decisions about the way in which cities are built will be decisive in addressing global climate change. Wheeler describes ways to think about the impact of cites on global climate change and practical measures cities are adopting to make things better.

The fate of the earth depends on how humans plan and manage the relationship between the human and built environments. More than half of the world's population now lives in cities, and cities cause a disproportionate share of humans' negative impact on the planet. Good planning theory and effective planning practices are essential.

"The City of Theory"

from *Cities of Tomorrow: An Intellectual History of Urban Planning and Design in the Twentieth Century*, 3rd edn (2001)

Peter Hall

Editors' Introduction

This selection from British geographer, planner and polymath Sir Peter Hall's magisterial intellectual history of urban planning and design in the twentieth century, *Cities of Tomorrow*, discusses the evolution of planning theory in the United States and the United Kingdom – the body of abstract philosophical writings that guide day-to-day urban planning practice.

While Hall does not use the term "paradigm" or couch his analysis in terms of "paradigm shifts", he describes a succession of planning theory paradigms that have evolved during the last hundred years. In the next selection, British planning theorist Nigel Taylor argues that significant developments in Anglo-American planning theory since 1945 do not amount to paradigm shifts (p. 386). After you have read both articles you can draw your own conclusions about whether the changes in planning theory Hall and Taylor describe amount to paradigm shifts or not.

According to Hall, the first planning paradigm of the modern era viewed urban planning as essentially architecture writ large. Hall points out that before World War II, urban planning was defined as the craft of physical planning. Professors who had been educated as architects dominated urban planning teaching and writing. The first generation of planning professors taught students to prepare architectural drawings extended to city scale – particularly self-contained, end-state physical plans like Raymond Unwin's plan for the Garden City of Letchworth, England. Planners were viewed as a privileged elite and were not expected to interact with the people they were planning for. They did not bring a gender perspective to their work. Once a well-crafted, aesthetically pleasing, functional plan was complete on paper, the early planning theorists believed that a city could be built just as a house is built from architectural drawings.

But, as Hall notes, few planners actually get to plan whole new towns from scratch. Rather, they have to decide how to integrate new housing, streets, commercial and retail districts, industrial areas, parks, open space, and infrastructure into existing cities and emerging suburbs. City planning does not end the way an architect's plan for a new house does with a final set of drawings that will be built exactly as drawn. Rather, city planning is an ongoing, fluid, messy process. Nor is the urban planning process a value-neutral, scientific process in which an educated elite can pick the best solution to a planning problem and have the client endorse it. As Paul Davidoff describes (p. 435) urban planning is a political process that requires planners to choose among conflicting values. John Forester points out that (p. 421) urban planning is often conflictual. Tidy ivory-tower plans have little chance of being implemented. Effective urban planning requires planners to interact with citizens, ideally along the higher rungs of Sherry Arnstein's ladder of citizen participation (p. 238).

During the 1960s, systems analysts who had been educated as computer scientists, quantitative geographers, engineers, and regional economists developed a competing view of what urban planning should be. They argued that urban planning should be a science, not a craft. They felt planners could and should base their plans on analysis of empirical data and use quantitative methodologies to plan transportation and other urban systems on an ongoing basis. The first generation of urban planners who viewed urban planning as a branch of systems analysis were pioneers in using computers at a time when computers were new, costly, and hard to use. The systems theorists thought plans should be mathematical models rather than end-state architectural drawings. A first-generation planner educated at the University of Liverpool in the 1920s who could produce a series of large, carefully drafted blueprints and design sketches for a new town would not know what to make of a systems planner educated at the University of Manchester in the 1960s presenting a series of equations and mathematical computer models showing how land should be subdivided, roads built, and parks dedicated over time as "the plan" for the same new town.

But planning for what kind of city and for whom? People hold strong and conflicting values of what makes a good city. Ebenezer Howard's vision of compact, human-scale garden cities (p. 328) is very different from Le Corbusier's vision of a radiant city for 3 million inhabitants consisting of massive high-rise buildings surrounded by parks and gardens (p. 321) or Frank Lloyd Wright's Broadacre City vision (p. 345) where every household would live in a single-family home on an acre of land per person.

In the 1960s and 1970s many planners shaken by urban racial and class conflict concluded that urban planning was too important to leave to either elitist designers following the craft approach to planning or technocratic planners following the systems-planning model. Liberal planners and academics of the 1960s and 1970s such as Paul Davidoff (p. 435) and Sherry Arnstein (p. 238) focused on planning outcomes and whom the city was being built for. While most cities were run by what Harvey Molotch (p. 251) calls "growth machines" that considered business interests most important, Davidoff, Arnstein, and other liberals saw urban planning as a vehicle that could benefit poor and powerless people rather than serve the interests of established business elites, pro-growth local governments and self-serving suburbanites intent on excluding low-income and minority households. Early in their academic careers, David Harvey (p. 230) and Manuel Castells (p. 572) developed neo-Marxist urban theory based on class conflict, though their writings in this book reflect interests developed later in their careers.

Twentieth-century neo-Marxists could not apply the categories Friedrich Engels applied to mid-nineteenth-century Manchester, England (p. 46). The stark class conflict between an oppressed industrial proletariat and wealthy owners of the means of production that characterized mid-nineteenth-century industrial cities had become far more complicated by the 1960s. Inequality, exclusion, racial segregation, and lack of opportunity were manifest in cities, but the categories had become more complex. Urban planning generally favored affluent, well-connected, white, male, straight, business-oriented residents rather than poor and working-class residents, racial and ethnic minorities, immigrants, women, and gays. Accordingly neo-Marxist theorists updated class analysis based on Karl Marx's theory. Neo-Marxist urban theory dominated much academic discourse throughout the 1970s.

Hall ends his tour of planning theory with some critical comments on the current divorce between planning theory and practice. He argues that as graduate programs in city and regional planning have grown in number and size, and as a formal body of planning theory has developed, too often academic planners today merely debate each other's academic theories with little attention to actual planning practice or the needs of planning practitioners on the front lines. Finding academic planning theory irrelevant, urban planning practitioners concern themselves only with the nuts-and-bolts of planning practice without the deeper understanding relevant theory could offer. Hall calls for an improved, reciprocal relationship between the two: theory that is informed by and relevant to planning practice and planning practice informed and improved by theory. John Forester's empirical research and practical theorizing (p. 421) is a good example of how Hall feels academic planning theorists might wed theory and practice. Forester interviewed dozens of practicing planners so that he was very knowledgeable about actual planning practice and then developed theory based on what they told him. This theory in turn has helped inform and improve planning practice as practitioners put into practice the theoretical insights that Forester developed.

Sir Peter Hall (b. 1932) is a professor of urban planning at the Bartlett School of Architecture and Planning, University College, London, where he has taught since 1994. Previously Hall shuttled between the University of California, Berkeley, Department of City and Regional Planning in the San Francisco Bay Area and the University of Reading's Geography Department in the United Kingdom. Hall has traveled widely and written prolifically on urban geography and urban planning. In addition to his academic accomplishments, Hall has worked on planning projects as varied as the English Channel tunnel link, bringing high-speed bullet trains to California, and revitalizing the depressed seaside town of Blackpool, England, where he grew up. Hall's theory is informed by the experience of actual planning. His numerous honors include the founder's medal of the Royal Geographical Society, membership of the British Academy, and the prestigious Balzan prize in 2005 for *Cities in Civilization* as well as his knighthood.

This selection is from *Cities of Tomorrow: An Intellectual History of Urban Planning and Design in the Twentieth Century*, 3rd edn (Oxford: Blackwell, 2001). Other of Peter Hall's books include *Urban and Regional Planning*, 5th edn (London: Routledge, 2009), *Cities in Civilization* (New York: Pantheon, 1998), *The World Cities*, 3rd edn (London: Weidenfeld & Nicolson; New York: St. Martin's, l984), and *Great Planning Disasters* (Berkeley, CA: University of California Press, 1982).

Section 2 of Eugenie Birch's *Urban and Regional Planning Reader* (London: Routledge, 2006) contains additional planning theory selections. Other overviews of planning theory include Philip Allmendinger, *Planning Theory* (New York: Palgrave Macmillan, 2002), Susan Fainstein and Scott Campbell, *Readings in Urban Theory* (London: Blackwell, 2001), Nigel Taylor, *Urban Planning Theory since 1945* (Thousand Oaks, CA: Sage, 1998), Seymour Mandelbaum, Luigi Mazza, and Robert Burchell (eds), *Explorations in Planning Theory* (New Brunswick, NJ: Center for Urban Policy Research, Rutgers University, 1996), and John Friedmann, *Planning in the Public Domain: From Knowledge to Action* (Princeton, NJ: Princeton University Press, l987).

PLANNING AND THE ACADEMY: PHILADELPHIA, MANCHESTER, CALIFORNIA, PARIS, 1955–1987

. . . about 1955 . . . city planning at last became legitimate; but in doing so, it began to sow the seeds of its own destruction. All too quickly, it split into two separate camps: the one, in the schools of planning, increasingly and exclusively obsessed with the theory of the subject; the other, in the offices of local authorities and consultants, concerned only with the everyday business of planning in the real world. That division was not at first evident; indeed, during the late 1950s and most of the 1960s, it seemed that at last a complete and satisfactory link had been forged between the world of theory and the world of practice. But all too soon, illusion was stripped aside: honeymoon was followed in quick succession during the 1970s by tiffs and temporary reconciliations, in the 1980s by divorce. And, in the process, planning lost much of its new-found legitimacy.

The prehistory of academic city planning 1930–1955

It was not that planning was innocent of academic influence before the 1950s. On the contrary: in virtually every urbanized nation, universities and polytechnics had created courses for the professional education of planners; professional bodies had come into existence to define and protect standards, and had forged links with the academic departments. Britain took an early lead when in 1909 . . . the soap magnate William Hesketh Lever, founder of Port Sunlight, won a libel action against a newspaper and used the proceeds to endow his local University of Liverpool with a Department of Civic Design. Stanley Adshead, the first professor, almost immediately created a new journal, the *Town Planning Review*, in which theory and good practice were to be firmly joined; its first editor was a young faculty recruit, Patrick Abercrombie, who was later to succeed Adshead in the chair first at Liverpool, then at Britain's second school of planning: University College London, founded in 1914. The Town Planning Institute – the Royal accolade was conferred only in 1959 – was founded in 1914 on the joint initiative of the Royal Institute of British Architects, the Institution of Civil Engineers and the Royal Institution of Chartered Surveyors; by the end of the 1930s, it had recognized seven schools whose examinations provided an entry to membership.

The United States was slower: though Harvard had established a planning course in 1909, neck and neck with Liverpool, it had no separate department until 1929. Nevertheless, by the 1930s America had schools also at MIT, Cornell, Columbia and Illinois, as well as courses taught in other departments at a great many universities across the country. And the American City Planning Institute, founded in 1917 as a breakaway from the National Conference on City Planning, ten years later became – mainly through the insistence of Thomas Adams – a full-fledged professional body on TPI lines, a status it retained when in 1938 it broadened to include regional planning and renamed itself the American Institute of Planners.

The important point about these, and other, initiatives was this: stemming as they did from professional needs, often through spin-offs from related professions like architecture and engineering, they were from the start heavily suffused with the professional styles of these design-based professions.

The job of the planners was to make plans, to develop codes to enforce these plans, and then to enforce those codes; relevant planning knowledge was what was needed for that job; planning education existed to convey that knowledge together with the necessary design skills. So, by 1950, the utopian age of planning . . . was over; planning was now institutionalized into comprehensive land-use planning. All this was strongly reflected in the curricula of the planning schools down to the mid-1950s, and often for years after that; and these in turn were reflected in the books and articles that academic planners wrote. Land-use planning, Keeble told his British audience in 1959 and Kent reminded the American counterpart in 1964, was a distinct and tightly bounded subject, quite different from social or economic planning. And these texts reflected the fact that "city planners early adopted the thoughtways and the analytical methods that engineers developed for the design of public works, and they then applied them to the design of cities."

The result, as Michael Batty has put it, was a subject that for the ordinary citizen was "somewhat mystical" or arcane, as law or medicine were, but that was – in sharp contrast to education for these older professions – not based on any consistent body of theory; rather, in it, "scatterings of social science bolstered the traditional architectural determinism." Planners acquired a synthetic ability not through abstract thinking, but by doing real jobs; in them, they used first creative intuition, then reflection. Though they might draw on bits and pieces of theory about the city – the Chicago school's sociological differentiation of the city, the land economists' theory of urban land rent differentials, the geographers' concepts of the natural region – these were employed simply as snippets of useful knowledge. In the important distinction later made by a number of writers, there was some theory in planning but there was no theory of planning. The whole process was very direct, based on a single-shot approach: survey (the Geddesian approach) was followed by analysis (an implicit learning approach), followed immediately by design.

True, as Abercrombie's classic text of 1933 argued, the making of the plan was only half the planner's job; the other half consisted of planning, that is implementation, but it was nowhere assumed that some kind of continuous learning process was needed. True, too, the 1947 Act provided for plans – and the surveys on which they were based – to be quinquennially updated; the assumption was still that the result would be a fixed land-use plan. And, a decade after that, though Keeble's equally classic text referred to the planning process, by this he simply meant the need for a spatial hierarchy of related plans from the regional to the local, and the need at each scale for survey before plan. Nowhere is found a discussion of implementation or updating. Thus – apart from extremely generalized statements like Abercrombie's famous triad of 'beauty, health and convenience' – the goals were left implicit; the planner would develop them intuitively from his own values, which by definition were "expert" and apolitical.

So, in the classic British land-use planning system created by the 1947 Town and Country Planning Act, no repeated learning process was involved, since the planner would get it right first time:

> The process was therefore not characterized by explicit feedback as the search "homed in" on the best plan, for the notion that the planner had to learn about the nature of the problem was in direct conflict with his assumed infallibility as an expert, a professional . . . The assumed certainty of the process was such that possible links back to the reality in the form of new surveys were rarely if ever considered . . . This certainty, based on the infallibility of the expert, reinforced the apolitical, technical nature of the process. The political environment was regarded as totally passive, indeed subservient to the "advice" of the planners and in practice, this was largely the case.
>
> (Batty, 1979)

It was, as Batty calls it, the golden age of planning: the planner, free from political interference, serenely sure of his technical capacities, was left to get on with the job. And this was appropriate to the world outside, with which planning had to deal: a world of glacially slow change – stagnant population, depressed economy – in which major planning interventions would come only seldom and for a short time, as after a major war. Abercrombie, in the plan for the West Midlands he produced with Herbert Jackson in 1948, actually wrote that a major objective of the plan should be to slow down the rate of urban change, thus reducing the rate at which built structures became obsolescent: the ideal city would be a static, stable city:

> Let us assume . . . that a maximum population has been decided for a town, arrived at after consideration of all the factors appearing to be relevant . . . Allowance has been made for proper space for all conceivable purposes in the light of present facts and the town planner's experience and imagination. Accordingly, an envelope or green belt has been prescribed, outside which the land uses will be those involving little in the way of resident population. The town planner is now in the happy position for the first time of knowing the limits of his problem. He is able to address himself to the design of the whole and the parts in the light of a basic overall figure for population. The process will be difficult enough in itself, but at least he starts with one figure to reassure him.
>
> (Abercrombie and Jackson, 1948)

American planning was never quite like that. Kent's text of 1964 on the urban general plan, though it deals with the same kind of land-use planning, reminds its students of end-directions which are continually adjusted as time passes. And, because the planner's basic understanding of the interrelationship between socio-economic forces and the physical environment was largely intuitive and speculative, Kent warned his student readers,

> In most cases it is not possible to know with any certainty what physical design measures should be taken to bring about a given social or economic objective, or what social and economic consequences will result from a given physical-design proposal. Therefore, the city council and the city-planning commission, rather than professional city planners, should make the final value judgements upon which the plan is based.
>
> (Kent, 1964)

But even Kent was certain that, despite all this, it was still possible for the planner to produce some kind of optimal land use plan; the problem of objectives was just shunted off.

The systems revolution

It was a happy, almost dream-like, world. But increasingly, during the 1950s, it did not correspond to reality. Everything began to get out of hand. In every industrial country, there was an unexpected baby boom, to which the demographers reacted with surprise, the planners with alarm; only its timing varied from one country to another, and everywhere it created instant demands for maternity wards and child-care clinics, only slightly delayed needs for schools and playgrounds. In every one, almost simultaneously, the great postwar economic boom got under way, bringing pressures for new investment in factories and offices. And, as boom generated affluence, these countries soon passed into the realms of high mass-consumption societies, with unprecedented demands for durable consumer goods: most notable among these, land-hungry homes and cars. The result everywhere – in America, in Britain, in the whole of western Europe – was that the pace of urban development and urban change began to accelerate to an almost superheated level. The old planning system, geared to a static world, was overwhelmed.

These demands in themselves would force the system to change; but, almost coincidentally, there were changes on the supply side too. In the mid-1950s there occurred an intellectual revolution in the whole cluster of urban and regional social studies, which provided planners with much of their borrowed intellectual baggage. A few geographers and industrial economists discovered the works of German theorists of location, such as Johann Heinrich von Thünen (1826) on agriculture, Alfred Weber (1909) on industry, Walter Christaller (1933) on central places, and August Lösch (1940) on the general theory of location; they began to summarize and analyse these works, and where necessary to translate them. In the United States, academics coming from a variety of disciplines began to find regularities in many distributions, including spatial ones. Geographers, beginning to espouse the tenets of logical positivism, suggested that their subject should cease to be concerned with descriptions of the detailed differentiation of the earth's surface, and should instead begin to develop general hypotheses about spatial distributions, which could then be rigorously tested against reality: the very approach which these German pioneers of location theory had adopted. These ideas, together with the relevant literature, were brilliantly synthesized by an American economist,

Walter Isard, in a text that became immediately influential. Between 1953 and 1957, there occurred an almost instant revolution in human geography and the creation, by Isard, of a new academic discipline uniting the new geography with the German tradition of locational economics. And, with official blessing – as in the important report of Britain's Schuster Committee of 1950, which recommended a greater social science content in planning education – the new locational analysis began to enter the curricula of the planning schools.

The consequences for planning were momentous: with only a short timelag, "the discipline of physical planning changed more in the 10 years from 1960 to 1970, than in the previous 100, possibly even 1000 years" (Batty, 1979).

The subject changed from a kind of craft, based on personal knowledge of a rudimentary collection of concepts about the city, into an apparently scientific activity in which vast amounts of precise information were garnered and processed in such a way that the planner could devise very sensitive systems of guidance and control, the effects of which could be monitored and if necessary modified. More precisely, cities and regions were viewed as complex systems – they were, indeed, only a particular spatially based subset of a whole general class of systems – while planning was seen as a continuous process of control and monitoring of these systems, derived from the then new science of cybernetics developed by Norbert Wiener.

There was thus, in the language later used in the celebrated work of Thomas Kuhn, a "paradigm shift." It affected city planning as it affected many other related areas of planning and design. Particularly, its main early applications – already in the mid-1950s – concerned defence and aerospace; for these were the Cold War years, when the United States was engaging in a crash programme to build new and complex electronically controlled missile systems. Soon, from that field, spun off another application. Already in 1954, Robert Mitchell and Chester Rapkin – colleagues of Isard at the University of Pennsylvania – had published a book suggesting that urban traffic patterns were a direct and measurable function of the pattern of activities – and thus land uses – that generated them. Coupled with earlier work on spatial interaction patterns, and using for the first time the data-processing powers of the computer, this produced a new science of urban transportation planning, which for the first time claimed to be able scientifically to predict future urban-traffic patterns. First applied in the historic Detroit Metropolitan Area transportation study of 1955, further developed in the Chicago study of 1956, it soon became a standardized methodology employed in literally hundreds of such studies, first across the United States, then across the world.

Heavily engineering-based in its approach, it adopted a fairly standardized sequence. First, explicit goals and objectives were set for the performance of the system. Then, inventories were taken of the existing state of the system: both the traffic flows, and the activities that gave rise to them. From this, models were derived which sought to establish these relationships in precise mathematical form. Then, forecasts were made of the future state of the system, based on the relationships obtained from the models. From this, alternative solutions could be designed and evaluated in order to choose a preferred option. Finally, once implemented the network would be continually monitored and the system modified as necessary.

At first, these relationships were seen as operating in one direction: activities and land uses were given; from these, the traffic patterns were derived. So the resulting methodology and techniques were part of a new field, transportation planning, which came to exist on one side of traditional city planning. Soon, however, American regional scientists suggested a crucial modification: the locational patterns of activities – commercial, industrial, residential – were in turn influenced by the available transportation opportunities; these relationships, too, could be precisely modelled and used for prediction; therefore the relationship was two-way, and there was a need to develop an interactive system of land-use–transportation planning for entire metropolitan or subregional areas. Now, for the first time, the engineering-based approach invaded the professional territory of the traditional land-use planner. Spatial interaction models, especially the Garin–Lowry model – which, given basic data about employment and transportation links, could generate a resulting pattern of activities and land uses – became part of the planner's stock in trade. As put in one of the classic systems texts:

> In this general process of planning we particularise in order to deal with more specific issues: that is, a specific real world system or subsystem must be represented by a specific conceptual system or subsystem within the general conceptual system. Such a particular representation of a system

> is called a *model* . . . the use of models is a means whereby the high variety of the real world is reduced to a level of variety appropriate to the channel capacities of the human being.
>
> (Chadwick, 1971)

This involved more than a knowledge of computer applications – novel as that seemed to the average planner of the 1960s. It meant also a fundamentally different concept of planning. Instead of the old master-plan or blueprint approach, which assumed that the objectives were fixed from the start, the new concept was of planning as a process, "whereby programmes are adapted during their implementation as and when incoming information requires such changes". And this planning process was independent of the thing that was planned; as Melvin Webber put it, it was "a special way of deciding and acting" which involved a constantly recycled series of logical steps: goal-setting, forecasting of change in the outside world, assessment of chains of consequences of alternative courses of action, appraisal of costs and benefits as a basis for action strategies, and continuous monitoring. This was the approach of the new British textbooks of systems planning, which started to emerge at the end of the 1960s, and which were particularly associated with a group of younger British graduates, many teaching or studying at the University of Manchester. It was also the approach of a whole generation of subregional studies, made for fast-growing metropolitan areas in Britain during that heroic period of growth and change, 1965–75: Leicester–Leicestershire, Nottinghamshire–Derbyshire, Coventry–Warwickshire–Solihull, South Hampshire. All were heavily suffused with the new approach and the new techniques; in several, the same key individuals – McLoughlin in Leicester, Batty in Notts–Derby – played a directing or a crucial consulting role.

But the revolution was less complete – at least, in its early stages – than its supporters liked to argue: many of these "systems" plans had a distinctly blueprint tint, in that they soon resulted in all-too-concrete proposals for fixed investments like freeway systems. Underlying this, furthermore, were some curious metaphysical assumptions, which the new systems planners shared with their blueprint elders: the planning system was seen as active, the city system as purely passive; the political system was regarded as benign and receptive to the planner's expert advice. In practice, the systems planner was involved in two very different kinds of activity: as a social scientist, he or she was passively observing and analysing reality; as a designer, the same planner was acting on reality to change it – an activity inherently less certain, and also inherently subject to objectives that could only be set through a complex, often messy, set of dealings between professionals, politicians and public.

The core of this problem was a logical paradox: despite the claims of the systems planners, the urban planning system was different from (say) a weapons system. In this latter kind of system, to which the "systems approach" had originally and successfully been applied, the controls were inside the system; but here, the urban-regional system was inside its own system of control. Related to this were other crucial differences: in urban planning, there was not just one problem and one overriding objective, but many, perhaps contradictory; it was difficult to move from general goals to specific operational ones; not all were fully perceived; the systems to be analysed did not self-evidently exist, but had to be synthesized; most aspects were not deterministic, but probabilistic; costs and benefits were difficult to quantify. So the claims of the systems school to scientific objectivity could not readily be fulfilled. Increasingly, members of the school came to admit that in such "open" systems, systematic analysis would need to play a subsidiary role to intuition and judgement; in other words, the traditional approach. By 1975 Britton Harris, perhaps the most celebrated of all the systems planners, could write that he no longer believed that the more difficult problems of planning could be solved by optimizing methods.

The search for a new paradigm

All this, in the late 1960s, came to focus in an attack from two very different directions, which together blew the ship of systems planning at least half out of the water. From the philosophical right came a series of theoretical and empirical studies from American political scientists, arguing that – at least in the United States – crucial urban decisions were made within a pluralist political structure in which no one individual or group had total knowledge or power, and in which, consequently, the decision-making process could best be described as "disjointed incrementalism" or "muddling through". Meyerson and Banfield's classic analysis of the Chicago Housing Authority concluded that it engaged in little real planning, and failed because it did not correctly identify the real power structure in the city; its elitist view of the public interest was totally opposed to the populist view of the ward politicians, which finally prevailed. Downs theorized about such a structure, suggesting that politicians buy votes by offering bundles of policies, rather as in a market. Lindblom contrasted the whole rational-comprehensive model of planning with what he found to be the actual process of policy development, which was characterized by a mixture of values and analysis, a confusion of ends and means, a failure to analyse alternatives, and an avoidance of theory. Altshuler's analysis of Minneapolis–St Paul suggested that the professional planner carried no clout against the political machine, which backed the highway-building engineers against him; they won by stressing expertise and concentrating on narrow goals, but theirs was a political game; the conclusion was that planners should recognize their own weakness, and devise strategies appropriate to that fact.

All these analyses arose from study of American urban politics, which is traditionally more populist, more pluralist, than most. Even there, Rabinowitz's study of New Jersey cities suggested that they varied greatly in style, from the highly fragmented to the very cohesive; while Etzioni, criticizing Lindblom, suggested that recent United States history showed several important examples of non-incremental decision-making, especially in defence. But, these reservations taken, the studies did at least suggest that planning in actuality was a very long way indeed from the cool, rational, Olympian style envisaged in the systems texts. Perhaps it might have been better if it had been closer; perhaps not. The worrisome point was that in practice, local democracy proved to be an infinitely messier business than the theory would have liked. Some theorists accordingly concluded that if this was the way planning was, this was the way it should be encouraged to be: partial, experimental, incremental, working on problems as they arose.

That emerged even more clearly, because – as so often seems to happen – in America the left-wing criticism was reaching closely similar conclusions. By the late 1960s, fuelled by the civil-rights movement and war on poverty, the protests against the Vietnam war and the campus free-speech movement, it was this wing that was making all the running. Underlying the general current of protest were three key themes, which proved fatal to the legitimacy of the systems planners. One was a widespread distrust of expert, top–down planning generally – whether for problems of peace and war, or for problems of the cities. Another, much more specific, was an increasing paranoia about the systems approach, which in its military applications was seen as employing pseudo-science and incomprehensible jargon to create a smokescreen, behind which ethically reprehensible policies could be pursued. And a third was triggered by the riots that tore through American cities starting with Birmingham, Alabama, in 1963 and ending with Detroit, in 1967. They seemed to prove the point: systems planning had done nothing to ameliorate the condition of the cities; rather, by assisting or at least conniving in the dismemberment of inner-city communities, it might actually have contributed to it. By 1967 one critic, Richard Bolan, could argue that systems planning was old-fashioned comprehensive planning, dressed up in fancier garb; both, alike, ignored political reality.

The immediate left-wing reaction was to call on the planners themselves to turn the tables, and to practise bottom–up planning by becoming advocate-planners. Particularly, in this way they would make explicit the debate about the setting of goals and objectives, which both the blueprint and systems approaches had bypassed by means of their comfortable shared assumption that this was the professional planner's job. Advocacy planners would intervene in a variety of ways, in a variety of groups; diversity should be their keynote. They would help to inform the public of alternatives; force public planning agencies to compete for support; help critics to generate plans that were superior to official ones; compel consideration of underlying values. The resulting structure was highly American: democratic, locally grounded, pluralistic, but also legalistic in being based on institutionalized conflict. But, interestingly, while demoting the planner in one respect, it enormously advanced his or her power in another: the planner was to take many of the functions that the locally elected official had previously exercised. And, in practice, it was not entirely clear how it would all work; particularly, how the process would resolve the very real conflicts of interest that could arise within communities, or how it could avoid the risk that the planners, once again, would become manipulators.

At any rate, there is more than a passing resemblance between the planner as a disjointed incrementalist, and the planner-advocate; and, indeed, between either of these and a third model set out in Bolan's paper of 1967, the planner as informal co-ordinator and catalyst, which in turn shades into a fourth: Melvin Webber's probabilistic planner, who uses new information systems to facilitate debate and improve decision-making. All are assumed to work within a pluralist world, with very many different competing groups and interests, where the planner has at most (and, further, should have) only limited power or influence; all are based, at least implicitly, on continued acceptance of logical positivism. As Webber put it, at the conclusion of his long two-part paper of 1968–9:

> The burden of my argument is that city planning failed to adopt the planning method, choosing instead to impose input bundles, including regulatory constraints, on the basis of ideologically defined images of goodness. I am urging, as an alternative, that planning tries out the planning idea and the planning method.

In turn, Webber's view of planning – which flatly denied the possibility of a stable predictable future or agreed goals – provided some of the philosophical underpinnings of the Social Learning or New Humanist approach of the 1970s, which stressed the importance of learning systems in helping cope with a turbulent environment. But finally, this approach divorced itself from logical positivism, returning to a reliance on personal knowledge which was strangely akin to old-style blueprint planning; and, as developed by John Friedmann of the University of California at Los Angeles, it finally resulted in a demand for all political activity to be decomposed into decision by minute political groups: a return to the anarchist roots of planning, with a vengeance.

So these different approaches diverged, sometimes in detailed emphasis, sometimes more fundamentally. What they shared was the belief that – at any rate in the American political system – the planner did not have much power and did not deserve to have much either; within a decade, from 1965 to 1975, these approaches together neatly stripped the planner of whatever priestly clothing, and consequent mystique, s/he may have possessed. Needless to say, this view powerfully communicated itself to the professionals themselves. Even in countries with more centralized, top–down political systems, such as Great Britain, young graduating planners increasingly saw their roles as rather like barefoot doctors, helping the poor down on the streets of the inner city, working either for a politically acceptable local authority, or, failing that, for community organizations battling against a politically objectionable one.

Several historical factors, in addition to the demolition job on planning by the American theorists, contributed to this change: planners and politicians belatedly discovered the continued deprivation of the inner-city poor; then, it was seen that the areas where these people lived were suffering depopulation and deindustrialization; in consequence, planners progressively moved away from the merely physical, and into the social and the economic. The change can be caricatured thus: in 1955, the typical newly graduated planner was at the drawing board, producing a diagram of desired land uses; in 1965, s/he was analysing computer output of traffic patterns; in 1975, the same person was talking late into the night with community groups, in the attempt to organize against hostile forces in the world outside.

It was a remarkable inversion of roles. For what was wholly or partly lost, in that decade, was the claim to any unique and useful expertise, such as was possessed by the doctor or the lawyer. True, the planner could still offer specialized knowledge on planning laws and procedures, or on how to achieve a particular design solution; though often, given the nature of the context and the changed character of planning education, s/he might not have enough of either of these skills to be particularly useful. And, some critics were beginning to argue, this was because planning had extended so thinly over so wide an area that it became almost meaningless; in the title of Aaron Wildavsky's celebrated paper, "If Planning is Everything, Maybe it's Nothing".

The fact was that planning, as an academic discipline, had theorized about its own role to such an extent that it was denying its own claim to legitimacy. Planning, Faludi pointed out in his text of 1973, could be merely *functional*, in that the goals and objectives are taken as given; or *normative*, in that they are themselves the object of rational choice. The problem was whether planning was really capable of doing that latter job. As a result, by the mid-1970s planning had reached the stage of a "paradigm crisis"; it had been theoretically useful to distinguish the planning process as something separate from what is planned, yet this had meant a neglect of substantive theory, pushing it to the periphery of the whole subject. Consequently, new theory is needed which attempts to bridge current planning strategies and the urban physical and social systems to which strategies are applied.

The Marxist ascendancy

That became ever clearer in the following decade, when the logical positivists retreated from the intellectual field of battle and the Marxists took possession. As the whole world knows, the 1970s saw a remarkable resurgence – indeed a veritable explosion – of Marxist studies. This could not fail to affect the closely related worlds of urban geography, sociology, economics and planning. True, like the early neo-classical economists, Marx had been remarkably uninterested in questions of spatial location – even though Engels had made illuminating comments on the spatial distribution of classes in mid-Victorian Manchester. The disciples now reverently sought to extract from the holy texts, drop by drop, a distillation that could be used to brew the missing theoretical potion. At last, by the mid-1970s, it was ready; then came a flood of new work. It originated in various places and in various disciplines: in England and the United States the geographers David Harvey and Doreen Massey helped to explain urban growth and change in terms of the circulation of capital; in Paris, Manuel Castells and Henri Lefebvre developed sociologically based theories. In the endless debates that followed among the Marxists themselves, a critical question concerned the role of the state. In France, Lokjine and others argued that it was mainly concerned, through such devices as macroeconomic planning and related infrastructure investment, directly to underpin and aid the direct productive investments of private capital. Castells, in contrast, argued that its main function had been to provide collective consumption – as in public housing, or schools, or transportation – to help guarantee the reproduction of the labour force and to dampen class conflict, essential for the maintenance of the system. Clearly, planning might play a very large role in both these state functions; hence, by the mid-1970s French Marxist urbanists were engaging in major studies of this role in the industrialization of such major industrial areas as Dieppe.

At the same time, a specifically Marxian view of planning emerged in the English-speaking world. To describe it adequately would require a course in Marxist theory. But, in inadequate summary, it states that the structure of the capitalist city itself, including its land-use and activity patterns, is the result of capital in pursuit of profit. Because capitalism is doomed to recurrent crises, which deepen in the current stage of late capitalism, capital calls upon the state, as its agent, to assist it by remedying disorganization in commodity production, and by aiding the reproduction of the labour force. It thus tries to achieve certain necessary objectives: to facilitate continued capital accumulation, by ensuring rational allocation of resources; by assisting the reproduction of the labour force through provision of social services, thus maintaining a delicate balance between labour and capital and preventing social disintegration; and by guaranteeing and legitimating capitalist social and property relations. As Dear and Scott put it: "In summary, planning is an historically-specific and socially-necessary response to the self-disorganizing tendencies of *privatized capitalist* social and property relations as these appear in urban space." In particular, it seeks to guarantee collective provision of necessary infrastructure and certain basic urban services, and to reduce negative externalities whereby certain activities of capital cause losses to other parts of the system.

But, since capitalism also wishes to circumscribe state planning as far as possible, there is an inbuilt contradiction: planning, because of this inherent inadequacy, always solves one problem only by creating another. Thus, say the Marxists, nineteenth-century clearances in Paris created a working-class housing problem; American zoning limited the powers of industrialists to locate at the most profitable locations. And planning can never do more than modify some parameters of the land development process; it cannot change its intrinsic logic, and so cannot remove the contradiction between private accumulation and collective action. Further, the capitalist class is by no means homogenous; different fractions of capital may have divergent, even contradictory interests, and complex alliances may be formed in consequence; thus, latter-day Marxist explanations come close to being pluralist, albeit with a strong structural element. But in the process, "the more that the State intervenes in the urban system, the greater is the likelihood that different social groups and fractions will contest the legitimacy of its decisions. *Urban life as a whole becomes progressively invaded by political controversies and dilemmas.*"

Because traditional non-Marxian planning theory has ignored this essential basis of planning, so Marxian commentators argue, it is by definition vacuous: it seeks to define what planning ideally ought to be, devoid of all context; its function has been to depoliticize planning as an activity, and thus to legitimate it.

It seeks to achieve this by representing itself as the force which produces the various facets of real-world planning. But in fact, its various claims – to develop abstract concepts that rationally represent real-world processes, to legitimate its own activity, to explain material processes as the outcome of ideas, to present planning goals as derived from generally shared values, and to abstract planning activity in terms of metaphors drawn from other fields like engineering – all these are both very large and quite unjustified. The reality, Marxists argue, is precisely the opposite: viewed objectively, planning theory is nothing other than a creation of the social forces that bring planning into existence.

It makes up a disturbing body of coherent criticism: yes, of course, planning cannot simply be an independent self-legitimating activity, as scientific inquiry may claim to be; yes, of course, it is a phenomenon that – like all phenomena – represents the circumstances of its time. As Scott and Roweis put it:

> . . . there is a definite mismatch between the world of current planning theory, on the one hand, and the real world of practical planning intervention on the other hand. The one is the quintessence of order and reason in relation to the other which is full of disorder and unreason. Conventional theorists then set about resolving this mismatch between theory and reality by introducing the notion that planning theory is in any case not so much an attempt to explain the world as it is but as it ought to be. Planning theory then sets itself the task of rationalizing irrationalities, and seeks to materialize itself in social and historical reality (like Hegel's World Spirit) by bringing to bear upon the world a set of abstract, independent, and transcendent norms.
>
> (Scott and Roweis, 1977)

It was powerful criticism. But it left in turn a glaringly open question, both for the unfortunate planner – whose legitimacy is now totally torn from him, like the epaulette from the shoulder of a disgraced officer – and, equally, for the Marxist critic: what, then, is planning theory about? Has it any normative or prescriptive content whatsoever? The answer, logically, would appear to be no. One of the critics, Philip Cooke, is uncompromising:

> The main criticism that tends to have been made, justifiably, of planning is that it has remained stubbornly normative . . . in this book it will be argued that [planning theorists] should identify mechanisms which cause changes in the nature of planning to be brought about, rather than assuming such changes to be either the creative idealizations of individual minds, or mere regularities in observable events.
>
> (Cooke, 1983)

This is at least consistent: planning theory should avoid all prescription; it should stand right outside the planning process, and seek to analyse the subject – including traditional theory – for what it is, the reflection of historical forces. Scott and Roweis, a decade earlier, seem to be saying exactly the same thing: planning theory cannot be normative, it cannot assume "transcendent operational norms". But then, they stand their logic on its own head, saying that "a viable theory of urban planning should not only tell us what planning is, but also what we can, and must, do as progressive planners".

This, of course, is sheer rhetoric. But it nicely displays the agony of the dilemma. Either theory is about unravelling the historical logic of capitalism, or it is about prescription for action. Since the planner-theorist – however sophisticated – could never hope to divert the course of capitalist evolution by more than a millimetre or a millisecond, the logic would seem to demand that s/he sticks firmly to the first and abjures the second. In other words, the Marxian logic is strangely quietist; it suggests that the planner retreats from planning altogether into the academic ivory tower.

Some were acutely conscious of the dilemma. John Forester tried to resolve it by basing a whole theory of planning action on the work of Jürgen Habermas. Habermas, perhaps the leading German social theorist of the post-World War Two era, had argued that latter-day capitalism justified its own legitimacy by spinning around itself a complex set of distortions in communication, designed to obscure and prevent any rational understanding of its own workings. Thus, he argued, individuals became powerless to understand how and why they act, and so were excluded from all power to influence their own lives,

> as they are harangued, pacified, mislead [*sic*], and ultimately persuaded that inequality, poverty, and ill-health are either problems for which the victim is responsible or problems so "political" and

> "complex" that they can have nothing to say about them. Habermas argues that democratic politics or planning requires the consent that grows from processes of collective criticism, not from silence or a party line.
>
> (Forester, 1980)

But, Forester argues, Habermas's own proposals for communicative action provide a way for planners to improve their own practice:

> By recognizing planning practice as normatively role-structured communication action which distorts, covers up, or reveals to the public the prospects and possibilities they face, a critical theory of planning aids us practically and ethically as well. This is the contribution of critical theory to planning: pragmatics with vision – to reveal true alternatives, to correct false expectations, to counter cynicism, to foster inquiry, to spread political responsibility, engagement, and action. Critical planning practice, technically skilled and politically sensitive, is an organizing and democratizing practice.
>
> (Forester, 1980)

Fine. The problem is that – stripped of its Germanic philosophical basis, which is necessarily a huge oversimplification of a very dense analysis – the practical prescription all comes out as good old-fashioned democratic common sense, no more and no less than Davidoff's advocacy planning of fifteen years before: cultivate community networks, listen carefully to the people, involve the less-organized groups, educate the citizens in how to join in, supply information and make sure people know how to get it, develop skills in working with groups in conflict situations, emphasize the need to participate, compensate for external pressures. True, if in all this planners can sense that they have penetrated the mask of capitalism, that may help them to help others to act to change their environment and their lives; and, given the clear philosophical impasse of the late 1970s, such a massive metaphysical underpinning may be necessary.

The world outside the tower: practice retreats from theory

Meanwhile, if the theorists were retreating in one direction, the practitioners were certainly reciprocating. Whether baffled or bored by the increasingly scholastic character of the academic debate, they lapsed into an increasingly untheoretical, unreflective, pragmatic, even visceral style of planning. That was not entirely new: planning had come under a cloud before, as during the 1950s, and had soon reappeared in a clear blue sky. What was new, strange, and seemingly unique about the 1980s was the divorce between the Marxist theoreticians of academe – essentially academic spectators, taking grandstand seats at what they saw as one of capitalism's last games – and the anti-theoretical, anti-strategic, anti-intellectual style of the players on the field down below. The 1950s were never like that; then, the academics were the coaches, down there with the team.

The picture is of course exaggerated. Many academics did still try to teach real-life planning through simulation of real-world problems. The Royal Town Planning Institute enjoined them to become ever more practice-minded. The practitioners had not all shut their eyes and ears to what comes out of the academy; some even returned there for refresher courses. And if all this was true in Britain, it was even more so of America, where the divorce had never been so evident. Yet the picture does describe a clear and unmistakable trend; and it was likely to be more than a cyclical one.

The reason is simple: as professional education of any kind becomes more fully absorbed by the academy, as its teachers become more thoroughly socialized within it, as careers are seen to depend on academic peer judgements, then its norms and values – theoretical, intellectual, detached – will become ever more pervasive; and the gap between teaching and practice will progressively widen. One key illustration: of the huge output of books and papers from the planning schools in the 1980s, there were many – often, those most highly regarded within the academic community – that were simply irrelevant, even completely incomprehensible, to the average practitioner.

Perhaps, it might be argued, that was the practitioner's fault; perhaps too we need fundamental science, with no apparent payoff, if we are later to enjoy its technological applications. The difficulty with that argument was to find convincing evidence that – not merely here, but in the social sciences generally – such payoff eventually comes. Hence the low esteem into which the social sciences had everywhere fallen, not least in Britain and the United States: hence too the diminished level of support for them, which – at any rate in Britain – had directly redounded on the planning schools. The relationship between planning and the academy had gone sour, and that is the major unresolved question that must now be addressed.

REFERENCES

Abercrombie, P. and Jackson, H. (1948) *West Midlands Plan*. Interim Confidential Edition. 5 vols. London: Ministry of Town and Country Planning.

Batty, M. (1979) "On Planning Processes", in: B. Goodall and A. Kirby (eds) *Resources and Planning*. Oxford: Pergamon.

Chadwick, G. (1971) *A Systems View of Planning: Towards a Theory of the Urban and Regional Planning Process*. Oxford: Pergamon.

Cooke, P.N. (1983) *Theories of Planning and Spatial Development*. London: Hutchinson.

Forester, J. (1980) "Critical Theory and Planning Practice", *Journal of the American Planning Association*, 46, 275–86.

Kent, T.J. (1964) *The Urban General Plan*. San Francisco: Chandler.

Scott, A.J. and Roweis, S.T. (1977) "Urban Planning in Theory and Practice: An Appraisal", *Environment and Planning*, 9, 1097–1119.

Webber, M.M. (1968–9) "Planning in an Environment of Change", *Town Planning Review*, 39, 179–95, 277–95.

"Anglo-American Town Planning Theory Since 1945: Three Significant Developments But No Paradigm Shifts"

Planning Perspectives (1999)

Nigel Taylor

Editors' Introduction

In the previous selection Sir Peter Hall described the evolution of modern planning theory from its origins through the early twenty-first century (p. 373). Most of the theorists Hall discussed are from the UK or United States. In this selection British planning theorist Nigel Taylor provides his own interpretation of Anglo-American planning theory from 1945 through 1999 when this article was written. Taylor argues that during this period there were significant developments in planning theory, but no true paradigm shifts.

The ideas of paradigms and revolutionary changes in scientific thinking were best articulated by physicist and historian of science Thomas Kuhn in an influential book titled *The Structure of Scientific Revolutions* (Chicago, IL: University of Chicago Press, 1962). Kuhn distinguished "scientific revolutions" in which revolutionary breakthroughs in thinking about the world occur from "normal science" in which knowledge advances slowly and incrementally. Sixteenth-century Italian astronomer Nicolaus Copernicus's theory placing a spherical sun revolving in space, rather than a stationary flat earth, at the center of our solar system, for example, would qualify as a revolutionary scientific breakthrough, while mapping craters on Mars would be normal science. The crater mapping is an addition to human knowledge consistent with established theory, but Copernicus's theory represents a whole new way of looking at the universe. Urban planning is a form of normative practice based on social science. A theory that urban planning should be done using computer models rather than architectural drawings might qualify as a paradigm shift, but John Forester's illuminating discussion of how urban planners resolve conflicts (p. 421) is a "normal science" extension of knowledge about urban planning.

Taylor discusses three significant shifts in the way urban planning has been conceived and defined that began in the 1960s: first, from the planners as a creative designer to the planner as a scientific analyst and rational decision-maker, second, from the planner as a technical expert to the planner as a manager and communicator, and third, from modernist to postmodernist ways of thinking about urban planning.

Taylor agrees with Peter Hall (p. 373) that initially urban planning theory viewed urban planning as essentially an exercise in physical design – architecture on the scale of a whole town, or part of a town – though he argues that this view remained dominant after World War II and that its continuing influence is greater than Hall describes. Taylor also agrees with Hall that in the mid-1960s some theorists challenged this conception based on systems theory and the rational planning model. In Taylor's formulation, two distinct theories of planning "the systems view" and the "rational process view" arose in the late 1960s. Both conceptions represented a radical departure from the then prevailing design-based view of planning. Cities were seen as systems of interrelated activities in a constant state of flux, whose ongoing processes needed to be analyzed scientifically rather than in static, end-state, aesthetically based blueprint-like drawings.

But is this change from design-based to systems-based planning a Kuhnian paradigm shift? Taylor thinks not. He points out that urban design continued to be an important feature of planning through the late 1960s and 1970s and that there was a revival of theoretical interest in urban design in the 1980s and 1990s. In Taylor's view, the systems and rational process views of planning added to rather than ousted the incumbent design-based view. They represented a significant change, but not a paradigm shift.

Taylor identifies the emergence of the view that planning is a value-laden political process in the 1960s as another significant development in planning theory. At that time neo-Marxists like Henri LeFevre, David Harvey, and Manuel Castells argued that planners should be concerned with social justice and equity. Paul Davidoff (p. 435) and Sherry Arnstein (p. 238) pressed for processes that would include the poor and powerless or make their interests know in planning processes.

Both the traditional design-based view of planning and the systems and rational-process views presume that planners possess specialized knowledge and expertise that laypeople do not. An extreme critic of this line of reasoning might argue that an angry neighborhood resident with no knowledge of design, economics, computers, or rational planning at all might be as (or more) qualified to articulate a plan for the neighborhood as a university-educated planner. Arguably an impassioned tirade against a redevelopment project that would displace most of the poor minority residents is a form of planning that might produce a better plan than an urban renewal plan based on lots of data, sophisticated computer modeling, and application of specialized design skills.

A less extreme theoretical position along the same lines – which Taylor notes has been widely adopted in planning theory – is that planners' expertise should lie mainly in eliciting opinions and managing planning processes. According to this theory of planning, the planner should "facilitate" planning, but not impose knowledge-based decisions. British planning theorist Patsy Healey has developed a theory viewing planners as both communicators and implementers: communicative action planning. US planners Judith Innes and David Booher also argued that collaborative planning is essential to solving complex planning problems. Is the shift to viewing planners as communicators and managers rather than technical experts a Kuhnian paradigm shift? Again, Taylor argues it is not. Taylor concedes that communicating and managing planning processes are important, but argues that planners can and should bring substantive knowledge to bear so that citizens, other stakeholders, and decision-makers understand the consequences of alternative courses of action.

The final significant development that started in the 1960s that Taylor describes involves a shift in planning from modernist to postmodernist approaches. The theory and practice of architecture and urban planning that Le Corbusier (p. 336), the International Congress of Modern Architecture (CIAM) and other modernists developed was at its height during from the late 1920s through the 1950s. The modernists found little of value in historic city designs, favored use of modern building materials such as steel and glass, liked aesthetically minimalist functional plans, advocated for rational, scientific problem-solving, and wanted big, efficient cities, built for speed. Since about 1960, modernism has been largely out of favor. David Harvey (p. 230), Mike Davis (p. 195), Michael Dear (p. 170) and others have developed a body of postmodernist urban theory.

Postmodernists value history and vernacular design. They like human-scale development. They are content with mixed-use projects and complexity. Efficiency and speed are not their dominant concerns. J.B. Jackson (p. 202) and Jane Jacobs (p. 105) reject modernist values. The urban design theory developed by Kevin Lynch (p. 499), William Whyte (p. 510), Allan Jacobs and Donald Appleyard (p. 518), and Jan Gehl (p. 530) reflects postmodernist values.

Taylor describes a postmodernist approach to planning developed by University of British Columbia planning professor Leonie Sandercock in some detail. Sandercock argues that planners should draw on experiential, grounded, contextual, intuitive knowledges manifested in speech, songs, stories, and various visual forms more than on rational scientific analysis. Normatively, Sandercock argues that today's diverse multicultural cities require postmodern community-based planning sensitive to cultural differences. Do the postmodern values and processes Sandercock and other postmodernist planners espouse amount to a paradigm shift? Again, Taylor says no. He argues that while multicultural, community-based planning may be desirable and "different ways of knowing" can contribute to understanding cities and making good plans, there is still a need for overarching normative values and governmental action to implement plans.

In summary, Taylor argues that while there have been significant shifts in planning theory since 1945, there have also been significant continuities. Overall, he concludes that changes in planning theory since 1945 have been developmental – filling out and enriching the rather primitive conception of planning that prevailed immediately after World War II.

Do you agree with Taylor that the change from the conception of planning as physical design to the systems and rational process views, while significant, did not amount to a paradigm shift? How about the shift from the planner as technical expert to the planner as a manager and communicator? The shift from modernist to postmodernist theories of urban planning?

Nigel Tayor is a principal lecturer in planning and architecture at the University of the West of England in Bristol and a visiting professor in the Department of Architecture and Planning at the University of Bologna, Italy. His teaching and research interests include planning theory, the history of town planning, aesthetics and urban design, and clear thinking and reasoning in policy-making and negotiation. He is on the editorial board and serves as book review editor of the journal *Planning Practice and Research*.

This selection – "Anglo-American Town Planning Theory Since 1945: Three Significant Developments But No Paradigm Shifts" – was published in *Planning Perspectives* 14(4) (1999). Taylor's planning theory book *Urban Planning Theory Since 1945* (Thousand Oaks, CA: Sage, 1998) contains a more extended discussion of the material summarized in this article. Other planning theory books are described in the introduction to Peter Hall's selection above (p. 373).

In recent times it has become fashionable to describe major changes in the history of ideas as 'paradigm' shifts, and some have described changes in town planning thought since the end of the Second World War in these terms. In this article I offer an overview of the history of town planning thought since 1945, and suggest that there have been three outstanding changes in planning thought over this period. These are, first, the shift in the 1960s from the view of town planning as an exercise in physical planning and urban design to the systems and rational process views of planning; second, the shift from the view of town planning as an activity requiring some technical expertise to the view of planning as a political process of making value-judgments about environmental change in which the planner acts as a manager and facilitator of that process; and third, the shift from 'modernist' to 'postmodernist' planning theory. I argue that none of these changes represents a paradigm change in anything like the strong sense of that term. Rather, they are better viewed as significant developments, which have 'filled out' and enriched the rather primitive town planning theory, which existed half a century ago.

INTRODUCTION

Over the fifty-year period since the end of the Second World War there have been a number of important shifts in town planning theory. But what have been the most significant changes, and how significant have these changes been? In this paper I offer a retrospective overview of the evolution of town planning thought since 1945, and an interpretation of the most significant shifts in planning thought over this period. My geographical focus will be on planning theory as it has developed in Britain and North America, though the developments I describe here have been influential elsewhere. My conceptual focus will be on those ideas or theories that have been concerned with clarifying what kind of an activity town planning is (and hence what skills are appropriate to its practice). In other words, I shall concentrate on changing conceptions of town planning itself over the last fifty years. But I shall also examine the modernist-postmodernist debate and its bearing on changing views about the purposes (and hence normative theory) of town planning.

In studies of the history of ideas, it has become fashionable to describe significant shifts in thought as 'paradigm' shifts, and some planning theorists have applied this concept to changes in town planning thought since 1945. Therefore, in addition to offering an account of the main shifts in (Anglo-American) town planning theory since 1945, I shall also assess whether it is appropriate to describe these changes as paradigm shifts. I begin, in the next section, by describing the concept of 'paradigms' as employed by the American historian of science, Thomas Kuhn. Following that, I offer, from a British perspective, an account of the three most significant shifts in the way the activity of town planning has been conceived since 1945. Though significant, I argue that, in anything like a strong sense of the term as it is used by Kuhn, it is inappropriate to describe these changes in planning thought as 'paradigm' shifts.

THE IDEA OF PARADIGMS AND PARADIGM SHIFTS

The use of the term 'paradigm' to describe major shifts in thought was first coined by Thomas Kuhn in his account of the history of scientific thought. Before Kuhn, it was widely assumed that scientific knowledge had grown steadily through history as more and more empirical evidence of phenomena had been accumulated. Kuhn's examination of the history of science led him to conclude that this gradualist, evolutionary view of the advance of scientific knowledge was misleading. For, according to Kuhn, if we examine any branch of science, we find that there are certain fundamental theories, conceptions or presuppositions which hold steady for very long periods – often for hundreds of years. These settled views of the world become so fundamental to people's whole conceptual scheme of reality that it is extremely difficult (and in some cases impossible) for most people to think of reality as being different; that, indeed, is why such views are fundamental. Because these fundamental theories constitute people's view of the world (or a significant part of it), they are, literally, 'world views', and it is these enduring world views which Kuhn describes as 'paradigms'. Examples of paradigms in the history of science would be the pre-Copernican view that the Earth was flat and at the centre of the Universe, the pre-Darwinian view that human beings had somehow been created on this planet separate from other species, and the Newtonian model of a mechanical Universe.

Kuhn's account of the history of science allows that, through the period in which any given paradigm prevails, advances in scientific theory still occur as a result of empirical research which uncovers fresh evidence about phenomena. Nevertheless, according to Kuhn, at any time most scientific research and theoretical development operates within the presuppositions of a prevailing, and more fundamental, world view or paradigm, and for the most part this latter goes unquestioned. In this respect, most scientific research amounts to filling in some of the details of, and so further refining, a given paradigm. Because most scientific research conforms to the norms of an established paradigm in this way, Kuhn also describes it as 'normal' science.

During the long periods of history in which a given paradigm prevails, scientists are often aware of empirical evidence which does not 'fit' the prevailing paradigm, which the prevailing paradigm seems unable satisfactorily to explain. However, according to Kuhn, most scientists do not allow this 'contrary' evidence to unseat their adherence to the paradigm on which they rely to explain phenomena in the world. Rather, they tend to 'turn a blind eye' to these puzzling phenomena, often in the belief that one day someone will succeed in explaining the seemingly 'anomalous' evidence within the framework of the given paradigm. The great scientists in history, however, have typically been curious about anomalous evidence which a prevailing paradigm is unable to explain and, as a result, have created a radically new account of the world which succeeds in explaining the hitherto puzzling evidence as well as the evidence previously explained by the 'old' paradigm. This new fundamental theory amounts to a whole new conceptual scheme, world view, or paradigm. In Kuhn's terms, then, a paradigm shift is a revolutionary shift in thought, because a whole way of perceiving some aspect of the world is overturned and replaced by a new theoretical perspective. Examples of such paradigm shifts noted by Kuhn were the shift from viewing the Earth as flat and at the centre of the Universe to seeing it as round and orbiting the Sun, the shift from a 'creationist' to a Darwinian evolutionary model of human origins and development, and the shift from a Newtonian to an Einsteinian view of space and time.

It should be clear from this account that, for Kuhn, paradigm shifts are fundamental theoretical changes. It is this, which explains why paradigm shifts typically occur infrequently in the history of science. Any given paradigm, once established, shapes the whole way a scientific community (and beyond that, the general public) views some aspect of the world, and tends to endure for centuries, not just decades. If, then, we adopt this 'strong' Kuhnian conception of paradigms, it would seem initially unlikely that there would be several paradigm shifts in the field of town planning theory over the short span of the last fifty years. Furthermore, Kuhn was describing changes in scientific thought – that is, major changes in the way people have described and explained some aspect of reality as a matter of fact – and town planning is not a science. Rather, it is a form of social action directed at shaping the physical environment to accord with certain valued ideals. In other words, town planning is a normative practice (although in seeking to realize certain valued ends town planning, like any normative practice, draws on relevant scientific understanding).

Of course, we are not compelled to adopt the strong, fundamentalist conception of paradigms described above. It is possible to use the concept in a weaker, more generous sense to describe shifts of thought which are significant, but not necessarily fundamental to people's world view or conceptual scheme, Moreover, although town planning is not strictly a science, the Kuhnian notion of paradigm shifts can also be extended to describe fundamental, or significant, shifts in values and ethical thinking. However, although there is nothing to stop us using the concept of paradigm shifts in these more liberal ways, we need to be alert to the dangers of over-using the concept. If every twist and turn in planning thought over the past fifty years is described as another paradigm shift, the very notion of a 'paradigm shift' becomes superfluous. I therefore favour the use of the term in its 'purer', more strict (and restricted) sense. Accordingly, it is in terms of this stronger conception of the term that I shall argue that, whilst there have been some significant changes in the way the activity of town planning has been conceived over the last half-century, none of these amounts to a paradigm shift.

My account of town planning theory since 1945 is organized as follows. In the next section I summarize what seems to me to be two significant innovations in the way the activity or 'discipline' of town planning has been conceived and defined, both of which emerged in Britain and the USA in the 1960s. The first was the shift from the urban design tradition of planning to the systems and rational process views of planning. The second was a shift from a substantive to a procedural conception of planning. This latter evolved further through the 1970s and 1980s, and eventually crystallized around the idea of the planner as a manager of the process of arriving at planning judgments, rather than someone who possesses a specialist expertise to make these judgments him- or herself. What is now termed 'communicative planning theory' is the latest version of this view of planning, under which the planner is seen as a kind of 'facilitator', drawing in other people's views and skills to the business of making planning judgments.

In the section that follows I examine a third significant change in post-war planning thought which some writers have identified – the alleged shift from 'modernist' to 'postmodernist' ways of thinking. For Sandercock, this shift from modern to postmodern planning theory is so fundamental as to lead her to claim that 'we are . . . living through a period of what Thomas Kuhn has called "paradigm shift" '.

The concluding section points out some of the continuities which run through the changes in town planning theory since 1945, and reiterates the thesis that none of the changes in planning thought over the past fifty years represents a paradigm shift in anything like the pure, or fundamental, Kuhnian sense.

TWO SIGNIFICANT SHIFTS IN THE WAY TOWN PLANNING HAS BEEN CONCEIVED

From the planner as a creative designer to the planner as a scientific analyst and rational decision-maker

For almost 20 years following the Second World War, town planning theory and practice was dominated by a conception which saw town planning essentially as an exercise in physical design. In fact, this view of town planning stretched back into history, arguably as far as the European Renaissance, arguably even further back than that. Its long historical lineage is shown by the fact that, for as far back as we can see, what came to be seen and described as town planning was assumed to be most appropriately carried out by architects. Indeed, such was the intimate connection between architecture and town planning that the two were not distinguished throughout most of human history. Thus what we call town planning was seen as architecture, its only distinctiveness being that it was architecture on the larger scale of a whole town, or at least part of a town, as distinct from individual buildings.

This conception of town planning as 'architecture writ large' persisted down to the 1960s, as was shown by the fact that most planners in the post-war years were architects by training, or 'architect-planners'. Indeed, because of this, the established professional body for architects in Britain, the Royal Institute of British Architects (RIBA), resisted the establishment of a separate professional body for town planning, arguing that town planning as a practice was already 'covered' by themselves. The close link in the post-war years between design and town planning, and hence between architecture and town planning, also explains why at this time aesthetic considerations were regarded as central to town planning. Like architecture, town planning was viewed as an 'art', albeit (again like architecture) an 'applied' or 'practical' art in which utilitarian or 'functional' requirements had to be accommodated . . .

Against this background, the bursting onto the scene in the 1960s of the systems and rational process views of planning represented a rupture with a centuries-old tradition, and so might well be viewed as a Kuhnian paradigm shift . . .

It is worth noting, in passing, that the systems and rational process views of planning are conceptually distinct, and so really two theories of planning, not one. Thus the systems view was premised on a view about the object that town planning deals with – towns, or regions, or the environment in general, were viewed as 'systems'. By contrast, the rational process view of planning was concerned with the method or process of planning itself, and in particular it advanced an 'ideal-type' conception of planning as a procedure for making instrumentally rational decisions. But setting aside this important distinction, both the systems and rational process views of planning, taken together, represented a radical departure from the then prevailing design-based view of town planning. This shift in planning thought can be summarized under four points.

- First, an essentially physical or morphological view of towns was to be replaced with a view of towns as systems of interrelated activities in a constant state of flux.
- Second, whereas town planners had tended to view and judge towns predominantly in physical and aesthetic terms, they were now to examine the town in terms of its social life and economic activities; in Harvey's terms, a sociological conception of space was to replace a geographical or morphological conception of space,
- Third, because the town was now seen as a 'live' functioning thing, this implied a 'process', rather than an 'end-state' or 'blueprint' approach to town planning and plan-making.
- Fourth, all these conceptual changes implied, in turn, a change in the kinds of skills, or techniques, which were appropriate to town planning. For if town planners were trying to control and plan complex, dynamic systems, then what seemed to be required were rigorously analytical, 'scientific' methods of analysis.

Overall, the shift in planning thought brought about by the systems and rational process views of planning can be summed up (albeit rather crudely) by saying that, whereas the design-based tradition saw town planning primarily as an art, the systems and rational process theorists suggested that town planning was a science.

For, on the one hand, the analysis of environmental systems (regions, cities, etc.) involved systematic empirical – and hence 'scientific' – investigation and analysis of interrelationships between activities at different locations. And, on the other hand, the conception of planning as a process of rational decision-making was also commonly equated with being 'scientific'. An indication of how significant this shift was from 'town planning as an art' to 'town planning as science' was that it was experienced as profoundly unsettling by many planners and planning students reared in the design tradition of town planning – understandably. Suddenly, within the space of a few years, town planners who had approached their task on the basis of an aesthetic appreciation of urban environments, and who saw themselves as creative, 'artistic' urban designers, were being told by a new generation of planning theorists that this conception of town planning was inappropriate, and that instead they should see themselves as – and so become – 'scientific' systems analysts.

However, significant though this shift in town planning thought undoubtedly was, it remains open to question whether it should be likened to a Kuhnian paradigm shift. Although the shift in planning thought described above did have the effect of marginalizing considerations of design and aesthetics in planning theory for about 20 years, in British planning practice (especially in the development control sections of planning authorities) planners still continued to evaluate development proposals partly in terms of their design quality and aesthetic impact . . . In planning practice, therefore, the design-based conception of town planning was not completely superseded by the change in theoretical perspective described above.

Admittedly, these observations are about planning practice, not planning theory. But the continuation of the physical design view of planning in planning practice was also theoretically significant. For it drew attention to the fact that, at the level of 'local' planning at least, many planners continued to believe that the physical form and aesthetic appearance of new development were important concerns of town planning. And although there were lessons for small area 'local' planning in systems and rational process thinking (e.g. in giving greater consideration to the social and economic effects of development proposals; in approaching local planning as a rational process; etc.), these lessons could be accommodated within an essentially traditional design-based conception of planning. At the level of local planning, therefore, the shift in planning thought described above did not replace the physical design view of planning; it was not a paradigmatic shift in that sense. Furthermore, the continuing relevance of the physical design-based conception of town planning to town planning theory has been shown by the revival of theoretical interest in questions of urban design in the 1980s and 1990s.

It was therefore primarily at the broader, more strategic level of planning that the design-based view of planning was supplanted by the changing conception of planning ushered in by the systems and rational process views of planning. In fact, the main shift in planning thought brought about by the systems and rational process views of planning was in clarifying a distinction between strategic and longer-term planning on the one hand, and 'local' and more short-term planning on the other. And it was at the former, strategic level of planning that the altered conception of town planning brought about by systems and rational process thinking was most relevant. In retrospect, then, the shift in town planning thought in the 1960s described above was not a wholescale revolution which completely ousted the incumbent design-based view of town planning; rather, the systems and rational process views of planning 'added' to the design-based view. The shift in planning thought described above was not therefore a revolution in thought comparable to the paradigm shifts in scientific theory described by Thomas Kuhn.

From the planner as technical expert to the planner as a manager and 'communicator'

Although there were marked differences between the traditional conception of town planning as an exercise in physical planning and design, and the conception of town planning as a rational process of decision-making directed at the analysis and control of urban systems, there was one thing that both these views had in common. Both presumed that the town planner was someone who possessed, or should possess, some specialist knowledge and skill – some substantive expertise – which the layperson did not possess. It was this, which qualified the planner to plan. And since a central condition of professionalism is the possession of some specialist knowledge or skill, it was this, too, which justified any claim town planners might make to constitute a distinct 'profession'.

Clearly, views about the content of the specialist skill appropriate to town planning varied according to which of the foregoing conceptions of planning were adopted. Under the traditional design-based view of town planning, the relevant skills were seen to be primarily those of aesthetic appreciation and urban design. Under the systems and rational process views, the skills were those of scientific analysis and rational decision-making. But still under both conceptions, the town planner was conceived as someone with a specialist knowledge, understanding, and/or skill. However, this whole idea of the planner as someone with some substantive expertise came in its turn to be challenged by an alternative view of town planning.

This challenge emerged again in the 1960s, when it came to be openly acknowledged that town-planning judgments were at root judgments of value (as distinct from purely scientific judgments) about the kinds of environments it is desirable to create or conserve. Once this view was taken, it naturally raised the question of whether town planners had any greater 'specialist' ability to make these judgments than the ordinary person in the street. Indeed, people's experience of much of the planning of the 1960s – such as comprehensive housing redevelopment or urban road planning – seemed to indicate not. The emergence of the view that town planning was a value-laden, political process therefore raised not so much the question of what the town planner's area of specialist expertise should be, but, more fundamentally, the question of whether town planning involved any such expertise at all.

From this radical questioning of the town planner's role, there developed a curious bifurcation in planning theory, which has persisted to this day. On the one hand, some planning theorists have continued to believe that the practice of town planning requires some specialist substantive knowledge or skills – be it about urban design, systems analysis, urban regeneration, sustainable development, or whatever. On the other hand, there has developed a tradition of planning thought, which openly acknowledges that town-planning judgments are value-laden and political. As noted above, one conclusion, which might have been drawn from this, would be to reject entirely the idea that town planning involves, or requires, some specialist expertise, and indeed, some 'radical' planning theorists have flirted with this view. However, most planning theorists who have openly acknowledged the value-laden and political nature of planning have developed an alternative line of thought. This rejects the idea that the town planner is someone who is specially qualified to make better planning decisions or recommendations – because what is 'better' is a matter of value, and (so the argument goes) planners have no superior expertise in making value-judgments over environmental options. However, the view is still taken that the town planner possesses (or should possess) some specialist skill, namely, a skill in managing the process of arriving at planning decisions and facilitating action to realize publicly agreed goals.

Through the 1970s and 1980s, therefore, a tradition of planning theory emerged which viewed the town planner's role (and hence his or her 'professional' expertise) as one of identifying and mediating between different interest groups involved in, or affected by, land development. In this way, the town planner was seen as someone who acts as a kind of cypher for other people's assessments of planning issues, rather than someone who is specially qualified to assess these issues him- or herself. The town planner was viewed as not so much a technical expert (i.e. as someone who possesses some superior skill to plan towns), but more as a 'facilitator' of other people's views about how a town, or part of a town, should be planned. To conceive of the town planner like this as a kind of manager of the process of making planning decisions could easily conjure up an image of the planner as a grey-suited chairperson of meetings. But tacked onto this view went a more particular ideological commitment which made it more appealing and inspiring to idealists in the profession, namely, a commitment to ensure that the process of planning was open and democratic, especially to disadvantaged or marginalized groups who tended to be ignored or overridden in decisions about land development.

An early version of this theory of town planning, and of the planner's role, was Paul Davidoff's 'advocacy' view of planning in the 1960s. The most recent version is the communicative planning theory inspired, particularly, by Habermas's theory of communicative action. In this, the skills of inter-personal communication and negotiation are seen as central to a non-coercive, 'facilitator' model of town planning. Indeed, in relation to involving the public in planning, it has even been suggested that the kinds of inter-personal skills needed by the communicative town planner are those of the listener and the counsellor:

> Meaningful dialogue – learning the language of the client – is at the heart of effective counselling. To counsel is not to give advice or push the client down a particular path, but to let the client see himself or herself fully and through this discovery achieve personal growth. As local government offices look for ways of including citizens in decision-making, they must adopt many counselling skills – active listening, non-judgmental acceptance, and the ability to empathize. How can people play a part in the decision-making process unless we 'enable' them to do so.

This view of the knowledge and skills relevant to town planning is a far cry from the view that the specialist skill of the town planner resides in being either an urban designer or a systems analyst.

In relation to this view of the planner as a manager, communicator, and 'facilitator' of planning decisions, it is also relevant to note here the emergence of a concern with implementation during the 1970s and 1980s. Perhaps the first planning theorist to articulate this concern was John Friedmann at the end of the 1960s, when he presented a critique of the rational process model of planning because it tended to emphasize the task of making decisions over that of taking action. This, together with the seminal work of Pressman and Wildavsky, drew attention to the much overlooked fact that, frequently, the most carefully thought through public decisions and policies did not actually result in the necessary action to realize their intentions. This was because, in attending primarily to the business of making decisions about appropriate policies and plans, insufficient attention was given to the problems of how these policies and plans might get implemented. A concern with implementation – with what John Friedmann termed 'action planning' – thus became a central preoccupation of some planning theorists in the 1970s and 1980s.

Although this concern with implementation had rather different roots from the view of planning as a political activity described above, it issued in much the same conception of planning and the skills appropriate for it. For the general conclusion of the implementation theory of the 1970s and 1980s was that, to be an effective implementer of public policies and plans, planners needed to become effective at networking, communicating and negotiating with other agents involved in the development process. In short, theoretical reflection on the problem of implementation led also to a view of the planner as one who should be a capable manager, 'networker' and communicator.

Nowadays, many planners do describe themselves as managers and facilitators of the process of planning rather than as people with a special expertise in planning towns. So the shift in town planning thought described here has undoubtedly been significant. Hence Judith Innes's description of 'communicative action and interactive practice' as the new 'paradigm' of planning theory. However, once again a word of caution is in order before it is too readily assumed that this change in planning thought represents a Kuhnian paradigm shift. For it is possible to imagine some kind of rapprochement between the two views described here. Thus one could adopt a view of the town planner as one whose role is primarily that of a communicator and negotiator, but where, in communicating and negotiating with others, the planner also brings to bear some specialist knowledge which, for example, would enable him or her to point out the likely consequences of development proposals on the form and functioning of a town. Such a model of the town planner would be akin to that of, say, civil servants who are experts in economic matters, and who impart their specialist economic understanding to those they advise who make decisions about economic policy. To be effective as an adviser, such a town planner would have to be skilled in communicating and negotiating with others, but he or she would also have to possess some specialist knowledge to bring to the communicating table to assist others in arriving at planning decisions. The alternative 'substantive' and 'procedural' conceptions of town planning described in this section are therefore not as fundamentally at odds, or 'incommensurable', as the paradigm shifts in the history of science described by Thomas Kuhn.

MODERNIST AND POSTMODERNIST PLANNING THEORY: A SHIFT IN NORMATIVE PLANNING THOUGHT

According to some commentators, since about the late 1960s there has been a significant shift in Western thought and culture from 'modernism' to 'postmodernism', and some view this as so fundamental as to constitute a shift in world view, or a paradigm shift. This alleged paradigm shift has a special bearing on town planning, because modern architecture and town planning have jointly provided one of the main 'sites' where the shift from modernism to postmodernism is supposed to have taken place. Indeed, according to Charles Jencks, the death-knell of modernism was sounded in July 1972, when the vandalized Pruitt-Igoe high-rise housing estate in St Louis (USA) – which had earlier won an award as an exemplar of modern architecture and town planning – was deemed uninhabitable and dynamited by the local city authority.

At one level, postmodernism can be viewed as a movement opposed to the styles of art and design associated with the modern movement. Thus in architecture, postmodernists rebelled against the aesthetic minimalism and anonymity of the plain geometrical buildings (and comprehensive planning schemes) of the modern movement, and against the modernist dogma of functionalism, which had legitimized this stripped-down architecture. Postmodern architects therefore sought to 'bring back style' to contemporary buildings to enrich their aesthetic content and give them 'meaning'. Thus in what is arguably the first text of postmodern architecture, Robert Venturi famously counterposed his preference for a stylistically more complex architecture over plain 'functional' modernism:

> I like complexity and contradiction in architecture – Architects can no longer afford to be intimidated by the puritanically moral language of orthodox Modern architecture. I like elements which are hybrid rather than 'pure', compromising rather than

> 'clean', distorted rather than 'straightforward', ambiguous rather than 'articulated', . . . inconsistent and equivocal rather than direct and clear. I am for messy vitality over obvious unity. . . . I am for richness of meaning rather than clarity of meaning . . .

In relation to town planning, Jane Jacobs [p. 105] and Christopher Alexander expressed a similar preference for complexity in the city, in opposition to the simplified order which modern town planning theorists like Ebenezer Howard [p. 328] and Le Corbusier [p. 336] had advocated. Jacobs berated the simple-mindedness of single use zoning and 'comprehensive' redevelopment in urban areas, both of which showed little understanding of, or regard for, the delicate social and economic fabric and vitality of existing city areas. Instead, she advocated mixing land uses, and leaving many so-called slum areas alone to 'unslum' themselves. Alexander similarly criticized modern town planning for seeking to impose a simplified 'tree-like' structure on urban areas (e.g. by planning for 'self-contained' neighborhoods), suggesting that successful cities contained within them complex and subtle 'semi-lattice' patterns of interrelationships.

These architectural and planning ideas certainly represented a departure from the prevailing 'modernist' orthodoxy. But, according to some accounts, the shift from modernism to postmodernism goes deeper than just a preference for greater 'complexity' in architecture and town planning. For, it is said, underpinning and driving the modern movement in architecture and town planning (and other developments in modern western culture) has been a more fundamental intellectual orientation, which has its roots in the European Enlightenment of the eighteenth century. This Enlightenment 'world view' has been characterized as an optimistic belief in human progress based on analytical reason and scientific understanding. Quite apart from its 'machine aesthetic', modern town planning thought and practice has been seen as expressing this more general Enlightenment world view or paradigm, and correspondingly, postmodern planning theory is seen by some as representing a break with this intellectual tradition.

Such a view is advanced, for example, by Leonie Sandercock. She emphasizes two important contrasts, in particular, between the modernist paradigm of town planning which she rejects and a postmodern paradigm which she endorses: one concerning the epistemological basis of planning, and the other concerning the values or normative theory of planning. On the first, Sandercock suggests that modernist planning has been 'concerned with making public/political decisions more rational' . . . As an adjunct to this, modernist planning has relied largely on 'a mastery of theory and methods in the social sciences', so that planning knowledge and expertise is 'grounded in positive science, with its propensity for quantitative modeling and analysis'. Second . . . 'the state is seen as possessing progressive, reformist tendencies'. Accordingly, the state is vested with the authority to undertake town planning . . . to realize what is in the overall public interest.

Against each of these (epistemological and normative) features of modernist planning, Sandercock counterposes postmodern alternatives. First, in contrast to the rationalist model of planning theory Sandercock urges that whilst 'Means–ends rationality may still be a useful concept . . . we also need greater and more explicit reliance on practical wisdom' . . . [S]uch practical wisdom derives from more varied kinds of knowledge and experience than just scientific knowledge; it can also include 'experiential, grounded, contextual, intuitive knowledges, which are manifested through speech, songs, stories, and various visual forms'. Planners should therefore 'learn to access these other ways of knowing'. Second, against the 'top–down' state-directed model of planning, Sandercock urges a 'move to community-based planning, from the ground up, geared to community empowerment'. Part of this involves a recognition that planning 'is no longer exclusively concerned with comprehensive, integrated, and coordinated action, . . . but more with negotiated, political and focused planning. This in turn makes it less document oriented and more people-centred'. Part of it also involves an acknowledgement that modern cities are increasingly inhabited by different ethnic and other social groups, with a diversity of cultures and interests. A postmodern community-based planning, which is sensitive to cultural differences, would therefore dispense with the idea of an overarching public interest because this 'tends to exclude difference'. Instead, 'planning in this multicultural area requires a new kind of multicultural literacy'.

The epistemological critique of modernism for its reliance on rationality and science has been central to some versions of postmodernism. However, many theorists have questioned whether the postmodern critique is, or can be made, coherent in its own terms. Consider, for example, the following statement of postmodern epistemology by Michael Dear:

> Postmodernism's principal target has been the rationality of the modern movement, especially its foundational character, its search for universal truth – The postmodern position is that all meta-narratives are suspect; that the authority claimed by any single explanation is ill-founded, and hence should be resisted. In essence, postmodernists assert that the relative merit of one meta-narrative over another is ultimately undecideable; and by extension, that any such attempts to forge intellectual consensus should be resisted.

Taken at face value (i.e. in terms of what it actually says), this statement implies a rejection of rational discourse altogether. For if postmodernists believe, as Dear here suggests, that there are literally no criteria against which we can judge the relative merits of different theoretical positions, then it would follow that there can be no reasoned debate about different theories at all. However, apart from the fact that such a position is intellectually hopeless (in the literal sense that there would be no point in hoping for greater enlightenment through rational discourse with others), it also seems to be self-defeating . . .

If, as these arguments suggest, the postmodern epistemological critique of 'modernist' rationality and science is itself incoherent, then the idea that postmodernism represents a paradigmatic break with Enlightenment reason turns out to be empty. This is not to deny Sandercock's suggestion that there may be 'different ways of knowing' which can supplement, 'fill out', or otherwise complement the understandings gained by reason and science. Certainly in relation to making practical planning judgments, the experience and knowledge of local communities, even if not strictly 'scientific', must be relevant to those judgments. However, if this is what the postmodern epistemological position comes down to, then for all its fruitfulness it does not represent an alternative to modernist 'Enlightenment' epistemology, but rather a supplement to it.

What, then, of the alternative values – or normative theory – which postmodernists advance in opposition to the values of modernism? Do these represent a paradigm shift in normative thinking?

Broadly, postmodernists argue that the world and our experience of it is far more complex and subtle than has typically been realized in the modern age. In relation to cities and the environment, postmodernists claim that people's experience of places, and from this the qualities of places, are much more diverse and 'open' than was implicit in many modernist schemes for improving the city – and especially in the bombastic simplicities of modernist architectural schemes for the ideal future city, such as Le Corbusier's radiant city. In place of the modernist architect's and planner's emphasis on simplicity, order, uniformity and tidiness, postmodernists typically celebrate complexity, diversity, difference, and pluralism. This echoes Jane Jacob and Christopher Alexander's celebration of urban complexity and diversity more than thirty years ago. Postmodernists therefore argue that there can be no one type of environment, which is ideal for everyone, no singular conception of environmental quality. Thus whilst some may continue to hold as a planning ideal Howard's vision of garden cities, others will prefer the buzz and excitement of what Elizabeth Wilson calls the 'teeming metropolis'. Hence, too, Sandercock's view that the ideal postmodern city is a multicultural city.

This postmodern emphasis on diversity, difference, pluralism and multiculturalism chimes with the political ideals of liberalism. For liberals also celebrate the plural society in which individuals have the opportunity to determine and 'realize' themselves in different ways through the exercise of free choice. From this point of view, the normative position of postmodernism might seem to accord with a liberal, market-sensitive system of planning in contrast to the 'statism' of socialist and social democratic forms of planning (which have been associated with modernism).

But do these postmodern values amount to a paradigmatic break with the values of so-called modernist planning? Two points can be made to dispute this. First, in relation to the general value stance of diversity and pluralism espoused by postmodernists, this can be taken to an extreme where any 'difference' is accepted or permitted; in other words, a position of complete moral and political relativism. And such an extreme ethical relativism is open to criticisms similar to those raised earlier against postmodern epistemological relativism. To be sure, we may endorse a more complex, variegated and 'multicultural' conception of environmental quality, but this need not exclude a commitment to some overarching – even universal – normative principles of planning. For example, shouldn't town planning, wherever it is practiced, do what it can to help bring about economically and environmentally sustainable development, development which is not socially divisive, and development which is experienced as an aesthetic delight? In fact, for all her celebration of difference and multiculturalism, even Sandercock acknowledges that her ideal 'cosmopolis' would require some overarching normative principles, such as some overall principles of social justice, and a democratic polity which fosters 'dialogue and negotiation as the habit of political participation'.

Second, with respect to the liberal inclinations of postmodernism, whilst the kind of centralized 'statist' planning of Soviet-style socialism and even post-war social democracy in the West may have been discredited, the idea that a liberal, market-supportive style of planning produces better environmental outcomes has by no means been universally accepted. Sandercock herself argues for a 'bottom–up' community-based style of planning as being the required antidote to 'top–down' statism. Yet even she insists that this

> is not to argue for the rejection of state-directed planning. There are transformative and oppressive possibilities in state planning, just as there are in community-based planning. Victories at the community level almost always need to be consolidated in some way through the state, through legislation and/or through the allocation of resources.

The idea, then, that either a kind of postmodern liberalism or some version of community planning constitute a paradigmatic break with social democratic politics is, to say the least, premature (consider, in this regard, Giddens's recent argument for a 'renewal' of social democracy).

CONCLUSION: CHANGE AND CONTINUITY IN TOWN PLANNING THOUGHT

Unquestionably, there have been significant changes in town planning thought since the end of the Second World War, and (for all the talk of the gulf between planning theory and practice), these changes have affected planning practice. Thus town planners now operate with a very different conception of town planning, and bring to it quite different skills, from the architect-planners of fifty years ago. The idea, for example, of town planning as an exercise in 'managing the process of arriving at planning decisions' and 'negotiating agreements for action', and hence of the town planner as someone who is primarily a communicator and a 'facilitator', would not really have occurred to early British post-war planners like Thomas Sharp, Frederick Gibberd or Lewis Keeble. Contrariwise, the idea that the prime task of town planners is to undertake surveys of the physical and aesthetic characteristics of towns, and then sit in front of drawing boards drawing up master plans, would strike most present-day planners as hopelessly limited, naive, and outmoded. Similarly, the 'modernist' ideas that better cities can be created by drastic surgery and 'comprehensive' development has long been discredited.

These are significant changes in people's conception of the activity of town and country planning, and the values, which should inform it. And there is no harm in describing these changes as 'paradigm shifts', so long as we appreciate that we are employing this term in a fairly loose or weak sense. However, in anything like the strong sense of the term as used by Thomas Kuhn to describe revolutionary changes in thought in the history of science, it is doubtful whether any of the changes in town planning thought since 1945 are appropriately described as paradigm shifts. For in the strong sense of the term, a paradigm shift is marked by a fundamental change in world view – a kind of Gestalt switch in people's whole conceptual scheme. And whilst there have been significant discontinuities in planning thought since 1945, there have also been continuities across these changes.

For example, whilst, on the face of it, the shift from the physical design view of town planning to the systems and rational process views of planning might look like a change of world view (and hence a Kuhnian paradigm shift), the systems and rational process views of planning did not completely supersede the urban design conception of town planning. Thus, at least at the 'local' level of small area or 'district' planning and the control of development, good urban design (and design control) is still regarded as central to good town planning; indeed, a concern with urban design within town planning has experienced something of a revival since the mid-1980s. Arguably, a more likely candidate for a Kuhnian paradigm shift has been the shift from a view of the town planner as an expert to the planner as a manager and facilitator – the shift, in other words, away from a view of the planner as the supplier of answers (in the form of 'master' plans) to that of the planner as someone skilled at eliciting other people's answers to urban problems and somehow 'mediating' between these. But even this view of town planning as a species of 'communicative action' is compatible with the view that town planners should possess at least some area of expertise, for example about the likely effects of proposed changes on urban environments. Similarly, the postmodern emphasis on complexity, difference and relative values is not completely incompatible with a commitment to some overarching, universal principles of environmental quality, still less with a reliance on (sophisticated) reason and scientific understanding.

Looking back over the changes in town planning theory since the end of the Second World War, some planning theorists have suggested that planning theory has fragmented into a plurality of diverse and even incompatible theoretical positions or 'paradigms'. However, this article has expressed scepticism with this idea. Although there have been significant shifts in planning thought since 1945, there have also been significant continuities. Indeed, the shifts in town planning thought over this period can be regarded as developmental rather than as ruptures between incompatible paradigms of planning. In other words, the changes to town planning theory described here can equally well be viewed as 'filling out', and thereby enriching, the rather primitive conception of planning which prevailed in the immediate post-war years. On this account, the story of town planning theory since 1945 has been one of developing sophistication as we have learnt more about the greater complexity of urban environments and the diverse values of different communities.

"Twentieth-Century Land Use Planning: A Stalwart Family Tree"

Journal of the American Planning Association (1995)

Edward J. Kaiser and David R. Godschalk

Editors' Introduction

Much of what city planning departments do is physical planning related to land use. Many urban planning programs offer graduate courses in "physical planning" or "land use planning," and physical planning is a common specialization within urban planning programs. Some geography departments also offer physical planning courses and specializations. The central focus of these courses is on how land – particularly urban land – should be used for housing, industry, retail, open space and other uses. Physical planning and land use planning are inextricably linked to transportation, capital improvements, and infrastructure.

In this selection, urban planning professors Edward J. Kaiser and David R. Godschalk describe how concepts of what land use plans should be like evolved during the twentieth century and what land use planning practice in the United States is like today. Their conclusions are based on their distinguished teaching careers at the University of North Carolina's Department of City and Regional Planning, a world center of excellence in land use planning. Kaiser and Godschalk are coauthors (with others) of different editions of the leading US land use planning text: *Urban Land Use Planning*.

Kaiser and Godschalk liken the development of the practice of land use planning to the growth of a tree. From disparate roots in planning theory and practice, the land use planning tree has grown from a slender set of ideas to a sturdy trunk based on the 1950s vision of a general plan for cities' long-term physical development. That vision has grown since the 1950s, with periodic branches, to a rich foliage of hybrid contemporary plans that typically blend aspects of design, policy, and management as well as prescribing where activities and land uses should be located.

Kaiser and Godschalk begin their account of twentieth-century land use planning by describing the elitist, architecturally based, and often unrealistic plans that Peter Hall (p. 373) and Nigel Taylor (p. 386) noted were the norm from the early twentieth century until after World War II. Many cities in late-nineteenth-century Germany did elaborate land use planning. New York, Chicago, other large US cities and many large cities in Europe also began to develop plans in the late nineteenth and early twentieth centuries. German cities pioneered zoning in the late nineteenth century and New York City adopted a comprehensive zoning ordinance in 1916.

Edward Bassett, the principal author of the New York city zoning law believed that cities should have what he called a "master plan" to guide long-term development. Bassett distinguished between master planning, which he thought should adopt a fifty-year time horizon and articulate a broad, non-binding vision, and zoning which he believe should a legally binding description of precisely what could be built where.

In the United States a 1926 supreme court decision upholding the constitutionality of zoning (Euclid v. Ambler Realty Company), a Model State Zoning Enabling Act (1926) and Standard City Planning Enabling Act (1928) recommended by the US Department of Commerce encouraged many states to authorize or require cities and counties to create planning commissions, adopt plans, and implement zoning. The term "master plan" was increasingly replaced by the politically less threatening term "general plan."

Kaiser and Godschalk trace the real roots of their tree to the middle of the twentieth century. By 1950 there was a consensus that city plans should be focused on long-term physical development – general plans based on Bassett's ideas. The early physically oriented general plans have become more sophisticated and have also evolved into contemporary hybrid urban land use plans today. Many still retain the solid trunk of physical planning practice, but modern plans incorporate elements of policy and management that make them far more realistic than plans developed during what Sir Peter Hall calls planning's comfortable, but ineffective golden age of general planning in the 1950s. Citizens are usually involved in formulating today's plans at the middle or high rungs of Sherry Arnstein's ladder of citizen participation (p. 238) with planners employing many of the consensus-building strategies that John Forester describes (p. 421). Modern general plans draw upon ideas proposed by University of California, Berkeley planning professor T.J. (Jack) Kent, who wrote a book titled *The Urban General Plan* in 1964.

In the United States, cities and counties are legal subdivisions of states and individual states may decide what planning to require or recommend. Sparsely populated states and states with a culture that is opposed to government interference with private property rights require little planning. Populous states, states with fragile ecosystems, and states whose political culture encourages more government regulation of private property require more planning. California, for example, requires every county and city to have a general plan, specifies mandatory elements the plans must contain, and requires local land use regulations to be consistent with the general plan.

In the 1950s, city general plans tended to be elitist, inspirational, long-range visions developed with little attention to implementation. In contrast, Kaiser and Godschalk argue that nowadays, urban land use plans have become frameworks for community consensus on future growth supported by fiscally grounded actions to manage change. Plans are becoming more sensitive to the green planning issues raised by Timothy Beatley (p. 446), sustainable urban development as proposed by the Brundtland Commission (p. 351), and strategies to reduce humans' carbon footprint and slow global warming discussed by Stephen Wheeler (p. 458).

Most modern city plans still contain maps showing projected long-range urban form for the city's land uses, transportation systems, community facilities, and other infrastructure. Present-day plans do not consist of architecturally complete renderings of an entire static new town or section of a city like many earlier plans. In addition to or in place of maps and drawings, the city plan's vision may be expressed in words and visual images. Maps in modern general plans are likely to be generated by computerized geographical information systems (GIS) software from census and other data rather than hand-colored approximations. Computerized statistical analysis of population, housing, and employment data ground the plans. Plan illustrations are often created using computer assisted design (CAD) software programs and illustration programs.

Urban planners also prepare regional plans. In the United States, Councils of Government (COGs) do some regional planning, but most are weak and their plans have limited impact. Metropolitan transportation planning agencies also do regionwide planning. The European Union requires member states to do spatial planning and in the United Kingdom and elsewhere this involves some level of regional planning. One metropolitan region in the United States (the Portland, Oregon, metropolitan area) has an elected regional government with substantial legal authority to do regional planning and an excellent track record. Portland State University urban historian Carl Abbott calls Portland the capital of good planning.

At the beginning of the twenty-first century it appears that the sturdy tree of urban land use planning will continue to grow. Kaiser and Godschalk anticipate that the next generation of development plans will mature and adapt without abandoning the urban physical plan heritage.

Edward J. Kaiser and David R. Godschalk are emeritus professors of city and regional planning at the University of North Carolina. They coauthored (with Stuart Chapin) the fourth edition of *Urban Land Use Planning* (Chicago, IL: University of Illinois Press, 1995), the leading American text on land use planning. David Godschalk is the coauthor with Philip Berke of the fifth edition of *Urban Land Use Planning* (2006).

Eugenie Birch (ed.), *The Urban and Regional Planning Reader* (London: Routledge, 2009) is an anthology of classic and contemporary writings on urban and regional planning. Other books on urban planning in the United States include the International City Management Association (ICMA), *Local Planning: Contemporary Principles and Practice* (Washington, DC: ICMA), John M. Levy, *Contemporary Urban Planning*, 8th edn (New York: Prentice-Hall, 2008), William Fulton and Paul Shigley, *Guide to California Planning* (Point Arena, CA: Solano Press, 2005), Eric Damian Kelley, *Community Planning: An Introduction to the Comprehensive Plan* (Washington, DC: Island Press, 2009), Alexander Garvin, *The American City: What Works, What Doesn't*, 2nd edn (New York: McGraw-Hill, 2002), and Jay Stein (ed.), *Classic Readings in Urban Planning* (New York: McGraw-Hill, 2004).

Leading books on European urban planning include J. Barry Cullingworth and Vincent Nadin, *Town and Country Planning in the UK*, 14th edn (London: Routledge, 2006), and Peter Hall and Mark Tewdyr-Jones, *Urban and Regional Planning*, 5th edn (London: Routledge, 2009). The latter has a good account of European spatial planning.

Sir Peter Hall, *Cities of Tomorrow*, 3rd edn (Oxford: Blackwell, 2002) is the definitive intellectual history of urban planning and design in the twentieth century that covers major trends in the United Kingdom, North America, and elsewhere. Mellior Scott, *American City Planning Since 1890* (Berkeley, CA: University of California Press, 1969) is a detailed history of the sixty years of American city planning from 1909 to 1969. Donald A. Krueckebert (ed.), *The American Planner: Biographies and Recollections* (New York: Methuen, 1983) provides additional insight on land use planning in the United States.

> How a city's land is used defines its character, its potential for development, the role it can play within a regional economy and how it impacts the natural environment.
>
> (Seattle Planning Commission, 1993)

During the twentieth century, community physical development plans have evolved from elite, City Beautiful designs to participatory, broad-based strategies for managing urban change. A review of land use planning's intellectual and practice history shows the continuous incorporation of new ideas and techniques. The traditional mapped land use design has been enriched with innovations from policy plans, land classification plans, and development management plans. Thanks to this flexible adaptation, local governments can use contemporary land use planning to build consensus and support decisions on controversial issues about space, development, and infrastructure. If this evolution persists, local plans should continue to be mainstays of community development policy into the twenty-first century.

Unlike the more rigid, rule-oriented modern architecture, contemporary local planning does not appear destined for deconstruction by a postmodern revolution. Though critics of comprehensive physical planning have regularly predicted its demise, the evidence demonstrates that spatial planning is alive and well in hundreds of United States communities. A 1994 tabulation found 2,742 local comprehensive plans prepared under state growth management regulations in twelve states. (See Table 1.) This figure of course significantly understates the overall nationwide total, which would include all those plans prepared in the other thirty-eight states and in the noncoastal areas of California and North Carolina. It is safe to assume that most, if not all, of these plans contain a mapped land use element. Not only do such plans help decision-makers to manage urban growth and change, they also provide a platform for the formation of community consensus about land use issues, now among the most controversial items on local government agendas.

This article looks back at the history of land use planning and forward to its future. It shows how planning ideas, growing from turn-of-the-century roots, culminated in a midcentury consensus on a general concept – the traditional land use design plan. That consensus was stretched as planning branched out to deal with public participation, environmental protection, growth management, fiscal responsibility, and effective implementation under turbulent conditions. To meet these new challenges, new types of plans arose: verbal policy plans, land classification plans, and growth management plans. These in turn became integrated into today's hybrid comprehensive plans, broadening and strengthening the traditional approach.

Future land use planning will continue to evolve in certain foreseeable directions, as well as in ways unforeseen. Among the foreseeable developments are even more active participation by interest groups, calling for planners' skills at consensus building and managing conflict; increased use of computers and electronic media, calling for planners' skills in information management and communication; and continuing

State	Number of comprehensive plans: Cities/towns	Counties	Regions	Total	Source
California (coastal)	97	7	0	104	Coastal Commission
Florida	377	49	0	426	Department of Community Affairs
Georgia	298	94	0	392	Department of Community Affairs
Maine	270	0	0	270	Dept. of Economic and Community Development
Maryland	1	1	0	2	Planning Office
New Jersey	567	0	0	567	Community Affairs Department
North Carolina (coastal)	70	20	0	90	Division of Coastal Management
Oregon	241	36	1	278	Department of Local Community Development
Rhode Island	39	0	1	40	Department of Planning and Development
Vermont	235	0	10	245	Department of Housing and Community Affairs
Virginia	211	94	0	305	Department of Housing and Community Affairs
Washington	23	9	9	23	Office of Growth Management
Total	**2429**	**301**	**12**	**2742**	

Table 1 Local comprehensive plans in growth-managing states and coastal areas as of 1994
Source: Compiled from telephone survey of state sources

concerns over issues of diversity, sustainability, and quality of life, calling for planners' ability to analyze and seek creative solutions to complex and interdependent problems.

THE LAND USE PLANNING FAMILY TREE

We liken the evolution of the physical development plan to a family tree. The early genealogy is represented as the roots of the tree (Figure 1). The general plan, constituting consensus practice at midcentury, is represented by the main trunk. Since the 1970s this traditional "land use design plan" has been joined by several branches – the verbal policy plan, the land classification plan, and the development management plan. These branches connect to the trunk although springing from different planning disciplines, in a way reminiscent of the complex structure of a Ficus tree. The branches combine into the contemporary, hybrid comprehensive plan integrating design, policy, classification, and management, represented by the foliage at the top of the tree.

As we discuss each of these parts of the family tree, we show how plans respond both to social climate changes and to "idea genes" from the literature. We also draw conclusions about the survival of the tree and the prospects for new branches in the future. The focus of the article is the plan prepared by a local government – a county, municipality, or urban region – for the long-term development and use of the land.

ROOTS OF THE FAMILY TREE: THE FIRST 50 YEARS

New World city plans certainly existed before this century. They included L'Enfant's plan for Washington, William Penn's plan for Philadelphia, and General Oglethorpe's plan for Savannah. These plans, however, were blueprints for undeveloped sites, commissioned

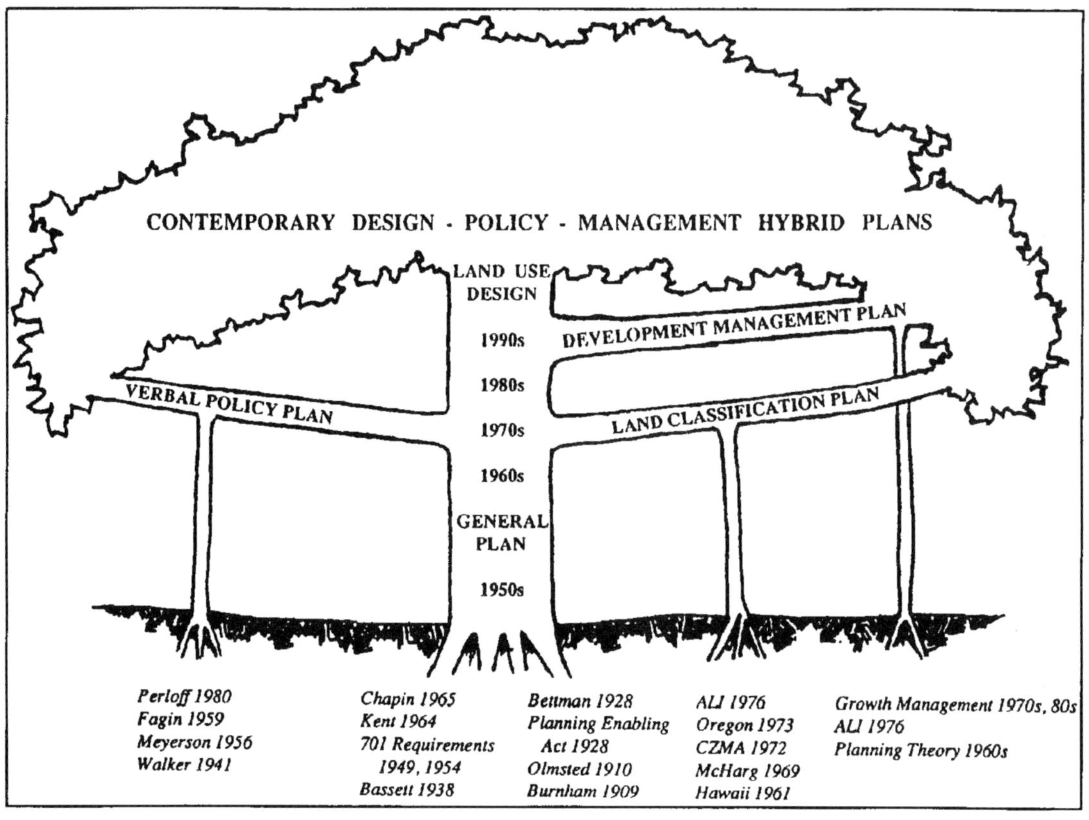

Figure 1 The family tree of the land use plan

by unitary authorities with power to implement them unilaterally.

In this century, perhaps the most influential early city plan was Daniel Burnham's plan for Chicago, published by the Commercial Club of Chicago (a civic, not a government entity) in 1909. The archetypical plan-as-inspirational-vision, it focuses only on design of public spaces as a City Beautiful effort.

The City Beautiful approach was soon broadened to a more comprehensive view. At the 1911 National Conference on City Planning, Frederick Law Olmsted, Jr., son of the famous landscape architect and in his own right one of the fathers of planning, defined a city plan as encompassing all uses of land, private property, public sites, and transportation. Alfred Bettman, speaking at the 1928 National Conference of City Planning, envisioned the plan as a master design for the physical development of the city's territory, including "the general location and extent of new public improvements . . . and in the case of private developments, the general distribution amongst various classes of land uses, such as residential, business, and industrial uses . . . designed for . . . the future, twenty-five to fifty years" (Black 1968, 352–3). Together, Olmsted and Bettman anticipated the development of the midcentury land use plan.

Another early influence, the federal Standard City Planning Enabling Act of 1928, shaped enabling acts passed by many states. However, the Act left many planners and public officials confused about the difference between a master plan and a zoning ordinance, so that hundreds of communities adopted "zoning plans" without having created comprehensive plans as the basis for zoning (Black 1968, 353). Because the Act also did not make clear the importance of comprehensiveness or define the essential elements of physical development, no consensus about the essential content of the plan existed.

Ten years later, Edward Bassett's book, *The Master Plan* (1938), spelled out the plan's subject matter and format – supplementing the 1928 Act, and consistent with it. He argued that the plan should have seven elements, all relating to land areas (not buildings) and capable of being shown on a map: streets, parks, sites for public buildings, public reservations, routes for public utilities, pier-head and bulkhead lines (all public facilities), and zoning districts for private lands. Bassett's views were incorporated in many state enabling laws.

The physical plans of the first half of the century were drawn by and for independent commissions, reflecting the profession's roots in the Progressive Reform movement, with its distrust of politics. The 1928 Act reinforced that perspective by making the planning commission, not the legislative body, the principal client of the plan, and purposely isolating the commission from politics. Bassett's book reinforced the reliance on an independent commission. He conceived of the plan as a "plastic" map, kept within the purview of the planning commission, capable of quick and easy change. The commission, not the plan, was intended to be the adviser to the local legislative body and to city departments.

By the 1940s, both the separation of the planning function from city government and the plan's focus on physical development were being challenged. Robert Walker, in *The Planning Function in Local Government*, argued that the "scope of city planning is properly as broad as the scope of city government." The central planning agency might not necessarily do all the planning, but it would coordinate departmental planning in the light of general policy considerations – creating a comprehensive plan but one without a physical focus. That idea was not widely accepted. Walker also argued that the independent planning commission should be replaced by a department or bureau attached to the office of mayor or city manager. That argument did take hold, and by the 1960s planning in most communities was the responsibility of an agency within local government, though planning boards still advised elected officials on planning matters.

This evolution of ideas over 50 years resulted at midcentury in a consensus concept of a plan as focused on long-term physical development; this focus was a legacy of the physical design professions. Planning staff worked both for the local government executive officer and with an appointed citizen planning board, an arrangement that was a legacy of the Progressive insistence on the public interest as an antidote to governmental corruption. The plan addressed both public and private uses of the land, but did not deal in detail with implementation.

THE PLAN AFTER MIDCENTURY: NEW GROWTH INFLUENCES

Local development planning grew rapidly in the 1950s, for several reasons. First, governments had to contend with the postwar surge of population and urban growth, as well as a need for the capital investment in infrastructure and community facilities that had been postponed during the depression and war years. Second, municipal legislators and managers became more interested in planning as it shifted from being the responsibility of an independent commission to being a function within local government. Third, and very important, Section 701 of the Housing Act of 1954 required local governments to adopt a long-range general plan in order to qualify for federal grants for urban renewal, housing, and other programs, and it also made money available for such comprehensive planning. The 701 program's double-barreled combination of requirements and financial support led to more urban planning in the United States in the latter half of the 1950s than at any previous time in history.

At the same time, the plan concept was pruned and shaped by two planning educators. T. J. Kent, Jr. was a professor at the University of California at Berkeley, a planning commissioner, and a city councilman in the 1950s. His book, *The Urban General Plan* (1964), clarified the policy role of the plan. F. Stuart Chapin, Jr. was a TVA planner and planning director in Greensboro, NC in the 1940s, before joining the planning faculty at the University of North Carolina at Chapel Hill in 1949. His contribution was to codify the methodology of land use planning in the various editions of his book, *Urban Land Use Planning* (1957, 1965).

What should the plan look like? What should it be about? What is its purpose (besides the cynical purpose of qualifying for federal grants)? The 701 program, Kent, and Chapin all offered answers.

The "701" program comprehensive plan guidelines

In order to qualify for federal urban renewal aid and, later, for other grants – a local government had to prepare a general plan that consisted of plans for physical development, programs for redevelopment, and administrative and regulatory measures for controlling and guiding development. The 701 program specified what the content of a comprehensive development plan should include:

- A land use plan, indicating the locations and amounts of land to be used for residential, commercial, industrial, transportation, and public purposes
- A plan for circulation facilities
- A plan for public utilities
- A plan for community facilities

T.J. Kent's urban general plan

Kent's view of the plan's focus was similar to that of the 701 guidelines: long-range physical development in terms of land use, circulation, and community facilities. In addition, the plan might include sections on civic design and utilities, and special areas, such as historic preservation or redevelopment areas. It covered the entire geographical jurisdiction of the community, and was in that sense comprehensive. The plan was a vision of the future, but not a blueprint; a policy statement, but not a program of action; a formulation of goals, but not schedules, priorities, or cost estimates. It was to be inspirational, uninhibited by short-term practical considerations.

Kent (1964, 65–89) believed the plan should emphasize policy, serving the following functions:

- Policy determination – to provide a process by which a community would debate and decide on its policy
- Policy communication – to inform those concerned with development (officials, developers, citizens, the courts, and others) and educate them about future possibilities
- Policy effectuation – to serve as a general reference for officials deciding on specific projects
- Conveyance of advice – to furnish legislators with the counsel of their advisors in a coherent, unified form

The format of Kent's proposed plan included a unified, comprehensive, but general physical design for the future, covering the whole community and represented by maps. (See Figure 2.) It also contained goals and policies (generalized guides to conduct, and the most important ingredients of the plan), as well as summaries of background conditions, trends, issues, problems, and assumptions. (See Figure 3.) So that the plan would be suitable for public debate, it was to be a complete, comprehensible document, containing factual data, assumptions, statements of issues, and goals, rather than merely conclusions and recommendations. The plan belonged to the legislative body and was intended to be consulted in decision-making during council meetings.

Kent recommended overall goals for the plan:

- Improve the physical environment of the community to make it more functional, beautiful, decent, healthful, interesting, and efficient
- Promote the overall public interest, rather than the interests of individuals or special groups within the community
- Effect political and technical coordination in community development
- Inject long-range considerations into the determination of short-range actions
- Bring professional and technical knowledge to bear on the making of political decisions about the physical development of the community

F. Stuart Chapin, Jr.'s urban land use plan

Chapin's ideas, though focusing more narrowly on the land use plan, were consistent with Kent's in both the 1957 and 1965 editions of *Urban Land Use Planning*, a widely used text and reference work for planners. Chapin's concept of the plan was of a generalized, but scaled, design for the future use of land, covering private land uses and public facilities, including the thoroughfare network.

Chapin conceived of the land use plan as the first step in preparing a general or comprehensive plan. Upon its completion, the land use plan served as a temporary general guide for decisions, until the comprehensive plan was developed. Later, the land use plan would become a cornerstone in the comprehensive plan, which also included plans for transportation, utilities, community facilities, and renewal, only the general rudiments of which are suggested in the land use plan. Purposes of the plan were to guide government decisions on public facilities, zoning, subdivision control, and urban renewal, and to inform private developers about the proposed future pattern of urban development.

The format of Chapin's land use plan included a statement of objectives, a description of existing conditions and future needs for space and services, and finally the mapped proposal for the future development

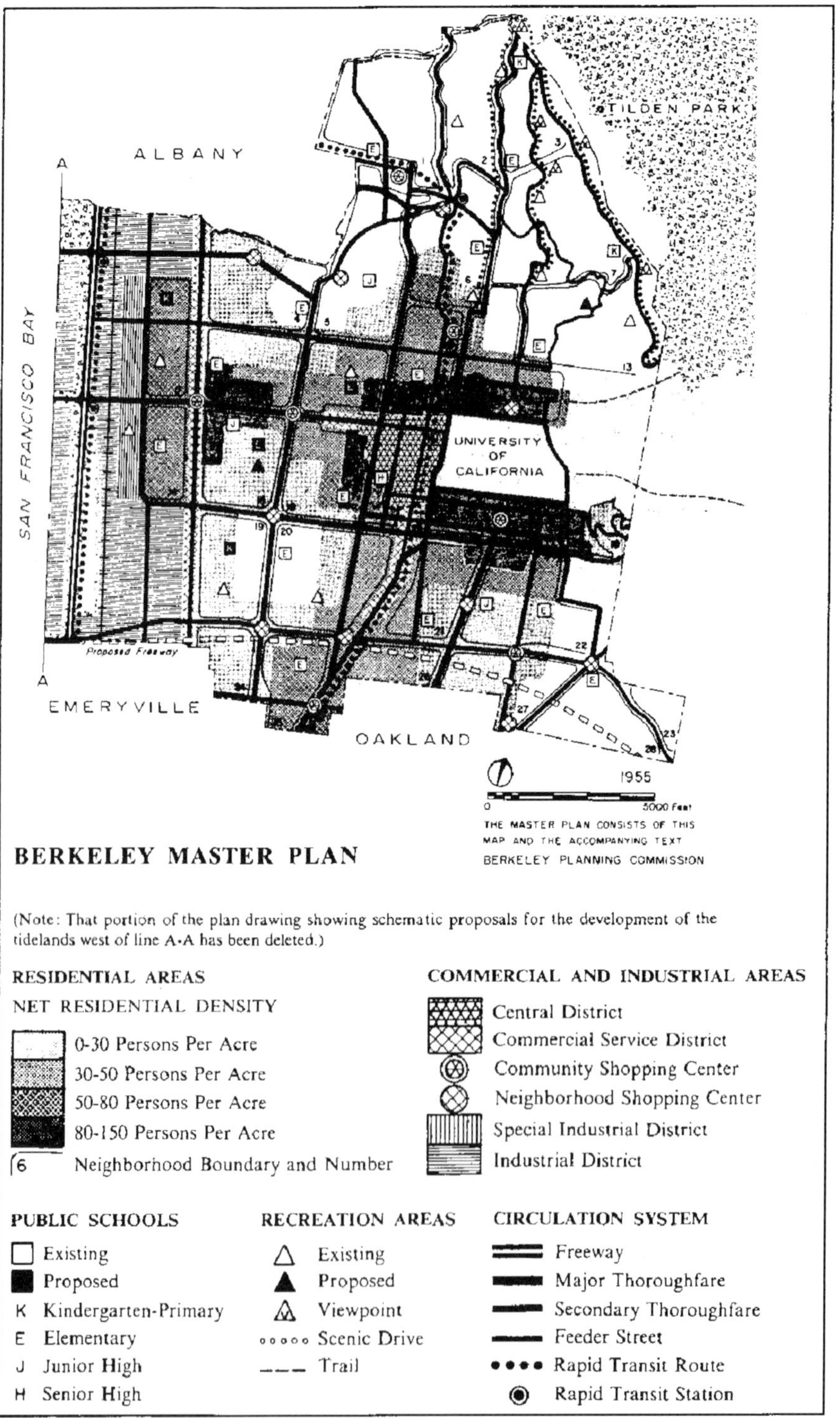

Figure 2 Example of the land use design map featured in the 1950s General Plan
Source: Kent 1991, 111

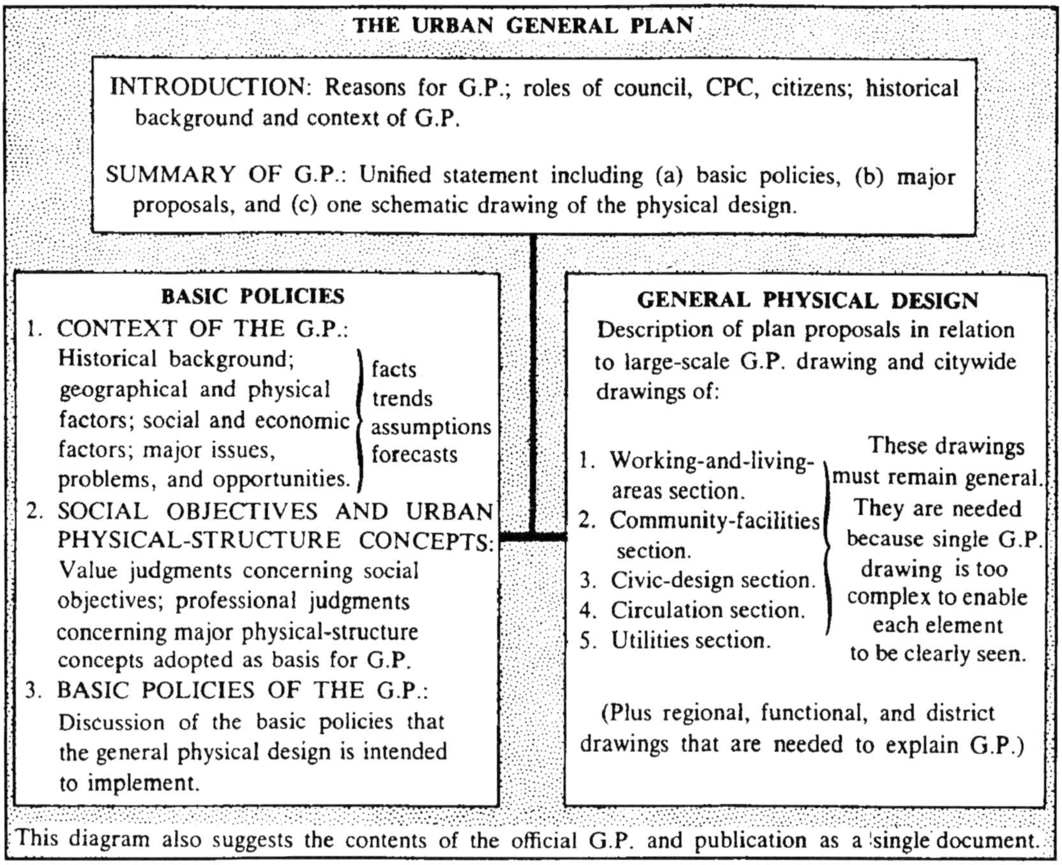

THE URBAN GENERAL PLAN

INTRODUCTION: Reasons for G.P.; roles of council, CPC, citizens; historical background and context of G.P.

SUMMARY OF G.P.: Unified statement including (a) basic policies, (b) major proposals, and (c) one schematic drawing of the physical design.

BASIC POLICIES

1. CONTEXT OF THE G.P.: Historical background; geographical and physical factors; social and economic factors; major issues, problems, and opportunities. } facts, trends, assumptions, forecasts
2. SOCIAL OBJECTIVES AND URBAN PHYSICAL-STRUCTURE CONCEPTS: Value judgments concerning social objectives; professional judgments concerning major physical-structure concepts adopted as basis for G.P.
3. BASIC POLICIES OF THE G.P.: Discussion of the basic policies that the general physical design is intended to implement.

GENERAL PHYSICAL DESIGN

Description of plan proposals in relation to large-scale G.P. drawing and citywide drawings of:

1. Working-and-living-areas section.
2. Community-facilities section.
3. Civic-design section.
4. Circulation section.
5. Utilities section.

} These drawings must remain general. They are needed because single G.P. drawing is too complex to enable each element to be clearly seen.

(Plus regional, functional, and district drawings that are needed to explain G.P.)

This diagram also suggests the contents of the official G.P. and publication as a single document.

Figure 3 Components of the 1950s–1960s General Plan
Source: Kent 1964, 93

of the community, together with a program for implementing the plan (customarily including zoning, subdivision control, a housing code, a public works expenditure program, an urban renewal program, and other regulations and development measures).

The typical general plan of the 1950s and 1960s

Influenced by the 701 program, Kent's policy vision, and Chapin's methods, the plans of the 1950s and 1960s were based on a clear and straightforward concept: The plans' purposes were to determine, communicate, and effectuate comprehensive policy for the private and public physical development and redevelopment of the city. The subject matter was long-range physical development, including private uses of the land, circulation, and community facilities. The standard format included a summary of existing and emerging conditions and needs; general goals; and a long-range urban form in map format, accompanied by consistent development policies. The coverage was comprehensive, in the sense of addressing both public and private development and covering the entire planning jurisdiction, but quite general. The tone was typically neither as "inspirational" as the Burnham plan for Chicago, nor as action-oriented as today's plans. Such was the well-defined trunk of the family tree in the 1950s and 1960s, in which today's contemporary plans have much of their origin.

CONTEMPORARY PLANS: INCORPORATING NEW BRANCHES

Planning concepts and practice have continued to evolve since midcentury, maturing in the process. By the 1970s, a number of new ideas had taken root. Referring back to the family tree in figure 1, we can see a trunk and several distinct branches:

- *The land use design*, a detailed mapping of future land use arrangements, is the most direct descendant of the 1950s plan. It still constitutes the trunk of the tree. However, today's version is more likely to be accompanied by action strategies, also mapped, and to include extensive policies.
- *The land classification plan*, a more general map of growth policy areas rather than a detailed land use pattern, is now also common, particularly for counties, metropolitan areas, and regions that want to encourage urban growth in designated development areas and to discourage it in conservation or rural areas. The roots of the land classification plan include McHarg's *Design With Nature* (1969), the 1976 American Law Institute (ALI) Model Land Development Code, the 1972 Coastal Zone Management Act, and the 1973 Oregon Land Use Law.
- *The verbal policy plan* de-emphasizes mapped policy or end-state visions and focuses on verbal action policy statements, usually quite detailed; sometimes called a strategic plan, it is rooted in Meyerson's middle-range bridge to comprehensive planning, Fagin's policies plan, and Perloff's strategies and policies general plan.
- *The development management plan* lays out a specific program of actions to guide development, such as a public investment program, a development code, and a program to extend infrastructure and services; and it assumes public sector initiative for influencing the location, type, and pace of growth. The roots of the development management plan are in the environmental movement, and the movements for state growth management and community growth control, as well as in ideas from Fagin (1959) and the ALI Code.

We looked for, but could not find, examples of land use plans that could be termed purely prototypical "strategic plans," in the sense of Bryson and Einsweller. Hence, rather than identifying strategic planning as a separate branch on the family tree of the land use plan, we see the influence of strategic planning showing up across a range of contemporary plans. We tend to agree with the planners surveyed by Kaufman and Jacobs that strategic planning differs from good comprehensive planning more in emphasis (shorter range, more realistically targeted, more market-oriented) than in kind.

The land use design plan

The land use design plan is the most traditional of the four prototypes of contemporary plans and is the most direct descendent of the Kent–Chapin–701 plans of the 1950s and 1960s. It proposes a long-range future urban form as a pattern of retail, office, industrial, residential, and open spaces, and public land uses and a circulation system. Today's version, however, incorporates environmental processes, and sometimes agriculture and forestry, under the "open space" category of land use. Its land uses often include a "mixed use" category, honoring the neotraditional principle of closer mingling of residential, employment, and shopping areas. In addition, it may include a development strategy map, which is designed to bring about the future urban form and to link strategy to the community's financial capacity to provide infrastructure and services. The plans and strategies are often organized around strategic themes or around issues about growth, environment, economic development, transportation, or neighborhood/community scale change.

Like the other types of plans in vogue today, the land use design plan reflects recent societal issues, particularly the environmental crisis, the infrastructure crisis, and stresses on local government finance. Contemporary planners no longer view environmental factors as development constraints, but as valuable resources and processes to be conserved. They also may question assumptions about the desirability and inevitability of urban population and economic growth, particularly as such assumptions stimulate demand for expensive new roads, sewers, and schools. While at midcentury plans unquestioningly accommodated growth, today's plans cast the amount, pace, location, and costs of growth as policy choices to be determined in the planning process.

The 1990 Howard County (Maryland) General Plan, winner of an American Planning Association (APA) award in 1991 for outstanding comprehensive planning, exemplifies contemporary land use design. (See Figure 4.) While clearly a direct descendent of the traditional general plan, the Howard County plan adds new types of goals, policies, and planning techniques. To enhance communication and public understanding, it is organized strategically around six themes/chapters

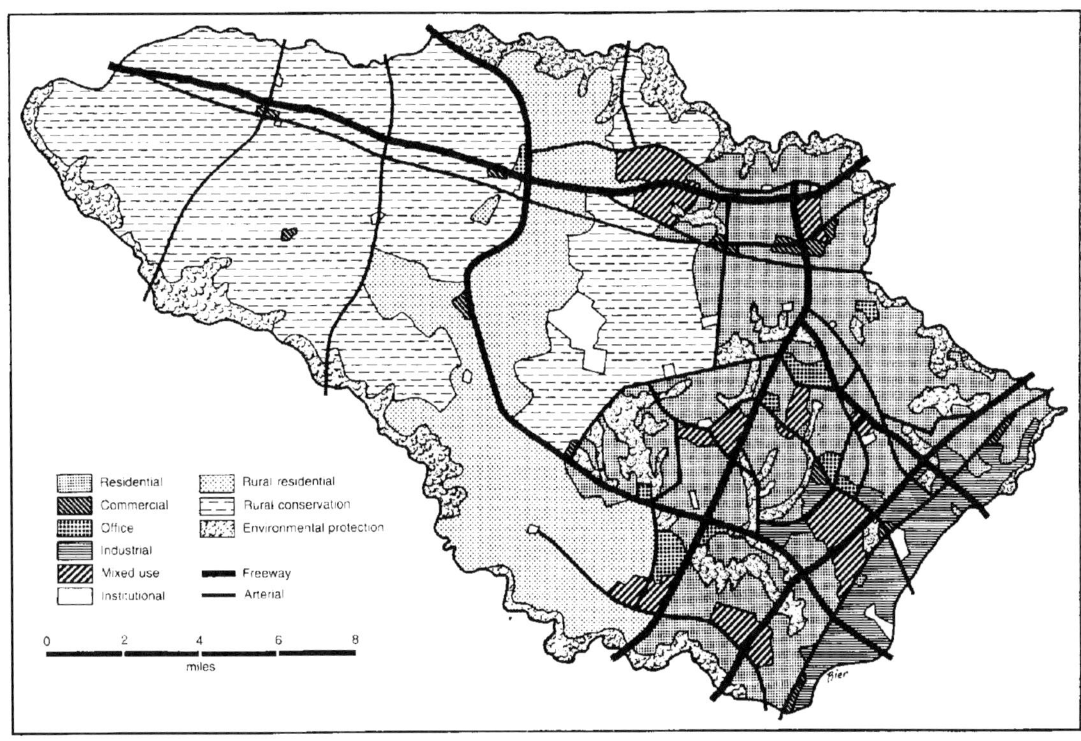

Figure 4 Howard County, Maryland, General Plan, Land Use 2010
Source: Adapted from Howard County 1990

(responsible regionalism, preservation of the rural area, balanced growth, working with nature, community enhancement, and phased growth), instead of the customary plan elements. Along with the traditional land use design, the plan includes a "policy map" (strategy map) for each theme and an overall policies map for the years 2000 and 2010. A planned service area boundary is used to contain urban growth within the eastern urbanized part of the county, home to the well-known Columbia New Town. The plan lays out specific next steps to be implemented over the next two years, and defines yardsticks for measuring success. An extensive public participation process for formulating the plan involved a 32-member General Plan Task Force, public opinion polling to discover citizen concerns, circulation of preplan issue papers on development impacts, and consideration of six alternative development scenarios.

The land classification plan

Land classification, or development priorities mapping, is a proactive effort by government to specify where and under what conditions growth will occur. Often, it also regulates the pace or timing of growth. Land classification addresses environmental protection by designating "nondevelopment" areas in especially vulnerable locations. Like the land use design, the land classification plan is spatially specific and map-oriented. However, it is less specific about the pattern of land uses within areas specified for development, which results in a kind of silhouette of urban form. On the other hand, land classification is more specific about development strategy, including timing. Counties, metropolitan areas, and regional planning agencies are more likely than cities to use a land classification plan.

The land classification plan identifies areas where development will be encouraged (called urban, transition, or development areas) and areas where development will be discouraged (open space, rural, conservation, or critical environmental areas). For each designated area, policies about the type, timing, and density of allowable development, extension of infrastructure, and development incentives or constraints apply. The planning principle is to concentrate financial resources, utilities, and services within a limited, prespecified area suitable for development, and to relieve pressure on nondevelopment areas by withholding facilities that accommodate growth.

Ian McHarg's (1969) approach to land planning is an early example of the land classification concept. He divides planning regions into three categories: natural use, production, and urban. Natural use areas, those with valuable ecological functions, have the highest priority. Production areas, which include agriculture, forestry, and fishing uses, are next in priority. Urban areas have the lowest priority and are designated after allocating the land suitable to the two higher-priority uses. McHarg's approach in particular, and land classification generally, also reflect the emerging environmental consciousness of the 1960s and 1970s.

As early as 1961, Hawaii had incorporated the land classification approach into its state growth management system. The development framework plan of the Metropolitan Council of the Twin Cities Area defined "planning tiers," each intended for a different type and intensity of development. The concepts of the "urban service area," first used in 1958 in Lexington, Kentucky, and the "urban growth boundary," used throughout Oregon under its 1973 statewide planning act, classify land according to growth management policy. Typically, the size of an urban growth area is based on the amount of land necessary to accommodate development over a period of ten or twenty years.

Vision 2005: A Comprehensive Plan for Forsyth County, North Carolina exemplifies the contemporary approach to land classification plans. The plan, which won honorable mention from APA in 1989, employs a six category system of districts, plus a category for activity centers. It identifies both short- and long-range growth areas (4A and 4B in Figure 5). Policies applicable to each district are detailed in the plan.

The verbal policy plan: shedding the maps

The verbal policy plan focuses on written statements of goals and policy, without mapping specific land use patterns or implementation strategy. Sometimes called a policy framework plan, a verbal policy plan is more easily prepared and flexible than other types of plans, particularly for incorporating nonphysical development policy. Some claim that such a plan helps the planner to avoid relying too heavily on maps, which are difficult to keep up to date with the community's changes in policy. The verbal policy plan also avoids falsely representing general policy as applying to specific parcels of property. The skeptics, however, claim that verbal statements in the absence of maps provide too little spatial specificity to guide implementation decisions.

The verbal policy plan may be used at any level of government, but is especially common at the state level, whose scale is unsuited to land use maps. The plan usually contains goals, facts and projections, and general policies corresponding to its purposes – to understand current and emerging conditions and issues, to identify goals to be pursued and issues to be addressed, and to formulate general principles of action. Sometimes communities do a verbal policy plan as an interim plan or a first step in the planning process. Thus, verbal policies are included in most land use design plans, land classification plans, and development management plans.

The Calvert County, MD, Comprehensive Plan (Calvert County 1983), winner of a 1985 APA award, exemplifies the verbal policy plan. Its policies are concise, easy to grasp, and grouped in sections corresponding to the six divisions of county government responsible for implementation. It remains a policy plan, however, because it does not specify a program of specific actions for development management. Though the plan clearly addresses physical development and discusses specific spatial areas, it contains no land use map. (See Figure 6 for an illustrative page from the Calvert County plan.)

The development management plan

The development management plan features a coordinated program of actions, supported by analyses and goals, for specific agencies of local government to undertake over a three-to-ten-year period. The program of actions usually specifies the content, geographic coverage, timing, assignment of responsibility, and coordination among the parts. Ideally, the plan includes most or all of the following components:

- Description of existing and emerging development conditions, with particular attention to development processes, the political-institutional context,

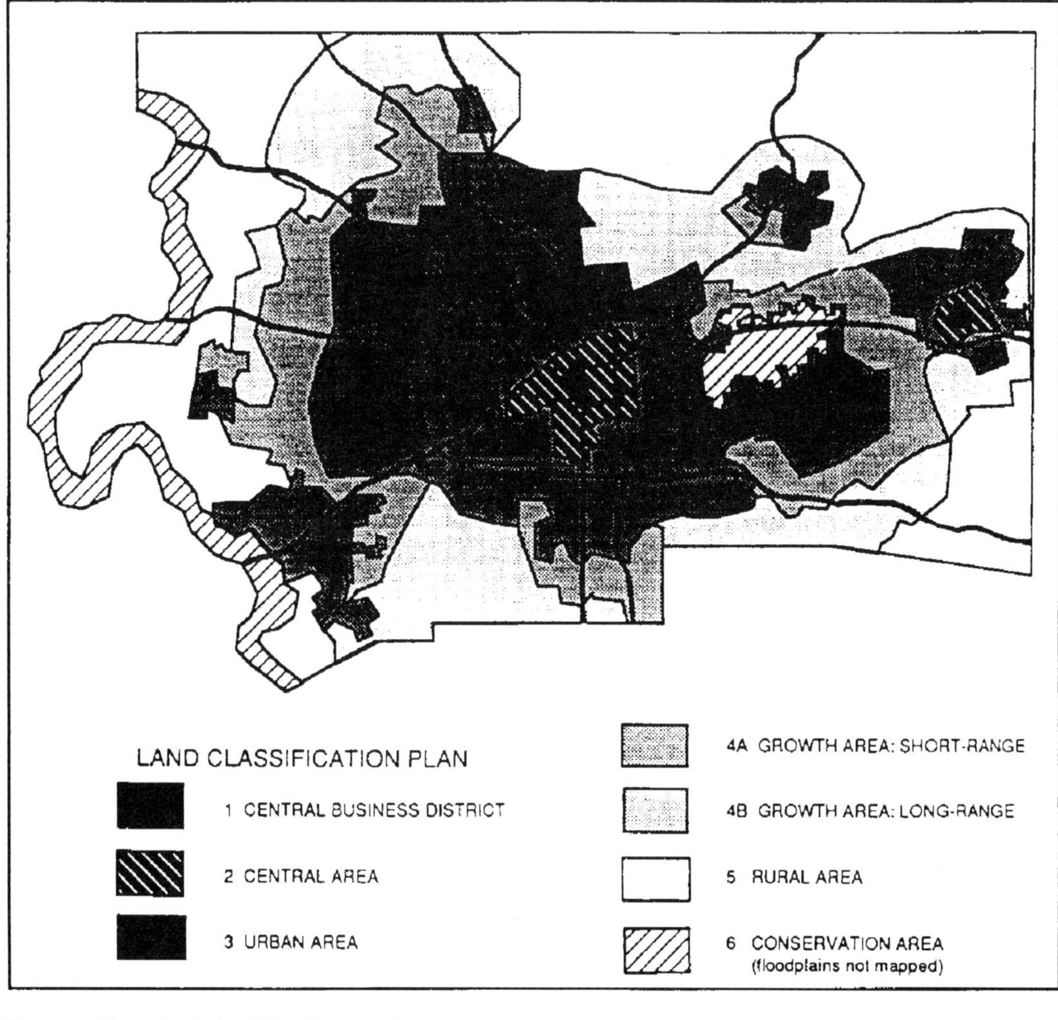

Figure 5 Example of a land classification plan
Source: Adapted from Forsyth County City–County Planning Board 1988

and a critical review of the existing systems of development management

- Statement of goals and/or legislative intent, including management-oriented goals
- Program of actions – the heart of the plan – including:
  1. Outline of a proposed development code, with: (a) procedures for reviewing development permits; (b) standards for the type of development, density, allowable impacts and/or performance standards; (c) site plan, site engineering, and construction practice requirements; (d) exactions and impact fee provisions and other incentives/disincentives; and (e) delineation of districts where various development standards, procedures, exactions, fees, and incentives apply
  2. Program for the expansion of urban infrastructure and community facilities and their service areas
  3. Capital improvement program
  4. Property acquisition program
  5. Other components, depending on the community situation, for example, a preferential taxation program, an urban revitalization program for specific built-up neighborhoods, or a historic preservation program
- Official maps, indicating legislative intent, which may be incorporated into ordinances, with force

INDUSTRIAL DISTRICTS

Industrial Districts are intended to provide areas in the county which are suitable for the needs of industry. They should be located and designed to be compatible with the surrounding land uses, either due to existing natural features or through the application of standards.

RECOMMENDATIONS:

1. Identify general locations for potential industrial uses.
2. Permit retail sales as an accessory use in the Industrial District.

SINGLE-FAMILY RESIDENTIAL DISTRICTS

Single-Family Residential Districts are to be developed and promoted as neighborhoods free from any land usage which might adversely affect them.

RECOMMENDATIONS:

1. For new development, require buffering for controlling visual, noise, and activity impacts between residential and commercial uses.
2. Encourage single-family residential development to locate in the designated towns.
3. Allow duplexes, triplexes, and fourplexes as a conditional use in the "R-1" Residential Zone so long as the design is compatible with the single-family residential development.
4. Allow home occupations (professions and services, but not retail sales) by permitting the employment of one full-time equivalent individual not residing on the premises.

MULTIFAMILY RESIDENTIAL DISTRICTS

Multifamily Residential Districts provide for townhouses and multifamily apartment units. Areas designated in this category are those which are currently served or scheduled to be served by community or multi-use sewerage and water supply systems.

RECOMMENDATIONS:

1. Permit multifamily development in the Solomons, Prince Frederick, and Twin Beach Towns.
2. Require multifamily projects to provide adequate recreational facilities—equipment, structures, and play surfaces.
3. Evaluate the feasibility of increasing the dwelling unit density permitted in the multifamily Residential Zone (R-2)

Figure 6 An excerpt from a verbal policy plan
Source: Calvert County, Maryland 1983

of law – among them, goal-form maps (e.g., land classification plan or land use design); maps of zoning districts, overlay districts, and other special areas for which development types, densities, and other requirements vary; maps of urban services areas; maps showing scheduled capital improvements; or other maps related to development management standards and procedures

The development management plan is a distinct type, emphasizing a specific course of action, nor general policy. At its extreme the management plan actually incorporates implementation measures, so that the plan becomes part of a regulative ordinance. Although the spatial specifications for regulations and other implementation measures are included, a land use map may not be.

One point of origin for development management plans is Henry Fagin's concept of the "policies plan," whose purpose was to coordinate the actions of line departments and provide a basis for evaluating their results, as well as to formulate, communicate, and implement policy (the traditional purpose). Such a plan's subject matter was as broad as the responsibilities of the local government, including but not limited to physical development. The format included a "state of the community" message, a physical plan, a financial plan, implementation measures, and detailed sections for each department of the government.

A more recent point of origin is *A Model Land Development Code* (American Law Institute 1976), intended to replace the 1928 Model Planning Enabling Act as a model for local planning and development management. The model plan consciously retains an emphasis on physical development (unlike Fagin's broader concept), but stresses a short-term program of action, rather than a long-term, mapped goal form. The ALI model plan contains a statement of conditions and problems; objectives, policies, and standards; and a short-term (from one to five years) program of specific public actions. It may also include land acquisition requirements, displacement impacts, development regulations, program costs and fund sources, and environmental, social, and economic consequences. More than other plan types, the development management plan is a "course of action" initiated by government to control the location and timing of development.

The Sanibel, Florida, Comprehensive Land Use Plan (1981) exemplifies the development management plan. The plan outlines the standards and procedures of regulations (i.e., the means of implementation), as well as the analyses, goals, and statements of intent normally presented in a plan. Thus, when the local legislature adopts the plan, it also adopts an ordinance for its implementation. Plan and implementation are merged into one instrument, as can be seen in the content of its articles:

Article 1: Preamble: including purposes and objectives, assumptions, coordination with surrounding areas, and implementation

Article 2: Elements of the Plan: Safety, Human Support Systems, Protection of Natural Environmental, Economic and Scenic Resources, Intergovernmental Coordination, and Land Use

Article 3: Development Regulations: Definitions, Maps, Requirements, Permitted Uses, Subdivisions, Mobile Home and Recreation Vehicles, Flood and Storm Proofing, Site Preparation, and Environmental Performance Standards

Article 4: Administrative Regulations (i.e., procedures): Standards, Short Form Permits, Development Permits, Completion Permits, Amendments to the Plan, and Notice, Hearing and Decision Procedures on Amendments

Figure 7 shows the Sanibel plan's map of permitted uses, which is more like a zoning plan than a land use design plan, because it shows where regulations apply, and boundaries are exact.

THE CONTEMPORARY HYBRID PLAN: INTEGRATING DESIGN, POLICY, AND MANAGEMENT

The rationality of practice has integrated the useful parts of each of the separate prototypes reviewed here into contemporary hybrid plans that not only map and classify land use in both specific and general ways, but also propose policies and management measures. For example, Gresham, Oregon, combined land use design (specifying residential, commercial, and industrial areas, and community facilities and public lands) with an overlay of land classification districts (developed, developing, rural, and conservation), and also included standards and procedures for issuing development permits (i.e., a development code). Prepared with considerable participation by citizens and interest groups, such plans usually reflect animated political debates about the costs and benefits of land use alternatives.

The states that manage growth have created new land use governance systems whose influence has broadened the conceptual arsenals of local planners. DeGrove identifies the common elements of these systems:

- Consistency – intergovernmentally and internally (i.e., between plan and regulations)
- Concurrency – between infrastructure and new development
- Compactness – of new growth, to limit urban sprawl affordability – of new housing
- Economic development, or "managing to grow"
- Sustainability – of natural systems

DeGrove attributes the changes in planning under growth management systems to new hard-nosed concerns for measurable implementation and realistic funding mechanisms. For example, Florida local governments must adopt detailed capital improvement programs as part of their comprehensive plans, and substantial state grants may be withheld if their plans do not meet consistency and concurrency requirements.

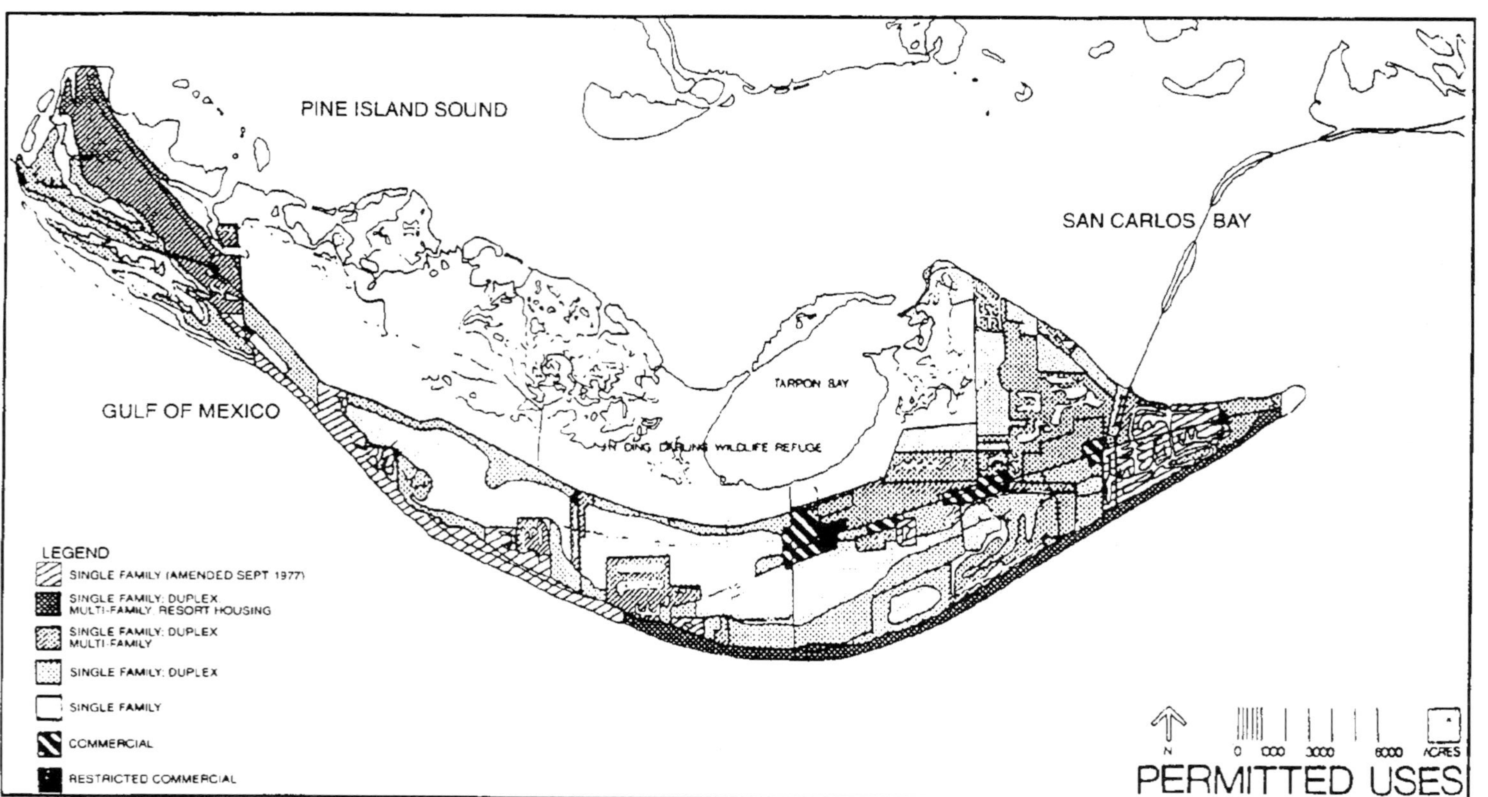

Figure 7 Map of permitted uses, Sanibel
Source: City of Sanibel 1981

Another important influence on contemporary plans is the renewed attention to community design. The neotraditional and transit-oriented design movements have inspired a number of proposals for mixed-use villages in land use plans. *Toward a Sustainable Seattle: A Plan for Managing Growth* (1994) exemplifies a city approach to the contemporary hybrid plan. Submitted as the Mayor's recommended comprehensive plan, it attempted to muster political support for its proposals. Three core values – social equity, environmental stewardship, and economic security and opportunity – underlie the plan's overall goal of sustainability. This goal is to be achieved by integrating plans for land use and transportation, healthy and affordable housing, and careful capital investment in a civic compact based on a shared vision. Citywide population and job growth targets, midway between growth completely by regional sprawl and growth completely by infill, are set forth within a 20-year time frame. The plan is designed to meet the requirements of the Washington State Growth Management Act.

The land use element designates urban center villages, hub urban villages, residential urban villages, neighborhood villages, and manufacturing/industrial centers, each with specific design guidelines (figure 8). The city's capacity for growth is identified, and then allocated according to the urban village strategy. Future development is directed to mixed-use neighborhoods, some of which are already established; existing single-family areas are protected. Growth is shaped to build community, promote pedestrian and transit use, protect natural amenities and existing residential and employment areas, and ensure diversity of people and activities. Detailed land use policies carry out the plan.

Loudoun County Choices and Changes: General Plan (1991), which won APA's 1994 award for comprehensive planning in small jurisdictions, exemplifies a county approach to the contemporary hybrid plan. Its goals are grouped into three categories:

1 Natural and cultural resources goals seek to protect fragile resources by limiting development or mitigating disturbances, while at the same time not unduly diminishing land values.
2 Growth management: goals seek to accommodate and manage the county's fair share of regional growth, guiding development into the urbanized eastern part of the county or existing western towns and their urban growth areas, and conserving agriculture and open space areas in the west. (See Figure 9.)
3 Community design goals seek to concentrate growth in compact, urban nodes to create mixed-use communities with strong visual identities, human-scale street networks, and a range of housing and employment opportunities utilizing neo-traditional design concepts (illustrated in Figure 9).

Three time horizons are addressed: the "ultimate" vision through 2040, the 20-year, long-range development pattern, and the five-year, short-range development pattern. The plan uses the concept of community character areas as an organizing framework for land use management. Policies are proposed for the overall county, as well as for the eastern urban growth areas, town urban growth areas, rural areas, and existing rural village areas. Implementation tools include capital facility and transportation proffers by developers, density transfers, community design guidelines, annexation guidelines, and an action schedule of next steps.

SUMMARY OF THE CONTEMPORARY SITUATION

Since midcentury, the nature of the plan has shifted from an elitist, inspirational, long-range vision that was based on fiscally innocent implementation advice, to a framework for community consensus on future growth that is supported by fiscally grounded actions to manage change. Subject matter has expanded to include the natural as well as the built environment. Format has shifted from simple policy statements and a single large-scale map of future land use, circulation, and community facilities, to a more complex combination of text, data, maps, and timetables. In a number of states plans are required by state law, and their content is specified by state agencies. Table 2 compares the general plan of the 1950s–1960s with the four contemporary prototype plans and the new 1990s hybrid design-policy management plan, which combines aspects of the prototype plans.

Today's prototype land use design continues to emphasize long-range urban form for land uses, community facilities, and transportation systems as shown by a map; but the design is also expressed in general policies. Land use design is still a common form of development plan, especially in municipalities.

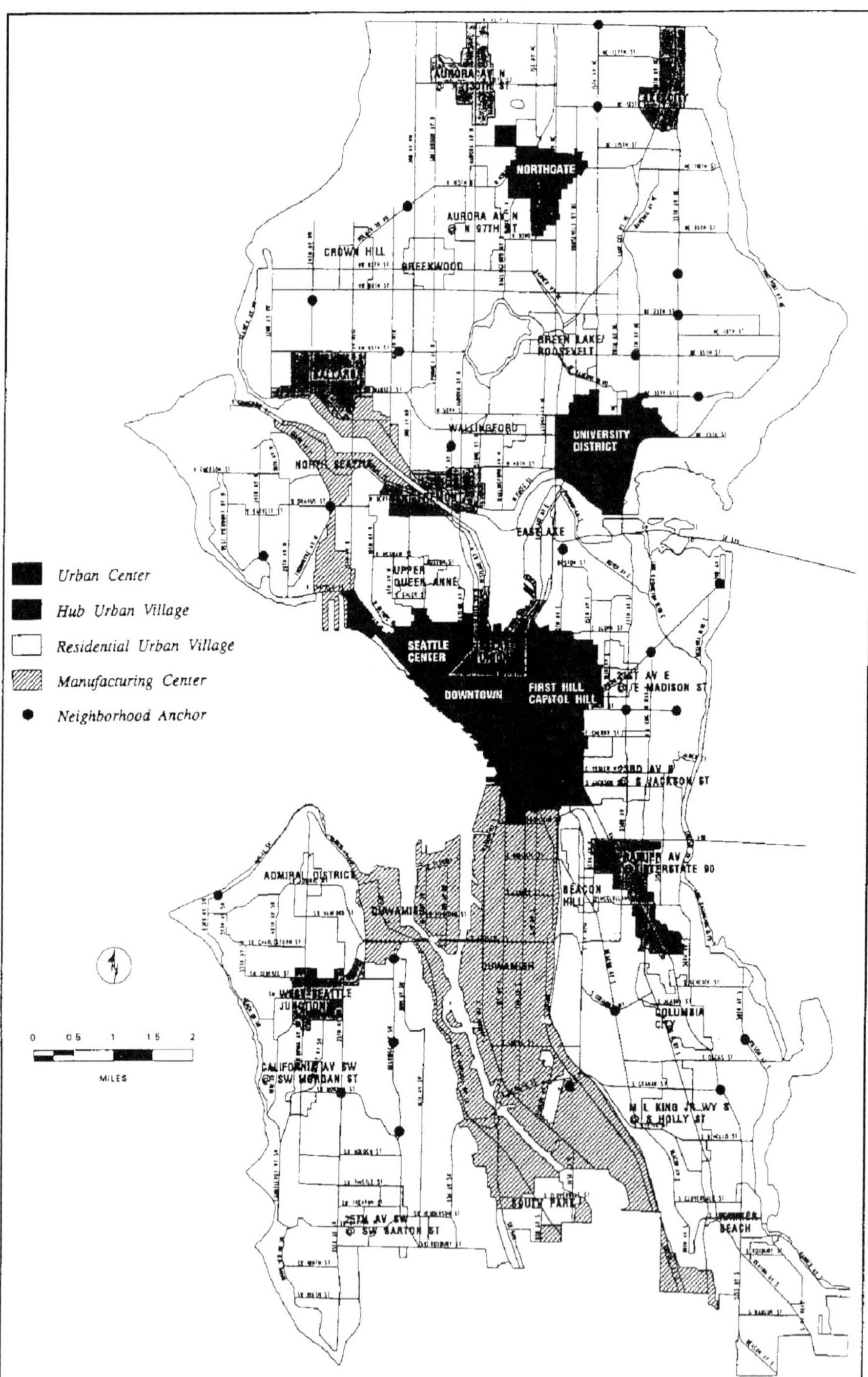

Figure 8 Seattle urban villages strategy
Source: Seattle Planning Department 1993

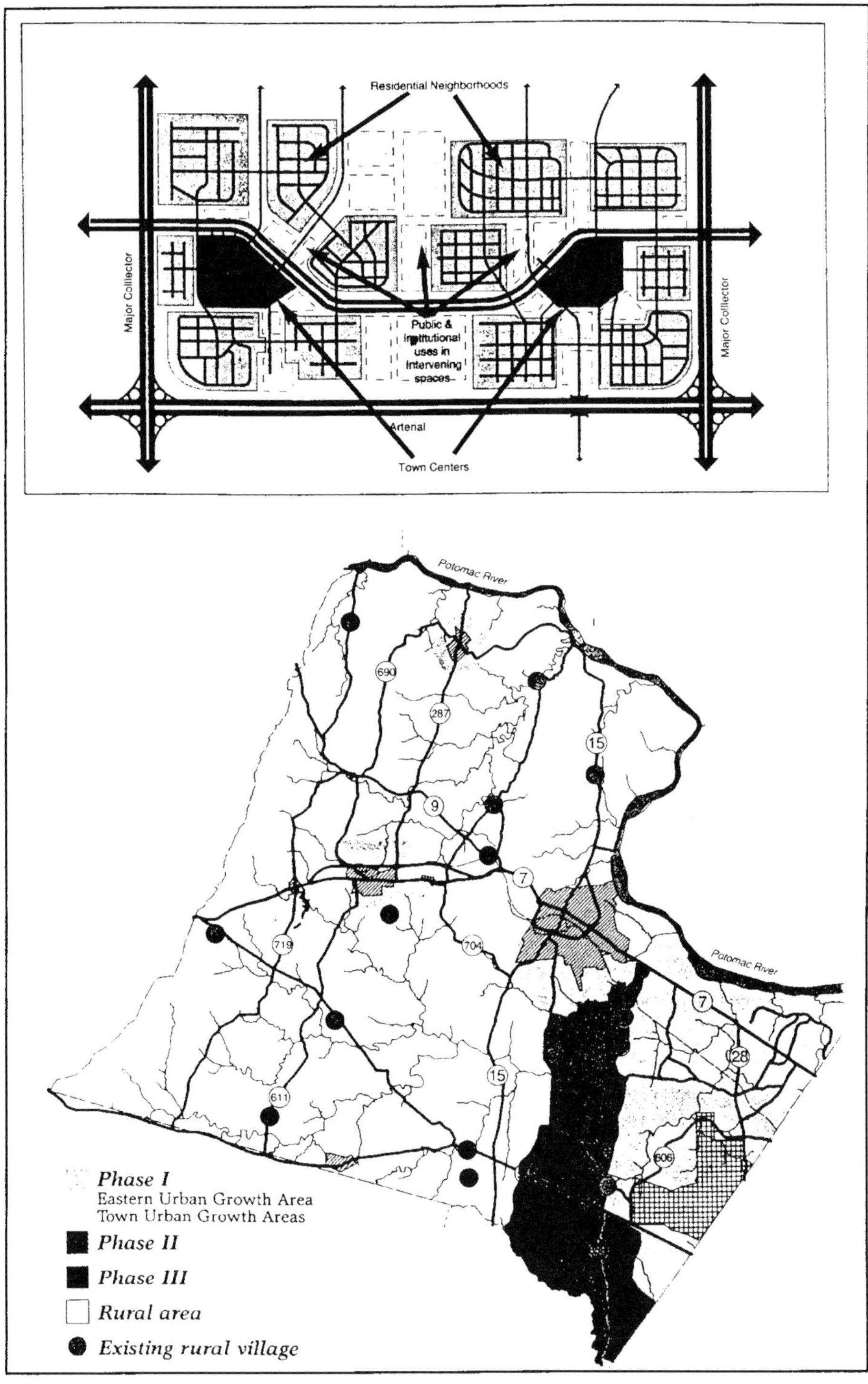

Figure 9 Neotraditional community schematic and generalized policy planning areas, Loudoun County, Virginia General Plan
Source: *Planning* 60, 3:10 (1994)

	Contemporary prototype plans					
Features of plans	1950s general plan	Land use design	Land classification plan	Verbal policy plan	Development management plan	1990s Hybrid design policy management plan
Land use maps	Detailed	Detailed	General	No	By growth areas	General *and* area specific
Nature of recommendations	General community goals	Land use policies & objectives	Growth locations & incentives	Variety of community policies	Specific management actions	Policy *and* actions
Time horizon	Long range	Long range	Long range	Intermediate range	Short range	Short *and* long range
Link to implementation	Very weak	Weak to moderate	Moderate	Moderate	Strong	Moderate to strong
Public participation	Pro-forma	Active	Moderate	Moderate	Active	Active
Capital improvements	Advisory	Recommended	Recommended	Recommended	Required	Recommended to required
Land use/transportation linkage	Moderate	Strong	Weak	Varies	Strong	Strong
Environmental protection	Weak	Moderate	Strong	Varies	Varies	Strong
Social policy linkage	Weak	Weak	Weak	Moderate to strong	Weak	Moderate

Table 2 Comparison of plan types

The land classification plan also still emphasizes mapping, but of development policy rather than policy about a pattern of urban land uses. Land classification is more specific about development management and environmental protection, but less specific about transportation, community facilities, and the internal arrangement of the future urban form. County and regional governments are more likely than are municipalities to use land classification plans.

The verbal policies plan eschews the spatial specificity of land use design and land classification plans and focuses less on physical development issues. It is more suited to regions and states, or may serve as an interim plan for a city or county while another type of plan is being prepared.

The development management plan represents the greatest shift from the traditional land use plan. It embodies a short-to-intermediate-range program of governmental actions for ongoing growth management rather than for long-range comprehensive planning.

In practice, these four types of plans are not mutually exclusive. Communities often combine aspects of them into a hybrid general plan that has policy sections covering environmental/social/economic/housing/infrastructure concerns, land classification maps defining spatial growth policy, land use design maps specifying locations of particular land uses, and development management programs laying out standards and procedures for guiding and paying for growth. Regardless of the type of plan used, the most progressive planning programs today regard the plan as but one part of a coordinated growth management program, rather than, as in the 1950s, the main planning product. Such a program incorporates a capital improvement program, land use controls, small area plans, functional plans, and other devices, as well as a general plan.

THE ENDURING LAND USE FAMILY TREE AND ITS FUTURE BRANCHES

For the first 50 years of this century, planning responded to concerns about progressive governmental reform, the City Beautiful, and the "City Efficient." Plans were advisory, specifying a future urban form, and were developed by and for an independent commission. By midcentury this type of plan, growing out of the design tradition, had become widespread in local practice. During the 1950s and 1960s the 701 program, T. J. Kent, and F. Stuart Chapin, Jr. further articulated the plan's content and methodology. Over the last 30 years, environmental and infrastructure issues have pushed planning toward growth management. As citizen activists and interest groups have taken more of a role, land use politics have become more heated. Planning theorists, too, have questioned the midcentury approach to planning, and have proposed changes in focus, process, subject matter, and format, sometimes challenging even the core idea of rational planning. As a result, practice has changed, though not to a monolithic extent and without entirely abandoning the traditional concept of a plan. Instead, at least four distinct types of plans have evolved, all descending more or less from the midcentury model, but advocating very different concepts of what a plan should be. With a kind of self-correcting common sense, the plans of the 1990s have subsequently incorporated the useful parts of each of these prototypes to create today's hybrid design/policy/management plans.

To return to our analogy of the plan's family tree: Roots for the physical development plan became well established during the first half of this century. By 1950, a sturdy trunk concept had developed. Since then, new roots and branches have appeared – land classification plans, verbal policy plans, and development management plans. Meanwhile development of the main trunk of the tree – the land use design – has continued. Fortunately, the basic gene pool has been able to combine with new genes in order to survive as a more complex organism – the 1990s design-policy management hybrid plan. The present family tree of planning reflects both its heredity and its environment.

The next generation of physical development plans also should mature and adapt without abandoning their heritage. We expect that by the year 2000, plans will be more participatory, more electronically based, and concerned with increasingly complex issues. An increase in participation seems certain, bolstered by interest groups' as well as governments' use of expert systems and computer databases. A much broader consideration of alternative plans and scenarios, as well as a more flexible and responsive process of plan amendment, will become possible. These changes will call upon planners to use new skills of consensus building and conflict management, as more groups articulate their positions on planning matters, and government plans and interest group plans compete, each backed by experts.

With the advent of the "information highway," plans are more likely to be drafted, communicated, and debated through electronic networks and virtual reality images. The appearance of plans on CD ROM and cable networks will allow more popular access and input, and better understanding of plans' three-dimensional consequences. It will be more important than ever for planners to compile information accurately and ensure it is fairly communicated. They will need to compile, analyze, and manage complex databases, as well as to translate abstract data into understandable impacts and images.

Plans will continue to be affected by dominant issues of the times: aging infrastructure and limited public capital, central city decline and suburban growth, ethnic and racial diversity, economic and environmental sustainability, global competition and interdependence, and land use/transportation/air quality spillovers. Many of these are unresolved issues from the last 30 years, now grown more complex and interrelated. Some are addressed by new programs like the Intermodal Surface Transportation Efficiency Act (ISTEA) and HUD's Empowerment Zones and Enterprise Communities. To cope with others, planners must develop new concepts and create new techniques.

One of the most troubling new issues is an attempt by conservative politicians (see the Private Property Protection Act of 1995 passed by the US House of Representatives) and "wise use" groups to reverse the precedence of the public interest over individual private property rights. These groups challenge the use of federal, state, and local regulations to implement land use plans and protect environmental resources when the result is any reduction in the economic value of affected private property. Should their challenge succeed and become widely adopted in federal and state law, growth management plans based on regulations could become toothless. Serious thinking by land use lawyers and planners would be urgently needed to create workable new implementation techniques, setting in motion yet another planning evolution.

We are optimistic, however, about the future of land use planning. Like democracy, it is not a perfect institution but works better than its alternatives. Because land use planning has adapted effectively to this century's turbulence and become stronger in the process, we believe that the twenty-first century will see it continuing as a mainstay of strategies to manage community change.

"Planning in the Face of Conflict"

Journal of the American Planning Association (1987)

John Forester

Editors' Introduction

Planning is never achieved without conflict. Planners, citizens, local elected officials, environmentalists, members of racial and ethnic groups, developers, and others invariably have different views on what a city should be like and how to plan and build it. In democracies, passions run high at important city planning commission meetings. One of the problems with the theoretical approach to early urban planning in the United Kingdom and United States was that it viewed planners as skilled technocrats who could produce good plans without much involvement with the people for whom they were planning. As Peter Hall (p. 373) and Nigel Taylor (p. 386) point out, the view that plans made in that way would be gratefully accepted by the public based on faith in the planner's superior knowledge proved unrealistic. Unless planners engage with the people they are planning for and confront the passions that different planning choices evoke, their plans will be deficient. Only a few urban planners comfortably bridge the gap between theory and practice. One theoretician who has directly confronted the conflictual nature of planning and has involved himself in mediating planning conflicts is Cornell planning professor John Forester.

Forester got down in the trenches with practicing city planners and other stakeholders involved in urban planning decisions to study what the practice of city planning is really like in the face of conflict. He has been involved in mediating planning conflicts in the area where he lives. The following selection summarizes what he learned about the planning process and his ideas on how planners can be effective in the face of conflict. It is valuable for the actual lessons Forester learned. But equally important, Forester helps point the way out of the impasse Peter Hall describes in much academic planning theory today (p. 373) between theorists who develop their ideas in a vacuum and practitioners who do not draw on planning theory to improve their work. Unlike ivory-tower theorists, Forester listened carefully to practicing city planners and learned from them. He learned from his practical experience mediating conflicts. Forester did not just write about conflict as an aspect of postmodern culture like David Harvey (p. 230) or throw up his hands in the face of conflict like Mike Davis (p. 195). Forester's work synthesizes what he learned and develops theoretical concepts that are highly relevant to planning practice.

Not surprisingly Forrester's ideas have been embraced by many practicing planners. Forester is notable among planners like Sherry Arnstein (p. 238) who believe citizen participation can be made to work. He has helped develop a body of knowledge about what British planning professor Patsy Healey calls collaborative planning.

Forester found that city planners use a variety of strategies to guide both developers and neighborhood residents through the complexities of the planning process. Successful planners handle conflicts through both formal and informal channels. Planners have to be attentive to timing. They must respond to complex and contradictory duties – tugged this way by local politicians, that way by legal mandates, and yet another way by citizen demands. Through all of this, successful city planners must be true to professional norms and hold fast to their own visions of high-quality city development. City planners who retreat to their offices to draw beautiful plans or create elegant computerized models of how they believe cities should develop without confronting the competing interest groups and conflicting ideas that all serious urban planning entails are doomed to fail. Planners need to wed professional expertise to practical skills for managing conflict among competing groups.

There are many lessons in Forester's work. People who want to be effective translating city plans into action need to expect opposition and should not be surprised or worn down by what often seems an endless and frustrating process. They need to be aware of their own power and also its limitations. They have to be sensitive to and understand the interests of the many different actors in the city development process.

Forester argues that city planners can self-consciously follow any of a number of strategies to keep projects on track and achieve success, as rule enforcers, negotiators and mediators, resource people, or shuttle diplomats.

Rule enforcers tell others in the city planning process what the law does and does not allow. They channel dialogue away from pie-in-the-sky discussions to options that are possible.

Negotiators and mediators use the kind of skills that family counselors employ with feuding domestic partners and labor mediators use trying to achieve consensus between labor unions and employers. This involves listening carefully to each side's demands, trying to make each side see what the other side is trying to accomplish, and getting both sides to compromise. Forester's analogy to shuttle diplomats invokes an image of the United Nations special envoy for the Mideast meeting with Palestinian leaders in Gaza to hear their views on a truce with Israel, then shuttling to Tel Aviv to share what they have learned, listen to Israeli views, pressure the Israelis to bend a little, and then shuttling back to Gaza to tell the Palestinians what they have learned, pressure them to bend a little, and listen to their views anew.

As resource people planners can provide information and interpretations that help with decision-making. For example if citizens want to build a green development, but do not know how to go about it, a planner who is familiar with solar panels, graywater systems, photovoltaic cells, composting, natural drainage and any of a host of other green planning concepts may inject ideas into the planning process for them to consider. This will empower citizens to make more informed choices based on information rather emotion.

Compare Forester's insights with David Harvey's description of social conflict around urban spatial issues (p. 230). Consider the kind of conflicts you would expect if you were trying to implement the different types of plans that Edward J. Kaiser and David R. Godschalk describe (p. 399).

John Forester (b. 1948) is a professor of city and regional planning at Cornell University. He is interested in the ways planners shape participatory processes and manage public disputes in diverse settings, planning ethics, and mediation. He is a mediator for the Community Dispute Resolution Center of Tompkins County, New York and has consulted on mediating urban planning disputes for the Consensus Building Institute, a Cambridge, Massachusetts nonprofit organization that seeks to improve the theory and practice of public consensus building and conflict resolution through training, capacity building, facilitation and mediation, and research.

Forester is the author of *Dealing with Differences: Dramas of Mediating Public Disputes* (New York: Oxford, 2009), *The Deliberative Practitioner: Encouraging Participatory Planning Processes* (Cambridge, MA: MIT Press, 1999), *Critical Theory, Public Choice, and Planning Practice: Towards a Critical Pragmatism* (Buffalo, NY: State University of New York Press, 1993) and *Planning in the Face of Power* (Berkeley, CA: University of California Press, 1988). He is the coauthor with former Cleveland city planning director Norman Krumholz of *Making Equity Planning Work* (Philadelphia, PA: Temple University Press, 1992), an account of Krumholz's experience implementing socially responsible equity planning in Cleveland. Forester is also the coauthor of *Israeli Planners and Designers: Profiles of Community Builders*, with Raphael Fischler and Deborah Shmueli (Albany, NY: SUNY Press, 2001) and the editor of *Critical Theory and Public Life* (Cambridge, MA: MIT Press, 1987), a book of essays about Jürgen Habermas's critical communications theory of society.

Another important article by Forester on how urban planners manage conflict is "Planning in the Face of Power," which appeared in the *Journal of the American Planning Association*, 48(1) (1982).

Other books on what city planning is actually like include Allan Jacobs, *Making City Planning Work* (Chicago, IL: American Society of Planning Officials, 1976) and Bruce W. McClendon (ed.), *Planners on Planning* (San Francisco, CA: Jossey-Bass, 1996). A description of how Norman Krumholz tried to make equity planning work as Cleveland, Ohio's planning director is Norman Krumholz, *Making Equity Planning Work* (Philadelphia, PA:

Temple University Press, 1990). Books on resolving urban planning conflicts include Patrick Field and Lawrence Susskind, *Dealing with an Angry Public: The Mutual Gains Approach to Resolving Disputes* (New York: Free Press, 1996), Lawrence Susskind and Jeffrey Cruikshank, *Breaking the Impasse: Consensual Approaches to Resolving Public Disputes* (New York: Basic Books, 1989), and Lawrence Susskind, Sarah McKearnan, and Jennifer Thomas-Larmer, *The Consensus Building Handbook: A Comprehensive Guide to Reaching Agreement* (Thousand Oaks, CA: Sage, 1999).

Patsy Healey, *Collaborative Planning: Shaping Places in Fragmented Societies* (Basingstoke, UK: Palgrave Macmillan, 2006), and Judith Innes and David Booher, *Planning with Complexity: An Introduction to Collaborative Rationality for Public Policy* (London: Routledge, 2010) are leading books on collaborative planning. Other books on collaborative planning include Robert J. Mason, *Collaborative Land Use Management: The Quieter Revolution in Place-Based Planning* (Lanham, MD: Rowman & Littlefield, 2008), and John McCarthy, *Partnership, Collaborative Planning and Urban Regeneration* (Aldershot, UK: Ashgate, 2007). The National Charrette Institute has produced a how-to-do-it book titled *The Charrette Handbook: The Essential Guide for Accelerated, Collaborative Community Planning* (Chicago, IL: American Planning Association, 2006).

In the face of local land-use conflicts, how can planners mediate between conflicting parties and at the same time negotiate as interested parties themselves? To address that question, this article explores planners' strategies to deal with conflicts that arise in local processes of zoning appeals, subdivision approvals, special permit applications, and design reviews.

Local planners often have complex and contradictory duties. They may seek to serve political officials, legal mandates, professional visions, and the specific requests of citizens' groups, all at the same time. They typically work in situations of uncertainty, of great imbalances of power, and of multiple, ambiguous, and conflicting political goals. Many local planners, therefore, may seek ways both to negotiate effectively, as they try to satisfy particular interests, and to mediate practically, as they try to resolve conflicts through a semblance of a participatory planning process.

But these tasks – negotiating and mediating – appear to conflict in two fundamental ways. First, the negotiator's interest in the subject threatens the independence and the presumed neutrality of a mediating role. Second, although a negotiating role may allow planners to protect less powerful interests, a mediating role threatens to undercut this possibility and thus to leave existing inequalities of power intact. How can local planners deal with these problems? I discuss their strategies in detail below.

This article first presents local planners' own accounts of the challenges they face as simultaneous negotiators and mediators in local land-use permitting processes. Planning directors and staff in New England cities and towns, urban and suburban, shared their viewpoints with me during extensive open-ended interviews. The evidence reported here, therefore, is qualitative, and the argument that follows seeks not generalizability but strong plausibility across a range of planning settings.

The article next explores a repertoire of mediated negotiation strategies that planners use as they deal with local land-use permitting conflicts. It assesses the emotional complexity of mediating roles and asks: What skills are called for? Why do planners often seem reticent to adopt face-to-face mediating roles?

Finally, the article turns to the implications of these discussions. How might local planning organizations encourage both effective negotiation and equitable, efficient mediation? How might mediated-negotiation strategies empower the relatively powerless instead of simply perpetuating existing inequalities of power?

ELEMENTS OF LOCAL LAND-USE CONFLICTS

Consider first the settings in which planners face local permitting conflicts. Private developers typically propose projects. Formal municipal boards – typically planning boards and boards of zoning appeals – have decision-making authority to grant variances, special permits, or design approvals. Affected residents often have a say – but sometimes little influence – in formal public hearings before these boards. Planning staff report to these boards with analyses of specific proposals. When the reports are positive, they often recommend conditions to attach to a permit or suggest design changes to improve the final project. When the reports are negative, there are arguments to be made, reasons to be given.

Some municipalities have elected permit-granting boards; some have appointed boards. Some municipal ordinances mandate design review; others do not. Some local by-laws call for more than one planning board hearing on "substantial" projects, but others do not. Nevertheless, for several reasons, planners' roles in these different settings may be more similar than dissimilar.

Common planning responsibilities

First, planners must help both developers and neighborhood residents to navigate a potentially complex review process; clarity and predictability are valued goods. Second, the planners need to be concerned with timing. When a developer or neighborhood resident is told about an issue may be even more important than the issue itself. Third, planners typically need to deal with conflicts between project developers and affected neighborhood residents that usually concern several issues at once: scale, the income of tenants, new traffic, existing congestion, the character of a street, and so on. Such conflicts simultaneously involve questions of design, social policy, safety, transportation, and neighborhood character as well. Fourth, how much planners can do in the face of such conflicts depends not only upon their formal responsibilities, but also upon their informal initiatives. A zoning by-law, for example, can specify a time by which a planning board is to hold a public hearing, but it usually will not tell a planner how much information to give a developer or a neighbor, when to hold informal meetings with either or both, how to do it, just whom to invite, or how to negotiate with either party. So within the formal guidelines of zoning appeals, special permit applications, site plan and design reviews, planning staff can exercise substantial discretion and exert important influence as a result.

Planners' influence

The complexity of permitting processes is a source of influence for planning staff. Complexity creates uncertainties for everyone involved. Some planners eagerly use the resulting leverage, as an associate planning director explains, beginning with a truism but then elaborating:

> Time is money for developers. Once the money is in, the clock is ticking. Here we have some influence. We may not be able to stop a project that we have problems with, but we can look at things in more or less detail, and slow them down. Getting back to [the developers] can take two days or two months, but we try to make it clear, "We're people you can get along with." So many developers will say, "Let's get along with these people and listen to their concerns . . ."

He continues,

> But we have influence in other ways too. There are various ways to interpret the ordinance, for example. Or I can influence the building commissioner. He used to work in this office and we have a good relationship . . . his staff may call us about a project they're looking over and ask, "Hey, do you want this project or not?"

Planners think strategically about timing not only to discourage certain projects but to encourage or capture others. The associate director explains,

> On another project, we waited before pushing for changes. We wanted to let the developer get fully committed to it; then we'd push. If we'd pushed earlier, he might have walked away . . .

A director in another municipality echoes the point:

> Take an initial meeting with the developer, the mayor, and me. Depending on the benefits involved – fiscal or physical – the mayor might kick me under the table; "Not now," he's telling me. He doesn't want to discourage the project . . . and so I'll be able to work on the problems later . . .

For the astute, it seems, the complexity of the planning process creates more opportunities than headaches. For the novice, no doubt, the balance shifts the other way.

But isn't everything, in the last analysis, all written down in publicly available documents for everyone to see? Hardly. Could all the procedures ever be made entirely clear? Consider the experience of an architect planner who grappled with these problems in several planning positions. The following conversation took place toward the end of my interview with this planner. The planner pulled a diagram from a folder and said, "Here's the new flow-chart I just drew up that shows how our design review process works. If you have any questions, let's talk. I think it's still pretty cryptic."

"If you think it's cryptic," interjected the zoning appeals planner, who was standing nearby and had overheard this, "just think what developers and neighborhood people will think!"

Both planners shook their heads and laughed, since the problem was all too plain: the arrows on the design review flow chart seemed to run everywhere. The chart was no doubt correct, but it did look complicated.

I recalled my first interview with the zoning appeals planner. Probing with a deliberately leading question, I had asked, "But what influence can you have in the process if everything's written down as public information, if it's all clear there on the page?"

The zoning appeals planner had grinned: "But that's just it! The process is not clear! And that's where I come in . . ." The architect-planner developed the point further:

> Where I worked before, the planning director wanted to adopt a new "policy and procedures" document that would have every last item defined. We were going to get it all clear. The whole staff spent a lot of time writing that, trying to get all the elements and subsections and so on clearly defined . . . But it was chaos. Once we had the document, everyone fought about what each item meant . . .

So clarity, apparently, has its limits!

Different actors, different strategies

Planning staff point almost poignantly to the different issues that arise as they work with developers and neighborhood residents. The candor of one planning director is worth quoting at length:

> It's easy to sit down with developers or their lawyers. They're a known quantity. They want to meet. There's a common language – say, of zoning – and they know it, along with the technical issues. And they speak with one voice (although that's not to say that we don't play off the architect and the developer at times – we'll push the developer, for example, and the architect is happy because he agrees with us) . . .
>
> But then there's the community. With the neighbors, there's no consistency. One week one group comes in, and the next week it's another. It's hard if there's no consistent view. One group's worried about traffic; the other group's not worried about traffic but about shadows. There isn't one point of view there. They also don't know the process (though there are cases where there are too many experts).
>
> So at the staff level (as opposed to planning board meetings) we usually don't deal with both developers and neighbors simultaneously.

Although these comments may distress advocates of neighborhood power, they say much about the practical situation in which the director finds himself.

All people may be created equal, but when they walk into the planning department, they are simply not all the same. This director suggests that getting all the involved parties together around the table in the planners' conference room is not an obviously good idea, for several reasons. (It is, however, an idea we shall consider more closely below.)

First, the director suggests, planners generally know what to expect from developers; the developers' interests are often clearer than the neighbors', and project proponents may actually want to meet with the staff. Neighborhood residents may be less likely to treat planners as potential allies; after all, the planners are not the decision makers, and the decision makers can often easily ignore the planners' recommendations. Because developers may cultivate good relations with planning staff (this is in part their business, after all), while neighborhood groups do not, local planning staff may find meetings with developers relatively cordial and familiar, but meetings with neighborhood activists more guarded and uncertain.

Second, the planning director suggests that planners and developers often share a common professional language. They can pinpoint technical and regulatory issues and know that both sides understand what is being said. But on any given project, he implies, he may need to teach the special terms of the local zoning code to affected neighbors before they can really get to the issues at hand.

The planning director makes a third point. Developers speak with one voice; neighbors do not. When planners listen to developers talk, they know whom they're listening to, and they know what they're likely to hear repeated, elaborated, defended, or qualified next week. When planners listen to neighborhood residents, though, this director suggests, they can't be so sure how strongly to trust what they hear. "Who really speaks for the neighborhood?" the director wonders.

Planners must make practical judgments about who represents affected residents and about how to interpret their concerns. This director implies, therefore, that until planners find a way to identify "the neighborhood's voice," the problems of conducting joint mediated negotiations between developers and neighbors are likely to seem insurmountable. We return to this issue of representation below.

Inequalities of information, expertise, and financing

What about imbalances of power? Developers, typically, initiate site developments. Planners respond. Neighbors, if they are involved at all, then try to respond to both. Developers have financing and capital to invest; neighbors have voluntary associations and not capital, but lungs. Developers hire expertise; neighborhood groups borrow it. Developers typically have economic resources; neighbors often have time, but not always the staying power to turn that time into real negotiating power.

Where power relations are unbalanced, must mediated negotiation simply lead to coopting the weaker party? No, because, as we shall see below, mediated negotiation is not a gimmick or a recipe; it is a practical and political strategy to be applied in ways that address the specific relations of power at hand.

When either developers or neighborhood groups are so strong that they need not negotiate, mediated negotiation is irrelevant, and other political strategies are more appropriate. But when both developers and neighbors want to negotiate, planners can act both as mediators, assisting the negotiations, and as interested negotiators themselves. But how is this possible? What strategies can planners use?

PLANNERS' STRATEGIES: SIX WAYS TO MEDIATE LOCAL LAND-USE CONFLICTS

Consider the following six mediated-negotiation strategies that planning staff can utilize in the face of local land-use conflicts. They are *mediated* strategies because planners employ them to assure that the interests of the major parties legitimately come into play. They are *negotiation* strategies because (except for the first) they focus attention on the informal negotiations that may produce viable agreements even before formal decision-making boards meet.

Strategy one: The facts! The rules! (The planner as regulator)

The first strategy is a traditional response, pristine in its simplicity, but obviously more complex in practice. A young planner who handles zoning appeals and design review says:

> I see my role often as a fact finder so that the planning board can evaluate this project and form a recommendation; whether it's design review, special permits, or variances, you still need lots of facts . . .

Here of course is the clearest echo of the planner as technician and bureaucrat; the planner processes information and someone else takes responsibility for making decisions. But the echo quickly fades. A moment later, this planner continues,

> Our role is to listen to the neighbors, to be able to say to the board, "Okay, this project meets the technical requirements but there will be impacts . . ." The relief will usually then be granted, but with conditions . . . We'll ask for as much in the way of conditions as we think necessary for the legitimate protection of the neighborhood. The question is, is there a legitimate basis for complaint? And it's not just a matter of complaint, but of the merits.

This planner's role is much more complex than that of fact finder; it is virtually judicial in character. He implies, essentially, "I'm not just a bureaucrat, I'm a professional. I need to think not only about the technical requirements, but about what's legitimate protection for the neighbors. Now I have to think about the merits!" Thinking about the merits, though, does not yet mean thinking about politics, the feelings of other agencies, the chaos at community meetings – it means making professional judgments and then recommending to the planning board the conditions that should be attached to the permits.

Consider now a slightly more complex strategy.

Strategy two: Pre-mediate and negotiate – representing concerns

When developers meet with planners to discuss project proposals, neighborhood representatives rarely join them. Yet planners might nevertheless speak *for* neighborhood concerns as well as *about* them. A planning director in a municipality where neighborhood groups are well-organized, vocal, and influential notes,

> We temper our recommendations to developers. While we might accept A, the neighbors want D, and so we'll tell the developers to think about something in the middle – if they can make it work.

Here, the planner anticipates the concerns of affected residents and changes the informal staff recommendation accordingly to search for an acceptable compromise with the developers. He explains,

> What we do is premediate rather than mediate after the fact. We project people's concerns and then raise them; so we do more before [explicit conflict arises] . . . The only other way we step in and mediate, later, is when we support changes to be made in a project, changes that consider the neighbors' views; but that's later, after the public hearing . . .

Unlike the planner-regulator quoted above, this planning director relies on far more than his professional judgment when he meets with a developer. He will negotiate to reach project outcomes that satisfy local statutes, professional standards, and the interests of affected residents as well. His calculation is not only judicial, but explicitly political. He anticipates the concerns of interested community members. So he seeks to represent neighborhood interests – without neighborhood representatives.

Such premediation – articulating others' concerns well before they can erupt into overt conflict – involves a host of political, strategic, and ethical issues. What relationships does the planner have with neighborhood groups? In what senses can the planner "know what the community wants"? To which "key actors" might the planner "steer" the developer? How much information and how much advice should the planner give, or withhold?

Such questions arise whether or not project developers ever meet with neighbors. In many cases, where "neighborhoods" are sprawling residential areas, and where "the interests of the neighbors" seem most difficult to represent through actual neighborhood representatives, the planners' premediation may be the only mediation that takes place.

Strategy three: Let them meet – the planner as a resource

The planner's influence might be used in still other ways. The director continues:

> Regardless of how our first meeting with a developer goes, we recommend to them that they meet with neighbors and the neighbors' representatives [on the permit-granting board]. We usually can give the developer a good inkling about what to expect both professionally and politically. The same elected representative might say that a project is "okay" professionally, but not "okay" for them in their elected capacity. We try to encourage back and forth meetings . . .

The director, then, regularly takes the pulse of neighborhood groups and elected representatives. Working in city hall has its advantages: "We'll discuss a project with the representatives; we see them so much here, just in the halls, and they ask us to let them know what's happening in their parts of the city." So the director listens to the developers, listens to the neighbors, and "encourages back and forth meetings."

A planning director who seldom met jointly with neighbors and developers had an acute sense of other strategies he used:

> We . . . urge the applicants, the developers, to deal directly with the neighborhood for several reasons: First, if the neighbors are confronted at a hearing with glossy plans, they'll think it's all a *fait accompli*;

> so they'll just adopt the "guns blazing, full charge ahead" strategy, since they think it'll just be a "yea" or "nay" decision. Second, we tell them to talk to the neighbors since if they can come up with something that the neighbors will "okay," it'll be easier at the board of appeals. Third, we try to get them to meet one on one, or maybe as a group, but in as deinstitutionalized a way as possible, informally. We try to get the developers to sell their case that way; it'll get a much better hearing than at the big formal public hearings.

But why should planners be reluctant to convene joint negotiating sessions between developers and neighbors, yet still be willing to encourage both parties to meet on their own? Why don't these planners embrace opportunities to mediate local land-use conflicts face to face? One planner could hardly imagine such a mediating role:

> Work as a neutral between developers and neighbors? I don't know how I'd approach it. I'd just answer questions, suggest what could be done, and so on. That's what our role should be – although we should reach compromises between developers and neighbors. But we have to work within the rules – that's my reference point – to say what the rules of the game are; that's the job.

This planner's image of a "neutral" between disputing parties is less that of a mediator facilitating agreement than it is of a referee in a boxing match. The referee assures that the rules are followed, but the antagonists might still kill each other. No wonder planners might find this image of mediation unattractive!

A senior planner envisions further complications:

> If I could be assured I could be wholly independent, then I could mediate – but I still have to pay my bills . . . The planning department always has some vested interests, as much as we try to stay objective, independent . . . I work for a mayor, for the elected representatives, for 14 committees . . . So there's always the question of compromise on my part: if the mayor says, "Tell me how to make this project work," for example. It took me a long time before I was able to say, "I'm going to have to say no." We have a very strong mayor . . .

Strategy four: Perform shuttle diplomacy – probe and advise both sides

A planning director proposes another way to facilitate developer–neighbor negotiations:

> I feel more comfortable in shuttle diplomacy, if you will; trying to get the neighbors' concerns on the table, to get the developers to deal with them . . . I'd rather bounce ideas off each side individually than be caught in the middle if they're both there. If both sides are there, I'm less likely to give my own ideas than if I'm alone with each of them.

Shuttle diplomacy, this director suggests, allows planners to address the concerns of each party in a professionally effective way. He explains:

> If I'm with the developer, I feel I can make a much more extreme proposal – "knock off three stories" – but I wouldn't dare say that if neighbors were there. The neighbors would be likely to pick up and run with it, and it could damage the negotiations rather than help them . . . I'm willing to back off on an issue if the developer has a good argument, but the neighborhood might not, and then they might use my point as a club to hit the developer with: "Well, the planning director suggested that; it must be a good idea" . . . and then I can't unsay it . . .

This planning director is as concerned about how his suggestions, proposals, queries, and arguments will be understood and used as he is about what ought to be altered in the project at hand. He recognizes clearly that when he talks he acts politically and inevitably fuels one argument or another. He not only conveys information in talking, but he acts practically, influentially. He focuses attention on specific problems, shapes future agendas, legitimates a point of view, and suggests lines of further argument.

The director continued,

> I might not want to concede to a developer that there won't be a traffic problem, because I want to push him to relieve a problem or a perceived problem . . . but I could say to the neighbors aside, "Look, this will be no big deal; it'll be five trips, not fifty." I can say that in a private meeting, but in a public meeting if I say it to a neighborhood representative I'm insulting him, even if the developer

> snickers silently . . . So I lose my ability to be frank with both sides if we're all together. Not that this should be completely shuttle diplomacy, but it has its place.

These comments suggest that planning staff can certainly mediate conflicts in local permitting processes, if not in ways that mediators are thought typically to act. The planners may not be independent third parties who assist developers and neighbors in face-to-face meetings to reach development agreements – but they might still mediate such conflicts as "shuttle diplomats."

Strategy five: Active and interested mediation – thriving as a non-neutral

We can consider a case that involves not a zoning appeal but a rezoning proposal. One planner, who had earlier worked as a community organizer, had convened a working group of five community representatives and five local business representatives to draft a rezoning proposal for a large stretch of the major arterial street in their municipality. She considers her work on that project a kind of mediation and reflects about how she as a planner acts as a mediator, dealing with substantive and affective issues alike:

> Am I in a position of having to think about everyone's interest and yet being trusted by no one? Sure, all the time. But I've been in this job for seven years, and I have a reputation that's good, fortunately . . . Trust is an issue of your integrity and planning process. I talk to people a lot; communication is a big part of it . . . My approach is to let people let off steam – let them say negative things about other people to me, and then in a different conversation at another time, I'll be sure to say something positive about that person – to try to let them feel that they can say whatever they want to me, and to try to confront them with the fact that the other person isn't just out to ruin the process. But I'd do that in another conversation; I let them let off steam if they're angry.

This planner is well aware that distrust on all sides is an abiding issue, so she tries to build trust as she works. She works to assure others that she will listen to them and more; that she will acknowledge and respect their thoughts and feelings, whatever they have to say. She pays attention first to the person, then to the words. Then, as she establishes trust with her committee members and with others, she can also make sure, carefully, that real evidence is not ignored.

She realizes that anger makes its own demands, so she responds with an interested patience. She seeks throughout to mediate the conflicting interests of the groups with whom she's working:

> I also make a point to tell each side the other's concerns – categorically, not with names, but all the other sides' concerns . . . Why's that important? I like to let people anticipate the arguments and prepare a defense, either to stand or fall on its own merits. For people to be surprised is unfortunate. It's better to let people know what's coming so they can build a case. They can hear an objection, if you can retain credibility, and absorb it; but in another setting they might not be able to hear it . . . If they hear an objection first as a surprise, you're likely to get blamed for it. If concerns are raised in an emotional setting, people concentrate more on the emotion than on the substance. This is a concern of mine. In emotional settings, lots gets thrown out, and lots is peripheral, but possibly also central later . . .

This planner is keenly aware that emotion and substance are interwoven, and that planners who focus only upon substance and try to ignore or wish away emotion do so at their own practical peril. Yet she is saying even more.

She knows that in some settings disputing groups can hear objections, understand the points at stake, and address them, while in other settings those points may be lost. She tries to present each side's concerns to the other so that they can be understood and addressed. Anticipating issues is central; learning of important objections late in the process will be mostly emotionally and financially, and planning staff are likely to share the blame. "Why didn't you tell us sooner . . .?" the refrain is likely to sound.

Consider next, then, this planner-mediator's thoughts about the sort of mediation role she is performing. She continues,

> But what I do is different from the independent mediator model. In a job like mine, you have an on-going relationship with parties in the city. You have more information than a mediator does about

> the history of various individuals, about participating organizations, about the political history of city agencies, and so on. You also have a vested interest in what happens. You want the process to be credible. You want the product to be successful; in my case I want the city council to adopt the committee's proposal. And you're invested . . . both professionally and emotionally. And then you have an opinion about particular proposals; you're a professional, you should have one, you should be able to look at a proposal and have an opinion.

Thus, she suggests, mediation has its place in local land-use conflicts, but the "rules of the game" will not be those that labor mediators follow. Indeed, planners who now mediate local land-use conflicts are not waiting for someone else to write the rules of the game, they are writing them themselves.

Strategy six: Split the job – you mediate, I'll negotiate

Consider finally a planning strategy that promotes face-to-face mediation with planning staff at the table – but as negotiators or advisors, not as mediators. A planning director explains:

> There's another way we deal with these conflicts; we might involve a local planning board member. For example, if there's a sophisticated neighborhood group that's well organized, we've brought in an architect from the board who's as good with words as he is with his pencil . . . The chair of the board might ask the board member to be a liaison to the neighborhood, say, and sometimes he'll talk just to the neighbors, sometimes with both . . .

Here the "process manager" comes from the planning board with highly developed "communications skills." How does the planner feel in these situations?

> It's more comfortable from my point of view, and the citizens', to have a board member in the convening role. I'm still a hired hand. It seems more appropriate in a negotiating situation to have a citizen in that role and not an employee . . . Since they've come from the neighborhoods, a board member is in a better position to bring neighbors and developers together – if they behave properly. Some board members are good communicators; some are more dynamic than others in pressing for specific solutions.

This planner identifies so strongly with the professional and political mandate of his position that he cannot imagine a role as neutral convener or mediator of neighborhood–developer negotiations. But that does not prevent mediation; it means rather that the planner retains a substantively interested posture while another party, here a planning board member, convenes informal, but organized, project negotiations between developers and neighbors. This planner's example makes the point:

> Take the example of the Mayfair Hospital site. The hospital was going to close, and the neighbors and the planning board were concerned about what might happen with the site. So Jan from the planning board got involved with the hospital and the neighborhood to look at the possibilities. Both the neighbors and the hospital set up re-use committees, and Jan and I went to the meetings. There was widespread agreement that the best use of the site would be residential – the neighbors definitely preferred that to an institutional use – but then there was a lot of haggling over scale, density, and so on. Ultimately, a special zoning district was proposed that included the site; the neighbors supported it, and it went to [the elected representatives] where they voted to rezone the several acres involved . . .

When local planners feel they cannot mediate disputes themselves, then one strategy may be to search for informal, most likely volunteer, mediators. These ad hoc mediators might be "borrowed" from respected local institutions, and their facilitation of meetings between disputing parties might allow planning staff to participate as professionally interested parties concerned with the site in question.

Table 1 summarizes the six approaches presented. Together, these approaches form a repertoire of strategies that land-use planners can use to encourage mediated negotiations in the face of conflicts in local zoning, special permit, and design review processes. To refine these strategies, local planning staff can build upon several basic theories and techniques of conflict resolution. Consider now the distinctive competences and sensitivities required by these strategies.

1 The Facts! The Rules! (The Planner as Regulator)
2 Pre-Mediate and Negotiate: Representing Concerns
3 Let Them Meet: The Planner as a Resource . . .
4 Perform Shuttle Diplomacy: Probe and Advise Both Sides
5 Active and Interested Mediation: Thriving as a Non-Neutral
6 Split the Job: You Mediate, I'll Negotiate

Table 1 A repertoire of mediated-negotiation strategies used by local land-use planners in permitting processes

THE EMOTIONAL COMPLEXITY OF MEDIATED-NEGOTIATION STRATEGIES

More than a lack of independence keeps planners from easily adopting roles as mediators. The emotional complexity of the mediating role makes quite different demands upon planners than those that they have traditionally been prepared to meet. The community-organizer-turned-planner makes the point brilliantly:

> In the middle, you get all the flak. You're the release valve. You're seen as having some power, and you do have some . . .Look, if you have a financial interest in a project, or an emotional one, you want the person in the middle to care about your point of view, and if you don't think they do, you'll be angry!
>
> ["So when planners try to be professional by appearing detached, objective, does it get people angry at them?" I asked.]
>
> Sure!

This comment cuts to the heart of planners' professional identities. Must "professional," "objective," and "detached" be synonymous? If so, this planner suggests, then planners' own striving for an independent professionalism will fuel the anger, resentment, and suspicion of the same people those planners presume to serve!

Thus we can understand the caution with which a planner speaks of his way of handling emotional participants in public hearings:

> How do I deal with people's anger? I try to keep cool, but occasionally I get irritated. But that's how we're expected to behave, to be rational. It's all right for citizens to be irrational, but not the staff!

How does one keep cool, be rational, and still respond to the claims of an emotional public at formal hearings? This planner elaborates:

> It's one thing to begin the discussion of a project [to present our analysis] and anticipate problems. But it's another thing to *rebut* a neighborhood resident in public in a gentle way . . . Part of the problem is that if you antagonize people it'll haunt you in the future . . . We're here for the long haul, and we have to try to maintain our credibility . . .

The planner's problem here is precisely *not* the facts of the case: the facts themselves may be clear enough. But how should the planner present the analysis that he feels must be made and how should he decide which arguments to make and which to hold back at a given time?

> The biggest problem I have in the board meetings is when to respond and when to keep quiet. In a hearing, for example, I can't possibly respond to all the accusations and issues that come up. So I have to pick a direction, to deal with a generally felt concern. It's just not effective to enter into a debate on each point in turn; it's better to clarify things, to explain what's misunderstood . . .

This planner does much more than simply recapitulate facts. He tries to avoid an adversarial posture, even when he feels the situation is quite conflictual. He listens as much to the individuals and their concern as he does to each point. He knows that points and demands and positions may change as issues are clarified, but that if he cannot respond to people's concerns, he's in some trouble. Because he and his staff are there "for the long haul," he wants to be able to work with neighbors, community leaders, and elected representatives alike not just now but in the future as well. How he relates to the parties involved in local disputes, he suggests, is as important as what he has to say.

Another planner points to the skills involved:

> Whom would I try to hire to deal with such conflicts? I'd look for someone who's a careful listener, someone who's good at explaining a position

SIX

coherently, succinctly, quietly, in a calm tone . . . someone who could hear a point, understand it whether he or she agreed with it or not, and then verbalize a clear, concise response. Most people though – myself included – try to jump the gun and answer before it's appropriate. So I want someone who's able to stay cool and stay on the issues . . .

A community development director first mentions "a good listener" and then elaborates:

> [To deal with these conflicting situations I'd want to hire staff] who won't say, "I know best," who won't get people's backs up just by their style. I'd want someone with some openness, with a sense of how things work who won't accept everything, but who won't offend people. They have to have critical judgment – to leave doors open, to give people a sense of involvement and a sense of the feasible – [someone who] can't be convinced of something that's not likely to work, just for the sake of getting agreement . . .

This planning director also points to the balance necessary between what planners say and how they say it. The "how" counts; he doesn't want staff who will "get people's backs up," "offend people," and not communicate an openness to others' concerns. Nor does he want someone who will sacrifice project viability for the temporary comfort of agreement. He asks for substantive judgment and the skills to manage a process.

Referring to the demands of working and negotiating with developers as they navigate the approval process, the director stresses the role of diplomacy:

> We [planners] have access to information, to resources, to skills . . . so developers usually want to work with us. They have certain problems getting through the process . . . so we'll go to them and ask, "What do you want?" and we'll start a process of meetings . . . It's diplomacy; that's the real work. You have to have the technical skill . . . but that's the first 25 percent. The next 75 percent is diplomacy, working through the process.

Percentages aside, the point remains. To the extent that planning practitioners and educators focus predominantly upon facts, rules, likely consequences, and mitigation measures, they may fail to attend to the pressing emotional and communicative dimensions of local land-use conflicts. Because the planning profession has not traditionally embraced the diplomat's skills, it should surprise no one that practicing planners envision mediating roles with more reticence than relish.

In the next section, we turn to administrative and political questions. What, initially, can be done in planning organizations to improve planners' abilities to mediate local land-use negotiations successfully? What about imbalances of power?

ADMINISTRATIVE IMPLICATIONS FOR PLANNING ORGANIZATIONS

What does this analysis imply for policymakers and planners who wish to build options for mediation into local review processes? Mediation may offer several opportunities, under conditions of interdependent power: a shift from adversarial to collaborative problem-solving; voluntary development controls and agreements; improved city–developer–neighborhood relationships enabling early and effective reviews of future projects; more effective neighborhood voice; and joint gains ("both gain" outcomes) for the municipality, neighbors, and developers alike. Such opportunities present themselves *only* when no single party is so dominant that it need not negotiate at all, that it is likely simply to get what it wants in any case.

Planners already use the strategies reviewed in diverse settings. Which strategy a planner uses, and at which times, depends largely on practical judgment: What skills does the planner have? How willing are developers or neighbors, or other agency staff, to meet jointly? Does enough time exist to allow early, joint meetings? Are the practical and political alternatives of any one party so attractive that they see no point in mediated negotiations?

No strategy is likely to be desirable in all circumstances, so no one approach will provide the model to formalize into new zoning or permitting procedures. But to say that we should not formalize these strategies does not mean that we cannot regularly use them. How, then, can planners apply the mediated-negotiation strategies in local zoning, permitting, and design review processes?

First, planning staff must distinguish clearly the two complementary but distinct mandates they typically must serve: to press professionally, and thus to negotiate, for particular substantive goals (design quality or affordable housing, for example), and to enable a participatory process that gives voice to affected parties; thus, like mediators, to facilitate negotiations between disputants.

Second, planning staff need to adopt, administratively if not formally, a goal of supplementing (not substituting for) formal permitting processes with mediated negotiations: attempting to craft workable and voluntary tentative agreements before formal hearing dates.

Third, planning staff should examine each of the strategies reviewed here. They need to determine how each could work, given the size of their agency, their zoning and related by-laws, the political and institutional history of elected officials, neighborhood groups, and other agencies. Planning staff must ask which skills and competencies they need to develop to employ each of these strategies appropriately.

Fourth, planning staff must be able to show others – developers, neighborhood groups, public works department staff, elected and appointed officials – how and when mediated negotiations can lead to "both gain" outcomes and so improve the local land-use planning and development process. Planners also have to be clear about what mediated negotiation will not do: it will not solve problems of radically unbalanced power, for example. It can, however, refine an adversarial process into a partially collaborative one. It will not solve problems of basic rights, but it can often expand the range of affected parties' interests that developers will take into account. Mediated negotiations will neither necessarily co-opt project opponents (as skeptical neighborhood residents might suspect) nor stall proposals and projects (as skeptical developers and builders might suspect). Yet when each side can effectively threaten the other, when each side's interests depend upon the other's actions, then mediated negotiations may enable voluntary agreements, incorporate measures of control on both sides, allow "both gain" trades to be achieved, and do so more efficiently for all sides than pursuing alternative strategies (e.g. going to court or, sometimes, community organizing).

Fifth, planners need administratively to create an organized process to match incoming projects with one or more of the mediated-negotiation strategies and to review their progress as they go along. With staff training in negotiation and mediation principles and techniques, planning departments would be better able to carry out these strategies effectively once they have organized administratively to promote them.

DEALING WITH POWER IMBALANCES: CAN THE SIX STRATEGIES MAKE A DIFFERENCE?

The six strategies we have considered are hardly "neutral." Planners who adopt them inevitably either perpetuate or challenge existing inequalities of information, expertise, political access, and opportunity. Consider each approach, briefly, in turn.

To provide only the facts, or information about procedures, to whomever asks for them seems to treat everyone equally. Yet where severe inequalities exist, to treat the strong and the weak alike only ensures that the strong remain strong, the weak remain weak. The planner who pretends to act as a neutral regulator may sound egalitarian but nevertheless act, ironically, to perpetuate and ignore existing inequalities.

The premediation strategy can involve substantial discretion on the part of the planning staff. If the staff fail to put the interests of weaker parties "on the negotiating table," then here, too, inequalities will be perpetuated, not mitigated. If the staff do defend neighborhood interests in the development negotiations, they may challenge existing inequalities. But which "neighborhood interests" should the planning staff identify? How should neighborhoods – especially weakly organized ones – be represented? These questions are both practical and theoretical and they have no purely technical, "recipe"-like answers.

At first glance, the strategy of letting developers and neighbors meet without an active staff presence seems only to reproduce the initial strengths of the parties. Yet depending on how the planning staff intervene, one party or another may be strengthened or weakened. At times planners have helped developers anticipate and ultimately evade the concerns of citizens who opposed projects. Yet planners may also provide expertise, access, information, and so on to strengthen weaker citizens' positions.

The same discretion exists for planning staff who act as shuttle diplomats. Here a planner may counsel weaker parties to help them both before and during actual negotiations by identifying concerns that might effectively be raised, experts or other influentials who might be called upon, prenegotiation strategies and tactics to be employed, and so on. The shuttle diplomat need not appear neutral to all parties but he or she does need to appear useful to, or needed by, those parties. Planners who act as "interested mediators" face many of the same problems and opportunities that shuttle diplomats confront. In addition, though, the activist mediator may risk being perceived by planning board members, officials, or elected representatives as making deals that preempt their own formal authority. Thus the invisibility of the shuttle diplomat has its advantages; the planners can give counsel discreetly, suggesting packages and "deals" but avoiding the glare – and the heat – of the limelight.

Finally, the strategy of separating mediation and negotiation functions also involves substantial staff discretion. Here, too, the ways that mediators and negotiators consider the interests and enable the voice of weaker parties will affect existing power imbalances.

Because negotiations always involve questions of relative power, they depend heavily upon the parties' *prenegotiation* work of marshalling resources, developing options, and organizing support. Thus politically astute planners need both organizing and mediated-negotiation skills if conflicts are to be addressed without pretending that structural power imbalances just do not exist. Finally, note that a planner who explicitly calls everyone's attention to class-based power imbalances, for example, may not obviously do better in any practical sense of the word than an activist mediator who knows the same thing and acts on it in just the same ways without explicitly framing the planning negotiations in those terms.

CONCLUSION

The repertoire of mediated-negotiation strategies inevitably requires that planners exercise practical judgment, both politically and ethically. These judgments involve who is and who is not invited to meetings; where, when, and which meetings are held; what issues should and should not appear on agendas; whose concerns are and are not acknowledged; how interventionist the planner's role is; and so on.

In local planning processes, then, planners often have the administrative discretion not only to mediate among conflicting parties, but to negotiate as interested parties themselves. Planning staff can routinely engage in the complementary tasks of supporting organizing efforts, negotiating, and mediating. In these ways, local planners can use a range of mediated-negotiation strategies to address practically existing power imbalances of access, information, class, and expertise that perpetually threaten the quality of local planning outcomes.

Mediated negotiations in local permitting processes will, of course, not resolve the structural problems of our society. Yet when local conflicts involve multiple issues, when differences in interests can be exploited by trading to achieve joint gains, and when diverse interests rather than fundamental rights are at stake, mediated-negotiation strategies for planners make good sense, politically, ethically, and practically.

"Advocacy and Pluralism in Planning"

Journal of the American Institute of Planners (1965)

Paul Davidoff

Editors' Introduction

Many different stakeholders and interest groups are typically involved in any significant urban planning decision. A single plan, usually produced by the local planning commission, emerges. Interest groups involved in the planning process may range in sophistication and influence from well-organized and well-funded industry groups with deep pockets to penniless ad hoc groupings of people affected by planning decisions who know little about planning procedures or plans. As John Forester describes (p. 421) city planning typically involves conflict between different groups and planners need to master conflict resolution skills.

Paul Davidoff, an activist lawyer and planner, felt that the conventional way of formulating city plans is deeply flawed. Davidoff believed that in democracies planning should be pluralistic, explicitly designed to incorporate the views of different groups. And since low-income and minority groups are not on an equal footing with the rich and powerful, they need advocates: planners acting in a capacity similar to lawyers representing clients. This selection – Davidoff's brilliant articulation of these views – is one of the classics in urban planning theory.

Unlike the advocacy planning Davidoff proposes, most city planning at the time he wrote this article in the late 1960s and still at the present time is performed by staff of local government planning departments working under the direction of planning commissions. The planning staff develop plans they feel will best serve the welfare of the whole community which are in turn reviewed, perhaps modified, and ultimately adopted by the planning commission and ultimately the city council. In theory the planning commission does not favor any particular interest group such as homeless people, merchants, environmentalists, or bicycle enthusiasts. Depending on the local political culture and the composition of the city planning commission, commissioners may be particularly sympathetic to some points of view and not to others. Many city planning commissions share the values of local "growth machines" that Harvey Molotch describes (p. 251).

Davidoff argues that since different groups in society have different interests, these interests will result in plans that are different from each other if they are recognized. For example, one planner might develop and advocate for a plan that would meet the needs of poor West Indian residents of London's Brixton neighborhood. Another planner might develop a different plan representing the point of view of shopkeepers in the same area. And a third might work with Brixton environmentalists to develop and advocate for a plan for the Brixton area incorporating the kind of sustainable urban development urged by the Brundtland Commission (p. 351) and Timothy Beatley (p. 446). Like a lawyer the advocate planner serves his or her client, not the public at large. The advocate planner leaves it to competing advocate planners to represent other interests, just as lawyers in legal cases leave it to opposing counsel to argue the other side of a legal case. Planning commissioners make the ultimate decisions about a plan's contents just as judges decide the outcome in legal cases.

Confronted with different proposed plans, the local planning commission would be forced to weigh the merits of the competing plans much as a court weighs competing evidence and opposing legal arguments in a legal case presented by competing lawyers. Davidoff believed plans that would emerge from such a process would be better than plans prepared by planning department staff without the interplay of competing advocacy planners. The justification for adversarial systems has been well developed by legal theorists. Law professors point out that conflict keeps people honest. It makes lawyers work hard because they know that their work will be critically scrutinized. And it gives judges competing points of view from which to choose. Davidoff reasoned that the needs of the poor and powerless would be better met in city plans if – a big if – they were adequately represented by advocacy planners speaking on their behalf. Davidoff was particularly concerned with low-income minority communities. Selections in Part Two of this reader on Urban Culture and Society suggest different kinds of community that might merit different plans. In addition to race and ethnicity, gender, age, sexual orientation, disability status, occupational structure and other characteristics may affect what kind of a plan is needed for an area.

Davidoff's view of how urban planning should be practiced profoundly influenced activist planners of the 1960s and 1970s, many of whom defined themselves as advocacy planners and developed plans to meet the needs and interests of underrepresented groups with some notable success. Some planners define themselves as advocacy or equity planners and continue this tradition today. While planning at the local level is still typically done by the staff of a city planning department, the planning profession has been sensitized to the importance of pluralism in planning and many planners and planning commissioners are far more open to advocacy by competing interest groups than they were before Davidoff's article was written. One former city planning director, Norman Krumholz, made social equity the fundamental principle in his tenure as director of the Cleveland, Ohio planning department and developed the theory and practice of what he calls equity planning, a theory of planning similar to, but distinct from, advocacy planning.

Review the accounts of the evolution of urban planning theory by Sir Peter Hall (p. 373) and Nigel Taylor (p. 386) and the description of mainstream physical planning by Edward Kaiser and David Godschalk (p. 399) to better understand the context within which Davidoff developed his critique. Compare Davidoff's humanistic, grassroots, pluralistic approach to city planning with Le Corbusier's brilliant but elitist vision of an elite cadre of CIAM architects imposing the forms they felt modern machine culture demanded on the fabric of cities (p. 336). Compare Davidoff's views with John Forester's comments on how planners working within the system can use their influence to empower stakeholders in the planning process (p. 421). Compare the advocacy planning approach to strategies to empower communities to reach the highest possible level on the ladder of citizen participation that Sherry Arnstein developed (p. 238).

Davidoff's advocacy planning model assumes that there will be planners available to represent underrepresented low-income and minority interests. For a brief time after Davidoff's article appeared, the US federal government funded Community Design Centers that provided some architectural and planning assistance along these lines. But this was not a popular use for taxpayer money. Most advocate planners for poor people have been dedicated volunteers. On the other hand planners representing growth machines, developers and private property owners are often well funded to develop and advocate for their visions.

Paul Davidoff (1930–1984) received degrees in urban planning (1956) and law (1961) from the University of Pennsylvania. He worked as a practicing planner in New York City and a number of other East Coast cities in the 1950s and 1960s. Davidoff taught urban planning at the University of Pennsylvania from 1958 to 1965 and Hunter College from 1965 to 1969, where he served as director of Hunter's graduate planning program. In 1969, with Neil Gold, he formed and became executive director of the Suburban Action Institute, a nonprofit organization that worked to get low-income housing built in suburbs. He had direct experience doing advocacy planning for low-income and minority groups around housing and other issues. The Suburban Action Institute became the Metropolitan Action Institute in 1980.

Each year at the annual meeting of the Association of Collegiate Schools of Planning (ACSP), North American professors of urban planning present the Paul Davidoff Award to a city planning scholar whose work exemplifies the practice and ideals of Paul Davidoff. It is an honor to receive the Davidoff award, because Davidoff exemplified professional commitment to vigorous advocacy on behalf of less fortunate members of society.

Cornell University library Collection Number 4250 contains Davidoff's papers from 1951 to 1985. A guide to these papers is available at http://rmc.library.cornell.edu/ead/htmldocs/RMM04250.html

Chester Hartman, *Cities for Sale: The Transformation of San Francisco* (Berkeley, CA: University of California Press, 2002) describes how advocacy planners and poverty lawyers fought to make urban renewal more responsive to residents of San Francisco's low-income South of Market Neighborhood. Former Cleveland, Ohio, planning director and Cleveland State University professor Norman Krumholz describes his experience practicing equity planning in *Making Equity Planning Work*, with John Forester (Philadelphia, PA: Temple University Press, 1990) and the experience of other equity planners in *Reinventing Cities: Equity Planners Tell Their Stories*, with Pierre Clavel (Philadelphia, PA: Temple University Press, 1994). Jacqueline Levitt discusses feminist advocacy planning in Barry Checkoway (ed.), *Strategic Perspectives on Planning Practice* (Lexington, MA: Lexington Books, 1986). Social work theorists Francis Fox Piven and Richard A. Cloward advanced a provocative radical critique of advocacy planning in a chapter titled "Who Does the Advocacy Planner Serve?" in Cloward and Piven, *The Politics of Turmoil* (New York: Vintage, 1970). Their answer was that by directing angry residents into planning processes for which they are ill equipped, well-intentioned advocacy planners actually undercut their political power and thereby serve establishment interests. This is similar to the argument some extreme postmodernists make nowadays that the best planners are the people themselves.

The present can become an epoch in which the dreams of the past for an enlightened and just democracy are turned into a reality. The massing of voices protesting racial discrimination have roused this nation to the need to rectify racial and other social injustices. The adoption by Congress of a host of welfare measures and the Supreme Court's specification of the meaning of equal protection by law both reveal the response to protest and open the way for the vast changes still required.

The just demand for political and social equality on the part of the Negro and the impoverished requires the public to establish the bases for a society affording equal opportunity to all citizens. The compelling need for intelligent planning, for specification of new social goals and the means for achieving them, is manifest. The society of the future will be an urban one, and city planners will help to give it shape and content.

The prospect for future planning is that of a practice which openly invites political and social values to be examined and debated. Acceptance of this position means rejection of prescriptions for planning which would have the planner act solely as a technician. It has been argued that technical studies to enlarge the information available to decision makers must take precedence over statements of goals and ideals:

> We have suggested that, at least in part, the city planner is better advised to start from research into the functional aspects of cities than from his own estimation of the values which he is attempting to maximize. This suggestion springs from a conviction that at this juncture the implications of many planning decisions are poorly understood, and that no certain means are at hand by which values can be measured, ranked, and translated into the design of a metropolitan system.

While acknowledging the need for humility and openness in the adoption of social goals, this statement amounts to an attempt to eliminate, or sharply reduce, the unique contribution planning can make: understanding the functional aspects of the city and recommending appropriate future action to improve the urban condition.

Another argument that attempts to reduce the importance of attitudes and values in planning and other policy sciences is that the major public questions are themselves matters of choice between technical methods of solution. Dahl and Lindblom put forth this position at the beginning of their important textbook *Politics, Economics, and Welfare*:

> In economic organization and reform, the "great issues" are no longer the great issues, if they ever were. It has become increasingly difficult for thoughtful men to find meaningful alternatives posed in the traditional choices between socialism and capitalism, planning and the free market, regulation and laissez faire, for they find their actual choices neither so simple nor so grand. Not so simple, because economic organization poses knotty problems that can only be solved by painstaking attention to technical details – how else, for example, can inflation be controlled? Nor so grand,

> because, at least in the Western world, most people neither can nor wish to experiment with the whole pattern of socio-economic organization to attain goals more easily won. If, for example, taxation will serve the purpose, why "abolish the wages system" to ameliorate income inequality?

These words were written in the early 1950s and express the spirit of that decade more than that of the 1960s. They suggest that the major battles have been fought. But the "great issues" in economic organization, those revolving around the central issue of the nature of distributive justice, have yet to be settled. The world is still in turmoil over the way in which the resources of nations are to be distributed. The justice of the present social allocation of wealth, knowledge, skill, and other social goods is clearly in debate. Solutions to questions about the share of wealth and other social commodities that should go to different classes cannot be technically derived; they must arise from social attitudes.

Appropriate planning action cannot be prescribed from a position of value neutrality, for prescriptions are based on desired objectives. One conclusion drawn from this assertion is that "values are inescapable elements of any rational decision-making process" and that values held by the planner should be made clear. The implications of that conclusion for planning have been described elsewhere and will not be considered in this article. Here I will say that the planner should do more than explicate the values underlying his prescriptions for courses of action; he should affirm them; he should be an advocate for what he deems proper.

Determinations of what serves the public interest, in a society containing many diverse interest groups, are almost always of a highly contentious nature. In performing its role of prescribing courses of action leading to future desired states, the planning profession must engage itself thoroughly and openly in the contention surrounding political determination. Moreover, planners should be able to engage in the political process as advocates of the interests both of government and of such other groups, organizations, or individuals who are concerned with proposing policies for the future development of the community.

The recommendation that city planners represent and plead the plans of many interest groups is founded upon the need to establish an effective urban democracy, one in which citizens may be able to play an active role in the process of deciding public policy. Appropriate policy in democracy is determined through a process of political debate. The right course of action is always a matter of choice, never of fact. In a bureaucratic age great care must be taken that choices remain in the area of public view and participation.

Urban politics, in an era of increasing government activity in planning and welfare, must balance the demands for ever-increasing central bureaucratic control against the demands for increased concern for the unique requirements of local, specialized interests. The welfare of all and the welfare of minorities are both deserving of support; planning must be so structured and so practiced as to account for this unavoidable bifurcation of the public interest.

The idealized political process in a democracy serves the search for truth in much the same manner as due process in law. Fair notice and hearings, production of supporting evidence, cross-examination, reasoned decision are all means employed to arrive at relative truth: a just decision. Due process and two- (or more) party political contention both rely heavily upon strong advocacy by a professional. The advocate represents an individual, group, or organization. He affirms their position in language understandable to his client and to the decision makers he seeks to convince.

If the planning process is to encourage democratic urban government then it must operate so as to include rather than exclude citizens from participating in the process. "Inclusion" means not only permitting the citizen to be heard. It also means that he be able to become well informed about the underlying reasons for planning proposals, and be able to respond to them in the technical language of professional planners.

A practice that has discouraged full participation by citizens in plan making in the past has been based on what might be called the "*unitary plan.*" This is the idea that only one agency in a community should prepare a comprehensive plan; that agency is the city planning commission or department. Why is it that no other organization within a community prepares a plan? Why is only one agency concerned with establishing both general and specific goals for community development, and with proposing the strategies and costs required to effect the goals? Why are there not plural plans?

If the social, economic, and political ramifications of a plan are politically contentious, then why is it that those in opposition to the agency plan do not prepare one of their own? It is interesting to observe that "rational" theories of planning have called for consideration of alternative courses of action by planning agencies. As a matter of rationality it has been argued that all of the alternative choices open as means to the ends ought be examined. But those, including myself, who have recommended agency consideration of alternatives have placed upon the agency planner the burden of inventing "a few representative alternatives." The agency planner has been given the duty of constructing a model of the political spectrum, and charged with sorting out what he conceives to be worthy alternatives. This duty has placed too great a burden on the agency planner, and has failed to provide for the formulation of alternatives by the interest groups who will eventually be affected by the completed plans.

Whereas in a large part of our national and local political practice contention is viewed as healthy, in city planning where a large proportion of the professionals are public employees, contentious criticism has not always been viewed as legitimate. Further, where only government prepares plans, and no minority plans are developed, pressure is often applied to bring all professionals to work for the ends espoused by a public agency. For example, last year a Federal official complained to a meeting of planning professors that the academic planners were not giving enough support to Federal programs. He assumed that every planner should be on the side of the Federal renewal program. Of course government administrators will seek to gain the support of professionals outside of government, but such support should not be expected as a matter of loyalty. In a democratic system opposition to a public agency should be just as normal and appropriate as support. The agency, despite the fact that it is concerned with planning, may be serving undesired ends.

In presenting a plea for plural planning I do not mean to minimize the importance of the obligation of the public planning agency. It must decide upon appropriate future courses of action for the community. But being isolated as the only plan maker in the community, public agencies as well as the public itself may have suffered from incomplete and shallow analysis of potential directions. Lively political dispute aided by plural plans could do much to improve the level of rationality in the process of preparing the public plan.

The advocacy of alternative plans by interest groups outside of government would stimulate city planning in a number of ways. First, it would serve as a means of better informing the public of the alternative choices open, *alternatives strongly supported by their proponents*. In current practice those few agencies which have portrayed alternatives have not been equally enthusiastic about each. A standard reaction to rationalists' prescription for consideration of alternative courses of action has been "it can't be done; how can you expect planners to present alternatives which they don't approve?" The appropriate answer to that question has been that planners, like lawyers, may have a professional obligation to defend positions they oppose. However, in a system of plural planning, the public agency would be relieved of at least some of the burden of presenting alternatives. In plural planning the alternatives would be presented by interest groups differing with the public agency's plan. Such alternatives would represent the deep-seated convictions of their proponents and not just the mental exercises of rational planners seeking to portray the range of choice.

A second way in which advocacy and plural planning would improve planning practice would be in forcing the public agency to compete with other planning groups to win political support. In the absence of opposition or alternative plans presented by interest groups the public agencies have had little incentive to improve the quality of their work or the rate of production of plans. The political consumer has been offered a yes–no ballot in regard to the comprehensive plan; either the public agency's plan was to be adopted or no plan would be adopted.

A third improvement in planning practice which might follow from plural planning would be to force those who have been critical of "establishment" plans to produce superior plans, rather than only to carry out the very essential obligation of criticizing plans deemed improper.

THE PLANNER AS ADVOCATE

Where plural planning is practiced, advocacy becomes the means of professional support for competing claims about how the community should develop. Pluralism in support of political contention describes the process; advocacy describes the role performed by the professional in the process. Where unitary planning prevails, advocacy is not of paramount importance, for there is little or no competition for the plan prepared by the public agency. The concept of advocacy as taken from legal practice implies the opposition of at least two contending viewpoints in an adversary proceeding.

The legal advocate must plead for his own and his client's sense of legal propriety or justice. The planner as advocate would plead for his own and his client's view of the good society. The advocate planner would be more than a provider of information, an analyst of current trends, a simulator of future conditions, and a detailer of means. In addition to carrying out these necessary parts of planning, he would be a *proponent* of specific substantive solutions.

The advocate planner would be responsible to his client and would seek to express his client's views. This does not mean that the planner could not seek to persuade his client. In some situations persuasion might not be necessary, for the planner would have sought out an employer with whom he shared common views about desired social conditions and the means toward them. In fact one of the benefits of advocate planning is the possibility it creates for a planner to find employment with agencies holding values close to his own. Today the agency planner may be dismayed by the positions affirmed by his agency, but there may be no alternative employer.

The advocate planner would be above all a planner. He would be responsible to his client for preparing plans and for all of the other elements comprising the planning process. Whether working for the public agency or for some private organization, the planner would have to prepare plans that take account of the arguments made in other plans. Thus the advocate's plan might have some of the characteristics of a legal brief. It would be a document presenting the facts and reasons for supporting one set of proposals, and facts and reasons indicating the inferiority of counter-proposals. The adversary nature of plural planning might, then, have the beneficial effect of upsetting the tradition of writing plan proposals in terminology which makes them appear self-evident.

A troublesome issue in contemporary planning is that of finding techniques for evaluating alternative plans. Technical devices such as cost–benefit analysis by themselves are of little assistance without the use of means for appraising the values underlying plans. Advocate planning, by making more apparent the values underlying plans, and by making definitions of social costs and benefits more explicit, should greatly assist the process of plan evaluation. Further, it would become clear (as it is not at present) that there are no neutral grounds for evaluating a plan; there are as many evaluative systems as there are value systems.

The adversary nature of plural planning might also have a good effect on the uses of information and research in planning. One of the tasks of the advocate planner in discussing the plans prepared in opposition to his would be to point out the nature of the bias underlying information presented in other plans. In this way, as critic of opposition plans, he would be performing a task similar to the legal technique of cross-examination. While painful to the planner whose bias is exposed (and no planner can be entirely free of bias) the net effect of confrontation between advocates of alternative plans would be more careful and precise research.

Not all the work of an advocate planner would be of an adversary nature. Much of it would be educational. The advocate would have the job of informing other groups, including public agencies, of the conditions, problems, and outlook of the group he represented. Another major educational job would be that of informing his clients of their rights under planning and renewal laws, about the general operations of city government, and of particular programs likely to affect them.

The advocate planner would devote much attention to assisting the client organization to clarify its ideas and to give expression to them. In order to make his client more powerful politically the advocate might also become engaged in expanding the size and scope of his client organization. But the advocate's most important function would be to carry out the planning process for the organization and to argue persuasively in favor of its planning proposals.

Advocacy in planning has already begun to emerge as planning and renewal affect the lives of more and more people. The critics of urban renewal have forced response from the renewal agencies, and the ongoing debate has stimulated needed self-evaluation by public agencies. Much work along the lines of advocate planning has already taken place, but little of it by professional planners. More often the work has been conducted by trained community organizers or by student groups. In at least one instance, however, a planner's professional aid led to the development of an alternative renewal approach, one which will result in the dislocation of far fewer families than originally contemplated.

Pluralism and advocacy are means for stimulating consideration of future conditions by all groups in society. But there is one social group which at present is particularly in need of the assistance of planners. This group includes organizations representing low-income families. At a time when concern for the condition of the poor finds institutionalization in community action programs, it would be appropriate for planners concerned with such groups to find means to plan with them. The plans prepared for these groups would seek to combat poverty and would propose programs affording new and better opportunities to the members of the organization and to families similarly situated. The difficulty in providing adequate planning assistance to organizations representing low-income families may in part be overcome by funds allocated to local antipoverty councils. But these councils are not the only representatives of the poor; other organizations exist and seek help. How can this type of assistance be financed? This question will be examined below, when attention is turned to the means for institutionalizing plural planning.

THE STRUCTURE OF PLANNING

Planning by special interest groups

The local planning process typically includes one or more "citizens'" organizations concerned with the nature of planning in the community. The Workable Program requirement for "citizen participation" has enforced this tradition and brought it to most large communities. The difficulty with current citizen participation programs is that citizens are more often *reacting* to agency programs than proposing their concepts of appropriate goals and future action.

The fact that citizens' organizations have not played a positive role in formulating plans is to some extent a result of both the enlarged role in society played by government bureaucracies and the historic weakness of municipal party politics. There is something very shameful to our society in the necessity to have organized "citizen participation." Such participation should be the norm in an enlightened democracy. The formalization of citizen participation as a required practice in localities is similar in many respects to totalitarian shows of loyalty to the state by citizen parades.

Will a private group interested in preparing a recommendation for community development be required to carry out its own survey and analysis of the community? The answer would depend upon the quality of the work prepared by the public agency, work which should be public information. In some instances the public agency may not have surveyed or analyzed aspects the private group thinks important; or the public agency's work may reveal strong biases unacceptable to the private group. In any event, the production of a useful plan proposal will require much information concerning the present and predicted conditions in the community. There will be some costs associated with gathering that information, even if it is taken from the public agency. The major cost involved in the preparation of a plan by a private agency would probably be the employment of one or more professional planners.

What organizations might be expected to engage in the plural planning process? The first type that comes to mind are the political parties; but this is clearly an aspirational thought. There is very little evidence that local political organizations have the interest, ability, or concern to establish well-developed programs for their communities. Not all the fault, though, should be placed upon the professional politicians, for the registered members of political parties have not demanded very much, if anything, from them as agents.

Despite the unreality of the wish, the desirability for active participation in the process of planning by the political parties is strong. In an ideal situation local parties would establish political platforms which would contain master plans for community growth and both the majority and minority parties in the legislative branch of government would use such plans as one basis for appraising individual legislative proposals. Further, the local administration would use its planning agency to carry out the plans it proposed to the electorate. This dream will not turn to reality for a long time. In the interim other interest groups must be sought to fill the gap caused by the present inability of political organizations.

The second set of organizations which might be interested in preparing plans for community development are those that represent special interest groups having established views in regard to proper public policy. Such organizations as chambers of commerce, real estate boards, labor organizations, pro- and anti-civil rights groups, and anti-poverty councils come to mind. Groups of this nature have often played parts in the development of community plans, but only in a very few instances have they proposed their own plans.

It must be recognized that there is strong reason operating against commitment to a plan by these organizations. In fact it is the same reason that in part limits the interests of politicians and which limits the potential for planning in our society. The expressed commitment to a particular plan may make it difficult for groups to find means for accommodating their various interests. In other terms, it may be simpler for professionals, politicians, or lobbyists to make deals if they have not laid their cards on the table.

There is a third set of organizations that might be looked to as proponents of plans and to whom the foregoing comments might not apply. These are the ad hoc protest associations which may form in opposition to some proposed policy. An example of such a group is a neighborhood association formed to combat a renewal plan, a zoning change, or the proposed location of a public facility. Such organizations may seek to develop alternative plans, plans which would, if effected, better serve their interests.

From the point of view of effective and rational planning it might be desirable to commence plural planning at the level of city-wide organizations, but a more realistic view is that it will start at the neighborhood level. Certain advantages of this outcome should be noted. Mention was made earlier of tension in government between centralizing and decentralizing forces. The contention aroused by conflict between the central planning agency and the neighborhood organization may indeed be healthy, leading to clearer definition of welfare policies and their relation to the rights of individuals or minority groups.

Who will pay for plural planning? Some organizations have the resources to sponsor the development of a plan. Many groups lack the means. The plight of the relatively indigent association seeking to propose a plan might be analogous to that of the indigent client in search of legal aid. If the idea of plural planning makes sense, then support may be found from foundations or from government. In the beginning it is more likely that some foundation might be willing to experiment with plural planning as a means of making city planning more effective and more democratic. Or the Federal Government might see plural planning, if carried out by local anti-poverty councils, as a strong means of generating local interest in community affairs.

Federal sponsorship of plural planning might be seen as a more effective tool for stimulating involvement of the citizen in the future of his community than are the present types of citizen participation programs. Federal support could only be expected if plural planning were seen, not as a means of combating renewal plans, but as an incentive to local renewal agencies to prepare better plans.

The public planning agency

A major drawback to effective democratic planning practice is the continuation of that non-responsible vestigial institution, the planning commission. If it is agreed that the establishment of both general policies and implementation policies are questions affecting the public interest and that public interest questions should be decided in accord with established democratic practices for decision making, then it is indeed difficult to find convincing reasons for continuing to permit independent commissions to make planning decisions. At an earlier stage in planning the strong arguments of John T. Howard and others in support of commissions may have been persuasive. But it is now more than a decade since Howard made his defense against Robert Walker's position favoring planning as a staff function under the mayor. With the increasing effect planning decisions have upon the lives of citizens the Walker proposal assumes great urgency.

Aside from important questions regarding the propriety of independent agencies which are far removed from public control determining public policy, the failure to place planning decision choices in the hands of elected officials has weakened the ability of professional planners to have their proposals effected. Separating planning from local politics has made it difficult for independent commissions to garner influential political support. The commissions are not responsible directly to the electorate and in turn the electorate is, at best, often indifferent to the planning commission.

During the last decade, in many cities power to alter community development has slipped out of the hands of city planning commissions, assuming they ever held it, and has been transferred to development coordinators. This has weakened the professional planner. Perhaps planners unknowingly contributed to this by their refusal to take concerted action in opposition to the perpetuation of commissions.

Planning commissions are products of the conservative reform movement of the early part of this century. The movement was essentially anti-populist and pro-aristocracy. Politics was viewed as dirty business. The commissions are relics of a not-too-distant past when it was believed that if men of good will discussed a problem thoroughly, certainly the right solution would be forthcoming. We know today, and perhaps it was always known, that there are no right solutions. Proper policy is that which the decision-making unit declares to be proper.

Planning commissions are responsible to no constituency. The members of the commissions, except for their chairman, are seldom known to the public. In general the individual members fail to expose their personal views about policy and prefer to immerse them in group decision. If the members wrote concurring and dissenting opinions, then at least the commissions might stimulate thought about planning issues. It is difficult to comprehend why this aristocratic and undemocratic form of decision making should be continued. The public planning function should be carried out in the executive or legislative office and perhaps in both. There has been some question about which of these branches of government would provide the best home, but there is much reason to believe that both branches would be made more cognizant of planning issues if they were each informed by their own planning staffs. To carry this division further, it would probably be advisable to establish minority and majority planning staffs in the legislative branch.

At the root of my last suggestion is the belief that there is or should be a Republican and Democratic way of viewing city development; that there should be conservative and liberal plans, plans to support the private market, and plans to support greater government control. There are many possible roads for a community to travel and many plans should show them. Explication is required of many alternative futures presented by those sympathetic to the construction of each such future. As indicated earlier, such alternatives are not presented to the public now. Those few reports which do include alternative futures do not speak in terms of interest to the average citizen. They are filled with professional jargon and present sham alternatives. These plans have expressed technical land use alternatives rather than social, economic, or political value alternatives. Both the traditional unitary plans and the new ones that present technical alternatives have limited the public's exposure to the future states that might be achieved. Instead of arousing healthy political contention as diverse comprehensive plans might, these plans have deflated interest.

The independent planning commission and unitary plan practice certainly should not co-exist. Separately they dull the possibility for enlightened political debate; in combination they have made it yet more difficult. But when still another hoary concept of city planning is added to them, such debate becomes practically impossible. This third of a trinity of worn-out notions is that city planning should focus only upon the physical aspects of city development.

AN INCLUSIVE DEFINITION OF THE SCOPE OF PLANNING

The view that equates physical planning with city planning is myopic. It may have had some historic justification, but it is clearly out of place at a time when it is necessary to integrate knowledge and techniques in order to wrestle effectively with the myriad of problems afflicting urban populations.

The city planning profession's historic concern with the physical environment has warped its ability to see physical structures and land as servants to those who use them. Physical relations and conditions have no meaning or quality apart from the way they serve their users. But this is forgotten every time a physical condition is described as good or bad without relation to a specified group of users. High density, low density, green belts, mixed uses, cluster developments, centralized or decentralized business centers are per se neither good nor bad. They describe physical relations or conditions, but take on value only when seen in terms of their social, economic, psychological, physiological, or aesthetic effects upon different users.

The profession's experience with renewal over the past decade has shown the high costs of exclusive concern with physical conditions. It has been found that the allocation of funds for removal of physical blight may not necessarily improve the overall physical condition of a community and may engender such harsh social repercussions as to severely damage both social and economic institutions. Another example of the deficiencies of the physical bias is the assumption of city planners that they could deal with the capital budget as if the physical attributes of a facility could be understood apart from the philosophy and practice of the service conducted within the physical structure. This assumption is open to question. The size, shape, and location of a facility greatly interact with the purpose of the activity the facility houses. Clear examples of this can be seen in public education and in the provision of low cost housing. The racial and other socioeconomic consequences of "physical decisions" such as location of schools and housing projects have been immense, but city planners, while acknowledging the existence of such consequences, have not sought or trained themselves to understand socioeconomic problems, their causes or solutions.

The city planning profession's limited scope has tended to bias strongly many of its recommendations toward perpetuation of existing social and economic practices. Here I am not opposing the outcomes, but the way in which they are developed. Relative ignorance of social and economic methods of analysis has caused planners to propose solutions in the absence of sufficient knowledge of the costs and benefits of proposals upon different sections of the population.

Large expenditures have been made on planning studies of regional transportation needs, for example, but these studies have been conducted in a manner suggesting that different social and economic classes of the population did not have different needs and different abilities to meet them. In the field of housing, to take another example, planners have been hesitant to question the consequences of locating public housing in slum areas. In the field of industrial development, planners have seldom examined the types of jobs the community needs; it has been assumed that one job was about as useful as another. But this may not be the case where a significant sector of the population finds it difficult to get employment.

"Who gets what, when, where, why, and how" are the basic political questions which need to be raised about every allocation of public resources. The questions cannot be answered adequately if land use criteria are the sole or major standards for judgment.

The need to see an element of city development, land use, in broad perspective applies equally well to every other element, such as health, welfare, and recreation. The governing of a city requires an adequate plan for its future. Such a plan loses guiding force and rational basis to the degree that it deals with less than the whole that is of concern to the public.

The implications of the foregoing comments for the practice of city planning are these. First, state planning enabling legislation should be amended to permit planning departments to study and to prepare plans related to any area of public concern. Second, planning education must be redirected so as to provide channels of specialization in different parts of public planning and a core focused upon the planning process. Third, the professional planning association should enlarge its scope so as to not exclude city planners not specializing in physical planning.

A year ago at the AIP convention it was suggested that the AIP Constitution be amended to permit city planning to enlarge its scope to all matters of public concern. Members of the Institute in agreement with this proposal should seek to develop support for it at both the chapter and national level. The Constitution at present states that the Institute's "particular sphere of activity shall be the planning of the unified development of urban communities and their environs and of states, regions and the nation *as expressed through determination of the comprehensive arrangement of land and land occupancy and regulation thereof.*" It is time that the AIP delete the words in my italics from its Constitution. The planner limited to such concerns is not a city planner, he is a land planner or a physical planner. A city is its people, their practices, and their political, social, cultural and economic institutions as well as other things. The city planner must comprehend and deal with all these factors.

The new city planner will be concerned with physical planning, economic planning, and social planning. The scope of his work will be no wider than that presently demanded of a mayor or a city councilman. Thus, we cannot argue against an enlarged planning function on grounds that it is too large to handle. The mayor needs assistance; in particular he needs the assistance of a planner, one trained to examine needs and aspirations in terms of both short- and long-term perspectives. In observing the early stages of development of Community Action Programs, it is apparent that our cities are in desperate need of the type of assistance trained planners could offer. Our cities require for their social and economic programs the type of long-range thought and information that have been brought forward in the realm of physical planning. Potential resources must be examined and priorities set.

What I have just proposed does not imply the termination of physical planning, but it does mean that physical planning be seen as part of city planning. Uninhibited by limitations on his work, the city planner will be able to add his expertise to the task of coordinating the operating and capital budgets and to the job of relating effects of each city program upon the others and upon the social, political, and economic resources of the community.

An expanded scope reaching all matters of public concern will make planning not only a more effective administrative tool of local government but it will also bring planning practice closer to the issues of real concern to the citizens. A system of plural city planning probably has a much greater chance for operational success where the focus is on live social and economic questions instead of rather esoteric issues relating to physical norms.

THE EDUCATION OF PLANNERS

Widening the scope of planning to include all areas of concern to government would suggest that city planners must possess a broader knowledge of the structure and forces affecting urban development. In general this would be true. But at present many city planners are specialists in only one or more of the functions of city government. Broadening the scope of planning would require some additional planners who specialize in one or more of the services entailed by the new focus.

A prime purpose of city planning is the coordination of many separate functions. This coordination calls for men holding general knowledge of the many elements comprising the urban community. Educating a man for performing the coordinative role is a difficult job, one not well satisfied by the present tradition of two years of graduate study. Training of urban planners with the skills called for in this article may require both longer graduate study and development of a liberal arts undergraduate program affording an opportunity for holistic understanding of both urban conditions and techniques for analyzing and solving urban problems.

The practice of plural planning requires educating planners who would be able to engage as professional advocates in the contentious work of forming social policy. The person able to do this would be one deeply committed to both the process of planning and to particular substantive ideas. Recognizing that ideological commitments will separate planners, there is tremendous need to train professionals who are competent to express their social objectives.

The great advances in analytic skills, demonstrated in the recent May issue of this journal [*Journal of the American Institute of Planners*] dedicated to techniques of simulating urban growth processes, portend a time when planners and the public will be better able to predict the consequences of proposed courses of action. But these advances will be of little social advantage if the proposals themselves do not have substance. The contemporary thoughts of planners about the nature of man in society are often mundane, unexciting or gimmicky. When asked to point out to students the planners who have a developed sense of history and philosophy concerning man's situation in the urban world one is hard put to come up with a name. Sometimes Goodman or Mumford might be mentioned. But planners seldom go deeper than acknowledging the goodness of green space and the soundness of proximity of linked activities. We cope with the problems of the alienated man with a recommendation for reducing the time of the journey to work.

CONCLUSION

The urban community is a system comprised of interrelated elements, but little is known about how the elements do, will, or should interrelate. The type of knowledge required by the new comprehensive city planner demands that the planning profession be comprised of groups of men well versed in contemporary philosophy, social work, law, the social sciences, and civic design. Not every planner must be knowledgeable in all these areas, but each planner must have a deep understanding of one or more of these areas and he must be able to give persuasive expression to his understanding. As a profession charged with making urban life more beautiful, exciting, and creative, and more just, we have had little to say. Our task is to train a future generation of planners to go well beyond us in its ability to prescribe the future urban life.

"Planning for Sustainability in European Cities: A Review of Practice in Leading Cities"

from *The Sustainable Urban Development Reader* (2003)

Timothy Beatley

Editors' Introduction

The urbanization of the human population described by Kingsley Davis (p. 20) and industrialization without regard for environmental consequences – described in an extreme form by Friedrich Engels in 1844 – raise serious concerns about the earth's capacity to sustain urban life, as the Bruntland Commission concluded (p. 351). Green urbanism and sustainable urban development are alternatives proposed to align human development with natural processes and assure that natural resources will be available to subsequent generations.

Most European cities take sustainable urban development more seriously than cities in the United States and elsewhere in the world. Vigorous green politics, participation in EU-sponsored information sharing, and hundreds of exemplary local projects are evidence of their concern. Because European cities have pioneered new policies, the many successful sustainability practices they have adopted and some shortcomings they have experienced offer important lessons for the United Kingdom, United States, China and cities in other countries worldwide.

Until recently, hardheaded decision-makers generally dismissed alternative green visions as hopelessly unrealistic. In their view it is easy to dream of cities where clean renewable solar and wind power produce much of the communities' energy needs rather than nonrenewable resources like coal that cause pollution and contribute to global warming. It is pleasant to fantasize about communities where people walk and bicycle to work rather than drive cars, recycle sewage sludge into biogas, and grow many of their vegetables in urban gardens. But, critics of green urbanism say, alternative energy sources can never produce enough energy to run cities. Bicycle and pedestrian paths are nice, but people need cars and super highways to get around. Agribusiness, not urban gardens, is the only way to grow enough food cheaply enough to feed the earth's 6 billion-plus inhabitants. Beatley disagrees. He produces hard evidence from Europe that compact, walkable, energy-efficient, clean, green communities can be created; cities that are economically viable as well as sustainable and livable.

Fundamental to many European sustainability practices is the way European cities have limited sprawl and encouraged compact development. Most Europeans reject the laissez-faire approach to sprawl that Robert Bruegmann advocates (p. 446). Accordingly most European cities have much higher average densities than their American counterparts, because their citizens accept much higher density development than suburb-loving, auto-dependent Americans. Paradoxically Americans who have chosen to live in low-density suburbs travel to Paris or Amsterdam to enjoy the energy and street life!

European cities have achieved relatively high average densities by restricting sprawl, building new areas adjacent to the existing city core at relatively high densities, and fostering urban development and industrial reuse. Beatley points out that higher density makes more efficient public transit and energy systems possible.

European cities generally invest much more per capita in public transportation systems such as high-speed rail lines (bullet trains), subways, and buses than cities in the United States. Some reduce auto-dependency through congestion pricing, imaginative incentive systems to reduce driving, car-sharing programs, and promotion of bicycle use. European transit systems are consistent with land use plans and well integrated with each other. This makes it possible for households to get around with a single car rather than one car for each driver in the household or even without a car at all. If there is a really good public transit system serving a compact city, as is the case in many European cities, people will use it rather than cars for many trips. This in turn generates revenue to keep the system viable. In contrast, if there is a terrible bus system that functions poorly as the transit system of last resort, it will not attract riders who have other alternatives. Without enough revenue the system will deteriorate further.

In many European countries gasoline taxes are two or three times as high as in the United States. This is a good example of Wilbur Thompson's ideas about pricing goods to achieve urban policies (p. 274). High gas taxes discourage driving and contribute to more compact cities, better and more used public transportation, and less air pollution. Revenue from gas taxes in Europe is often used to fund public transit systems.

Amsterdam and other European cities work hard to make bicycling possible. Some provide systems for people to use bicycles on a pay-for-use basis or free public bicycles.

Not only are European cities generally better served by public transit, but also they are often more pedestrian-friendly than sprawling, low-density cities in the United States and elsewhere in the world. Greater density can make it easier for people to get around on foot if planners and architects follow Jan Gehl's principles (p. 530). Conscious policies to promote attractive, exciting pedestrian areas reinforce walking as an alternative. Sustainability can contribute to livability.

European cities have made impressive strides in increasing energy efficiency and reducing waste. Stockholm, for example, has reorganized its government departments so that the city offices dealing with waste, water, and energy are all grouped together. Sewage sludge in Stockholm is used for fertilizer and – hard-headed critics notwithstanding – to produce biogas to fuel the cities' buses and a local power plant.

Green politics has now established itself as a worldwide political movement. In Europe green parties have elected enough representatives to be political forces in the governing coalitions of a number of countries. In the United States green politics is stimulating new thinking about the nature of urban development.

Green urbanism is not a new idea. Before World War I, eccentric Scottish biologist Patrick Geddes had classified the environmental needs of different ecological systems and developed a systematic approach to building cities that respect natural systems. Ian McHarg, another Scot, wrote a seminal book titled *Design with Nature* (1969) that inspired the environmentally conscious generation of the 1960s. There is now a movement in ecological design among architects, and respect for the natural environment is a cornerstone of the Charter of the New Urbanism (p. 356).

Timothy Beatley (b. 1957) is the Theresa Heinz Professor of Urban and Environmental Planning at the University of Virginia, Charlottesville. His primary teaching and research interests are in environmental planning and policy, with special emphasis on coastal and natural hazards planning, environmental values and ethics, and biodiversity conservation.

The practices reviewed in this selection are discussed in greater detail in Beatley, *Green Urbanism: Learning from European Cities* (Washington, DC: Island Press, 1999). Professor Beatley's other books include *Native to Nowhere: Sustaining Home and Community in a Global Age* (Washington, DC: Island Press, 2005), *Natural Hazard Mitigation*, with David Godschalk and others (Washington, DC: Island Press, 1998), *The Ecology of Place*, with Kristy Manning (Washington, DC: Island Press, 1997), *After the Hurricane: Linking Recovery To Sustainable Development in the Caribbean*, with Philip Berke (Baltimore, MD: Johns Hopkins University Press, 1997), *An Introduction to Coastal Zone Management*, 2nd edn (Washington, DC: Island Press, 1994), *Ethical Land Use: Principles of Policy and Planning* (Baltimore, MD: Johns Hopkins University Press, 1994), and *Habitat Conservation Planning: Endangered Species and Urban Growth* (Austin, TX: University of Texas Press, 1994).

Key classic and contemporary writings on sustainable urban development and green urbanism are contained in Stephen Wheeler and Timothy Beatley (eds), *The Sustainable Urban Development Reader*, 2nd edn (London: Routledge, 2008).

For descriptions of the destructive effects of urbanization on the natural environment, see William Cronon, *Nature's Metropolis* (New York: Norton, 1992), which describes how nineteenth-century Chicago prospered from exploitation of natural resources, and Mark Reisner, *Cadillac Desert* (New York: Penguin, 1993), which describes destructive development in the American Southwest.

Important writings on sustainable urban development include Peter Newman and Isabella Jennings, *Cities as Sustainable Ecosystems: Principles and Practices* (Washington, DC: Island Press, 2008), Douglas Farr, *Sustainable Urbanism: Urban Design with Nature* (Hoboken, NJ: Wiley, 2007), Richard Register, *EcoCities: Rebuilding Cities in Balance with Nature*, revised edn (Gabriola Island, BC: New Society Publishers, 2006), and Mike Jencks and Nicola Dempsey, *Future Forms and Design for Sustainable Cities* (Oxford: Architectural Press, 2005).

Ian McHarg's classic *Design with Nature* (Garden City, NY: Doubleday, 1969) and Patrick Geddes, *Cities in Evolution* (London: Williams & Norgate, 1915), reprinted in Richard T. LeGates and Frederic Stout (eds), *Early Urban Planning, 1870–1940* (London: Routledge/Thoemmes, 1998) are essential foundational works that anticipate present-day sustainable development, regional planning, and ecological design debates.

For more on green politics in Europe, see Michael Dobson, *Green Political Thoughts*, 4th edn (London: Routledge, 2007), Andrew Dobson and Robyn Eckersley (eds), *Political Theory and the Ecological Challenge* (Cambridge: Cambridge University Press, 2006), John S. Dryzek (ed.), *Green States and Social Movements: Environmentalism in the United States, United Kingdom, Germany and Norway* (London: Oxford University Press, 2002), and Michael O'Neill, *Green Parties and Political Change in Contemporary Europe: New Politics, Old Predicaments* (Aldershot, UK: Ashgate, 1997). For green politics in the United States, see John Rensenbrink, *Against All Odds: The Green Transformation of American Politics* (Raymond, ME: Leopold Press, 1999).

INTRODUCTION: LEARNING FROM EUROPEAN CITIES

In few other parts of the world is there as much interest in sustainability as in Europe, especially northern and northwestern Europe, and as much tangible evidence of applying this concept to cities and urban development. For approximately the last six years this author has been researching innovative urban sustainability practice in European cities. The findings from the first phase of this work are presented in the book *Green Urbanism: Learning from European Cities* (Island Press, 2000). What follows is a summary of some of the key themes and most promising ideas and strategies found in the 30 or so cities, in 11 countries, described in this book, as well as more recent case studies and field work.

An initial observation from this work is just how important sustainability is at the municipal level in Europe, especially evident in the cities chosen. "Sustainable cities" resonates well and has important political meaning and significance in these cities, and on the European urban scene generally. One measure of this is the success of the Sustainable Cities and Towns campaign, an EU-funded informal network of communities pursuing sustainability begun in 1994. Participating cities have signed the so-called Aalborg Charter (from Aalborg, Denmark, the site of the first campaign conference), and more than 1800 cities and towns have done so. Among the activities of this organization are the publication of a newsletter, networking between cities, and initiation of conferences and workshops. The organization has also created the annual European Sustainable City award (with the first of these awards issued in 1996), and it is clear that they have been coveted and highly valued by politicians and city officials.

Many European cities have also gone through, or are currently going through, some form of local Agenda 21 process (including many of the same cities that have signed the Aalborg charter), and this is another important indicator of the relevance of local sustainability. Indeed, in the countries studied, high percentages of municipal governments are participating (for instance, in Sweden 100 percent of all local governments are at some stage in the local Agenda 21 process). Often these programs represent tremendous local efforts to engage the community in a dialogue about sustainability, and typically involve the creation of a local sustainability forum, sustainability indicators, local state-of-the-environment reports, and the preparation of comprehensive local sustainability action plans. European cities and towns demonstrate serious commitment to environmental and sustainability values and what follows are a few of the more important ways in which these concerns are being addressed.

Compact cities and regions

Urban form and land use patterns are primary determinants of urban sustainability. While European cities have been experiencing considerable decentralization pressures, they are typically much more compact and dense than American cities. Peter Newman and Jeffrey Kenworthy have monitored and tracked average density in a number of cities throughout the world. Western European cities like Amsterdam and Paris have substantially higher densities, as measured in persons per hectare, than typical American cities. Overall or whole-city densities for European cities are typically in the 40–60 persons per hectare range; American cities are much lower, commonly under 20 persons per hectare. Even American cities that we tend to think of as particularly dense, for example New York, are comparatively less dense when the entire metropolitan wide pattern is considered. Density and compactness directly translate into much lower energy use, per capita, and lower carbon emissions, air and water pollution, and other resource demands compared with less dense, less compact cities.

Many of these European examples, moreover, show that compactness and density need not translate to skyscrapers and excessive high-rise. Density and compactness in cities like Amsterdam happens through a building pattern of predominately low-rise structures. While many sustainability proponents advocate the need for the green high-rise development (e.g. see Ken Yeang's designs for bio-climatic skyscrapers), these European cities demonstrate convincingly that tremendous compactness and density can be accomplished at a clearly human scale. The European model is appealing to many precisely because of its more traditional form of density and compactness, and many believe its more human scale.

These characteristics of urban form make many other dimensions of local sustainability more feasible, of course (e.g. public transit, walkable places, energy efficiency). There are many factors that explain this urban form, including an historic pattern of compact villages and cities, a limited land base in many countries, and different cultural attitudes about land. Nevertheless in the cities studied there are conscious policies aimed at strengthening a tight urban core. Indeed, the major new growth areas in almost every city studied are situated in locations within or adjacent to existing developed areas, and are designed generally at relatively high densities.

Exemplary and for the most part effective efforts at maintaining the traditional tight urban form can be seen in many cities. Cities like Amsterdam are actively promoting urban redevelopment and industrial reuse (e.g. through its eastern docklands redevelopment). Berlin's plan calls for most future growth to be accommodated with its urbanized area through a variety of infill and reurbanization strategies. Freiburg, Germany, has been able to effectively steer relatively compact, high-density new growth along the main corridors of its tram system, as well as to protect existing housing supply in the center (there is now a prohibition on the conversion of housing to offices and other uses).

European cities are utilizing a variety of planning strategies to promote compactness and to maintain a tight urban form. These include strict limits on building outside of designated development areas, a strong role for municipal governments in designing and developing new growth areas, extensive public acquisition and ownership of land (especially in Scandinavian cities like Stockholm), and a willingness to make significant transportation and other infrastructure investments that facilitate and support compactness.

GREEN URBANISM: COMPACT AND ECOLOGICAL URBAN FORM

Growth areas and redevelopment districts in these European cities are incorporating a wide range of ecological design and planning concepts, from solar energy to natural drainage to community gardens, and effectively demonstrate that *ecological* and *urban* can go together. Good examples of this compact green growth can be seen in the new development districts planned for or recently completed in Utrecht (Leidsche Rijn), Freiburg (Rieselfeld), Amsterdam (e.g. IJburg), Copenhagen (Orestad), Helsinki (Viikki), and Stockholm (Hammerby Sjostad).

Leidsche Rijn, for example, is an innovative new growth district in the Dutch city of Utrecht. In addition to incorporating a mixed-use design, and a balance of jobs and housing (30,000 dwelling units and 30,000 new jobs), it will include a number of ecological design features. Much of the area will be heated through district heating supplied from the waste energy of a nearby power plant, a double-water system which will provide recycled water for non-potable uses, and a storm water management through a system of natural swales (what the Dutch call "wadies"). Higher-density uses will be clustered around several new train stations and bicycle-only and bicycle/pedestrian-only bridges will provide fast, direct connections to the city center. Homes and buildings will meet a low-energy standard and only certified sustainably harvested wood will be allowed.

European cities also provide excellent and generally successful examples of redevelopment and adaptive reuse of older, deteriorated areas within the center-city. Good examples include Amsterdam's eastern docklands, where 8000 new homes have been accommodated on recycled land. In *Java-eiland*, one major piece of this project, an overall plan (prepared by urban designer Sjoerd Soeters) lays out broad density, massing, and circulation for the district. Diversity and distinctiveness in actual design of the buildings, however, was encouraged through a restriction on the number of buildings that could be designed by a single architect. The result is a stimulating community where buildings have been created by scores of different designers. This island district successfully balances connection to the past (a series of canals and building scale reminiscent of historic Amsterdam) with unique modern design (each of the pedestrian bridges crossing the canals offers a distinctive look and design). *Java-eiland* demonstrates that city building can occur in ways that create interesting and organically evolved places, and which also acknowledge and respect history and context, overcoming sameness.

European cities on the whole (and especially the cities examined in this study) have been able to maintain and strengthen their center cities and urban cores. In no small part this is a function of historic density and compactness, but they are also the result of numerous efforts to maintain and enhance the quality and attractiveness of the city-center. In the cities studied, the center has remained a mixed-use zone, with a significant residential population. *Groningen*, for instance, has undertaken a host of actions to improve its center including the creation of new pedestrian-only shopping areas (creating a system of two linked circles of pedestrian areas), and installation of (yellow) brick surfaces and new street furniture in walking areas, among other actions.

Committed to a policy of compact urban form, Groningen has also made a strong effort to keep all major new public buildings and public attractions close-in. As one example, a new modern art museum has been sited and designed to provide an important pedestrian link between the city's main train station and the town center.

SUSTAINABLE MOBILITY

Achieving a more sustainable mix of mobility options is a major challenge, and in almost all of the cities studied in *Green Urbanism* a very high level of priority is given to building and maintaining a relatively fast, comfortable and reliable systems of public transport.

There are impressive examples of cities that have been working hard to expand and enhance transit, in the face of rising auto use in many areas. Zurich implements an aggressive set of measures to give priority to its transit on streets. Trams and buses travel on protected, dedicated lanes. A traffic control system gives trams and buses green lights at intersections and numerous changes and improvements have been made to reduce the interference of autos with transit movement (e.g. bans on left turns on tram line roads; prohibiting stopping or parking in certain areas; building pedestrian islands; etc.) A single ticket is good for all modes of transit in the city (including buses, trams, and a new underground regional metro system). The frequency of service is high and there are few areas in the canton that are not within a few hundred meters of a station of stop. Cities like Freiburg and Copenhagen have made similar strides.

In these European cities transit modes are integrated to an impressive degree. This means coordination of investments and routes so that transit modes complement each other. In most of the cities studied, for instance, regional and national trains systems are fully integrated with local routes. It is easy, as well, to shift from one mode to another. Local transit centers are viewed in these cities as multi-modal, mixed-use centers of activity. Arnhem's new central train station in the Netherlands is a case in point. It integrates in a single location high-speed and conventional train service, local transit, bicycle parking, rental, and repair, as well as shops, offices and housing. These uses are all within a few hundred meters of the city center.

The ease of traveling throughout Europe is aided tremendously by the commitment on this continent to high-speed rail. Cross-national movement by high-speed train is increasingly comfortable and easy, and investments in dedicated tracks and infrastructure reflect impressive forward thinking on this issue. And increasingly it is not just the northern and northwestern European nations leading the way. Major new high-speed rail systems are under construction in Italy and Spain for instance. Overall, plans are on the books to double the length of dedicated high-speed rail track in Europe over the next eight years. And, the newest generation of trains will travel faster – on average 300 kph or higher.

Importantly, investments in transit complement, and are coordinated with, important land use decisions. Virtually all the major new growth areas identified in this study have good public transit service as a basic, underlying design assumption. The cities studied here do not wait until after the housing is built, but rather the lines and investments occur contemporaneously with the projects. The new community growth area *Rieselfeld* in Freiburg, for instance, has a new tram line even before the project has been fully built. In Amsterdam, as a further example, at the new neighborhood of *Nieuw Sloten*, tram service began when the first homes were built. In the new ecological housing district *Kronsberg*, in Hannover, three new tram stops ensure that no resident is further than 600 meters away from a station. There is an recognition in these cities of the importance of providing new residents with options, and establishing mobility patterns early.

Car sharing has become a viable and increasingly popular option in Europe cities. Here, by joining a car sharing company or organization residents have access to neighborhood-based cars, on an hourly or per-kilometer cost. There are now some 100,000 members served by car sharing companies or organizations in 500 European cities. Some of the newest car sharing companies, such as *GreenWheels* in the Netherlands, are also pursuing creative strategies for enticing new customers. This company has been developing strategic alliances, for example with the national train company, to provide packages of benefits at reduced prices. One of the key issues for the success of car sharing is the availability of convenient spaces, and a number of cities, including Amsterdam and Utrecht, have been setting-aside spaces for this purpose. In cities such as Hannover, Germany, the car sharing organization there (a non-profit called Okostadt) has strategically placed cars at the stations of the Stadtbahn, or city tram, furthering enhancing their accessibility.

Thinking beyond the automobile

Many of these cities are in the vanguard of new mobility ideas and concepts and are working hard to incorporate them into new development areas. Amsterdam, for example, has taken an important strategy in developing *Jjburg*. It is working to develop a comprehensive mobility package that all new residents will be offered and which includes, among other things, a free transit pass (for a certain specified period) and discounted membership in local car sharing companies. Minimizing from the beginning the reliance on automobiles, and giving residents more mobility options, are the goals. Eventually this new area will be served both by an extension of the city's underground metro and fast tram.

An increasing number of carfree housing estates are also being developed in these cities, as a further reflection of the commitment to minimizing auto-dependence. The *GWL-Terrein* project, also in Amsterdam, built on the city's old waterworks site, incorporates only very limited peripheral parking. An on-site car sharing company, in combination with good tram service, are part of what makes this concept work there. The interior of the project incorporates extensive gardens (and 120 community gardens available to residents) and pedestrian environment, with key-lock access for fire and emergency vehicles.

Another carfree experiment is the new ecological district *Vauban*, in Freiburg. Built on the former site of an army barracks, this project is unique because it gives new residents the opportunity to declare their intentions to be carfree, and rewards them financially for doing so. Specifically, if residents choose to have a car, they must pay approximately $13,000 for the cost of a space in the nearby parking garage (a bit less than one-tenth the cost of the housing units). In this way there is a strong financial incentive to choose to be carfree and so far about half the residents have taken the carfree path. Projects like *Vauban* challenge new residents to think and act more sustainably and reward them for doing so.

Bicycles are an impressive mobility option in almost all of the cities studied in *Green Urbanism*, and many of these cities have taken tremendous efforts to expand bicycle facilities and to promote bicycle use. Berlin has 800 km of bike lanes, and Vienna has more than doubled its bicycle network since the late 1980s. Copenhagen now has a policy of installing bike lanes along all major streets, and bicycle use in that city has risen substantially. Few have gone as far, of course, as the Dutch cities, with cities like Groningen, where more than half of the daily trips are made on bicycles. In virtually all new growth areas in the Dutch cities, as well as many Scandinavian and German cities, bicycle mobility is an essential design feature, including providing important connections to existing city bicycle networks.

A number of actions have been taken by these cities to promote bicycle use. These include separated bike lanes with separate signaling, separate signaling and priority at intersections, signage and provision of extensive bicycle parking facilities (e.g. especially at train stations, public buildings), and minimum bicycle storage and parking standards for new development. Many cities are gradually converting spaces for auto parking to spaces to bicycles. Utrecht has discovered that it can fit 6–10 bicycles in the same space it takes to park one automobile. Tilburg, in the Netherlands, has recently built an underground valet bicycle parking facility in the heart of that city's shopping district. Freiburg's mobility center combines two levels of bicycle parking, with car-sharing cars on the ground level, a café, travel agency, and office of the Deutsche Bahn (and the structure has a green roof and a photovoltaic array generating electricity!).

These cities are also innovating in the area of public bikes. The most impressive program is Copenhagen's "City Bikes," which now makes available more than 2000 public bicycles throughout the center of the city. The bikes are brightly painted (companies sponsor and purchase the bikes in exchange for the chance to advertise on their wheels and frames), and can be used by simply inserting a coin as a deposit. The bikes are geared in such a way that the pedaling is difficult enough to discourage their theft. The program has been a success, and the number of bikes has been expanding. These sustainable European cities have discovered that bicycles are an important and legitimate alternative mode of transport to the car and with modest planning and investments substantial ridership can be achieved.

BUILDING PEDESTRIAN CITIES; EXPANDING THE PUBLIC REALM

European cities represent, as well, exemplary efforts at creating walkable, pedestrian urban environments. Relatively compact, dense, and mixed-use urban environments make cities much more walkable, of course. And most European cities and regions benefit from having a compact historic core, designed and evolved around walking and face-to-face commerce. The vitality, beauty, and attraction of European cities is in no small part a function of the impressive public and pedestrian spaces. Cities like Barcelona and Venice remain positive and compelling models of pedestrian urban society. The uses of these spaces are varied and many: they are outdoor stages, the "living rooms" in which citizens socialize, interact, and come together, places where political events occur and democracy plays itself out. These areas are now the social heart of these communities – places where children play, casual conversations and unexpected meetings take place, and people come to watch and be seen.

The overall land use pattern in these cities, and the priority given to maintaining their compact form, certainly make a walking culture more feasible. What is especially impressive, however, is the continued attention given to this issue and the continued expanding of pedestrian areas and the strengthening of the public and pedestrian realm. Cities like Copenhagen have set the stage, beginning in the early 1960s, gradually taking back their urban centers from cars. That city pedestrianized the Stroget, one of its main downtown streets, in 1962. Copenhagen continues this pedestrianizing in a gradual way each year. The city has adopted the policy of converting 2–3 percent of its downtown parking to pedestrian space each year, to dramatic effect over a 20–40 year period. Today the amount of pedestrian space is tremendous. Eighteen pedestrian squares have been created in Copenhagen where there was once auto parking – some 100,000 square meters in all. Had proponents of public space in Copenhagen attempted to convert this amount of space all at once it would have been very politically difficult to do so.

Many other cities have followed suit, especially Dutch and German cities, but examples can be found throughout Europe. Cities like Vienna and Groningen have pedestrianized much of their centers, creating delightful, highly functional public spaces. Groningen's compact city policy ensures that major new public buildings and facilities are kept in the center, and accessible through walking – it is a compact city of "short distances." In cities like Leiden, emphasis has been given to installing new pedestrian bridges over canals connecting major streets, and every new residential area is designed to include a grocery, post office, and other shops within an easy walk. The greater mixing of uses means that residents of these cities typically have many shops, services, cafés within a walkable range.

The experience of these European cities in pedestrianizing much of their urban centers has been a positive one, both economically and in terms of quality of life. The spaces created commonly contain fountains, sculptures and public art, extensive seating and, of course, many reasons for being there – restaurants, cafés, shops. Each city has its own unique history and features that can be used to strengthen the unique character of its pedestrian environment. Freiburg's "backle," or urban streams that run through the streets of its old center, as well as its pebble mosaics are delightful and special and this city has done an excellent job expanding and adding onto to these unique qualities of place.

Good public transit appears a major factor strengthening the pedestrian realm in these cities, as well as commitments to bicycles, as in the case of Copenhagen. Extensive efforts to calm urban traffic, to restrict auto access, and to raise the cost of parking and auto mobility are also important elements. A number of European cities have experimented with or are anticipating some form of road pricing. The City of London is the most recent notable example, now charging a fee of five pounds for cars wishing to enter central London (and already resulting in a significant reduction in car traffic there). These European experiences support that a pedestrian culture and community life is indeed possible, even where the climate may be harsh and that these spaces serve an incredible range of social, cultural, and economic functions.

GREENING THE URBAN ENVIRONMENT

Ensuring that compact cities are also green cities is a major challenge, and there are a number of impressive greening initiatives among the study cities. First, in many of these cities there is an extensive greenbelt and regional open space structure, with a considerable amount of natural land actually owned by the cities. Extensive tracts of forest and open lands are owned by cities such as Vienna, Berlin, and Graz, among others. Cities such as Helsinki and Copenhagen are spatially structured so that large wedges of green nearly penetrate the center for these cities. Helsinki's large *Keskuspuisto* central park extends in an almost unbroken wedge from the center to an area of old growth forest to the north of city. It is 1000 hectares is size and 11 km long.

In Hannover an extensive system of protected greenspaces exists, including the *Eilenriede*, a 650 hectare dense forest located in the center of the city. Hannover has also recently completed a 80-kilometre long *green ring* (der grune ring) which circles the city, providing a continuous hiking and biking route, and exposing residents to a variety of landscape types, from hilly Borde to the river valleys of the Leineaue river.

There is a trend in the direction of creating and strengthening ecological networks within and between urban centers. This is perhaps most clearly evident in Dutch cities, where extensive attention to ecological networks has occurred at the national and provincial levels. Under the national government's innovative Nature Policy Plan, a national ecological network has been established consisting of core areas, nature development areas, and corridors, which must be more specifically elaborated and delineated at the provincial level. Cities in turn are attempting to tie into this network and build upon it. At a municipal level, such networks can consist of ecological waterways (e.g. canals), tree corridors, and green connections between parks and open space systems. Dutch cities like Groningen, Amsterdam, and Utrecht have full time urban ecology staff, and are working to create and restore these important ecological connections and corridors.

Many examples exist of efforts to mandate or subsidize the greening of existing urban areas. There is a continuing trend, for instance, towards installation of ecological or green rooftops, especially in German, Austrian, and Dutch cities. Linz, Austria, for instance, has one of the most extensive green roof programs in Europe. Under this program, the city frequently requires building plans to compensate for the loss of green space taken by a building. Creation of green roofs has frequently been the response. Also since the late

1980s the city has subsidized the installation of green roofs – specifically, it will pay up to 35 percent of the costs. The program has been quite successful and there are now some 300 green roofs scattered around the city. They have been incorporated into many different types of buildings including a hospital, a kindergarten, a hotel, a school, a concert hall, and even the roof of a gas station. Green roofs have been shown to provide a number of important environmental benefits, and to accommodate a surprising amount of biological diversity. Many other innovative urban greening strategies can be found in these cities from green streets, to green bridges, to urban stream daylighting.

RENEWABLE ENERGY AND CLOSED-LOOP CITIES

A number of the cities have taken action to promote more closed-loop urban metabolism, in which, as in nature, wastes represent inputs or "food," for other urban processes. The city of Stockholm has made some of the most impressive progress in this area, and has even administratively reorganized its governmental structure so that the departments of waste, water, and energy are grouped within an eco-cycles division. A number of actions in support of ecocycle balancing have already occurred. These include, for instance: the conversion of sewage sludge to fertilizer and its use in food production, and the generation of biogas from sludge. The biogas is used to fuel public vehicles in the city, and to fuel a combined heat and power plant. In this way, wastes are returned to residents in the form of district heating. Another powerful example of the closed-loop concept can be seen in Rotterdam, [where] the Roca3 power plant supplies [*sic*] district heating and carbon dioxide to 120 greenhouses in the area. A waste product becomes a useful input, and in this case prevents some 130,000 metric tonnes of carbon emissions annually.

Energy is very much on the planning agenda, and these exemplary cities are taking a host of serious measures to conserve energy and to promote renewable sources. The heavy use of combined heat and power (CHP) generation, and district heating, especially in northern European cities, is one reason for typically lower per capita levels of CO_2 production here. Helsinki, for instance, has one of the most extensive district heating systems: more than 91 percent of the city's buildings are connected to it. The result is a substantial increase in fuel efficiency, and significant reductions in pollution emissions. District heating and decentralized combined heat and power plants are now commonly integrated into new housing districts in these cities. In *Kronsberg*, in Hannover, for instance, heat is provided by two CHP plants, one of which, serving about 600 housing units and a small school, is actually located in the basement of a building of flats.

Many cities, including Heidelberg and Freiburg, have set ambitious maximum energy consumption standards for new construction projects. Heidelberg has recently sponsored a low-energy social housing project, to demonstrate the feasibility of very low-energy designs (specifically a standard of 47 kwh/m^2 per year). The Dutch are promoting the concept of energy-balanced housing – housing that will over the course of a year produce as much energy as it uses – and the first two of these units have been completed in the *Nieuwland* district in Amersfoort.

Many cities such as Heidelberg have undertaken programs to evaluate and reduce energy consumption in schools and other public buildings. Incentive programs have been established which allow schools to keep a certain percentage of the savings from energy conservation and retrofitting investments. Heidelberg has engaged in an innovative system of performance contracts, in which private retrofitting companies get to keep a certain share of the conservation benefits.

There is an explosion of interest in solar and other renewable energy sources in these cities (and countries). Cities like Freiburg and Berlin have been competing for the label "solar city," with each providing significant subsidies for solar installations. In the Netherlands, major new development areas, such as *Nieuwland* in Amersfoort and *Nieuw Sloten* in Amsterdam, are incorporating solar energy, both passive and active, into their designs. In Nieuwland, described as a "solar suburb," there are more than 900 homes with rooftop photovoltaics, 1100 homes with thermal solar units, and a number of major public buildings producing power from solar (including several schools, a major sports hall, and a childcare facility). What is particularly exciting is to see the effective integration of solar into the architectural design of homes, schools and other buildings.

The degree of public and governmental support in these European cities, financial and technical, for renewable energy developments is truly impressive.

Reflecting a generally overall level of concern for global warming issues and energy self-sufficiency, significant production subsidies and consumer subsidies have both been given. The degree of creativity in incorporating renewable energy ideas and technologies in many of these cities is also quite impressive. Oslo's new international airport, for example, provides heating through a bark/wood bio-energy district heating system. This system provides heat for buildings through 8 km of pipes, as well as the airport's de-icing system. The moist bark fuel is a local product, and costs only one-third as much as fuel oil. In Sundsvall, Sweden, snow is collected, stored, and used as a major cooling source for the city's main hospital. In Copenhagen, twenty 2 MW wind turbines have been installed offshore which will together generate enough energy for about 30,000 homes.

Green cities, green governance

Many of these cities are taking a hard look at ways their own operations and management can become more environmentally responsible. As a first step, many local governments have undertaken some form of internal environmental audit. Variously called green audits or environmental audits, they represent attempts to study comprehensively the environmental implications of a city's policies and governance structure. A number of local governments are now going through the process of becoming certified (the London borough of Sutton being the first) under the EU's Eco-Management and Audit Scheme (EMAS), an environmental management system more commonly applied to private companies. Several German cities are preparing environmental budgets, under a pilot program. The cities of Den Haag and London have calculated their ecological footprints and are using these measures as policy guideposts. Albertslund, Denmark, has developed an innovative system of "green accounts," used to track and evaluate key environmental trends at city and district levels, and many of the study cities have developed sustainability indicators (e.g. Leicester, London and Den Haag). Cities like Lahti, Helsinki, and Bologna have gone through extensive in-house education and involvement of city personnel, often as part of the local Agenda 21 process, in examining environmental impacts and in identifying ways that personnel and city departments can reduce waste, energy, and environmental impacts.

Municipal governments have taken a variety of measures to reduce the environmental impacts of their actions. A number of communities have adopted environmental purchasing and procurement policies. Cities like Alberstlund have adopted policies mandating that only organic food can be served in schools and child care facilities, and restricting use of pesticides in public parks and grounds. Other cities are aggressively promoting the development of environmental vehicles. Stockholm's environmental vehicles program is one of the largest (a pilot program under the EU-funded initiative ZEUS), with over 300 vehicles. A number of cities have sought to modify the mobility patterns of employees, for instance by creating financial incentives for the use of transit or bicycles. Cities like Saarbrucken, Germany, have made great strides in reducing energy, waste, and resource consumption in public buildings.

Communities have also engaged in extensive public involvement and outreach on sustainability matters. A variety of creative approaches have been taken. Leicester, for instance, has developed alliances with the local media and has sponsored a series of educational campaigns on particular community issues. As a further example, it has established (with its NGO partner Environ) an environmental center and cyber-cafe called the Ark, as well as a demonstration ecological home. Officials in these exemplary cities often express the belief that it is essential to set a positive example for the community and that before they could ask citizens to change their behaviors and lifestyles, the municipal government must have its environmental house in order.

Understanding European cities: some concluding thoughts

To be sure, many European cities are facing some serious problems and trends working against sustainability, in particular a dramatic rise in automobile ownership and use, and a continuing pattern of de-concentration of people and commerce. And, with their relatively affluent populations consuming substantial amounts of resources, European cities exert a tremendous ecological footprint on the world. Yet, these most exemplary cities provide both tangible examples of sustainable practice, and inspiration that progress can be made in the face of these difficult pressures.

"Urban Planning and Global Climate Change"

Stephen Wheeler

Editors' Introduction

In this selection, newly written for this edition of *The City Reader*, Stephen Wheeler, an associate professor of landscape architecture and environmental design at the University of California, Davis, addresses one of the most important threats to cities and towns around the world and what many consider the greatest challenge humanity has ever faced – global climate change. As Wheeler points out, carbon from greenhouse gases (GHG) is the biggest culprit, so planning low-emission or carbon-neutral cities for the future is a major urban planning challenge everywhere in the world. Wheeler argues that even meeting that very difficult urban planning goal will not be enough. Harmful effects of global climate change are already inevitable and in the coming decades urban planners must help communities mitigate the effects of global climate change that have already occurred or will occur despite best efforts now.

Wheeler's selection provides a theoretical framework for urban planning to address climate change and suggests a range of practical planning practices at every scale to deal with this large and complex problem.

The Brundtland Commission's conclusion in 1987 that world development practices were not sustainable (p. 351) was much disputed at the time, but is now accepted by all but a handful of climate change skeptics and special interest groups. The narrow window for action that the Brundtland Commission identified in the late 1980s to slow or stop enormously destructive development without huge, costly, and wrenching programs has now closed. No matter how effectively urban planners change plans for cities of the future, so much damage has now been done to the earth that world cities will experience severe climate-change-related problems. The scientific consensus is that temperatures will rise by about 2°C (3.6°F) by mid-century. These projections are based on complex computer models and assumptions that are open to debate. But changes approaching this magnitude will have enormous impacts. Heat waves will likely increase mortality among people and animals. Climate change will affect agriculture and food availability. Water scarcity will become a problem as mountain snowpacks and glaciers melt. Shifting global air circulation patterns will cause droughts in many parts of the world. Storm surges and sea level rise will require costly flood protection systems and may flood cities built near sea level regardless. Global climate change is likely to produce excessive rainfall in some areas of the world and drought in other areas. It will have complex effects on ecosystems, agriculture, and health. These changes will likely require the relocation of millions of people and in hard-hit areas may produce political instability and even provoke wars.

Wheeler points out that greenhouse gases seep into the atmosphere from sources as varied as smokestacks, automobile tailpipes, livestock manure, and air conditioning equipment. They come from transportation, electricity and heat used in buildings, industry, land use changes, agriculture, and landfills and many other sources. Accordingly any effort to reduce global warming must consider many sources and a new and extremely holistic way of thinking about human actions.

How did we get into this situation? The enormous and exponential growth of the world's population described by Kingsley Davis in Part One on The Evolution of Cities (p. 20) helps explain our predicament. The emergence of enormous mega-city regions described by Tingwei Zhang (p. 590) also illuminates this important question.

Both population increase and urbanization have had profound and often catastrophic effects on the natural environment of the planet. Nonrenewable energy sources have been consumed, forests cleared, and species extinguished. Technology, how humans produce material goods, and how they move around also contribute to global climate change. More and more people everywhere in the world participate in the auto-centered culture Kenneth Jackson describes (p. 65), depleting nonrenewable energy sources and contributing to GHG emissions. Low density, sprawling land-use patterns that Bruegmann (p. 211) and Calthorpe and Fulton (p. 360) describe increase auto dependency.

Wheeler's selection is an excellent example of relating urban planning theory and practice. He describes important theories about how to address global climate change. Princeton professors Stephen Pacala and Robert Socolow, for example, have proposed a "wedges" strategy calling for the world's nations to identify and pursue "wedge" policies such as improved motor vehicle fuel efficiency, and substitution of wind, photovoltaic, and nuclear technologies for coal power that could each reduce global carbon emissions by one gigaton annually. Wheeler describes alternative theoretical frameworks to address global climate change that Lester Brown, of the Worldwatch Institute and Earth Policy Institutes, James Hansen or the US National Aeronautics and Space Administration (NASA), Stanford professor Paul Ehrlich, and Harvard professor John Holdren, have proposed.

At the level of urban planning practice, Wheeler points out that planning to address global climate change is a complex task that requires holistic and interdisciplinary approaches connecting the insights of biologists and transit planners, agronomists, economists, and many other disciplines.

Wheeler feels that governments at every level have been much too slow in planning for GHG reductions. But some progress has been made. Wheeler provides many examples of best practices to address global climate change and Timothy Beatley (p. 446) enumerates many more.

At the national and multinational scales, important, albeit insufficient, multinational agreements have been made. Some regions, such as the European Union, and many of the world's countries have set goals for reducing GHG emissions and are pursuing strategies to meet those goals. At regional and subnational scales, Wheeler describes plans many governments and agencies have developed for reducing GHG emissions. A majority of US states now have some sort of climate change plan. Metropolitan regional agencies are also adapting to the new reality. Many are reworking plans for public transit systems to promote transit-oriented development and ensure more compact urban form.

A growing number of cities, towns, and rural communities have adopted climate change plans and policies to reduce GHG emissions. Some require public buildings to be certified under the US Green Building Council's Leadership in Energy and Environmental Development (LEED) rating system or encourage "passive houses" that use little or no off-site energy. Other local approaches that Wheeler describes include investing in bicycle and pedestrian transportation systems, creating district heating and cooling systems, promoting small-scale "community energy systems" using wind and solar power, reducing the land area covered by dark asphalt (which absorbs heat) and painting roofs light colors (which reflect heat), and many more.

Wheeler concludes that global climate change is likely to be the greatest challenge this generation of urban planners face. But it is also an opportunity – a chance to create far more livable, equitable, and sustainable communities and lifestyles. Climate change may finally provide the impetus for sustainable development.

Stephen Wheeler is an associate professor of landscape architecture and environmental design at the University of California, Davis. Previously he taught in the Department of Community and Regional Planning at the University of New Mexico. Prior to his academic career, Wheeler was a lobbyist for Friends of the Earth, a board member of Urban Ecology, and, for eight years, the editor of *The Urban Ecologist*. He is the author of *Planning for Sustainability: Creating Livable, Equitable, and Ecological Communities* (London: Routledge, 2004) and the coeditor, with Timothy Beatley, of *The Sustainable Urban Development Reader*, 2nd edn (London: Routledge, 2008).

Among the most important books on global climate change are *Our Choice: A Plan to Solve the Climate Crisis* (Emmaus, PA: Rodale, 2009) by former US vice-president and Nobel peace prize winner Albert Gore and *The Report of the Intergovernmental Panel on Climate Change* (Geneva: IPCC, 2007).

Other books on global climate change include Arnold Bloom, *Global Climate Change* (Basingstoke, UK: Sinauer, 2008), Andrew Dessler and Edward A. Parson, *The Science and Politics of Global Climate Change*

(Cambridge: Cambridge University Press, 2006), Tim Flannery, *The Weather Makers: How Man is Changing the Climate and What It Means for Life on Earth* (New York: Grove, 2005), Diane Dumanoski, *The End of the Long Summer: Why We Must Remake Our Civilization to Survive on a Volatile Earth* (New York: Crown, 2009), Mark Lynas, *Six Degrees: Our Future on a Hotter Planet* (Washington, DC: National Geographic, 2008), Elizabeth Kolbert, *Field Notes from a Catastrophe: Man, Nature, and Climate Change* (New York: Bloomsbury, 2006), Karen McGlothlin, *Global Climate Change* (Lanham, MD: Rowman & Littlefield, 2006), Thomas R. Karl, Jerry M. Melillo, Thomas C. Peterson, and Susan J. Hassol, *Global Climate Change Impacts in the United States* (Cambridge: Cambridge University Press, 2009), George Monbiot, *Heat: How to Stop the Planet from Burning* (Cambridge, MA: South End Press, 2007), Fred Pearce, *With Speed and Violence: Why Scientists Fear Tipping Points in Climate Change* (Boston, MA: Beacon, 2006), Joseph Romm, *Hell and High Water: Global Warming – the Solution and the Politics – and What We Should Do* (New York: Morrow, 2007), Michael E. Schlesinger et al. (eds) *Human Induced Climate Change: An Interdisciplinary Assessment* (Cambridge: Cambridge University Press, 2007).

For contrarian views, see Patrick J. Michaels and Robert C. Balling, Jr., *Climate of Extremes: Global Warming Science They Don't Want You to Know* (Washington, DC: Cato Institute, 2010) and Bjørn Lomborg, *The Skeptical Environmentalist: Measuring the Real State of the World* (Cambridge: Cambridge University Press, 2001).

For a critique of climate change skeptics, see James Hoggan and Richard Littlemore, *Climate Change Coverup: The Crusade to Deny Global Warming* (Petersburg, VA: Graystone, 2009).

■

Perhaps the largest planning challenge humanity has ever faced – and one of the preeminent threats to cities and towns around the world – is climate change. Within the lifetimes of many of those living today, societies will need to become virtually carbon-neutral, a huge challenge given that our transportation systems, buildings, agriculture, and industries are currently extremely fossil fuel-dependent. And then there will be the enormous task of adapting to climate change impacts that to a certain extent are now inevitable. These coming effects include a temperature rise of at least 2 degrees Celsius (3.6 degrees Fahrenheit), drought or flooding in many parts of the world, and a sea level rise of several meters within a century and much more after that.

But this challenge is also an opportunity, a chance to create far more livable, equitable, and sustainable communities and lifestyles. Climate change may finally provide the impetus for sustainable development, something many environmentalists and social justice advocates have been urging since the 1960s, but which global capitalism and allied political and cultural forces have resisted.

Humankind has in fact known that it may be changing the earth's climate since the nineteenth century. In 1859 John Tyndall discovered that gases such as carbon dioxide and water vapor create a greenhouse effect for the Earth, and in the 1890s Svante Arrhenius calculated with a surprising degree of accuracy the amount that a doubling of atmospheric carbon dioxide would heat the planet. Additional elements of the science of climate change were put in place in the 1950s, when scientists discovered that the oceans would not be able to absorb nearly as much carbon dioxide as previously thought, and when the first annual documentation of rising atmospheric carbon dioxide levels began. US national science agencies issued authoritative reports on the likelihood of climate change in the 1970s, and top US climate scientist James Hansen announced in 1988 that global warming could then be seen against the background "noise" of annual variations. However, it is only now, in the early twenty-first century, that our species is coming to grips with the fact that this change is actually happening, and that it must take rapid action to reduce emissions and prepare for a changed world.

Responding to global warming will require an evolutionary step forward in humanity's ability to plan for and manage its own future. As part of this process, people and societies will need to get better at understanding complex issues, governing themselves, collaborating across cultures and jurisdictions, and regulating the destructive or self-interested forces that tend to undermine collective welfare and environmental health. Global warming is the most sweeping planning challenge humanity has ever faced, and moreover one that cannot be avoided. But by the same token addressing climate change can be an exciting and creative task, one that can lead towards much healthier communities and give great meaning to the work of current and future generations.

THE MITIGATION CHALLENGE

The most pressing need, in the view of many, has been to reduce human emissions of greenhouse gases (GHGs), a process that has become known by the somewhat wonkish term "mitigation." Greenhouse gases seep into the atmosphere from many sources: carbon dioxide from smokestacks, tailpipes, and forest fires; methane from the manure and belches of livestock as well as from the decomposition of organic matter in landfills; nitrous oxides from farm fertilizers and cattle; hydrofluorocarbons and other chemicals from refrigeration and air conditioning equipment; and other trace gases from industrial processes. In addition, other human pollutants such as soot affect climate change (in this case, by darkening snow and causing it to absorb more of the sun's energy). Any effort to reduce global warming must consider all of these sources – a complex task that requires a new and extremely holistic way of thinking about human actions.

Globally, according to the World Resources Institute, about 13 percent of GHG emissions come from transportation, 34 percent from electricity and heat used in buildings and the energy industry itself, 18 percent from other forms of industry, 18 percent from land use changes including rainforest destruction and the draining of wetlands, 13 percent from agriculture, and 4 percent from landfills and other waste processes. Since the sources of GHGs are so broad based, it is not as though putting scrubbers on smokestacks in a particular industry can address the problem (as it has done for some local air pollutants). *Every* human process must be examined for its climate change impacts.

Governments at every level have been slowly – much too slowly – attempting to plan for GHG reductions. At the international scale, a United Nations-backed process of international conferences and agreements was established in the early 1990s. The Kyoto Protocol of 1997, which set GHG mitigation targets for 30 industrial countries to reach by the 2008–2012 period, was one product of this process. Many European nations and Japan did in fact meet their Kyoto goals, but other nations including the United States did not (the United States signed but never ratified the treaty). Developing nations such as China and India were for the most part left out of the Kyoto framework. The 2009 Copenhagen Climate Conference was intended to produce a stronger treaty that could be agreed to by all the world's nations, but unfortunately failed to do so. Less broad-based international agreements to mitigate climate change, such as the Copenhagen Accord between the United States, China, India, Brazil, and South Africa, have been developed instead. Much is being done at the international level, certainly, but overall such planning efforts have been far from adequate to reduce global warming emissions.

At national or multinational scales, many of the world's countries have set goals for reducing GHG emissions and are pursuing strategies to meet those targets. For example the European Union, representing twenty-seven nations, has an overall goal of reducing emissions 20 percent below 1990 levels by 2020 and has considered targets of as much as 95 percent below 1990 levels by 2050. To achieve these goals, the EU and its member nations have established a variety of product regulations, carbon taxes, and incentives for wind and solar power, as well as a continent-wide "cap-and-trade" system. The latter sets allowable emissions levels for different industries, lowered over time, and then allows companies that do more than their share to sell credits for the foregone emissions to firms not able to reduce GHG pollution so quickly. Such a framework provides an incentive to businesses that make rapid progress, while accommodating those not able to change so quickly.

At state, regional, and local scales, many governments and agencies have also developed plans for reducing emissions. States frequently regulate electric utilities, and so have often set "renewable portfolio standards" requiring those firms to produce a certain percentage of their power from renewable sources. States can also toughen their building codes to require more energy-efficient construction, as California has done repeatedly since 1978, and can establish frameworks for better land use planning, as Oregon has done for a variety of reasons since the 1960s. As of 2010, about thirty US states had developed overall climate change plans incorporating dozens of potential actions. Those plans were only a beginning, though; most lacked the funding and regulatory power to bring about the necessary emissions reductions.

Metropolitan regional agencies – powerful sometimes in Canada, the United Kingdom, and continental

Europe but less so in the United States and Australia – can play a role in reducing greenhouse gases through actions such as improving public transit systems, ensuring compact urban form, and promoting transit-oriented development. For example the San Diego Association of Governments regional agency, known as SANDAG, has developed an ambitious plan to expand the area's light rail and bus system and to cluster new housing, office, and commercial development around stations. Regional agencies in Toronto, London, and Portland have historically sought similar goals.

At the lowest end of the governance scale, a growing number of local cities, towns, and rural communities have adopted climate change plans and policies to reduce emissions in those areas that they have control over. Bicycle and pedestrian transportation systems, for example, are most appropriately developed at the local level. Detailed urban design and land use changes are likewise best implemented locally, following goals set by higher levels of government. Green economic development programs, recycling, ecological education, and many social services are primarily the responsibility of local government. Nonprofit organizations such as ICLEI – Local Governments for Sustainability have assisted local governments worldwide in developing emissions inventories and climate change plans.

Cities, indeed, have often been at the forefront of pushing the world toward climate change action. Local planning for greenhouse gas mitigation has been underway in some communities since the late 1980s, and got a major boost following the 1992 United Nations "Earth Summit" conference in Rio de Janeiro. One typical local action has been to require public buildings and motor vehicle fleets to be energy-efficient. Many American cities and towns now require public buildings to be certified under the US Green Building Council's Leadership in Energy and Environmental Development (LEED) rating system. By producing a sizeable crop of green buildings, this step in turn has helped to reduce costs of green design and technology. But public buildings are a small fraction of overall construction, and so it is essential to change building codes to require low energy buildings in the private sector as well. Since most structures built now will be around in 2050, it is important to require carbon-neutral buildings almost immediately. Such structures would generate some of their own power on-site to offset their own, very low energy needs. Already, there are examples worldwide of what the Germans call "passive houses," buildings that use little or no off-site energy.

More proactive energy planning of all sorts is needed in cities. District heating and cooling systems – which very efficiently provide those services to an entire neighborhood – have been pursued in sustainable community projects such as the Hammarby district of Stockholm and the South Coast city of Southampton, UK (which has met many of its energy needs through geothermal facilities since the late 1980s). Cogeneration, in which both heat and electric power are produced together, promises further energy efficiencies. Meanwhile, small-scale wind power and solar energy systems can be integrated into rooftops, parking lots, and other urban spaces. Although large wind and solar farms will also be needed, these small-scale "community energy systems" can help integrate sustainable energy production into every neighborhood and reduce the need for long-distance transmission lines.

In recent decades cities and towns worldwide have also taken steps to reduce the amount that people drive, and thus the carbon dioxide emissions from motor vehicle use. This is generally done through initiatives in three areas: programs to promote alternatives to driving (train, bus, bike, and walking); economic incentives to reduce driving, such as higher parking charges; and policies to change land use so that destinations such as homes, workplaces, and shops are closer to one another and to public transit.

The problem has been that many communities have continued at the same time to allow forms of development that increase driving, for example, widening roads; approving outlying malls, office parks, and housing tracts; and permitting low-density exurban development in the countryside. Unfortunately, communities cannot have things both ways. Continued suburban and rural sprawl undermines the market for compact, low-emission neighborhoods and generates political demand for more roads and motor vehicle subsidies. Communities urgently need to find the political will to outlaw types of development that produce high levels of GHG emissions.

Planning to reduce emissions in the countryside is also needed. By preserving and expanding forests, which take carbon dioxide from the atmosphere and sequester it in wood and soils for centuries, societies can partially offset their own urban and suburban emissions, while preventing the emissions that result from deforestation. Preserving wetlands, especially peat bogs, is also essential, since when waterlogged areas are drained or burned they tend to release large amounts of methane and carbon dioxide into the atmosphere. This source of emissions is a special problem in developing nations such as Indonesia.

Farming practices will need to change to mitigate emissions, and also to reduce dependency on fossil fuels that will be in increasingly short supply as world oil supplies decline. Deep plowing of soils releases nitrous oxide, a powerful greenhouse gas, so low-till or no-till farming practices are desirable. Synthetic nitrogen-based fertilizers also produce nitrous oxide emissions, which argues for using organic practices to maintain soil fertility. Rice growing often leads to methane emissions, as does the raising of livestock, especially ruminants (animals such as cows that digest plant matter in multiple stomachs). Farm machinery and irrigation release GHG emissions by burning fossil fuels. The exact mechanisms by which agricultural systems produce GHGs are complex and vary depending on crops, soils, local climate and ecology, production techniques, and distances to markets. But many food production practices will certainly need to change. These reforms can affect urban dwellers in positive ways, for example by having agricultural production located in and around cities, providing easier access to healthy food, and by having fewer synthetic chemicals used to produce food.

The overall process of mitigating greenhouse gas emissions will lead to large changes in landscapes and lifestyles. It is likely that several centuries from now, humans will live in much more compact cities than currently – especially in North America – and in smaller dwellings than the huge 2000+ square foot homes currently being built there (since big houses contain a great deal of embodied energy and take more energy to heat and cool). It is also likely that people will use motor vehicles far less, and will buy far fewer material goods, since even with the most efficient technologies these vehicles and products will produce substantial carbon emissions. Every resource will be carefully recycled to further reduce emissions. Diets will have to change, for example to eliminate meats produced in ways that release greenhouse gases. If the problem were only reducing emissions 10, 20, or even 50 percent it might be possible to get away with smaller changes. But the need is for 80 to 100 percent reduction. So mitigating climate change will require planning for radical changes in urban form, function, and lifestyle.

THE ADAPTATION CHALLENGE

Early climate change plans focused almost entirely on mitigating emissions, with little attention to how nations or communities are to adapt to a changing climate. But as science provides ever more proof that substantial changes are already occurring, adaptation has become a topic of increasing attention.

Heat is of course one main cause for concern. Models show average temperatures over much of the continental United States and Europe rising 2 °C (3.6 °F) by 2050 and 3 °C to 5 °C (5.4 °F to 9 °F) by 2100 (the exact amount depends in part on how successful mitigation efforts are between now and then). Warmer temperatures will have a wide range of effects that in turn require many different adaptation strategies.

For one thing, heat waves may lead to greater mortality, especially among poor and elderly people without sufficient cooling. According to the French government, a 2003 heat wave killed 14,800 people in France; a total of 37,000 died across Europe. To prevent such waves of death on very hot days, local government programs might identify beforehand and cool or insulate the homes of at-risk individuals. Human health effects also include the possibility that diseases may spread into new geographic regions when their animal or insect hosts migrate due to changing climate. Proactive scientific analysis of changing disease trends can help identify such risks, leading to appropriate public health measures.

Apart from health effects, people's basic comfort levels are likely to decline during hotter summers. This problem is worsened by urban heat islands – city landscapes dominated by pavement and buildings that can raise urban temperatures by as much as 5–10 degrees Fahrenheit. Simply increasing air conditioning is not a desirable response, since this would probably lead to more energy consumption and greenhouse gas emissions. Many cities instead have initiated tree-planting programs. Already New York, Denver, and Los Angeles have undertaken "million tree" programs to increase urban forests. Also, use of building overhangs, covered walkways, and thermal mass within buildings to retain nighttime coolness can help improve comfort as temperatures rise. Vernacular architecture in hot climates such as around the Mediterranean has long used such strategies to make hot summers endurable. Reducing land area covered by dark asphalt (which absorbs heat), painting roofs light colors (which reflect heat), or creating vegetated "green roofs" can also help.

Hotter temperatures will affect agriculture in many ways, in turn affecting food availability. Some places will become too hot to grow current crops, since each species has maximum temperatures beyond which it does not flower, fruit, or even survive. Crops such as peaches and plums whose reproductive cycles depend on winter cold may not be viable either. And new crop pests will appear in many places due to the changed climate. Extensive scientific studies are needed to refine adaptation strategies related to agriculture. In some places new heat-tolerant strains of crops can be grown, but many other places face a permanent change in what can be grown locally.

Quite apart from heat, changes in precipitation will require a wide range of adaptation responses. As global air circulation patterns shift, many parts of the Earth will experience greater drought. These areas include the Mediterranean basin, the American Southwest, northern and southern Africa, parts of Brazil, and Australia. In these places communities will need to plan even more vigorously to conserve water. Conversely, some parts of the world are likely to become wetter. These areas include northern North America, Europe, and Asia, and potentially East Africa. Here, flood prevention programs will be necessary. In many parts of the world storm intensities are likely to increase, since hot weather systems contain more energy. Communities will need to undertake programs to protect against flood and wind damage, not just from hurricanes, but from many lesser storms which are likely to become more intense.

Even in areas where rainfall remains constant or increases, water scarcity may become a problem because mountain snowpacks – which currently store water and release it slowly throughout the year – will diminish or vanish. Such problems are likely to particularly affect communities in California, which depends on the Sierra Nevada snowpack to store water, in Bolivia, which uses water from the Andes mountains, and in India and Pakistan, which rely on rivers originating in the Himalaya and other central Asian ranges. Here there is no easy adaptation solution except increased conservation. No amount of dam-building and lake creation is likely to replace the vast quantity of water stored by these mountain ranges.

In the long run – beginning as early as the middle of the current century – sea level rise is likely to be one of the biggest threats to cities. Many of the world's largest urban areas are built virtually at sea level, including New York, London, Amsterdam, Dhaka, and Shanghai. Even a rise of 2–3 meters, likely this century, will threaten them. Drastic measures such as large floodgates on rivers are being contemplated; London has already constructed floodgates on the Thames River south of the city to protect urban areas from storm surges. The Netherlands has also undertaken many protection measures. But as seas keep rising century after century, up to a maximum of perhaps 40 meters above current sea levels, such defenses will eventually fail. (Examination of the geologic past has shown scientists that this amount of sea level rise has occurred when the planet has warmed significantly, as ice sheets have melted in places such as Greenland and Antarctica.) Along the way there will be no recourse except to move populations inland. Already in low-lying areas such as Florida and the Netherlands, much discussion is taking place of how to change land use planning so as to discourage development in areas likely to be flooded, and how to move infrastructure inland as seas rise.

Many secondary effects of climate change will require adaptive responses. If global food production falls because of drought, floods, pests, or changes in temperatures, programs will be needed to ensure that the most vulnerable populations still have enough to eat. If water crises or famines lead to warfare, international diplomacy or peacekeeping will be required. If populations need to be resettled – for example, a 2 meter sea level rise will make much of Bangladesh uninhabitable – then humane and equitable ways to assist refugees will be needed.

Adaptation has been relatively little emphasized in climate change plans to date. But more attention is being paid to this topic by the year, and it is likely to become a central element of city and regional planning. A large unanswered question is how societies, especially in developing nations, will pay for adaptation programs, or for green development in general. Even though energy efficiency programs often pay for themselves in the long run, up-front costs are needed for conversion. Many in developing nations feel that industrialized nations such as the United States, which have been responsible for a disproportionate share of greenhouse gases historically, should help them pay for adaptation. This responsibility has been acknowledged to some extent through the Clean Development Mechanism created at the 1997 Kyoto conference and a pledge by the United States and others at the 2009 Copenhagen conference to raise $100 billion a year for assistance to the developing world. But the first of these mechanisms has been widely criticized as inadequate, and it remains to be seen whether subsequent financial pledges will materialize.

THE SOCIOPOLITICAL CHALLENGE

Although communities and nations around the world are increasingly taking action to mitigate emissions and adapt to climate change, such actions have fallen short of what is needed. Instead of declining, actual global greenhouse gas emissions rose 26 percent between 1990 and 2008, according to data from the US National Oceanic and Atmospheric Administration. The inadequacy of adaptation plans has been shown by disasters such as 2005's Hurricane Katrina, and recurrent crises from flood, drought, and famine in many other parts of the world.

Potential ways to take stronger action are well known. Environmentalists have called for energy conservation and renewable energy programs since the 1970s. Strategies to reduce driving and nonrenewable resource consumption are well known as well. Many creative proposals have been made for stronger climate change action. For example, Lester Brown, founder of the Worldwatch Institute and Earth Policy Institute, has proposed an extensively detailed "Plan B" through which the world's societies would move rapidly toward renewable energy, conservation, sustainable agriculture, stabilized population, and improved social equity. James Hansen, head of NASA's Goddard Institute for Space Studies, has suggested a "fee-and-dividend" system under which energy producers would pay a gradually rising fee for each ton of carbon dioxide in their fuel, with the proceeds equitably refunded to the public. Unlike cap-and-trade systems, which put economic pressure only on some, relatively inefficient companies – and then only if the price of emissions permits is high enough – this system would create a powerful economic incentive for every possible action to reduce emissions.

Perhaps the most influential proposal for stronger action has been the "wedges" strategy laid out in 2004 by Princeton professors Stephen Pacala and Robert Socolow. This approach calls for the world's nations to identify and pursue a handful of "wedge" policies that could each reduce global carbon emissions by one gigaton annually. Wedges are categories of action such as improved motor vehicle fuel efficiency, reduced vehicle use, more efficient buildings, conservation tillage of agricultural land, and substitution of wind, photovoltaic, and nuclear technologies for coal power. The proposal is conceptually simple and elegant. However, each wedge would require enormous effort. The wind power wedge, for example, requires erecting 1 million one-megawatt wind turbines, an undertaking perhaps on a par with the US mobilization for World War II.

The problem is not that humanity could not take such steps. It could, and probably at a net financial savings due to greater energy efficiency, according to research by McKinsey & Company and the separate Stern Review by the British government. The problem instead is one of political will. Entrenched fossil fuel industries and their allies oppose action of any sort. (ExxonMobil has been a leading funder of climate change denial organizations.) Political parties, politicians, and most media commentators in countries such as the United States have not been willing to push for drastic action. And the public within industrialized nations has not been willing to change its own lifestyles or elect politicians who would change the emissions trajectory known as "business as usual."

A number of key strategies needed to prevent climate change are not even on the table for discussion, and if pursued would fundamentally challenge current mainstream beliefs and lifestyles. Stabilizing and actually reducing the world's population is one such action. Global population is expected to level off during the twenty-first century at approximately 10 billion people – since as societies become more affluent birthrates decline. However, this is probably far more than the planet can support, either in terms of GHG emissions or other resource use, at anything approaching an industrialized lifestyle. Yet few people want to talk about limiting or reducing population. Discussing population in many nations also runs into complex and difficult debates about immigration, and cultural and religious traditions advocating large family sizes.

Another unacknowledged need is to reduce consumption. Industrial economies depend to a large extent on continual increases in production and consumption of material goods. On a finite planet this is by definition unsustainable. It also makes reducing GHG emissions extremely difficult. Economist Herman Daly has called since the 1970s for a steady-state economy that emphasizes quality of life rather than quantity of material goods, but this possibility has been ignored. Despite occasional recessions, the current juggernaut of poorly regulated, consumption-oriented capitalism continues unabated. The Buddhist kingdom of Bhutan in the Himalayas appears to be the only country in the world that measures success in terms of what it calls "gross national happiness" instead of "gross national product."

A third off-the-table issue is mobility. We have grown used to driving increasingly long distances each year and flying frequently by jet airplane. Yet only human-powered travel is emissions free. Air travel, in particular, generates extremely large quantities of emissions – flying from New York to Los Angeles produces about as many GHGs as driving one's car for a year. Unfortunately, no good technological substitutes exist for the jet engine. In a carbon-neutral society we will have to travel much less, and live much more locally. This can have many advantages; we may then more actively care for our cities and towns, and learn to appreciate local culture and ecology. We will also spend less time commuting or sitting in traffic. But this alternative direction is so far from current lifestyles that few people yet contemplate it.

A final subject that no one wants to talk about is equity. It is profoundly unfair that a small percentage of the world's people have produced the majority of emissions to date, and burdened the rest with the costs of adaptation. It would also be greatly unfair if, having brought the planet to an ecological precipice, the rich told the poor that it was not possible for them to enjoy the same affluent lifestyles. Somehow a sustainable standard of living has to be found that can be shared equally by all the world's people. But this will inevitably mean the rich giving up many of their overconsumptive ways, and few in industrial countries wish to consider that.

One way to understand these linked issues at the core of the climate change challenge is through the formula "I=PATE," which is a version of one originally developed in the early 1970s by Paul Ehrlich and John Holdren to describe the resource crises of that era. "I" stands for the global warming impact of humans on the planet. "P" stands for Population, the sheer number of human beings on the Earth. "A" stands for Affluence, which here includes both material consumption and mobility, since travel is a form of consumption. "T" stands for Technology, in other words the efficiency and carbon intensity of technologies employed for human lifestyles. And "E" stands for Equity, the extent to which consumptive lifestyles are shared around the world.

Humanity's global warming impact, in other words, is a function of population, affluence, technology, and equity. If any of these factors is high, then it becomes very difficult for the species to have a low climate impact. For example, if the world's population is high, then affluence (consumption) must be very low, technology must be very low carbon, and/or equity must be very low.

To date, "T" (technology) has been the usual focus of climate change debates. "P," "A," and "E" are rarely discussed. The reason has to do with the extent to which meaningful action related to these factors would challenge existing economic, political, and cultural systems. What is needed is to break through the constraints of those institutions, so as to be able to look at what is really necessary to reduce GHG emissions and adapt to global warming.

Planning to address climate change, in other words, requires not just steps toward greener buildings, renewable energy, and better public transit systems, but planning to improve democracy and capitalism. Only if our fundamental social ecology is strengthened – so that, for example, we have wiser and more educated publics that will elect strong and creative leaders, or we have strong regulations to prevent large corporations from skewing public debates – are we likely to be able to move towards a more sustainable, carbon neutral world.

CONCLUSION

The world's cities, towns, and suburbs have a pivotal role to play in climate change planning. They are responsible for perhaps the majority of greenhouse gas emissions, and need to plan immediately for carbon-neutral lifestyles for their inhabitants. Urban communities are also frequently at risk for the impacts of climate change, and need to prepare to adapt to a changed world. Climate change will test urban and regional planning, in other words, on a scale never before seen.

But neither mitigation nor adaptation planning can be effective unless societies examine and address the underlying reasons they have gotten into the current fix. This means developing new economies that do not depend on continual increases in material production and resource consumption, while generating large amounts of pollution. It means developing forms of governance that continually help their publics understand the complexity and interdependency of the current world, in turn producing an electorate and leaders that will support constructive action. It means developing cultures that do not glorify consumption, violence, or national self-interest. And it means rethinking lifestyles so to live much more lightly on the planet.

These things may sound impossible. However, during the past century science and psychology have made it clear that societies can to a large extent shape human nature. This is the biggest planning challenge that we have for ourselves – one of proactively shaping social ecologies in such a way as to bring out the best potential of human beings, individually and collectively. It is time, in short, for our species to take an evolutionary step forward to address the challenges of living on a small planet. Climate change is forcing our hand. We have no choice but to radically change our current institutions. Though daunting, this process can also be immensely rewarding, leading to much improved lives, communities, and societies as well as a healthy earth.

Perspectives on urban design

INTRODUCTION TO PART SEVEN

Part Seven focuses on urban design – the way in which humans actually shape the built environment. Urban designers are usually trained as architects with further training in urban design, city and regional planning, or both. They focus on the design of sites larger than individual buildings – site plans for one or more buildings, blocks, neighborhoods, park systems, highway corridors, or even entire new towns. Professionals from the related field of landscape architecture are educated to manage the relationship between the natural environment and the built environment. As in other areas where academics study cities or professionals work to build cities, material from many disciplines and professional fields is relevant to urban design. Urban designers blend artistic right-brain approaches and rational left-brain functions. The best designers are also social scientists who study how people use the environments they are designing. They may draw on psychology to understand how people perceive the space around them and interact with other people, history to understand how the physical form of a place evolved, and anthropology and sociology to create places that meet the needs of different social groups. Urban designers use computer assisted design (CAD) and illustration software. They use the full gamut of qualitative and quantitative research methods social scientists and urban planners use.

Urban designers may disagree on what makes for a good design, but they share a belief in the value of design itself. They believe professionals should consciously think about physical relationships in the creation of urban space. They may respect and even emulate the vernacular design elements that J.B. Jackson so admires (p. 202), but work as self-conscious design professionals. Urban designs are often expressed through drawings – generated by hand or computer – but design ideas may be expressed in words, photographs, maps, and other media.

The selections in this part by architects Camillo Sitte and Jan Gehl, urban designer Kevin Lynch, sociologist William Whyte, and urban planners Clarence Perry, Allan Jacobs and Donald Appleyard illustrate how different academic disciplines and professional fields can contribute to good urban design that is sensitive to human needs.

Good urban design usually begins with intensive observation. Late-nineteenth-century Austrian architect Camillo Sitte (p. 476) observed squares and plazas in cities all over Europe in order to develop his artistic principles for designing cities. Massachusetts Institute of Technology urban design professor Kevin Lynch and his students (p. 499) surveyed people to see how they perceived the cities where they lived and asked them to draw mental maps to help understand what city elements were and were not clear to them. Sociologist William Whyte (p. 510) and his students from Hunter College spent hundreds of hours watching people use New York City parks and plazas – filming them, analyzing the films, and quantifying what they observed about how people used parks and plazas in order to specify principles for good park and plaza design. Planners Allan Jacobs and Donald Appleyard's urban design manifesto (p. 512) grew out of many hours surveying, sketching, counting, mapping, photographing, measuring, and simply walking and looking at San Francisco.

The design of the built environment may not determine human behavior, but implicit in the following selections is the notion that bad design can numb the human spirit and good design can have powerful, positive influences on human beings. Ugly, impersonal, dirty, dangerous, economically depressed, unsustainable, dysfunctional, race- and gender-segregated areas are now prevalent in many large cities. Urban designers may have different priorities with respect to the need to improve traffic flow versus making pedestrian-friendly streets, economically revitalizing an area versus retaining historic buildings, or protecting the natural environment versus keeping the city competitive in the global economy. But whatever their priorities they bring a distinct self-conscious design approach to their work.

Part Seven begins with Austrian architect Camillo Sitte's theories on the art of building cities "according to artistic principles," written at the end of the nineteenth century. Sitte is often viewed as the father of modern urban design. Dissatisfied with the impersonality of building projects in his own city of Vienna, Sitte set out to rediscover the artistic principles that guided classical Greek and Roman city builders and their medieval and Renaissance successors by carefully observing public spaces throughout Europe. Sitte emphasizes the aesthetic, artistic character of city design. Underlying Sitte's theories is a profound belief in the importance of public places as venues to celebrate civic life. Like Lewis Mumford (p. 91), Sitte values human-scale development, the retention of historical elements, the joys of irregularity, and cities as theaters for the display of culture and civilization. Sitte's writings ran counter to the grand building schemes the new industrial wealth in Europe was creating during his own day. He touched a responsive chord and quickly developed followers throughout the world.

In the 1930s, modernists such as Le Corbusier (p. 336) and his followers in the CIAM, enamored of industrial technology and efficiency, dismissed Sitte. But in the present-day postmodern era his ideas are enjoying a resurgence of interest. Sitte's perspective – the aesthetic character of cities – did not deal with many of the difficult and important issues raised in prior parts of this reader. Issues of class and race, economics and governance, and how to implement city plans are nowhere discussed in his work. He did not bring a gender perspective to his work, but within the sphere of aesthetics and design he made an important contribution.

Three decades after Sitte's seminal work, architect Clarence Perry produced one of the most influential writings on urban design ever written (p. 486). The Russell Sage Foundation provided generous funding for a massive plan for the New York City Region. As part of this important project, the foundation commissioned background papers by many of the United States' best known planners. Perry was assigned the task of developing design standards for new neighborhoods. The final Plan for the New York Region proved controversial. Lewis Mumford dismissed the plan as timid and uninspired. But Mumford and others immediately recognized Perry's short piece as a brilliant contribution to planning. It has had enormous influence worldwide ever since 1929.

Perry's answer to the growth of enormous, impersonal metropolitan regions such as the New York Region and to the sprawl and potential destruction of community that mass auto ownership was already bringing by the 1920s was to design human-scale neighborhood units organized around an educational and cultural complex that would bring people together and maintain community. Education, Perry reasoned, was a fundamental value of American families. If a neighborhood was designed around an area large enough to support an elementary school, the school would provide the glue to get neighbors to interact with each other. Better yet, if the school building and grounds were used on evenings and weekends for adult education classes, sports and cultural events, neighbors would get to know one another even better. The anomie that Louis Wirth deplored in "Urbanism as a Way of Life" (p. 96) could be overcome. A family-oriented, school-centered neighborhood where children could walk to school and neighborhood cultural activities brought people together would help build the kind of social capital Robert Putnam (p. 134) considers so essential. Perry proposed a whole series of design ideas that would help knit the neighborhood unit together and keep automobiles from destroying the urban fabric – separation of through and local traffic, culs-de-sac to prevent through traffic, overpasses to eliminate messy intersections, and of course a school complex at the center of the neighborhood. Perry's writing was the most influential urban design writing in the world until a slim volume on urban design by Massachusetts Institute of Technology planning professor Kevin Lynch appeared in 1961.

Lynch's *The Image of the City* (p. 499) quickly established itself as the foundation of contemporary urban design. Lynch asked basic questions: how do people perceive the built environment? What are the underlying elements common to human perception of the city? What aspects of the built environment disorient people and which help them to grasp their surroundings? Armed with a better theoretical understanding of how people perceive the city image, what can urban designers actually do to design better cities? The selection from *The Image of the City* reprinted in Part Seven describes Lynch's core findings about how people perceive the city image and summarizes the five elements of urban form that Lynch considers to be most fundamental: paths, edges, districts, nodes, and landmarks. It is rich in suggestions about how these findings can shape better urban design. Urban planners like Allan Jacobs and Donald Appleyard (p. 518) have put Lynch's ideas into practice all over the world.

Sociologist William Whyte's writing on the design of spaces summarizes ideas he developed studying the way in which New Yorkers use urban parks and plazas. Whyte was a sharp observer and a fine writer. His writings are exemplary of the way in which social scientists can use understanding of human behavior to produce excellent urban designs. Whyte's work illustrates how urban research can lead directly to changes in city policy. The New York City Planning Department and organizations in New York involved in planning and managing parks have incorporated Whyte's ideas directly into policy. As a result of Whyte's ideas, iron fences isolating New York's Bryant Park from the street were torn down. Food stalls were expanded. Moveable chairs were brought in. Nowadays, Bryant Park is a vibrant, well-used urban park rather than a sinister and dangerous park that New Yorkers shunned. Other cities have used Whyte's ideas to improve parks and plazas.

Urban planners Allan Jacobs and Donald Appleyard also illustrate the connection between theory and practice. Jacobs is an emeritus professor of city planning at the University of California, Berkeley, where Appleyard also taught until his death. Both were students of Kevin Lynch. Jacobs served as the director of the San Francisco City Planning Department, and Appleyard worked with him on notable studies of street livability and urban design. Under Jacobs's leadership, the San Francisco City Planning Department produced an award-winning urban design plan, which draws heavily on Lynch's ideas and the insights of Appleyard's studies. That work has profoundly shaped the development of San Francisco for more than a quarter century and serves as a model for other cities.

The final selection in Part Seven presents some of Danish architect and educator Jan Gehl's ideas on public spaces. Like the other authors represented in Part Seven, Jan Gehl developed his theories and practical principles of urban design from careful observation. Gehl chose to focus his attention on the way in which people conduct ordinary outdoor day-to-day activities such as walking to the bus stop, going to do a local grocery shopping, washing the car, playing with their children, or walking the dog. He observed ordinary "spaces between buildings" – front yards, neighborhood streets, doorways of public buildings, vacant lots, local parks, and bus stops. His observations convinced him that people will choose to engage in optional outdoor activities if they find the physical space inviting, but avoid them if they do not. Even necessary outdoor activities will be prolonged and enriched if people enjoy being outside. Gehl believes that people have a basic need for human contact and that well designed ordinary outdoor spaces will bring more people in contact with each other. In addition to meeting a basic individual psychological need, even outdoor encounters at a very low level of intensity – just passing others on the street, or choosing to take a bus ride – provide a measure of satisfaction and can stimulate other, more important connections between people.

Gehl's careful observations of spaces that people in his native Copenhagen did and did not like produced many practical insights about better design. He noticed, for example, that people enjoy observing other people. Where benches along paths in Danish parks were placed back-to-back people almost always chose the bench facing the path so they could enjoy contact with – even mere observation of – other people. Gehl was an early advocate of pedestrian-only streets, traffic calming, and designs to accomodate bicycles. He played a significant role in the design of Copenhagen's very successful Strøget, the longest pedestrianized street in the world.

Designing the urban environment requires sensitivity to human needs and the natural environment. Urban planners, architects, landscape architects, and other design professions must be sensitive to biological and other natural systems, physical form and function, the history and culture of the areas they are designing, aesthetics, economics, transportation systems, and all the myriad ways in which people use urban space. Each part of this reader can contribute to sensitive urban design.

"Author's Introduction," "The Relationship Between Buildings, Monuments, and Public Squares," and "The Enclosed Character of the Public Square"

from *The Art of Building Cities* (1889)

Camillo Sitte

Editors' Introduction

Nineteenth-century Austrian architect, Camillo Sitte (1843–1903) felt that much was lost in the transformation of his native city of Vienna in the latter part of the nineteenth century. Sitte witnessed the old city walls – no longer useful against modern artillery – torn down, a ring road (Ringstrasse) with new electric streetcars built to encircle the city, and old areas in the city leveled for monumental boulevards and impressive new buildings. Modernization eliminated slums, replaced medieval alleyways with much wider streets, improved sanitation and public health, increased mobility, and provided beautiful new buildings and public spaces. But Sitte felt nostalgia for the oddly shaped cathedral squares and narrow streets of Vienna, Salzburg, and other European cities that had evolved over time. He mourned the loss of structures built to human scale and public spaces embellished with statues, fountains, and other municipal art that adorned cites in classical Greece and Rome, the Middle Ages, and the Renaissance.

Sitte knew instinctively that he enjoyed qualities in cities he had visited that had been built before the modern era. He admired the urban form of Greek and Roman cities of classical antiquity discernible in ruins and fragments of existing cities. But what exactly were these qualities? And how might modern-day designers, architects, and city planners conserve the best aspects of historic buildings and public spaces and incorporate the principles that Sitte and so many others enjoyed into new building projects? To answer these questions, Sitte embarked on a careful study of the built environment of notable European cities. Armed with a sketchbook he visited Athens, Rome, Florence, Venice, Paris, Pisa, Salzburg, Rothenburg on the Tauber, Dresden, and dozens of other European cities. Everywhere he went, Sitte carefully sketched the physical form of squares and plazas, the building footprints of cathedrals and public buildings, street patterns, and the location of statues and fountains. He thought about scale and building materials, views and elevations, the integration of ornamental features with functional buildings. He imagined what civic life in these urban spaces must have been like at the time of Pericles and Julius Caesar, of Lorenzo the Magnficent, and Louis XI of France. Sitte reflected on how architects and city planners designed aesthetically pleasing spaces to reinforce civic culture. The result was a masterful little book that set in motion the modern study of urban design. Sitte produced a volume that was passionate in its advocacy of human-scale building and consideration of artistic principles in city building. Good design involves more than attention to cosmetic aspects of surface features. While Sitte couched his critique and suggestions in terms of art, he was attentive to the complexity of urban form and human interaction.

Sitte was the first in a long line of urban designers who used simple observation as their research method of choice and a principal source of inspiration for their design concepts. Later urban designers like Kevin Lynch (p. 499) and Allan Jacobs (p. 518) also sketched parts of the built environment they studied. The act of drawing forced them to concentrate on details and understand the physical spaces. Indeed all of the authors in this section were astute observers of how people use urban space. William Whyte (p. 510) and a team of students from Hunter College in New York developed principles of park and plaza design by looking carefully at how people responded to such everyday artifacts as benches, food carts, grass, ledges, and water. Kevin Lynch and his students from the Massachusetts Institute of Technology (p. 499) observed different parts of Boston and other cities and asked passersby to sketch maps showing their image of the city. Jan Gehl spent hours watching how people in Danish cities interacted or failed to interact in the space between buildings. Despite his busy schedule during eight years as city planning director of San Francisco, Allan Jacobs (p. 518) made time to walk the city streets, take photos, and make sketches. One of Jacobs' books is titled simply *Looking at Cities*.

Sitte celebrated public space and was particularly enamored of public squares and plazas. His sketches showed that many of the best loved public spaces were irregular in shape and had a complex jumble of features that had built up over time rather than being designed all at one time as a completely thought out whole. He applauded the practice in ancient Greece and Rome and during the Italian Renaissance of concentrating civic buildings around public squares and plazas and ornamenting the resulting centers of community life with fountains, monuments, and statues – again often reflecting the taste and mores of different historic periods.

Sitte had limited influence on the rebuilding of his native Vienna, but enormous and continuing impact elsewhere. Sitte Schülen (Sitte schools) sprang up all over Europe as young architects and planners read his book and discussed how to implement his ideas. *The Art of Building Cities* was translated into other languages. In the United States, it was the Bible of the turn-of-the-century municipal arts movement. Dozens of local committees inspired by Sitte and the American writer Charles Mulford Robinson formed committees to "embellish and adorn" American cities.

Sitte fell out of favor in the interwar period when Le Corbusier and the insurgent young architects of the Congrès International d'Architecture Moderne (CIAM) developed plans to raze and rebuild what they saw as the obsolete cities using modern materials, monumental scale, and designs inspired by industrial society (p. 336). In *The City of Tomorrow and its Planning* (1929) Le Corbusier dismissed the process of allowing cities to grow organically, which he labeled "the pack-donkey's way" of designing cities in contrast to rational modernist designs, which he called "man's way"! There is now a renewed interest in human-scale postmodernist designs, and New Urbanist architects and planners associated with the Congress of the New Urbanism (p. 356) have rediscovered Sitte's writings and find much of value in the principles he developed more than a century ago.

While Sitte limited his observations and the examples in his book to Europe, city planning in other parts of the world raised and continue to raise similar issues. In China, India, and other rapidly developing countries rapidly rising land values and the desire to modernize cities have created enormous pressure to raze historic districts and built modern new buildings at an enormous scale.

Tensions exist between historic preservationists and neighborhood residents, who value physical reminders of the past and the human-scale aspects of their neighborhoods, and modernizers who value modern, efficient cities built at a scale appropriate to megacities. As J.B. Jackson (p. 202) and Harvey Molotch (p. 251) point out, development-oriented small town chambers of commerce and urban growth machines often prevail.

Note the connections between Sitte's celebration of plazas and public squares in classical Greece and Rome and in medieval and Renaissance European cities with Lewis Mumford's notion that cities, above all, should be theaters in which humans can display their culture (p. 91) and William Whyte's views on how urban parks and plazas contribute to urban life (p. 510). Note the similarities between his appreciation for irregular and surprising spaces to Jane Jacobs' love of disorderly street life (p. 105).

Camillo Sitte's treatise is available in many languages and editions. Christiane Crasemann Collins and George R. Collins, *Camillo Sitte: The Birth of Modern City Planning* (Mineola, NY: Dover, 2006) includes a translation of the 1889 Austrian edition. Another English language edition was published by Hyperion Press in 1979.

An exhaustive study of the form of European (and non-European) cities before the Industrial Revolution is A.E.J. Morris, *The History of Urban Form Before the Industrial Revolution*, 3rd edn (New York: Prentice Hall, 1996).

SEVEN

Another study of public spaces in European cities is Paul Zucker, *Town and Square from the Agora to the Village Green*, 2nd edn (Cambridge, MA: MIT Press, 1970). Spiro Kostof's monumental studies of urban form, *The City Shaped* (Boston, MA: Little, Brown, 1991) and *The City Assembled* (Boston, MA: Little, Brown, 1992), contain illustrations of many of the features Sitte discusses. Leonardo Benevolo, *The History of the City* (Cambridge, MA: MIT Press) has more than a thousand pages with hundreds of illustrations of features of European cities. Erwin Anton Gutkind's eight-volume *International History of City Development* (New York: Free Press, 1964) also has hundreds of photographs and line drawings of public spaces in European cities. Suzanne H. Crowhurst and Henry L. Lennard, *Genius of the European Square* (Portland, OR: International Making Cities Livable, 2008) has descriptions and illustrations of contemporary European squares. An account of the transformation of Vienna at the time that Sitte lived and wrote is contained in Carl Shorske, *Fin de Siècle Vienna: Politics and Culture* (New York: Vintage, 1981).

AUTHOR'S INTRODUCTION

Memory of travel is the stuff of our fairest dreams. Splendid cities, plazas, monuments, and landscapes thus pass before our eyes, and we enjoy again the charming and impressive spectacles that we have formerly experienced. If we could but stop again at those places where beauty never satiates, we could bear many dreary hours with a light heart and pursue life's long struggle with new energies. Assuredly the imperturbable lightheartedness of the South, on the Hellenic coast, in lower Italy and other favored climes, is above all a gift of nature. And the old cities of these countries, built after the beauty of nature itself, continue to augment nature's gentle and irresistible influence upon the soul of man. Only the person who has never understood the beauty of an ancient city could contradict this assertion.

Let him go ramble on the ruins of Pompeii to convince himself of it. If, after a day of patient investigation there, he walks across the bare Forum, he will be drawn, in spite of himself, to the summit of the monumental staircase toward the terrace of Jupiter's temple. On this platform, which dominates the entire place, he will sense, rising within him, waves of harmony like the pure, full tones of sublime music. Under this influence he will truly understand the words of Aristotle, who thus summarized all principles of city building: "A city should be built to give its inhabitants security and happiness."

The science of the technician will not suffice to accomplish this. We need, in addition, the talent of the artist. Thus it was in ancient times, in the Middle Ages, and in the Renaissance, wherever fine arts were held in esteem. It is only in our mathematical century that the construction and extension of cities has become a purely technical matter. Perhaps, then, it is not beside the point to recall that these problems have diverse aspects, and that he who has been given the least attention in our time is perhaps not the least important.

The object of this study, then, is clear. It is not our purpose to republish ancient and trite ideas, nor to reopen sterile complaints against the already proverbial banality of modern streets. It is useless to hurl general condemnations and to put everything that has been done in our time and place once more to the pillory. That kind of purely negative effort should be left to the critic who is never satisfied and who can only contradict. Those who have enough enthusiasm and faith in good causes should be convinced that our own era can create works of beauty and worth. We shall examine the plan of a number of cities, but neither as historian nor as critic. We wish to seek out, as technician and artist, the elements of composition which formerly produced such harmonious effects, and those which today produce only loose and dull results. Perhaps this study will permit us to find the means of satisfying the three principal requirements of practical city building: to rid the modern system of blocks and regularly aligned houses; to save as much as possible of that which remains from ancient cities; and in our creation to approach more closely the ideal of the ancient models.

This standpoint of practical art will lead us to consider especially the cities of the Middle Ages and the Renaissance. We shall be content, on recalling examples from Greek and Roman conceptions, either to explain the creations of following epochs, or to support the ideas that we propose to develop. For the principal architectural elements of cities have greatly changed since antiquity. Public squares (Forum, market, etc.) are used in our times not so much for great popular festivals or for the daily needs of our life. The sole reason for their existence is to provide more air and light, and to break the monotony of oceans of houses. At times they also enhance a monumental edifice by freeing its walls. It was quite different in ancient times. Public squares, or plazas, were then of prime necessity, for they were theaters for the principal scenes of public life, which today take place in enclosed halls. Under the open sky, on the agora, the council of the ancient Greeks gathered.

The market place, a further center of activity for our ancestors, has persisted, it is true, to the present time, but more and more it is being replaced by vast enclosed halls. And how many other scenes of public life have totally disappeared? Sacrifices before the temples, games, and theatrical presentations of all kinds. The temples themselves were scarcely covered, and the principal part of dwellings, around which were grouped large and small rooms, consisted of an open court. In a word, the distinction between the public square and other structures was so slight that it is amazing to our modern minds, accustomed to a very different state of things.

A review of the writings of the period proves to us that the ancients themselves sensed this similarity. Thus Vitruvius does not discuss the Forum in connection with the placement of public buildings or the arrangement of streets in his account of Dinocrates and his plan of Alexandria. But he does mention it in the same chapter which discusses the Basilica, and in the same book (1, 5) he deals with the theaters, palaces, the circus, and the baths. That is to say, all gathering places under the open sky constituted architectural works. The ancient Forum corresponds exactly to this definition, and Vitruvius logically places it in this group. This close relationship between the Forum and a public hall enhanced architecturally by statues and paintings is brought out clearly by the Latin writer's description, and more clearly still by an examination of the Forum of Pompeii. Vitruvius writes again on this subject:

> The Greeks arrange their market places in the form of a square and surround them by vast double column supporting stone or marble architraves above which run the promenades. In Italian cities the Forum takes another aspect, for from time immemorial it has been the theater of gladiatorial combats. The columns, therefore, must be less densely grouped. They shelter the stalls of the silversmiths, and their upper floors have projections in the form of balconies which are advantageously placed for frequent use and for public revenue.

This description illustrates well the correspondence between theater and Forum. This relationship appears still more striking when we examine the plan of the Forum of Pompeii (Figure 1). The square is surrounded on all sides by public buildings. The temple of Jupiter alone rises in isolation. And the two-story colonnade which surrounds the entire space is interrupted only by the peristyle of the temple of the household gods, which makes a greater projection than the other buildings. The center of the Forum remains free, but its periphery is occupied by numerous monuments, the pedestals of which, covered with inscriptions, are still visible.

What a grandiose impression this place must have made! To our modern point of view its effect is like that of a great concert hall without ceiling. In every direction the eye fell upon edifices which in no respect resembled our files of modern houses, and there were far fewer streets opening directly on the plaza. Streets ran behind buildings III, IV, and V, but they did not extend as far as the Forum. Streets C, D, E, and F were closed by grilles, and even those on the north side passed under the monumental portals, A and B.

Forum Romanum (Figure 2) was conceived according to the same principles. It is surrounded, of course, by buildings more varied in type but all monumental. The streets which open onto it were arranged to avoid too frequent openings in the frame of the plaza. Monuments are located around its sides rather than in its center. In brief, the place of the forum in cities corresponds to that of the principal room of a house. It is to the city, so to speak, the principal hall, as well arranged as it is richly furnished. There stand assembled in immense bulk the columns, the statues, the monuments, and everything that can contribute to the splendor of the place. The art treasures of some of them were said to be numbered in hundreds and thousands. As they did not encumber the midst of the plaza, but were always located at the periphery, it was possible to encompass them all with a single glance; and the spectacle must have been imposing. This concentration of plastic and architectural masterpieces at a single point was a stroke of genius. Aristotle had taught it. He advocated grouping the temples of the gods with public buildings. Pausanias wrote similarly, "A city without public edifices and squares is not worthy of its name."

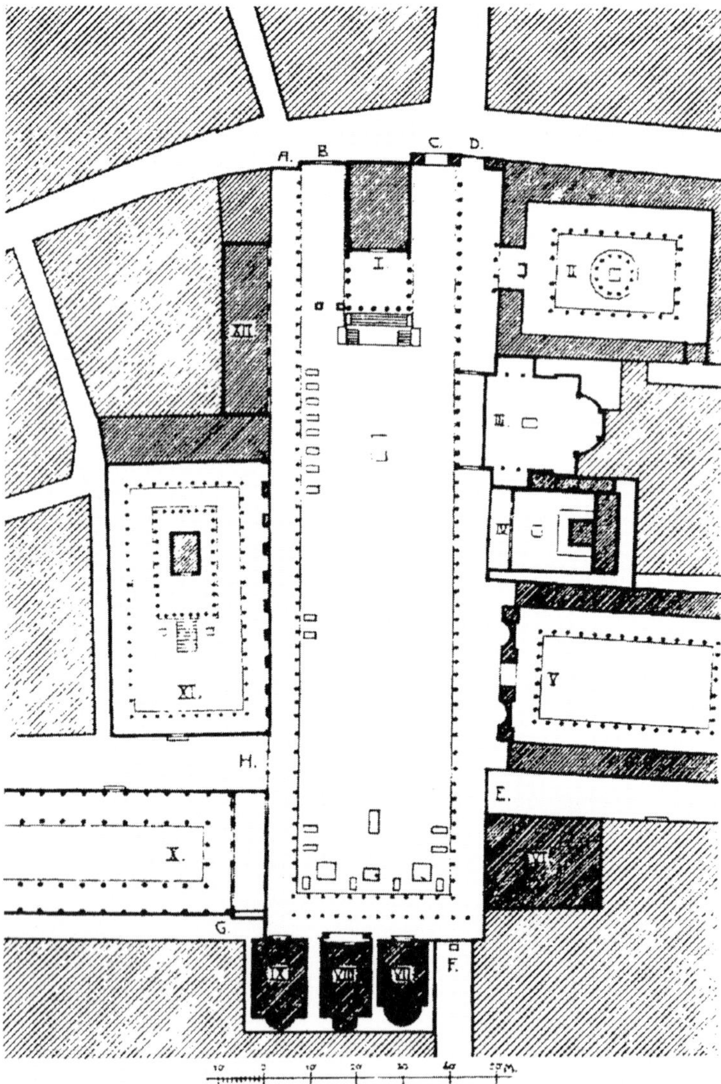

Figure 1 Forum of Pompeii
I Temple of Jupiter, II Enclosed Market, III Temple of Household Gods, IV Temple of Vespasianus, V Eumachia, VI Comitium, VII–IX Public Buildings, X Basilica, XI Temple of Apollo, XII Market Hall

The market place of Athens is arranged in its principal features according to the same rules, as well as may be judged from the restoration projects. They are applied on a still grander scale in the consecrated cities of Hellenic antiquity (Olympia, Delphi, Eleusis) (Figure 3). Masterpieces of architecture, painting, and sculpture are found there in a superb and imposing union capable of rivaling the most powerful tragedies and the most majestic symphonies. The Acropolis of Athens (Figure 4) is the most finished creation of this character. A high plateau surrounded by high walls is the base of it. The lower entrance portal, the enormous flight of steps, and monumental vestibules constitute the first phrase of this symphony in marble, gold, ivory, bronze, and color. The interior temples and monuments are the stone myths of the Greek people. The highest poetry and thought are embodied in them. It is truly the center of a considerable city, an expression of the feelings of a great people. It is no longer a simple square in the ordinary sense of the term, but the work of several centuries grown to the maturity of pure art.

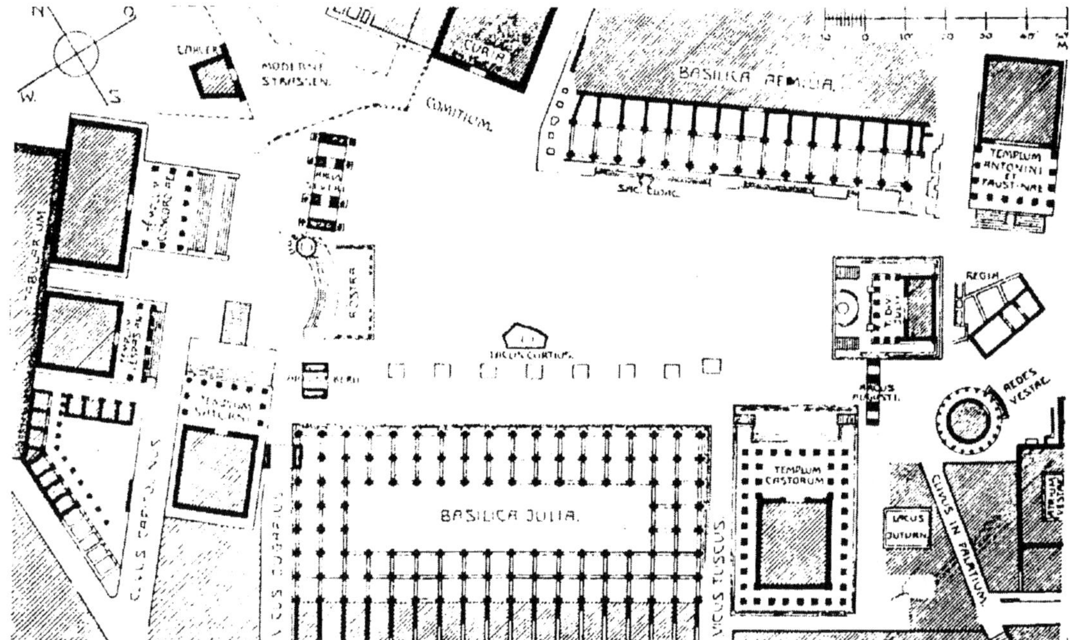

Figure 2 Forum Romanum

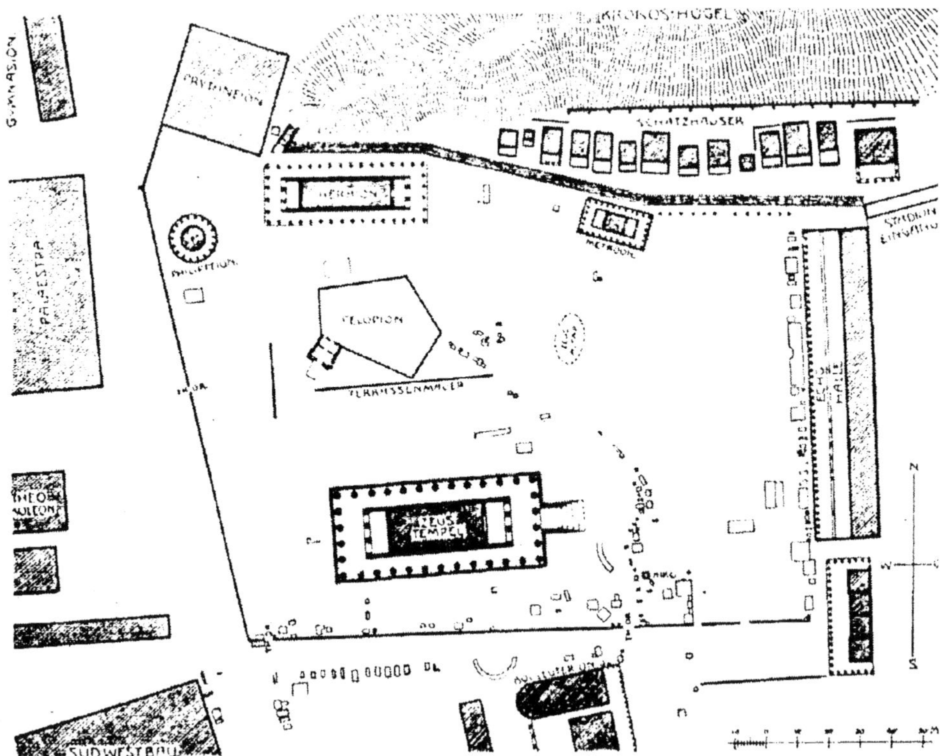

Figure 3 The Temple of Zeus and Plaza of Olympia

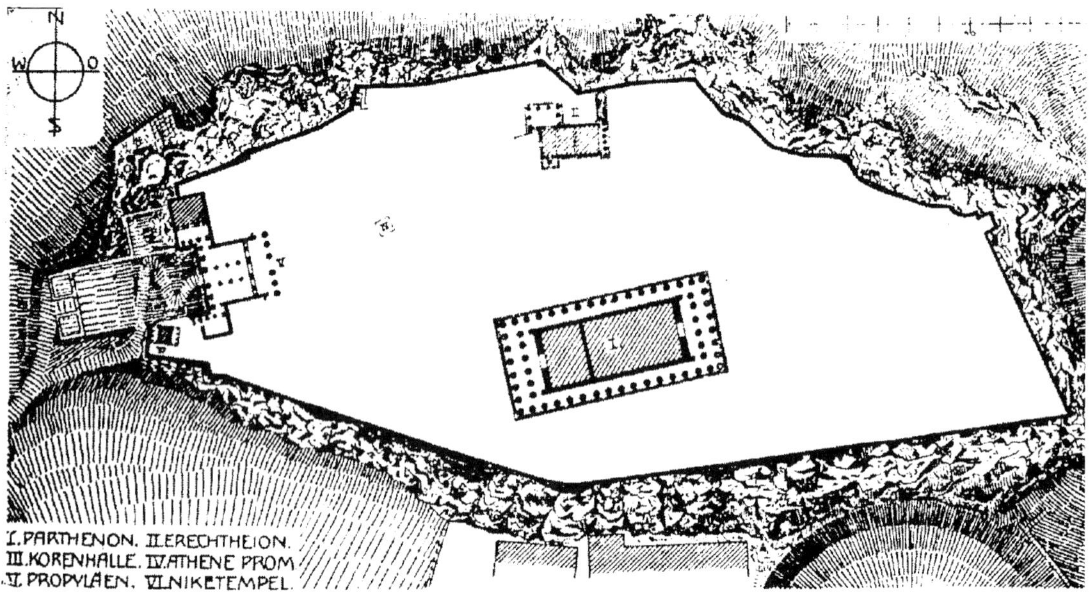

Figure 4 The Acropolis of Athens in the age of Pericles

It is impossible to establish a higher aim in this style, and it is difficult to imitate successfully this splendid model, but it should always remain before our eyes in all our works as the most sublime ideal to attain. In the progress of our study we shall see that the principles which have inspired such building are not entirely lost, but that they remain to us.

THE RELATIONSHIP BETWEEN BUILDINGS, MONUMENTS, AND PUBLIC SQUARES

In the South of Europe, and especially in Italy, where ancient cities and ancient public customs have remained alive for ages, even to the present in some places, public squares still follow the type of the ancient forum. They have preserved their role in public life. Their natural relationships with the buildings which enclose them may still be readily discerned. The distinction between the forum, or agora, and the market place also remains. As before, we find the tendency to concentrate outstanding buildings at a single place, and to ornament this center of community life with fountains, monuments, and statues which can bring back historical memories and which, during the Middle Ages and the Renaissance, constituted the glory and pride of each city.

It was there that traffic was most intense. That is where public festivals and theatrical presentations were held. There it was that official ceremonies were conducted and laws promulgated. In Italy, according to varying circumstances, two or three public places, rarely a single one, served these practical purposes.

The existence of two powers, temporal and spiritual, required two distinct centers: one, the cathedral square (Figure 5) dominated by the campanile, the baptistry, and the palace of the bishop; the other, the Signoria, or manor place, which is a kind of vestibule to a royal residence. It is enclosed by houses of the country's great and adorned with monuments. Sometimes we see there a loggia, or open gallery, used by a military guard, or a high terrace from which laws and public statements were promulgated. The Signoria of Florence (Figure 6) is the finest example of this. The market square, rarely lacking even in cities of northern Europe, is the meeting place of the citizens. There stand the City Hall and the more or less richly decorated traditional fountain, the sole vestige of the past that has been conserved since the lively activity of merchants and traders has been moved within to iron cages and glass market places.

The important function of the public square in the community life of past ages is evident. The period of the Renaissance saw the birth of masterpieces in the manner of the Acropolis of Athens, where everything

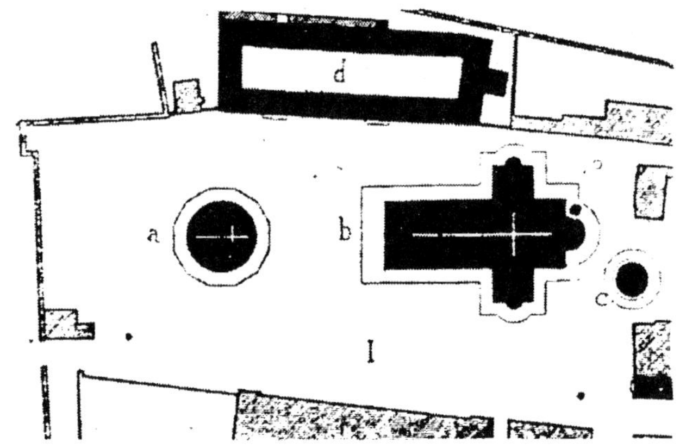

Figure 5 Pisa: Cathedral Square
Key
a. Saint-Jean b. Cathedral c. Campanile d. Campo Santo (Cemetery)

Figure 6 Florence: Piazza of the Signoria

concurred to produce a finished artistic effect. The cathedral place at Pisa, an Acropolis of Pisa (Figure 5), is the proof of this. It includes everything that the people of the City have been able to create in building religious edifices of unparalleled richness and grandeur. The splendid cathedral, the campanile, the baptistry, the incomparable Campo Santo are not depreciated by profane or banal surroundings of any kind. The effect produced by such a place, removed from the world of baseness while rich in the noblest works of the human spirit, is overpowering. Even those with a poorly developed sensitiveness to art are unable to escape the power of this impression. There is nothing there to distract our thoughts or to intrude our daily affairs. The esthetic enjoyment of those who look upon the noble facade of the Cathedral is not spoiled by the sight of a modern haberdashery, by the cries of drivers and porters, or by the tumult of a cafe. Peace reigns over the place. It is thus possible to give full attention to the artwork assembled there.

This situation is almost unique, although that of Saint Francis of Assisi and the arrangement of the Certosa de Pavia closely approach it. In general, the modern period does not encourage the formation of such perfect groupings. Cities, even in the fatherland of art, undergo the fate of palaces and dwellings. They no longer have distinct character. They present a mixture of motifs borrowed as much from the architecture of the north as from that of the southern countries. Ideas and tastes have been mingled as the people themselves have been interchanged. Local characteristics are gradually disappearing. The market place alone, with its City Hall and fountain, has here and there remained intact.

In passing we should like to remark that our intention is not to suggest a sterile imitation of the beauties spoken of as "picturesque" in the ancient cities for our present needs. The proverb, "Necessity breaks even iron," is fully applicable here. Changes made necessary by hygiene or other requirements must be carried out, even if the picturesque suffers from it. But that does not prevent us from examining the work of our forebears at close range to determine how much of it may be adapted to modern conditions. In this way alone can we resolve the esthetic part of the practical problem of city building, and determine what can be saved from the heritage of our ancestors.

Before determining the question in a positive manner, we state the principle that during the Middle Ages and Renaissance public squares were often used for practical purposes, and that they formed an entirety with the buildings which enclosed them. Today they serve at best as places for stationing vehicles, and they have no relation to the buildings which dominate them. Our parliament buildings have no agora enclosed by columns. Our universities and cathedrals have lost their atmosphere of peace. Surging throngs no longer circulate on market days before our City Halls. In brief, activity is lacking precisely in those places where, in ancient times, it was most intense – near public structures. Thus, to a great extent, we have lost that which contributed to the splendor of public squares.

And the fabric of their very splendor, the numerous statues, is almost entirely lacking today. What have we to compare to the richness of ancient forums and to works of majestic style like the Signoria of Florence and its Loggia dei Lanzi?

A few years ago there flourished at Vienna a remarkable school of sculpture whose works of merit cannot be scorned. They were generally used to adorn buildings. In only a few exceptional cases were their works used in public squares. Statues adorn the two museums, the palace of Parliament, the two Court theaters, the City Hall, the new university, the Votive Church. But there is no interest in adorning public open spaces. And that is true not only in Vienna, but nearly everywhere.

Buildings lay claim to so many statues that commissions are needed to find new subjects to be represented. It is often necessary to wait for years to find a suitable place for a statue although many appropriate places remain empty in the meantime. After long efforts we have reconciled ourselves to modern public squares as vast as they are deserted, and the monument, without a place of refuge, becomes stranded on some small and ancient space. That is even more strange, yet true. After much groping about, this fortunate result occurs, for it is thus that a work of art derives its value and produces a more powerful impression. Indifferent artists who neglect to provide for such effects must bear the entire responsibility of it.

The story of Michelangelo's *David* at Florence shows how mistakes of this kind are perpetrated in modern times. This gigantic marble statue stands close to the walls of the Palazzo Vecchio, to the left of its principal entrance, in the exact place chosen by Michelangelo. The idea of erecting a statue on this place of ordinary appearance would have appeared to moderns as absurd if not insane. Michelangelo chose it, however, and without doubt deliberately; for all those who have seen the masterpiece in this place testify to the extraordinary impression that it makes. In contrast to the relative scantiness of the place, affording an easy comparison with human stature, the enormous statue seems to swell even beyond its actual dimensions. The sombre and uniform, but powerful, walls of the palace provide a background on which we could not wish to improve to make all the lines of the figure stand out.

Today the David is moved into one of the academy's halls under a glass cupola in the midst of plaster reproductions, photographs, and engravings. It serves as a model for study and an object of research for historians and critics. A special mental preparation is needed now to resist the morbid influences of an art prison that we call a museum, and to have the ability to enjoy the imposing work. Moreover, the spirit of the times, which believed that it was perfecting art, and which was still not satisfied with this innovation, had a bronze cast made of the David in its original grandeur and put it up on a vast plaza (naturally in its mathematical center) far from Florence at the Via dei Colli. It has a superb horizon before it; behind it, cafes; on one side, a carriage station, a corso; and from all sides the murmurs of Baedeker readers ascend to it. In this setting the statue produces no effect at all. The opinion that its dimensions do not exceed human stature is often heard. Michelangelo thus understood best the kind of placement that would be suitable for his work, and, in general, the ancients were abler than we are in these matters.

The fundamental difference between the procedures of former times and those of today rests in the fact that we constantly seek the largest possible space for each little statue. Thus we diminish the effect that it could produce, instead of augmenting it with the assistance of a neutral background such as painters have used in their portraits.

This explains why the ancients erected their monuments by the sides of public places, as is shown in the view of the Signoria of Florence. In this way, the number of statues could increase indefinitely without obstructing the circulation of traffic, and each of them had a fortunate background. Contrary to this, we hold the middle of a public place as the sole spot worthy to receive a monument. Thus no esplanade, however magnificent, can have more than one. If by misfortune it is irregular and if its center cannot be located geometrically we become confused and allow the space to remain empty for eternity.

THE ENCLOSED CHARACTER OF THE PUBLIC SQUARE

The old practice of setting churches and palaces back against other buildings brings to mind the ancient forum and its unbroken frame of public buildings. In examining the public squares that came into being during the Middle Ages and the Renaissance, especially in Italy, it is seen that this pattern has been retained for ages by tradition. The old plazas produce a collective harmonious effect because they are uniformly enclosed. In fact, the public square owes its name to this characteristic in an expanse at the center of a city. It is true that we now use the term to indicate any parcel of land bounded by four streets on which all construction has been renounced.

That can satisfy the public health officer and the technician, but for the artist these few acres of ground are not yet a public square. Many things must be done to embellish the area to give it character and importance. For just as there are furnished and unfurnished rooms, we could speak of complete and incomplete squares. The essential thing of both room and square is the quality of enclosed space. It is the most essential condition of any artistic effect, although it is ignored by those who are now elaborating on city plans.

The ancients, on the contrary, employed the most diverse methods of fulfilling this condition under the most diverse circumstances. They were, it is true, supported by tradition and favored by the usual narrowness of streets and less active traffic movement. But it is precisely in cases where these aids were lacking that their talent and artistic feeling is displayed most conspicuously.

A few examples will assist in accounting for this. The following is the simplest. Directly facing a monumental building a large gap was made in a mass of masonry, and the square thus created, completely surrounded by buildings, produced a happy effect. Such is the Piazza S. Giovanni at Brescia. Often a second street opens on to a small square, in which case care is taken to avoid an excessive breach in the border, so that the principal building will remain well enclosed. The methods used by the ancients to accomplish this were so greatly varied that chance alone could not have guided them. Undoubtedly they were often assisted by circumstance, but they also knew how to use circumstances admirably.

Today in such cases all obstructions would be taken down and large breaches in the border of the public place would be opened, as is done when we decide to "modernize" a city. Ancient streets would be found to open on the square in a manner precisely contrary to the methods of modern city builders, and mere chance would not account for this. Today the practice is to join two streets that intersect at right angles at each corner of the square, probably to enlarge as much as possible the opening made in the enclosure and to destroy every impression of cohesion. Formerly the procedure was entirely different. There was an effort to have only one street at each angle of the square. If a second artery was needed in a direction at right angles to the first it was designed to terminate at a sufficient distance from the square to remain out of view from the square. And better still, the three or four streets which came in at the corners each ran in a different direction. This interesting arrangement was reproduced so frequently, and more or less completely, that it can be considered as one of the conscious or subconscious principles of ancient city building.

Careful study shows that there are many advantages to an arrangement of street openings in the form of turbine arms. From any part of the square there is but one exit on the streets opening into it, and the enclosure of buildings is not broken. It even seems to enclose the square completely, for the buildings set at an angle conceal each other, thanks to perspective, and unsightly impressions which might be made by openings are avoided. The secret of this is in having streets enter the square at right angles to the visual lines instead of parallel to them. Joiners and carpenters have followed this principle since the Middle Ages when, with subtle art, they sought to make joints of wood and stone inconspicuous if not invisible.

The Cathedral Square at Ravenna shows the purest type of the arrangement just described. The square of Pistoia (Figure 7) is in the same manner; as is . . . the Piazza Grande at Parma (Figure 8).

The ancients had recourse to still other means of closing in their squares. Often they broke the infinite perspective of a street by a monumental portal or by several arcades of which the size and number were determined by the intensity of traffic circulation. This splendid architectural pattern has almost entirely disappeared, or, more accurately, it has been suppressed. Again Florence gives us one of the best examples in the portico of the Uffizi with its view of the Arno in the distance. Every Italian city of average importance has its portico, and this is also true north of the Alps. We mention only the Langasser Thor at Danzig, the entrance portal of the City Hall and Chancellery at Bruges, the Kerkboog at Nimeguen, the great Bell Tower at Rouen, the monumental Portals of Nancy, and the windows of the Louvre.

More or less ornate portals like those that simply but effectively frame the Piazza dei Signori at Verona (Figure 9) are to be found in all the royal residences, in the chateaux and city halls, and they are used as much for vehicular traffic as by pedestrians. While ancient architects used this pattern wherever possible with infinite variations, our modern builders seem to ignore its existence. Let us recall, to demonstrate again the persistence of ancient traditions, that at Pompeii, too, there is an Arc de Triomphe at the entrance to the Forum.

Columns were used with porticos to form enclosures for public squares. Saint Peter's in Rome is the best example of this . . .

Arcades were used to embellish monumental buildings more frequently in former times than at present, either on the higher stories, as in the City Halls of Halle (1548) and Cologne (1568), or on the ground level . . .

All of these above-mentioned architectural forms in former times made up a complete system of enclosing public squares. Today there is a contrary tendency to open them on all sides. It is easy to describe the results that have come about. It has tended to destroy completely the old public squares. Wherever these openings have been made the cohesive effect of the square has been completely nullified.

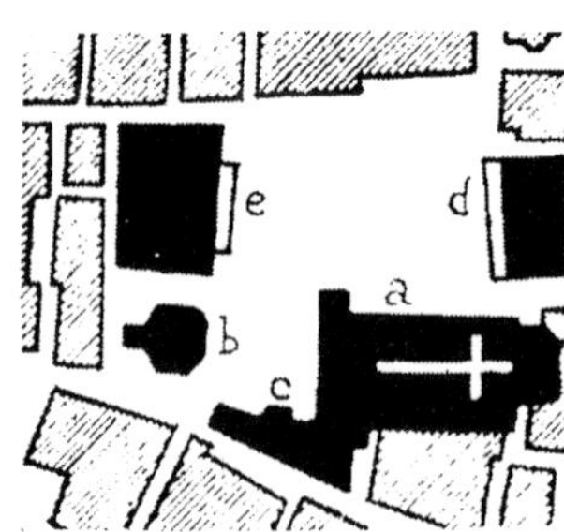

Figure 7 Pistoia: Cathedral Square
Key
a. Cathedral b. Baptistery
c. Residence d. Palais de la Commune
e. Palais du Podestat

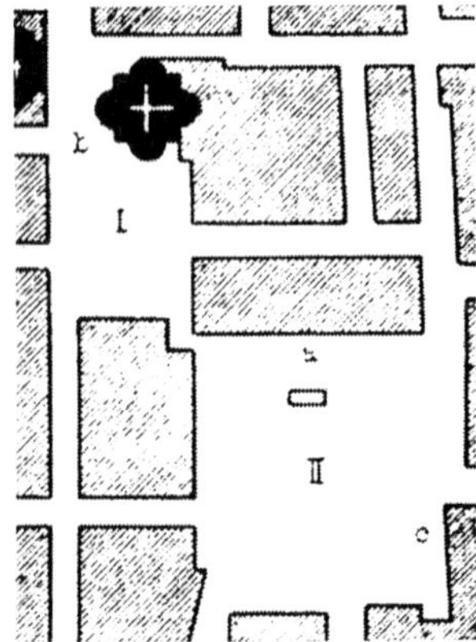

Figure 8 Parma
Key
a. Pal. del Commune b. Madonna della Steccata
c. Pal. della Podesteria
I. Piazza d. Steccata II. Piazza Grande

Figure 9 Verona: Piazza dei Signori

"The Neighborhood Unit"

from *The Regional Plan of New York and its Environs* (1929)

Clarence Perry

Editors' introduction

In the 1920s, early in the automobile age, American architect Clarence Perry thought deeply about the way in which the growth of cities and the rise of the automobile were affecting neighborhoods and characteristics that make good neighborhoods. He articulated a philosophy for maintaining human-scale neighborhoods in the modern world that has had a profound impact on twentieth-century urban planning and remains extremely relevant nowadays.

Every great city, Perry argued, is a conglomerate of smaller communities. The "cellular city" is the inevitable product of the automobile age. It is the quality of life within these smaller communities that will most shape individuals' experiences. New York and Paris routinely turn up in the top popular rankings of the best places in the world to live. But someone experiencing exclusion in an all-immigrant Parisian suburb Ali Madanipour deplores (p. 186) or the mean ghetto streets that Elijah Anderson describes (p. 127) may experience these great world cities as dirty, noisy, crowded, and dangerous. Residents may be unhappy with their neighborhood in a great city if they do not dare let their children play outside because of fast-moving traffic, their children cannot get to school without crossing a freeway, there is no convenience store nearby to buy groceries, or they do not have access to parks and playgrounds.

Perry noted that in the past many people felt a strong identity with villages and small towns, a perception shared by the new urbanists (p. 356). These places had a distinct spatial structure and culture. But by the time Perry wrote this selection in 1929, express highways were cutting up residential areas into small islands separated from each other by raging streams of traffic. As the growing population filled in the interstices between villages, there was what Perry called "a growing attenuation of community characteristics." While residents of some newly developed areas continued to associate with their neighbors, in many of these interstitial areas they did not. It was the amount and kind of association among their residents that Perry felt would distinguish good neighborhoods from bad ones.

Perry related the need for identity to a geographic neighborhood community to the human life cycle. Young singles often enjoy the relative anonymity of city living. But, he noted, when they marry and have children they "long for a detached house and yard and the social benefit of a congenial neighborhood." Thus, for Perry, a primary challenge was to create spaces that would best suit families with children. His solution was "the neighborhood unit."

Perry noted that the primary school was the central institution to which nuclear families with young children related. The quality of the school was the most important factor in deciding which school district to live in and the location of a house in relation to school affected home-buying choices. Every weekday during the school year one parent (usually the mother) took one or more children to and from school. Family members went to school plays, sporting events, and other events at the school. Many of the families' friends were parents of children in their children's classes. One or both parents were often active in parent-teacher associations and other school institutions. For these, and other, reasons Perry argued that neighborhood units should be built around schools.

In "The Neighborhood Unit" he introduces the idea – developed more fully in other of his writings – that the school should be a community center with adult education classes and cultural events in the evening.

Since educators at that time felt that 800–1500 students was an appropriate size for a primary school, Perry argued that the residential land in a neighborhood unit should be large enough to house families with that number of children. At prevailing densities, a five-acre primary school site in the center of a circle with a half-mile radius would work well. That would allow children to walk to school, so long as they did not have to cross busy streets. Of course variations in density, the average number of primary-age school children per household, and geographical particularities would affect this idealized model.

A graded street system was central to Perry's plan. Streets would serve two different groups: people passing by the neighborhood unit and the residents themselves. Perry placed arterials along which through traffic could move rapidly at the boundaries of the neighborhood unit. Unless there were (expensive) bridges or tunnels, Perry knew it would be dangerous for children to cross highways to get from home to school so he opposed arterials between residences and schools. Residential streets, designed primarily for use by neighborhood unit residents would be in the interior. Long before Jan Gehl (p. 530) proposed traffic calming to improve life between buildings, Perry proposed residential street widths and designs to ensure that traffic would move slowly enough that pedestrians would be safe within the residential areas.

Most of the neighborhood would be residential – mainly single-family detached houses on separate yards. But Perry argued that neighborhood residents would want easy access to grocery stores and other neighborhood-serving retail stores. He proposed locating a business district on the edge of the neighborhood unit so that neighborhood residents could reach it on interior streets and through traffic could reach it on arterials. In addition to the school and playground, street system, and residential areas, Perry was an advocate of parks and open space.

Politically, Perry was a conservative pragmatist. Writing just as the Great Depression was beginning, Perry did not believe that much government intervention or taxpayer subsidies could realistically be set aside for neighborhood units. He felt that private developers had to be convinced his neighborhood unit would attract private-market buyers. Accordingly Perry described how private developers could use his ideas for new developments on raw land (greenfield sites), redevelopment of blighted areas, and to improve already built up areas with poorly located schools, traffic problems and lack of open space.

Perry's prescriptions focused on a limited segment of society. His interest was to create functional, safe and attractive neighborhoods with a sense of community for middle and upper-income nuclear families with children. He did not value the kind of messy, mixed, urban neighborhood that Jane Jacobs celebrates (p. 105). While he focused on predominantly white, middle-class, nuclear families, his ideas on education, neighborhood design, and community are extremely relevant for residents of Black neighborhoods that concern W.E.B. DuBois (p. 114), William Julius Wilson (p. 117), and Elijah Anderson (p. 127); immigrant neighborhoods that concerned Louis Wirth (p. 96); single women and female-headed households that are the focus of Daphne Spain's analysis (p. 176), or the urban poor that Michael Porter hopes to help through the private sector (p. 282).

Note the similarity between Perry's observations of the interstitial areas that he observed growing up between villages in 1929 to Thomas Sievert's concept of the Zwischenstadt (literally in-between city) proposed in 1977 discussed in Bruegmann (p. 211). Robert Fishman discusses similar patches of Technoburbia (p. 75).

Do Perry's ideas, developed for middle- and upper-income, nuclear families with children, still apply today as fewer and fewer households are like that? Does he have an anti-urban bias? Will his ideas just help promote urban sprawl? Do they support Bruegmann's argument that sprawl exists because it provides the type of housing people in affluent democracies want? As more and more people worldwide can afford to live in low density developments, are Perry's prescriptions an effective antidote to the kind of alienation that Louis Wirth described in "Urbanism as a Way of Life" (p. 96)? Will twenty-first-century neighborhood units counteract the lack of civic engagement that Robert Putnam deplores (p. 134)?

Clarence Arthur Perry (1872–1944) was an architect and planner. Perry wrote a series of reports on education and the use of schools for community centers for the Russell Sage Foundation including *Wider Use of the School Plant* (1911), *Community Center Activities* (1916), *Educational Extension* (1916), and *The Extension of Public Education* (1915). He lived in Forest Hills Garden, a garden suburb the Russell Sage Foundation had supported, at the time he wrote "The Neighborhood Unit".

This selection is the second part of a 118-page monograph titled "The Neighborhood Unit: A Scheme of Arrangement for the Family-Life Community" published in Volume VII of *The Regional Plan of New York and its Environs* titled *Neighborhood and Community Planning* (New York: Russell Sage Foundation, 1929, reprinted New York: Arno Press, 1974).

Perry's neighborhood unit ideas were further disseminated in *Housing for the Mechanical Age* (New York: Russell Sage Foundation, 1933) and *Housing for the Machine Age* (New York: Russell Sage Foundation, 1939).

Other books on neighborhood planning and design include Tridib Bannerjee and William C. Baer, *Beyond the Neighborhood Unit: Residential Environments and Public Policy* (New York: Springer, 1984), Frederick D. Jarvis, *Site Planning and Community Design for Great Neighborhoods* (Washington, DC: Home Builder Press, 1993), Randolph Hester, *Neighborhood Space* (Stroudsburg, PA: Dowden, Hutchinson & Ross; New York: Halsted Press, 1975), Randolph Hester, *Planning Neighborhood Space with People* (New York: Van Nostrand Reinhold, 1982), Sidney Bower, *Good Neighborhoods: A Study of In-Town and Suburban Residential Environments* (Westport, CT: Praeger, 2000), and Urban Design Associates, *The Architectural Pattern Book: A Tool for Building Great Neighborhoods* (New York: Norton, 2004).

AUTHOR'S INTRODUCTION

What is known as a neighborhood, and what is now commonly defined as a region, have at least one characteristic in common – they possess a certain unity which is quite independent of political boundaries. The area with which the Regional Plan of New York is concerned, for instance, has no political unity, although it is possessed of other unifying characteristics of a social, economic and physical nature. Within this area there are definite political entities, such as villages, counties and cities, forming suitable divisions for sub-regional planning, and within those units there are definite local or neighborhood communities which are entirely without governmental limits and sometimes overlap into two or more municipal areas. Thus, in the planning of any large metropolitan area, we find that three kinds of communities are involved:

1. The regional community, which embraces many municipal communities and is, therefore, a family of communities;
2. The village, county or city community;
3. The neighborhood community.

Only the second of these groups has any political framework, although all three have an influence upon political life and development. While the neighborhood community has no political structure, it frequently has greater unity and coherence than are found in the village or city and is, therefore, of fundamental importance to society.

THE NEIGHBORHOOD UNIT

The above title is the name which, to facilitate discussion, has been given to the scheme of arrangement for a family-life community that has evolved as the main conclusion of this study. Our investigations showed that residential communities, when they meet the universal needs of family life, have similar parts performing similar functions. In the neighborhood-unit system those parts have been put together as an organic whole. The scheme is put forward as the framework of a model community and not as a detailed plan. Its actual realization in an individual real-estate development requires the embodiment and garniture which can be given to it only by the planner, the architect, and the builder.

The underlying principle of the scheme is that an urban neighborhood should be regarded both as a unit of a larger whole and as a distinct entity in itself. For government, fire and police protection, and many other services, it depends upon the municipality. Its residents, for the most part, find their occupations outside of the neighborhood. To invest in bonds, attend the opera or visit the museum, perhaps even to buy a piano, they have to resort to the "downtown" district. But there are certain other facilities, functions or aspects which are strictly local and peculiar to a well-arranged residential community. They may be classified under four heads: (1) the elementary school, (2) small parks and playgrounds, (3) local shops, and (4) residential environment. Other neighborhood institutions and services are sometimes found, but these are practically universal.

Parents have a general interest in the public school system of the city, but they feel a particular concern regarding the school attended by their children. Similarly, they have a special interest in the playgrounds where their own and their neighbors' children spend so many formative hours. In regard to small stores, the main concern of householders is that they be accessible but not next to their own doors. They should also be concentrated and provide for varied requirements.

Under the term "residential environment" is included the quality of architecture, the layout of streets, the planting along curbs and in yards, the arrangement and set-back of buildings, and the relation of shops, filling stations and other commercial institutions to dwelling places – all the elements which go into the environment of a home and constitute its external atmosphere. The "character" of the district in which a person lives tells something about him. Since he chose it, ordinarily, it is an extension of his personality. One individual can do but little to create it. It is strictly a community product.

It is with the neighborhood itself, and not its relation to the city at large, that this study is concerned. If it is to be treated as an organic entity, then it logically follows that the first step in the conversion of unimproved acreage for residential purposes will be its division into unit areas, each one of which is suitable for a single neighborhood community. The next step consists in the planning of each unit so that adequate provision is made for the efficient operation of the four main neighborhood functions. The attainment of this major objective – as well as the securing of safety to pedestrians and the laying of the structural foundation for quality in environment – depends, according to our investigations, upon the observance of the following requirements.

Neighborhood-unit principles

1. *Size* – A residential unit development should provide housing for that population for which one elementary school is ordinarily required, its actual area depending upon population density.
2. *Boundaries* – The unit should be bounded on all sides by arterial streets, sufficiently wide to facilitate its by-passing by all through traffic.
3. *Open Spaces* – A system of small parks and recreation spaces, planned to meet the needs of the particular neighborhood, should be provided.
4. *Institution Sites* – Sites for the school and other institutions having service spheres coinciding with the limits of the unit should be suitably grouped about a central point or common area.
5. *Local Shops* – One or more shopping districts, adequate for the population to be served, should be laid out in the circumference of the unit, preferably at traffic junctions and adjacent to similar districts of adjoining neighborhoods.
6. *Internal Street System* – The unit should be provided with a special street system, each highway being proportioned to its probable traffic load, and the street net as a whole being designed to facilitate circulation within the unit and to discourage its use by through traffic.

[For] each of these principles [. . .], it is desirable [. . .] to obtain a clearer picture of them, and for that purpose a number of plans and diagrams in which they have been applied will now be presented.

Low-cost suburban development

Character of district

[The plan shown in Figure 1] is based upon an actual tract of land in the outskirts of the Borough of Queens. The section is as yet entirely open and exhibits a gently rolling terrain, partly wooded. So far, the only roads are of the country type, but they are destined some day to be main thoroughfares. There are no business or industrial establishments in the vicinity.

Complete unit	160 acres	100 per cent
Dwelling-house lots	86.5	54.0
Apartment-house lots	3.4	2.1
Business blocks	6.5	4.1
Market squares	1.2	0.8
School and church sites	1.6	1.0
Parks and playgrounds	13.8	8.6
Greens and circles	3.2	2.0
Streets	43.8	27.4

Table 1 Area relations of the plan

SEVEN

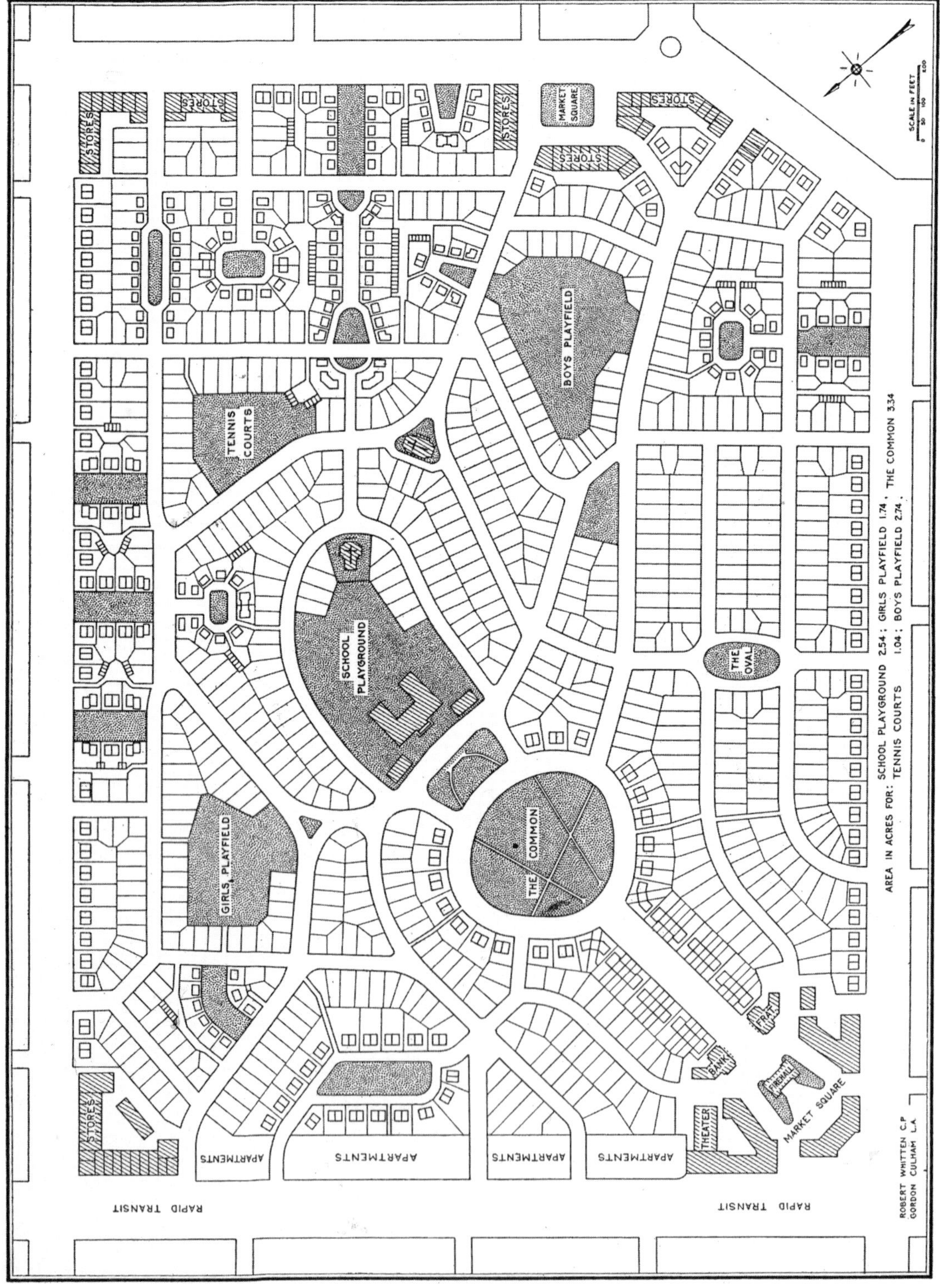

Figure 1 A Subdivision for Modest Dwellings Planned as a Neighborhood Unit

Population and housing

The lot subdivision provides 822 single-family houses, 236 double houses, 36 row houses and 147 apartment suites, accommodations for a total of 1,241 families. At the rate of 4.93 persons per family, this would mean a population of 6,125 and a school enrollment of 1,021 pupils. For the whole tract the average density would be 7.75 families per gross acre.

Open spaces

The parks, playgrounds, small greens and circles in the tract total 17 acres, or 10.6 per cent of the total area. If there is included also the 1.2 acres of market squares, the total acreage of open space is 18.2 acres. The largest of these spaces is the common of 3.3 acres.

This serves both as a park and as a setting or approach to the school building. Back of the school is the main playground for the small children, of 2.54 acres, and near it is the girls' playfield of 1.74 acres. On the opposite side of the schoolyard, a little farther away, is the boys' playground of 2.7 acres. Space for tennis courts is located conveniently in another section of the district. At various other points are to be found parked ovals or small greens which give attractiveness to vistas and afford pleasing bits of landscaping for the surrounding homes.

Community center

The pivotal feature of the layout is the common, with the group of buildings, which face upon it. These consist of the schoolhouse and two lateral structures facing a small central plaza. One of these buildings might be devoted to a public library and the other to any suitable neighborhood purpose. Sites are provided for two churches, one adjoining the school playground and the other at a prominent street intersection. The school and its supporting buildings constitute a terminal vista for a parked main highway coming up from the market square. In both design and landscape treatment the common and the central buildings constitute an interesting and significant neighborhood community center.

Shopping districts

Small shopping districts are located at each of the four corners of the development. The streets furnishing access to the stores are widened to provide for parking, and at the two more important points there are small market squares, which afford additional parking space and more opportunity for unloading space in the rear of the stores. The total area devoted to business blocks and market plazas amounts to 7.7 acres. The average business frontage per family provided by the plan is about 2.3 feet.

Street system

In carrying out the unit principle, the boundary streets have been made sufficiently wide to serve as main traffic arteries. One of the bounding streets is 160 feet wide, and the other three have widths of 120 feet. Each of these arterial highways is provided with a central roadway for through traffic and two service roadways for local traffic separated by planting strips. One-half of the area of the boundary streets is contributed by the development. This amounts to 15.3 acres, or 9.5 per cent of the total area, which is a much larger contribution to general traffic facilities than is ordinarily made by the commercial subdivision, but not greater than that which is required by present-day traffic needs. The interior streets are generally 40 or 50 feet in width and are adequate for the amount of traffic, which will be developed in a neighborhood of this single- family density. By the careful design of blocks, the area devoted to streets is rather lower than is usually found in a standard gridiron subdivision. If the bounding streets were not over 50 feet wide, the per cent of the total street area would be reduced from the 27.4 per cent to about 22 per cent. It will be observed that most of the streets opening on the boundary thoroughfares are not opposite similar openings in the adjacent developments. There are no streets which run clear through the development without being interrupted . . .

A neighborhood unit for an industrial section

[Figure 2] is presented as a sketch of the kind of layout which might be devised for a district in the vicinity of factories and railways. Many cities possess somewhat

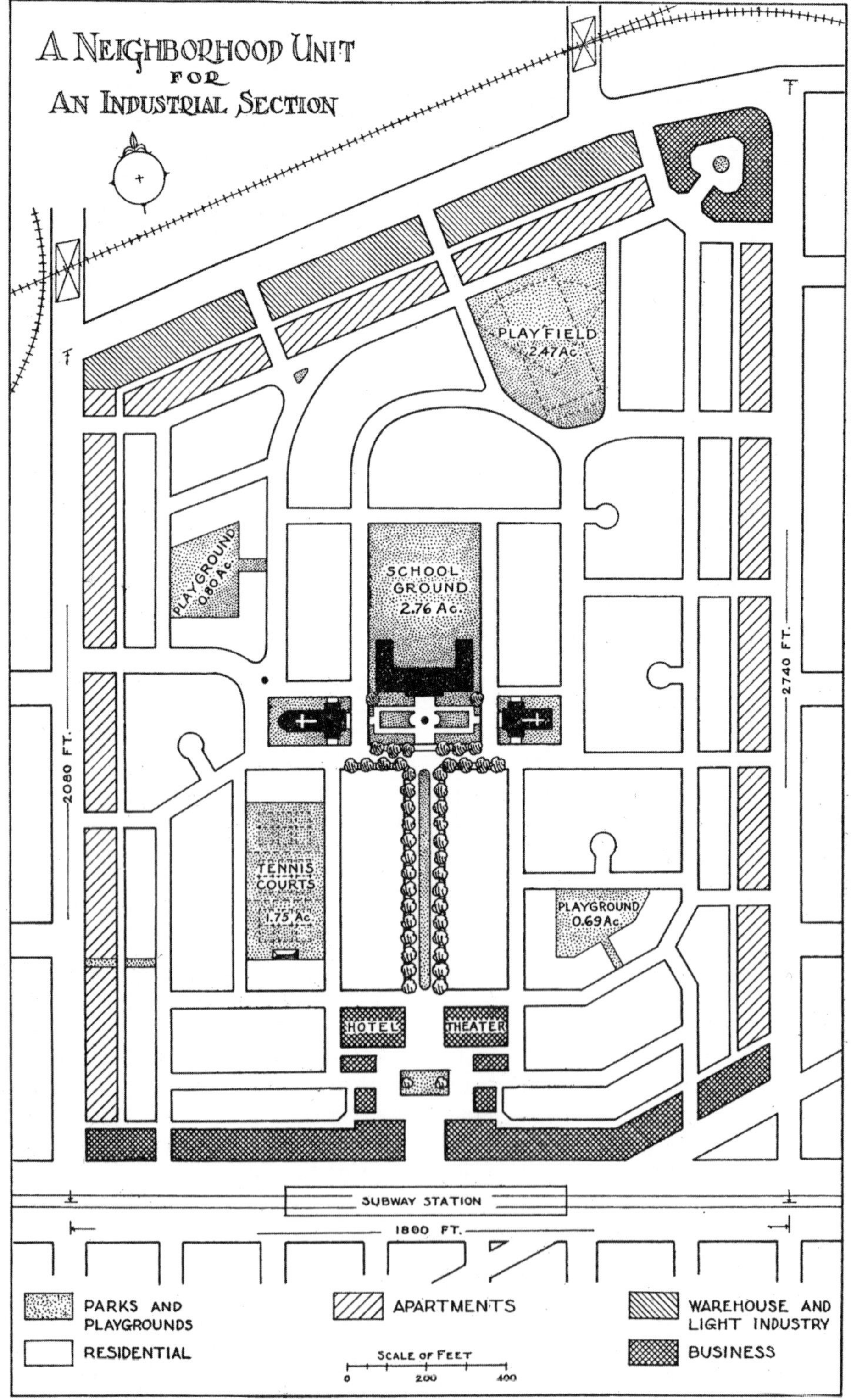

Figure 2 Suggested Treatment for a Denser and More Central District

Complete unit	101.4 acres	100 per cent
Residences—houses	37.8	37.3
Residences—apartments	8.4	8.3
Parks and play spaces[1]	10.8	10.6
Business	5.2	5.1
Warehouses	3.2	3.2
Streets	36.0	35.5

Table 2 Distribution of area in Fig. 2

central areas of this character, which have not been pre-empted by business or industry but which are unsuitable for high-cost housing and too valuable for a low-cost development entirely of single-family dwellings.

Economically, the only alternative use for such a section is industrial. If it were built up with factories, however, the non-residential area thereabouts would be increased and the daily travel distance of many workers would be lengthened. One of the main objectives of good city planning is therefore attained when it is made available for homes.

Along the northern boundary of the tract illustrated lie extensive railroad yards, while its southern side borders one of the city's main arteries, affording both an elevated railway and wide roadbeds for surface traffic. An elevated station is located at a point opposite the center of the southern limit, making that spot the main portal of the development.

The functional dispositions

The above features dictated the employment of a tree-like design for the street system. Its trunk tests upon the elevated station, passes through the main business district, and terminates at the community center. Branches, covering all sections of the unit, facilitate easy access to the school, to the main street stem, and to the business district.

Along the northern border, structures suitable for light industry, garages, or warehouses have been designated. These are to serve as a buffer both for the noises and the sights of the railway yards. Next to them, separated only by a narrow service street, is a row of apartments, whose main outlooks will all be directed toward the interior of the unit and its parked open spaces.

The apartments are assigned to sites at the sides of the unit that they may serve as conspicuous visible boundaries and enable the widest possible utilization of the attractive vistas which should be provided by the interior features – the ecclesiastical architecture around the civic center and the park-like open spaces.

Housing density

The above diagram is intended to suggest mainly an arrangement of the various elements of a neighborhood and is not offered as a finished plan. The street layout is based upon a housing scheme providing for 2,000 families, of which 68 per cent are allotted to houses, some semi-detached and some in rows; and 32 per cent to apartments averaging 800 square feet of ground area per suite. On the basis of 4.5 persons in houses and 4.2 in suites, the total population would be around 8,800 people and there would be some 1,400 children of elementary school age, a fine enrollment for a regulation city school. The average net ground area per family amounts to 1,003.7 square feet. If the parks and play areas are included, this figure becomes 1,216 square feet.

Recreation spaces

These consist of a large schoolyard and two playgrounds suitable for the younger children, grounds accommodating nine tennis courts, and a playfield adapted either for baseball or soccer football. In distributing these spaces regard was had both to convenience and to their usefulness as open spaces and vistas for the adjacent homes. All should have planting around the edges, and most of them could be seeded, thus avoiding the barren aspect so common to city playgrounds.

Community center

The educational, religious and civic life of the community is provided for by a group of structures, centrally located and disposed so as to furnish an attractive vista for the trunk street and a pivotal point for the whole layout. A capacious school is flanked by two churches, and all face upon a small square which might be embellished with a monument, fountain, or other ornamental feature. The auditorium, gymnasium, and library of the school, as well as certain other rooms, could be used for civic, cultural and recreational activities of the neighborhood. With such equipment and an environment possessing so much of interest and service to all the residents, a vigorous local consciousness would be bound to arise and find expression in all sorts of agreeable and useful face-to-face associations.

Shopping districts

The most important business area is, of course, around the main portal and along the southern arterial highway. For greater convenience and increased exposures a small market square has been introduced. Here would be the natural place for a motion-picture theatre, a hotel, and such services as a branch post office and a fire-engine house. Another and smaller shopping district has been placed at the northeast corner to serve the needs of the homes in that section.

Economic aspects

While this development is adapted to families of moderate means, comprehensive planning makes possible an intensive and profitable use of the land without the usual loss of a comfortable and attractive living environment. The back and side yards may be smaller, but pleasing outlooks and play spaces are still provided. They belong to all the families in common and the unit scheme preserves them for the exclusive use of the residents.

While this is primarily a housing scheme, it saves and utilizes for its own purposes that large unearned increment, in business and industrial values, which rises naturally out of the mere aggregation of so many people. The community creates that value and while it may apparently be absorbed by the management, nevertheless, some of it goes to the individual householder through the improved home and environment which a corporation, having that value in prospect, is able to offer.

The percentage of area devoted to streets (35.5) is higher than is usually required in a neighborhood-unit scheme. In this case the proportion is boosted by the generous parking space provided in the market square and by the adjoining 200-foot boulevard, one-half of whose area is included in this calculation. Ordinarily the unit scheme makes possible a saving in street area that is almost, if not quite, equal to the land devoted to open spaces. The school and church sites need not be dedicated. They may simply be reserved and so marked in the advertising matter with full confidence that local community needs and sentiment will bring about their ultimate purchase by the proper bodies. If either or both of the church sites should not be taken, their very location will ensure their eventual appropriation for some public, or semi-public, use.

Apartment-house unit

Population

On the basis of five-story and basement buildings and allowing 1,320 square feet per suite, this plan would accommodate 2,381 families. Counting 4.2 persons per family, the total population would number 10,000 individuals, of whom about 1,600 would probably be of elementary school age, a number which could be nicely accommodated in a modern elementary school.

Environment

The general locality is that section where downtown business establishments and residences begin to merge. One side of the unit faces on the principal street of the city and this would be devoted to general business concerns. A theatre and a business block, penetrated by an arcade, would serve both the residents of the unit and the general public.

Street system

The unit is bounded by wide streets, while its interior system is broken up into shorter highways that give easy circulation within the unit but do not run uninterruptedly through it. In general they converge upon the community center. Their widths are varied to fit probable traffic loads and parking needs.

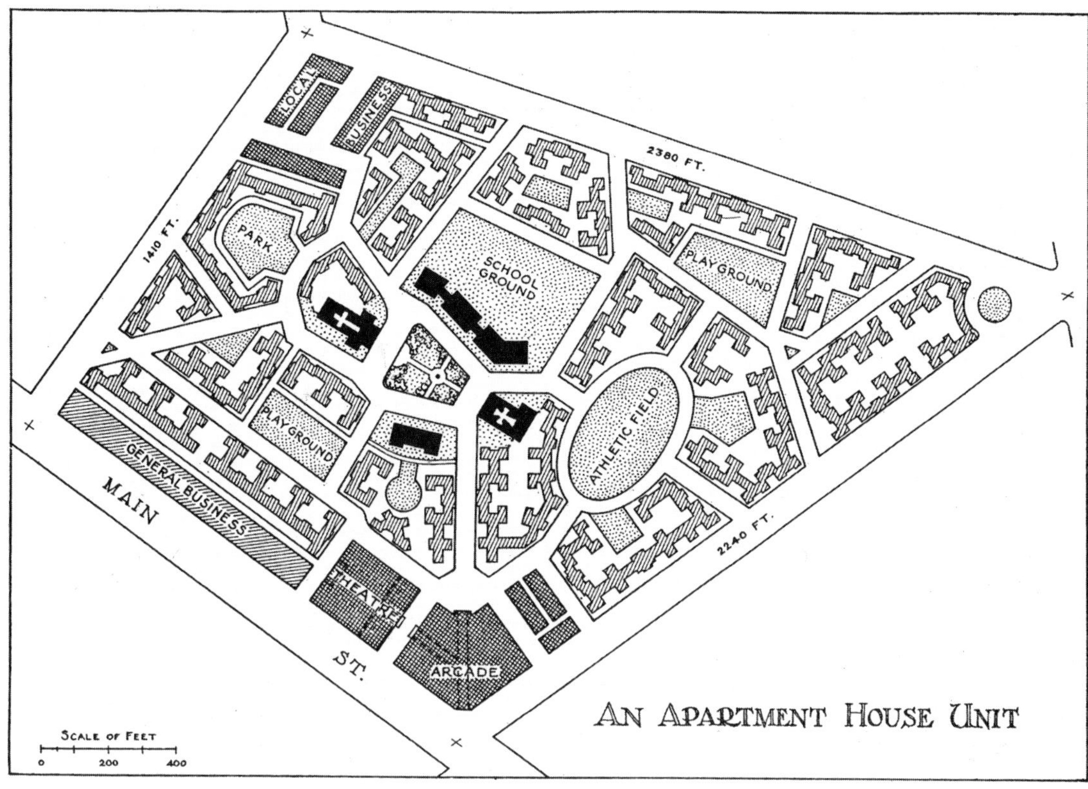

Figure 3 A Method of Endowing a Multiple Family District with Interesting Window Vistas, Greater Street Safety, More Liberal Open Spaces, and a Neighborhood Character

Total area of unit	75.7 acres	100 per cent
Apartment buildings	12.0	15.9
Apartment yards	21.3	28.0
Parks and playgrounds	10.4	13.8
Streets	25.3	33.4
Local business	4.9	6.5
General business	1.8	2.4

Table 3 Distribution of area in Fig. 3

Kind	Acres
School grounds	3.27
Athletic field	1.85
Common	.81
Park	.61
Playground	1.03
Playground	.81
Circle	.18
Small greens	1.86
Total	10.42

Table 4 Area of open spaces in Fig. 3

Open spaces

The land devoted to parks and playgrounds averages over one acre per 1,000 persons. If the space in apartment yards is also counted, this average amounts to 3.17 acres per 1,000 persons. The distribution is (shown in Table 4).

For 1,600 children the space in the school yard provides an average of 89 square feet per pupil, which is a fair allowance considering that all the pupils will seldom be in the yard at the same time. The athletic field is large enough for baseball in the spring and summer, and football in the fall. By flooding it with a hose in the winter time it can be made available for skating.

On the smaller playground it will be possible, if desired, to mark off six tennis courts. The bottle-neck park is partly enclosed by a group of apartments, but it is also accessible to the residents in general.

The recreation spaces should be seeded and have planting around the edges, thus adding attractiveness to the vistas from the surrounding apartments.

Community center

Around a small common are grouped a school, two churches, and a public building. The last might be a branch public library, a museum, a "little theatre," or a fraternal building. In any case it should be devoted to a local community use.

The common may exhibit some kind of formal treatment in which a monument and perhaps a band-stand may be elements of the design. The situation is one that calls for embellishment, by means of both architecture and landscaping, and such a treatment would contribute greatly to local pride and the attractiveness of the development. The ground plan of the school indicates a type in which the auditorium, the gymnasium and the classrooms are in separate buildings, connected by corridors. This arrangement greatly facilitates the use of the school plant by the public in general and permits, at the same time, an efficient utilization of the buildings for instruction purposes.

Apartment pattern

The layout of the apartment structures follows quite closely an actual design employed by Mr. Andrew J. Thomas for a group of "garden apartments" now being constructed for Mr. John D. Rockefeller, Jr., in New York City. The suites are of four, five, six and seven rooms and, in the case of the larger ones, two bath-rooms. Light comes in three sides of a room as a rule and, in some cases, from four sides. All rooms enjoy cross-ventilation.

In the Rockefeller plan every apartment looks out upon a central garden, which is ornamented with a Japanese rookery and a foot-bridge over running water. The walks are to be lined with shrubbery and the general effect will be park-like and refreshing.

Similar treatments could be given to the various interior spaces of the unit layout. Here, however, due to the short and irregular streets and the odd positions of the buildings, the charm of a given court would be greatly extended because, in many cases, it would constitute a part of the view of not merely one, but several, apartments.

Five-block apartment-house unit

Locality

The plan shown in Figure 4 is put forward as a suggestion of the type of treatment which might be given to central residential areas of high land values destined for rebuilding because of deterioration or the sweep of a real estate movement. The blocks chosen for the ground site are 200 feet wide and 670 feet long, a length which is found in several sections on Manhattan. In this plan, which borders a river, two streets are closed and two are carried through the development as covered roadways under terraced central courts.

Ground plan

The dimensions of the plot between the boundary streets are 650 feet by 1,200 feet, and the total area is approximately 16 acres. The building lines are set back from the streets 30 feet on the northern and southern boundaries. Both of the end streets, which were originally 60 feet, have been widened to 80 feet, the two 20-foot extra strips being taken out of the area of the development. The western boundary has been enlarged from 80 to 100 feet. The area given to street widening and to building set-back amounts to 89,800 square feet, or 11,800 square feet more than the area of the two streets which were appropriated.

It will be observed that the plan of buildings encloses 53 per cent of the total area devoted to open space in the form of central courts. The main central court is about the size of Gramercy Park, Manhattan, with its surrounding streets. Since this area would receive an unusual amount of sunlight, it would be susceptible to the finest sort of landscape and formal garden treatment.

Both of the end courts are on a level 20 feet higher than the central space and cover the two streets which

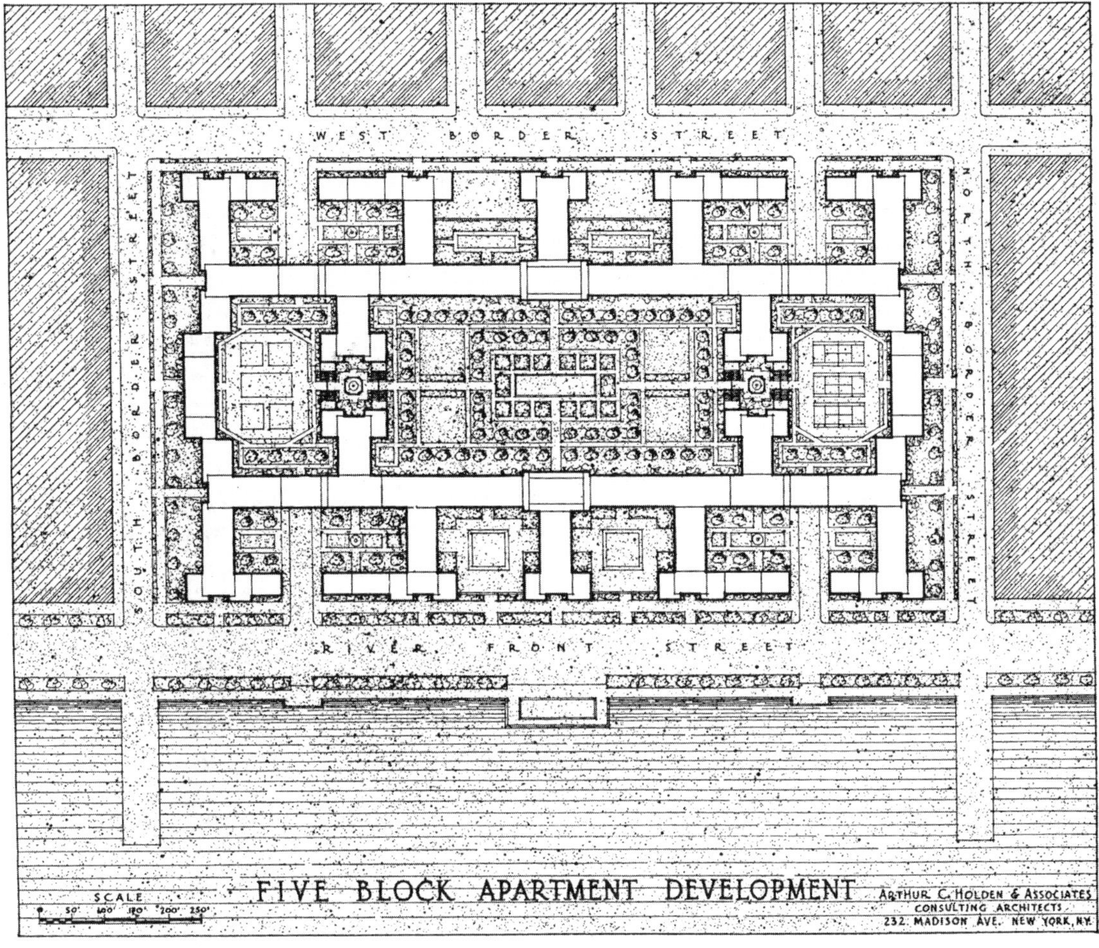

Figure 4 How a Slum District Might be rehabilitated

Five blocks and four cross streets	19.07 acres
Two cross streets taken	78,000 sq. ft.
Given to boundary streets	50,800 sq. ft.
Area of set-backs	39,000 sq. ft.
Land developed	16.4 acres
Covered by buildings	6.5 acres
Coverage	40.0 per cent
Three central courts	5.3 acres

Table 5 Area relations

are carried through the development. Underneath these courts are the service areas for the buildings. At one end of the central space there is room for tennis courts and, at the other, a children's playground of nearly one acre. By reason of the large open spaces and the arrangement of the buildings, the plan achieves an unusual standard as to light in that there is no habitable room that has an exposure to sunlight of less than 45 degrees. The width of all the structures is 50 feet, so that apartments of two-room depth are possible throughout the building, while the western central rib, being 130 feet from a 100-foot street, will never have its light unduly shut off by buildings on the adjacent blocks.

Accommodations

The capacity of the buildings is about 1,000 families, with suites ranging from three to fourteen rooms in size, the majority of them suitable for family occupancy. In addition there would be room for a hotel for transients, an elementary school, an auditorium, a gymnasium, a swimming pool, handball courts, locker rooms and other athletic facilities. The first floors of certain buildings on one or more sides of the unit could be devoted to shops. The auditorium could be suitable for motion pictures, lectures, little-theatre performances, public meetings, and possibly for public worship. Dances could be easily held in the gymnasium. In the basement there might be squash courts.

Height

The buildings range in height from two and three stories on the boundary streets to ten stories in the abutting ribs, fifteen stories in the main central ribs, and thirty-three stories in the two towers. Many of the roofs could be given a garden-like treatment and thus contribute to the array of delightful prospects which are offered by the scheme.

This plan, though much more compact than the three others, nevertheless observes all of the unit principles. Neither the community center nor the shopping districts are conspicuous, but they are present. Children can play, attend school, and visit stores without crossing traffic ways.

"The City Image and its Elements"

from *The Image of the City* (1960)

Kevin Lynch

Editors' Introduction

Kevin Lynch (1918–1984), a professor of urban design at the Massachusetts Institute of Technology, is the towering figure of twentieth-century urban design. *The Image of the City*, from which this selection is taken, is the most widely read urban design book of all time.

Drawing widely on material from psychology and the humanities, Lynch sought to understand how people perceive urban environments and how design professionals can respond to the deepest human needs. Lynch's rambling, profoundly humane writings weave together a unique blend of theory and practical design suggestions drawn from his voluminous reading in history, anthropology, architecture, art, psychology, literature, and a host of other areas.

This influential chapter on "The City Image and its Elements" presents Lynch's best known concepts on how people perceive cities. Lynch argues that people perceive cities as consisting of underlying city form elements such as "paths" (along which people and goods flow), "edges" (which differentiate one part of the urban fabric from another), "landmarks" (which stand out and help orient people), "districts" (perceived as physically or culturally distinct even if their boundaries are fuzzy), and "nodes" where activities, and often paths, meet. Lynch believes that humans have an innate desire to understand their surroundings and do this best if a clear city image is discernible from these elements. If urban designers understand how people perceive these elements and design to make cities more imageable, Lynch argues, urban designers can create more psychologically satisfying urban environments.

Lynch's research is a good example of creative qualitative research. Rather than starting from a theory and reasoning from it about how people perceive cities (deductive logic), Lynch started by gathering empirical information from people themselves and then constructing theory that explained the broad patterns he found (inductive logic). This is a common research strategy. Planners, for example, often conduct surveys to get people's opinions on a particular planning issue. What distinguishes Lynch's work from a run-of-the-mill survey report is his skill at generalizing what he found into high order conceptual categories rather than just reporting survey results. Lynch pioneered the technique of having people draw "mental maps" and analyzing them to understand how people perceived their surroundings. Lynch came to his conclusions about the basic elements of city form that shape perception largely by observing recurring patterns in the maps. Almost everyone drew streets and other geographical features along which people and goods moved. Lynch generalized these features into a category he named "paths". He also noticed that boundaries between parts of the city were often clear in the maps – and named this urban form element an "edge". Similarly Lynch created categories he called landmarks, nodes, and districts by generalizing repeating patterns he discovered in the maps ordinary citizens drew.

Influenced by Lynch's work, urban designers in cities as diverse as San Francisco, Cairo, Havana, and Ciudad Guyana, Venezuela have sketched out the elements of cities or parts of cities they are designing as paths, edges, nodes, landmarks, and districts – the underlying city form elements that Lynch identified – and have used his theories and practical suggestions about good city form to strengthen the image of the cities in which they practice.

Lynch used qualitative methods such as observation, interviews, and analysis of conceptual maps. Like many architects and designers, Lynch was a good artist, who sketched aspects of the built environment both in schematic conceptual ways and as detailed line drawings. Urban design is a science as well as an art, and computers and computer assisted design (CAD) software have revolutionized the way in which urban design is done today. CAD software allows designers to create two- and three-dimensional renderings of urban space. Using CAD software urban designers can see and show others how a site, neighborhood or an entire city might be designed with the volumes and dimensions to scale and surfaces that look like the real thing. Three dimensional virtual reality software makes it possible to "walk through" virtual designs and experience what they will be like.

Like Lynch, other notable urban designers ground their practical design suggestions in underlying theory about human psychological needs. Compare Lynch's ideas about what people find psychologically satisfying and aesthetically appealing about cities with Camillo Sitte's ideas on planning public spaces according to artistic principles (p. 476) so that the entire city is a kind of outdoor art gallery. Note the similarities to Jan Gehl's emphasis on people's psychological need for contact with other people and his ideas about how thoughtful design of the space between buildings can encourage people to engage in more outdoor activities, increase social contact, and satisfy basic human needs (p. 530). Observe how William Whyte grounds his imaginative applied principles for park and plaza design (p. 510) in theories about how people respond to the built environment around them even at a subconscious level. Consider how Frederick Law Olmsted based his practical suggestions about urban park design in a broader vision that urban parks should provide oases of tranquility in which city dwellers could escape the stress of city life and be spiritually refreshed (p. 321).

Urban design draws heavily on the social science discipline of psychology, particularly the subfields social psychology and urban psychology. Urban psychology was pioneered by German psychologist George Simmel, who wrote a seminal essay titled "The Metropolis and Mental Life" in 1905. Lewis Wirth's essay on "Urbanism as a Way of Life" (p. 96) builds on Simmel's work. Studies of how people perceive the personal space around them by University of California, Davis, psychology professor emeritus Robert Sommer are widely used by architects and urban designers to inform them work.

Unlike urban geography, urban politics, urban economics, and urban history, which are mainstream subfields regularly taught in their respective departments, urban psychology remains a specialty taught in a minority of psychology departments.

Culture is a particular concern of anthropologists. Some urban anthropologists study the way in which culture affects perceptions of space. Anthropologist Edward T. Hall developed the field of proxemics – how people in different cultures respond to the space immediately around them.

Kevin Lynch studied architecture at Yale University and apprenticed himself to Frank Lloyd Wright, the brilliant and opinionated architect who envisioned Broadacre City (p. 345) and was eventually recognized as one of the United States' greatest architects. Lynch received a bachelor's degree in urban planning from the Massachusetts Institute of Technology (MIT) in 1947 and joined its faculty a year later. At MIT Lynch taught courses in urban design and site planning. He maintained an active urban design practice and, after publication of *The Image of the City*, cemented his reputation, lectured and consulted worldwide.

In addition to *The Image of the City* (Cambridge, MA: MIT Press, 1960), Lynch's many books include a textbook on site design coauthored with Gary Hack, *Site Planning*, 3rd edn (Cambridge, MA: MIT Press, 1984), and books on historic preservation, *What Time Is This Place* (Cambridge, MA: MIT Press, 1979), regional planning, *Managing the Sense of a Region* (Cambridge, MA: MIT Press, 1976), and his magnum opus, *Good City Form* (Cambridge, MA: MIT Press, 1991). Other of Lynch's writings are contained in Kevin Lynch, Tridib Banerjee, and Michael Southworth (eds), *City Sense and City Design: Writings and Projects of Kevin Lynch* (Cambridge, MA: MIT Press, 1995). Lynch's papers are in the Massachusetts Institute of Technology Library Archives and Special Collections, MC 208, Box X in Cambridge, Massachusetts.

Classic and contemporary writings on urban design are collected in Michael Larice and Elizabeth Macdonald (eds), *The Urban Design Reader* (London: Routledge, 2006). Other books on urban design include Alex Krieger and William S. Saunders (eds), *Urban Design* (Minneapolis, MN: University of Minnesota Press, 2009), Lance Jay

Brown, David Dixon, and Oliver Gillham, *Urban Design for an Urban Century: Placemaking for People* (New York: Wiley, 2009), Doug Kelbaugh, *Common Place: Toward Neighborhood and Regional Design* (Seattle, WA: University of Washington Press, 1997), Mike Greenberg, *The Poetics of Cities: Designing Neighborhoods that Work* (Columbus, OH: Ohio State University Press, 1995), Spiro Kostof, *The City Shaped* (Boston, MA: Little, Brown, 1991) and *The City Assembled* (Boston, MA: Little, Brown, 1992), and Edmund Bacon, *The Design of Cities* (New York: Viking, 1976). A practitioner's book on urban design is Ray Gindoz, Karen Levine, and Urban Design Associates, *The Urban Design Handbook: Techniques and Working Methods* (New York: Norton, 2003).

Psychologist Robert Sommers' book on urban psychology is *Personal Space* (Bristol: Bosko, 2008). Anthony Hiss, *The Experience of Place* (New York: Knopf, 1990) also explores the psychology of how people perceive urban space.

Anthropologist Edward T. Hall's books on proxemics (the study of how people respond to the space immediately around them) are *The Silent Language* (Garden City, NY: Doubleday, 1959) and *The Hidden Dimension* (Garden City, NY: Doubleday, 1966).

Books on the relationship between gender, design, and space include Daphne Spain, *Gendered Spaces* (Chapel Hill, NC: University of North Carolina Press, 1992), Doreen Massey, *Space, Place, and Gender*, and Linda Mcdowell, *Gender, Identity, and Place: Understanding Feminist Geographies* (Minneapolis, MN: University of Minnesota Press, 1999).

■

There seems to be a public image of any given city which is the overlap of many individual images. Or perhaps there is a series of public images, each held by some significant number of citizens. Such group images are necessary if an individual is to operate successfully within his environment and to cooperate with his fellows. Each individual picture is unique, with some content that is rarely or never communicated, yet it approximates the public image, which, in different environments, is more or less compelling, more or less embracing.

This analysis limits itself to the effects of physical, perceptible objects. There are other influences on imageability, such as the social meaning of an area, its function, its history, or even its name. These will be glossed over, since the objective here is to uncover the role of form itself. It is taken for granted that in actual design form should be used to reinforce meaning, and not to negate it.

The contents of the city images so far studied, which are referable to physical forms, can conveniently be classified into five types of elements: paths, edges, districts, nodes, and landmarks.

Indeed, these elements may be of more general application, since they seem to reappear in many types of environmental images. . . . These elements may be defined as follows:

1 *Paths.* Paths are the channels along which the observer customarily, occasionally, or potentially moves. They may be streets, walkways, transit lines, canals, railroads. For many people, these are the predominant elements in their image. People observe the city while moving through it, and along these paths the other environmental elements are arranged and related.

2 *Edges.* Edges are the linear elements not used or considered as paths by the observer. They are the boundaries between two phases, linear breaks in continuity: shores, railroad cuts, edges of development, walls. They are lateral references rather than coordinate axes. Such edges may be barriers, more or less penetrable, which close one region off from another; or they may be seams, lines along which two regions are related and joined together. These edge elements, although probably not as dominant as paths, are for many people important organizing features, particularly in the role of holding together generalized areas, as in the outline of a city by water or wall.

3 *Districts.* Districts are the medium-to-large sections of the city, conceived of as having two-dimensional extent, which the observer mentally enters "inside of," and which are recognizable as having some common, identifying character. Always identifiable from the inside, they are also used for exterior reference if visible from the outside. Most people structure their city to some extent in this way, with individual differences as to whether paths or districts are the dominant

elements. It seems to depend not only upon the individual but also upon the given city.

4 *Nodes.* Nodes are points, the strategic spots in a city into which an observer can enter, and which are the intensive foci to and from which he is traveling. They may be primarily junctions, places of a break in transportation, a crossing or convergence of paths, moments of shift from one structure to another. Or the nodes may be simply concentrations, which gain their importance from being the condensation of some use or physical character, as a street-corner hangout or an enclosed square. Some of these concentration nodes are the focus and epitome of a district, over which their influence radiates and of which they stand as a symbol. They may be called cores. Many nodes, of course, partake of the nature of both junctions and concentrations. The concept of node is related to the concept of path, since junctions are typically the convergence of paths, events on the journey. It is similarly related to the concept of district, since cores are typically the intensive foci of districts, their polarizing center. In any event, some nodal points are to be found in almost every image, and in certain cases they may be the dominant feature.

5 *Landmarks.* Landmarks are another type of point-reference, but in this case the observer does not enter within them, they are external. They are usually a rather simply defined physical object: building, sign, store, or mountain. Their use involves the singling out of one element from a host of possibilities. Some landmarks are distant ones, typically seen from many angles and distances, over the tops of smaller elements, and used as radial references. They may be within the city or at such a distance that for all practical purposes they symbolize a constant direction. Such are isolated towers, golden domes, great hills. Even a mobile point, like the sun, whose motion is sufficiently slow and regular, may be employed. Other landmarks are primarily local, being visible only in restricted localities and from certain approaches. These are the innumerable signs, storefronts, trees, doorknobs, and other urban detail, which fill in the image of most observers. They are frequently used clues of identity and even of structure, and seem to be increasingly relied upon as a journey becomes more and more familiar.

The image of a given physical reality may occasionally shift its type with different circumstances of viewing. Thus an expressway may be a path for the driver, and edge for the pedestrian. Or a central area may be a district when a city is organized on a medium scale, and a node when the entire metropolitan area is considered. But the categories seem to have stability for a given observer when he is operating at a given level.

None of the element types isolated above exist in isolation in the real case. Districts are structured with nodes, defined by edges, penetrated by paths, and sprinkled with landmarks. Elements regularly overlap and pierce one another. If this analysis begins with the differentiation of the data into categories, it must end with their reintegration into the whole image . . .

PATHS

For most people interviewed, paths were the predominant city elements, although their importance varied according to the degree of familiarity with the city. People with least knowledge of Boston tended to think of the city in terms of topography, large regions, generalized characteristics, and broad directional relationships. Subjects who knew the city better had usually mastered part of the path structure; these people thought more in terms of specific paths and their interrelationships. A tendency also appeared for the people who knew the city best of all to rely more upon small landmarks and less upon either regions or paths.

The potential drama and identification in the highway system should not be underestimated. One Jersey City subject, who can find little worth describing in her surroundings, suddenly lit up when she described the Holland Tunnel. Another recounted her pleasure:

> You cross Baldwin Avenue, you see all of New York in front of you, you see the terrific drop of land [the Palisades] . . . and here's this open panorama of lower Jersey City in front of you and you're going down hill, and there you know: there's the tunnel, there's the Hudson River and everything. . . . I always look to the right to see if I can see the . . . Statue of Liberty. . . . Then I always look up to see the Empire State Building, see how the weather is. . . . I have a real feeling of happiness because I'm going someplace, and I love to go places.

Particular paths may become important features in a number of ways. Customary travel will of course be one of the strongest influences, so that major access lines, such as Boylston Street, Storrow Drive, or Tremont Street in Boston, Hudson Boulevard in Jersey City, or the freeways in Los Angeles, are all key image features. . . .

Concentration of special use or activity along a street may give it prominence in the minds of observers. Washington Street is the outstanding Boston example: subjects consistently associated it with shopping and theatres. . . . People seemed to be sensitive to variations in the amount of activity they encountered, and sometimes guided themselves largely by following the main stream of traffic. Los Angeles' Broadway was recognized by its crowds and its street cars; Washington Street in Boston was marked by a torrent of pedestrians. . . .

Characteristic spatial qualities were able to strengthen the image of particular paths. In the simplest sense, streets that suggest extremes of either width or narrowness attracted attention. Cambridge Street, Commonwealth Avenue, and Atlantic Avenue are all well known in Boston, and all were singled out for their great width. . . . Spatial qualities of width and narrowness derived part of their importance from the common association of main streets with width and side streets with narrowness. Looking for, and trusting to the "main" (i.e., wide) street becomes automatic, and in Boston the real pattern usually supports this assumption. . . . Some of the orientation difficulties in Boston's financial district, or the anonymity of the Los Angeles grid, may be due to this lack of spatial dominance.

Special façade characteristics were also important for path identity. Beacon Street and Commonwealth Avenue were distinctive partly because of the building façades that line them. . . .

Proximity to special features of the city could also endow a path with increased importance. In this case the path would be acting secondarily as an edge. Atlantic Avenue derived much importance from its relation to the wharves and the harbor, Storrow Drive from its location along the Charles River.

[. . .]

Where major paths lacked identity, or were easily confused one for the other, the entire city image was in difficulty. . . . Many of the paths in Jersey City were difficult to find, both in reality and in memory.

That the paths, once identifiable, have continuity as well, is an obvious functional necessity. People regularly depended upon this quality.

[. . .]

People tended to think of path destinations and origin points: they liked to know where paths came from and where they led. Paths with clear and well-known origins and destinations had stronger identities, helped tie the city together, and gave the observer a sense of his bearings whenever he crossed them. . . .

[. . .]

A few important paths may be imaged together as a simple structure, despite any minor irregularities, as long as they have a consistent general relationship to one another. The Boston street system is not conducive to this kind of image, except perhaps for the basic parallelism of Washington and Tremont Streets. But the Boston subway system, whatever its involutions in true scale, seemed fairly easy to visualize as two parallel lines cut at the center by the Cambridge–Dorchester line, although the parallel lines might be confused one with the other, particularly since both go to North Station. The freeway system in Los Angeles seemed to be imaged as a complete structure. . . .

[. . .]

A large number of paths may be seen as a total network, when repeating relationships are sufficiently regular and predictable.

The Los Angeles grid is a good example. Almost every subject could easily put down some twenty major paths in correct relation to each other. At the same time, this very regularity made it difficult for them to distinguish one path from another.

Boston's Back Bay is an interesting path network. Its regularity is remarkable in contrast to the rest of the central city, an effect that would not occur in most American cities. But this is not a featureless regularity. The longitudinal streets were sharply differentiated from the cross streets in everyone's mind, much as they are in Manhattan. The long streets all have individual character – Beacon Street, Marlboro Street, Commonwealth Avenue, Newbury Street, each one is different – while the cross streets act as measuring devices. The relative width of the streets, the block lengths, the building frontages, the naming system, the relative length and number of the two kinds of streets, their functional importance, all tend to reinforce this differentiation. Thus a regular pattern is given form and character. The alphabet formula for naming the cross streets was frequently used as a location device, much as the numbers are used in Los Angeles.

[. . .]

The frequent reduction of the South End to a geometrical system was typical of the constant tendency of the subjects to impose regularity on their surroundings. Unless obvious evidence refuted it, they tried to organize paths into geometrical networks, disregarding curves and non-perpendicular intersections. The lower area of Jersey City was frequently drawn as a grid, even though it is one only in part. Subjects absorbed all of central Los Angeles into a repeating network, without being disturbed by the distortion at the eastern edge. Several subjects even insisted on reducing the street maze of Boston's financial district to a checkerboard! . . .

EDGES

Edges are the linear elements not considered as paths: they are usually, but not quite always, the boundaries between two kinds of areas. They act as lateral references. They are strong in Boston and Jersey City but weaker in Los Angeles. Those edges seem strongest which are not only visually prominent, but also continuous in form and impenetrable to cross movement. The Charles River in Boston is the best example and has all of these qualities.

The importance of the peninsular definition of Boston has already been mentioned. It must have been much more important in the 18th century, when the city was a true and very striking peninsula. Since then the shore lines have been erased or changed, but the picture persists. One change, at least, has strengthened the image: the Charles River edge, once a swampy backwater, is now well defined and developed. It was frequently described, and sometimes drawn in great detail. Everyone remembered the wide open space, the curving line, the bordering highways, the boats, the Esplanade, the Shell.

The water edge on the other side, the harborfront, was also generally known, and remembered for its special activity. But the sense of water was less clear, since it was obscured by many structures, and since the life has gone out of the old harbor activities. . . .

[. . .]

In Jersey City, the waterfront was also a strong edge, but a rather forbidding one. It was a no-man's land, a region beyond the barbed wire. Edges, whether of railroads, topography, throughways, or district boundaries, are a very typical feature of this environment and tend to fragment it. Some of the most unpleasant edges, such as the bank of the Hackensack River with its burning dump areas, seemed to be mentally erased.

[. . .]

While continuity and visibility are crucial, strong edges are not necessarily impenetrable. Many edges are uniting seams, rather than isolating barriers, and it is interesting to see the differences in effect. Boston's Central Artery seems to divide absolutely, to isolate. Wide Cambridge Street divides two regions sharply but keeps them in some visual relation. Beacon Street, the visible boundary of Beacon Hill along the Common, acts not as a barrier but as a seam along which the two major areas are clearly joined together. Charles Street at the foot of Beacon Hill both divides and unites, leaving the lower area in uncertain relation to the hill above. Charles Street carries heavy traffic but also contains the local service stores and special activities associated with the Hill. It pulls the residents together by attracting them to itself. It acts ambiguously either as linear node, edge, or path for various people at various times.

Edges are often paths as well. Where this was so, and where the ordinary observer was not shut off from moving on the path . . . then the circulation image seemed to be the dominant one. The element was usually pictured as a path, reinforced by boundary characteristics.

[. . .]

It is difficult to think of Chicago without picturing Lake Michigan. It would be interesting to see how many Chicagoans would begin to draw a map of their city by putting down something other than the line of the lake shore. Here is a magnificent example of a visible edge, gigantic in scale, that exposes an entire metropolis to view. Great buildings, parks, and tiny private beaches all come down to the water's edge, which throughout most of its length is accessible and visible to all. The contrast, the differentiation of events along the line, and the lateral breadth are all very strong. The effect is reinforced by the concentration of paths and activities along its extent. The scale is perhaps unrelievedly large and coarse, and too much open space is at times interposed between city and water, as at the Loop. Yet the façade of Chicago on the Lake is an unforgettable sight.

DISTRICTS

Districts are the relatively large city areas which the observer can mentally go inside of, and which have some common character. They can be recognized internally, and occasionally can be used as external reference as a person goes by or toward them. Many persons interviewed took care to point out that Boston, while confusing in its path pattern even to the experienced inhabitant, has, in the number and vividness of its differentiated districts, a quality that quite makes up for it. As one person put it: Each part of Boston is different from the other. You can tell pretty much what area you're in.

Jersey City has its districts too, but they are primarily ethnic or class districts with little physical distinction. Los Angeles is markedly lacking in strong regions, except for the Civic Center area. The best that can be found are the linear, street-front districts of Skid Row or the financial area. . . .

Subjects, when asked which city they felt to be a well-oriented one, mentioned several, but New York (meaning Manhattan) was unanimously cited. And this city was cited not so much for its grid, which Los Angeles has as well, but because it has a number of well-defined characteristic districts, set in an ordered frame of rivers and streets. Two Los Angeles subjects even referred to Manhattan as being "small" in comparison to their central area! Concepts of size may depend in part on how well a structure can be grasped.

In some Boston interviews, the districts were the basic elements of the city image. One subject, for example, when asked to go from Faneuil Hall to Symphony Hall, replied at once by labeling the trip as going from North End to Back Bay. But even where they were not actively used for orientation, districts were still an important and satisfying part of the experience of living in the city. Recognition of distinct districts in Boston seemed to vary somewhat as acquaintance with the city increased. People most familiar with Boston tended to recognize regions but to rely more heavily for organization and orientation on smaller elements. A few people extremely familiar with Boston were unable to generalize detailed perceptions into districts: conscious of minor differences in all parts of the city, they did not form regional groups of elements.

The physical characteristics that determine districts are thematic continuities which may consist of an endless variety of components: texture, space, form, detail, symbol, building type, use, activity, inhabitants, degree of maintenance, topography. In a closely built city such as Boston, homogeneities of façade – material, modeling, ornament, color, skyline, especially fenestration – were all basic clues in identifying major districts. Beacon Hill and Commonwealth Avenue are both examples. The clues were not only visual ones: noise was important as well. At times, indeed, confusion itself might be a clue, as it was for the woman who remarked that she knows she is in the North End as soon as she feels she is getting lost.

Usually, the typical features were imaged and recognized in a characteristic cluster, the thematic unit. The Beacon Hill image, for example, included steep narrow streets; old brick row houses of intimate scale; inset, highly maintained, white doorways; black trim; cobblestones and brick walks; quiet; and upper-class pedestrians. The resulting thematic unit was distinctive by contrast to the rest of the city and could be recognized immediately. . . .

A certain reinforcement of clues is needed to produce a strong image. All too often, there are a few distinctive signs, but not enough for a full thematic unit. Then the region may be recognizable to someone familiar with the city, but it lacks any visual strength or impact. Such, for example, is Little Tokyo in Los Angeles, recognizable by its population and the lettering on its signs but otherwise indistinguishable from the general matrix. Although it is a strong ethnic concentration, probably known to many people, it appeared as only a subsidiary portion of the city image.

Yet social connotations are quite significant in building regions. A series of street interviews indicated the class overtones that many people associate with different districts. Most of the Jersey City regions were class or ethnic areas, discernible only with difficulty for the outsider. Both Jersey City and Boston have shown the exaggerated attention paid to upper-class districts, and the resulting magnification of the importance of elements in those areas. District names also help to give identity to districts even when the thematic unit does not establish a striking contrast with other parts of the city, and traditional associations can play a similar role.

When the main requirement has been satisfied, and a thematic unit that contrasts with the rest of the city has been constituted, the degree of internal homogeneity is less significant, especially if discordant elements occur in a predictable pattern. Small stores on street corners establish a rhythm on Beacon Hill that one subject perceived as part of her image. These stores in no way weakened her non-commercial image of Beacon Hill but merely added to it. Subjects could pass over a surprising amount of local disagreement with the characteristic features of a region.

Districts have various kinds of boundaries. Some are hard, definite, precise. Such is the boundary of the Back Bay at the Charles River or at the Public Garden. All agreed on this exact location. Other boundaries may be soft or uncertain, such as the limit between downtown shopping and the office district, to whose existence and approximate location most people would testify. Still other regions have no boundaries at all, as did the South End for many of our subjects. . . .

These edges seem to play a secondary role: they may set limits to a district, and may reinforce its identity, but they apparently have less to do with constituting it. Edges may augment the tendency of districts to fragment the city in a disorganizing way. A few people sensed disorganization as one result of the large number of identifiable districts in Boston: strong edges, by hindering transitions from one district to another, may add to the impression of disorganization.

That type of district which has a strong core, surrounded by a thematic gradient which gradually dwindles away, is not uncommon. Sometimes, indeed, a strong node may create a sort of district in a broader homogeneous zone, simply by "radiation," that is, by the sense of proximity to the nodal point. These are primarily reference areas, with little perceptual content, but they are useful organizing concepts, nevertheless.

Some well-known Boston districts were unstructured in the public image. The West End and North End were internally undifferentiated for many people who recognized these regions. Even more often, thematically vivid districts such as the market area seemed confusingly shapeless, both externally and internally. The physical sensations of the market activity are unforgettable. Faneuil Hall and its associations reinforce them. Yet the area is shapeless and sprawling, divided by the Central Artery, and hampered by the two activity centers which vie for dominance: Faneuil Hall and Haymarket Square. Dock Square is spatially chaotic. The connections to other areas are either obscure or disrupted by the Artery. Thus the market district simply floated in most images. Instead of fulfilling its potential role as a mosaic link at the head of the Boston peninsula, as does the Common farther down, the district, while distinctive, acted only as a chaotic barrier zone. Beacon Hill, on the other hand, was very highly structured, with internal subregions, a node at Louisburg Square, various landmarks, and a configuration of paths.

Again, some regions are introvert, turned in upon themselves with little reference to the city outside them, such as Boston's North End or Chinatown. Others may be extrovert, turned outward and connected to surrounding elements. The Common visibly touches neighboring regions, despite its inner path confusions. Bunker Hill in Los Angeles is an interesting example of a district of fairly strong character and historical association, on a very sharp topographical feature lying even closer to the city's heart than does Beacon Hill. Yet the city flows around this element, buries its topographic edges in office buildings, breaks off its path connections, and effectively causes it to fade or even disappear from the city image. Here is a striking opportunity for change in the urban landscape.

Some districts are single ones, standing alone in their zone. The Jersey City and Los Angeles regions are practically all of this kind, and the South End is a Boston example. Others may be linked together, such as Little Tokyo and the Civic Center in Los Angeles, or West End–Beacon Hill in Boston. In one part of central Boston, inclusive of the Back Bay, the Common, Beacon Hill, the downtown shopping district, and the financial and market areas, the regions are close enough together and sufficiently well joined to make a continuous mosaic of distinctive districts. Wherever one proceeds within these limits, one is in a recognizable area. The contrast and proximity of each area, moreover, heightens the thematic strength of each. The quality of Beacon Hill, for example, is sharpened by its nearness to Scollay Square, and to the downtown shopping district.

NODES

Nodes are the strategic foci into which the observer can enter, typically either junctions of paths, or concentrations of some characteristic. But although conceptually they are small points in the city image, they may in reality be large squares, or somewhat extended linear shapes, or even entire central districts when the city is being considered at a large enough level. Indeed, when conceiving the environment at a national or international level, then the whole city itself may become a node.

The junction, or place of a break in transportation, has compelling importance for the city observer. Because decisions must be made at junctions, people heighten their attention at such places and perceive nearby elements with more than normal clarity. This tendency was confirmed so repeatedly that elements located at junctions may automatically be assumed to derive special prominence from their location. The perceptual importance of such locations shows in another way as well. When subjects were asked where on a habitual trip they first felt a sense of arrival in downtown Boston, a large number of people singled out break-points of transportation as the key places. In a number of cases, the point was at the transition from a highway (Storrow Drive or the Central Artery) to a city street; in another case, the point was at the first railroad stop in Boston (Back Bay Station) even though the subject did not get off there. Inhabitants of Jersey City felt they had left their city when they had passed through the Tonnelle Avenue Circle. The transition from one transportation channel to another seems to mark the transition between major structural units.

[. . .]

The subway stations, strung along their invisible path systems, are strategic junction nodes. Some, like Park Street, Charles Street, Copley, and South Station, were quite important in the Boston map, and a few subjects would organize the rest of the city around them. . . .

Major railroad stations are almost always important city nodes, although their importance may be declining. Boston's South Station was one of the strongest in the city, since it is functionally vital for commuter, subway rider, and intercity traveler, and is visually impressive for its bulk fronting on the open space of Dewey Square. The same might have been said for airports, had our study areas included them. . . .

The other type of node, the thematic concentration, also appeared frequently. Pershing Square in Los Angeles was a strong example, being perhaps the sharpest point of the city image, characterized by highly typical space, planting, and activity. . . .

Louisburg Square is another thematic concentration, a well-known quiet residential open space, redolent of the upper-class themes of the Hill, with a highly recognizable fenced park. It is a purer example of a concentration than is the Jordan–Filene corner, since it is no transfer point at all, and was only remembered as being "somewhere inside" Beacon Hill. Its importance as a node was out of all proportion to its function.

Nodes may be both junctions and concentrations, as is Jersey City's Journal Square, which is an important bus and automobile transfer and is also a concentration of shopping. Thematic concentrations may be the focus of a region, as is the Jordan–Filene corner, and perhaps Louisburg Square. Others are not foci but are isolated special concentrations, such as Olvera Street in Los Angeles.

A strong physical form is not absolutely essential to the recognition of a node: witness Journal Square and Scollay Square. But where the space has some form, the impact is much stronger. The node becomes memorable.

[. . .]

Nodes, like districts, may be introvert or extrovert. Scollay Square is introverted, it gives little directional sense when one is in it or its environs. The principal direction in its surroundings is toward or away from it; the principal locational sensation on arrival is simply "here I am." Boston's Dewey Square, on the other hand, is extroverted. General directions are explained, and connections are clear to the office district, the shopping district, and the waterfront. . . .

Many of these qualities may be summed up by the example of a famous Italian node: the Piazza San Marco in Venice. Highly differentiated, rich and intricate, it stands in sharp contrast to the general character of the city and to the narrow, twisting spaces of its immediate approaches. Yet it ties firmly to the major feature of the city, the Grand Canal, and has an oriented shape that clarifies the direction from which one enters. It is within itself highly differentiated and structured: into two spaces (Piazza and Piazzetta) and with many distinctive landmarks (Duomo, Palazzo Ducale, Campanile, Libreria). Inside, one feels always in clear relation to it, precisely micro-located, as it were. So distinctive is this space that many people who have never been to Venice will recognize its photograph immediately.

LANDMARKS

Landmarks, the point references considered to be external to the observer, are simple physical elements which may vary widely in scale. There seemed to be a tendency for those more familiar with a city to rely increasingly on systems of landmarks for their guides – to enjoy uniqueness and specialization, in place of the continuities used earlier.

Since the use of landmarks involves the singling out of one element from a host of possibilities, the key physical characteristic of this class is singularity, some aspect that is unique or memorable in the context.

Landmarks become more easily identifiable, more likely to be chosen as significant, if they have a clear form; if they contrast with their background; and if there is some prominence of spatial location. Figure–background contrast seems to be the principal factor. The background against which an element stands out need not be limited to immediate surroundings: the grasshopper weathervane of Faneuil Hall, the gold dome of the State House, or the peak of the Los Angeles City Hall are landmarks that are unique against the background of the entire city.

In another sense, subjects might single out landmarks for their cleanliness in a dirty city (the Christian Science buildings in Boston) or for their newness in an old city (the chapel on Arch Street). The Jersey City Medical Center was as well known for its little lawn and flowers as for its great size. The old Hall of Records in the Los Angeles Civic Center is a narrow, dirty structure, set at an angle to the orientation of all the other civic buildings, and with an entirely different scale of fenestration and detail. Despite its minor functional or symbolic importance, this contrast of siting, age, and scale makes it a relatively well-identified image, sometimes pleasant, sometimes irritating. It was several times reported to be "pie-shaped," although it is perfectly rectangular. This is evidently an illusion of the angled siting.

Spatial prominence can establish elements as landmarks in either of two ways: by making the element visible from many locations (the John Hancock Building in Boston, the Richfield Oil Building in Los Angeles), or by setting up a local contrast with nearby elements, i.e., a variation in setback and height. In Los Angeles, on 7th Street at the corner of Flower Street, is an old, two-story gray wooden building, set back some ten feet from the building line, containing a few minor shops. This took the attention and fancy of a surprising number of people. One even anthropomorphized it as the "little gray lady." The spatial setback and the intimate scale is a very noticeable and delightful event, in contrast to the great masses that occupy the rest of the frontage.

[. . .]

Distant landmarks, prominent points visible from many positions, were often well known, but only people unfamiliar with Boston seemed to use them to any great extent in organizing the city and selecting routes for trips. It is the novice who guides himself by reference to the John Hancock Building and the Custom House.

Few people had an accurate sense of where these distant landmarks were and how to make one's way to the base of either building. Most of Boston's distant landmarks, in fact, were "bottomless"; they had a peculiar floating quality. The John Hancock Building, the Custom House, and the Court House are all dominant on the general skyline, but the location and identity of their base is by no means as significant as that of their top.

The gold dome of Boston's State House seems to be one of the few exceptions to this elusiveness. Its unique shape and function, its location at the hill crest and its exposure to the Common, the visibility from long distances of its bright gold dome, all make it a key sign for central Boston. It has the satisfying qualities of recognizability at many levels of reference, and of coincidence of symbolic with visual importance.

People who used distant landmarks did so only for very general directional orientation, or, more frequently, in symbolic ways. For one person, the Custom House lent unity to Atlantic Avenue because it can be seen from almost any place on that street. For another, the Custom House set up a rhythm in the financial district, for it can be seen intermittently at many places in that area.

The Duomo of Florence is a prime example of a distant landmark: visible from near and far, by day or night; unmistakable; dominant by size and contour; closely related to the city's traditions; coincident with the religious and transit center; paired with its campanile in such a way that the direction of view can be gauged from a distance. It is difficult to conceive of the city without having this great edifice come to mind.

But local landmarks, visible only in restricted localities, were much more frequently employed in the three cities studied. They ran the full range of objects available. The number of local elements that become landmarks appears to depend as much upon how familiar the observer is with his surroundings as upon the elements themselves. Unfamiliar subjects usually mentioned only a few landmarks in office interviews, although they managed to find many more when they went on field trips. Sounds and smells sometimes reinforced visual landmarks, although they did not seem to constitute landmarks by themselves.

[. . .]

Element interrelations

These elements are simply the raw material of the environmental image at the city scale. They must be patterned together to provide a satisfying form. The preceding discussions have gone as far as groups of similar elements (nets of paths, clusters of landmarks, mosaics of regions). The next logical step is to consider the interaction of pairs of unlike elements.

Such pairs may reinforce one another, resonate so that they enhance each other's power: or they may conflict and destroy themselves. A great landmark may dwarf and throw out of scale a small region at its base.

Properly located, another landmark may fix and strengthen a core; placed off center, it may only mislead, as does the John Hancock Building in relation to Boston's Copley Square. . . .

[. . .]

We are continuously engaged in the attempt to organize our surroundings, to structure and identify them. Various environments are more or less amenable to such treatment. When reshaping cities it should be possible to give them a form which facilitates these organizing efforts rather than frustrates them.

"The Design of Spaces"

from *City: Rediscovering the Center* (1988)

William H. Whyte

Editors' Introduction

Puzzled by why some of New York's parks and plazas were well used while others were almost always nearly empty, the New York City Planning Commission asked sociologist William Whyte to study park and plaza use and help draft a comprehensive design plan to improve New York City's parks and plazas.

While he had no formal training in urban planning, landscape architecture, or design, Whyte's lucid writings in these areas gave him great credibility. Hunter College (City University of New York) appointed him a distinguished professor, and the National Geographic Society gave him the first domestic expedition grant it had ever made to investigate what makes for good urban parks and plazas. He named his study "The Street Life Project."

Whyte worked with Hunter College students, bright young designers and planners at the New York City Planning Department, and other talented people he drew to the Street Life Project. This team produced an exceptional study of how people use urban space and a set of urban design guidelines for New York that have been widely praised and used in New York and many other cities.

The Street Life Project is an excellent example of how to do urban research. Whyte formed hypotheses about how people use urban space. Then, like Camillo Sitte (p. 476) and Jan Gehl (p. 530), he and his team observed how people used urban spaces. Whyte tested his hypotheses by filming people using different plazas and parks in New York City and carefully analyzing the films, having his researchers note down who sat where during different times of the day and year, and personally watching people interact with each other and the physical spaces around them.

Whyte's results were often surprising. Some initial hypotheses were validated, but Whyte had to reject or modify others, including many that seem intuitively obvious. Whyte hypothesized, for example, that the number of people using a plaza would be strongly influenced by the amount of space or its shape. More people should use big parks than little ones. But Chart 1 in Figure 1 shows that is not the case. New Yorkers use tiny Greenacre Park much more than J.C. Penny Park, which is much larger. Intuitively more people should use a park that is about as wide as it is long than a long skinny park. But Whyte found that one of New York's most popular parks is just a long, narrow indentation in a building. What did attract people to parks and plazas? Whyte eventually concluded that the amount of sitting space in a park or plaza was much more important than either the total space or its shape. The presence of food vendors, an open relationship to the street, the presence of water, movable chairs, sunlight (even reflected sunlight), and many other influences that were rarely included in formal park and plaza designs before Whyte's groundbreaking research turned out to make a big difference.

Daphne Spain (p. 176) notes that most writing about the spatial organization of cities ignores gender issues. Most writings on urban design either ignore gender differences or are written from a male perspective with, at best, a separate section on design implications for women. Whyte included women researchers in the Street Life Project and listened carefully to what they told him about how women used the spaces the team was observing. The Street Life Project was pioneering in its conscious observation of gender difference in how men and women use urban space. Whyte found that women are more discriminating than men as to where they will sit and are more sensitive to annoyances. He concluded that if a plaza has a high proportion of women, it is probably a good and well-managed one.

William Whyte's ideas have had a wide impact. The New York City Planning Commission held hearings on his recommendations and, after much debate, adopted many of his suggestions as requirements or guidelines for new park and plaza development. Whyte helped develop a restoration plan for Bryant Park, and a public–private partnership inspired by Whyte's ideas tore down walls isolating Bryant Park from the street, put in sitting space and food, and transformed a dangerous, little-used park into one bustling with life (while at the same time evicting many homeless park residents and raising troubling questions about who should control public space and prescribe how it is to be used).

Note the importance of good public spaces in Camillo Sitte's work on "the art of building cities" (p. 476) and in H.D.F. Kitto's description of the bustling street life in the agora and other public spaces in the Greek polis (p. 40). Observe the similarity between Whyte's description of Seagram's Plaza as the best of stages and Lewis Mumford's emphasis on the city as theater (p. 91). As Whyte comments, it takes real work to create a lousy place, but Mike Davis (p. 195) found designers in Los Angeles worked hard to design public spaces to keep homeless people and other undesirables away.

The way in which New York City encouraged parks and plazas is an interesting example how cities can use the economic incentives Wilbur Thompson (p. 274) suggests to encourage the type of development they want. There is a strong market for additional office space in the central business districts of many cities. Zoning ordinances set limits on the height and bulk of office buildings. Permission to build more office space than zoning allows is worth money to developers. New York City awarded developers "density bonuses" allowing them to build more office space if the private developers agreed to provide park and plaza space at street level. They granted enough additional square footage that almost all office developers participated in the program. While some developers worked hard to design attractive parks and plazas, other just wanted to build something that would get them the density bonus. That helps explain the huge difference in the quality of urban parks and plazas in New York City (and other cities with similar zoning laws and policies). Requiring the developers to meet design guidelines was intended to produce good parks and plazas.

William Hollingsworth "Holly" Whyte (1918–1999) was a sociologist and journalist who first achieved prominence with a study of corporate culture titled *The Organization Man* that became a best-seller in the United States in 1956, selling more than 2 million copies. Whyte served as an advisor to Laurence S. Rockefeller on environmental issues and as a planning consultant for major US cities. He was a trustee of the American Conservation Association, and was active in the Municipal Art Society, the Hudson River Valley Commission and President Lyndon B. Johnson's Task Force on Natural Beauty. But he remains best known for his work with the Street Life Project. The Project for Public Spaces, a New York City nonprofit organization, continues to promote Whyte's vision.

This selection is from *City: Rediscovering the Center* (New York: Anchor, 1988). In other chapters of *City*, Whyte explores water, wind, trees, light, steps and entrances, undesirables, walls, sun and shadows, and many other factors that affect public spaces. Whyte produced a delightful film titled *Public Spaces/Human Places* based on his research (available from Direct Cinema Limited in Los Angeles). The original report of Whyte's classic street life project has been reprinted as *The Social Life of Small Urban Spaces* (New York: Project for Public Spaces, 2001). An anthology of William Whyte's most important writings is Albert LaFarge (ed.), *The Essential William Whyte* (New York: Fordham University Press, 2000). Whyte's most widely known book is *The Organization Man* (New York: Simon & Schuster, 1956).

A book by the Project for Public Spaces, *How To Turn a Place Around: A Handbook for Creating Successful Public Spaces* (New York: Project for Public Spaces, 2000), further develops Whyte's ideas. A similar book is Jay Walljasper, *The Great Neighborhood Book: A Do-It-Yourself Guide to Placemaking* (New York: New Society, 2007). Oscar Newman's influential *Defensible Space* (New York: Macmillan, 1972) is a study of the way in which low-rent public housing project residents use space, with suggestions to architects and planners on how to meet their security concerns. Clare Cooper and Wendy Sarkissien, *Housing as if People Mattered* (Berkeley, CA: University of California Press, 1986) provides practical suggestions for designing housing responsive to the needs of all its residents, particularly working women and children. Allan Jacobs, *Looking at Cities* (Cambridge, MA: MIT

Press, 1985) provides a stimulating discussion of how close observation such as Whyte undertook can inform city planning, and how to do it. John Lofland, David A. Snow, Leon Anderson, and Lyn H. Lofland, *Analyzing Social Settings: A Guide to Qualitative Observation and Analysis*, 4th edn (Belmont, CA: Wadsworth, 2005) is a standard text on observation and research method.

. . . Since 1961 New York City had been giving incentive bonuses to developers who would provide plazas . . . Every new office building qualified for the bonus by providing a plaza or comparable space; in total, by 1972 some twenty acres of the world's most expensive open space.

Some plazas attracted lots of people . . .

But on most plazas there were few people. In the middle of the lunch hour on a beautiful day the number of people sitting on plazas averaged four per thousand square feet of space – an extraordinarily low figure for so dense a center . . .

. . . The city was being had. For the millions of dollars of extra floor space it was handing out to developers, it had every right to demand much better spaces in return.

I put the question to the chairman of the city planning commission, Donald Elliott . . . He felt tougher zoning was in order. If we could find out why the good places worked and the bad ones didn't and come up with tight guidelines, there could be a new code . . .

We set to work. We began studying a cross section of spaces – in all, sixteen plazas, three small parks, and a number of odds and ends of space . . .

[. . .]

We started by charting how people used plazas. We mounted time-lapse cameras at spots overlooking the plaza . . . and recorded the dawn-to-dusk patterns. We made periodic circuits of the plazas and noted on sighting maps where people were sitting, their gender, and whether they were alone or with others . . . We also interviewed people and found where they worked, how frequently they used the plaza, and what they thought of it. But mostly we watched what they did.

Most of them were young office workers from nearby buildings. Often there would be relatively few from the plaza's own building. As some secretaries confided, they would just as soon put a little distance between themselves and the boss come lunchtime. In most cases the plaza users came from a building within a three-block radius. Small parks, such as Paley and Greenacre, had a somewhat more varied mix of people – with more upper-income older people – but even here office workers predominated.

This uncomplicated demography underscores an elemental point about good spaces: supply creates demand. A good new space builds a new constituency. It gets people into new habits – such as alfresco lunches – and induces them to use new paths . . .

The best-used plazas are sociable places, with a higher proportion of couples and groups than you will find in less-used places. At the plazas in New York, the proportion of people in twos or more runs about 50–62 percent; in the least-used, 25–30 percent. A high proportion is an index of selectivity. If people go to a place in a group or rendezvous there, it is most often because they decided to beforehand. Nor are these places less congenial to the individual. In absolute numbers, they attract more individuals than do the less-used spaces. If you are alone, a lively place can be the best place to be.

The best-used places also tend to have a higher than average proportion of women. The male–female ratio of a plaza reflects the composition of the work force and this varies from area to area. In midtown New York it runs about 60 percent male, 40 percent female. Women are more discriminating than men as to where they will sit, they are more sensitive to annoyances, and they spend more time casing a place. They are also more likely to dust off a ledge with their handkerchief.

The male–female ratio is one to watch. If a plaza has a markedly low proportion of women, something is wrong. Conversely, if it has a high proportion, the plaza is probably a good and well-managed one and has been chosen as such.

The rhythms of plaza life are much alike from place to place. In the morning hours, patronage will be sporadic . . .

Around noon the main clientele begins to arrive. Soon activity will be near peak and will stay there until a little before two . . .

Some 80 percent of the people activity on plazas comes during the lunchtime, and there is very little of any kind after five-thirty . . .

During the lunch period, people will distribute themselves over space with considerable consistency, with some sectors getting heavy use day in and day out, others much less so. We also found that off-peak use often gives the best clues to people's preferences. When a place is jammed, people sit where they can; this may or may not be where they most want to. After the main crowd has left, however, the choices can be significant. Some parts of the plaza become empty; others continue to be used . . .

Men show a tendency to take the front row seats and if there is a kind of gate they will be the guardians of it. Women tend to favor places slightly secluded. If there are double-sided ledges parallel to the street, the inner side will usually have a higher proportion of women; the outer, of men.

Of the men up front the most conspicuous are the girl watchers. As I have noted, they put on such a show of girl watching as to indicate that their real interest is not so much the girls as the show. It is all machismo. Even in the Wall Street area, where girl watchers are especially demonstrative you will hardly ever see one attempt to pick up a girl.

Plazas are not ideal places for striking up acquaintances. Much better is a very crowded street with lots of eating and quaffing going on. An outstanding example is the central runway of the South Street Seaport. At lunch sometimes, one can hardly move for the crush. As in musical chairs, this can lead to interesting combinations. On most plazas, however, there isn't much mixing. If there are, say, two smashing blondes on a ledge, the men nearby will usually put on an elaborate show of disregard. Look closely, however, and you will see them giving away the pose with covert glances.

Lovers are to be found on plazas, but not where you would expect them. When we first started interviewing, people would tell us to be sure to see the lovers in the rear places. But they weren't usually there. They would be out front. The most fervent embracing we've recorded on film has taken place in the most visible of locations, with the couple oblivious of the crowd. (In a long clutch, however, I have noted that one of the lovers may sneak a glance at a wristwatch.)

Certain locations become rendezvous points for groups of various kinds. The south wall of the Chase Manhattan Plaza was, for a while, a gathering point for camera bugs, the kind who are always buying new lenses and talking about them. Patterns of this sort may last no more than a season – or persist for years. A black civic leader in Cincinnati told me that when he wants to make contact, casually, with someone, he usually knows just where to look at Fountain Square . . .

Standing patterns on the plazas are fairly regular. When people stop to talk they will generally do so athwart one of the main traffic flows, as they do on streets. They also show an inclination to station themselves near objects, such as a flagpole or a piece of sculpture. They like well-defined places, such as steps or the border of a pool. What they rarely choose is the middle of a large space.

There are a number of explanations. The preference might be ascribed to some primeval instinct: you have a full view of all comers but your rear is covered. But this doesn't explain the inclination men have for lining up at the curb. Typically, they face inward, with their backs exposed to the vehicle traffic of the street.

Whatever their cause, people's movements are one of the great spectacles of a plaza. You do not see this in architectural photographs, which are usually devoid of human beings and are taken from a perspective that few people share. It is a misleading one. Looking down on a bare plaza, one sees a display of geometry, done almost in monochrome. Down at eye level the scene comes alive with movement and color – people walking quickly, walking slowly, skipping up steps, weaving in and out on crossing patterns, accelerating and retarding to match the moves of others. Even if the paving and the walls are gray, there will be vivid splashes of color – in winter especially, thanks to women's fondness for red coats and colored umbrellas.

There is a beauty that is beguiling to watch, and one senses that the players are quite aware of this themselves. You can see this in the way they arrange themselves on ledges and steps. They often do so with a grace that they must appreciate themselves. With its brown-gray setting, Seagram is the best of stages – in the rain, too, when an umbrella or two puts spots of color in the right places, like Corot's red dots.

Let us turn to the factors that make for such places. The most basic one is so obvious it is often overlooked: people. To draw them, a space should tap a strong flow of them. This means location, and, as the old adage has it, location and location. The space should be in the heart of downtown, close to the 100 percent corner – preferably right on top of it.

Because land is cheaper further out, there is a temptation to pick sites away from the center. There may also be some land for the asking – some underused spaces, for example, left over from an ill-advised civic center campus of urban renewal days. They will be poor bargains. A space that is only a few blocks too far might as well be ten blocks for all the people who will venture to walk to it.

People *ought* to walk to it, perhaps; the exercise would do them good. But they don't. Even within the core of downtown the effective radius of a good place is about three blocks. About 80 percent of the users will have come from a place within that area. This does indicate a laziness on the part of pedestrians and this may change a bit, just as the insistence on close-in parking may. But there is a good side to the constrained radius. Since usage is so highly localized, the addition of other good open spaces will not saturate demand. They will increase it.

Given a fine location, it is difficult to design a space that will not attract people. What is remarkable is how often this has been accomplished. Our initial study made it clear that while location is a prerequisite for success, it in no way assures it. Some of the worst plazas are in the best spots . . .

All of the plazas and small parks that we studied had good locations; most were on the major avenues, some on attractive side-streets. All were close to bus-stops or subway stations and had strong pedestrian flows on the sidewalks beside them. Yet when we rated them according to the number of people sitting at peak time, there was a wide range: from 160 people at 77 Water Street to 17 at 280 Park Avenue (Figure 1).

How come? The first factor we studied was the sun. We thought it might well be the critical one, and our first time-lapse studies seemed to bear this out. Subsequent studies did not. As I will note later they show that sun was important but did not explain the differences in popularity of plazas.

Nor did aesthetics . . . The elegance and purity of a complex's design, we had to conclude, had little relationship to the usage of the spaces around it.

[. . .]

Another factor we considered was the shape of spaces. Members of the commission's urban design group believed this was very important and hoped our findings would support tight criteria for proportions and placement. They were particularly anxious to rule out strip plazas: long, narrow spaces that were little more than enlarged sidewalks, and empty of people more times than not . . .

Our data did not support such criteria. While it was true that most strip plazas were little used, it did not follow that their shape was the reason. Some squarish plazas were little used too, and, conversely, several of the most heavily used spaces were in fact long, narrow strips. One of the five most popular sitting places in New York is essentially an indentation in a building, long and narrow. Our research did not prove shape unimportant or designers' instincts misguided. As with the sun, however, it proved that other factors were more critical.

If not the shape of the space, what about the *amount* of it? Some conservationists believed this would be the key factor. In their view, people seek open space as a relief from overcrowding and it would follow that

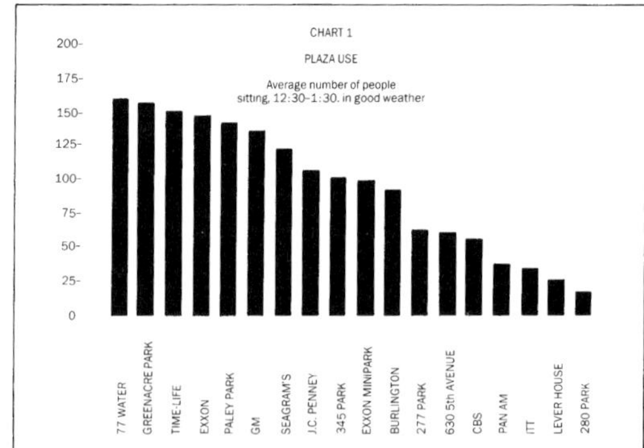

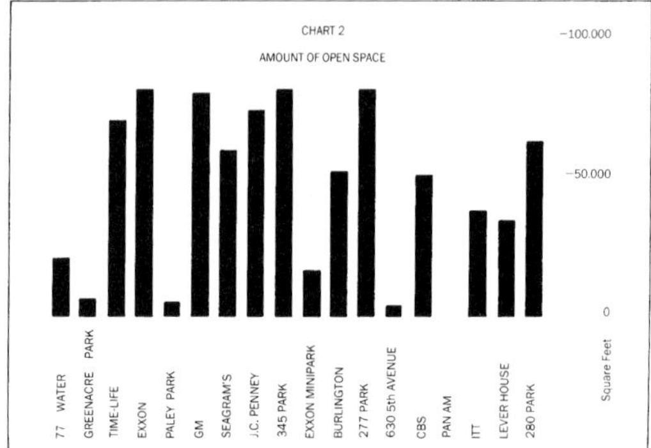

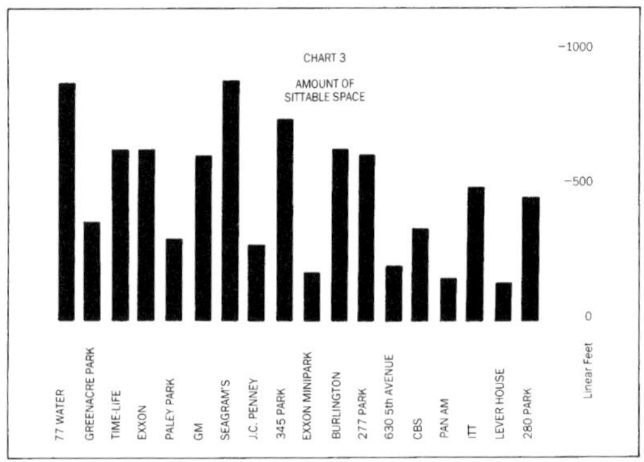

Figure 1

places with the greatest sense of space and light and air would draw the best. If we ranked plazas by the amount of space they provided, there surely would be a positive correlation between space and people.

Once again we found no clear relationship. Several of the smallest spaces had the largest number of people, and several of the largest spaces had the least number of people . . .

What about the amount of *sittable* space? Here we began to get close. As we tallied the number of linear feet of sitting space, we could see that the plazas with the most tended to be among the most popular . . .

. . . No matter how many other variables we checked, one basic point kept coming through. We at last recognized that it was the major one.

People tend to sit most where there are places to sit.

This may not strike the reader as an intellectual bombshell, and now that I look back on our study I wonder why it was not more apparent to us from the beginning . . . Whatever the attractions of a space, it cannot induce people to come and sit if there is no place to sit.

INTEGRAL SEATING

The most basic kind of seating is the kind that is built into a place, such as steps and ledges. Half the battle is seeing to it that these features are usable by people. And there is a battle. Another force has been diligently at work finding ways to deny these spaces. Here are some of the ways:

Horizontal metal strip with sawtooth points.
Jagged rocks set in concrete (Southbridge House, New York City).
Spikes imbedded in ledges (Peachtree Plaza Hotel).
Railing placed to hit you in small of back (GM Plaza, New York City).
Canted ledges of slippery marble (Celanese Building, New York City).

It takes real work to create a lousy place. In addition to spikes and metal objects, there are steps to be made steep, additional surveillance cameras to be mounted, walls to be raised high. Just not doing such things can produce a lot of sitting space.

It won't be the most comfortable kind but it will have the great advantage of enlarging choice. The more sittable the inherent features are made, the more freedom people have to sit up front, in the back, to the side, in the sun, or out of it. This means designing ledges and parapets and other flat surfaces so they can do double duty as seating, tables, and shelves. Since most building sites have some slope in them, there are bound to be opportunities for such features, and it is no more trouble to leave them sittable than not.

[. . .]

SITTING HEIGHT

One guideline we thought would be easy to establish was for sitting heights. It seemed obvious enough that somewhere around sixteen to seventeen inches would probably be the optimum. But how much higher or lower could a surface be and still be sittable? Thanks to slopes, several of the most popular ledges provided a range of continuously variable heights. The front ledge at Seagram, for example started at seven inches at one corner and rose to forty-four inches at the other. Here was an opportunity for a definitive study, we thought; by recording over time how many people sat at what heights, we would get a statistical measure of preferences.

We didn't . . . We had to conclude that people will sit almost anywhere between a height of one foot and three, and this was the range that was to be specified in the zoning. People will sit lower or higher, of course, but there are apt to be special conditions – a wall too high for most adults to mount but just right for teenagers.

A dimension that is truly important is the human backside. It is a dimension many architects ignore. Rarely will you find a ledge or bench that is deep enough to be sittable on both sides . . . Most frustrating are the ledges just deep enough to tempt people to sit on both sides, but too shallow to let them do so comfortably. At peak times people may sit on both sides but they won't be comfortable doing it. They will be sitting on the forward edge, awkwardly.

Thus to another of our startling findings: ledges and spaces two backsides deep seat more people than those that are not as deep . . .

[. . .]

Steps work for the same reason. They afford an infinity of possible groupings, and the excellent sight lines make all the seating great for watching the theatre of the street . . .

[. . .]

Circulation and sitting, in sum, are not antithetical but complementary. I stress this because a good many planners think that the two should be kept separate. More to the point, so do some zoning codes. New York's called for "pedestrian circulation areas" separate from "activity areas" for sitting. People ignore such boundaries.

We felt that pedestrian circulation through and within plazas should be encouraged. Plazas that are sunken or elevated tend to attract low flows, and for that reason the zoning specifies that plazas be not more than three feet below street level or above it. The easier the flow between street and plaza the more likely they are to come in and tarry and sit.

This is true of the handicapped also. If a place is planned with their needs in mind, the place is apt to function more easily for everyone. Drinking fountains that are low enough for wheelchair users are low enough for children. Walkways that are made easier for the handicapped by ramps, handrails, and steps of gentle pitch are easier for all. The guidelines make such amenities mandatory . . . For the benefit of the handicapped, it is required that at least 5 percent of the seating spaces have backrests. These are not segregated for the handicapped. No facilities are segregated. The idea is to make all of the place useful for everyone.

BENCHES

Benches are design artifacts the purpose of which is to punctuate architectural photographs. They are most often sited in modular form, spaced equidistant from one another in a symmetry that is pleasing in plan view. They are not very good, however, for sitting. There are usually too few of them; they are too short and too narrow; they are isolated from other benches and from what action there is to look at.

[. . .]

Watch how benches fill up. The first arrival will usually take the end of a bench, not the middle. The next arrival will take the end of another bench. Subsequent arrivals head for whatever end spots are not taken. Only when there are few other places left will people sit in the middle of the bench, and some will elect to stand.

Since it's the ends of the benches that do most of the work, it could be argued that benches ought to be shortened so they're all end and no middle. But the unused middles are functional for not being used. They provide buffer space. They also provide choice, and if it is the least popular choice, that does not negate its utility.

[. . .]

CHAIRS

We come now to a wonderful invention: the movable chair. Having a back, it is comfortable, and even more so if it has armrests as well. But the big asset is movability. Chairs enlarge choice: to move into the sun, out of it; to move closer to someone, further away from another.

The possibility of choice is as important as the exercise of it. If you know you can move if you want to, you can feel all the more comfortable staying put. This is why, perhaps, people so often approach a chair and then, before sitting on it, move the chair a few inches this way or that, finally ending up with the chair just about where it was in the first place. These moves are functional. They are a declaration of one's free will to oneself, and rather satisfying. In this one small matter you are the master of your fate.

Small moves can say things to other people. If a newcomer chooses a chair next to a couple in a crowded situation, he may make several moves with the chair. He is conveying a message: Sorry about the closeness, but it can't be helped and I am going to respect your privacy as you will mine. A reciprocal shift of a chair may signal acknowledgment.

Chair arranging by groups is a ritual worth watching. In a group of three or four women, one may be dominant and direct the sitting, including the fetching of an extra chair. More times, the members of the group work it out themselves, often with false starts and second choices. The chair arranging can take quite a bit of time on occasion – it is itself a form of recreation – but people enjoy it. Watching these exercises in civility is one of the pleasures of a good place.

Fixed individual seats deny choice. They may be good to look at, and in the form of stools, metal love seats, granite cubes, and the like, they make interesting decorative elements. That is their primary function. For sitting, however, they are inflexible and socially uncomfortable.

[. . .]

Where space is at a premium – in theatres, stadia – fixed seats are a necessity. In open spaces, however, they are uncalled for; there is so much space around them that the compression makes for awkward sitting . . . On one campus a group of metal love seats was cemented to the paving with epoxy glue; in short order they were wrenched out of position by students. The designer is unrepentant. His love seats have won several design awards.

[. . .]

A salute to grass is in order. It is a wonderfully adaptable substance, and while it is not the most comfortable seating, it is fine for napping, sunbathing, picnicking, and Frisbee throwing. Like movable chairs, it also has the great advantage of offering people the widest possible choice of sitting arrangements. There are an infinity of possible groupings, but you will note that the most frequent has people self-positioned at oblique angles from each other.

Grass offers a psychological benefit as well. A patch of green is a refreshing counter to granite and concrete, and when people are asked what they would like to see in a park, trees and grass usually are at the top of the list . . .

RELATIONSHIP TO THE STREET

Let us turn to a more difficult consideration. With the kind of amenities we have been discussing, there are second chances. If the designers have goofed on seating, more and better seating can be provided. If they have been too stingy with trees, more trees can be planted. If there is no food, a food cart can be put in – possibly a small pavilion or gazebo. If there is no water feature, a benefactor might be persuaded to donate a small pool or fountain. Thanks to such retrofitting, spaces regarded as hopeless dogs have been given new life.

What is most difficult to change, however, is what is most important: the location of the space and its relationship to the street. The real estate people are right about location, location, location. For a space to function truly well it must be central to the constituency it is to serve – and if not in physical distance, in visual accessibility . . .

The street functions as part of the plaza or square; indeed, it is often hard to tell where the street leaves off and the plaza begins. The social life of the spaces flows back and forth between them. And the most vital space of all is often the street corner. Watch one long enough and you will see how important it is to the life of the large spaces. There will be people in 100 percent conversations or prolonged goodbyes. If there is a food vendor at the corner, like Gus at Seagram, people will be clustered around him, and there will be a brisk traffic between corner and plaza.

It is a great show, and one of the best ways to make the most of it is, simply, not to wall off the plaza from it. Frederick Law Olmsted spoke of an "interior park" and an "outer park," and he argued that the latter – the surrounding streets – was vital to the enjoyment of the former. He thought it an abomination to separate the two with walls or, worse yet, with a spiked iron fence. "In expression and association," he said, "it is in the most distinct contradiction and discord with all the sentiment of a park. It belongs to a jail or to the residence of a despot who dreads assassination."

But walls are still being put up, usually in the mistaken notion that they will make the space feel safer. They do not . . . they make a space feel isolated and gloomy. Lesser defensive measures can work almost as much damage. The front rows of a space – whether ledges or steps or benches – are the best of sitting places, yet they are often modified against human use. At the General Motors Building on Fifth Avenue, the front ledges face out on one of the greatest of promenades. But you cannot sit on the ledges for more than a minute or so. There is a fussy little railing that catches you right in the small of your back. I do not think it was deliberately planned to do so. But it does and you cannot sit for more than a few moments before your back hurts. Another two inches of clearance for the railing and you would be comfortable. But day after day, year after year, one of the great front rows goes scarcely used, for want of two inches. Canted ledges, especially ones of polished marble, are another nullifying feature. You can almost sit on them if you keep pressing down on your heel hard enough.

[. . .]

A good space beckons people in, and the progression from street to interior is critical in this respect. Ideally, the transition should be such that it's hard to tell where one ends and the other begins. You shouldn't have to make a considered decision to enter; it should be almost instinctive . . .

[. . .]

"Toward an Urban Design Manifesto"

Journal of the American Planning Association (1987)

Allan Jacobs and Donald Appleyard

Editors' Introduction

Allan Jacobs and Donald Appleyard deplore many of the same aspects of Los Angeles, London, New York, and other large cities that Mike Davis (p. 195), Ali Madanipour (p. 186), and David Harvey (p. 230) criticize: vast anonymous areas developed by giant public and private developments; dangerous, polluted, noisy, anonymous living environments; fortress-like buildings that present windowless facades to the street; and a pervasive semiotics that tells outsiders they are not welcome in subtle and not-so-subtle ways. But "Toward an Urban Design Manifesto" moves beyond observation and critique to set out goals for urban life and advance ideas for how the urban fabric of cities might be designed for livable urban environments.

Jacobs and Appleyard title their piece a manifesto and model it on the celebrated Charter of Athens adopted by the International Congress of Modern Architecture (CIAM), the organization that advanced ideas for building contemporary cities based on Le Corbusier's principles (p. 336). But the values they espouse are opposed to modernist principles. This is essentially an anti-modernist manifesto. The modernist approach has been out of favor for half a century, so it is not surprising that many of the design suggestions that the Congress of the New Urbanism (p. 356), Kevin Lynch (p. 499), Jan Gehl (p. 530), and William Whyte (p. 510) advance are consistent with Jacobs and Appleyard's approach.

Jacobs and Appleyard do not like the vast clearance projects, highways, and high-rise buildings surrounded by enormous open space that have resulted from the CIAM's design ideology. Both authors loved the human qualities of European cities as well as their native San Francisco. Before his untimely death in 1982, Appleyard spent time exploring and observing European cities. For many years Allan Jacobs has spent summers in Europe and South America engaged in the same kind of observation that Camillo Sitte (p. 476) pioneered. An accomplished artist and photographer, Jacobs has spent countless hours photographing and sketching European and American cities. His important book on great streets is filled with illustrations of humanistic qualities of the world's most exciting and beloved streets. These are very different concerns from efficiency, speed, and use of modern building materials that preoccupied Le Corbusier (p. 336) and other modernists.

Jacobs and Appleyard acknowledge that the garden city ideas of Ebenezer Howard (p. 328) have produced some pleasant communities, but dismiss garden cities as more like suburbs than true cities. Their manifesto suggests an approach that is more subtle and humane than Le Corbusier's (p. 336) and more truly urban than Howard's (p. 328). They want to create truly urban communities at densities great enough to sustain street life and urban vitality.

Jacobs and Appleyard's manifesto is grounded in both a command of academic theory and their own practical experience in urban design. In this manifesto they propose urban development at densities higher than Howard proposed for Garden City designs – high enough to qualify as truly urban. But they do not endorse urban densities nearly as great as the CIAM theorists do, particularly in megastructures surrounded by parks. (Le Corbusier deliberately shocked the architectural establishment by producing a plan to tear down much of historic Paris and replace it with modern concrete and steel high-rise buildings!)

While Jacobs and Appleyard favor reasonable engineering standards for decibel levels and street widths, they oppose excessive standards that destroy the texture of urban life. Like Jane Jacobs (p. 105), William Whyte (p. 510), and Lewis Mumford (p. 91), they relish some of the disorder that makes urban life enjoyable, including noise, smells, and jumbled land uses that some engineers and many modernist architects wanted to separate into orderly zones. Like Jan Gehl (p. 530), Jacobs and Appleyard value pedestrians and public space that promotes human interaction. They argue that participatory planning of the kind Sherry Arnstein (p. 238) and John Forester (p. 421) describe is essential, unlike the elitist CIAM theorists. Both Allan Jacobs and Donald Appleyard were students of and worked closely with Kevin Lynch, whose ideas on the elements that make up the city image (p. 499) strongly influenced their work.

Allan Jacobs (b. 1928) is an emeritus professor of city and regional planning at the University of California, Berkeley, and was San Francisco's planning director for eight years. Jacobs alternated between careers as a practicing city planner in Pittsburgh, Philadelphia, New Delhi, and San Francisco and teaching urban planning and urban design at the University of Pennsylvania and the University of California, Berkeley. While he was San Francisco's City Planning Director, Jacobs enlisted Appleyard to work on studies of street livability in San Francisco and to help develop an award-winning citywide urban design plan reflecting Lynch's ideas.

Donald Appleyard (1928–1982) also taught urban planning and urban design at Berkeley. Tragically, Appleyard was killed by a speeding automobile shortly before this selection was published. Appleyard's emphasis on the importance of streets is consistent with Jane Jacobs description of the street ballet (p. 105) and Clarence Perry's (p. 428) and Jan Gehl's (p. 530) emphasis on street design as a major determinant neighborhood quality and the life between buildings. Planners and urban designers have given street design a great deal of attention because streets are the most significant publicly owned and funded component of most cities. If a city chooses to follow an urban designer's recommendation for streets that can have an enormous impact on the design of the city.

Like Jane Jacobs (p. 105), Appleyard emphasized that streets perform many functions in addition to serving as conduits for cars. Just as Jacobs described how street designs that permit residents to keep their eyes on the street can reduce crime, Appleyard concluded that street design could help or hamper neighborliness. In one study Appleyard found that residents of San Francisco streets with light traffic had, on average, three times as many friends and twice as many acquaintances on the opposite site of the street from where they lived as people on streets with heavy traffic. Accordingly Appleyard was a vigorous advocate for reducing street widths where possible and traffic calming to slow down cars and allow people to safely use streets.

In *Making City Planning Work* (Chicago, IL: Planner's Press, 1976), Jacobs alternates chapters describing the practical aspects of a city planning director's job with case studies on successful and not so successful projects he undertook during his tenure as San Francisco's city planning director. Allan Jacobs's book *Looking at Cities* (Cambridge, MA: MIT Press, 1985) grew out of a class he taught at Berkeley in which students took him to an unfamiliar neighborhood, left him to observe it carefully, and then compared what he found out from observation with what they learnt by examining data and city planning reports on the same neighborhood. *Looking at Cities* reminds professionals to follow in the footsteps of Sitte, Lynch, Gehl, Whyte, Appleyard, and Jacobs himself and carefully observe the areas they are planning. It outlines a methodology for reading clues in the built environment that can improve urban planning practice. Jacobs's most recent books are *Great Streets* (Cambridge, MA: MIT Press, 1995) and *The Boulevard Book: History, Evolution, Design of Multiway Boulevards*, with Elizabeth MacDonald and Yodan Rofe (Cambridge, MA: MIT Press, 2001).

Donald Appleyard, *The View from the Road*, with Kevin Lynch and John Myer (Cambridge, MA: MIT Press, 1963) and *Livable Streets* (Berkeley, CA: University of California Press, 1981) show how ideas can be translated into action in street design. A new edition of *Livable Streets* with supplementary material by Appleyard's son, Bruce Appleyard – an urban planner and designer who has revisited sites his father studied and supplemented and extended his work – will be published by Routledge in 2011.

Two books with the same title, *The Urban Design Reader*, are anthologies of key writings on urban design. Elizabeth McDonald and Michael Larice (eds), *The Urban Design Reader* (London: Routledge, 2006) is a companion volume in the Routledge Urban Reader Series. It contains classic and contemporary writings on urban design with editors' introductions to the book, sections, and selections in a format similar to *The City Reader fifth edition*. Matthew Carmona and Steve Tiesdell (eds), *The Urban Design Reader* (Oxford: Architectural Press, 2007) places greater emphasis on contemporary urban design writings from the UK.

Other overviews of the field of urban design include Lance Jay Brown, David Dixon, and Oliver Gillham, *Urban Design for an Urban Century: Placemaking for People* (Hoboken, NJ: Wiley, 2009), Peter Bosselmann, *Urban Transformation: Understanding City Form and Design* (Washington, DC: Island Press, 2008), and Jonathan Barnett, *An Introduction to Urban Design* (New York: Harper & Row, 1982). Alex Krieger and William S. Saunders (eds), *Urban Design* (Minneapolis, MN: University of Minnesota Press, 2009) contains essays by urban designers reflecting on the evolution of the field since the 1950s.

Edmund N. Bacon, *Design of Cities* (New York: Penguin, 1976) is a pioneering book by the influential architect and planner who served as executive director of the Philadelphia City Planning Commission from 1949 to 1970 and whose visionary leadership helped transform Philadelphia.

We think it's time for a new urban design manifesto. Almost 50 years have passed since Le Corbusier and the International Congress of Modern Architecture (CIAM) produced the Charter of Athens, and it is more than 20 years since the first Urban Design Conference, still in the CIAM tradition, was held (at Harvard in 1957). Since then the precepts of CIAM have been attacked by sociologists, recently by architects themselves. But it is still a strong influence, and we will take it as our starting point. Make no mistake: the charter was, simply, a manifesto – a public declaration that spelled out the ills of industrial cities as they existed in the 1930s and laid down physical requirements necessary to establish healthy, humane, and beautiful urban environments for people. It could not help but deal with social, economic, and political phenomena, but its basic subject matter was the physical design of cities. Its authors were (mostly) socially concerned architects, determined that their art and craft be responsive to social realities as well as to improving the lot of man. It would be a mistake to write them off as simply elitist designers and physical determinists.

So the charter decried the medium-size (up to six storys) high-density buildings with high land coverage that were associated so closely with slums. Similarly, buildings that faced streets were found to be detrimental to healthy living. The seemingly limitless horizontal expansion of urban areas devoured the countryside, and suburbs were viewed as symbols of terrible waste. Solutions could be found in the demolition of unsanitary housing, the provision of green areas in every residential district, and new high-rise, high-density buildings set in open space. Housing was to be removed from its traditional relationship facing streets, and the whole circulation system was to be revised to meet the needs of emerging mechanization (the automobile). Work areas should be close to but separate from residential areas. To achieve the new city, large land holdings, preferably owned by the public, should replace multiple small parcels (so that projects could be properly designed and developed).

Now thousands of housing estates and redevelopment projects in socialist and capitalist countries the world over, whether built on previously undeveloped land or developed as replacements for old urban areas, attest to the acceptance of the charter's dictums. The design notions it embraced have become part of a world design language, not just the intellectual property of an enlightened few, even though the principles have been devalued in many developments.

Of course, the Charter of Athens has not been the only major urban philosophy of this century to influence the development of urban areas. Ebenezer Howard, too, was responding to the ills of the nineteenth-century industrial city, and the Garden City movement has been at least as powerful as the Charter of Athens. New towns policies, where they exist, are rooted in Howard's thought. But you don't have to look to new towns to see the influence of Howard, Olmsted, Wright, and Stein. The superblock notion, if nothing else, pervades large housing projects around the world, in central cities as well as suburbs. The notion of buildings in a park is as common to garden city designs as it is to charter-inspired development. Indeed, the two movements have a great deal in common: superblocks, separate paths for people and cars, interior common spaces, housing divorced from streets, and central ownership of land. The garden city-inspired communities place greater emphasis on private outdoor space. The most significant difference, at least as they have evolved, is in density and building type: the garden city people preferred to accommodate people in row houses, garden apartments, and maisonettes, while Corbusier and the CIAM designers went for high-rise buildings and, inevitably, people living in flats and at significantly higher densities.

We are less than enthralled with what either the Charter of Athens or the Garden City movement has produced in the way of urban environments. The emphasis of CIAM was on buildings and what goes on within buildings that happen to sit in space, not on the public life that takes place constantly in public spaces. The orientation is often inward. Buildings tend to be islands, big or small. They could be placed anywhere. From the outside perspective, the building, like the work of art it was intended to be, sits where it can be seen and admired in full. And because it is large it is best seen from a distance (at a scale consistent with a moving auto). Diversity, spontaneity, and surprise are absent, at least for the person on foot. On the other hand, we find little joy or magic or spirit in the charter cities. They are not urban, to us, except according to some definition one might find in a census. Most garden cities, safe and healthy and even gracious as they may be, remind us more of suburbs than of cities. But they weren't trying to be cities. The emphasis has always been on "garden" as much as or more than on "city."

Both movements represent overly strong design reactions to the physical decay and social inequities of industrial cities. In responding so strongly, albeit understandably, to crowded, lightless, airless, "utilitiless," congested buildings and cities that housed so many people, the utopians did not inquire what was good about those places, either socially or physically. Did not those physical environments reflect (and maybe even foster) values that were likely to be meaningful to people individually and collectively, such as publicness and community? Without knowing it, maybe these strong reactions to urban ills ended up by throwing the baby out with the bathwater.

In the meantime we have had a lot of experience with city building and rebuilding. New spokespeople with new urban visions have emerged. As more CIAM-style buildings were built people became more disenchanted. Many began to look through picturesque lenses back to the old preindustrial cities. From a concentration on the city as a kind of sculpture garden, the townscape movement, led by the *Architectural Review*, emphasized "urban experience." This phenomenological view of the city was espoused by Rasmussen, Kepes, and ultimately Kevin Lynch and Jane Jacobs. It identified a whole new vocabulary of urban form – one that depended on the sights, sounds, feels, and smells of the city, its materials and textures, floor surfaces, facades, style, signs, lights, seating, trees, sun, and shade all potential amenities for the attentive observer and user. This has permanently humanized the vocabulary of urban design, and we enthusiastically subscribe to most of its tenets, though some in the townscape movement ignored the social meanings and implications of what they were doing.

The 1960s saw the birth of community design and an active concern for the social groups affected, usually negatively, by urban design. Designers were the "soft cops," and many professionals left the design field for social or planning vocations, finding the physical environment to have no redeeming social value. But at the beginning of the 1980s the mood in the design professions is conservative. There is a withdrawal from social engagement back to formalism. Supported by semiology and other abstract themes, much of architecture has become a dilettantish and narcissistic pursuit, a chic component of the high art consumer culture, increasingly remote from most people's everyday lives, finding its ultimate manifestation in the art gallery and the art book. City planning is too immersed in the administration and survival of housing, environmental, and energy programs and in responding to budget cuts and community demands to have any clear sense of direction with regard to city form.

While all these professional ideologies have been working themselves out, massive economic, technological, and social changes have taken place in our cities. The scale of capitalism has continued to increase, as has the scale of bureaucracy, and the automobile has virtually destroyed cities as they once were.

In formulating a new manifesto, we react against other phenomena than did the leaders of CIAM 50 years ago. The automobile cities of California and the Southwest present utterly different problems from those of nineteenth-century European cities, as do the CIAM-influenced housing developments around European, Latin American, and Russian cities and the rash of squatter settlements around the fast-growing cities of the Third World. What are these problems?

PROBLEMS FOR MODERN URBAN DESIGN

Poor living environments

While housing conditions in most advanced countries have improved in terms of such fundamentals as light, air, and space, the surroundings of homes are still frequently dangerous, polluted, noisy, anonymous wastelands. Travel around such cities has become more and more fatiguing and stressful.

Giantism and loss of control

The urban environment is increasingly in the hands of the large-scale developers and public agencies. The elements of the city grow inexorably in size, massive transportation systems are segregated for single travel modes, and vast districts and complexes are created that make people feel irrelevant.

People, therefore, have less sense of control over their homes, neighborhoods, and cities than when they lived in slower-growing locally based communities. Such giantism can be found as readily in the housing projects of socialist cities as in the office buildings and commercial developments of capitalist cities.

Large-scale privatization and the loss of public life

Cities, especially American cities, have become privatized, partly because of the consumer society's emphasis on the individual and the private sector, creating Galbraith's "private affluence and public squalor," but escalated greatly by the spread of the automobile. Crime in the streets is both a cause and a consequence of this trend, which has resulted in a new form of city: one of closed, defended islands with blank and windowless facades surrounded by wastelands of parking lots and fast-moving traffic. As public transit systems have declined, the number of places in American cities where people of different social groups actually meet each other has dwindled. The public environment of many American cities has become an empty desert, leaving public life dependent for its survival solely on planned formal occasions, mostly in protected internal locations.

Centrifugal fragmentation

Advanced industrial societies took work out of the home, and then out of the neighborhood, while the automobile and the growing scale of commerce have taken shopping out of the local community. Fear has led social groups to flee from each other into homogeneous social enclaves. Communities themselves have become lower in density and increasingly homogeneous. Thus the city has spread out and separated to form extensive monocultures and specialized destinations reachable often only by long journeys – a fragile and extravagant urban system dependent on cheap, available gasoline, and an effective contributor to the isolation of social groups from each other.

Destruction of valued places

The quest for profit and prestige and the relentless exploitation of places that attract the public have led to the destruction of much of our heritage, of historic places that no longer turn a profit, of natural amenities that become overused. In many cases, as in San Francisco, the very value of the place threatens its destruction as hungry tourists and entrepreneurs flock to see and profit from it.

Placelessness

Cities are becoming meaningless places beyond their citizens' grasp. We no longer know the origins of the world around us. We rarely know where the materials and products come from, who owns what, who is behind what, what was intended. We live in cities where things happen without warning and without our participation. It is an alien world for most people. It is little surprise that most withdraw from community involvement to enjoy their own private and limited worlds.

Injustice

Cities are symbols of inequality. In most cities the discrepancy between the environments of the rich and the environments of the poor is striking. In many instances the environments of the rich, by occupying and dominating the prevailing patterns of transportation and access, make the environments of the poor relatively worse. This discrepancy may be less visible in the low-density modern city, where the display of affluence is more hidden than in the old city; but the discrepancy remains.

Rootless professionalism

Finally, design professionals today are often part of the problem. In too many cases, we design for places and people we do not know and grant them very little power or acknowledgment. Too many professionals are more part of a universal professional culture than part of the local cultures for whom we produce our plans and products. We carry our "bag of tricks" around the world and bring them out wherever we land. This floating professional culture has only the most superficial conception of particular place. Rootless, it is more susceptible to changes in professional fashion and theory than to local events. There is too little inquiry, too much proposing. Quick surveys are made, instant solutions devised, and the rest of the time is spent persuading the clients. Limits on time and budgets drive us on, but so do lack of understanding and the placeless culture. Moreover, we designers are often unconscious of our own roots, which influence our preferences in hidden ways.

At the same time, the planning profession's retreat into trendism, under the positivist influence of social science, has left it virtually unable to resist the social pressures of capitalist economy and consumer sovereignty. Planners have lost their beliefs. Although we believe citizen participation is essential to urban planning, professionals also must have a sense of what we believe is right, even though we may be vetoed.

GOALS FOR URBAN LIFE

We propose, therefore, a number of goals that we deem essential for the future of a good urban environment: livability; identity and control; access to opportunity, imagination, and joy; authenticity and meaning; open communities and public life; self-reliance; and justice.

Livability

A city should be a place where everyone can live in relative comfort. Most people want a kind of sanctuary for their living environment, a place where they can bring up children, have privacy, sleep, eat, relax, and restore themselves. This means a well-managed environment relatively devoid of nuisance, overcrowding, noise, danger, air pollution, dirt, trash, and other unwelcome intrusions.

Identity and control

People should feel that some part of the environment belongs to them, individually and collectively, some part for which they care and are responsible, whether they own it or not. The urban environment should be an environment that encourages people to express themselves, to become involved, to decide what they want and act on it. Like a seminar where everybody has something to contribute to communal discussion, the urban environment should encourage participation. Urbanites may not always want this. Many like the anonymity of the city, but we are not convinced that the freedom of anonymity is a desirable freedom. It would be much better if people were sure enough of themselves to stand up and be counted. Environments should therefore be designed for those who use them or are affected by them, rather than for those who own them. This should reduce alienation and anonymity (even if people want them); it should increase people's sense of identity and rootedness and encourage more care and responsibility for the physical environment of cities.

Respect for the existing environment, both nature and city, is one fundamental difference we have with the CIAM movement. Urban design has too often assumed that new is better than old. But the new is justified only if it is better than what exists. Conservation encourages identity and control and, usually, a better sense of community, since old environments are more usually part of a common heritage.

Access to opportunity, imagination, and joy

People should find the city a place where they can break from traditional molds, extend their experience, meet new people, learn other viewpoints, have fun. At a functional level, people should have access to alternative housing and job choices; at another level, they should find the city an enlightening cultural experience. A city should have magical places where fantasy is possible, a counter to and an escape from the mundaneness of everyday work and living. Architects and planners take cities and themselves too seriously; the result too often is deadliness and boredom, no imagination, no humor, alienating places. But people need an escape from the seriousness and meaning of the everyday. The city has always been a place of excitement; it is theater, a stage upon which citizens can display themselves and see others. It has magic, or should have, and that depends on a certain sensuous, hedonistic mood, on signs, on night lights, on fantasy, color, and other imagery. There can be parts of the city where belief can be suspended, just as in the experience of fiction. It may be that such places have to be framed so that people know how to act. Until now such fantasy and experiment have been attempted mostly by commercial facilities, at rather low levels of quality and aspiration, seldom deeply experimental. One should not have to travel as far as the Himalayas or the South Sea Islands to stretch one's experience. Such challenges could be nearer home. There should be a place for community utopias; for historic, natural, and anthropological evocations of the modern city, for encounters with the truly exotic.

Authenticity and meaning

People should be able to understand their city (or other people's cities), its basic layout, public functions, and institutions; they should be aware of its opportunities. An authentic city is one where the origins of things and places are clear. All this means an urban environment should reveal its significant meanings; it should not be dominated only by one type of group, the powerful; neither should publicly important places be hidden. The city should symbolize the moral issues of society and educate its citizens to an awareness of them.

That does not mean everything has to be laid out as on a supermarket shelf. A city should present itself as a readable story, in an engaging and, if necessary, provocative way, for people are indifferent to the obvious, overwhelmed by complexity. A city's offerings should be revealed or they will be missed. This can affect the forms of the city, its signage, and other public information and education programs.

Livability, identity, authenticity, and opportunity are characteristics of the urban environment that should serve the individual and small social unit, but the city has to serve some higher social goals as well. It is these we especially wish to emphasize here.

Community and public life

Cities should encourage participation of their citizens in community and public life. In the face of giantism and fragmentation, public life, especially life in public places, has been seriously eroded. The neighborhood movement, by bringing thousands, probably millions of people out of their closed private lives into active participation in their local communities, has begun to counter that trend, but this movement has had its limitations. It can be purely defensive, parochial, and self-serving. A city should be more than a warring collection of interest groups, classes, and neighborhoods; it should breed a commitment to a larger whole, to tolerance, justice, law, and democracy. The structure of the city should invite and encourage public life, not only through its institutions, but directly and symbolically through its public spaces. The public environment, unlike the neighborhood, by definition should be open to all members of the community. It is where people of different kinds meet. No one should be excluded unless they threaten the balance of that life.

Urban self-reliance

Increasingly cities will have to become more self-sustaining in their uses of energy and other scarce resources. "Soft energy paths" in particular not only will reduce dependence and exploitation across regions and countries but also will help reestablish a stronger sense of local and regional identity, authenticity, and meaning.

An environment for all

Good environments should be accessible to all. Every citizen is entitled to some minimal level of environmental livability and minimal levels of identity, control, and opportunity. Good urban design must be for the poor as well as the rich. Indeed, it is more needed by the poor.

We look toward a society that is truly pluralistic, one where power is more evenly distributed among social groups than it is today in virtually any country, but where the different values and cultures of interest- and place-based groups are acknowledged and negotiated in a just public arena.

These goals for the urban environment are both individual and collective, and as such they are frequently in conflict. The more a city promises for the individual, the less it seems to have a public life; the more the city is built for public entities, the less the individual seems to count. The good urban environment is one that somehow balances these goals, allowing individual and group identity while maintaining a public concern, encouraging pleasure while maintaining responsibility, remaining open to outsiders while sustaining a strong sense of localism.

AN URBAN FABRIC FOR AN URBAN LIFE

We have some ideas, at least, for how the fabric or texture of cities might be conserved or created to encourage a livable urban environment. We emphasize the structural qualities of the good urban environment – qualities we hope will be successful in creating urban experiences that are consonant with our goals.

Do not misread this. We are not describing all the qualities of a city. We are not dealing with major transportation systems, open space, the natural environment, the structure of the large-scale city, or even the structure of neighborhoods, but only the grain of the good city.

There are five physical characteristics that must be present if there is to be a positive response to the goals and values we believe are central to urban life. They must be designed, they must exist, as prerequisites of a sound urban environment. All five must be present, not just one or two. There are other physical characteristics that are important, but these five are essential: livable streets and neighborhoods; some minimum density of residential development as well as intensity of land use; an integration of activities – living, working, shopping – in some reasonable proximity to each other; a manmade environment, particularly buildings, that defines public space (as opposed to buildings that, for the most part, sit in space); and many, many separate, distinct buildings with complex arrangements and relationships (as opposed to few, large buildings).

Let us explain, keeping in mind that all five of the characteristics must be present. People, we have said, should be able to live in reasonable (though not excessive) safety, cleanliness, and security. That means livable streets and neighborhoods: with adequate sunlight, clean air, trees, vegetation, gardens, open space, pleasantly scaled and designed buildings; without offensive noise; with cleanliness and physical safety. Many of these characteristics can be designed into the physical fabric of the city.

The reader will say, "Well of course, but what does that mean?" Usually it has meant specific standards and requirements, such as sun angles, decibel levels, lane widths, and distances between buildings. Many researchers have been trying to define the qualities of a livable environment. It depends on a wide array of attributes, some structural, some quite small details. There is no single right answer. We applaud these efforts and have participated in them ourselves. Nevertheless, desires for livability and individual comfort by themselves have led to fragmentation of the city. Livability standards, whether for urban or for suburban developments, have often been excessive.

Our approach to the details of this inclusive physical characteristic would center on the words "reasonable, though not excessive . . ." Too often, for example, the requirement of adequate sunlight has resulted in buildings and people inordinately far from each other, beyond what demonstrable need for light would dictate. Safety concerns have been the justifications for ever wider streets and wide, sweeping curves rather than narrow ways and sharp corners. Buildings are removed from streets because of noise considerations when there might be other ways to deal with this concern. So although livable streets and neighborhoods are a primary requirement for any good urban fabric – whether for existing, denser cities or for new development – the quest for livable neighborhoods, if pursued obsessively, can destroy the urban qualities we seek to achieve.

A *minimum density* is needed. By density we mean the number of people (sometimes expressed in terms of housing units) living on an area of land, or the number of people using an area of land.

Cities are not farms. A city is people living and working and doing the things they do in relatively close proximity to each other.

We are impressed with the importance of density as a perceived phenomenon and therefore relative to the beholder and agree that, for many purposes, perceived density is more important than an "objective" measurement of people per unit of land. We agree, too, that physical phenomena can be manipulated so as to render perceptions of greater or lesser density. Nevertheless, a narrow, winding street, with a lot of signs and a small enclosed open space at the end, with no people, does not make a city. Cities are more than stage sets. Some minimum number of people living and using a given area of land is required if there is to be human exchange, public life and action, diversity and community.

Density of people alone will account for the presence or absence of certain uses and services we find important to urban life. We suspect, for example, that the number and diversity of small stores and services – for instance, groceries, bars, bakeries, laundries and cleaners, coffee shops, secondhand stores, and the like – to be found in a city or area is in part a function of density. That is, that such businesses are more likely to exist, and in greater variety, in an area where people live in greater proximity to each other ("higher" density). The viability of mass transit, we know, depends partly on the density of residential areas and partly on the size and intensity of activity at commercial and service destinations. And more use of transit, in turn, reduces parking demands and permits increases in density. There must be a critical mass of people, and they must spend a lot of their time in reasonably close proximity to each other, including when they are at home, if there is to be an urban life. The goal of local control and community identity is associated with density as well. The notion of an optimum density is elusive and is easily confused with the health and livability of urban areas, with lifestyles, with housing types, with the size of area being considered (the building site or the neighborhood or the city), and with the economics of development. A density that might be best for child rearing might be less than adequate to support public transit. Most recently, energy efficiency has emerged as a concern associated with density, the notion being that conservation will demand more compact living arrangements.

Our conclusion, based largely on our experience and on the literature, is that a minimum net density (people or living units divided by the size of the building site, excluding public streets) of about 15 dwelling units (30–60 people) per acre of land is necessary to support city life. By way of illustration, that is the density produced with generous town houses (or row houses). It would permit parcel sizes up to 25 feet wide by about 115 feet deep. But other building types and lot sizes also would produce that density. Some areas could be developed with lower densities, but not very many. We don't think you get cities at 6 dwellings to the acre, let alone on half-acre lots. On the other hand, it is possible to go as high as 48 dwelling units per acre (96 to 192 people) for a very large part of the city and still provide for a spacious and gracious urban life. Much of San Francisco, for example, is developed with three-story buildings (one unit per floor) above a parking story, on parcels that measure 25 feet by 100 or 125 feet. At those densities, with that kind of housing, there can be private or shared gardens for most people, no common hallways are required, and people can have direct access to the ground. Public streets and walks adequate to handle pedestrian and vehicular traffic generated by these densities can be accommodated in rights-of-way that are 50 feet wide or less. Higher densities, for parts of the city, to suit particular needs and lifestyles, would be both possible and desirable. We are not sure what the upper limits would be but suspect that as the numbers get much higher than 200 people per net residential acre, for larger parts of the city, the concessions to less desirable living environments mount rapidly.

Beyond residential density, there must be a minimum intensity of people using an area for it to be urban, as we are defining that word. We aren't sure what the numbers are or even how best to measure this kind of intensity. We are speaking here, particularly, of the public or "meeting" areas of our city. We are confident that our lowest residential densities will provide most meeting areas with life and human exchange, but are not sure if they will generate enough activity for the most intense central districts.

There must be an *integration of activities* – living, working, and shopping as well as public, spiritual, and recreational activities – reasonably near each other.

The best urban places have some mixtures of uses. The mixture responds to the values of publicness and diversity that encourage local community identity. Excitement, spirit, sense, stimulation, and exchange are more likely when there is a mixture of activities than when there is not. There are many examples that we all know. It is the mix, not just the density of people and uses, that brings life to an area, the life of people going about a full range of normal activities without having to get into an automobile.

We are not saying that every area of the city should have a full mix of all uses. That would be impossible. The ultimate in mixture would be for each building to have a range of uses from living, to working, to shopping, to recreation. We are not calling for a return to the medieval city. There is a lot to be said for the notion of "living sanctuaries," which consist almost wholly of housing. But we think these should be relatively small, of a few blocks, and they should be close and easily accessible (by foot) to areas where people meet to shop or work or recreate or do public business. And except for a few of the most intensely developed office blocks of a central business district or a heavy industrial area, the meeting areas should have housing within them. Stores should be mixed with offices. If we envision the urban landscape as a fabric, then it would be a salt-and-pepper fabric of many colors, each color for a separate use or a combination. Of course, some areas would be much more heavily one color than another, and some would be an even mix of colors. Some areas, if you squinted your eyes, or if you got so close as to see only a small part of the fabric, would read as one color, a red or a brown or a green. But by and large there would be few if any distinct patterns, where one color stopped and another started. It would not be patchwork quilt, or an even-colored fabric. The fabric would be mixed.

In an urban environment, *buildings* (and other objects that people place in the environment) *should be arranged in such a way as to define and even enclose public space, rather than sit in space.* It is not enough to have high densities and an integration of activities to have cities. A tall enough building with enough people living (or even working) in it, sited on a large parcel, can easily produce the densities we have talked about and can have internally mixed uses, like most "mixed use" projects. But that building and its neighbors will be unrelated objects sitting in space if they are far enough apart, and the mixed uses might be only privately available. In large measure that is what the Charter of Athens, the garden cities, and standard suburban development produce.

Buildings close to each other along a street, regardless of whether the street is straight, or curved, or angled, tend to define space if the street is not too wide in relation to the buildings. The same is true of a plaza or a square. As the spaces between buildings become larger (in relation to the size of the buildings, up to a point), the buildings tend more and more to sit in space. They become focal points for few or many people, depending on their size and activity. Except where they are monuments or centers for public activities (a stadium or meeting hall), where they represent public gathering spots, buildings in space tend to be private and inwardly oriented. People come to them and go from them in any direction. That is not so for the defined outdoor environment. Avoiding the temptation to ascribe all kinds of psychological values to defined spaces (such as intimacy, belonging, protection – values that are difficult to prove and that may differ for different people), it is enough to observe that spaces surrounded by buildings are more likely to bring people together and thereby promote public interaction. The space can be linear (like streets) or in the form of plazas of myriad shapes. Moreover, interest and interplay among uses is enhanced. To be sure, such arrangements direct people and limit their freedom – they cannot move in just any direction from any point – but presumably there are enough choices (even avenues of escape) left open, and the gain is in greater potential for sense stimulation, excitement, surprise, and focus. Over and over again we seek out and return to defined ways and spaces as symbolic of urban life emphasizing the public space more than the private building.

It is important for us to emphasize *public places* and a *public* way system. We have observed that the central value of urban life is that of publicness, of people from different groups meeting each other and of people acting in concert, albeit with debate. The most important public places must be for *pedestrians*, for no public life can take place between people in automobiles. Most public space has been taken over by the automobile, for travel or parking. We must fight to restore more for the pedestrian. Pedestrian malls are not simply to benefit the local merchants. They have an essential public value. People of different kinds meet each other directly. The level of communication may be only visual, but that itself is educational and can encourage tolerance. The revival of street activities, street vending, and street theater in American cities may be the precursor of a more flourishing public environment, if the automobile can be held back.

There also must be symbolic, public meeting places, accessible to all and publicly controlled. Further, in order to communicate, to get from place to place, to interact, to exchange ideas and goods, there must be a healthy public circulation system. It cannot be privately controlled. Public circulation systems should be seen as significant cultural settings where the city's finest products and artifacts can be displayed, as in the piazzas of medieval and renaissance cities.

Finally, *many different buildings and spaces with complex arrangements and relationships* are required. The often elusive notion of human scale is associated with this requirement – a notion that is not just an architect's concept but one that other people understand as well.

Diversity, the possibility of intimacy and confrontation with the unexpected, stimulation, are all more likely with many buildings than with few taking up the same ground areas.

For a long time we have been led to believe that large land holdings were necessary to design healthy, efficient, aesthetically pleasing urban environments. The slums of the industrial city were associated, at least in part, with all those small, overbuilt parcels. Socialist and capitalist ideologies alike called for land assembly to permit integrated, socially and economically useful developments. What the socialist countries would do via public ownership the capitalists would achieve through redevelopment and new fiscal mechanisms that rewarded large holdings. Architects of both ideological persuasions promulgated or were easily convinced of the wisdom of land assembly. It's not hard to figure out why. The results, whether by big business or big government, are more often than not inward-oriented, easily controlled or controllable, sterile, large-building projects, with fewer entrances, fewer windows, less diversity, less innovation, and less individual expression than the urban fabric that existed previously or that can be achieved with many actors and many buildings. Attempts to break up facades or otherwise to articulate separate activities in large buildings are seldom as successful as when smaller properties are developed singly.

Health, safety, and efficiency can be achieved with many smaller buildings, individually designed and developed. Reasonable public controls can see to that. And, of course, smaller buildings are a lot more likely if parcel sizes are small than if they are large. With smaller buildings and parcels, more entrances must be located on the public spaces, more windows and a finer scale of design diversity emerge. A more public, lively city is produced. It implies more, smaller groups getting pieces of the public action, taking part, having a stake. Other stipulations may be necessary to keep public frontages alive, free from the deadening effects of offices and banks, but small buildings will help this more than large ones. There need to be large buildings, too, covering large areas of land, but they will be the exception, not the rule, and should not be in the centers of public activity.

ALL THESE QUALITIES . . . AND OTHERS

A good city must have all those qualities. Density without livability could return us to the slums of the nineteenth century. Public places without small-scale, fine-grain development would give us vast, overscale cities. As an urban fabric, however, those qualities stand a good chance of meeting many of the goals we outlined. They directly attend to the issue of livability though they are aimed especially at encouraging public places and a public life. Their effects on personal and group identity are less clear, though the small-scale city is more likely to support identity than the large-scale city. Opportunity and imagination should be encouraged by a diverse and densely settled urban structure. This structure also should create a setting that is more meaningful to the individual inhabitant and small group than the giant environments now being produced. There is no guarantee that this urban structure will be a more just one than those presently existing. In supporting the small against the large, however, more justice for the powerless may be encouraged.

Still, an urban fabric of this kind cannot by itself meet all these goals. Other physical characteristics are important to the design of urban environments. Open space, to provide access to nature as well as relief from the built environment, is one. So are definitions, boundaries if you will, that give location and identity to neighborhoods (or districts) and to the city itself. There are other characteristics as well: public buildings, educational environments, places set aside for nurturing the spirit, and more. We still have work to do.

MANY PARTICIPANTS

While we have concentrated on defining physical characteristics of a good city fabric, the process of creating it is crucial. As important as many buildings and spaces are many participants in the building process. It is through this involvement in the creation and management of their city that citizens are most likely to identify with it and, conversely, to enhance their own sense of identity and control.

AN ESSENTIAL BEGINNING

The five characteristics we have noted are essential to achieving the values central to urban life. They need much further definition and testing. We have to know more about what configurations create public space: about maximum densities, about how small a community can be and still be urban (some very small Swiss villages fit the bill, and everyone knows some favorite examples), about what is perceived as big and what small under different circumstances, about landscape material as a space definer, and a lot more. When we know more we will be still further along toward a new urban design manifesto.

We know that any ideal community, including the kind that can come from this manifesto, will not always be comfortable for every person. Some people don't like cities and aren't about to. Those who do will not be enthralled with all of what we propose.

Our urban vision is rooted partly in the realities of earlier, older urban places that many people, including many utopian designers, have rejected, often for good reasons. So our utopia will not satisfy all people. That's all right. We like cities. Given a choice of the kind of community we would *like* to live in – the sort of choice earlier city dwellers seldom had – we would choose to live in an urban, public community that embraces the goals and displays the physical characteristics we have outlined. Moreover, we think it responds to what people want and that it will promote the good urban life.

"Three Types of Outdoor Activities," "Life Between Buildings," and "Outdoor Activities and the Quality of Outdoor Space"

from *Life Between Buildings: Using Public Space* (1987)

Jan Gehl

Editors' Introduction

With its long winters and reserved residents, Copenhagen, Denmark, seems an unlikely setting for vibrant new uses of outdoor urban spaces. When Copenhagen created one of Europe's first pedestrian-free zones – Strøget (Literally, "the sweep") – in 1962, skeptics predicted the experiment would fail. But now the Strøget carfree zone is the longest pedestrian shopping area in the world, swarming with people shopping, walking, sitting, chatting, playing, drawing, eating, making and listening to music, people-watching, and simply being with other people.

Danish architect Jan Gehl and his followers have been at the forefront of innovative designs to promote the "life between buildings" that Strøget exemplifies. As Strøget and other innovative Danish designs for space between buildings succeeded, Gehl's ideas have been embraced by architects, urban designers and urban planners throughout the world. There is no one-size-fits-all prescription for pedestrian-only streets. Some, like New York's Times Square or the area around the Acropolis in Greece, have been extremely successful. Others have failed to attract the expected pedestrian flow and failed.

It is the millions of day-to-day interactions in ordinary neighborhoods that determine the quality of life for most of humanity: walking the dog, taking chicken soup next door to a sick neighbor, washing the car, puttering in a front yard garden, leaning over a fence to gossip with a friend, just going outside for the joy of it. Gehl argues that designs that encourage people to spend time outdoors and make interacting with other people outdoors enjoyable can make a big difference in city dwellers' quality of life.

Gehl notes that some outdoor activities – like delivering the mail and going to work or school – have to take place regardless of the quality of the built environment or how people feel about being outside. Good design will have a negligible impact on whether or not these activities take place, though they will affect how enjoyable they are and may affect how much time people choose to spend doing them. But, Gehl notes, many outdoor activities that take place in the space between buildings – taking a walk, chatting with a neighbor, sunbathing – are optional. If the physical environment makes them pleasant, people will engage in them; if it does not they won't. Gehl feels that designs that encourage contact among people at any level, from very simple and noncommittal contacts such as seeing, hearing, and being among other people to complex and emotionally involved connections, enrich people's lives.

Outdoor social interactions result from both necessary and optional activities. Since the extent to which people choose to engage in optional activities depends on how enjoyable they find them, designers can help create lively cities by designing good outdoor space, particularly ones that will encourage optional time spent outside.

The heart of Gehl's theory involves four dualities: designs that assemble or disperse, integrate or segregate, invite or repel, and open up or close in. Gehl advocates designs that assemble, integrate, invite, and open up.

Gehl likes designs that assemble. The idea of how design can assemble people is well illustrated by an everyday example. Shopping mall designers usually design mall stores to be narrow and deep so that people will pass many different store windows as they walk through the mall – a design that assembles people. Gehl made a brilliant connection. Narrower residential lots (and the houses on them) will result in more housing units per linear foot of street frontage and more people walking along the streets. People walking along streets with narrow lots will pass more of their neighbors on the way to the store, school, or bus stop than they would if house were the same size, but built on wider, shallower lots. Accordingly Gehl advocates narrow residential lots in order to assemble people and increase social contact.

Gehl favors designs that integrate. Good design can bring people in contact with one another regardless of gender, age, income, sexual orientation, occupation, and ethnic group. Gehl likes the sprawling University of Denmark campus that developed piecemeal and is mixed into Copenhagen's downtown area. He deplores the sterile campus of the newer Technical University of Denmark, built on the outskirts of the city. Students at the University of Denmark mix with other city residents, patronize public cafés, and can enjoy Copenhagen's amenities. Students at the Technical University of Denmark mix only with faculty and other students, eat in the university cafeteria, and remain separate from the life of the city. Like Jane Jacobs (p. 105), Gehl thinks a little bit of urban disorder can be a good thing.

Gehl likes designs that open up. A library with windows directly on the street, for example, will be open to passersby who can participate vicariously in the library experience by watching the librarians and browsers even if they do not go in.

Contrast Gehl's view that even fleeting, anonymous contact with other human beings is innately satisfying with Louis Wirth's view in "Urbanism as a Way of Life" (p. 96) that the transitory, impersonal contacts between people characteristic of modern cities illustrates just how disconnected people become when they move from small rural communities to large, anonymous cities.

Gehl blames the well-intentioned ideas of modernists like Le Corbusier (p. 336) for destruction of livable streets and a thinning of cities that make human contact difficult. Modernists sought to bring light, air, sun, and ventilation into residential and commercial areas. But big modernist multistory residential urban areas with long distances between different land uses destroy street life and eliminate intimate squares. Similarly the wide dispersal of people and events in low, open, single-family areas in suburbs has reduced outdoor communal activities.

At the core of Gehl's philosophy is the belief that people need and want human contact in outdoor public spaces. Is that necessarily so? Some people (illegal immigrants, runaway teenagers, people who simply want to be alone) may not want to come in contact with other people. Is the space between buildings the most important space for human contact? What about the home? The workplace? Other public spaces? Ray Oldenburg argues that "third places" like cafés, coffee shops, bookstores, bars, and hair salons are more important venues for human contact and socializing than outdoor space between buildings.

Jan Gehl (b. 1936) is a Danish architect and urban designer based in Copenhagen. He received a Masters of Architecture degree from the Royal Danish Academy of Fine Arts in 1960. The first (Danish) edition of *Life Between Buildings* was published in 1971 and subsequent revised editions have been published regularly since that time, most recently in 2008. The first English language edition of *Life Between Buildings* was published in 1987 and the most recent English language edition was published in 2008.

The verb "Copenhagenize" is not yet in common parlance, but Gehl uses it to describe the design principles he hopes to export from his native city. In addition to many projects in Denmark and other Scandinavian countries, Gehl has designed projects in London, Stoke-on-Trent, and Brighton in England; Melbourne, Perth, Adelaide, Wellington, and Sydney in Australia; Cork and Dublin in Ireland; New York and Pittsburgh in the United States; Belgrade in Serbia; Prague in the Czech Republic; and Rabat in Morocco. The selection here is from *Life Between Buildings: Using Public Space* translated by Jo Koch (New York: Van Nostrand Reinhold, 1987). A coauthored book by Jan Gehl and Lars Gemzoe is *New City Spaces, Strategies and Projects* (Copenhagen: Danish Architectural Press, 2008).

University of California, Berkeley, professor of architecture emeritus Clare Cooper Marcus has written and edited excellent books about the way in which people use both public and private spaces. Her books provide extensive practical guidelines for architects and planners. *People Places: Design Guidelines for Urban Open Space*, coedited with Carolyn Francis (Hoboken, NJ: Wiley, 1997) is an anthology that nicely complements *Life Between Buildings.* Her first book, *Easter Hill Village: Some Social Implications of Design* (New York: Free Press, 1975) documents what residents themselves liked and disliked about the design of a low-rent housing project in Richmond, California. *Housing as if People Mattered: Site Design Guidelines for the Planning of Medium-Density Family Housing*, with Wendy Sarkissien (Berkeley, CA: University of California Press, 1988) is filled with examples and principles for designing moderate income housing, particularly for single parents with children.

Other books about the design of public spaces include Lorna McNeur, *Theatre of the City: Interpreting Public Space* (London: Routledge, 2010), William Whyte, *City: Rediscovering the Center* (Philadelpha, PA: University of Pennsylvania Press, 2009), Sharon Zukin, *Naked City* (Oxford: Oxford University Press, 2009), Roger Yee, *Public Spaces* (New York: Visual Reference Publications, 2009), Sarah Gaventa, *New Public Spaces* (London: Mitchell Beazley, 2006), Raymond Gastil and Zoë Ryan, *Open: New Designs for Public Space* (New York: Princeton Architectural Press, 2006), Matthew Carmona, Tim Heath, Taner Oc, and Steve Tiesdell, *Public Places, Urban Spaces* (Oxford: Architectural Press, 2003), Doug Kelbaugh, *Common Place: Toward Neighborhood and Regional Design* (Seattle, WA: University of Washington Press, 1997), Stephen Carr, Mark Francis, Leanne G. Rivlin, and Andew M. Stone *Public Space* (Cambridge: Cambridge University Press, 1993), and Michael Sorkin, *Variations on a Theme Park: The New American City and the End of Public Space* (New York: Hill & Wang, 1992). A classic early critique of placenessness is Edward Relph, *Place and Placelessness.* (London: Pion, 1976).

Ray Oldenburg, *The Great Good Place: Cafés, Coffee Shops, Bookstores, Bars, Hair Salons and Other Hangouts at the Heart of a Community* (New York: Marlowe, 1999) analyzes interior public spaces. Clare Cooper Marcus, *House as a Mirror of Self: Exploring the Deeper Meaning of Home* (Lake Worth, FL: Nicholas-Hays, 2006) explores the symbolic meaning of space in private homes.

THREE TYPES OF OUTDOOR ACTIVITIES

An ordinary day on an ordinary street. Pedestrians pass on the sidewalks, children play near front doors, people sit on benches and steps, the postman makes his rounds with the mail, two passersby greet on the sidewalk, two mechanics repair a car, groups engage in conversation. This mix of outdoor activities is influenced by a number of conditions. Physical environment is one of the factors: a factor that influences the activities to a varying degree and in many different ways. Outdoor activities, and a number of the physical conditions that influence them, are the subject of this book.

Greatly simplified, outdoor activities in public spaces can be divided into three categories, each of which places very different demands on the physical environment: *necessary activities, optional activities*, and *social activities.*

Necessary activities include those that are more or less compulsory – going to school or to work, shopping, waiting for a bus or a person, running errands, distributing mail – in other words, all activities in which those involved are to a greater or lesser degree required to participate.

In general, everyday tasks and pastimes belong to this group. Among other activities, this group includes the great majority of those related to walking.

Because the activities in this group are necessary, their incidence is influenced only slightly by the physical framework. These activities will take place throughout the year, under nearly all conditions, and are more or less independent of the exterior environment. The participants have no choice.

Optional activities – that is, those pursuits that are participated in if there is a wish to do so and if time and place make it possible – are quite another matter.

This category includes such activities as taking a walk to get a breath of fresh air, standing around enjoying life, or sitting and sunbathing.

These activities take place only when exterior conditions are optimal, when weather and place invite them. This relationship is particularly important in connection with physical planning because most of the recreational activities that are especially pleasant to pursue outdoors are found precisely in this category of activities. These activities are especially dependent on exterior physical conditions.

When outdoor areas are of poor quality, only strictly necessary activities occur.

When outdoor areas are of high quality, necessary activities take place with approximately the same frequency – though they clearly tend to take a longer time, because the physical conditions are better. In addition, however, a wide range of optional activities will also occur because place and situation now invite people to stop, sit, eat, play, and so on.

In streets and city spaces of poor quality, only the bare minimum of activity takes place. People hurry home.

In a good environment, a completely different, broad spectrum of human activities is possible.

Social activities are all activities that depend on the presence of others in public spaces. Social activities include children at play, greetings and conversations, communal activities of various kinds, and finally – as the most widespread social activity – passive contacts, that is, simply seeing and hearing other people.

Different kinds of social activities occur in many places: in dwellings; in private outdoor spaces, gardens, and balconies; in public buildings; at places of work; and so on; but in this context only those activities that occur in publicly accessible spaces are examined.

These activities could also be termed "resultant" activities, because in nearly all instances they evolve from activities linked to the other two activity categories. They develop in connection with the other activities because people are in the same space, meet, pass by one another, or are merely within view.

Social activities occur spontaneously, as a direct consequence of people moving about and being in the same spaces. This implies that social activities are indirectly supported whenever necessary and optional activities are given better conditions in public spaces.

The character of social activities varies, depending on the context in which they occur. In the residential streets, near schools, near places of work, where there are a limited number of people with common interests or backgrounds, social activities in public spaces can be quite comprehensive: greetings, conversations, discussions, and play arising from common interests and because people "know" each other, if for no other reason than that they often see one another.

In city streets and city centers, social activities will generally be more superficial, with the majority being passive contacts – seeing and hearing a great number of unknown people. But even this limited activity can be very appealing.

Very freely interpreted, a social activity takes place every time two people are together in the same space. To see and hear each other, to meet, is in itself a form of contact, a social activity, The actual meeting, merely being present, is furthermore the seed for other, more comprehensive forms of social activity.

This connection is important in relation to physical planning. Although the physical framework does not have a direct influence on the quality, content, and intensity of social contacts, architects and planners can affect the possibilities for meeting, seeing, and hearing

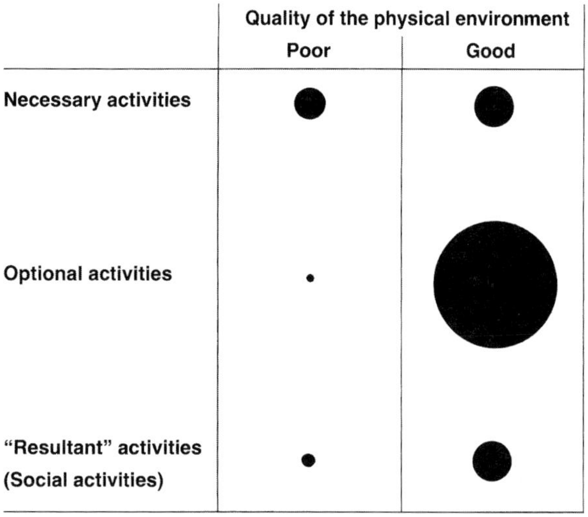

Graphic representation of the relationship between the quality of outdoor spaces and the rate of occurrence of outdoor activities.

When the quality of outdoor areas is good, optional activities occur with increasing frequency. Furthermore, as levels of optional activity rise, the number of social activities usually increases substantially.

	Quality of the physical environment	
	Poor	**Good**
Necessary activities	●	●
Optional activities	•	⬤
"Resultant" activities (Social activities)	•	●

people – possibilities that both take on a quality of their own and become important as background and starting point for other forms of contact.

This is the background for the investigation . . . of meeting possibilities and opportunities to see and hear other people. Another reason for a comprehensive review of these activities is that precisely the presence of other people, activities, events, inspiration, and stimulation comprise one of the most important qualities of public spaces altogether.

If we look back at the street scene that was the starting point for defining the three categories of outdoor activities, we can see how necessary, optional, and social activities occur in a finely interwoven pattern. People walk, sit, and talk. Functional, recreational, and social activities intertwine in all conceivable combinations. Therefore, this examination of the subject of outdoor activities does not begin with a single, limited category of activities. Life between buildings is not merely pedestrian traffic or recreational or social activities. Life between buildings comprises the entire spectrum of activities, which combine to make communal spaces in cities and residential areas meaningful and attractive.

Both necessary, functional activities and optional, recreational activities have been examined quite thoroughly over the years in different contexts. Social activities and their interweaving to form a communal fabric have received considerably less attention.

This is the background for the following, more detailed examination of social activities in public spaces.

LIFE BETWEEN BUILDINGS

It is difficult to pinpoint precisely what life between buildings means in relation to the *need for contact*.

Opportunities for meetings and daily activities in the public spaces of a city or residential area enable one to be among, to see, and to hear others, to experience other people functioning in various situations.

These modest "see and hear contacts" must be considered in relation to other forms of contact and as part of the whole range of social activities, from very simple and noncommittal contacts to complex and emotionally involved connections.

The concept of varying-degrees of contact intensity is the basis of the following simplified outline of various contact forms.

High intensity	↑	Close friendships
		Friends
		Acquaintances
		Chance contacts
Low intensity		Passive contacts ("see and hear" contacts)

In terms of this outline, life between buildings represents primarily the low-intensity contacts located at the bottom of the scale. Compared with the other contact forms, these contacts appear insignificant, yet they are valuable both as independent contact forms and as prerequisites for other, more complex interactions.

Opportunities related to merely being able to meet, see, and hear others include:

- contact at a modest level
- a possible starting point for contact at other levels
- a possibility for maintaining already established contacts
- a source of information about the social world outside
- a source of inspiration, an offer of stimulating experience.

The possibilities related to the low-intensity contact forms offered in public spaces perhaps can best be described by the situation that exists if they are lacking.

If activity between buildings is missing, the lower end of the contact scale also disappears. The varied transitional forms between being alone and being together have disappeared. The boundaries between isolation and contact become sharper – people are either alone or else with others on a relatively demanding and exacting level.

Life between buildings offers an opportunity to be with others in a relaxed and undemanding way. One can take occasional walks, perhaps make a detour along a main street on the way home or pause at an inviting bench near a front door to be among people for a short while. One can take a long bus ride every day, as many retired people have been found to do in large cities. Or one can do daily shopping, even though it would be more practical to do it once a week. Even looking out of the window now and then, if one is fortunate enough to have something to look at, can be rewarding. Being among others, seeing and hearing others, receiving impulses from others, imply positive experiences, alternatives to being alone. One is not necessarily with a specific person, but one is, nevertheless, with others.

As opposed to being a passive observer of other people's experiences on television or video or film, in public spaces the individual himself is present, participating in a modest way, but most definitely participating.

Low-intensity contact is also a situation from which other forms of contact can grow. It is a medium for the unpredictable, the spontaneous, the unplanned.

These opportunities can be illustrated by examining how play activities among children get started.

Such situations can be arranged. Formalized play occurs at birthday parties and arranged play groups in schools. Generally, however, play is not arranged. It evolves when children are together, when they see others at play, when they feel like playing and "go out to play" without actually being certain that play will get started. The first prerequisite is being in the same space. Meeting.

Contacts that develop spontaneously in connection with merely being where there are others are usually very fleeting – a short exchange of words, a brief discussion with the next man on the bench, chatting with a child in a bus, watching somebody working and asking a few questions, and so forth. From this simple level, contacts can grow to other levels, as the participants wish. Meeting, being present in the same space, is in each of these circumstances the prime prerequisite.

The possibility of meeting neighbors and co-workers often in connection with daily comings and goings implies a valuable opportunity to establish and later maintain acquaintances in a relaxed and undemanding way.

Social events can evolve spontaneously. Situations are allowed to develop. Visits and gatherings can be arranged on short notice, when the mood dictates. It is equally easy to "drop by" or "look in" or to agree on what is to take place tomorrow if the participants pass by one another's front doors often and, especially, meet often on the street or in connection with daily activities around the home, place of work, and so on.

Frequent meetings in connection with daily activities increase chances of developing contacts with neighbors, a fact noted in many surveys. With frequent meetings friendships and the contact network are maintained in a far simpler and less demanding way than if friendship must be kept up by telephone and invitation. If this is the case, it is often rather difficult to maintain contact, because more is always demanded of the participants when meetings must be arranged in advance.

This is the underlying reason why nearly all children and a considerable proportion of other age groups maintain closer and more frequent contact with friends and acquaintances who live or work near them – it is the simplest way to stay "in touch."

The opportunity to see and hear other people in a city or residential area also implies an offer of valuable information, about the surrounding social environment in general and about the people one lives or works with in particular.

This is especially true in connection with the social development of children, which is largely based on observations of the surrounding social environment, but all of us need to be kept up-to-date about the surrounding world in order to function in a social context.

Through the mass media we are informed about the larger, more sensational world events, but by being with others we learn about the more common but equally important details. We discover how others work, behave, and dress, and we obtain knowledge about the people we work with, live with, and so forth. By means of all this information we establish a confidential relationship with the world around us. A person we have often met on the street becomes a person we "know."

In addition to imparting information about the social world outside, the opportunity to see and hear other people can also provide ideas and inspiration for action.

We are inspired by seeing others in action. Children, for example, see other children at play and get the urge to join in, or they get ideas for new games by watching other children or adults.

The trend from living to lifeless cities and residential areas that has accompanied industrialization, segregation of various city functions, and reliance on the automobile also has caused cities to become duller and more monotonous. This points up another important need, namely *the need for stimulation*.

Experiencing other people represents a particularly colorful and attractive opportunity for stimulation. Compared with experiencing buildings and other inanimate objects, experiencing people, who speak and move about, offers a wealth of sensual variation. No moment is like the previous or the following when people circulate among people. The number of new situations and new stimuli is limitless. Furthermore, it concerns the most important subject in life: people.

Living cities, therefore, ones in which people can interact with one another, are always stimulating because they are rich in experiences, in contrast to lifeless cities, which can scarcely avoid being poor in experiences and thus dull, no matter how many colors and variations of shape in buildings are introduced.

If life between buildings is given favorable conditions through sensible planning of cities and housing areas alike, many costly and often stilted and strained attempts to make buildings "interesting" and rich by using dramatic architectural effects can be spared.

Life between buildings is both more relevant and more interesting to look at in the long run than are any combination of colored concrete and staggered building forms.

The value of the many large and small possibilities that are attached to the opportunity of being in the same space as and seeing and hearing other people is underlined by a series of observations investigating people's reaction to the presence of other people in public spaces.

Wherever there are people – in buildings, in neighborhoods, in city centers, in recreational areas, and so on – it is generally true that people and human activities attract other people. People are attracted to other people. They gather with and move about with others and seek to place themselves near others. New activities begin in the vicinity of events that are already in progress.

In the home we can see that children prefer to be where there are adults or where there are other children, instead of, for example, where there are only toys. In residential areas and in city spaces, comparable behavior among adults can be observed. If given a choice between walking on a deserted or a lively street, most people in most situations will choose the lively street. If the choice is between sitting in a private backyard or in a semiprivate front yard with a view of the street, people will often choose the front of the house where there is more to see.

In Scandinavia an old proverb tells it all: "people come where people are."

A series of investigations illustrates in more detail the interest in being in contact with others. Investigations of children's play habits in residential areas show that children stay and play primarily where the most activity is occurring or in places where there is the greatest chance of something happening.

Both in areas with single-family houses and in apartment house surroundings, children tend to play more on the streets, in parking areas, and near the entrances of dwellings than in the play areas designed for that purpose but located in backyards of single-family houses or on the sunny side of multi-story buildings, where there are neither traffic nor people to look at.

Corresponding trends can be found regarding where people choose to sit in public spaces. Benches that provide a good view of surrounding activities are used more than benches with less or no view of others.

An investigation of Tivoli Garden in Copenhagen, carried out by the architect John Lyle, shows that the most used benches are along the garden's main path, where there is a good view of the particularly active areas, while the least used benches are found in the quiet areas of the park. In various places, benches are arranged back to back, so that one of the benches faces a path while the other "turns its back." In these instances it is always the benches facing the path that are used.

Comparable results have been found in investigations of seating in a number of squares in central Copenhagen. Benches with a view of the most trafficked pedestrian routes are used most, while benches oriented toward the planted areas of the squares are used less frequently.

At sidewalk cafés, as well, the life on the sidewalk in front of the café is the prime attraction. Almost without exception café chairs throughout the world are oriented toward the most active area nearby. Sidewalks are, not unexpectedly, the very reason for creating sidewalk cafés.

The opportunity to see, hear, and meet others can also be shown to be one of the most important attractions in city centers and on pedestrian streets. This is illustrated by an attraction analysis carried out on Strøget, the main pedestrian street in central Copenhagen, by a study group from the School of Architecture at the Royal Danish Academy of Fine Arts. The analysis was based on an investigation of where pedestrians stopped on the walking street and what they stopped to look at.

Fewest stops were noted in front of banks, offices, showrooms, and dull exhibits of, for example, cash registers, office furniture, porcelain, or hair curlers.

Conversely, a great number of stops were noted in front of shops and exhibits that had a direct relationship to other people and to the surrounding social environment, such as newspaper kiosks, photography exhibits, film stills outside movie theaters, clothing stores, and toy stores.

Even greater interest was shown in the various human activities that went on in the street space itself. All forms of human activity appeared to be of major interest in this connection.

Considerable interest was observed in both the ordinary, everyday events that take place on a street – children at play, newlyweds on their way from the photographers, or merely people walking by – and in the more unusual instance – the artist with his easel, the street musician with his guitar, street painters in action, and other large and small events.

It was obvious that human activities, being able to see other people in action, constituted the area's main attraction.

The street painters collected a large crowd as long as their work was in progress, but when they left the area, pedestrians walked over the paintings without hesitation. The same was true of music. Music blaring out on the street from loudspeakers in front of record shops elicited no reaction, but the moment live musicians began to play or sing, there was an instantaneous show of lively interest.

The attention paid to people and human activities was also illustrated by observations made in connection with the expansion of a department store in the area. While excavation and pouring of foundations were in progress, it was possible to see into the building site through two gates facing the pedestrian street. Throughout this period more people stopped to watch the work in progress on the building site than was the case for stops in front of all the department store's fifteen display windows together.

In this case, too, it was the workers and their work, not the building site itself, that was the object of interest. This was demonstrated further during lunch breaks and after quitting time – when no workers were on the site, practically nobody stopped to look.

A summary of observations and investigations shows that people and human activity are the greatest object of attention and interest. Even the modest form of contact of merely seeing and hearing or being near to others is apparently more rewarding and more in demand than the majority of other attractions offered in the public spaces of cities and residential areas.

Life in buildings and between buildings seems in nearly all situations to rank as more essential and more relevant than the spaces and buildings themselves.

OUTDOOR ACTIVITIES AND THE QUALITY OF OUTDOOR SPACE

Life between buildings is discussed here because the extent and character of outdoor activities are greatly influenced by physical planning. Just as it is possible through choice of materials and colors to create a certain palette in a city, it is equally possible through planning decisions to influence patterns of activities, to create better or worse conditions for outdoor events, and to create lively or lifeless cities.

The spectrum of possibilities can be described by two extremes. One extreme is the city with multistory buildings, underground parking facilities, extensive automobile traffic, and long distances between buildings and functions. This type of city can be found in a number of North American and "modernized" European cities and in many suburban areas.

In such cities one sees buildings and cars, but few people, if any, because pedestrian traffic is more or less impossible, and because conditions for outdoor stays in the public areas near buildings are very poor. Outdoor spaces are large and impersonal. With great distances in the urban plan, there is nothing much to experience outdoors, and the few activities that do take place are spread out in time and space. Under these conditions most residents prefer to remain indoors in front of the television or on their balcony or in other comparably private outdoor spaces.

Another extreme is the city with reasonably low, closely spaced buildings, accommodation for foot traffic, and good areas for outdoor stays along the streets and in direct relation to residences, public buildings, places of work, and so forth. Here it is possible to see buildings, people coming and going, and people stopping in outdoor areas near the buildings because the outdoor spaces are easy and inviting to use. This city is a living city, one in which spaces inside buildings are supplemented with usable outdoor areas, and where public spaces are allowed to function.

It has already been mentioned that the outdoor activities that are particularly dependent on the quality of the outdoor spaces are the optional, recreational activities, and by implication, a considerable part of the social activities.

It is these specifically attractive activities that disappear when conditions are poor and that thrive where conditions are favorable.

The significance of quality improvement to daily and social activities in cities can be observed where pedestrian streets or traffic-free zones have been established in existing urban areas. In a number of examples, improved physical conditions have resulted in a doubling of the number of pedestrians, a lengthening of the average time spent outdoors, and a considerably broader spectrum of outdoor activities.

In a survey recording all activities occurring in the center of Copenhagen during the spring and summer of 1986, it was found that the number of pedestrian streets and squares in the city center had tripled between 1968 and 1986. Parallel to this improvement of the physical conditions, a tripling in the number of people standing and sitting was recorded.

In cases where neighboring cities offer varying conditions for city activities, great differences can also be found.

In Italian cities with pedestrian streets and automobile-free squares, the outdoor city life is often much more pronounced than in the car-oriented neighboring cities, even though the climate is the same.

A 1978 survey of street activities in both trafficked and pedestrian streets in Sydney, Melbourne, and Adelaide, Australia, carried out by architectural students from the University of Melbourne and the Royal Melbourne Institute of Technology found a direct connection between street quality and street activity. In addition, an experimental improvement of increasing the number of seats by 100 percent on the pedestrian street in Melbourne resulted in an 88 percent increase in seated activities.

William H. Whyte, in his book *The Social Life of Small Urban Spaces*, describes the close connection between qualities of city space and city activities and documents how often quite simple physical alterations can improve the use of the city space noticeably.

Comparable results have been achieved in a number of improvement projects executed in New York and other US cities by the Project for Public Spaces.

In residential areas as well, both in Europe and the United States, traffic reduction schemes, courtyard clearing, laying out of parks, and comparable outdoor improvements have had a marked effect.

Conversely, the effect of the deterioration of quality on activities in ordinary residential streets is illustrated by a study of three neighboring streets in San Francisco carried out in 1971 by [Donald] Appleyard and [Mark] Lintell.

The study shows the dramatic effect of increased traffic in two of the streets, all of which formerly had a modest rate of traffic.

In the street where there was only little traffic (2,000 vehicles per day), a great number of outdoor activities were registered. Children played on sidewalks and in the streets. Entranceways and steps were used widely for outdoor stays, and an extensive network of neighbor contacts was noted.

In one of the other streets, where the traffic volume was greatly increased (16,000 vehicles per day), outdoor activities became practically nonexistent. Comparable, neighbor contacts in this street were poorly developed.

In the third street, with middle to high traffic intensity (8,000 vehicles per day), a surprisingly great reduction in outdoor activities and neighbor contacts was noted, emphasizing that even a relatively limited deterioration of the quality of the outdoor environment can have a disproportionately severe negative effect on the extent of outdoor activities.

In summarizing the studies, a close relationship between outdoor quality and outdoor activities can be noted.

In at least three areas, it appears possible, in part through the design of the physical environment, to influence the activity patterns in public spaces in cities and residential areas. Within certain limits – regional, climatic, societal – it is possible to influence *how many* people and events use the public spaces, *how long* the individual activities last, and *which* activity types can develop.

The fact that a marked increase of outdoor activities is often seen in connection with quality improvements emphasizes that the situation found in a specific area at a certain time frequently gives an incomplete indication of the need for public spaces and outdoor activities, which can indeed exist in the area. The establishment of a suitable physical framework for social and recreational activities has time after time revealed a suppressed human need that was ignored at the outset.

When the main street in Copenhagen was converted to a pedestrian street in 1962 as the first such scheme in Scandinavia, many critics predicted that the street would be deserted because "city activity just doesn't belong to the northern European tradition."

Today this major pedestrian street, plus a number of other pedestrian streets later added to the system, are filled to capacity with people walking, sitting, playing music, drawing, and talking together. It is evident that the initial fears were unfounded and that city life in Copenhagen had been so limited because there was previously no physical possibility for its existence.

In a number of new Danish residential areas as well, where physical possibilities for outdoor activity have been established in the form of high-quality public spaces, activity patterns that no one had believed possible in Danish residential areas have evolved.

Just as it has been noted that automobile traffic tends to develop concurrently with the building of new roads, all experience to date with regard to human activities in cities and in proximity to residences seems to indicate that were a better physical framework is created, outdoor activities tend to grow in number, duration, and scope.

Plate 33 **Central Park, New York, 1863.** Frederick Law Olmsted and his partner Calvert Vaux conceived and executed a park, which not only was a masterpiece of design excellence, but also articulated a philosophy about what urban parks were for. Central Park provided the illusion of nature in the city. It was an oasis of calm and an intended meeting place for different classes. Central Park provided areas for quiet contemplation, boating, strolling, riding, baseball, Sunday school picnics and countless other activities. (Copyright © Museum of the City of New York).)

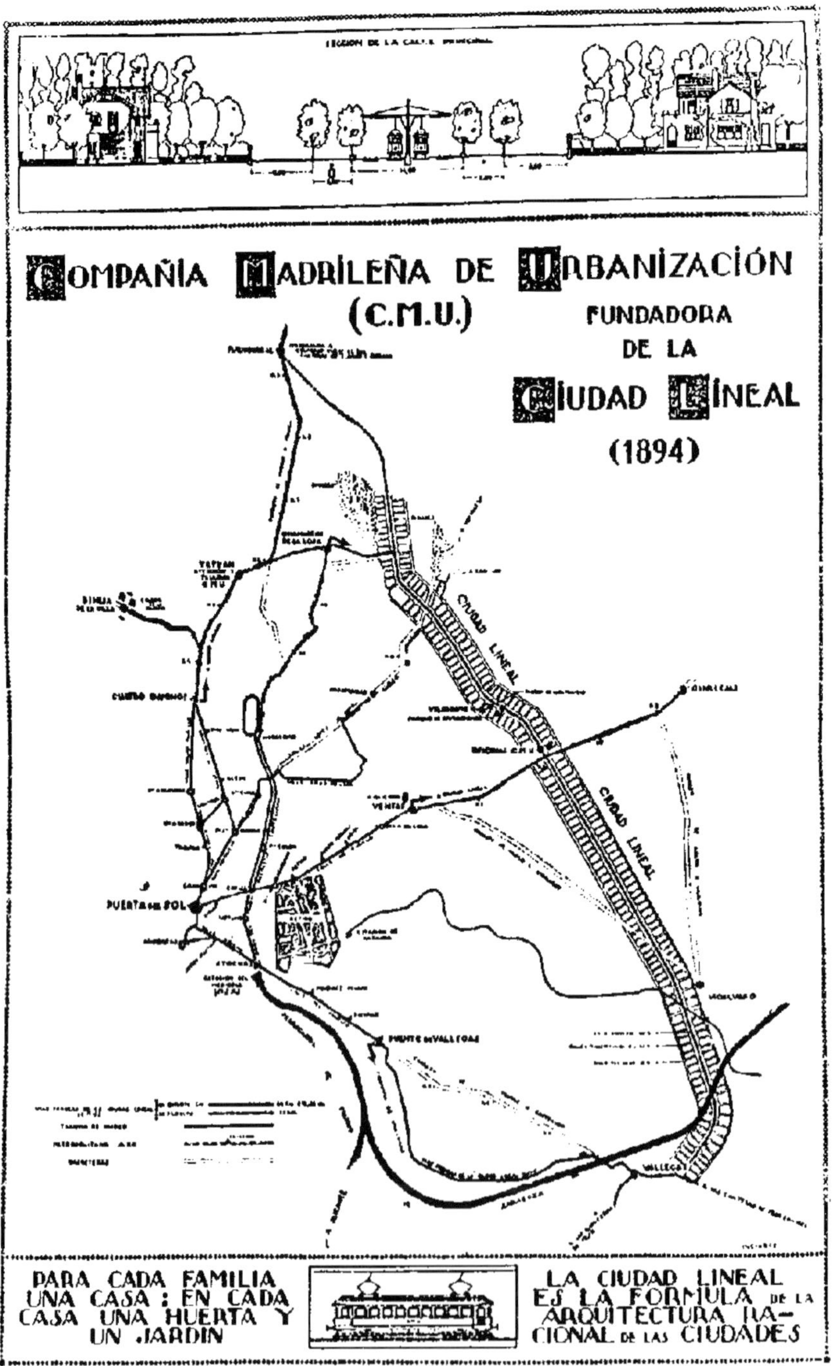

Plate 34 **Arturo Soria y Mata's plan for a linear city around Madrid, 1894.** Spanish engineer and planner Arturo Soria y Mata envisaged the liberating force of new transportation technology. He conceived of "linear cities" built along electric street car lines which would provide for access to nature, large and relatively inexpensive lots, quick transportation, and efficient provision of infrastructure. Portions of the linear city around Madrid pictured here were built.

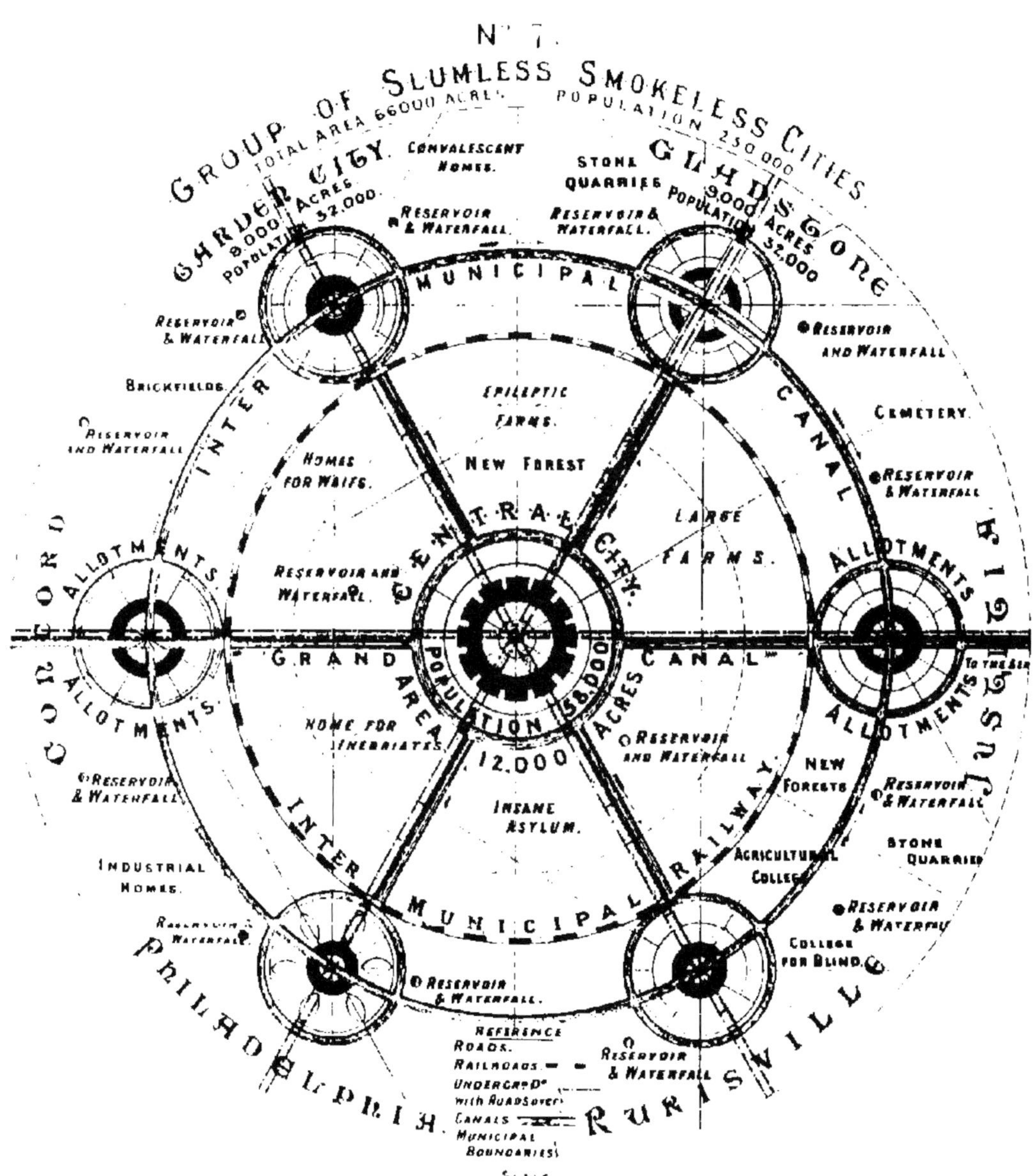

Plate 35 **Ebenezer Howard's plan for a Garden City, 1898.** Reacting to the squalor of the nineteenth-century city, Ebenezer Howard proposed self-sufficient "Garden Cities" of about 32,000 people, carefully planned and surrounded by a permanent green belt. The Garden City movement spread worldwide and continues to inspire city planners. Illustration from Ebenezer Howard, *To-morrow: A Peaceful Path to Real Reform* (London: Swan Sonnenschein, 1898)

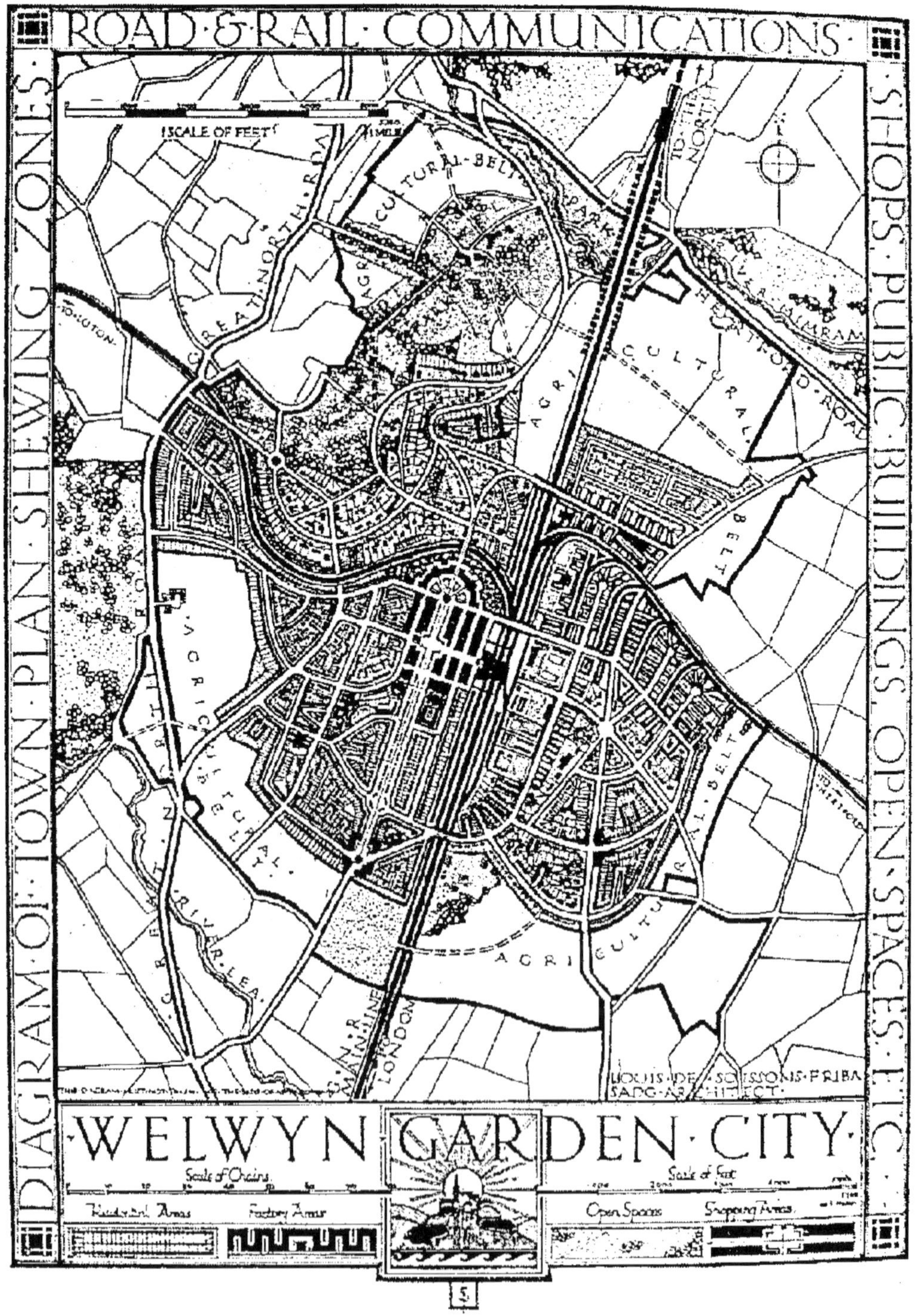

Plate 36 **Plan for Welwyn Garden City, 1909.** Welwyn, the second of Britain's Garden Cities, closely followed Ebenezer Howard's vision. Howard lived to see Welwyn built and spent the last years of his life living there.

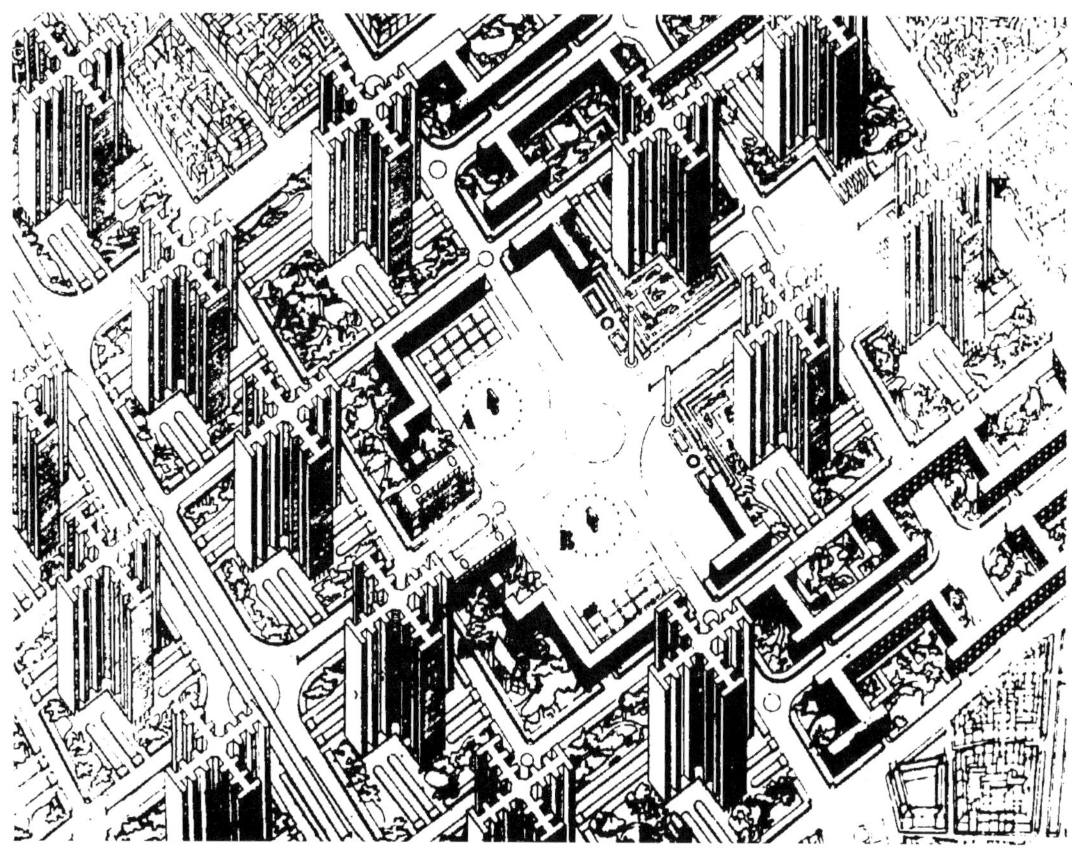

Plate 37 **Le Corbusier's "Plan Voisin" for a city of 3 million people, 1925.** Visionary modernist Le Corbusier planned huge new cities of steel and concrete dominated by highways and large modern high-rise buildings in park-like settings. This 1925 plan contemplates an entire new city of 3 million people built on these principles.

Plate 40 **Paseo del Rio, San Antonio, Texas.** Good city planning can turn the most mundane landscape into a good urban space. Through adopting a bold vision and seeing it through to completion, San Antonio, Texas, turned a blighted riverfront into a magnificent area of parkland, water-oriented activities, and retail shopping. (Copyright © Alexander Garvin)

Plate 41 **Quincy Market, Boston, Massachusetts.** Suburban malls and shopping centers may be the "new downtown" in many metropolitan areas, but some cities like Boston, Massachusetts, are rebuilding their urban core into attractive shopping and entertainment areas. Boston's Quincy Market has been transformed from a seedy and economically marginal market area into a bustling and successful commercial area. (Copyright © Alexander Garvin)

Plate 42 **Peter Calthorpe's plan for a transit-oriented development: "The Crossings," Mountain View, California.** Architects and planners are paying increased attention to developing commercial properties and housing in relation to transit nodes. Commuters in this new California development can walk to a light rail line that connects to work sites and shopping. (Image courtesy of Peter Calthorpe Associates)

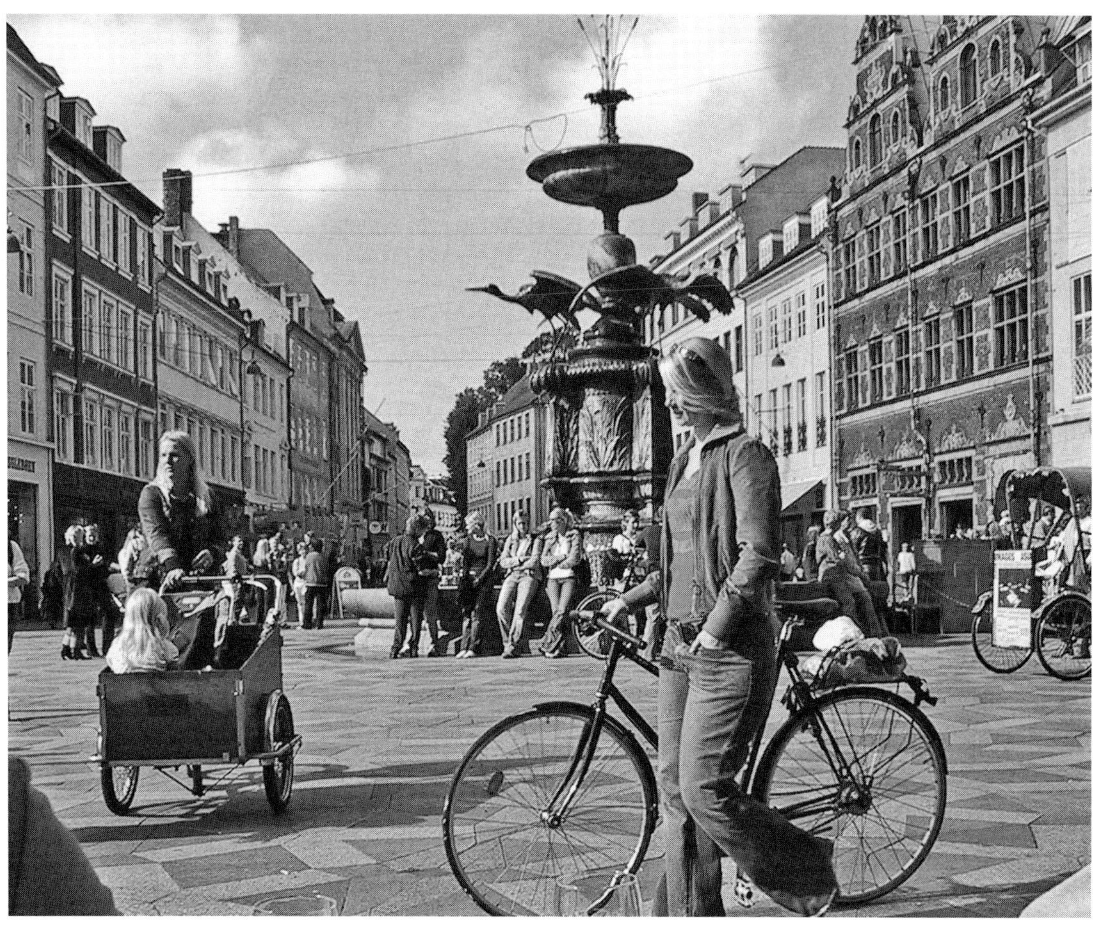

Plate 43 **Strøget. A pedestrian-only street in Copenhagen, Denmark.** Strøget exemplifies Danish architect and planner Jan Gehl's theories of how good design of the space between buildings can encourage people to interact with each other outside of their homes and other buildings.

PART EIGHT

Cities in a global society

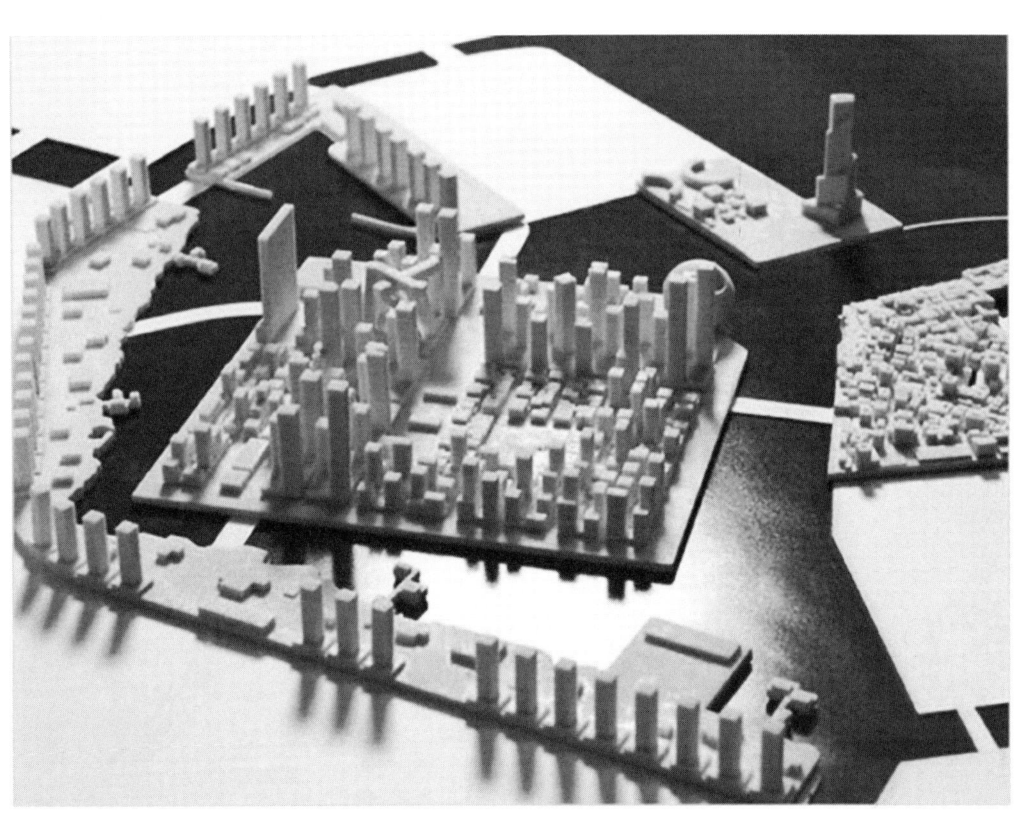

INTRODUCTION TO PART EIGHT

It is customary for books about cities to conclude – as earlier editions of this book did – with sections on the future of the urban world. And for the past several decades, speculations about the urban future have often referred to an emerging postmodern, post-industrial world order as different from the world of modernism and industrialization as that world was from the medieval or the ancient. A new "revolution" in the history of human development would be added to those posited by V. Gordon Childe – the agricultural, the urban, and the industrial (p. 31) – and a new urban future would unfold. Now, a global society *has* emerged. The urban future has arrived, the future is now!

Of course, the desire to peer into the future is a human trait as old as the biblical prophets and the oracle at Delphi, and projecting *urban* futures is at least as old as the biblical New Jerusalem or Plato's description of the ideal city-state in *The Republic*. But the pace of futurist predictions quickens at times when great cultural and historic shifts are taking place. Such was the case during the Industrial Revolution of the nineteenth century when fantasists like Jules Verne and political idealists like Edward Bellamy (author of *Looking Backward, 2000–1887*) captured the popular imagination or when the utopian visionaries discussed in Part Five of this volume helped to establish the theoretical basis of urban planning practice. And such is the case nowadays as the realization becomes every day more clear that the advanced economies of the world are entering a major new global restructuring and an information-based, post-industrial stage of development that promises to reveal new forms of urban civilization and human community. The new urban reality is the city of globalization or, more properly, an interlocking system of global urbanism. Writing about Friedrich Engels and his description of Manchester, England, in the nineteenth century (p. 46), cultural historian Steven Marcus wrote that the "historical experience of industrialization is not to be separated from that of urbanization." Today one would say that the experience of the new urban order is not to be separated from globalization.

The variety of possible globalized futures is extraordinarily diverse. Each option mirrors some of our deepest hopes and fears. Marshall McLuhan, the 1960s guru of communications theory, suggested that the whole world would one day become a "global village," with every member of humanity interacting with every other in a real-time simulacrum of the neolithic community. Some radical environmentalists who used to suggest going "back to nature" by establishing rural communes along the fringes of urbanized civilization now advocate for "urban farming," even in high-rise office buildings. Other equally radical social activists strive to establish "urban kibbutzim" and other forms of alternative community in our inner cities and suburbs. Global techno-optimists see earth's future in space colonization projects and the widespread utilization of nuclear energy, while techno-pessimists envision post-apocalyptic cities like the ones depicted in "Terminator," "Blade Runner," "The Matrix," and other popular science fiction entertainments. But three broad socio-economic forces have always propelled fundamental urban change: the demographic destinies of urbanization tied to rapid population growth; the expanding and transformative possibilities of new technologies; and the environmental inevitabilities of climate, resources, and carrying capacity. The new urban paradigm of globalism responds to all three.

In order to predict the shape of the newly emerging city, one must first have a clear sense of the probable direction of the world urbanization process described by Kingsley Davis in "The Urbanization of the

Human Population," the essay with which Part One, The Evolution of Cities, begins (p. 20). In 1900, the world population was about 2 billion. In 2010, it is nearly 7 billion. How fast and to what extent will the world's population continue to grow? Will the percentage of the world's total population living in cities – now about 50 percent – continue to increase in the decades and centuries to come? Will the "townward drift," as Frederick Law Olmsted (p. 321) called it, continue? To what eventual level: 70 percent, 80, 100 percent? Or has urbanization reached its peak, ready to stabilize at more or less the present level? Or perhaps the S curve on the chart representing world urbanization (Plate 1) will prove to be a bell-shaped curve, with the percentage of the human population living in cities going into a long, gradual decline until only a small number of the total population remains urbanized in the traditional, place-based sense of the term.

This last possibility has spawned an intriguing, if highly controversial body of literature. Some, pondering the possible effects of modern transportation and telecommunications technologies, see a gradual withering away of cities. As early as 1845, a newspaper editorialist proclaimed that Samuel F.B. Morse's new invention, the telegraph, would "annihilate distance." Others, particularly environmentalists, talk about urbanization reaching its natural limits and beginning to reverse in an age of ecological and economic constraints, or suggest that unrestrained growth is sure to result in widespread economic and environmental collapse. All of this is, of course, conjectural, and both recent experience and the most widely accepted projections from various United Nations agencies tell a different story. The Population Division of the UN Department of Economic and Social Affairs predicts that world population will reach 9 billion by the year 2050, and the United Nations Population Fund in its 2007 State of World Population report sees rural populations dwindling and urbanization continuing, especially in the developing world, for at least the next several decades to the point that "*all* future population growth will thus be in towns and cities."

One possible urban future based on the idea of declining urbanization might be called "post-urbanism." In "The Post-City Age" (1968), Melvin Webber, a distinguished professor of urban planning, argued that certain technological developments would necessarily result in an end to traditional cities and the emergence of a post-urban period of human development. When Webber wrote his prophetic essay in the 1960s, the technologies he referred to were air travel and telephones, and the post-urban society he foresaw was similar to Frank Lloyd Wright's Broadacre City of the 1930s (p. 345). Now, the advent of computers and telecommunications would seem to make those predictions even more plausible. And even if we recognize that his basic idea of a reversal of the course of urban development has so far proven to be wrong – as it surely has – Webber's insights into the way the very definition of urban may be changing and how the new technological world will result in what many call a "digital divide" that leaves some people informationally rich and others informationally poor are troubling realities that should not be ignored.

Webber was something of a utopian visionary engaged in brilliant, if somewhat mystical, speculation about humankind's future. It is perhaps more useful, when considering the future of the city, to come down to earth, to shorten the futurist perspective to the near term, and to project immediate futures based on present and observable trends. Among the most important current trends are the parallel emergence of, first, a global, post-industrial, telecommunications-based economy, second, a new, far-flung suburban ring of development that journalist Joel Garreau has dubbed "Edge City" and that Robert Fishman has analyzed as "Technoburbia" (p. 75), third, transnational environmental concerns such as sustainability, global warming and atmospheric pollution described by Timothy Beatley (p. 446) and Stephen Wheeler (p. 458), and fourth, the persistence, even intensification, of the slum conditions and social conflicts along racial, ethnic, and class lines that have been a feature of urban life for hundreds of years. These are the influences that are likely to determine the course of early-twenty-first-century urban development.

Professor Saskia Sassen, writing in the 1990s with the benefit of real information about the urban future of globalism – not just the speculations that Webber entertained – comments in "The Impact of the New Technologies and Globalization on Cities" that the "geography of globalism contains both a dynamic of dispersal and of centralization" (p. 554). She freely admits that "according to the standard conceptions of information industries, the rapid growth and disproportionate concentration of producer services in central cities should not have happened." But cities have not disappeared in the era of globalism. Rather, they have become bigger, denser, and more critically important to the emerging global society and economy because cities still offer "agglomeration economies and highly innovative environments." This last point – global cities as centers of innovation – adds emphasis to the importance of developments that Richard Florida describes in "The Creative Class" (p. 143).

Sassen notes that there is "no longer a simple straightforward relation between centrality and . . . the central business district." Yes, New York's Manhattan still operates as a global-city CBD, but the changing spatial pattern of other global cities suggests "a reconstitution of the concept of region" that can extend into "a metropolitan area in the form of a grid of business nodes." Cities in the new global society are indeed different from what has come before, but Sassen persuasively argues the case for continuity as well as discontinuity. Economies, societies, and technologies all change but, she writes, "What stands out . . . is the extent to which the city remains an integral part of these new configurations."

When Sassen notes that new forms of urban centrality "are being constituted in electronically generated spaces," she calls to mind the work of one of the most influential theorists of global urbanism, Manuel Castells, author of "Space of Flows, Space of Places: Materials for a Theory of Urbanism in the Information Age" (p. 572).

"We have entered," Castells writes, "a new age, the Information Age." This is not only a statement about new global economic relationships but also a vision of a new "network society" in which people live, work, and are defined by their simultaneous inhabitation of both a "space of flows" – the online global telecommunications grid – and a "space of place" – their local neighborhoods and communities. For Castells, this new twenty-first-century urban reality can be analyzed along three conceptual axes: function, meaning, and form. Functionally, the "dominant processes in the economy . . . are organized in global nets," but "day-to-day, private life and cultural identity are essentially local." At the human level of meaning, this suggests that individual personal identities are shared and that communal identities continually overlap. And at the spatial level of the city, "the space of flows is folded into the space of places . . . yet their logics are distinct." These insights have profound implications for urban planning, architecture, urban design, and urban governance for the cities of our global society. Planners must understand that a sense of "urbanity, street life, and civic culture" are as important as "economic competitiveness" in connecting the space of flows and the space of places on a metropolitan scale. Architects and designers must connect urban communities through the "sharing of public spaces" that can be "sites of spontaneous social interaction," and they must work to "restore symbolic meaning" to cities through projects like Dubai's Burj Khalifa or the Guggenheim Museum in Bilbao. And politicians and policy makers must seek new modes of effective governance at the national and international levels to accommodate the emergence of the new "network state."

In assessing the impacts of telecommunications and globalization on urban space, Saskia Sassen tends to focus on specific cities and metropolitan regions, but she also refers to an emerging world dominated by what she calls "a series of transnational networks of cities." This idea of urban networks – or even a single interconnected global urban network – immediately brings to mind one of the most important contributions to recent urban theory, "World-City Network: A New Metageography?" by Jonathan Beaverstock, Richard Smith, and Peter Taylor (p. 563). The question posed by the title of this seminal article suggests the breadth and depth of the concept. Can traditional notions of urban space continue to have any relevance in the era of globalization? Or must they be swept aside by a totally new "metageography" of global urbanism? These are among the questions posed by these members of the Globalization and World Cities Research Network (known as the GaWC), founded and directed by Professor Taylor at the University of Loughborough in the UK.

The new conception of urban geography that the GaWC group call for is made necessary by the insufficiency of earlier urban geographic models – for example, the Burgess model based on industrial Chicago (p. 161) or the diagrammatic representation of functional zones common to the utopian visionary planners Howard, Le Corbusier, and Wright. It builds on German geographer Walter Christaller's central place theory and ideas about urban hierarchy within systems of cities. Today, under the conditions of post-modern global urbanism, the issues are not zoning or individual city-suburb-hinterland relationships but nothing less than the global interrelationship between all the major global cities and metropolitan regions that increasingly operate independently of their nation-state locations. This is now a world not of separate and competing cities but of a global urban network of cities, an interconnected world of electronic information flows combining to create "a new functional space that will be crucial to understanding in the new millennium." Within this new urban space are what Beaverstock, Smith, and Taylor term "alpha cities" like London, New York, Paris, and Hong Kong; secondary world cities like Los Angeles, Frankfurt, and Tokyo; and still other globally linked cities such as Chicago, São Paulo, Amsterdam and Bangkok or Minneapolis, Warsaw and Shanghai. The result is an emerging hierarchy of urban power in the process of active reorganization, perhaps even a global "dystopia in the making." But as these new theorists of globalism argue, no one can "afford to ignore this new metageography, the world-city network."

The work of Sassen, Castells, and the GaWC group makes it clear that cities everywhere have been affected – to one degree or another, for good or for ill – by the dynamics of globalization. But China presents an exceptionally important case for the workings of this new phenomenon. In "Chinese Cities in a Global Society," a contribution specially commissioned for this edition of *The City Reader* (p. 590), Chinese American scholar Tingwei Zhang outlines the extraordinary transformation of Chinese cities in the new globalized society and makes the case that the changes that have taken place there are of world-historical significance.

Cities first arose in the river valleys of China as early as the second millennium BCE, and Chinese cities have been among the world's largest and most splendid for as far back as antiquity and the European Middle Ages (as the writings of Marco Polo attest). But for at least five thousand years, China has been a largely agricultural society with a strong cultural bias against cities. In part, this is because all peasant societies tend to see cities as the strongholds of oppressive rulers. In greater part, especially in the nineteenth century, Chinese saw many of their largest cities – Shanghai, Canton, Hong Kong – become colonized "concessions" ruled by outside imperialist powers. Urbanization increased significantly after the foundation of the People's Republic in 1949, but even under Mao Zedong China remained a rural-majority country. Both the Great Leap Forward policies and the Cultural Revolution period actually resulted in a slight decline of urbanization as many city dwellers were relocated to the rural countryside. But in the late 1970s, under the leadership of Deng Xiaoping, the Communist Party of China announced major structural reforms, including the historic market reforms of 1978 that ended the central political authority's long-standing hostility to capitalist enterprise and freed local entrepreneurs to innovate and foreign capitalists to invest in new enterprises. It was a dramatic new approach – one in which the citadel retained strong political powers of central control while the market was given near complete independence to produce, trade, and create new wealth. The effects were to raise millions out of poverty and to encourage the slow development of a new middle class (while enriching a small minority to extravagant levels of income). It also rapidly increased the pace of urban growth based on unprecedented rural-to-urban migration.

As Zhang reports, China in the 2000s has an urbanization rate of nearly 46 percent as millions stream from the rural countryside, mostly in the interior of the country, into the mostly coastal cities and metropolitan regions. Zhang notes that China has long been the world's largest country in terms of population (20 percent of the total world population). In 2010, China's 655 cities contain a total population of over 830 million (including recent migrants) and account for some 75 percent of "all urban population in the developed world including North America, the EU, Japan, and Australia."

As a result of China's recent explosion of urban growth, four Chinese cities – Shanghai, Beijing, Guangzhou, and Shenzhen – will likely be among the world's twenty largest cities in 2025. This growth has been facilitated by both market forces and government policies related to the creation of Economic Development Zones (EDZs) and the encouragement of Foreign Direct Investments (FDIs). Urbanization under the conditions of globalization has not been without negative consequences. More rural migrants have moved to the cities than there are jobs and housing, resulting in peripheral slum conditions for many. The disproportionally rapid growth of a small number of coastal metropolitan regions has created a pattern of "uneven development" nationwide. And the widespread use of the internet has radically changed traditional Chinese culture and sometimes threatened the central control of the political authorities. All these developments represent major planning and policy challenges for the Chinese government and for the new private sector as well. Inevitably, conflicts between the citadel and the market arise and need to be worked out to the satisfaction of both the government and the business interests – and of the larger community as well. China is and will continue to be a test case for the long-term success or failure of the emerging global society.

In China and elsewhere, much of the literature on global urbanism focuses on the revolutionary changes in economic and social life brought about by technology and a general reorganization of worldwide relationships between cities, nation-states, and transnational forces of investment and production. Visually, the icons of such changes are the gleaming new towers that define the skylines of global cities from London and Los Angeles to Shanghai and Dubai. What is often overlooked is the extreme poverty that continues to be a feature of urban life, particularly in the developing world. Not only have the skyscrapers grown bigger, so too have the slums.

In 2003, the United Nations Human Settlements Programme (UN-HABITAT) published *The Challenge of Slums* (p. 583), a report that estimated that the number of people living in urban slums was already 1 billion, nearly one-third of the total urban population and predicted that the number would likely increase to 2 billion by the year 2020. Some of these slum areas are in the inner cities, but more are arrayed in vast peripheral areas surrounding the rapidly enlarging global centers. Here, living conditions are almost unimaginable, with open sewers and overcrowding reminiscent of the reality described by Freidrich Engels in the mid-nineteenth century (p. 46) but on a vastly larger scale. Clearly, one of the major challenges facing the new cities of globalism will be the need to develop innovative planning policies that will progressively alleviate the horrors of slums conditions and provide a full range of health, educational, and employment opportunities for slum dwellers.

World cities – with their vast wealth and equally vast slums – are not an entirely new phenomenon. Patrick Geddes wrote about a "world league of cities" as early as 1924, and cities like imperial Rome, Ottoman Istanbul, the capitals of European imperialist powers, and many others have projected power globally in times past. But the extent and depth of globalism's urban influence nowadays is truly a new reality, and it has given rise to an entire body of scholarly literature that seeks to describe, analyze, and theorize about the ways cities have developed and will continue to develop in the age of globalization. That body of literature is vast, and few scholars are prepared to summarize its entirety. But two who can are Neil Brenner and Roger Keil, the co-editors of *The Global Cities Reader* (2006) in the Routledge Urban Reader Series.

Neil Brenner is a professor of sociology and metropolitan studies at New York University, and Roger Keil is a professor of environmental studies at York University in Toronto where he also serves as the director of the Canadian Centre for German and European Studies. In "From Global Cities to Globalized Urbanization," a contribution specially commissioned for this edition of *The City Reader* (p. 599), Brenner and Keil review the history of global cities and global city networks as they have developed over the past several decades and provide insightful commentaries on the academic literature and schools of thought that they have spawned.

Brenner and Keil carefully evaluate the ground-breaking work of scholars like Saskia Sassen, Manuel Castells, and Peter Taylor and the GaWC group, all represented in selections in this part of *The City Reader*. They also take note of the work of many others: Sir Peter Hall's *The World Cities* of 1966, the historical perspective of Janet Abu-Lughod, the work on inequality in global cities by Susan Fainstein, Stephen Graham's insights on the role of telecommunications, the postmodern analyses of Edward Soja and Mike Davis – and many more. Brenner and Keil give special attention to two seminal scholars who have deeply influenced their own work: Henri LeFebvre, whose prophetic *The Urban Revolution* of 1970 predicted what he called the "generalization" of world capitalism, and John Friedmann, whose "world city hypothesis" of 1986 crystallized much of the thinking in the field and whose deeply humanistic program was based not on brilliant new policies or massive building megaprojects but rather on "people. Their habitat and quality of life, the claims of invisible migrant citizens and now, in yet another turn, the concept of civil society."

In the introduction to Part One of this book, we noted that urban history has progressed from its Mesopotamian origins to its globalized present with much fundamental continuity punctuated by several heightened moments of change and discontinuity such as the paradigm-shift "revolutions" that V. Gordon

Childe identified (p. 31). The emergence of global cities, global city networks, and the fundamental restructuring of political and economic reality that characterize the contemporary age of globalization certainly constitute one of those historic moments of radical discontinuity. But although a revolution in urban history has taken place as a result of globalization, some long-wave continuities remain: global internet connections have not entirely replaced local attachments, ethnic and class identities have shifted but not disappeared. And perhaps most important of all, Brenner and Keil recognize that social inequality, poverty, and degrading slum conditions remain as a reality of urban life even – perhaps particularly – in the new global society. That is why they end their review of globalized urbanization and the literature it has produced with a call for "research – and action," and we, the editors of *The City Reader*, join them in that call. Brenner and Keil see the new global society as "profoundly authoritarian" and in need of "radical or progressive social change." Whether one agrees with that perspective or with a more moderate, perhaps even optimistic view that global cities need only a reinvigoration of the ongoing process of incremental reform to achieve their future promise, the challenge that lies at the heart of urban studies remains both a call to research *and* a call to action: not only to strive always for a better understanding of urban processes, but also to seek a higher, better form of urban community that will enlarge the lives of the individuals it supports.

"The Post-City Age"

Daedalus (1968)

Melvin M. Webber

Editors' Introduction

In the introduction to Kingsley Davis's "The Urbanization of the Human Population" (p. 20), we asked what the future of urbanization might be. Would the urban population of the earth stabilize at 50 percent or 70 percent? Might it increase to 100 percent? Or might the S curve of historic urbanization (see Plate 1) become a standard bell curve in the future, charting a steady decline in the proportion of the total human population that lives in cities? As long ago as 1968, Melvin Webber thought through the possibility of a gradual decline in urbanization in his brilliant essay "The Post-City Age." Prophetic? Maybe, maybe not. Major urban centers have not decreased in density and population since the early 1980s. To the contrary, major global cities – as well as newly emerging mega-cities in Asia, Africa, and Latin America – have grown tremendously in the decades since Webber made his bold and counter-intuitive prediction. Admitting as much in a later essay entitled "Tenacious Cities," Webber wrote that "despite growing ease of interaction over distance and the eroding requirements for propinquity . . . metropolitan areas have not disappeared. Indeed, they continue to grow." Still, Webber was prescient in pointing to two major technological developments – air transportation and the telephone – that even in 1968 were eliminating the traditional space-time constraints on human interaction and bringing about stunning cultural changes in the nature of an increasingly global human civilization.

Webber argues that the widespread availability of commercial air transportation and global telephonic communication permits an educated and affluent class of people to live anywhere – in suburbs, in rural districts, on mountain-tops, for that matter – and still be thoroughly "urban," participating fully in intellectual, professional, and economic life. At the same time, it is precisely the poor populations that are trapped in inner cities and deprived of access to technology that are becoming increasingly "rural" in the sense that they are non-participants in global community affairs.

Are these insights still applicable, perhaps even more applicable, now that networks of instantaneous global telecommunications exist to serve the needs of technologically advanced individuals and populations? Is the society that Webber foresaw in 1968 something like the contemporary "technoburbs" (p. 75) that Robert Fishman describes? Might not computers and modern telecommunications make the Broadacre City that Frank Lloyd Wright prophesied (p. 345) even more possible now and in the future? And might Webber's vision of a "post-city age" prefigure the kind of "informational city" that Manuel Castells describes in "Space of Flows, Space of Places" (p. 572)?

Melvin Webber (1920–2006) was a professor of city and regional planning at the University of California, Berkeley, where he was the director of the Institute of Urban and Regional Development and the University of California Transportation Center. He is the author of *Explorations into Urban Structure* (Philadelphia, PA: University of Pennsylvania Press, 1964) and wrote extensively on issues of urban transit and cyberspace citizenship. He was a winner of the Distinguished Planning Education Award of the American Institute of Planners.

THE POST-CITY AGE

We are passing through a revolution that is unhitching the social processes of urbanization from the locationally fixed city and region. Reflecting the current explosion in science and technology, employment is shifting from the production of goods to services; increasing ease of transportation and communication is dissolving the spatial barriers to social intercourse; and Americans are forming social communities comprised of spatially dispersed members. A new kind of large scale urban society is emerging that is increasingly independent of the city. In turn, the problems of the city place generated by early industrialization are being supplanted by a new array different in kind. With but a few remaining exceptions (the new air pollution is a notable one), the recent difficulties are not place type problems at all. Rather, they are the transitional problems of a rapidly developing society-economy-and-polity whose turf is the nation. Paradoxically, just at the time in history when policy-makers and the world press are discovering the city, "the age of the city seems to be at an end."

Our failure to draw the rather simple conceptual distinction between the spatially defined city or metropolitan area and the social systems that are localized there clouds current discussions about the "crisis of our cities." The confusion stems largely from the deficiencies of our language and from the anachronistic thoughtways we have carried over from the passing era. We still have no adequate descriptive terms for the emerging social order, and so we use, perforce, old labels that are no longer fitting. Because we have named them so, we suppose that the problems manifested inside cities are, therefore and somehow, "city problems." Because societies in the past had been spatially and locally structured, and because urban societies used to be exclusively city-based, we seem still to assume that territorial is a necessary attribute of social systems.

The error has been a serious one, leading us to seek local solutions to problems whose causes are not of local origin and hence are not susceptible to municipal treatment. We have been tempted to apply city-building instruments to correct social disorders, and we have then been surprised to find that they do not work. (Our experience with therapeutic public housing, which was supposed to cure "social pathologies," and urban renewal, which was supposed to improve the lives of the poor, may be our most spectacular failures.) We have lavished large investments on public facilities, but neglected the quality and the distribution of the social services. And we have defended and reinforced home-rule prerogatives of local and state governments with elaborate rhetoric and protective legislation.

Neither crime-in-the-streets, poverty, unemployment, broken families, race riots, drug addiction, mental illness, juvenile delinquency, nor any of the commonly noted "social pathologies" marking the contemporary city can find its causes or its cure there. We cannot hope to invent local treatments for conditions whose origins are not local in character, nor can we expect territorially defined governments to deal effectively with problems whose causes are unrelated to territory or geography. The concepts and methods of civil engineering and city planning suited to the design of unitary physical facilities cannot be used to serve the design of social change in a pluralistic and mobile society. In the novel society now emerging – with its sophisticated and rapidly advancing science and technology, its complex social organization, and its internally integrated societal processes – the influence and significance of geographic distance and geographic place are declining rapidly.

This is, of course, a most remarkable change. Throughout virtually all of human history, social organization coincided with spatial organization. In preindustrial society, men interacted almost exclusively with geographic neighbors. Social communities, economies, and polities were structured about the place in which interaction was least constrained by the frictions of space. With the coming of large-scale industrialization during the latter half of the nineteenth century, the strictures of space were rapidly eroded, abetted by the new ease of travel and communication that the industrialization itself brought.

The initial counterparts of industrialization in the United States were, first, the concentration of the nation's population into large settlements and, then, the cultural urbanization of the population. Although these changes were causally linked, they had opposite spatial effects. After coming together at a common place, people entered larger societies tied to no specific place. Farming and village peoples from throughout the continent and the world migrated to the expanding cities, where they learned urban ways, acquired the occupational skills that industrialization demanded, and became integrated into the contemporary society.

In recent years, rising societal scale and improvements in transportation and communications systems have loosed a chain of effects robbing the city of its once unique function as an urbanizing instrument of society. Farmers and small-town residents, scattered throughout the continent, were once effectively removed from the cultural life of the nation. City folks visiting the rural areas used to be treated as strangers, whose styles of living and thinking were unfamiliar. News of the rest of the world was hard to get and then had little meaning for those who lived the local life. Country folk surely knew there was another world out there somewhere, but little understood it and were affected by it only indirectly. The powerful anti-urban traditions in early American thought and politics made the immigrant city dweller a suspicious character whose crude ways marked him as un-Christian (which he sometimes was) and certainly un-American. The more sophisticated urban upper classes – merchants, landowners, and professional men – were similarly suspect and hence rejected. In contrast, the small-town merchant and the farmer who lived closer to nature were the genuine Americans of pure heart who lived the simple, natural life. Because the contrasts between the rural and the urban ways-of-life were indeed sharp, antagonisms were real, and the differences became institutionalized in the conduct of politics. America was marked by a diversity of regional and class cultures whose followers interacted infrequently, if ever.

By now this is nearly gone. The vaudeville hick-town and hayseed characters have left the scene with the vaudeville act. Today's urbane farmer watches television documentaries, reads the national news magazines, and manages his acres from an office (maybe located in a downtown office building), as his hired hands ride their tractors while listening to the current world news broadcast from a transistor. Farming has long since ceased to be a handicraft art; it is among the most highly technologized industries and is tightly integrated into the international industrial complex.

During the latter half of the nineteenth century and the first third of the twentieth, the traditional territorial conception that distinguished urbanites and ruralites was probably valid: The typical rural folk lived outside the cities, and the typical urbanites lived inside. By now this pattern is nearly *reversed*. Urbanites no longer reside exclusively in metropolitan settlements, nor do ruralites live exclusively in the hinterlands. Increasingly, those who are least integrated into modern society – those who exhibit most of the attributes of rural folk – are concentrating within the highest-density portions of the large metropolitan centers. This profoundly important development is only now coming to our consciousness, yet it points up one of the major policy issues of the next decades.

Cultural diffusion is integrating immigrants, city residents, and hinterland peoples into a national urban society, but it has not touched all Americans evenly. At one extreme are the intellectual and business elites, whose habitat is the planet; at the other are the lower-class residents of city and farm who live in spatially and cognitively constrained worlds. Most of the rest of us, who comprise the large middle class, lie somewhere in-between, but in some facets of our lives we all seem to be moving from our ancestral localism toward the unbounded realms of the cosmopolites.

High educational attainments and highly specialized occupations mark the new cosmopolites. As frequent patrons of the airlines and the long-distance telephone lines, they are intimately involved in the communications networks that tie them to their spatially dispersed associates. They contribute to and consume the specialized journals of science, government, and industry, thus maintaining contact with information resources of relevance to their activities, whatever the geographic sources or their own locations. Even though some may be employed by corporations primarily engaged in manufacturing physical products, these men trade in information and ideas. They are the producers of the information and ideas that fuel the engines of societal development. For those who are tuned into the international communications circuits, cities have utility precisely because they are rich in information. The way such men use the city reveals its essential character most clearly, for to them the city is essentially a massive communications switchboard through which human interaction takes place.

Indeed, cities exist *only* because spatial agglomeration permits reduced costs of interaction. Men originally elected to locate in high-density settlements precisely because space was so costly to overcome. It is still cheaper to interact with persons who are nearby, and so men continue to locate in such settlements. Because there *are* concentrations of associates in city places, the new cosmopolites establish their offices there and then move about from city to city conducting their affairs. The biggest settlements attract the most long-distance telephone and airline traffic and have undergone the most dramatic growth during the era of city-building.

The recent expansion of Washington, DC is the most spectacular evidence of the changing character of metropolitan development. Unlike the older settlements whose growth was generated by expanding manufacturing activities during the nineteenth and early-twentieth centuries, Washington produces almost no goods whatsoever. Its primary products are information and intelligence, and its fantastic growth is a direct measure of the predominant roles that information and the national government have come to play in contemporary society.

This terribly important change has been subtly evolving for a long time, so gradually that it seems to have gone unnoticed. The preindustrial towns that served their adjacent farming hinterlands were essentially alike. Each supplied a standardized array of goods and services to its neighboring market area. The industrial cities that grew after the Civil War and during the early decades of this century were oriented to serving larger markets with the manufacturing products they were created to produce. As their market areas widened, as product specialization increased, and as the information content of goods expanded, establishments located in individual cities became integrated into the spatially extensive economies. By now, the large metropolitan centers that used to be primarily goods-producing loci have become interchange junctions within the international communications networks. Only in the limited geographical, physical sense is any modern metropolis a discrete, unitary, identifiable phenomenon. At most, it is a localized node within the integrating international networks, finding its significant identity as contributor to the workings of that larger system. As a result, the new cosmopolites belong to none of the world's metropolitan areas, although they use them. They belong, rather, to the national and international communities that merely maintain information exchanges at these metropolitan junctions.

Their capacity to interact intimately with others who are spatially removed depends, of course, upon a level of wealth adequate to cover the dollar costs of long-distance intercourse, as well as upon the cognitive capacities associated with highly skilled professional occupations. The intellectual and business elites are able to maintain continuing and close contact with their associates throughout the world because they are rich not only in information, but also in dollar income.

As the costs of long-distance interaction fall in proportion to the rise in incomes, more and more people are able and willing to pay the transportation and communication bills. As expense-account privileges are expanded, those costs are being reduced to zero for ever larger numbers of people. As levels of education and skill rise, more and more people are being tied into the spatially extensive communities that used to engage only a few.

Thus, the glue that once held the spatial settlement together is now dissolving, and the settlement is dispersing over ever widening terrains. At the same time, the pattern of settlement upon the continent is also shifting (moving toward long strips along the coasts, the Gulf, and the Great Lakes). These trends are likely to be accelerated dramatically by cost-reducing improvements in transportation and communications technologies now in the research-and-development stages. (The SST, COMSAT communications, high-speed ground transportation with speeds up to 500 mph, TV and computer-aided educational systems, no-toll long-distance telephone service, and real-time access to national computer-based information systems are likely to be powerful ones.) Technological improvements in transport and communications reduce the frictions of space and thereby ease long-distance intercourse. Our compact, physical city layouts directly mirror the more primitive technologies in use at the time these were built. In a similar way, the locational pattern of cities upon the continent reflects the technologies available at the time the settlements grew. If currently anticipated technological improvements prove workable, each of the metropolitan settlements will spread out in low-density patterns over far more extensive areas than even the most frightened future-mongers have yet predicted. The new settlement-form will little resemble the nineteenth-century city so firmly fixed in our images and ideologies. We can also expect the large junction points will no longer have the communications advantage they now enjoy, and smaller settlements will undergo a major spurt of growth in all sorts of now isolated places where the natural amenities are attractive.

Moreover, as ever larger percentages of the nation's youth go to college and thus enter the national and international cultures, attachments to places of residence will decline dramatically. This prospect, rather than the spatial dispersion of metropolitan areas, portends the functional demise of the city. The signs are already patently clear among those groups whose worlds are widest and least bounded by parochial constraints.

Consider the extreme cosmopolite, if only for purposes of illustrative cartooning. He might be engaged in scientific research, news reporting, or international business, professions exhibiting critical common traits. The astronomer, for example, maintains instantaneous contact with his colleagues around the world; indeed, he is a day-to-day collaborator with astronomers in all countries. His work demands that he share information and that he and his colleagues monitor stellar events jointly, as the earth's rotation brings men at different locales into prime viewing position. Because he is personally committed to their common enterprise, his social reference group is the society of astronomers. He assigns his loyalties to the community of astronomers, since their work and welfare matter most to him.

To be sure, as he plays out other roles – say, as citizen, parent, laboratory director, or grocery shopper – he is a member of many other communities, both interest-based and place-defined ones. But the striking thing about our astronomer, and the millions of people like him engaged in other professions, is how little of his attention and energy he devotes to the concerns of place-defined communities. Surely, as compared to his grandfather, whose life was largely bound up in the affairs of his locality, the astronomer, playwright, newsman, steel broker, or wheat dealer lives in a life-space that is not defined by territory and deals with problems that are not local in nature. For him, the city is but a convenient setting for the conduct of his professional work; it is not the basis for the social communities that he cares most about.

"The Impact of the New Technologies and Globalization on Cities"

from Arie Graafland and Deborah Hauptmann (eds), *Cities in Transition* (2001)

Saskia Sassen

Editors' Introduction

Sociologist Saskia Sassen taught for many years at the University of Chicago and is now the Robert S. Lynd Professor of Sociology and a member of the Committee on Global Thought at Columbia University. Sassen has an extraordinarily international background. She was born in The Hague in 1949, and grew up in Argentina and Italy, learning to speak five languages. Before earning her PhD at Notre Dame University in Indiana, she studied and taught at the University of Poitiers, the University of Buenos Aires, and the Università degli Studidi Roma. While teaching at Chicago and Columbia, she has also frequently been a visiting professor at the London School of Economics. It is perhaps not surprising that her work has analyzed data on information technology, the economies, and the organization of physical space in the most advanced cities and metropolitan regions in the world and that she originally coined the term "global city."

In this selection Sassen describes how globalization and information technology are changing relationships among cities and reconfiguring the physical arrangement of activities within metropolitan space. The global cities that Sassen is most interested in – as in her ground-breaking book *The Global City: New York, London, Tokyo* (Princeton, NJ: Princeton University Press, 1991) – are places where international financial functions are concentrated and whose economies are most closely integrated with the world economy. She argues that globalization is both concentrating and simultaneously dispersing activity at the global, national, and metropolitan levels. At the global scale, economic power is increasingly *concentrated* in cities like New York, London, and Tokyo. But cities as dispersed as Mexico City, Taipei, Bangkok, Buenos Aires, São Paulo, Frankfurt, Zurich, and Sydney may also be characterized as global cities – not merely megacities – in that they serve as focal points and operation centers of the global economy.

As economic activity is increasingly globalized and as industries need more specialized services, Sassen believes a new world "system of cities" is emerging unlike anything that has previously existed. What Sassen terms "corporate service complexes" – sophisticated networks of high-level financiers, lawyers, accountants, advertising professionals, and other skilled professionals serving international corporations – are clustered in global cities. Decisions made by joint headquarters and corporate services complexes in London, New York, Tokyo, and other global cities affect not only the residents of these cities, but also jobs, wages, and the economic health of cities as dispersed around the globe as Kuala Lumpur, Malaysia, Saigon (now Ho Chi Minh City), Vietnam, and Santiago, Chile. If financial analysts advise their corporate clients that Argentina's economy is weakening and lawyers inform them that legal reforms in China present new opportunities for super profits, the corporations may pull billions of dollars out of Argentina and redirect the funds to Shanghai – assisted by a small army of advertising executives touting their products in China's vast emerging markets.

A traditional reason why businesses cluster in large cities has been to be in touch with other businesses and with lawyers, accountants, bankers, advertising firms and other specialized service providers that help them do business. It is efficient to walk next door to a lawyer's office and down the block to do business with a major business partner. Cities had what economists call "agglomeration economies." Now in the age of digital telecommunications, as Manuel Castells (p. 572) makes clear, information can travel almost instantly to anywhere in the world. Business professionals no longer need to walk next door to communicate with their lawyers or down the street to meet with a business partner: they can just phone, fax, email, or video conference next door or to a remote location anywhere in the world, so long as the telecommunications infrastructure permits. Highly developed telecommunications infrastructure in global cities facilitates transmission of information in staggering quantities at lightning speed. International banks in Rio de Janeiro can bounce a year's worth of financial records to a New York bank via satellite in seconds.

Anticipating these trends, in 1968 Melvin Webber (p. 549) prophesied that information technology would make space increasingly irrelevant and, as a consequence, cities would diminish in importance. But Sassen makes clear that has not happened. Surveying the data, she notes that global cities such as New York, London, and Tokyo have actually grown in population, wealth, and power since the information revolution. And their status and importance continues to grow, not decline. On the other hand, many cities that historically once served as secondary corporate command and control centers are in economic decline as corporate power continues to concentrate in the most advanced global cities. Paris is growing in economic power and wealth; Marseilles is declining.

Sassen points out that production and retailing are becoming more dispersed. Many corporations design products at a headquarters location – perhaps (but not necessarily) in a global city like London. They contract with firms in a developing country like Malaysia to produce the products they have designed, and then they market the finished products worldwide. This kind of production process requires sophisticated support to manage dispersed and rapidly changing operations all over the world – lawyers familiar with British law, accountants who understand Malaysian accounting practices, and advertising executives sensitive to the cultural preferences of German consumers. Corporations rarely have the internal capacity to do all that. Instead they turn to networks of specialized firms located in global cities to provide the services they need.

Sassen questions the whole notion of "rich" countries and "rich" cities; places central to the world economy, and those that exist at the margin. She argues that economic inequality is sharply increasing – particularly in global cities at the center of the world economy. The opportunity for hyper-profits in international finance is creating extraordinary wealth for Wall Street bankers. But many of the low-paid janitors and file clerks who work on Wall Street were born in less developed countries and live in ethnic New York neighborhoods just a short distance away. In São Paulo, wealthy Brazilian nationals and expatriates earn salaries and participate in lifestyles more like those of wealthy New Yorkers than those of the people in São Paulo neighborhoods a few blocks from where they work. An important public policy question facing countries all over the world is how to promote economic equality and help more of their citizens benefit from the new world economic order.

Urban regions are also changing as a result of globalization and information technology. Sassen argues that economic activity within the metropolitan areas of global cities manifests a dynamic of both concentration and dispersion just as is occurring in the world system of cities. At the time that Ernest W. Burgess developed his concentric zone theory (p. 161), Chicago and many other cities had a distinct central business district (CBD) where intense economic activity was concentrated. In advanced metropolitan areas today there is often no longer a single, clearly demarcated CBD. Sassen believes four different models are emerging. In some cities something close to a classic CBD still exists. New York City's Wall Street area is an example. In others there is a new pseudo-CBD just outside the historic city center, such as the massive planned office complex named La Défense just outside the center of Paris. In other regions Sassen sees nodes of intense business activity emerging along "cyber routes" or "digital highways." Sassen points out that these spaces along which information flows often follow historic infrastructure for highways, high-speed rail lines, and airports. Twentieth-century infrastructure appears to be shaping the spatial organization of twenty-first-century regions. Sassen also discerns agglomeration and centralization across physical space and within cyber-space – transterritorial centers of intense economic activity and centrality in electronically generated space.

In addition to *The Global City*, Sassen's books include *Territory, Authority, Rights: From Medieval to Global Assemblages* (Princeton, NJ: Princeton University Press, 2006), *Denationalization: Economy and Polity in a Global Digital Age* (Princeton, NJ: Princeton University Press, 2003), *Global Networks/Linked Cities* (New York: Routledge, 2002), *Guests and Aliens* (New York: New Press, 1999), *Globalization and its Discontents*, with Anthony Appiah (New York: New Press, 1998), *Losing Control? Sovereignty in an Age of Globalization* (New York: Columbia University Press, 1996), *Cities in a World Economy*, 3rd edn (Thousand Oaks, CA: Pine Forge Press, 2006) and *The Mobility of Labor and Capital: A Study in International Investment and Labor Flows* (New York: Cambridge University Press, 1988).

For an excellent set of readings on the impact of globalization on cities, see Neil Brenner and Roger Keil (eds) *The Global Cities Reader* (London: Routledge, 2005). Jan Lin and Christopher Lee (eds), *The Urban Sociology Reader* (London: Routledge, 2005) contains important materials on urbanization and global change. Part III of Nicholas Fyfe and Judith Kenny (eds), *The Urban Geography Reader* (London: Routledge, 2005) discusses the impact of global economic and cultural restructuring on cities. J. John Palen, *The Urban World*, 8th edn (Boulder, CO: Paradigm, 2008) is a useful overview, and Stephen Graham (ed.) *The Cybercities Reader* (London: Routledge, 2003) contains additional writings on how information technology is impacting cities.

■

Telecommunications and globalization have emerged as major forces shaping the organization of urban space. This reorganization ranges from the spatial virtualization of a growing number of social and economic activities to the reconfiguration of the geography of the built environment for these activities. Whether in electronic space or in the geography of the built environment, this reorganization involves a repositioning of the urban and of urban centrality in particular.

The growth of global markets for finance and specialized services, the need for transnational servicing networks due to sharp increases in international investment, the reduced role of the government in the regulation of international economic activity and the corresponding ascendance of other institutional arenas, notably global markets and corporate headquarters—all these point to the existence of a series of transnational networks of cities. We can see here the formation, at least incipient, of transnational urban systems. To a large extent it seems to me that the major business centers in the world today draw their importance from these transnational networks. There is no such thing as a single global city—and in this sense there is a sharp contrast with the erstwhile capitals of empires.

[. . .]

WORLDWIDE NETWORKS AND CENTRAL COMMAND FUNCTIONS

The geography of globalization contains both a dynamic of dispersal and of centralization, the latter a condition that began receiving recognition only recently. The massive trends towards the spatial dispersal of economic activities at the metropolitan, national and global level which we associate with globalization have contributed to a demand for new forms of territorial centralization of top-level management and control operations. The rapid growth of affiliates illustrates this dynamic. By 1998 firms had about half a million affiliates outside their home countries. The sheer number of dispersed factories and service outlets that are part of a firm's integrated operation creates massive new needs for central coordination and servicing. Thus the spatial dispersal of economic activity made possible by telecommunications and the new legal frameworks for globalization contribute to an expansion of central functions if this dispersal is to take place under the continuing concentration in control, ownership and profit appropriation that characterizes the current economic system.

Another instance today of this negotiation between a global cross-border dynamic and territorially specific sites is that of the global financial markets. The orders of magnitude in these transactions have risen sharply, as illustrated by the $75 U.S. trillion in turnover in the global capital market, a major component of the global economy. These transactions are partly embedded in telecommunications systems that make possible the instantaneous transmission of money/information around the globe. Much attention has been given to the capacity for instantaneous transmission of the new technologies. But the other half of the story is the extent to which the global financial markets are located in particular cities in the highly developed countries; indeed, the degrees of concentration are unexpectedly high, a subject I discuss empirically in a later section.

Stock markets worldwide have become globally integrated. Besides deregulation in the 1980s in all the major European and North American markets, the late 1980s and early 1990s saw the addition of such markets as Buenos Aires, São Paulo, Mexico City, Bangkok, Taipei, etc. The integration of a growing number of stock markets has contributed to raise the capital that can be mobilized through stock markets. Worldwide market value reached over 20 trillion dollars in 1998. . . .

The specific forms assumed by globalization over the last decade [i.e. since 1990] have created particular organizational requirements. The emergence of global markets for finance and specialized services, the growth of investment as a major type of international transaction, all have contributed to the expansion in command functions and in the demand for specialized services for firms.

By central functions I do not only mean top level headquarters; I am referring to all the top level financial, legal, accounting, managerial, executive, planning functions necessary to run a corporate organization operating in more than one country, and increasingly in several countries. These central functions are partly embedded in headquarters, but also in good part in what has been called the corporate services complex, that is, the network of financial, legal, accounting, advertising firms that handle the complexities of operating in more than one national legal system, national accounting system, advertising culture, etc. and do so under conditions of rapid innovations in all these fields. Such services have become so specialized and complex, that headquarters increasingly buy them from specialized firms rather than producing them in-house. These agglomerations of firms producing central functions for the management and coordination of global economic systems, are disproportionately concentrated in the highly developed countries—particularly, though not exclusively, in the kinds of cities I call global cities. . . . Such concentrations of functions represent a strategic factor in the organization of the global economy, and they are situated right here, in New York, in Paris, in Amsterdam.

[. . .]

NEW FORMS OF CENTRALITY

Today there is no longer a simple straightforward relation between centrality and such geographic entities as the downtown, or the central business district. In the past, and up to quite recently in fact, the center was synonymous with the downtown or the CBD. Today, the spatial correlate of the center can assume several geographic forms. It can be the CBD, as it still is largely in New York City, or it can extend into a metropolitan area in the form of a grid of nodes of intense business activity, as we see in Frankfurt and Zurich. The center has been profoundly altered by telecommunications and the growth of a global economy, both inextricably linked; they have contributed to a new geography of centrality (and marginality). Simplifying, one could identify four forms assumed by centrality today. First, while there is no longer a simple straightforward relation between centrality and such geographic entities as the downtown, or the central business district as was the case in the past, the CBD remains a key form of centrality. But the CBD in major international business centers is one profoundly reconfigured by technological and economic change.

We may be seeing a difference in the pattern of global city formation in parts of the United States and in parts of Western Europe. In the United States, major cities such as New York and Chicago have large centers that have been rebuilt many times, given the brutal neglect suffered by much urban infrastructure and the imposed obsolescence so characteristic of U.S. cities. This neglect and accelerated obsolescence produce vast spaces for rebuilding the center according to the requirements of whatever regime of urban accumulation or pattern of spatial organization of the urban economy prevails at a given time. In Europe, urban centers are far more protected and they rarely contain significant stretches of abandoned space; the expansion of workplaces and the need for intelligent buildings necessarily will have to take place partly outside the old centers. One of the most extreme cases is the complex of La Défense, the massive, state-of-the-art office complex developed right outside Paris to avoid harming the built environment inside the city. This is an explicit instance of government policy and planning aimed at addressing the growing demand for central office space of prime quality. Yet another variant of this expansion of the 'center' onto hitherto peripheral land can be seen in London's Docklands. Similar projects for recentralizing peripheral areas were launched in several major cities in Europe, North America, and Japan during the 1980s.

Second, the center can extend into a metropolitan area in the form of a grid of nodes of intense business activity. One might ask whether a spatial organization characterized by dense strategic nodes spread over a broader region does or does not constitute a new form of organizing the territory of the 'center', rather than, as in the more conventional view, an instance of suburbanization or geographic dispersal. Insofar as these various nodes are articulated through cyber-routes or digital highways, they represent a new geographic correlate of the most advanced type of 'center'. The places that fall outside this new grid of digital highways, however, are peripheralized. This regional grid of nodes represents, in my analysis, a reconstitution of the concept of region. Far from neutralizing geography the regional grid is likely to be embedded in conventional forms of communications infrastructure, notably rapid rail and highways connecting to airports. Ironically perhaps, conventional infrastructure is likely to maximize the economic benefits derived from telematics. I think this is an important issue that has been lost somewhat in discussions about the neutralization of geography through telematics.

Third, we are seeing the formation of a trans-territorial 'center' constituted via telematics and intense economic transactions. . . . The most powerful of these new geographies of centrality at the inter-urban level binds the major international financial and business centers: New York, London, Tokyo, Paris, Frankfurt, Zurich, Amsterdam, Los Angeles, Sydney, Hong Kong, among others. But this geography now also includes cities such as São Paulo and Mexico City. The intensity of transactions among these cities, particularly through the financial markets, trade in services, and investment has increased sharply, and so have the orders of magnitude involved. At the same time, there has been a sharpening inequality in the concentration of strategic resources and activities between each of these cities and others in the same country. For instance, Paris now concentrates a larger share of leading economic sectors and wealth in France than it did fifteen years ago, while Marseilles, once a major economic hub, has lost its share and is suffering severe decline.

[. . .]

Fourth, new forms of centrality are being constituted in electronically generated spaces. Electronic space is often read as a purely technological event and in that sense a space of innocence. But if we consider for instance that strategic components of the financial industry operate in such space we can see that these are spaces where profits are produced and power is thereby constituted. Insofar as these technologies strengthen the profit-making capability of finance and make possible the hyper-mobility of finance capital, they also contribute to the often devastating impacts of the ascendance of finance on other industries, on particular sectors of the population, and on whole economies. Cyberspace, like any other space can be inscribed in a multiplicity of ways, some benevolent or enlightening; others, not. My argument is that structures for economic power are being built in electronic space and that their highly complex configurations contain points of coordination and centralization.

[. . .]

A CONCENTRATION AND THE REDEFINITION OF THE CENTER: SOME EMPIRICAL REFERENTS

The trend towards concentration of top-level management, coordination and servicing functions is evident at the national and international scales in all highly developed countries. For instance, the Paris region accounts for over 40% of all producer services in France, and over 80% of the most advanced ones. New York City is estimated to account for between a fourth and a fifth of all U.S. producer services exports though it has only 3% of the U.S. population. London accounts for 40% of all exports of producer services in the U.K. Similar trends are also evident in Zurich, Frankfurt, and Tokyo, all located in much smaller countries.

[. . .]

In the financial district in Manhattan, the use of advanced information and telecommunication technologies has had a strong impact on the spatial organization of the district because of the added spatial requirements of 'intelligent' buildings. A ring of new office buildings meeting these requirements was built over the last decade immediately around the old Wall Street core, where the narrow streets and lots made this difficult; furthermore, renovating old buildings in the Wall Street core is extremely expensive and often not possible. The new buildings in the district were mostly corporate headquarters and financial services industry facilities. These firms tend to be extremely intensive users of telematics, and the availability of the most advanced forms typically is a major factor in their real estate and locational decisions. They need complete redundancy of telecommunications systems, high carrying capacity, often their own private branch exchange, etc. With this often goes a need for large spaces. For instance, the technical installations backing a firm's trading floor are likely to require additional space equivalent to the size of the trading floor itself.

The case of Sydney illuminates the interaction of a vast, continental economic scale and pressures towards spatial concentration. Rather than strengthening the multipolarity of the Australian urban system, the developments of the 1980s—increased internationalization of the Australian economy, sharp increases in foreign investment, a strong shift towards finance, real estate and producer services—contributed to a greater concentration of major economic activities and actors in Sydney. This included a loss of share of such activities and actors by Melbourne, long the center of commercial activity and wealth in Australia.

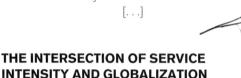

[. . .]

THE INTERSECTION OF SERVICE INTENSITY AND GLOBALIZATION

To understand the new or sharply expanded role of a particular kind of city in the world economy since the early 1980s, we need to focus on the intersection of two major processes. The first is the sharp growth in the globalization of economic activity; this has raised the scale and the complexity of transactions, thereby feeding the growth of top-level multinational headquarter functions and the growth of advanced corporate services. It is important to note that even though globalization raises the scale and complexity of these operations, they are also evident at smaller geographic scales and lower orders of complexity, as is the case with firms that operate regionally. Thus while regionally oriented firms need not negotiate the complexities of international borders and the regulations of different countries, they are still faced with a regionally dispersed network of operations that requires centralized control and servicing.

The second process we need to consider is the growing service intensity in the organization of all industries. This has contributed to a massive growth in the demand for services by firms in all industries, from mining and manufacturing to finance and consumer services. Cities are key sites for the production of services for firms. Hence the increase in service intensity in the organization of all industries has had a significant growth effect on cities in the 1980s. It is important to recognize that this growth in services for firms is evident in cities at different levels of a nation's urban system. Some of these cities cater to regional or sub-national markets; others cater to national markets and yet others cater to global markets. In this context, globalization becomes a question of scale and added complexity.

The key process from the perspective of the urban economy is the growing demand for services by firms in all industries and the fact that cities are preferred production sites for such services, whether at the global, national, or regional level. As a result we see in cities the formation of a new urban economic core of banking and service activities that comes to replace the older typically manufacturing oriented core.

In the case of cities that are major international business centers, the scale, power, and profit levels of this new core suggest that we are seeing the formation of a new urban economy. This is so in at least two regards. First, even though these cities have long been centers for business and finance, since the late 1970s there have been dramatic changes in the structure of the business and financial sectors, as well as sharp increases in the overall magnitude of these sectors and their weight in the urban economy. Second, the ascendance of the new finance and services complex, particularly international finance, engenders what may be regarded as a new economic regime, that is, although this sector may account for only a fraction of the economy of a city, it imposes itself on that larger economy. Most notably, the possibility for super-profits in finance has the effect of devalorizing manufacturing insofar as the latter cannot generate the super-profits typical in much financial activity. This is not to say that everything in the economy of these cities has changed. On the contrary, they still show a great deal of continuity and many similarities with cities that are not global nodes. Rather, the implantation of global processes and markets has meant that the internationalized sector of the economy has expanded sharply and has imposed a new valorization dynamic—that is, a new set of criteria for valuing or pricing various economic activities and outcomes. This has had devastating effects on large sectors of the urban economy. High prices and profit levels in the internationalized sector and its ancillary activities, such as top-of-the-line restaurants and hotels, have made it increasingly difficult for other sectors to compete for space and investments. Many of these other sectors have experienced considerable downgrading and/or displacement, as, for example, neighborhood shops tailored to local needs are replaced by upscale boutiques and restaurants catering to new high-income urban elites.

Though at a different order of magnitude, these trends also became evident during the late 1980s in a number of major cities in the developing world that have become integrated into various world markets: São Paulo, Buenos Aires, Bangkok, Taipei, and Mexico City are only a few examples. Also here the new urban core was fed by the deregulation of financial markets, ascendance of finance and specialized services, and integration into the world markets. The opening of stock markets to foreign investors and the privatization of what were once public sector firms have been crucial institutional arenas for this articulation. Given the vast size of some of these cities, the impact of this new core on the broader city is not always as evident as in central London or Frankfurt, but the transformation is still very real.

[. . .]

The formation of a new production complex

According to standard conceptions about information industries, the rapid growth and disproportionate concentration of producer services in central cities should not have happened. Because they are thoroughly embedded in the most advanced information technologies, producer services could be expected to have locational options that bypass the high costs and congestion typical of major cities. But cities offer agglomeration economies and highly innovative environments. The growing complexity, diversity, and specialization of the services required have contributed to the economic viability of a freestanding specialized service sector.

The production process in these services benefits from proximity to other specialized services. This is especially the case in the leading and most innovative sectors of these industries. Complexity and innovation often require multiple highly specialized inputs from several industries. The production of a financial instrument, for example, requires inputs from accounting, advertising, legal expertise, economic consulting, public relations, designers, and printers. The particular characteristics of production of these services, especially those involved in complex and innovative operations, explain their pronounced concentration in major cities. The commonly heard explanation that high-level professionals require face-to-face interactions needs to be refined in several ways. Producer services, unlike other types of services, are not necessarily dependent on spatial proximity to the consumers, i.e. firms, served. Rather, economies occur in such specialized firms when they locate close to others that produce key inputs or whose proximity makes possible joint production of certain service offerings. The accounting firm can service its clients at a distance, but the nature of its service depends on proximity to specialists, lawyers, programmers. Moreover, concentration arises out of the needs and expectations of the people likely to be employed in these new high-skill jobs, who tend to be attracted to the amenities and lifestyles that large urban centers can offer. Frequently, what is thought of as face-to-face communication is actually a production process that requires multiple simultaneous inputs and feedbacks. At the current stage of technical development, immediate and simultaneous access to the pertinent experts is still the most effective way, especially when dealing with a highly complex product. The concentration of the most advanced telecommunications and computer network facilities in major cities is a key factor in what I refer to as the production process of these industries.

[. . .]

This combination of constraints suggests that the agglomeration of producer services in major cities actually constitutes a production complex. This producer services complex is intimately connected to the world of corporate headquarters; they are often thought of as forming a joint headquarters-corporate services complex. But in my reading, we need to distinguish the two. Although it is true that headquarters still tend to be disproportionately concentrated in cities, over the last two decades [i.e. since 1980] many have moved out. Headquarters can indeed locate outside cities, but they need a producer services complex somewhere in order to buy or contract for the needed specialized services and financing. Further, headquarters of firms with very high overseas activity or in highly innovative and complex lines of business tend to locate in major cities. In brief, firms in more routinized lines of activity, with predominantly regional or national markets, appear to be increasingly free to move or install their headquarters outside cities. Firms in highly competitive and innovative lines of activity and/or with a strong world market orientation appear to benefit from being located at the center of major international business centers, no matter how high the costs.

[. . .]

The region in the global information age

The massive use of telematics in the economy and the corresponding possibility for geographic dispersal and mobility of firms suggest that the whole notion of regional specialization and of the region may become obsolete. But there are indications that, as is the case for large cities, so also for regions the hypermobility of information industries and the heightened capacity for geographic dispersal may be only part of the story. The evidence on regional specialization in the U.S. and in other highly developed countries along with new insights into the actual work involved in producing these services point to a different set of outcomes.

What is important from the perspective of the region is that the existence of, for instance, a producer services complex in the major city or cities in a region creates a vast concentration of communications infrastructure which can be of great use to other economic nodes in that region. Such nodes can (and do) connect with the major city or cities in a region and thereby to a worldwide network of firms and markets. The issue from the regional perspective is, then, that somewhere in its territory the region connects with state-of-the-art communication facilities which connect it with the world and which bring foreign firms from all over the world to the region. Given a regional grid of economic nodes, the benefits of this concentration in the major city or cities are no longer confined only to firms located in those cities.

Secondly, given the nature of the production process in advanced information industries, as described in the preceding section, the geographic dispersal of activities has limits. The importance of actual face-to-face transactions means that a metropolitan or regional network of firms will need conventional communications infrastructure, e.g. highways or rapid rail, and locations not farther than something like two hours. One of the ironies of the new information technologies is that to maximize their use we need access to conventional infrastructure. In the case of international networks it takes airports and planes; and in the case of metropolitan or regional networks, trains and cars.

The importance of conventional infrastructure in the operation of economic sectors that are heavy users of telematics has not received sufficient attention. The dominant notion seems to be that telematics obliterates the need for conventional infrastructure. But it is precisely the nature of the production process in advanced industries, whether they operate globally or nationally, which contributes to explain the immense rise in business travel we have seen in all advanced economies over the last decade [i.e. since 1990], the new electronic era. The virtual office is a far more limited option than a purely technological analysis would suggest. Certain types of economic activities can be run from a virtual office located anywhere. But for work processes requiring multiple specialized inputs, considerable innovation and risk taking, the need for direct interaction with other firms and specialists remains a key locational factor. Hence the metropolitanization and regionalization of an economic sector has boundaries that are set by the time it takes for a reasonable commute to the major city or cities in the region. The irony of today's electronic era is that the older notion of the region and older forms of infrastructure re-emerge as critical for key economic sectors. This type of region in many ways diverges from older forms of region. It corresponds rather to the second form of centrality posited above in this paper—a metropolitan grid of nodes connected via telematics. But for this digital grid to work, conventional infrastructure—ideally of the most advanced kind—is also a necessity.

[. . .]

THE INTERSECTION BETWEEN ACTUAL AND DIGITAL SPACE

There is a new topography of economic activity, sharply evident in this subeconomy. This topography weaves in and out between actual and digital space. There is today no fully virtualized firm or economic sector. Even finance, the most digitalized, dematerialized and globalized of all activities has a topography that weaves back and forth between actual and digital space. To different extents in different types of sectors and different types of firms, a firm's tasks now are distributed across these two kinds of spaces; further the actual configurations are subject to considerable transformation as tasks are computerized or standardized, markets are further globalized, etc. More generally, telematics and globalization have emerged as fundamental forces reshaping the organization of economic space.

The question I have for architects here is whether the point of intersection between these two kinds of spaces in a firm's or a dynamic's topography of activity, is one worth thinking about, theorizing, exploring. This intersection is unwittingly, perhaps, thought of as a line that divides two mutually exclusive zones. I would propose, again, to open up this line into an 'analytic borderland' which demands its own empirical specification and theorization, and contains its own possibilities for architecture. The space of the computer screen, which one might posit as one version of the intersection, will not do, or is at most a partial enactment of this intersection.

What does contextuality mean in this setting? A networked subeconomy that operates partly in actual space and partly in globe-spanning digital space cannot easily be contextualized in terms of its surroundings. Nor can the individual firms. The orientation is simultaneously towards itself and towards the global. The intensity of its internal transactions is such that it overrides all considerations of the broader locality or region within which it exists. On another, larger scale, in my research on global cities I found rather clearly that these cities develop a stronger orientation towards the global markets than to their hinterlands. Thereby they override a key proposition in the urban systems literature, to wit, that cities and urban systems integrate, articulate national territory. This may have been the case during the period when mass manufacturing and mass consumption were the dominant growth machines in developed economies and thrived on the possibility of a national scale. But it is not today with the ascendance of digitalized, globalized, dematerialized sectors such as finance. The connections with other zones and sectors in its 'context' are of a special sort—one that connects worlds that we think of as radically distinct. For instance, the informal economy in several immigrant communities in New York provides some of the low-wage workers for the 'other' jobs on Wall Street, the capital of global finance. The same is happening in Paris, London, Frankfurt, Zurich . . .

[. . .]

CONCLUSION

Economic globalization and telecommunications have contributed to produce a spatiality for the urban that pivots on cross-border networks and territorial locations with massive concentrations of resources. This is not a completely new feature. Over the centuries cities have been at the crossroads of major, often worldwide, processes. What is different today is the intensity, complexity and global span of these networks, the extent to which significant portions of economies are now dematerialized and digitalized and hence the extent to which they can travel at great speeds through some of these networks, and, thirdly, the numbers of cities that are part of cross-border networks operating at vast geographic scales.

The new urban spatiality thus produced is partial in a double sense: it accounts for only part of what happens in cities and what cities are about, and it inhabits only part of what we might think of as the space administrative boundaries or in the sense of a city's public imaginary. What stands out, however, is the extent to which the city remains an integral part of these new configurations.

"World-City Network: A New Metageography?"

Annals of the Association of American Geographers (2000)

Jonathan V. Beaverstock, Richard G. Smith, and Peter J. Taylor

Editors' Introduction

In the wake of Saskia Sassen's research on the producer and financial services complex in global cities (p. 554), a number of scholars began to explore alternative approaches to the study of the global urban system. One of the most sophisticated efforts to analyze the global city hierarchy has been developed by a group of scholars based at Loughborough University in the UK, through an innovative research network known as the Globalization and World Cities Research Network, or "GaWC". Led by Peter J. Taylor, an economic geographer and urbanist, the GaWC group has generated a variety of new data sources for the study of global city formation and some extremely innovative methodological strategies for analyzing that data. Taylor, along with fellow geographers Richard G. Smith and Jonathan V. Beaverstock, have also grappled with the problem of *theorizing* about global cities and global urban networks.

The selection reprinted here represents a relatively early, programmatic statement by the three core members of the GaWC group that was published in the *Annals of the Association of American Geographers* in 2000. The piece opens by questioning the inherited, territorialist "metageography" of mainstream social science which conceives the world in terms of bordered, state-defined containers of political-economic life. By "metageography," the authors mean a totally new conception of urban geography that goes well beyond the traditional conceptions of urban space, even theoretical urban spatial patterns such as the famous Burgess model (p. 161), to a new vision that encompasses the interrelationships of cities and global metropolitan regions throughout the world. The authors argue for greater attention to inter-city networks – which they, like Manuel Castells (p. 572), say are based on flows, linkages, connections and relations between cities globally. Such inter-city networks, they argue, have acquired unprecedented importance under contemporary globalization; yet our knowledge of their developmental trajectories and geographical contours remains seriously underdeveloped. While the GaWC researchers acknowledge their debt to the first wave of global cities research, they also criticize established approaches for a tendency to focus on the fixed *attributes* of global cities (such as the number of transnational headquarter locations) rather than examining the changing *relations* between globally interlinked urban centers. In "World-City Network: A New Metageography?" the authors introduced an approach to the study of the global urban system that emphasizes relationships between cities not only by examining the worldwide office networks of seventy-four major producer and financial services firms but also through a paradigm case study of the global networks of major producer and financial services firms in London. Their quantitative-descriptive methodology opened up a radically new theoretical and empirical perspective on the global urban system and has had an immediate impact on the study of world cities and globalization generally.

The founder and director of the GaWC Research Network is Peter J. Taylor (b. 1944), Professor of Geography at Loughborough. Taylor is the editor of *Political Geography Quarterly* and the *Review of International Political*

Economy and the author of some three hundred publications, sixty of which have been translated into as many as twenty-three languages. His research falls into three broad categories: analysis of the contemporary world city network; comparative historical studies of urban networks as far back as the sixteenth century; and theories of the generic process of urban social transformation. Among his many influential books are *World City Network: A Global Urban Analysis* (London: Routledge, 2004), *Political Geography: World-Economy, Nation-State Locality* (Englewood, NJ: Prentice-Hall, 2006), and *Cities in Globalization*, coedited with Ben Derudder, Pieter Saey, and Frank Witlox (London: York: Routledge, 2006).

Since 2007, Jonathan V. Beaverstock has been Professor of Economic Geography at the University of Nottingham. Formerly, he was the head of the Department of Geography at Loughborough and co-director of the GaWC. He is an editor of *Geoforum* and the *Journal of Contemporary European Studies*. His research interests include globalization and world cities, the relational geographies of international financial centers (especially London, Frankfort, and Singapore), the globalization of banking, and patterns of skilled labor migration in the world economy. Richard G. Smith is a Senior Lecturer at Swansea University where he specializes in the urban theory of globalization and world cities.

"Global City Network: A New Metageography?" builds upon two other articles written by the same team of Beaverstock, Smith and Taylor: "A Roster of World Cities," *Cities*, 16 (1999) and "The Long Arm of the Law: London's Law Firms in a Globalizing World-Economy," *Environment and Planning*, 31 (1999). For further information on the research projects of the GaWC, see the group's website (www.lboro.ac.uk/gawc/). The literature on globalization and world cities is vast. The best introduction to the field as a whole is Neil Brenner and Roger Keil (eds), *The Global Cities Reader* (London: Routledge, 2006), Peter Taylor, *World City Network: A Global Urban Analysis* (London: Routledge, 2004), Saskia Sassen, *Cities in a World Economy*, 3rd edn (Thousand Oaks, CA: Pine Forge Press, 2008), Manuel Castells, *The Rise of the Network Society* (Oxford: Blackwell, 1996), and J. John Palen, *The Urban World*, 8th edn (Boulder, CO: Paradigm, 2008).

During the Apollo space flights, it was reported that one of the astronauts, looking back to Earth, expressed amazement that he could see no boundaries. This new view of our world as the "blue planet" contradicted the taken-for-granted, state-centric Ptolemaic model or image of world-space that most modern people carry around in their heads: a world of grids, graticule, and territorial boundaries. As a further jolt to the arrogance of modernity, it was soon accepted as a truism that the only "man-made" artifact visible from space was the ancient Great Wall of China. Interestingly, however, the Great Wall is not the only visible feature: at night, modern settlements are clearly visible as pin-pricks of electric light on a black canvas. The globality of modern society is clear for all to see in the photo prints, communicated back to Earth, of lights delimiting a global pattern of cities, consisting of a broad swath girdling the mid-latitudes of the northern hemisphere plus many oases of light elsewhere.

The fact that these "outside views" of Earth identified a world-space of settlements rather than the more familiar world-space of countries has contributed to the growth of contemporary "One World" rhetoric (also "Spaceship Earth" or "Whole Earth") which has culminated in "borderless world" theories of globalization. Of course, geographies do not depend solely upon visibility or metaphors. The fact that state boundaries are missing from space-flight photographs tells us nothing, therefore, about the current power of states to affect world geography. The photographs can, however, influence "metageography," or the "spatial structures through which people order their knowledge of the world." In the modern world, this has been notably Eurocentric and state-based in character. It is this mosaic spatial structure that the night-time photographs challenge since, first and foremost, people live in settlements. [Here] we consider the largest pinpricks of light, "the world cities" whose transnational functions materially challenge states and their territories. These cities exist in a world of flows, linkages, connections, and relations. World cities represent an alternative metageography, one of networks rather than the mosaic of states.

Historically, cities have always existed in environments of linkages, both material flows and information transfers. They have acted as centers from where their hinterlands are serviced and connected to wider realms. This is reflected in how economic geographers have treated economic sectors: primary and secondary activities are typically mapped as formal agricultural or industrial regions, tertiary activities as functional regions, epitomized by central-place theory. Why is our concern for contemporary cities in a world of flows any different from this previous tertiary activity and its study? First, the twentieth century has witnessed a remarkable sectoral turnabout in advanced economies: originally defined by their manufacturing industry, economic growth has become increasingly dependent on service industries. Second, this trend has been massively augmented by more recent developments in information technology that has enabled service and control to operate not only more rapidly and effectively, but crucially on a global scale. Contemporary world cities are an outcome of these economic changes. The large electric pinpricks of light on space photos are actually connected by massive electronic flows of information, a new functional space that will be crucial to geographical understanding in the new millennium.

[We] report preliminary research on the empirical groundwork required for describing the new metageography of relations between world cities. Such a modest goal is made necessary because of a critical empirical deficit within the world-city literature on intercity relations: studies of world cities are generally full of information that facilitates evaluations of individual cities and comparative analyses of several cities; yet, the data upon which these analyses are based has been overwhelmingly derived from measures of city attributes. Such information is useful for estimating the general importance of cities and for studying intra-city processes, but it tells us nothing directly about relations between cities. Hence cities can be ranked by attributes, but a hierarchical ordering aimed at uncovering flows or networks requires a different type of data based upon measures of relations between cities. It is the dearth of relational data that [has been called] the "dirty little secret" of this research area. In other words, we know about the nodes but not the links in this new metageography.

Our particular solution to this data problem . . . is to focus on the global office-location strategies of major corporate-service firms. After outlining this data-collection exercise, we analyze the resultant data in two ways: the first defines a network; the second deals with the global relations of a single city, London. We claim both of these analyses to be unique, first empirical studies of their kind. In a brief conclusion, we consider the future implications of this new metageography: are we witnessing a dystopia in the making?

GLOBAL OFFICE LOCATION STRATEGIES

The only published data available for studying relations between cities at a global scale are international airline-passenger statistics. Not surprisingly, therefore, empirical studies that present *networks* of world cities have focused upon this source. There are, however, serious limitations to these statistics as descriptions of relations between world cities: first, the information includes much more than trips associated with world-city processes (e.g., tourism), and second, important intercity trips within countries are not recorded in international data (e.g., New York–Toronto does feature in the data, New York–Los Angeles does not). While the latter can be overcome by augmenting the data with domestic flight statistics, the particularities of hub-and-spoke systems operated by airlines create another important caveat to using this data to describe the world-city network.

Studying the global location strategies of advanced producer-service firms is an alternative approach for describing world-city networks, one which overcomes these problems. Firms that provide business services on a global scale have to decide on the distribution of their practitioners and professionals across world cities. Setting up an office is an expensive undertaking, but a necessary investment if the firm believes that a particular city is a place where it must locate in order to fulfill its corporate goals. Hence the office geographies of advanced producer firms provide a strategic insight into world-city processes by interpreting intrafirm office networks as intercity relations. In this argument, world-city network formation consists of the aggregate of the global location strategies of major, advanced producer-service firms.

Information on the office networks of firms can be obtained by investigating a variety of sources, such as company web sites, internal directories, handbooks for customers, and trade publications. We have collected data on the distributions of offices for 74 companies (covering accountancy, advertising, banking/finance, and commercial law) in 263 cities. An initial analysis of this data identified the 143 major office centers in these cities, and 55 of these were designated world cities on the basis of the number, size, and importance of their offices. No other such roster of world cities exists; it is used here as the basic framework for studying the world-city network.

AN INTERCITY GLOBAL NETWORK

The roster of 55 world cities is divided into three levels of service provision comprising 10 Alpha cities, 10 Beta cities, and 35 Gamma cities. Only the Alpha cities – Chicago, Frankfurt, Hong Kong, London, Los Angeles, Milan, New York, Paris, Singapore, and Tokyo – are used in this section to illustrate how office geographies can define intercity relations. Note the geographical spread of these top 10 world cities; they are distributed relatively evenly across three regions we have previously identified as the major "globalization arenas": the U.S., Western Europe, and Pacific Asia. World-city network patterns are constructed for these Alpha world cities, using simple presence/absence data for the largest 46 firms in the data (all of these firms have offices in 15 or more different cities).

Shared presences are shown in Table 1. Each cell in this intercity matrix indicates the number of firms with offices in both cities. Thus, London and New York "share" 45 of the 46 firms in the data; only one firm in the data does not have offices in both of these cities. Obviously these two cities are the places to be for a corporate-service firm with serious global pretensions. This finding is not, of course, at all surprising; interest comes when lower levels of intercity relations are explored.

In Figure 1, the highest 20 shared presences are depicted at two levels of relation.

The higher level picks out Saskia Sassen's trio of global cities – London, New York, and Tokyo – as a triangular relationship (but note that, in addition, Hong Kong has such a relationship with London and New York). Bringing in the lower level of relations, London and New York have shared presences with eight other cities in all, but note again the high Pacific Asia profile in this data: Singapore joins with Tokyo and Hong Kong in showing relations with five other cities, the same level as Paris. This contrasts with the U.S. world cities below New York; Los Angeles is in the next-to-bottom class of shared presences with Frankfurt and Milan, and Chicago stands alone, with no intercity relations at the minimum level for inclusion in the diagram. This pattern can be interpreted in terms of the different degrees of political fragmentation in the three major globalization arenas. In the most fragmented, Pacific Asia, there is no dominant world city, so that presences are needed in at least three cities to cover the region: Hong Kong for China, Singapore for southeast Asia, and Tokyo for Japan. In contrast, the U.S. consists of a single state such that one city can suffice for a presence in that market. The result is that New York throws a shadow effect over other U.S. cities. In between, Western Europe is becoming more unified politically, but numerous national markets remain so that London does not dominate its regional hinterland to the same degree as New York.

Shared presences define a symmetric matrix that shows sizes, but not the direction, of intercity relations. By contrast, Table 2 is an asymmetric matrix showing probabilities of connections.

	CH	FF	HK	LN	LA	ML	NY	PA	SG	TK
Chicago										
Frankfurt	21									
Hong Kong	21	30								
London	23	32	38							
Los Angeles	21	23	29	33						
Milan	19	28	29	32	22					
New York	23	32	38	45	32	32				
Paris	21	30	32	35	27	28	34			
Singapore	20	30	34	35	26	29	35	32		
Tokyo	23	30	34	37	30	29	37	32	32	

Table 1 Relations between Alpha world cities: shared firm presences, number of firms with offices in both cities

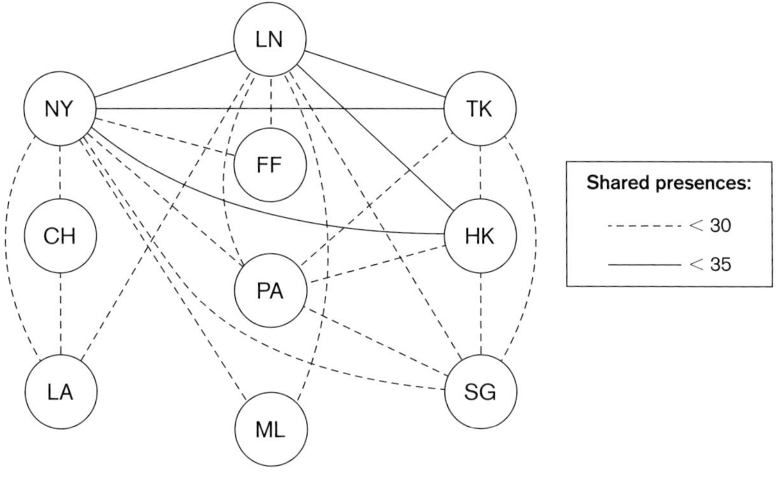

Figure 1

	Linkage to									
	CH	FF	HK	LN	LA	ML	NY	PA	SG	TK
Chicago		89	89	100	91	79	100	89	83	100
Frankfurt	67		93	100	72	87	100	95	94	95
Hong Kong	60	82		100	80	80	100	85	92	90
London	59	77	87		78	78	98	83	83	86
Los Angeles	67	73	89	100		70	97	84	81	89
Milan	59	88	93	100	67		100	88	91	93
New York	59	77	87	98	77	77		79	83	85
Paris	64	85	90	100	80	81	97		90	90
Singapore	60	87	98	100	78	83	100	92		95
Tokyo	64	84	93	100	83	81	100	87	88	

Table 2 Matrix of office-presence linkage indices for Alpha world cities

Each cell contains the probability that a firm in city A will have an office in city B. Thus, Table 2 shows that if you do business with a Chicago-based firm, then there is a 0.91 probability that that firm will also have an office in Frankfurt. On the other hand, go to a Frankfurt-based firm, and the probability of it having an office in Chicago is only 0.66. Such asymmetry is represented by vectors 2 and 3

Primary vectors are defined by probabilities above 0.95. Note that all cities connect to London and New York at this level (Figure 2). As in Figure 1, only Tokyo and Hong reach this highest category of connection, but each with only one link. Again, it is also interesting to look at the lower level relations, and these are shown in Figure 3. This diagram reinforces the interpretation concerning the three globalization arenas presented above: Chicago and Los Angeles have no inward vectors from the other arenas in what is largely a Eurasian pattern of connections. Vectors to the Pacific Asian cities dominate, but Frankfurt and Paris also have a reasonable number of inward vectors.

This is the first time intercity relations on a global scale have been studied in this way. As expected of such initial research, several opportunities for further

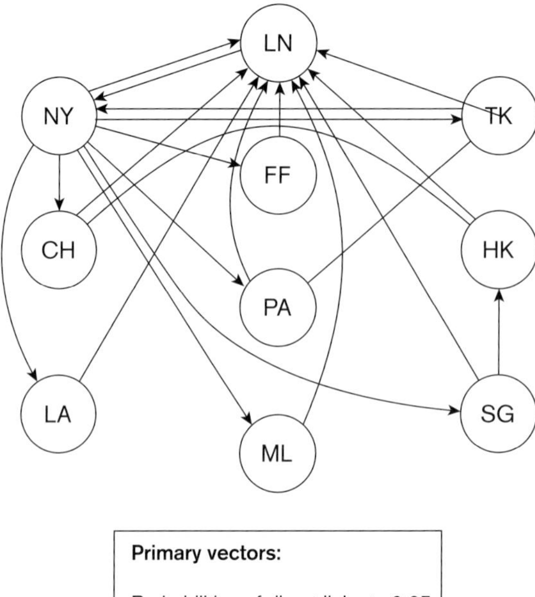
LN
NY
TK
FF
CH
HK
PA
LA
SG
ML
Primary vectors:
Probabilities of direct links > 0.95

Figure 2

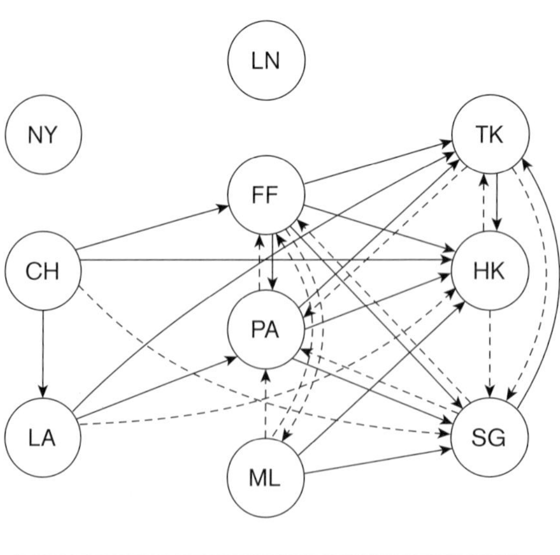
LN
NY
TK
FF
CH
HK
PA
LA
SG
ML

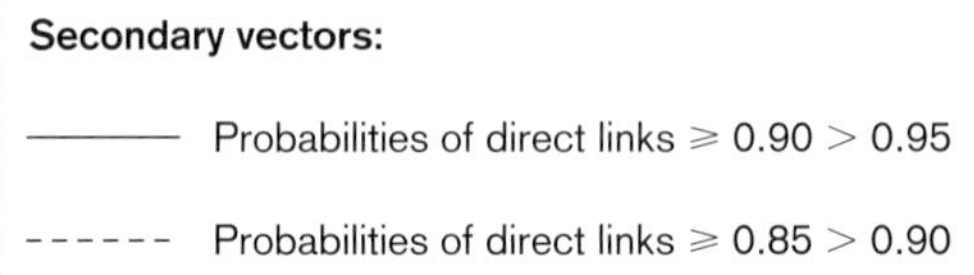
Secondary vectors:
Probabilities of direct links ≥ 0.90 > 0.95
Probabilities of direct links ≥ 0.85 > 0.90

Figure 3

investigations are suggested, not least using more cities and more sophisticated network analysis to tease out further features of the contemporary world-city network. But the most urgent task is to go beyond this cross-sectional analysis and study changes over time in order to delineate the evolution of world-city network formation. Only in this way will we be able to make informed assessments of how the network will develop in the new millennium and how this will affect different cities. For instance, is the New York shadow effect growing or declining? We simply do not know.

CASE STUDY: LONDON'S GLOBAL REACH

There is no published study assessing the global capacity of a world city in terms of its relations with other world cities. The producer-service-office geography dataset is particularly suited for such an exercise; here we illustrate this with a brief case study of London (Figure 4).

The data we employ for London differs from that used in the last section in three ways. First, it is obvious that since we will consider only London-based firms, one of the firms used previously is dropped. In addition, we add data for smaller London-based firms, creating a total of 69. Second, we consider all 55 world cities in our roster. Third, for many firms, there is richer information than simply whether they are present or absent in a city. Further information provides interval-level measurements on the numbers of practitioners or professionals employed by a firm across all its offices, as well as ordinal-level measurements in which the importance of offices was allocated to ranked classes on the basis of given functions. In order to combine this data into a single, comparable set of measures, all three levels – interval, ordinal, and nominal (presence/absence) – were combined as a single ordinal scale. For every world city, each firm is scored as one of the following: (0) indicating absence; (1) indicating presence, or where additional information is available, indicating only minor presence; (2) when additional information indicates a medium presence in a city; and

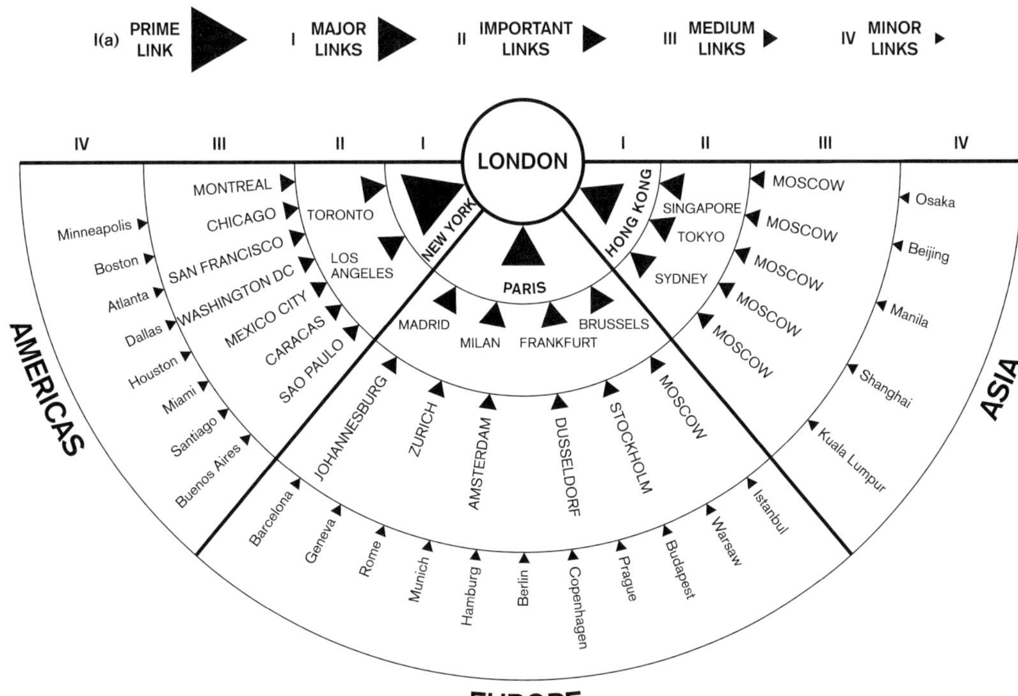

Figure 4

(3) when additional information indicates a major presence. In delineating these data when additional information is available, we were careful to be sensitive to the range of data; for large accountancy firms, for example, "minor presence" was defined as fewer than 20 practitioners in a city, while "major presences" required more than 50 professionals; yet, for law firms, the equivalent figures were 10 and 20. Through this approach, we can move beyond simple geographies of presence to geographies of the *level* of producer services available in a city.

For each of the producer-service sectors represented in our data, levels of service are summed for London-based firms in each of the other 54 world cities. This provides an estimate of the level of external service that can be expected when doing business in another world city from London. In Table 3, the top 10 world cities are ranked in terms of service available for each of the producer services represented in our data. As would be expected, the Alpha cities identified in the last section figure prominently in these rankings, with New York first or first equal in all four sectors. Yet, other world cities now make an appearance: the notable examples are the key political cities of Washington, DC, and Brussels featuring prominently in law, Dusseldorf easily outranking Germany's Alpha world city Frankfurt in accountancy, and Britain's imperial links represented by Sydney and Toronto in several lists. Average levels of linkage have been computed from standardized sector scores (city totals as percentage of maximum possible; i.e., 3 [maximum score] X number of firms in a given sector), showing all the Alpha cities in London's top 10 except Chicago. Given also Los Angeles's bottom ranking in the list, this can be interpreted as the New York shadow effect operating even from London.

Average levels of linkage with London, when computed for all other world cities, provide an illustration of London's global reach within the world-city network. These average percentages range from a top score of 87 for New York, followed by Paris (68) and Hong Kong (64), to the lowest score for Minneapolis of only 15, with Osaka (21) and Munich (22) just above the bottom. Using these averages, world cities can be divided into five groups in terms of the intensity of their relations with London. Out on its own is New York, the "prime link," followed by Paris and Hong Kong as the other two "major links." Below these three, and all with scores over 50, come nine "important links." The remainder of the cities are divided between 18 "medium links" (36–50) and the remaining 24 "minor links." These links are arrayed in Figure 3, showing a relatively even distribution across the three globalization arenas and their adjacent regions, the New York shadow effect notwithstanding.

The big question for London, as we enter the next millennium, is whether it can retain its position as Europe's leading world city. With the European Central

Accountancy	Advertising	Finance	Law services	Average
1. Düsseldorf, New York, Paris, Tokyo, Toronto	1. New York 2. Brussels, Madrid, Sydney, Toronto	1. New York 2. Singapore 3. Hong Kong, Tokyo 5. Frankfurt	1. New York 2. Washington 3. Brussels, Hong Kong 5. Paris	1. New York 2. Paris 4. Hong Kong
6. Chicago, Milan, Sydney, Washington, DC	6. Milan, Paris	6. Paris, Zurich	6. Los Angeles	5. Tokyo, Brussels
10. Atlanta, Brussels, Frankfurt, San Francisco	8. Los Angeles, Singapore, Stockholm	8. Sydney 9. Madrid 10. Milan, Taipei	7. Tokyo 8. Singapore 9. Moscow 10. Frankfurt	6. Singapore 7. Sydney 8. Milan 9. Frankfurt, Los Angeles, Toronto

Table 3 Top ten office linkages to London by advanced producer services

Bank located in Frankfurt, the scene is set for some intense intercity competition. Currently, as we showed in the previous section, London is clearly preeminent in relation to all its European rivals but, as before, in order to see which way this competition is moving, we need to supplement our cross-sectional analysis with evolutionary data and analysis, and repeat the exercise across several cities.

CONCLUSION: METAGEOGRAPHIC DYSTOPIA?

Riccardo Petrella, sometimes referred to as the "official futurist of the European Union," has warned of the rise of a "wealthy archipelago of city regions . . . surrounded by an impoverished *lumpenplanet*". He envisages a scenario in which the 30 most powerful city regions (the CR-30) will replace the G-7 (the seven most powerful states), presiding over a new global governance by 2025. Such a scenario is given credence by the fact that contemporary world cities are implicated in the current polarization of wealth and wages accompanying economic globalization. World-city practitioners and professionals operating in a global labor market have demanded and received "global wages" (largely in the form of bonuses) to create a new income category of the "waged rich."

Petrella sets out his global apartheid dystopia as a warning about current trends so as to alert us to the dangers ahead. But cities do not have to play the *bête noir* role of the future. It is within cosmopolitan cities that cultural tensions can best be managed and creatively developed. Certainly modern states, in their ambition to be nation-states, have an appalling record in dealing with matters of cultural difference. But the key point is that this is not a simple matter of cities versus states. World cities are not eliminating the power of states, they are part of a global restructuring which is "rescaling" power relations, in which states will change and adapt as they have done many times in previous restructurings. The "renegotiations" going on between London's world role and the nation's economy, between New York's world role and the U.S. economy, and with all other world cities and their encompassing territorial "home" economies, are part of a broader change affecting the balance between networks and territories in the global space-economy. In this chapter, we have illustrated how empirical analysis of city economic networks might be undertaken to complement traditional economic geography's concern for comparative advantage between states. Our one firm conclusion is that in the new millennium, we cannot afford to ignore this new metageography, the world-city network.

"Space of Flows, Space of Places: Materials for a Theory of Urbanism in the Information Age" (2001, 2002)

Manuel Castells

Editors' Introduction

For postmodern theorist Manuel Castells, the concept of *space* is a fundamental dimension that "expresses" urban society. According to Castells, the spatial experience of contemporary city life is expressed both through what he calls "the space of flows" – the electronic, computerized network of telecommunications through which most of the work of the new global economy is conducted – and through "the space of places", that is, the physical world of neighborhoods and local business nodes within metropolitan regions where people live their day-to-day lives and develop personal, familial relationships and individual identities.

Castells begins with the observation that the twenty-first-century world has entered "a new age, the Information Age" characterized by a new "network society" and the "informational city," the postmodern equivalent of the "industrial city" of the nineteenth and twentieth centuries. He then reviews the changes that have taken place in this postmodern urban world. Some are merely continuations of earlier developments: for example, rapid urbanization and metropolitan regionalism, a breakdown of the patriarchal family, increasingly multiethnic urban communities, and a social segregation spurred in part by a growing criminal culture of urban violence and paranoia. Others are totally new developments: a "new geography of networks and urban nodes" based on telecommunications technologies; social relationships that pit virtual communities against physical communities which nonetheless "develop in close interaction"; and new forms of local governance based on the emergence of "the network state." Based on this survey of the new urban reality, Castells builds his theoretical approach to cities along three axes that he calls function, meaning, and form. By *function* he means the dynamic opposition between the electronic global and the face-to-face local. By *meaning* he implies a complex relationship between "individuation" (personal identity) and "communalism" (the shared identities of ethnicity, social class, and culture). And *form* is the conflict and interaction between the two spatial dimensions of online and physical. The "space of flows is folded into the space of places," he writes, "yet their logics are distinct."

Castells argues that the organization of society in the information age involves centralization as dominant processes in the economy, technology, media, and institutionalized authority are organized in global networks. Important aspects of the world economy constitute transterritorial networks as described by Jonathan Beaverstock, Richard Smith, and Peter Taylor (p. 563) and Saskia Sassen (p. 554). On the other hand, Castells observes exclusion by spatial separation that leads to the fortress quality of Los Angeles as described by Mike Davis (p. 195) and the forms of social exclusion that Ali Madanipour (p. 186) describes. Like Neil Brenner and Roger Keil (p. 599), Castells is committed to the idea that the social inequalities of the new global cities must be challenged by new forms of social and political activism.

The theory that Castells constructs has direct application to issues of urban planning, architecture, urban design, and governance. Some of his themes are new wine in old bottles. Castells's argument that "urbanity, street life, [and] civic culture" should be as important to urban planners as "economic competitiveness" echoes Lewis Mumford's idea of "the urban drama" (p. 91), and Castells himself notes that his emphasis on the social importance of public spaces restates and updates the ideas of Kevin Lynch (p. 499) and Allan Jacobs (p. 518) and is perfectly congruent with the ideas of Camillo Sitte (p. 474), William H. Whyte (p. 510), and Jan Gehl (p. 530). But Castells recognizes that the new urban world of networks and spatial oppositions calls for a complete recasting of our ideas about cities and urban life. Cities can now only be understood on the scale of metropolitan regions, and the challenge of urban planning, design, and governance is to create a meaningful and effective "connectivity" between the very different urban worlds in which people now live their lives: the electronic space of flows and the physical space of places.

Manuel Castells is a true internationalist. Born in Spain in 1942, Castells fled the country while still a teenager because of his anti-Franco political activities and studied (and later taught) sociology at the University of Paris. In 1979, he came to the University of California, Berkeley, where he taught in the departments of city and regional planning and sociology for twenty-four years. He now divides his time between Barcelona, where he is a research professor at the Open University, and Los Angeles, where he holds the Wallis Annenberg Chair of Communication Technology and Society at the University of Southern California. He is also the Marvin and Joanne Grossman Distinguished Professor of Technology and Society at the Massachusetts Institute of Technology and a distinguished visiting professor of internet studies at Oxford University. In the 1980s, the period that Peter Hall (p. 373) terms "the Marxist ascendancy" in urban planning theory, Castells crafted sophisticated neo-Marxian theories on the role of the capitalist state, grassroots urban protest movements, and urban planning. More recently, he has turned his attention to the implications of high technology and the information revolution for community life and urban development. On the basis of this work, Castells has been dubbed a "cyber-culture theorist" and "the first great philosopher of cyberspace."

As the author of some twenty books, fifteen coauthored books, and over one hundred journal articles, Castells is one of the most influential and frequently cited scholars in the world. The selection reprinted here is an expanded and revised version of two lectures that Castells delivered in 2001 and 2002. His magnum opus is the *Information Age* trilogy: *The Rise of the Network Society* (Oxford: Blackwell, 1996), *The Power of Identity* (Oxford: Blackwell, 1997) and *The End of the Millennium* (Oxford: Blackwell, 1998). Other books on information technology and cities by Castells include *The Informational City: Information Technology, Economic Restructuring, and the Urban-Regional Process* (Oxford: Blackwell, 1991), an edited anthology *High Tech, Space, and Society* (Beverly Hills, CA: Sage, 1985), *Technopoles of the World*, coauthored with Sir Peter Hall (London: Routledge, 1994), *Dual City: Restructuring New York*, coedited with John Mollenkopf (New York: Russell Sage Foundation, 1991), *The City and the Grassroots* (Berkeley, CA: University of California Press, 1983), and *The Urban Question* (London: Edward Arnold, 1977). An excellent anthology of works from throughout his career is Ida Susser (ed.), *The Castells Reader on Cities and Social Theory* (Oxford: Blackwell, 2002). For a complete overview of the literature of cyber-urbanism and the information age city, see Steven Graham (ed.), *The Cybercities Reader* (London: Routledge, 2005).

■

We have entered a new age, the Information Age. Spatial transformation is a fundamental dimension of the overall process of structural change. We need a new theory of spatial forms and processes, adapted to the new social, technological, and spatial context where we live. I will attempt here to propose some elements of this theory, a theory of urbanism in the Information Age. . . .

I will not build theory from other theories, but from the observation of social and spatial trends in the world at large. Thus, I will start with a summary characterization of the main spatial trends at the onset of the 21st century. Then I will propose a tentative theoretical interpretation of observed spatial trends. Subsequently I will highlight the main issues arising in cities in the information age, with particular emphasis on the crisis of the city as a socio-spatial system of cultural communication. I will conclude by drawing some of the implications of my analysis for planning, architecture and urban design.

THE TRANSFORMATION OF URBAN SPACE IN THE EARLY 21ST CENTURY

Spatial transformation must be understood in the broader context of social transformation: space does not reflect society, it expresses it, it is a fundamental dimension of society, inseparable of the overall process of social organization and social change. Thus, the new urban world arises from within the process of formation of a new society, the network society, characteristic of the Information Age. The key developments in spatial patterns and urban processes associated with these macro-structural changes, can be summarized under the following headings:

- Because commercial agriculture has been, by and large, automated, and a global economy has integrated productive networks throughout the planet, the majority of the world's population is already living in urban areas, and this will be increasingly the case: we are heading towards a largely urbanized world, which will comprise between two-thirds and three-quarters of the total population by the middle of the century;
- This process of urbanization is concentrated disproportionately in metropolitan areas of a new kind: urban constellations scattered throughout huge territorial expanses, functionally integrated and socially differentiated, around a multi-centered structure. I call these new spatial forms metropolitan regions;
- Advanced telecommunications, Internet, and fast, computerized transportation systems allow for simultaneous spatial concentration and decentralization, ushering in a new geography of networks and urban nodes throughout the world, throughout countries, between and within metropolitan areas;
- Social relationships are characterized simultaneously by individuation and communalism, both processes using, at the same time, spatial patterning and on-line communication. Virtual communities and physical communities develop in close interaction, and both processes of aggregation are challenged by increasing individualization of work, social relationships, and residential habits;
- The crisis of the patriarchal family, with different manifestations depending on cultures and levels of economic development, gradually shifts sociability from family units to networks of individualized units (most often, women and their children, but also individualized co-habiting partnerships), with considerable consequences in the uses and forms of housing, neighborhoods, public space, and transportation systems;
- The emergence of the network enterprise as a new form of economic activity, with its highly decentralized, yet coordinated, form of work and management, tends to blur the functional distinction between spaces of work and spaces of residence. The work–living arrangements characteristic of the early periods of industrial craft work are back, often taking over the old industrial spaces, and transforming them into informational production spaces. This is not just New York's Silicon Alley or San Francisco's Multimedia Gulch, but a phenomenon that also characterizes London, Tokyo, Beijing, Taipei, Paris, or Barcelona, among many other cities. Transformation of productive uses becomes more important than residential succession to explain the new dynamics of urban space;
- Urban areas around the world are increasingly multi-ethnic, and multi-cultural. An old theme of the Chicago School, now amplified in terms of its extremely diverse racial composition;
- The global criminal economy is solidly rooted in the urban fabric, providing jobs, income, and social organization to a criminal culture, which deeply affects the lives of low-income communities, and of the city at large. It follows rising violence and/or widespread paranoia of urban violence, with the corollary of defensive residential patterns;
- Breakdowns of communication patterns between individuals and between cultures, and the emergence of defensive spaces, leads to the formation of sharply segregated areas: gated communities for the rich, territorial turfs for the poor;
- In a reaction against trends of suburban sprawl and the individualization of residential patterns, urban centers and public space become critical expressions of local life, benchmarking the vitality of any given city. Yet, commercial pressures and artificial attempts at mimicking urban life often transform public spaces into theme parks where symbols rather than experience create a life-size, urban virtual reality, ultimately destined to mimic the real virtuality projected in the media. It follows increasing individualization, as urban places become consumption items to be individually appropriated;
- Overall, the new urban world seems to be dominated by the double movement of inclusion into

transterritorial networks, and exclusion by the spatial separation of places. The higher the value of people and places, the more they are connected into interactive networks. The lower their value, the lower their connection. In the limit, some places are switched off, and bypassed by the new geography of networks, as it is the case of depressed rural areas and urban shanty towns around the world. Splintering urbanism operates on the basis of segregated networks of infrastructure . . .;

- The constitution of mega-metropolitan regions, without a name, without a culture, and without institutions, weakens the mechanism of political accountability, of citizen participation, and of effective administration. On the other hand, in the age of globalization, local governments emerge as flexible institutional actors, able to relate at the same time to local citizens and to global flows of power and money. Not because they are powerful, but because most levels of government, including the nation states, are equally weakened in their capacity of command and control if they operate in isolation. Thus, a new form of state emerges, the network state, integrating supra-national institutions made up of national governments, nation-states, regional governments, local governments, and even non-governmental organizations. Local governments become a node of the chain of institutional representation and management, able to input the overall process, yet with added value in terms of their capacity to represent citizens at a closer range. Indeed in most countries, opinion polls show the higher degree of trust people have in their local governments, relative to other levels of government.
- Urban social movements have not disappeared, by any means. But they have mutated. In an extremely schematic representation they develop along two main lines. The first is the defense of the local community, affirming the right to live in a particular place, and to benefit from adequate housing and urban services in their place. The second is the environmental movement, acting on the quality of cities within the broader goal of achieving quality of life: not only a better life but a different life. Often, the broader goals of environmental mobilizations become translated into defensive reactions to protect one specific community, thus merging the two trends. Yet, it is only by reaching out to the cultural transformation of urban life as proposed by ecological thinkers and activists that urban social movements can transcend their limits of localism. Indeed, enclosing themselves in their communities, urban social movements may contribute to further spatial fragmentation, ultimately leading to the breakdown of society.

It is against the background of these major trends of urban social change that we can understand new spatial forms and processes, thus re-thinking architecture, urban design and planning in the 21st century.

A THEORETICAL APPROACH TO SPATIAL TRANSFORMATION

To make the transition from the observation of urban trends to the new theorization of cities, we need to grasp, at a more analytical level, the key elements of socio-spatial change. I think the transformation of cities in the information age can be organized around three bipolar axes. The first relates to function, the second to meaning, the third to form.

Function

Functionally speaking the network society is organized around the opposition between the global and the local. Dominant processes in the economy, technology, media, institutionalized authority are organized in global networks. But day-to-day work, private life, cultural identity, political participation, are essentially local. Cities, as communication systems, are supposed to link up the local and the global, but this is exactly where the problems start since these are two conflicting logics that tear cities from the inside when they try to respond to both, simultaneously.

Meaning

In terms of meaning, our society is characterized by the opposing development of individuation and communalism. By individuation I understand the enclosure of meaning in the projects, interests, and representations of the individual, that is a biologically embodied personality system (or, if you want, translating from French structuralism, a person). By communalism

I refer to the enclosure of meaning in a shared identity, based on a system of values and beliefs to which all other sources of identity are subordinated. Society, of course, exists only in-between, in the inter-face between individuals and identities mediated by institutions, at the source of the constitution of civil society. . . .

Trends I observe in the formative stage of the network society indicate the increasing tension and distance between personality and culture, between individuals and communes. Because cities are large aggregates of individuals, forced to coexist, and communes are located in the metropolitan space, the split between personality and commonality brings extraordinary stress upon the social system of cities as communicative and institutionalizing devices. The problematic of social integration becomes again paramount, albeit under new circumstances and in terms radically different from those of early industrial cities. This is mainly because of the role played in urban transformation by a third, major axis of opposing trends, this one concerning spatial forms.

Forms

There is a growing tension and articulation between the space of flows and the space of places.

The space of flows links up electronically separate locations in an interactive network that connects activities and people in distinct geographical contexts. The space of places organizes experience and activity around the confines of locality. Cities are structured, and destructured simultaneously by the competing logics of the space of flows and the space of places. Cities do not disappear in the virtual networks. But they are transformed by the interface between electronic communication and physical interaction, by the combination of networks and places. . . . The informational city is built around this double system of communication. Our cities are made up, at the same time, of flows and places, and of their relationships. Two examples will help to make sense of this statement, one from the point of view of the urban structure, another in terms of the urban experience.

Turning to urban structure, the notion of global cities was popularized in the 1990s. Although most people assimilate the term to some dominant urban centers, such as London, New York and Tokyo, the concept of global city does not refer to any particular city, but to the global articulation of segments of many cities into an electronically linked network of functional domination throughout the planet. The global city is a spatial form rather than a title of distinction for certain cities, although some cities have a greater share of these global networks than others. In a sense, most areas in all cities, including New York and London, are local, not global. And many cities are sites of areas, small and large, which are included in these global networks, at different levels. This conception of global city as a spatial form resulting from the process of globalization is closer to the pioneering analysis by Saskia Sassen than to its popularized version by city marketing agencies. Thus, from the structural point of view, the role of cities in the global economy depends on their connectivity in transportation and telecommunication networks, and on the ability of cities to mobilize effectively human resources in this process of global competition. As a consequence of this trend, nodal areas of the city, connecting to the global economy, will receive the highest priority in terms of investment and management, as they are the sources of value creation from which an urban node and its surrounding area will make their livelihood. Thus, the fate of metropolitan economies depends on their ability to subordinate urban functions and forms to the dynamics of certain places that ensure their competitive articulation in the global space of flows.

From the point of view of the urban experience, we are entering a built environment that is increasingly incorporating electronic communication devices everywhere. Our urban life fabric . . . becomes an *e-topia*, a new urban form in which we constantly interact, deliberately or automatically, with on-line information systems, increasingly in the wireless mode. Materially speaking the space of flows is folded into the space of places. Yet, their logics are distinct: on-line experience and face-to-face experience remain specific, and the key question then is to assure their articulation in compatible terms.

* * *

THE URBAN THEMES OF THE INFORMATION AGE

The issue of social integration comes again at the forefront of the theory of urbanism, as was the case during the process of urbanization in the industrial era. Indeed, it is the very existence of cities as communication artifacts that is called into question, in spite of the fact that we live in a predominantly urban world. But what is at stake is a very different kind of integration. In the early 20th century the quest was for assimilation of urban subcultures into the urban culture. In the early 21st century the challenge is the sharing of the city by irreversibly distinct cultures and identities. There is no more dominant culture, because only global media have the power to send dominant messages, and the media have in fact adapted to their market, constructing a kaleidoscope of variable content depending on demand, thus reproducing cultural and personal diversity rather than overimposing a common set of values. The spread of horizontal communication via the Internet accelerates the process of fragmentation and individualization of symbolic interaction. Thus, the fragmented metropolis and the individualization of communication reinforce each other to produce an endless constellation of cultural sub-sets. The nostalgia of the public domain will not be able to countervail the structural trends towards diversity, specification, and individualization of life, work, space and communication, both face to face, and electronic. On the other hand, communalism adds collective fragmentation to individual segmentation. Thus, in the absence of a unifying culture, and therefore of a unifying code the key question is not the sharing of a dominant culture but the communicability of multiple codes.

The notion of communication protocols is central here. Protocols may be physical, social, and electronic, with additional protocols being necessary to relate these three different planes of our multidimensional experience.

Physically, the establishment of meaning in these nameless urban constellations relates to the emergence of new forms of symbolic nodality which will identify places, even through conflictive appropriation of their meaning by different groups and individuals.

The second level of urban interaction refers to social communication patterns. Here, the diversity of expressions of local life, and their relationship to media culture, must be integrated into the theory of communication by doing rather than by saying. In other words, how messages are transmitted from one social group to another, from one meaning to another in the metropolitan region requires a redefinition of the notion of public sphere moving from institutions to the public place. . . . Public places, as sites of spontaneous social interaction, are the communicative devices of our society, while formal, political institutions have become a specialized domain that hardly affects the private lives of people, that is what most people value most. Thus, it is not that politics, or local politics, does not matter. It is that its relevance is confined to the world of instrumentality, while expressiveness, and thus communication, refers to social practice, outside institutional boundaries. Therefore, in the practice of the city, its public spaces, including the social exchangers (or communication nodes) of its transportation networks become the communicative devices of city life. How people are, or are not, able to express themselves, and communicate with each other, outside their homes and off their electronic circuits, that is, in public places, is an essential area of study for urbanism. I call it the sociability of public places in the individualized metropolis.

The third level of communication refers to the prevalence of electronic communication as a new form of sociability. Studies by a growing legion of social researchers have shown the density and intensity of electronic networks of communication, providing evidence to sustain the notion that virtual communities are often communities, albeit of a different kind than face to face communities. Here again, the critical matter is the understanding of the communication codes between various electronic networks, built around specific interests or values, and between these networks and physical interaction. There is no established theory yet on these communication processes, as the Internet as a widespread social practice is still in its infancy. But we do know that on-line sociability is specified, not downgraded, and that physical location does contribute, often in unsuspected ways, to the configuration of electronic communication networks. Virtual communities as networks of individuals are transforming the patterns of sociability in the new metropolitan life, without escaping into the world of electronic fantasy.

Fourth, the analysis of code sharing in the new urban world requires also the study of the inter-face between physical layouts, social organisation, and electronic networks. It is this interface that William Mitchell considers to be at the heart of the new urban form, what he calls e-topia. . . . In other words, we must understand at the same time the process of communication and that of in-communication.

The contradictory and/or complementary relationships between new metropolitan centrality, the practice of public space, and new communication patterns emerging from virtual communities, could lay the foundations for a new theory of urbanism – the theory of cyborg cities or hybrid cities made up by the intertwining of flows and places.

Let us go farther in this exploration of the new themes for urban theory. We know that telecommuting – meaning people working full time on-line from their home – is another myth of futurology. Many people, including you and me, work on-line from home part of the time, but we continue to go to work in places, as well as moving around (the city or the world) while we keep working, with mobile connectivity to our network of professional partners, suppliers, and clients. The latter is the truly new spatial dimension of work. This is a new work experience, and indeed a new life experience. Moving physically while keeping the networking connection to everything we do is a new realm of the human adventure. . . . The analysis of networked spatial mobility is another frontier for the new theory of urbanism. To explore it in terms that would not be solely descriptive we need new concepts. The connection between networks and places has to be understood in a variable geometry of these connections. The places of the space of flows, that is the corridors and halls that connect places around the world, will have to be understood as exchangers and social refuges, as homes on the run, as much as offices on the run. The personal and cultural identification with these places, their functionality, their symbolism, are essential matters that do not concern only the cosmopolitan elite. Worldwide mass tourism, international migration, transient work, are experiences that relate to the new huddled masses of the world. How we relate to airports, to train and bus stations, to freeways, to customs buildings, are part of the new urban experience of hundreds of millions. We can build on an ethnographic tradition that addressed these issues in the mature industrial society. But here again, the speed, complexity, and planetary reach of the transportation system have changed the scale and meaning of the issues. Furthermore, the key reminder is that we move physically while staying put in our electronic connection. We carry flows and move across places.

Urban life in the 21st century is also being transformed by the crisis of patriarchalism. This is not a consequence of technological change, but I have argued in my book *The Power of Identity* that it is an essential feature of the Information Age. To be sure, patriarchalism is not historically dead. Yet, it is contested enough, and overcome enough so that everyday life for a large segment of city dwellers has already been redefined vis-à-vis the traditional patterns of an industrial society based on a relatively stable patriarchal nuclear family. Under conditions of gender equality, and under the stress suffered by traditional arrangements of household formation, the forms and rhythms of urban life are dramatically altered. Patterns of residence, transportation, shopping, education, and recreation evolve to adjust to the multi-directionality of individual needs that have to share household needs. This transformation is mediated by variable configurations of state policies. For instance, how child care is handled by government, by firms, by the market, or by individual networking largely conditions the time and space of daily lives, particularly for children.

We have documented how women are discriminated against in the patriarchal city. We can empirically argue that women's work makes possible the functioning of cities – an obvious fact rarely acknowledged in the urban studies literature. Yet, we need to move forward, from denunciation to the analysis of specific urban contradictions resulting from the growing dissonance between the de-gendering of society and historical crystallization of patriarchalism in the patterns of home and urban structure. How do these contradictions manifest themselves as people develop strategies to overcome the constraints of a gendered built environment? How do women, in particular, re-invent urban life, and contribute to re-design the city of women, in contrast to the millennial heritage of the city of men? These are the questions to be researched, rather than stated, by a truly post-patriarchal urban theory.

Grass-roots movements continue to shape cities, as well as societies at large. They come in all kinds of formats and ideologies, and one should keep an open mind on this matter, not deciding in advance which ones are progressive, and which ones are regressive, but taking all of them as symptoms of society in the making. We should also keep in mind the most fundamental rule in the study of social movements. They are what they say they are. They are their own consciousness. We can study their origins, establish their rules of engagement, explore the reasons for their victories and defeats, link their outcomes to overall social transformation, but not to interpret them, not to explain to them what they really mean by what they say. Because, after all, social movements are nothing else than their own symbols and stated goals, which ultimately means their words.

Based on the observation of social movements in the early stage of the network society, two kinds of issues appear to require privileged attention from urban social scientists. The first one is what I called some time ago the grass-rooting of the space of flows, that is the use of Internet for networking in social mobilization and social challenges. This is not simply a technological issue, because it concerns the organization, reach, and process of formation of social movements. Most often these on-line social movements connect to locally based movements, and they converge, physically, in a given place at a given time. A good example was the mobilization against the World Trade Organization meeting in Seattle in December 1999, and against subsequent meetings of globalizing institutions, which, arguably, set a new trend of grass-roots opposition to uncontrolled globalization, and redefined the terms of the debate on the goals and procedures of the new economy. The other major issue in the area of social movements is the exploration of the environmental movement, and of an ecological view of social organization, as urban areas become the connecting point between the global issues posed by environmentalism and the local experience through which people at large assess their quality of life. To redefine cities as eco-systems, and to explore the connection between local eco-systems and the global eco-system lays the ground for the overcoming of localism by grass-roots movements.

On the other hand, the connection cannot be operated only in terms of ecological knowledge. Implicit in the environmental movement, and clearly articulated in the deep ecology theory . . . is the notion of cultural transformation. A new civilization, and not simply a new technological paradigm, requires a new culture. This culture in the making is being fought over by various sets of interests and cultural projects. Environmentalism is the code word for this cultural battle, and ecological issues in the urban areas constitute the critical battleground for such struggle.

Besides tackling new issues, we still have to reckon in the 21st century with the lingering questions of urban poverty, racial and social discrimination, and social exclusion. In fact, recent studies show an increase of urban marginality and inequality in the network society. Furthermore, old issues in a new context, become in fact new. Thus, Ida Susser . . . has shown the networking logic underlying the spread of AIDS among New York's poor along networks of destitution, stigma, and discrimination. Eric Klinenberg, in his social anatomy of the devastating effects of the 1995 heat wave in Chicago, shows why dying alone in the city, the fate of hundreds of seniors in a few days, was rooted in the new forms of social isolation emerging from people's exclusion from networks of work, family, information and sociability. The dialectics between inclusion and exclusion in the network society redefines the field of study of urban poverty, and forces us to consider alternative forms of inclusion (e.g. social solidarity or, else, the criminal economy), as well as new mechanisms of exclusion technological apartheid in era of Internet.

The final frontier for a new theory of urbanism, indeed for social sciences in general, is the study of new relationships between time and space in the Information Age. In my analysis of the new relationships of time and space I proposed the hypothesis that in the network society, space structures time, in contrast to the time-dominated constitution of the industrial society, in which urbanization, and industrialization were considered to be part of the march of universal progress, erasing place-rooted traditions and cultures. In our society, the network society, where you live determines your time frame of reference. If you are an inhabitant of the space of flows, or if you live in a locality that is in the dominant networks, timeless time (epitomized by the frantic race to beat the clock) will be your time as in Wall Street or Silicon Valley. If you are in a Pearl River Delta factory town, chronological time will be imposed upon you as in the best days of Taylorism in Detroit. And if you live in a village in Mamiraua, in Amazonia, biological time, usually a much shorter life-span, will still rule your life. Against this spatial determination of time, environmental movements assert the notion of slow-motion time, the time of the long now, in the words of Stewart Brand, by broadening the spatial dimension to its planetary scale in the whole complexity of its interactions thus including our great-grand children in our temporal frame of reference.

Now, what is the meaning of this multidimensional transformation for planning, architecture, and urban design?

PLANNING, ARCHITECTURE, AND URBAN DESIGN IN THE RECONSTRUCTION OF THE CITY

The great urban paradox of the 21st century is that we could be living in a predominantly urban world without cities – that is without spatially based systems of cultural communication and sharing of meaning, even conflictive sharing. Signs of the social, symbolic, and functional disintegration of the urban fabric multiply around the world. So do the warnings from analysts and observers from a variety of perspectives.

But societies are produced, and spaces are built, by conscious human action. There is no structural determinism. So, together with the emphasis on the economic competitiveness of cities, on metropolitan mobility, on privatization of space, on surveillance and security, there is also a growing valuation of urbanity, street life, civic culture, and meaningful spatial forms in the metropolitan areas around the world The process of reconstruction of the city is under way. And the emphasis of the most advanced urban projects in the world is on communication, in its multidimensional sense: restoring functional communication by metropolitan planning; providing spatial meaning by a new symbolic nodality created by innovative architectural projects; and re-instating the city in its urban form by the practice of urban design focused on the preservation, restoration, and construction of public space as the epitome of urban life.

However, the defining factor in the preservation of cities as cultural forms in the new spatial context will be the capacity of integration between planning, architecture, and urban design. This integration can only proceed through urban policy influenced by urban politics. Ultimately, the management of metropolitan regions is a political process, made of interests, values, conflicts, debates, and options that shape the interaction between space and society. Cities are made by citizens, and governed on their behalf. Only when democracy is lost can technology and the economy determine the way we live. Only when the market overwhelms culture and when bureaucracies ignore citizens can spatial conurbations supersede cities as living systems of multidimensional communication.

Planning

The key endeavor of planning in the metropolitan regions of the information age is to ensure their connectivity, both intra-metropolitan and inter-metropolitan. Planning has to deal with the ability of the region to operate within the space of flows. The prosperity of the region and of its dwellers will greatly depend on their ability to compete and cooperate in the global networks of generation/appropriation of knowledge, wealth, and power. At the same time planning must ensure the connectivity of these metropolitan nodes to the space of places contained in the metropolitan region. In other words, in a world of spatial networks, the proper connection between these different networks is essential to link up the global and the local without opposing the two planes of operation.

This means that planning should be able to act on a metropolitan scale, ensuring effective transportation, accepting multinodality, fighting spatial segregation by acting against exclusionary zoning, providing affordable housing, and desegregated schooling. Ethnic and social diversity is a feature of the metropolitan region, and ought to be protected. Planning should seek the integration of open space and natural areas in the metropolitan space, going beyond the traditional scheme of the greenbelt. The new metropolitan region embraces a vast territorial expanse, where large areas of agricultural land and natural land should be preserved as a key component of a balanced metropolitan territory. The new metropolitan space is characterized by its multifunctionality, and this is a richness that supersedes the functional specialization and segregation of modernist urbanism. New planning practice induces a simultaneous process of decentering and recentering of population and activities, leading to the creation of multiple subcenters in the region.

The social and functional diversity of the metropolitan region requires a multimodal approach to transportation, by mixing the private automobile/highway system with public metropolitan transportation (railways, subways, buses, taxis), and with local transportation (bicycles, pedestrian paths, specialized shuttle services). Furthermore, in a post-patriarchal world, childcare becomes a critical urban service, and therefore must be integrated in the schemes of metropolitan planning. In the same way that some cities require additional housing and transportation investment per each new job created in certain areas, child care provision should be included in these planning standards.

Overall, most metropolitan planning nowadays is geared towards the adaptation of the space of places of the metropolitan region to the space of flows that conditions the economic competitiveness of the region. The challenge would be to use planning, instead, to structure the space of places as a living space, and to ensure the connection and complementarity between the economy of the metropolitan region and the quality of life of its dwellers.

Architecture

Restoring symbolic meaning is a most fundamental task in a metropolitan world in crisis of communication. This is the role that architecture has traditionally assumed. It is more important than ever. Architecture, of all kinds, must be called to the rescue in order to recreate symbolic meaning in the metropolitan region, marking places in the space of flows. In recent years, we have observed a substantial revival of architectural meaningfulness that in some cases has had a direct impact in revitalizing cities and regions, not only culturally but economically as well. To be sure, architecture per se cannot change the function, or even the meaning, of a whole metropolitan area. Symbolic meaning has to be inserted in the whole fabric of the city, and this is, as I will argue below, the key role of urban design. But we still need meaningful forms, resulting from architectural intervention, to stir a cultural debate that makes space a living form. Recent trends in architecture signal its transformation from an intervention on the space of places to an intervention on the space of flows, the dominant space of the Information Age by acting on spaces dedicated to museums, convention centers, and transportation nodes. These are spaces of cultural archives, and of functional communication that become transformed into forms of cultural expression and meaningful exchange by the act of architecture.

The most spectacular example is Frank Gehry's Guggenheim Museum in Bilbao, that symbolized the way of life of a city immersed in a serious economic crisis and a dramatic political conflict. . . . Ricardo Bofill's Barcelona airport, Rafael Moneo's AVE railway station in Madrid and Kursaal Convention Center in San Sebastian, Richard Meier's Modern Art Museum in Barcelona, or Rem Koolhaas's Lille Grand Palais, are all examples of these new cathedrals of the Information Age, where the pilgrims gather to search for the meaning of their wandering. Critics point at the disconnection between many of these symbolic buildings and the city at large. The lack of integration of this architecture of the space of flows into the public space would be tantamount to juxtapose symbolic punctuation and spatial meaninglessness. This is why it is essential to link up architecture with urban design, and with planning. Yet, architectural creation has its own language, its own project that cannot be reduced to function or to form. Spatial meaning is still culturally created. But their final meaning will depend on its interaction with the practice of the city organized around public space.

Urban design

The major challenge for urbanism in the Information Age is to restore the culture of cities. This requires a socio-spatial treatment of urban forms, a process that we know as urban design. But it must be an urban design able of connecting local life, individuals, communes, and instrumental global flows through the sharing of public places. Public space is the key connector of experience, opposed to private shopping centers as the spaces of sociability.

It is public space that makes cities as creators of culture, organizers of sociability, systems of communication, and seeds of democracy, by the practice of citizenship. This is in opposition to the urban crisis characterized by the dissolution, fragmentation, and privatization of cities. . . .

This is in fact a long tradition in urban design, associated with the thinking and practice of Kevin Lynch, and best represented nowadays by Allan Jacobs. Jacobs' work on streets, and, with Elizabeth McDonald, on boulevards as urban forms able to integrate transportation mobility and social meaning in the city, shows that there is an alternative to the edge city, beyond the defensive battles of suburbanism with a human face. The success of the Barcelona model of urban design is based on the ability to plan public squares, even mini-squares in the old city, that bring together social life, meaningful architectural forms (not always of the best taste, but this does not matter), and the provision of open space for people's use. That is, not just open space, but marked open space, and street life induced by activities, such as the tolerance of informal trade, street musicians etc.

The reconquest of public space operates throughout the entire metropolitan region, highlighting particularly the working class peripheries, those that need the most attention at socio-spatial reconstruction. Some times the public space is a square, some times a park, some times a boulevard, some times a few square meters around a fountain or in front of a library or a museum. Or an outdoor café colonizing the sidewalk. In all instances what matters is the spontaneity of uses, the density of the interaction, the freedom of expression, the multifunctionality of space, and the multiculturalism of the street life. This is not the nostalgic reproduction of the medieval town. In fact, examples of public space (old, new, and renewed) dot the whole planet. . . . It is the dissolution of public space under the combined pressures of privatization of the city and the rise of the space of flows that is an historical oddity. Thus, it is not the past versus the future, but two forms of present that fight each other in the battleground of the emerging metropolitan regions. And the fight, and its outcome, is of course, political, in the etymological sense: it is the struggle of the polis to create the city as a meaningful place.

THE GOVERNMENT OF CITIES IN THE INFORMATION AGE

The dynamic articulation between metropolitan planning, architecture, and urban design is the domain of urban policy. Urban policy starts with a strategic vision of the desirable evolution of the metropolitan space in its double relationship to the global space of flows and to the local space of places. This vision, to be a guiding tool, must result from the dynamic compromise between the contradictory expression of values and interests from the plurality of urban actors. Effective urban policy is always a synthesis between the interests of these actors and their specific projects. But this synthesis must be given technical coherence and formal expression, so that the city evolves in its form without submitting the local society to the imperatives of economic constraints or technological determinism.

The constant adjustment between various structural factors and conflictive social processes is implemented by the government of cities. This is why good planning or innovative architecture cannot do much to save the culture of cities unless there are effective city governments, based on citizen participation and the practice of local democracy. Too much to ask for? Well, in fact, the planet is dotted with examples of good city government that make cities livable by harnessing market forces and taming interest groups on behalf of the public good. Portland, Toronto, Barcelona, Birmingham, Bologna, among many other cities, are instances of the efforts of innovative urban policy to manage the current metropolitan transformation. However, innovative urban policy does not result from great urbanists (although they are indeed needed), but from courageous urban politics able to mobilize citizens around the meaning of their environment.

IN CONCLUSION

The new culture of cities is not the culture of the end of history. Restoring communication may open the way to restore meaningful conflict. Currently, social injustice and personal isolation combine to induce alienated violence. So, the new culture of urban integration is not the culture of assimilation into the values of a single dominant culture, but the culture of communication between an irreversibly diverse local society connected/disconnected to global flows of wealth, power, and information.

Architecture and urban design are sources of spatio-cultural meaning in an urban world in dramatic need of communication protocols and artifacts of sharing. It is commendable that architects and urban designers find inspiration in social theory, and feel as concerned citizens of their society. But first of all, they must do their job as providers of meaning by the cultural shaping of spatial forms. Their traditional function in society is more critical than ever in the Information Age, an age marked by the growing gap between splintering networks of instrumentality and segregated places of singular meaning. Architecture and design may bridge technology and culture by creating shared symbolic meaning and reconstructing public space in the new metropolitan context. But they will only be able to do so with the help of innovative urban policy supported by democratic urban politics.

"Key Findings and Messages"

from *The Challenge of Slums: Global Report on Human Settlements 2003*

United Nations Human Settlements Programme (UN-HABITAT)

Editors' Introduction

As all the selections in this part on Cities in a Global Society suggest, the very nature of urbanization considered as a whole and the conditions of urban life considered in specific detail is in the process of transformation. New technologies lead to new social relationships, new global economies, and new challenges of local, regional, and global governance. But within the context of global urban change, no issue is more important than the persistence of an age-old urban problem: the complex of poverty, social inequality, and communities plagued by slum conditions of almost unimaginable proportions.

Some degree of economic inequality has existed in all cities and in every historical period. The homes and neighborhoods of the very poor have always been markedly different from the palaces of the rich and the comfortable precincts of the urban middle class. But the problem of urban slums – areas either in center-cities or on their peripheries where masses of the disenfranchised live hand-to-mouth lives and cope with terrible living conditions – is peculiarly an issue of global urbanism. Nowadays, whether in run-down inner-city ghettoes and barrios or in ramshackle shantytowns and *favelas* on the outskirts of global metropolitan regions, extreme poverty and slum conditions represent the urban reality of hundreds of millions of people – perhaps as many as a billion people, nearly one-third of the urban population worldwide.

Slums have always posed challenges for policy makers and growing economies. The slums that Friedrich Engels (p. 46) described in the 1840s during the rise of industrial urbanism were surely bleak, but the wealth created by industrial progress helped better housing and community conditions to emerge for the industrial working class over the course of a century. Even the strictly segregated Black ghettoes of America described by W.E.B. Du Bois (p. 110) were multi-class communities – although disproportionally poor – with some chance of upward mobility. But the vast urban slums of the present day are the creation of new global economies that offer little promise for advancement through education and employment. As United Nations Secretary General Kofi Annan noted in his foreword to *The Challenge of Slums: Global Report on Human Settlements 2003*, "the locus of global poverty is moving to the cities, a process now recognized as the 'urbanization of poverty'."

The principal findings of *The Challenge of Slums* are that the majority of slum dwellers are in the developing regions of the world, that their numbers have increased dramatically during the 1990s, and that they will most likely double (to 2 billion) by the year 2020. Surprisingly, slum dwellers are not all poor, but most slum dwellers – those who are not utterly destitute – earn their livings in what are called "informal sector" activities: that is, off-the-books and unregulated trades that are sometimes clearly illegal but which are nonetheless in demand within the larger global urban economy. Local and regional authorities urgently need to implement urban planning and economic development policies designed to prevent the emergence of new slums and must institute, as much as possible, *in situ* "slum upgrading" policies – not devastating slum clearance projects – to ameliorate the living conditions within existing slums. These findings are echoed in the call of Neil Brenner and Roger Keil for "research – and action" moving toward "radical, progressive" social change in their essay "From Global Cities to Globalized Urbanization" (p. 599).

The Challenge of Slums (London: Earthscan, 2003) was prepared by the United Nations Settlements Programme, commonly known as UN-HABITAT, an agency that was established in 1978 and headquartered in Nairobi, Kenya. When the General Assembly promulgated the United Nations Millennium Declaration in September of 2000 – a sweeping set of global development goals aimed at achieving world peace, human rights, universal education, environmental sustainability, the elimination of HIV-AIDS, and the eradication of poverty – UN-HABITAT was tasked with reporting on issues of human settlements and urban development worldwide in much the same way as the World Commission on Environment and Development was charged with reporting on global sustainable development in the Bruntland Report of 1987 (see p. 351).

Other UN-HABITAT publications include *Slums of the World: The Face of Urban Poverty in the New Millennium?* (2003), *The State of the World's Cities* (2008), a series of reports on urban water and sanitation issues, and several nation-specific housing finance strategy papers. *The Challenge of Slums* itself contains the highlights of some twenty-nine city case studies – from Cairo and Lusaka to São Paulo and Los Angeles – as well as a useful statistical index. Another important analysis of global slums is Mark Kamer, *Dispossessed: Life in the World's Urban Slums* (Maryknoll, NY: Orbis, 2006). Mike Davis, *Planet of Slums* (London: Verso, 2006) is written as a direct response to *The Challenge of Slums* and is a passionate diatribe against the "neo-liberal" world order that, he argues, permits and profits from human degradation. Robert Neuwirth, *Shadow Cities: A Billion Squatters, a New Urban World* (London: York: Routledge, 2005) addresses the same material but sees the slums as "squatter settlements" and is sharply critical of UN-HABITAT.

See also John Hagedorn and Mike Davis, *A World of Gangs: Young Men and Gangista Culture* (Minneapolis, MN: University of Minnesota Press, 2008). For a longer-term perspective on the history of slums, consult the Engels and Du Bois texts cited above as well as William Julius Wilson, *The Truly Disadvantaged: The Inner City, the Underclass, and Public Policy* (Chicago, IL: University of Chicago Press, 1987) and *When Work Disappears: The World of the New Urban Poor* (New York: Knopf, 1996) as well as Elijah Anderson, *Code of the Street: Decency, Violence, and the Moral Life of the Inner City* (New York: W.W. Norton, 1999). For an even deeper historical view, consult Robert Roberts, *The Classic Slum: Salford Life in the First Quarter of the Century* (Manchester: University of Manchester Press, 1971), which describes one of the neighborhoods visited by Engels, and Tyler Anbinder's entertaining *Five Points: The 19th-Century New York City Neighborhood that Invented Tap Dance, Stole Elections, and Became the World's Most Notorious Slum* (New York: Free Press, 2001).

Following the adoption of the Millennium Declaration by the United Nations General Assembly in 2000, a Road Map was established identifying the Millennium Development Goals and Targets for combating poverty, hunger, disease, illiteracy, environmental degradation and discrimination against women and for improving the lives of slum dwellers. *The Challenge of Slums: Global Report on Human Settlements 2003* presents the first global assessment of slums. Starting from a newly accepted operational definition of slums, the report first presents global estimates of the number of urban slum dwellers, followed by an examination of the global, regional and local factors underlying the formation of slums, as well as the social, spatial and economic characteristics and dynamics of slums. Finally, it identifies and assesses the main slum policies and approaches that have guided responses to the slum challenge in the last few decades.

From this assessment, the immensity of the challenge posed by slums is clear and daunting. Without serious and concerted action on the part of municipal authorities, national governments, civil society actors and the international community, the numbers of slum dwellers are likely to increase in most developing countries. In pointing the way forward, the report identifies recent promising approaches to slums, including scaling up of participatory slum upgrading programmes that include, within their objectives, urban poverty reduction. In light of this background, the key findings and messages of this issue of the

Global Report on Human Settlements are presented below.

THE MAIN FINDINGS

In 2001, 924 million people, or 31.6 per cent of the world's urban population, lived in slums. The majority of them were in the developing regions, accounting for 43 per cent of the urban population, in contrast to 6 per cent in more developed regions. Within the developing regions, sub-Saharan Africa had the large proportion of the urban population resident in slums in 2001 (71.9 per cent) and Oceania had the lowest (24.1 per cent). In between these were South-central Asia (58 per cent), Eastern Asia (36.4 per cent), Western Asia (33.1 per cent), Latin America and the Caribbean (31.9 per cent), Northern Africa (28.2 per cent) and Southeast Asia (28 per cent).

With respect to absolute numbers of slum dwellers, Asia (all of its sub-regions combined) dominated the global picture, having a total of 554 million slum dwellers in 2001 (about 60 per cent of the world's total slum dwellers). Africa had a total of 187 million slum dwellers (about 20 per cent of the world's total), while Latin America and the Caribbean had 128 million slum dwellers (about 14 per cent of the world's total) and Europe and other developed countries had 54 million slum dwellers (about 6 per cent of the world's total).

It is almost certain that slum dwellers increased substantially during the 1990s. It is further projected that in the next 30 years, the global number of slum dwellers will increase to about 2 billion, if no firm and concrete action is taken. The urban population in less developed regions increased by 36 per cent in the last decade. It can be assumed that the number of urban households increased by a similar ratio. It seems very unlikely that slum improvement or formal construction kept pace to any degree with this increase, as very few developing countries had formal residential building programmes of any size, so it is likely that the number of households in informal settlements increased by more than 36 per cent. However, it is clear that trends in different parts of the world varied from this overall pattern.

In Asia, general urban housing standards improved during the decade, and formal building kept pace with urban growth, until the financial crisis of 1997. Even after the crisis, some countries like Thailand continued to improve their urban conditions. In India, economic conditions also improved in some cities such as Bangalore. However, it is generally considered that urban populations grew faster than the capacity of cities to support them, so slums increased, particularly in South Asia.

In some countries of Latin America, there was a wholesale tenure regularization and a large drop in numbers of squatter households, which would reduce the number of slums under most definitions. Also, urbanization reached saturation levels of 89 per cent, so that slum formation slowed. Still, housing deficits remain high and slums are prominent in most cities.

Most cities in sub-Saharan Africa and some in Northern Africa and Western Asia showed considerable housing stress, with rents and prices rising substantially while incomes fell, probably corresponding to higher occupancy rates. In addition, slum areas increased in most cities, and the rate of slum improvement was very slow or negligible in most places. In South Africa, a very large housing programme reduced the numbers in informal settlements significantly.

More than half of the cities on which case studies were prepared for this Global Report indicated that slum formation will continue (Abidjan, Abmedabad, Beirut, Bogota, Cairo, Havana, Jakarta, Karachi, Kolkata, Los Angeles, Mexico City, Nairobi, Newark, Rabat-Sale, Rio de Janeiro and Sao Paulo). A few (Bangkok, Chengdu, Colombo and Naples) reported decreasing slum formation, while the rest reported no or insufficient data on this topic (Durban, Ibadan, Lusaka, Manila, Moscow, Phnom Penh, Quito and Sydney).

There is growing global concern about slums, as manifested in the recent United Nations Millennium Declaration and subsequent identification of new development priorities by the international community. In light of the increasing number of urban slum dwellers, governments have recently adopted a specific target on slums . . . which aims to significantly improve the lives of at least 100 million slum dwellers by the year 2020. Given the enormous scale of predicted growth in the number of people living in slums (which might rise to about 2 billion in the next 30 years), the Millennium Development target on slums should be considered as the bare minimum that the international community should aim for. Much more will need to be done if 'cities without slums' are to become a reality.

Slums are a physical and spatial manifestation of urban poverty and intra-city inequality. However, slums do not accommodate all of the urban poor, nor are all slum dwellers always poor. Based on the World Bank poverty definitions, it is estimated that half the world – nearly 3 billion people – lives on less than US$2 per day. About 1.2 billion people live in extreme poverty, that is on less than US$1 per day. The proportion of people living in extreme poverty declined from 29 per cent in 1990 to 23 per cent in 1999, mostly due to a large decrease of 140 million people in East Asia during the period 1987 to 1998. However, in absolute terms, global numbers in extreme poverty increased up until 1993, and were back to about 1988 levels in 1998.

Despite well-known difficulties in estimating urban poverty, it is generally presumed that urban poverty levels are less than rural poverty and that the rate of growth of the world's urban population living in poverty is considerably higher than that in rural areas. The absolute number of poor and undernourished in urban areas is increasing, as is the share of urban areas in overall poverty and malnutrition. In general, the locus of poverty is moving to cities, a process now recognized as the 'urbanization of poverty'.

Slums and poverty are closely related and mutually reinforcing, but the relationship is not always direct or simple. On the one hand, slum dwellers are not a homogeneous population, and some people of reasonable incomes live within or on the edges of slum communities. Even though most slum dwellers work in the informal economy, it is not unusual for them to have incomes that exceed the earnings of formal sector employees. On the other hand, in many cities, there are more poor people outside slum areas than within them. Slum areas have the most visible concentrations of poor people and the worst shelter and environmental conditions, but even the most exclusive and expensive areas will have some low-income people. In some cities, slums are so pervasive that rather than designate residential areas for the poor, it is the rich who segregate themselves behind gated enclaves.

The majority of slum dwellers in developing country cities earn their living from informal sector activities located either within or outside slum areas, and many informal sector entrepreneurs whose operations are located within slums have clienteles extending to the rest of the city. Most slum dwellers are in low paying occupations such as informal jobs in the garment industry, recycling of solid waste, a variety of home-based enterprises and many are domestic servants, security guards, piece rate workers and self-employed hair dressers and furniture makers. The informal sector is the dominant livelihood source in slums. However, information on the occupations and income generating activities of slum dwellers from all over the world emphasizes the diversity of slum populations, who range from university lecturers, students and formal sector employees, to those engaged in marginal activities bordering on illegality, including petty crime. The main problems confronting the informal sector at present are lack of formal recognition, as well as low levels of productivity and incomes.

National approaches to slums, and to informal settlements in particular, have generally shifted from negative policies such as forced eviction, benign neglect and involuntary resettlement, to more positive policies such as self-help and in situ upgrading, enabling and rights-based policies. Informal settlements, where most of the urban poor in developing countries live, are increasingly seen by public decision-makers as places of opportunity, as 'slums of hope' rather than 'slums of despair'. While forced evictions and resettlement still occur in some cities, hardly any governments still openly advocate such repressive policies today.

There is abundant evidence of innovative solutions developed by the poor to improve their own living environments, leading to the gradual consolidation of informal settlements. Where appropriate upgrading policies have been put in place, slums have become increasingly socially cohesive, offering opportunities for security of tenure, local economic development and improvement of incomes among the urban poor. However, these success stories have been rather few, in comparison to the magnitude of the slum challenge, and have yet to be systematically documented.

With respect to the issue of crime, which has long been associated with slums and has accounted for much of the negative views of slums by public policy-makers, there is an increasing realization that slum dwellers are not the main source of crime. Instead, slum dwellers are now seen as more exposed to organized crime than non-slum dwellers as a result of the failure of public housing and other policies that have tended to exclude slum dwellers, including in matters of public policing. The result is a growing belief that most slum dwellers are more victims than perpetrators of crime. While some slums (especially traditional inner-city slums) may be more exposed to crime and violence, and may be characterized by transient households and 'counter-culture' social patterns, many are generally not socially dysfunctional.

THE MAIN MESSAGES

In facing the challenge of slums, urban development policies should more vigorously address the issue of livelihoods of slum dwellers and urban poverty in general, thus going beyond traditional approaches that have tended to concentrate on improvement of housing, infrastructure and physical environmental conditions. Slums are, to a large extent, a physical and spatial manifestation of urban poverty, and the fundamental importance of this fact has not always been recognized by past policies aimed at either the physical eradication or the upgrading of slums. Future policies should go beyond the physical dimension of slums by addressing problems underlying urban poverty. Slum policies should seek to support the livelihoods of the urban poor, by enabling urban informal sector activities to flourish, linking low-income housing development to income generation, and ensuring easy access to jobs through pro-poor transport and low-income settlement location policies.

In general, slum policies should be integrated with, or should be seen as part of, broader, people-focused urban poverty reduction policies that address the various dimensions of poverty, including employment and incomes, food, health and education, shelter and access to basic urban infrastructure and services. It should be recognized, however, that improving incomes and jobs for slum dwellers requires robust growth of the national economy, which is itself dependent upon effective and equitable national and international economic policies, including trade.

Up-scaling and replication of slum upgrading is among the most important of the strategies that have received greater emphasis in recent years, though it should be recognized that slum upgrading is only one solution among several others. The failure of past slum upgrading and low-income housing development has, to a large extent, been a result of inadequate allocation of resources, accompanied by ineffective cost-recovery strategies. Future slum upgrading should be based on sustained commitment of resources sufficient to address the existing slum problem in each city and country. Proper attention should also be paid to the maintenance and management of the existing housing stock, both of which require the consistent allocation of adequate resources. Slum upgrading should be scaled up to cover the whole city, and replicated to cover all cities. Up-scaling and replication should therefore become driving principles of slum upgrading, in particular, and of urban low-income housing policies in general. Some countries have made significant strides by consistently allocating modest percentages of their national annual budgets to low-income housing development, for example Singapore, China and, more recently, South Africa.

For slum policies to be successful, the kind of apathy and lack of political will that has characterized both national and local levels of government in many developing countries in recent decades needs to be reversed. Recent changes in the global economic milieu have resulted in increased economic volatility, decreasing levels of formal urban employment (especially in developing countries) and growing levels of income inequality both between and within cities. At the same time, economic structural adjustment policies have required, among other conditionalities, the retreat of the state from the urban scene, leading to the collapse of low-income housing programmes. Much more political will is needed at both the national and local levels of government to confront the very large scale of slum problems that many cities face today and will continue to face in the foreseeable future. With respect to urban poverty and slums, greater state involvement is, in fact, necessary now more than ever, especially in developing countries, given increasing levels of urban poverty, decreasing levels of formal employment and growing levels of income inequality and vulnerability of the urban poor.

There is great potential for enhancing the effectiveness of slum policies by fully involving the urban poor and those traditionally responsible for investment in housing development. This requires urban policies to be more inclusive and the public sector to be much more accountable to all citizens. It has long been recognized that the poor play a key role in the improvement of their own living conditions and that their participation in decision-making is not only a right, thus an end in itself, but is also instrumental in achieving greater effectiveness in the implementation of public policies.

Slum policies should seek to involve the poor in the formulation, financing and implementation of slum upgrading programmes and projects, building on the logic of the innovative solutions developed by the poor themselves to improve their living conditions. Such involvement, or participation of the poor, should also extend to the formal recognition of the non-governmental organizations (NGOs) working with the urban poor at both the community and higher levels, and their formal incorporation within the mechanisms of urban governance. Further, slum solutions should build on the experience of all interested parties, that is informal sector landlords, land owners and the investing middle class. This should be done in ways that encourage investment in low income housing, maximize security of tenure and minimize financial exploitation of the urban poor.

Many poor slum dwellers work in the city, ensuring that the needs of the rich and other higher income groups are met; the informal economic activities of slums are closely intertwined with the city's formal economy; and informal services located in slums often extend to the whole city in terms of clientele. Clearly, the task is how to ensure that slums become an integral, creative and productive part of the city. The broader context, therefore, has to be good, inclusive and equitable urban governance. But inclusive and equitable urban governance requires greater, not less, involvement of the state at both the national and local levels. Particularly needed in this respect are equitable policies for investment in urban infrastructure and services.

It is now recognized that security of tenure is more important for many of the urban poor than home ownership, as slum policies based on ownership and large-scale granting of individual land titles have not always worked. A significant proportion of the urban poor, may not be able to afford property ownership, or may have household priorities more pressing than home ownership, so that rental housing is the most logical solution for them – a fact not always recognized by public policy-makers. Slum policies have therefore started placing greater emphasis on security of tenure (for both owner-occupied and rental accommodation) and on housing rights for the urban poor, especially their protection from unlawful eviction. There is also increasing focus on the housing and property rights of women. Improving security of tenure and housing rights of slum dwellers lie at the heart of the norms of the Global Campaign for Secure Tenure (GCST), although several international organizations, especially bilateral, still place emphasis on formal access to home ownership and titling. However, it is clear that future policies should incorporate security of tenure and enhance housing rights of the poor, with specific provisions for poor women. For the poorest and most vulnerable groups unable to afford market-based solutions, access to adequate shelter for all can only be realized through targeted subsidies.

To improve urban inclusiveness, urban policies should increasingly aim at creating safer cities. This could be achieved through better housing policies for the urban low-income population (including slum dwellers), effective urban employment generation policies, more effective formal policing and public justice institutions, as well as strong community-based mechanisms for dealing with urban crime. Evidence from some cities, especially in Latin America and the Caribbean, points to the need to confront the underlying causes of urban crime and violence and making slums safer for habitation. During the 1960s and 1970s, the greatest fear among slum dwellers in some Latin American cities, especially those in squatter settlements or *favelas*, was of eviction either by government or private landowners. Today, this has been replaced by fear of violence and crime, including shootings related to drug trafficking. While more globally representative empirical evidence on the linkages between crime and slums is needed, some recent analyses suggest that slum dwellers are not a threat to the larger city, but are themselves victims of urban crime and related violence, often organized from outside slum areas. Slum dwellers are, in fact, more vulnerable to violence and crime by virtue of the exclusion of slums from preventive public programmes and processes, including policing.

To attain the goal of cities without slums, developing country cities should vigorously implement urban planning and management policies designed to prevent the emergence of slums, alongside slum upgrading and within the strategic context of poverty reduction. The problem of urban slums should be viewed within the broader context of the general failure of both welfare oriented and market-based low-income housing policies and strategies in many (though not all) countries. Slums develop because of a combination of rapid rural-to-urban migration, increasing urban poverty and inequality, marginalization of poor neighbourhoods, inability of the urban poor to access affordable land for housing, insufficient investment in new low-income housing and poor maintenance of the existing housing stock.

Upgrading of existing slums should be combined with clear and consistent policies for urban planning and management, as well as for low-income housing development. The latter should include supply of sufficient and affordable serviced land for the gradual development of economically appropriate low-income housing by the poor themselves, thus preventing the emergence of more slums. At the broader national scale, decentralized urbanization strategies should be pursued, where possible, to ensure that rural-to-urban migration is spread more evenly, thus preventing the congestion in primate cities that accounts, in part, for the mushrooming of slums. This is a more acceptable and effective way of managing the problem of rapid rural-to-urban migration than direct migration control measures. However, decentralized urbanization can only work if pursued within the framework of suitable national economic development policies, inclusive of poverty reduction.

Investment in city-wide infrastructure is a pre-condition for successful and affordable slum upgrading, as the lack of it is one strong mechanism by which the urban poor are excluded, and also by which improved slum housing remains unaffordable for them. At the core of efforts to improve the environmental habitability of slums and to enhance economically productive activities is the provision of basic infrastructure, especially water and sanitation, but also including electricity, access roads, footpaths and waste management. Experience has shown the need for significant investment in city-wide trunk infrastructure by the public sector if housing in upgraded slums is to be affordable to the urban poor and if efforts to support the informal enterprises run by poor slum-dwellers are to be successful. Future low-income housing and slum upgrading policies therefore need to pay greater attention to the financing of city-wide infrastructure development.

Experience accumulated over the last few decades suggests that in-situ slum upgrading is more effective than resettlement of slum dwellers and should be the norm in most slum-upgrading projects and programmes. Forced eviction and demolition of slums, as well as resettlement of slum dwellers create more problems than they solve. Eradication and relocation destroys, unnecessarily, a large stock of housing affordable to the urban poor and the new housing provided has frequently turned out to be unaffordable, with the result that relocated households move back into slum accommodation. Resettlement also frequently destroys the proximity of slum dwellers to their employment sources. Relocation or involuntary resettlement of slum dwellers should, as far as possible, be avoided, except in cases where slums are located on physically hazardous or polluted land, or where densities are so high that new infrastructure (especially water and sanitation) cannot be installed. In-situ slum upgrading should therefore be the norm, with justifiable involuntary or voluntary resettlement being the exception. Easy access to livelihood opportunities is one of the main keys to the success of slum upgrading programmes.

Chinese Cities in a Global Society

Tingwei Zhang

Editors' Introduction

Globalization has affected every country in the world, but perhaps the cities of no other country on the planet have been so dramatically transformed as those of China. Chinese-born scholar Tingwei Zhang notes that China is the largest country on earth in terms of population and that the sweeping market reforms carried out by the Communist Party leadership beginning in 1978 transformed not only China's cities but also China's economic relationship with the rest of the world. Nowadays, China's economy is the second largest in the world, and China's burgeoning cities are deeply embedded in both the world city network and the technological space of flows.

In "Chinese Cities in a Global Society," written specially for this edition of *The City Reader*, Zhang notes that China's rapid urbanization since the early 1980s has resulted in a geographically "uneven distribution pattern," with coastal cities favored over interior regions of the country. He also notes that most of the new urbanization has been driven by unprecedented rural-to-urban migration that present extraordinary challenges to Chinese city planners and policy makers. It has been difficult for the provision of housing and jobs to meet the demand of former villagers flowing into the cities, and the migration itself has seemed sometimes to go against the grain of traditional Chinese culture patterns rooted in village life. For all this, Zhang writes, the case of China's response to urban globalism has "lessons that can be shared with other developing countries." First, most of the population and economic growth of the foreseeable future will be urban. Second, urbanization has had many positive economic effects and does not necessarily result in "a burden to a nation's economy or a negative impact on cities." Third, "active public–private partnerships" are key to the success of economic and urban development. And fourth, important choices must be made in terms of the development policies that cities and central authorities embrace. Great progress can be achieved, but it will be necessary to take measures that avoid "increased social stratification and a widening gap between the urban poor and the new rich."

Tingwei Zhang received both his BA degree in architecture and his MA in urban planning from Tongji University in Shanghai. After being certificated as a United Nations Senior Planner at the University of Leuven in Belgium and further studies at both Tongji and the University of North Carolina at Chapel Hill, Zhang received his PhD in public policy analysis at the University of Illinois, Chicago, in 1992. He is currently professor of urban planning and policy and director of the Asia and China Research Program in the Great Cities Institute in the College of Urban Planning and Public Affairs at the University of Illinois, Chicago. He is also guest professor of urban planning at Tongji University and was president of International Association of City Planners (2005–2007), a member of the Global Planning Committee of the American Institute of City Planning (2001–2005). He is also a member of China National Planning Expert Committee, and as planning adviser to several Chinese cities including Wuhan, Shenzhen, and Shanghai. He serves on the editorial boards of several academic journals including *City Planning Review, Urban Planning Forum, Planners, Urban Planning International*, and *Time and Architecture*. His research interests cover planning theory, China's transition and urban policy, and urban development in American cities. He has published over a hundred articles and book chapters in China, France, the United Kingdom and United State and authored and coauthored seven books published in China and Switzerland.

Among Tingwei Zhang's principal publications are *Design and Development of Waterfront Areas*, with Feng Hui and Peng Zhiquang (Shanghai: Tongji University Press, 2002), *Citizen, Local Government and the Development of Chicago's Near South Side*, with David Ranney and Pat Wright (Geneva: United Nations Research Institute for Social Development, 1997), *Principles of City Construction and City Planning*, with Ran Yishan, (Tianjin: Tianjin Sciences and Technology Press, 1993), and *Urban Planning for Small Towns* (Beijing: China Construction Industry Press, 1986).

There are many books on the history of Chinese urbanization and urban culture. Among the best are Patricia Buckley Ebrey and Kwang-Ching Liu, *The Cambridge Illustrated History of China* (Cambridge: Cambridge University Press, 1999), John King Fairbank and Merle Goldman, *China: A New History*, 2nd edn (Cambridge, MA: Belknap/Harvard University Press, 2006), and J.A.G. Roberts, *A Concise History of China* (Cambridge, MA: Harvard University Press, 1999). Good sources on contemporary urban development in China include Jieming Zhu, *The Transition of China's Urban Development: From Plan-Controlled to Market-Led* (Westport, CT: Praeger, 1999), John Friedman, *China's Urban Transition* (Minneapolis, MN: University of Minnesota Press, 2005), and Larry Ma and Fulong Wu, *Restructuring the Chinese City: Changing Society, Economy, and Space* (London: Routledge, 2005). John R. Logan (ed.), *The New Chinese City: Globalization and Market Reform* (Oxford: Blackwell, 2002) provides a useful compilation of readings. Other useful sources are Fulong Wu, *Globalisation and the Chinese City* (London: Routledge, 2006), and Joseph Esherick (ed.), *Remaking the Chinese City: Modernity and the National Identity, 1900–1950* (Honolulu, HI: University of Hawaii Press, 2002). There are also many interesting and useful studies of individual Chinese cities. For Shanghai, Stella Dong, *Shanghai: The Rise and Fall of a Decadent City, 1842–1949* (New York: Morrow, 2000) is fascinating. For Beijing, consult Jasper Becker, *The City of Heavenly Tranquility: Beijing in the History of China* (Oxford: Oxford University Press, 2008), and Stephen Haw, *Beijing: A Concise History* (London: Routledge, 2008). For Hong Kong, see Steve Tsang, *A Modern History of Hong Kong* (London: I.B. Tauris, 2007), and John M. Carroll, *A Concise History of Hong Kong* (Lanham, MD: Rowman & Littlefield, 2007).

It is well known that China is the largest country in the world in terms of population. China's total population of 1.31 billion in 2005 was 20.6 percent of the world population. That means that one out of every five human beings is Chinese. This huge population size suggests the importance and the complicated relation of China to the rest of the world. Because about one-half of the Chinese now live in cities, this contribution focuses on China's urbanization in the context of globalization since the early 1980s. It discusses the history and development trends of Chinese cities, painting a holistic picture of Chinese cities, and exploring internal driving forces to China's urbanization including the nation's history, culture, and urban policy, as well as external forces such as global capital mobilization, in an approach that compares the Chinese case to urbanization in other countries.

The first section reviews China's urbanization trajectory, highlighting the huge urban population size, the rapid urbanization speed, a fluctuating urbanization in trajectory, and an uneven distribution pattern of cities. In the second section, both internal and external factors will be examined to explore influential driving forces to China's rapid urbanization since the early 1980s. A brief conclusion on lessons learned from the China case will be provided in the third section.

CHINA'S URBANIZATION: SIZE, TRAJECTORY AND DISTRIBUTION PATTERNS

China's urbanization has four notable characteristics: huge size, a fluctuating trajectory, a rapid growth rate since the 1980s, and an uneven distribution pattern. These characteristics may be attributed largely to China's history, culture and urban policy (Table 1).

According to UN-HABITAT, China's urban population was 561.6 million in 2005, which is 74.5 percent of all urban population in the developed world including North America, the EU, Japan and Australia. China's urban population alone (831 million in 2008) is 2.7 times the total population of the United States (304 million in 2008). In 2007, China had two cities (Shanghai and Beijing) among the twenty largest cities in the world; the number will increase to four (the two

Urban population			Number of cities
Urban residents	Migrants	Total population in urban area	
606 million (45.7% of total population)	225 million	831 million	655*

Table 1 Basic information about China's urbanization (2008)

Source: China POPIN (China Population Information Network), 2009, National Bureau of Statistic of China, 2009

Note: * Four state-administrated cities (Beijing, Shanghai, Tianjin, Chongqing; a status similar to Washington, DC in the United States) plus 651 other cities

already mentioned plus Guangzhou and Shenzhen) in 2025, according to UN-HABITAT predictions.

In the history of the world, China had the largest cities in the Tang Dynasty through the Qing Dynasty. Chang'An (Xi'An today), the capital of the Tang Empire, had a population of over 1 million in 500 BCE; as the capital of Qing Empire, Beijing had a population of over 1 million in the 1700s to 1800s. The cultural tradition of capital cities with the largest population as administration centers under government's rigid control, rather than trading and commercial centers in hands of merchandisers like European cities, has left a legacy to Chinese urban development and urban policy up to the 1950s (to a much less extent, even nowadays).

China's urbanization has experienced a fluctuating trajectory. Despite the fact that China had the largest population in the world ever since the Zhou Dynasty (1066 BCE), most Chinese people lived in the countryside for thousands of years. Traditionally, Chinese culture is basically an agriculture culture respecting farm activity and despising commerce and trading. Early in the *Spring and Autumn Era* in 770 BCE, the Confucians ranked all citizens as "official, farmer, handcraft producer, and merchandiser," a categorization that lays the foundation of the anti-urban tradition. Throughout Chinese history, the urban population concentrates in several administrative cities with a low urbanization rate until the 1980s.

The urbanization process since the early 1950s has been especially fluid, although the main trend is a continuous urban population growth (Figure 1 and Table 2). From 1951 to 1965 under the Soviet-style planned economy, urban population increased consistently with industrialization, except in a short period from 1960 to 1962 when a recession caused a forced urban–rural relocation to reduce pressure on urban services. The resident registration policy that requires all urban residents to register at urban police stations was put in place in the 1950s aiming to control rural–urban migration, which reflects the limited urban resources to be shared with migrants at that time. In 1965, the urbanization level reached 18 percent, a lower figure compared to now but the peak of the pre-reform era. The political movement of the Cultural Revolution from 1966 to 1976 again forced city residents to move to the countryside, making the urbanization level drop to 17.3 percent in 1975.

The steady increase of urban population since the sweeping and historic reforms of 1978 under Deng Xiaoping that introduced the market mechanism into the formerly planned economy and supported a decentralization of decision-making power from the central to the local level is a result of combined internal and external forces. Urbanization rates increased rapidly especially since the 1990s when China was becoming the "world's manufacturing plant," a by-product of the globalization process. Urban population increased by 18.2 million annually from 1995 to the 2000s, and the urbanization level reached 45.7 percent in 2008, the highest in China's history.

Due to both economic and social considerations, the resident registration system has been relaxed since the 2000s, and it will probably be abolished in the near future. The elimination of the migrant registration system known as *Hukou* removes the critical obstacle to rural–urban migration so it could be expected that even more of the rural population will move to cities. Moreover, urbanization has been officially employed as a means to promote economic growth. In a national debate in the 1990s about "the importance of urbanization to China's economy," supporters argued that urbanization had fallen behind China's industrialization and economic development so it should be

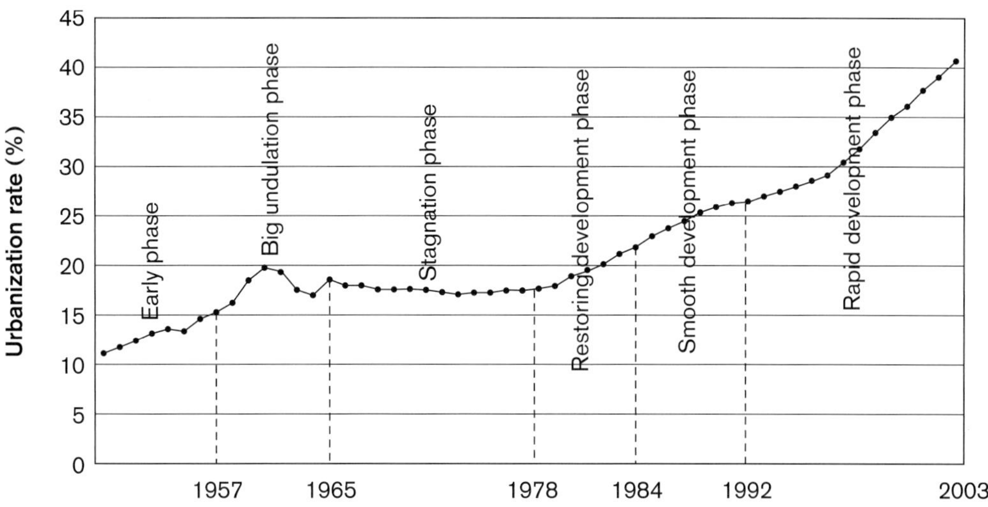

Figure 1 Trajectory of China's urbanization

Source: based on National Statistics Bureau of China, 2004 (urban population refers to "officially registered urban population" excluding migrants)

Year	1951	1965	1975	1978	1995	2000	2002	2008
Urbanization	11.8	18.0	17.3	17.9	29.0	36.2	39.1	45.7

Table 2 China's urbanization level (%, in selected years)

Source: www.stats.gov.ch (urban population refers to "officially registered urban population" excluding migrants)

encouraged. At that time, the central government adopted the policy of "speeding up urbanization to stimulate economy growth" by promoting urbanization at a rate of 1.5–2 percent annually. The average urban growth rate in all developing nations was 1.83 percent from 1990 to 2000, so China's urban growth seems matching the world trend (UN-HABITAT, *State of the World's Cities 2008*). In practice, expanding existing cities by creating new urban districts to absorb new factories and employees (migrants as majority), and reducing the amount of peasants by merging rural townships and villages (*che xiang bing cun*) were the main strategies. From 1995 to 2002, the average urbanization level increased by 1.44 percent annually. Some Chinese scholars warn that given China's huge population baseline, the urbanization growth rate is too high to be sustainable. They suggest that an annual growth rate of 0.8 percent, or at most 1 percent will be more appropriate and rational.

The post-reform urbanization has been facilitated not only by the government but also by the market-place. Foreign and domestic investment in real estate and manufacturing stimulated the demand of urban land and urban labor. Most urban expansion or the so-called "Chinese version of sprawl" has been a result of the establishment of suburban Economic Development Zones (EDZs) where local government attracts Foreign Direct Investment (FDI) by providing low-priced land and free infrastructure. Researchers reported that 34–46 percent of urban land expansion was caused by the government-led land acquisition in EDZ. The immature land and housing market often fueled irrational land acquisition via large-scale land use conversion from agriculture to urban uses.

In *State of the World's Cities 2008*, UN-HABITAT scholars discuss the uneven distribution of the world's cities. Most cities are located in coastal regions – either near seas or oceans, or by rivers or lakes. In 2000, 65 percent of the populations in the world's waterfront areas were urban residents. The same pattern is found in China. With only 2 percent of the land territory, China's coast area contains 23 percent of the urban

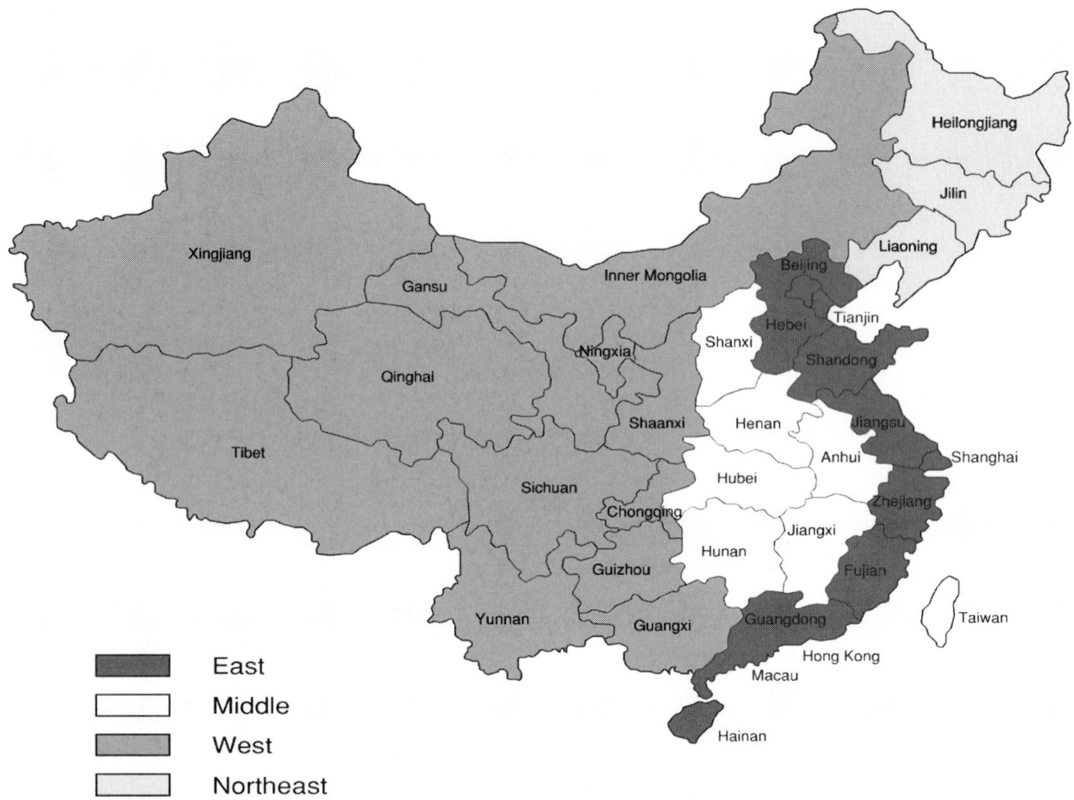

Figure 2 China's urban development pattern: the four zones

Source: National Statistics Bureau of China, 2005

population, or 14 percent of China's total population. Chinese cities concentrate in the rich area of the east coast region. This region has three state-managed cities (Beijing, Shanghai, and Tianjin) and 146 cities, comprising 22.7 percent of all Chinese cities (Figures 2 and 3). Shanghai, China's economic hub with a population of 1,578.9 million, and Beijing, the political and culture center with a population of 1,174.1 million in 2010, together with Hong Kong, an international trading and financing hub, play key roles in China's political and economic life. The other regions, especially the west region, lag behind economically and demographically, although their territories are very large. Since 2003, the central government has announced a set of policies to promote economic development of the vast west and the middle China region, aiming to reach a "balanced development" across the nation. But the outcome has not been as much of a success as expected, although statistics do show that the annual GDP growth rate in the west region has been faster than that in the east region in recent years. In part, the higher urban economic growth rate should be attributed to the lower baseline of the west region, rather than to the policy of improving life quality in the west, which is evidenced by people still moving to the coast region from the west. It seems that this trend will continue, and more urban residents will be moving from west and central-region cities to coast cities in addition to rural–urban migrants, a phenomenon UN-HABITAT scholars observed also in many Latin American countries.

China now has four metropolitan areas, or so-called "city-regions": the Yangtze River Delta led by Shanghai; the Jing-Jin-Tang region centered at Beijing and Tianjin; the Pearl River Delta led by Hong Kong, Shenzhen, and Guangzhou; and the South Liaoning region centered in Shenyang and Dalian. With the exception of the South Liaoning region which still remains as a domestic regional economic center, the city-regions are all recognized as main engines of

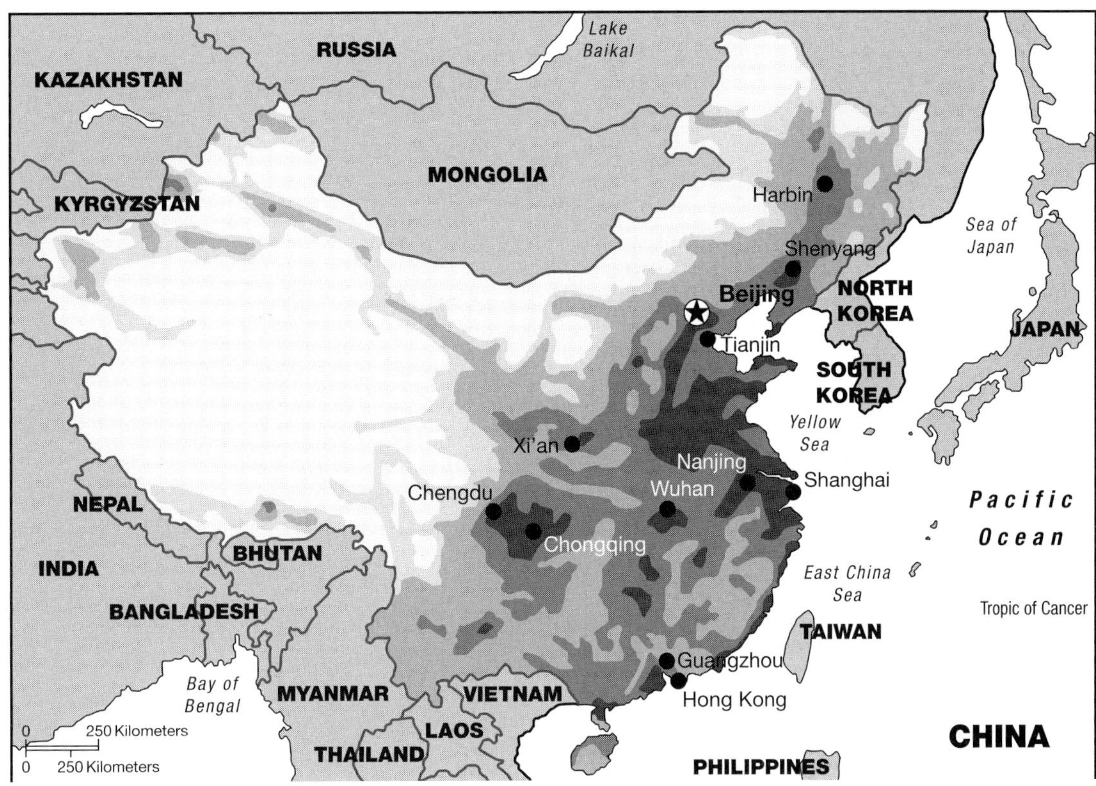

Figure 3 China's population density

Source: www.travelchinaguide.com/images/map/

China's rapid economic growth serving to realize the nation's ambitions as a global power.

Hong Kong, Shanghai, and Beijing are all striving to become international service centers in the next stage of globalization. Hong Kong is already a mature international trading and financing hub and is listed together with several top developed nations as a "developed economy" with a per capita GDP of $34,552 in 2008. Its goal is to retain the status of international financing hub and Asia's main container-shipping transfer center. Shanghai, with a per capita GDP of $10,754 in 2008, is competing with Hong Kong, aiming to be a new global financing and logistics center from China's most important manufacturing center. The central government announced full support to Shanghai's development goal at the national congress in the spring of 2009. Beijing's per capita GDP also reached $9,269 in 2008, which exceeds the international standard for middle to high income nations according to a 2008 World Bank report. In addition to being China's political and culture center, Beijing is actively seeking opportunities for high-tech and financing industry development and intends to become another of China's major economic centers.

There are a number of second-tier cities, located basically in the middle-China region, such as Wuhan of Hubei Province, Zhengzhou of Henan Province, and Changsha of Hunan Province. A few second-tier cities are in the vast west region, including Chongqing, Xi'an of Shaanxi Province, and Chengdu in Sichuan Province. These cities are regional economic, political and culture centers, and a few of them – for example, Chongqing and Xi'an – have experienced significant urban growth. All of the second tier cities are considered "super-big" cities each with a population of over 1 million, especially Chongqing with a total metropolitan population of 28.4 million (6.9 million urban residents).

It is important to understand the Chinese urban classification system. There are four classes of cities based on population size. A city with a population of less than 200,000 is considered a small town in China, although it could be a big city in other contexts. A population of 200,000 to 500,000 is a medium-size city in China, which differs significantly to the US city hierarchy system in which a city of 500,000 is recognized as a big city. From half a million to 1 million population is a big city, and over 1 million population is a "super-big city" (*teda cheng shi*). The number of big and super-big cities is increasing at a rate faster than that of small towns, which echoes UN-HABITAT's finding in Asian countries other than India. China's big and super-big cities are increasing at a rate of 3.9 percent annually from 1990 to 2000, which is two times faster than the world average rate for big cities (UN-HABITAT, *State of the World's Cities 2008*). It is foreseeable that more big and super-big cities will appear in China by 2050.

INTERNAL AND EXTERNAL FORCES ON URBAN GROWTH SINCE THE REFORM

It is obvious that China's urbanization level displays different increase rates throughout its history, especially in the pre- and post-reform era. As an agriculture society for about 5,000 years, China had a low urbanization rate with most urban residents living in only a few big cities. Entering modern times, China suffered by civil wars and foreign invasions, so its urbanization level remained low. The urbanization trajectory since the early 1950s (see Figure 1) suggests that urbanization in the planned economy was largely under the control of the central government and heavily influenced by central policies and various political movements. At that period, a very weak marketplace had almost no impact on population allocation and urban development. The urban resident registration policy which reflected the anti-urbanism ideological bias as a legacy of China's pro-agriculture culture (and numerous peasant-rebel political traditions) a distorted economic strategy of "recession recovery through reducing urban population and services," a weak national economy all contributed to low urbanization level, and resulted in the "controlled urbanization" before the 1980s. Isolated from the west since the 1949 communist revolution, internal forces were the only factor influencing urbanization in the pre-reform era.

The 1978 reform brought China into a new age. The reform was a fundamental paradigm shift from an ideology-led to an economy-led society and government. In general, the reform consists of two stages: reforming the agriculture sector in rural areas from 1978 to early 1980s; followed by reforms in manufacturing and trading sectors in cities in the 1980s to 1990s. With the implementation of the urban reform, three main policies were formulated: introducing a market mechanism to replace the planned system in all economic realms; decentralizing decision-making power on urban development issues from the central to the local government; and establishing the urban land and housing market to materialize the market value of urban land. China's constitution has been revised several times to reflect the new foundation of the nation – a free market economically with an authoritarian government politically. For example, the 1988 constitution separates urban land ownership (state-owned) from land-use right (a commodity) which lays the foundation for the property law by which a partial property right is created and protected.

By introducing market forces – foreign investments as well as western entrepreneur ideas – the reform created China's economic miracle and urban boom since the 1980s. The 1978 "Open door to outside world" reform was implemented at an early stage of the globalization era. It was historically perfect timing. As western investors and cross-national companies looked for growth opportunities globally, an open China provided the ideal place due to its vast market potential, cheap labor, inexpensive land and energy, fewer regulations on environmental and labor protection, and, more importantly, a market-friendly government providing almost free infrastructure and favorable tax regulations. From then on, both the internal and external factors have contributed to the rapid urban growth.

As an external force, globalization impacts Chinese cities in various ways. To multinational companies globalizing their supply chains, reducing production costs – the labor cost in particular is one of the key motivations to invest in China. Research found that 68 percent of China's manufacturing workers and 80 percent of construction workers are rural–urban migrants, and their average wage is only about $100 to $120 per month. Foreign direct investment (FDI) creates jobs that attract migrants to cities, which in turn stimulates booming urban economies since the 1980s. International managers and visitors generate huge demands for high-end hotels, housing and commercial development; these projects compose the major part of development activity in old downtowns and suburban new towns. Once China becomes a key hub in the global supply chain, urban growth led by output-oriented manufacturing developments takes place in cities and town all over China, the east coast region in particular.

Domestically, continuous economic growth provides important legitimacy to the non-elected government, especially at the local level. In addition, GDP growth rate may decide a mayor's fate under the Chinese promotion mechanism which gives great weight to a city's economic performance in making promotion decisions. Therefore, to ensure a city's attractiveness to FDI becomes a common practice in cities of all sizes. Declining profits from agricultural activity and a shrinking job pool in village- and township-owned enterprises in rural areas also push farmers to leave the countryside and seek their fortunes in cities. The outcome of the combination of the pulling and pushing forces is more migrants heading to the cities.

With the decentralization policy transferring decision power on urban development to local governments, two key elements for development projects – development funds and land – are now in the hands of mayors. The Chinese constitution divides land into two classes: all urban land is owned by the state, and rural land is owned collectively by farmers. Selling urban land (more accurately, leasing the land-use right) becomes the main source of local revenue. Researchers found that about 40 percent of local budgets are generated from land leasing; and most land-generated income goes to infrastructure and various public projects such as new sport and culture facilities. A new type of development company has thus emerged: the City Investment Corporation (CIC). CICs are owned, funded and supervised by the local municipality, but they operate as for-profit private companies. This is true not only at the municipal level, but also at urban district – even street office – level (the lowest level of administrative agency in Chinese cities to take care of routine maintenance and management business in a community). The neo-liberalist idea of "selling the place" has been employed fully in Chinese cities. City Beautiful-style projects targeting the international market can be easily seen in all cities.

Transferring rural land to urban uses is another key element to urban development. Chinese law allows rural land to be converted to urban uses only by municipal governments. Land acquisition as a public action applies to all kinds of projects, from manufacturing to housing. This power was held by the central government before the reform, and the new policy of diverting land-use decisions to localities implies considerable benefits to cities and stimulates rapid urban growth, even urban sprawl. Acquiring rural land at a lower price and leasing it for urban uses at a higher price helps cities to accumulate funds for development projects, at a cost of causing potential unrest among rural farmers.

Scholars have pointed out that removing the obstacles on rural–urban migrants imposed via the registration system (*Hukou*) has released the huge pressure on labor markets in the coast region where FDI-invested factories demand more workers, especially the low skilled and low paid. Therefore, globalization in the form of FDI has created a local–global interaction: Chinese cities offering cheap labor and land to meet the needs of global capital; multinational companies bringing capital and technology to Chinese cities to take the advantage of low production costs and taxes. As the vehicles of profit generation and transfer, Chinese cities are booming.

A typical case is the growth of Shenzhen, a super-big city in southern China situated immediately north of Hong Kong. Shenzhen had over 12 million urban residents in 2007, although the official statistics list the population as being 8.6 million, which excludes as many as 4 million temporary residents. The super-big city was developed from a small town of less than 20,000 in 1980 when it was officially announced as China's first Special Economic Zone (SEZ) where national taxation and trading regulations were reduced to a very low level. Between 1980 and 2010, its population increased by 600 times. The economic success has stimulated both the city's prosperity and growth problems as well. According to the latest survey in 2008, Shenzhen's population density tops large and medium cities in the Chinese mainland at 3,597 people per square kilometer. That density is growing 15.32 percent per year. In comparison, the population density in Beijing is 881 people per square kilometer, 2,902 in Shanghai, and 975 in Guangzhou. It is estimated that by 2025, the city will run out of land while education and hospital sectors have been overloaded for years.

Unlike other cities, where population growth is a result of natural birth of permanent residents, Shenzhen's population growth has been driven by an inflow of floating residents. According to the

2000–2001 Census by the Shenzhen Population and Family Planning Bureau, Shenzhen's temporary population grew 22.11 percent to 6.77 million. Another concern is that 53.05 percent of the floating population are female and most are of child-bearing age, which has made it more difficult for the local government to control and improve the population. A most recent solution is to expand the city's jurisdictional territory to five times what it is today. The Hong Kong authority would be happy to see the expansion, because Shenzhen is viewed as the "business backyard" to Hong Kong. It is very possible that a megacity will appear at the boundary of the mainland and Hong Kong in the near future.

To summarize, China has experienced significant urban growth since the 1978 reforms, although its urban civilization has existed for over 5,000 years. Both internal and external forces contributed to the rapid urbanization between 1980 and 2010. While globalization provides China a unique opportunity to connect to the outside world and join the global economy, the reform policy makes Chinese cities ready to catch global capital and advanced technology which will eventually extend the range of China's economic miracle as well as its urban development.

CONCLUSION: LESSONS AND OBSERVATIONS

Reviewing China's urbanization since the early 1950s, we may draw some lessons that can be shared with other developing countries.

First, China's rapid urban growth since the early 1980s, measured both by the increasing number of urban population and cities and by the rate of urbanization, displays an impressive record. The case of China's urbanization supports UN-HABITAT's finding that the urban population increase in developing countries is experiencing its fastest growth period – one that differs significantly from the slower urban population growth in developed nations – and that the trend will continue. In fact, the increase of urban population in developing nations contributes 95 percent of urban population growth in the world, and China plays a critical role in that growth.

Second, while many developing countries, Latin American and African countries in particular, face severe challenges in urbanization due to poorer economic performance and lack of urban resources for migrants, the case of China may provide lessons to these developing nations. China's rapid urbanization has had a positive relationship with its unprecedented economic growth. Economic development in China is the driving force as well as the product of its rapid urbanization since the economic and political reforms of the 1980s. Urban population increase, therefore, does not necessary mean a burden to a nation's economy or a negative impact on cities.

Third, an active public–private partnership has been a key factor to China's success, and the building of such a partnership could help other developing nations to rise out of urban poverty. China had an authoritarian regime from 1949 to 1976, and the heavy-handed government failed in economic development and urbanization largely because of its anti-market ideology and anti-urban policy. With globalization came the opportunity for reform and real progress for China in the 1980s, and those reforms allowed cities to use international capital resources and exploit opportunities for global manufacturing relocation. Public investment in infrastructure and communication projects also has important impacts on localities' output-oriented strategy. UN scholars found that 40 percent of world cities benefit from transportation improvement projects funded by central governments. The economic success of cities in China's east coast region, where highways, airports and harbors are well developed, can be positive role-models for cities throughout the developing world.

Fourth and finally, globalization can have positive and/or negative impacts on cities, largely depending on the development policies the cities embrace. The China case reveals many complicated lessons – particularly on the way economic success and urban growth can be achieved, but often at a cost of increased social stratification and a widening gap between the urban poor and the new rich.

"From Global Cities to Globalized Urbanization"

Neil Brenner and Roger Keil

Editors' Introduction

As with the emergence of the modern industrial city and the twentieth-century metropolis, the rise of a new kind of urban reality in the age of globalism has spawned an enormous body of descriptive, analytical, and theoretical literature that has led – and continues to lead – to a fuller understanding of the still-emerging urban future. No scholars have studied that literature more carefully and persuasively than Neil Brenner and Roger Keil, the co-editors of *The Global Cities Reader* (2006) in the Routledge Urban Reader Series.

Neil Brenner is an expert on urban political economy, urban geography, and urban theory who studied at Yale and the University of California, Los Angeles, before receiving his PhD from the University of Chicago in 1999. He is now associate professor of sociology and metropolitan studies at New York University and serves on the editorial boards of *European Urban and Regional Studies* and *Antipode: A Radical Journal of Geography.* He is the author of the seminal article "Global Cities, 'Glocal' States: Global City Formation and State Territorial Restructuring in Contemporary Europe" (*Review of International Political Economy*, 1998). Roger Keil received his doctorate from the University of Frankfurt and is the director of the City Institute at York University in Toronto, where he is also professor of environmental studies and director of the Canadian Centre for German and European Studies. Keil is the author of *Los Angeles: Urbanization, Globalization and Social Struggles* (Chichester: Wiley, 1998), *Nature and the City: Making Environmental Policy in Toronto and Los Angeles*, with Gene Desfor (Tucson, AZ: University of Arizona Press, Nature and Society Series, 2004) and *Networked Disease: Emerging Infections and the Global City* (Oxford: Wiley-Blackwell, 2008). He is the co-editor of the *International Journal of Urban and Regional Research* (IJURR) and a co-founder of the International Network for Urban Research and Action (INURA).

In "From Global Cities to Globalized Urbanization," specially commissioned for this edition of *The City Reader*, Brenner and Keil begin by stating that currently "all major indicators suggest that urbanization rates across the world economy are now higher and more rapid than ever before in human history." That revolutionary new reality, they argue, was prophesied by the French philosopher of urbanism Henri Lefebvre in his book *The Urban Revolution* (1970) where he "anticipated the 'generalization' of capitalist urbanization processes through the establishment of a planetary 'fabric' or 'web' of urbanized spaces." Today, they note, Lefebvre's "prediction is no longer futurist speculation" and that urbanization has now "come to condition all major aspects of planetary social existence and . . . the fate of human social life."

Very different from the realities analyzed by the Chicago school of urban researchers, and even from the visions of pioneers like Patrick Geddes who used the term "world cities" as early as 1924, Brenner and Keil argue that the contemporary urban world reveals "new forms of global connectivity – along with new patterns of disconnection, peripheralization, exclusion and vulnerability – among and within urbanizing regions across the globe." Examining the new urbanization as an expression of global capitalism in the post-World War II and post-Cold War contexts, they see new global cities that are increasingly detached from nation-states and subject to "supranational or global forces" that have been explored by Neo-Marxists like Lefebvre, David Harvey, and Manuel Castells (p. 572). In the eyes of these theorists, they observe, urbanization has now become "an active moment within the ongoing production and transformation of capitalist sociospatial configurations."

Turning their attention to global interurban networks and the ground-breaking work of Saskia Sassen (p. 554), Doreen Massey, Ananya Roy, Jennifer Robinson, and especially Peter Taylor and the Globalization and World Cities (GaWC) group at the University of Loughborough in the UK (p. 563), Brenner and Keil argue that world cities are not just major corporate headquarters locations nor even global command and control centers. Rather, the new global cities raise questions about "restructuring urban governance and the new contexts for urban social struggles." Increasingly, the process of studying these cities must engage "a broad range of globalized or globalizing vectors" that include not just "economic flows" but "the crystallization of new social, cultural, political, ecological, media and diasporic networks as well." In the end, the authors issue an "invitation to research – and action" to a new generation of urbanist scholars who, they hope, are reading this book. Building on the work and example of John Friedmann and others, Brenner and Keil challenge us to think, and act, more clearly about the realities of globalization that they regard as "a fundamentally disjointed, yet profoundly authoritarian, new world order." Whether this will lead to new "possibilities for radical or progressive social change," they write, "is ultimately a political question that can only be decided through ongoing social mobilizations and struggles."

For further reading about global cities and global urban networks, the best introductions are Neil Brenner and Roger Keil (eds), *The Global Cities Reader* (New York: Routledge, 2006) and the bibliographies attached to each selection in this part of *The City Reader*. Peter Taylor, *World City Network: A Global Urban Analysis* (London: Routledge, 2004) is fundamental to the study of global urbanism. Also important are Saskia Sassen, *The Global City: New York, London, Tokyo* (Princeton, NJ: Princeton University Press, 1991), *Globalization and its Discontents* (New York: New Press, 1998), *Global Networks/Linked Cities* (London: Routledge, 2002), and *Cities in a World Economy*, 3rd edn (Thousand Oaks, CA: Pine Forge Press, 2006); and Manuel Castells, *The Informational City: Information Technology, Economic Restructuring, and the Urban-Regional Process* (Oxford: Blackwell, 1991) and his magisterial Information Age Trilogy, especially *The Rise of the Network Society* (Oxford: Blackwell, 1996).

Other important sources include Henri Lefebvre, *The Urban Revolution* (Minneapolis, MN: University of Minnesota Press, 1970), Peter Marcuse and Ronald van Kempen (eds), *Globalizing Cities: A New Spatial Order?* (Oxford: Blackwell, 2000), Doreen Massey, *World City* (London: Polity, 2007), and J. John Palen, *The Urban World*, 8th edn (Boulder, CO: Paradigm, 2008).

Of special importance to the study of cities in a globalizing society are the works of Mike Davis (p. 195), especially *City of Quartz* (London: Verso, 1990) and *Planet of Slums* (London: Verso, 2006). Other useful overviews of the field include Fu-Chen Lo and Yue-Man Yeung (eds), *Globalization and the World of Large Cities* (Tokyo: United Nations University Press, 1998), John R. Logan (ed.), *The New Chinese City: Globalization and Market Reform* (Oxford: Blackwell, 2002), and Mark Abrahamson, *Global Cities* (Oxford: Oxford University Press, 2004).

■

INTRODUCTION

Urbanization is rapidly accelerating, and extending ever more densely, if unevenly, across the earth's surface. The combined demographic, economic, socio-technological, material-metabolic and sociocultural processes of urbanization have resulted in the formation of a globalized network of spatially concentrated human settlements and infrastructural configurations in which major dimensions of modern capitalism are at once concentrated, reproduced and contested. This pattern of increasingly globalized urbanization contradicts earlier predictions, in the waning decades of the twentieth century, that the era of urbanization was nearing its end due to new information technologies (such as the internet), declining transportation costs and new, increasingly dispersed patterns of human settlement. Despite these trends, all major indicators suggest that urbanization rates across the world economy are now higher and more rapid than ever before in human history.

Four decades ago, in his pioneering book, *The Urban Revolution* [1970], the French philosopher Henri Lefebvre anticipated the "generalization" of capitalist urbanization processes through the establishment of a planetary "fabric" or "web" of urbanized spaces. Today, Lefebvre's prediction is no longer a futuristic speculation, but instead provides a realistic starting point for inquiry into our global urban reality. This is not to suggest that the entire world has become a single, densely concentrated city; on the contrary, uneven spatial development, sociospatial polarization and territorial inequality remain pervasive, endemic features of modern capitalism. Rather, Lefebvre's prediction was that the process of urbanization would increasingly come to condition all major aspects of planetary social existence and, in turn, that the fate of human social life – indeed, that of the earth itself – would subsequently hinge upon the discontinuous dynamics and uneven trajectories of urbanization.

The urban revolution poses major challenges for the field of urban studies. As other contributions to *The City Reader* demonstrate, the origins of this research field lie in the concern to investigate relatively bounded urban settlements, understood as internally differentiated, self-contained "worlds," in isolation from surrounding networks of economic, political and environmental relationships – as, for instance, in the concentric ring model developed in the work of Chicago school of urban sociology. Today, however, it is not the internal differentiation of urban worlds within neatly contained ecologies of settlement, or the extension of such urbanized settlements into rural hinterlands, that constitutes the central focal point for urban studies. Instead, in conjunction with the uneven yet worldwide generalization of urbanization, we are confronted with new forms of global connectivity – along with new patterns of disconnection, peripheralization, exclusion and vulnerability – among and within urbanizing regions across the globe. How to decipher these transformations, their origins, and their consequences? What categories and models of urbanization are most appropriate for understanding them, and for coming to terms with their wide-ranging implications?

Since the early 1980s, critical urban researchers have devoted intense energies to precisely these questions: on the one hand, by analyzing emergent forms of globalized urbanization and their impacts upon social, political and economic dynamics within and beyond major cities; on the other hand, by introducing a host of new methods and conceptualizations intended to grasp the changing realities of planetary urbanization under late-twentieth and early-twenty-first-century capitalism. The resultant literatures on "world", "global" and "globalizing" cities contain fascinating, provocative and often controversial insights. Meanwhile, ongoing debates on the missing links and open questions within these literatures continue to inspire new generations of urban researchers as they work to decipher the urbanizing world in which we are living. In this brief chapter, we cannot attempt to survey the intricacies of these diverse research traditions (for a detailed introduction, overview and suggestions for further reading, see *The Global Cities Reader*, Routledge, 2006). Instead, we outline some of the methodological foundations and major lines of investigation within research on globalizing cities, while also alluding to several emergent debates and agendas that are currently animating this field, with specific reference to the conceptualization and investigation of global interurban networks. In so doing, we hope to stimulate readers of this book, the next generation of urban researchers, to contribute their own critical energies to the tasks of understanding and shaping the future dynamics and trajectories of planetary urbanization.

URBANIZATION AND GLOBAL CAPITALISM

Although the notion of a world city has a longer historical legacy, it was consolidated as a core concept for urban studies during the 1980s, in the context of interdisciplinary attempts to decipher the crisis-induced restructuring of global capitalism following the collapse of the post-World War II political-economic and spatial order. Until this period, the dominant approaches to urban studies tended to presuppose that cities were neatly enclosed within national territories and nationalized central place hierarchies. Thus, for example, postwar regional development theorists viewed the nation-state as the basic container of spatial polarization between core urban growth centers and internal peripheries. Similarly, postwar urban geographers generally assumed that the national territory was the primary scale upon which rank-size urban hierarchies and city-systems were organized. Indeed, even early uses of the term "world city" by famous twentieth-century urbanists such as Patrick Geddes and Peter Hall likewise expressed this set of assumptions. In their work, the cosmopolitan character of world cities was interpreted as an outgrowth of their host states' geopolitical power. The possibility that urban development or the formation of urban hierarchies might be conditioned by supranational or global forces was not systematically explored.

This nationalized vision of the urban process was challenged as of the late 1960s and early 1970s, with the rise of radical approaches to urban political economy. The seminal contributions of Neo-Marxist urban theorists such as Henri Lefebvre, David Harvey and Manuel Castells generated a wealth of new categories and methods through which to analyze the specifically capitalist character of modern urbanization processes. From this perspective, contemporary cities were viewed as spatial materializations of the core social processes associated with the capitalist mode of production, including, in particular, capital accumulation and class struggle. While these new approaches did not, at that time, explicitly investigate the global parameters for contemporary urbanization, they did suggest that cities had to be understood within a macrogeographical context defined by the ongoing development and restless spatial expansion of capitalism. In this manner, radical urbanists elaborated an explicitly spatialized and reflexively multiscalar understanding of capitalist urbanization. Within this new conceptual framework, the spatial and scalar parameters for urban development could no longer be taken for granted, as if they were pre-given features of the social world. Instead, urbanization was now increasingly viewed as an active moment within the ongoing production and transformation of capitalist sociospatial configurations.

Crucially, these new approaches to urban political economy were consolidated during a period in which, throughout the older industrialized world, cities, regions and national economies were undergoing any number of disruptive sociospatial transformations associated with the crisis of North Atlantic Fordism and the consolidation of a new international division of labor dominated by transnational corporations. Fordism was the accumulation regime that prevailed in much of the Western industrialized world during the post-World War II period through the early 1970s. Productivity increases in the Fordist model were grounded upon mass production technologies and tied closely to a class compromise between capital and labour that contributed to relatively collaborative industrial relations and rising working class incomes; the latter were in turn reinforced through an expanding welfare state apparatus that stabilized domestic demand for consumer goods. Internationally, Fordism was regulated and reproduced through American cultural, financial and military hegemony and was rooted in the impressive dynamism of large-scale industrial regions across the older industrialized world. This sociospatial formation was widely superseded, after the 1970s, due to the consolidation of increasingly flexible, specialized models of production, industrial organization and inter-firm relations, a tendential liberalization of various inherited institutional restraints upon market competition, a creeping commodification of social reproduction, generally weaker welfare states, and the emergence of new patterns of regional growth and decline across the world economy. In the global North, older industrial regions such as Detroit, Chicago, the English Midlands, the German Ruhr district and parts of northern Italy underwent major economic crises characterized by plant closings, high unemployment rates and infrastructural decay. Meanwhile, new industrial districts generally located outside the traditional heartlands of Fordism – for instance, in Silicon Valley, southern California, parts of Southern Germany, Emilia-Romagna and parts of southern France – were experiencing unprecedented industrial dynamism and growth. Outside of the global core zones of capitalism, new forms of industrialization were emerging in key manufacturing regions within late developing states, for instance in Mexico, Brazil, South Korea, Taiwan and India. These transformations were accompanied by an increasingly prominent role for transnational corporations in all zones of the world economy.

Following the crisis of Fordism, extensive research emerged among urban scholars on topics such as industrial decline, urban property markets, territorial polarization, regionalism, collective consumption, local state intervention, the politics of place and urban social movements. Among many other, more specific insights, these research initiatives indicated that the sources of contemporary urban transformations could not be understood in purely local, regional or national terms. Rather, the post-1970s restructuring of cities and regions had to be understood as an expression and outcome of worldwide economic, political and sociospatial transformations. Thus, for instance, plant closings and workers' struggles in older industrial cities such as Chicago, Detroit, Liverpool, Dortmund or Turin could not be explained simply in terms of local, regional or even national developments, but had to be analyzed in relation to broader secular trends within the world economy that were fundamentally reworking the conditions for profitable capital accumulation and reconstituting the global geographies of industrial production. Analogous arguments regarding the significance of global context were meanwhile articulated regarding other major aspects of urban and regional restructuring, for instance, the crystallization of new patterns of intra-national spatial inequality, the emergence of new, place- and region-specific forms of economic and social policy, and the activities of new territorially based social and political movements.

In opening up their analyses to the global dimensions of urban restructuring, critical urban political economists in the 1970s and early 1980s also began to draw upon several newly consolidated approaches to the political economy of capitalism that likewise underscored its intrinsically globalizing dimensions. Foremost among these was the model of world system analysis developed by Immanuel Wallerstein and others, which explored the worldwide polarization of economic development and living conditions under capitalism among distinct core, semi-peripheral and peripheral zones. World system theorists insisted that capitalism could be understood adequately only on the largest possible spatial scale, that of the world economy, and over a very long temporal period spanning many centuries. World system theorists thus sharply criticized the methodologically nationalist assumptions of mainstream social science, arguing instead for an explicitly globalist, long-term understanding of modern capitalism. The rise of world system theory during the 1970s resonated with a more general resurgence of Neo-Marxian approaches to geopolitical economy during this period. In the context of diverse studies of transnational corporations, underdevelopment, dependency, class formation, crisis theory and the internationalization of capital, these new approaches to radical political economy likewise explored the global parameters of capitalism both in historical and contemporary contexts.

It is against this background that the emergence of the research field that has today come to be known as global cities research must be contextualized. Like the other critical analyses of urban restructuring that were being pioneered during the 1980s, global city theorists built extensively upon the analytical foundations that had been established by Neo-Marxist urban political economists, world system theorists and other radical analysts of global capitalism during the preceding decade.

GLOBAL CITIES AND URBAN RESTRUCTURING

According to Peter Taylor, "The world city literature as a cumulative and collective enterprise begins only when the economic restructuring of the world-economy makes the idea of a mosaic of separate urban systems appear anachronistic and frankly irrelevant." During the course of the 1980s and 1990s, the latter assumption was widely abandoned among critical urban researchers, leading to a creative outpouring of research on the interplay between urban restructuring and various worldwide economic – and, subsequently, political, cultural and environmental – transformations. Numerous scholars contributed key insights to this emergent research agenda, but the most influential, foundational statements were presented by John Friedmann and Saskia Sassen. To date, the work of these authors is associated most closely with the global city concept, and is routinely cited in studies of the interplay between globalization and urban development.

During the course of the late 1980s and into the 1990s, global city theory was employed extensively in studies of the role of major cities as global financial centers, as headquarters locations for TNCs and as agglomerations for advanced producer and financial services industries. During this time, much research was conducted on several broad issues:

- *The formation of a global urban hierarchy.* Global city theory postulates the formation of a worldwide urban hierarchy in and through which transnational corporations coordinate their production and investment activities. The geography, composition and evolutionary tendencies of this hierarchy have been a topic of intensive research and debate since the 1980s. Following the initial interventions of Sassen and Friedmann, subsequent scholarship has explored a variety of methodological strategies and empirical data sources through which to map this hierarchy (see the work of the GaWC research team at Loughborough University – www.lboro.ac.uk/gawc/, and the concept of a "new metageography" developed by Beaverstock, Smith, and Taylor in this part of *The City Reader*). However, whatever their differences of interpretation, most studies of the global urban system have conceptualized this grid of cities simultaneously not only as a fundamental spatial infrastructure for the accelerated and intensified globalization of capital, including finance capital, but also as a medium and expression of the new patterns of global polarization that have emerged during the post-1970s period.

- *The contested restructuring of urban space.* The consolidation of global cities is understood, in this literature, not only with reference to the global scale, on which new, worldwide linkages among cities are being established. Just as importantly, researchers in this field have suggested that the process of global city formation also entails significant social, technological and spatial transformations at the urban scale, within cities themselves, as well as within their surrounding metropolitan regions. According to global cities researchers, the globalization of urban development has generated powerful expressions in the built and sociospatial environment. In Castells' influential terminology, the construction of a global "space of flows" necessarily entails major transformations in the "space of places." For example, the intensified clustering of transnational corporate headquarters and advanced corporate services firms in the city core overburdens inherited land use infrastructures, leading to new, often speculative, real estate booms as new office towers and high-end residential, infrastructural, cultural and entertainment spaces are constructed both within and beyond established downtown areas. Meanwhile, the need for new socio-technological infrastructures and the rising cost of office space in the global city core may generate massive spillover effects on a regional scale, as small- and medium-sized agglomerations of corporate services and back offices crystallize throughout the urban region. Finally, the consolidation of such headquarters economies may also generate significant shifts within local housing markets as developers attempt to transform once-devalorized inner city properties into residential space for corporate elites and other members of the putative "creative class." Consequently, gentrification ensues in formerly working-class neighbourhoods and deindustrialized spaces, and considerable residential and employment displacement may be caused in the wake of rising rents and housing prices. Global cities researchers have tracked these and many other spatial transformations at some length: the urban built environment is viewed as an arena of contestation in which competing social forces and interests, from transnational firms, developers and corporate elites to workers, residents and social movements – struggle over issues of urban design, land use and public space. Of course, such issues are hotly contested in nearly all contemporary cities. Global cities researchers acknowledge this, but were particularly concerned in the 1980s and 1990s to explore their distinctive forms and outcomes in cities that had come to serve key command and control functions in the global capitalist system.
- *The transformation of the urban social fabric.* One of the most provocative, if also controversial, aspects of global cities research during its initial phase involved claims regarding the effects of global city formation upon the urban social fabric. Friedmann and Sassen, in particular, suggested that the emergence of a global city hierarchy would generate a "dualized" urban labor market structure dominated, on the one hand, by a high-earning corporate elite and, on the other hand, by a large mass of workers employed in menial, low-paying and/or informalized jobs. For many, at the time, the so-called *Blade-Runner*-scenario, named after the famous futuristic movie directed by Ridley Scott in 1982, provided a fitting set of images for these new patterns of sociospatial polarization within globalizing cities. Based on an imaginary Los Angeles, the film expressed what many social scientists saw as a possible future in which most urban inhabitants would be migrants, many of them poor and often spatially sequestered in residential enclaves and ghettos. John Carpenter's film *Escape from New York* (1981) developed a similarly grim prognosis for the future of New York, representing all of Manhattan as a high-security prison. For Sassen, this "new class alignment in global cities" emerged in direct conjunction with the downgrading of traditional manufacturing industries and the emergence of the advanced producer and financial services complex. Her work on London, New York and Tokyo suggested that broadly analogous, if place-specific patterns of social polarization were emerging in these otherwise quite different cities, as a direct consequence of their new roles as global command and control centers. This "polarization thesis" has attracted considerable discussion and debate. Whereas some scholars have attempted to apply their argument to a range of globalizing cities, other analysts, for example Peter Marcuse

and Ronald van Kempen, have questioned its logical and/or empirical validity.

In close conjunction with the consolidation of global cities research around the above-mentioned themes, many critical urban scholars began to extend the empirical scope of the theory beyond the major urban command and control centers of the world economy – that is, cities such as New York, London, Tokyo; as well as various supraregional centers in East Asia (Singapore, Seoul, Hong Kong), North America (Los Angeles, Chicago, Miami, Toronto) and Western Europe (Paris, Frankfurt, Amsterdam, Zurich, Milan). In this important line of research, the basic methodological impulses of global city theory were applied to diverse types of cities around the world, but particularly in the global North, that were undergoing processes of economic and sociospatial restructuring that had been induced through geoeconomic transformations. Here, the central analytical agenda was to relate the dominant socioeconomic trends within particular cities – for instance, industrial restructuring, changing patterns of capital investment, processes of labor-market segmentation, sociospatial polarization and class and ethnic conflict – to the emergence of a worldwide urban hierarchy and the global economic forces that underlie it. In this manner, analysts demonstrated the usefulness of global city theory not simply for analyzing the transnational command and control centers that had been investigated in the first wave of research in this field, but for exploring a broad range of urban transformations – also now including questions about the restructuring urban governance and the new contexts for urban social struggles – that were unfolding in conjunction with the post-1970s wave of geoeconomic restructuring. They thus signaled a significant reorientation of the literature away from "global cities" as such, to what Marcuse and van Kempen famously labeled "globalizing cities," a term intended to underscore the diversity of pathways and the place-specific patterns in and through which processes of globalization and urban restructuring were being articulated.

GLOBAL INTERURBAN NETWORKS – DEBATES AND HORIZONS

The debate on global city formation thus no longer focuses primarily on the headquarters locations for transnational capital, the associated agglomeration of specialized producer and financial services, and the resultant transformation of urban and regional spaces. Increasingly, work on globalizing cities engages with a broad range of globalized or globalizing vectors – including not only economic flows, but also the crystallization of new social, cultural, political, ecological, media and diasporic networks. In this context, scholars have begun to reflect more systematically on the nature of the very network connectivities that link cities together across the world system. Such explorations have animated various strands of empirical research on cities, as well as ongoing debates about the nature of globalized urbanization itself. The contours of research on global cities are now increasingly differentiated as the field expands and advances, but certain shared concerns have nonetheless emerged. Accordingly, we summarize here four major dimensions of global interurban connectivity that have, in recent years, been inspiring both research and debate among contemporary urbanists.

- *Types of interurban networks.* In the 1980s and 1990s, scholars tended to assume that a single global urban hierarchy existed; debates focused on how to map it, and on what empirical indicators were most appropriate for doing so. However, the discussion has shifted considerably during the last decade [since 2000], as researchers now argue that the world system is composed of multiple, interlocking interurban networks. While the question of transnational corporate command and control remains central, there is now an equal interest in global cultural flows, political networks, media cities and other modalities of interurban connectivity, including those associated with large-scale infrastructural configurations. For instance, the cases of Washington DC, Geneva, Brussels, Nairobi and other bureaucratic headquarters of the global diplomatic and NGO communities point towards a network of global political centers. Religious centers such as Mecca, Rome and Jerusalem, among many others, constitute yet another such network. Moreover, in some cases, places that ostensibly lack strategic economic assets nonetheless acquire global significance through their role in the worldwide networks of social movement activism. Porto Alegre, Brazil, where the World Social Forum has been based, and Davos, Switzerland, where the World Economic Forum takes place every January, are cases in point. This line of investigation suggests that, interwoven around the structures of capital that underpin the world urban system, there also exists a complex lattice-work of interurban linkages that are constituted around a broad range of interconnectivities.

- *The spatiality of interurban networks.* In contrast to the somewhat simplistic understanding of global cities as neatly bounded, local places in which transnational capital could be anchored, several scholars have suggested alternative understandings of the geographies produced through the processes of globalized urbanization. Doreen Massey, for instance, argues against the notion that global cities contain distinct properties that make them inherently global. Instead, she suggests an understanding of the global cities network as a set of dialectical relationships that connect actors in cities, and cities as collective actors, through a variety of simultaneously globalized and localized streams. Thus, the space of global cities is "*relational*, not a mosaic of simply juxtaposed differences" and the global city "has to be conceptualized, not as a simple diversity, but as a meeting place, of jostling, potentially conflicting, trajectories." Other scholars have explored the ways in which processes of global city formation have been connected to rescaling processes that rework inherited configurations of global, national, regional and local relations, often in unpredictable, unexpected ways. Newer research explores the methodological and empirical implications of these interventions with reference to diverse aspects of globalized urbanization, from urban political ecologies and governance realignments to new social movement mobilizations. Each breaks in important ways with inherited, relatively place-bound conceptualizations of global cities, pointing instead towards new concepts of relationality, topology and rescaling as bases for understanding the dynamics of globalized urbanization.
- *The scope of interurban networks.* Much global cities research in the 1980s and 1990s focused on major cities and city-regions in the global North. More recently, several scholars have questioned this focus, and explored some of its problematic implications for the conceptualization of global city formation itself. For instance, in an influential intervention, Jennifer Robinson criticized the project of classifying cities by their alleged importance in a single global hierarchy or network, arguing instead for a broader understanding of the diverse, often rather "ordinary" ways in which the globality of cities might be constituted and reproduced. While directing attention back towards locally embedded and place-based social relations, Robinson's work also advocates a reconceptualization of transnational flows and interconnectivities themselves, from points of view that are not focused one-sidedly on the logics of capital investment and finance. An analogous idea is taken up by Ananya Roy in her plea for a rethinking of the theoretical geographies of urban studies. She suggests

> a rather paradoxical combination of specificity and generalizability: that theories have to be produced *in* place (and it matters *where* they are produced), but that they can then be appropriated, borrowed, and remapped. In this sense, the sort of theory being urged is simultaneously located and dislocated.

In practical terms, the dynamic relationships between specificity and generalizability, expounded forcefully by Robinson, Roy and others, refer back, to some degree, to the necessity for all cities under contemporary capitalism to manage two divergent dynamics: their internal contradictions and their external integration. More generally, though, this line of research and theory suggests some highly productive ways in which cities throughout the world system – including those located outside of the economic "heartlands" of the global North – might also be investigated through the tools of a critical revised approach to globalized urbanization.

- *The dangers of interurban networks.* Although critical of them, most global city research in the 1980s and 1990s emphasized the newly emergent strategic connectivities of capital, labor and information across the world economy, which were widely viewed as the preconditions for local economic development. In that context, foreign direct investment and thick webs of interfirm relationships were seen as the "stuff" of which global city relationships were made. Of course, as noted earlier, such "positive" connectivities were seen as being deeply contradictory insofar as they intensified polarization and sociospatial inequalities both within and among cities. Yet, aside from this emphasis on the problem of polarization *in situ*, the downsides of interurban connectivity itself and failures in the network have only recently been recognized among critical urban researchers. There has always been a sharp divide between optimistic, normative versions of global city parlance and the often dystopic, critical or analytical uses of concepts such as global city or world city. Among the former, we can count the boosterist, hyperbolic attempts by city governments to rank a particular place among the top tier global cities that everyone talks about and that apparently attract all attention and investment. In recent years, the attention on mega-infrastructures such as airports and convention centres has been supplemented by an obsession with "human capital" and creativity. Yet in both the boosterist and the critical literatures, little has been said specifically about the pitfalls and vulnerabilities that lie *within* the global interurban network itself. It is only recently that scholars have begun to track some of the dangers that lie in being networked per se. However, as a new strand of scholarship on networked vulnerabilities indicates, globalizing cities today find themselves increasingly confronted with challenges that lie beyond their control. First, in the wake of the global economic crisis of 2008–2010, the limits and contradictions of market-based, competition-oriented forms of urban governance are becoming more pervasive across the worldwide interurban network: crisis-tendencies and socio-ecological disruptions are no longer contained within particular niches within the network, but spread increasingly rapidly across its various conduits. Second, the worldwide urban political ecology that emerges through such crisis-tendencies is characterized and structured by rising vulnerabilities within the network as a whole. Such vulnerabilities are articulated not only through the traditional network of global economic centers, but also through international networks of infectious disease transmission and attainment, as well as through metropolitan infrastructural networks.

AN INVITATION TO RESEARCH – AND ACTION

What we know now about global cities in a world system has confirmed some and contradicted other predictions that were made in the 1980s. At the time, the world was still in the midst of the Cold War, and the so-called "Third World" was little more than an afterthought in much social research and theorizing. We live in a different world now. Moscow is not behind an "iron curtain", Berlin is unified, South Africa has overcome apartheid and hosted the 2010 World Cup, Brazilian cities are players in the global game, Shanghai, Dubai, Mumbai and Lagos have become household names not only in specialized urban lexica but also in popular discourse, film and musical imagination. Bollywood movie production has transgressed the boundaries of the Indian subcontinent, hiphop music is the vernacular of an urban and suburban youth around the globe, and the American coffee multinational Starbucks has captured the street corners of cities around the world and has changed the way those who can afford it consume coffee, whether in Romania, China or Peru. If anything, the post-Cold War world has become more tightly connected through a range of overlapping global urban networks. Hong Kong, London and Vancouver exist on a tangible map in which plausible connectivities exist that are lived and sustained across three continents through complex and expanding family and business relationships. While geographical proximities among cities and their inhabitants have increased, social distancing inside cities and across networks has often increased dramatically. Although the much touted *Blade Runner* scenario has not materialized in most cities of the West, internal sociospatial divisions have, and have led to new forms of exclusivity, ghettoization, gated communities and the like. On a global scale, the "planet of slums" predicted by Mike Davis in the early 2000s has indeed emerged and stands in contrast to the shining citadels of banking, culture and entertainment centres in Europe, Asia and North America. Across urban regions themselves, the tendency of the 100-mile city has dramatically intensified, as rapid urbanization in most parts of the world continues to push into the ever more distant hinterlands of erstwhile "rural" zones. New forms of politics have also emerged as globalized and diversified urban communities lay claim to the right to the city in new, potentially revolutionary ways. And as the consequences of the global economic crisis of 2008–2010 continue to be felt around the globe, we can anticipate new alignments and realignments of political-economic power relations and socio-natural metabolisms. All of this (and more) has necessarily challenged the assumptions and agendas associated with the first generation of global cities research. Yet, despite these transformations, the classic texts of global city theory remain a foundational reference point today due to their salient emphasis on the major role of globally networked city-regions in the making (and unmaking) of globalizing capitalism.

One of the more persistent criticisms that has been leveled at global city researchers is that their work serves to glorify the status of particular cities in worldwide interurban competition, and thus represents an uncritical affirmation of global neoliberalism. Relatedly, it has also been insinuated, at times, that research on global cities tends to affirm the policies of municipal boosters concerned to acquire distinction for their cities on the world stage. In our view, the misunderstanding that underlies these criticisms is based on a mistaken identification of the colloquial notion of the global/world city with the scholarly concept developed in the literatures we have discussed above. While the former is a descriptive, affirmative notion often used by municipal power brokers to draw attention to specific places, the latter is a polysemic analytical term that has been employed by critical urbanists concerned to decipher the globalizing dimensions of contemporary urbanization.

Still, some of the confusion around the notion of the global city may also be attributed to the substantive content of social science research on this topic. In some cases, such as Los Angeles, it would appear that the "hype" generated through studies of the purported "globality" of a particular place actually permits academic researchers to be enlisted, often unwittingly, as "mercenaries" into the camp of global city boosterism. In this context, it is crucial to recall that John Friedmann and Goetz Wolff's first foray into global cities research contained the programmatic subtitle, "an agenda for research and *action*" (our emphasis). For Friedmann and many of his colleagues, the analysis and description of the global city was meant to be a first step in actively effecting positive, progressive and even radical social change. Thus, data on the formation of global urban hierarchies and on the intensification of sociospatial polarization within global cities was clearly understood as a call to arms for progressive planners. Their role, in Friedmann's view, was to mobilize new public policies designed to reduce the suffering of the global city's increasingly impoverished internationalized working classes and migrant populations and, more ambitiously still, to subject the apparently deterritorialized operations of transnational capital to localized, democratic political control. For others, of course, this call to action was interpreted as an imperative to establish the positive business climate and general investment conditions that were deemed necessary for world city formation. However, in an incisive intervention into the public policy debate in East Asian city states craving world city status in the 1990s, Friedmann reminded his audience:

> [U]rban outcomes are to a considerable extent the result of *public policies*. They are, in part, what we choose them to be. The cities of the next century will thus be a result of planning in the broadest sense of that much abused term. This is not to fall into the naïve belief that all we need to do is to draw a pretty picture of the future, such as a master plan, or adopt wildly ambitious regulatory legislation as a template for future city growth. . . . Instead of waxing enthusiastic about megaprojects – bridges, tunnels, airports, and the cold beauty of glass-enclosed skyscrapers – which so delight the heart of big-city mayors, I am talking about people, their habitat and quality of life, the claims of invisible migrant citizens and now, in yet another turn, the concept of civil society.

What, then, can research on world cities/global cities teach us about the situation and prospects of contemporary capitalism? Beyond its significance to urban specialists, does research on global cities make a more general contribution to our understanding of contemporary social life, and to our ability to shape the latter in progressive, emancipatory ways? Global city research, in our view, offers us some bearings, some intellectual and political grounding, as we attempt to orient ourselves within a fundamentally disjointed, yet profoundly authoritarian, new world order. Whether or not this intellectual perspective can help open up possibilities for radical or progressive social change is ultimately a political question that can only be decided through ongoing social mobilizations and struggles.

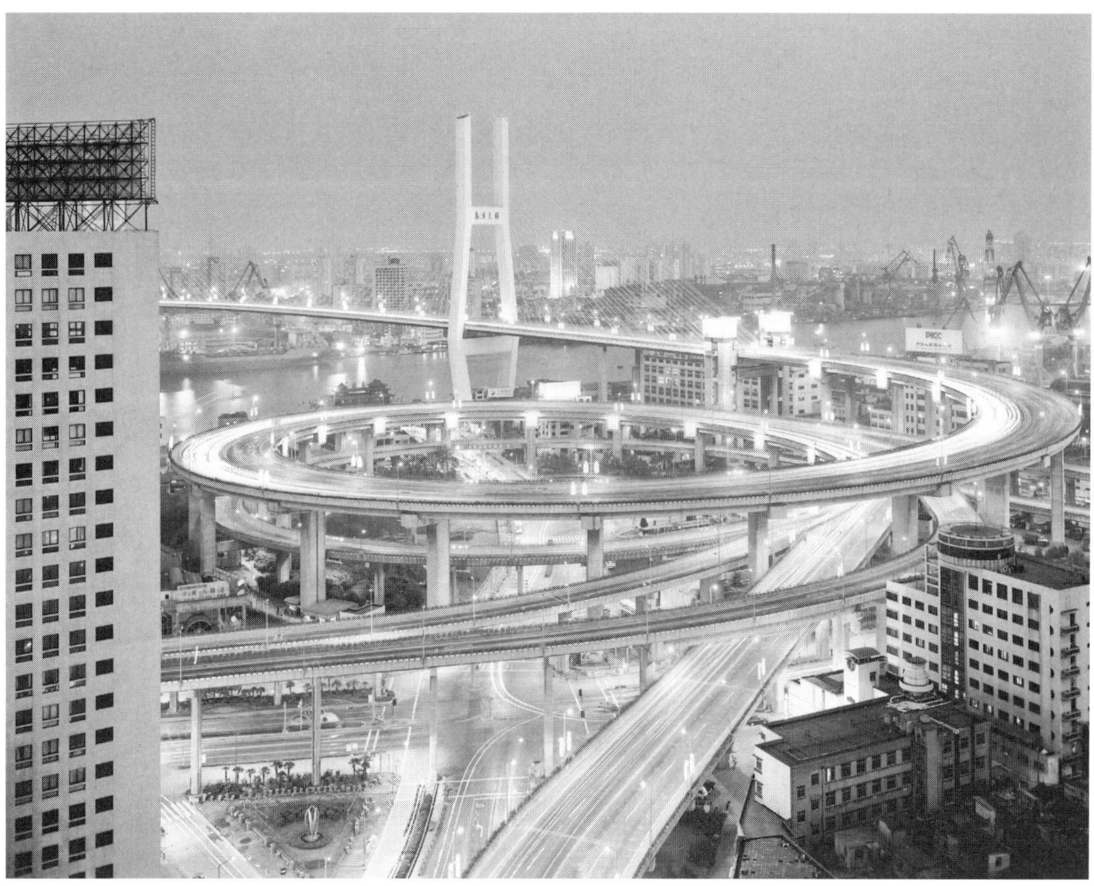

Plate 44 **An Asian megacity: Shanghai, China.** As the world's population increases and higher percentages of the population live in cities, there are increasing numbers of megacities of 10 million residents and more.

Plate 45 **Work in a contemporary Chinese factory.** New, efficient, and highly regimented workplaces characterize the new industrial leadership of contemporary China. Factories like this one account for China's extraordinary economic growth after the freeing of market forces brought on by the market reforms of the 1970s and 1980s. This image is by Edward Burtnysky from Jennifer Baichwal's film *Manufactured Landscapes* (2006).

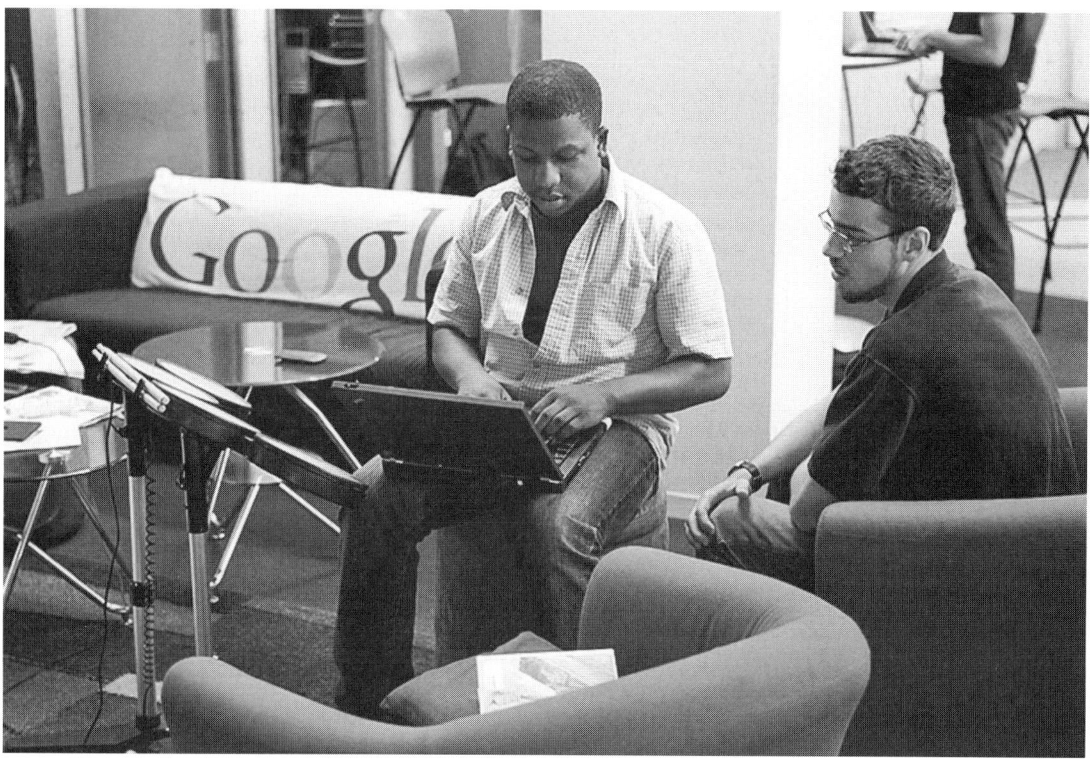

Plate 46 **Work at Google.com**. Young, diverse, and relaxed, the highly educated (and highly paid) workers at telecommunications and computer industries like Google enjoy remarkable freedom as they pioneer new ways to exploit the potential of digital technologies. (Photograph courtesy of Google.com)

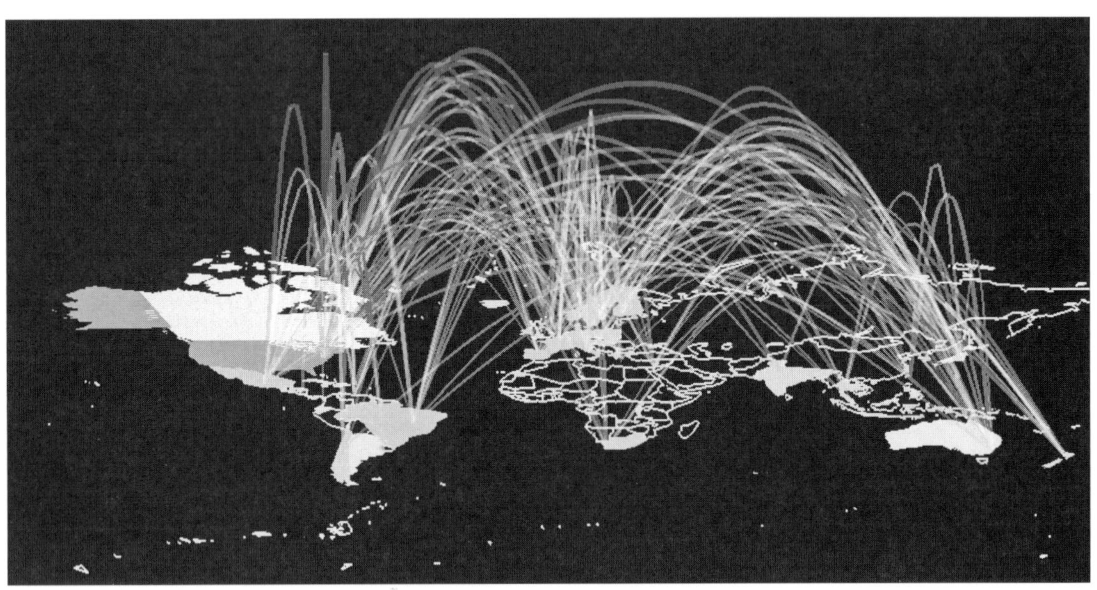

Plate 47 **Visualizing the digital interconnections of the global cities network.** Some realities are best explained visually. One can intellectually grasp the way telecommunications technologies connect global urban centers, but this "visualization" by Stephen Eick of VisTracks, LLC, is breathtakingly iconic in the way that it illustrates the new global networks.

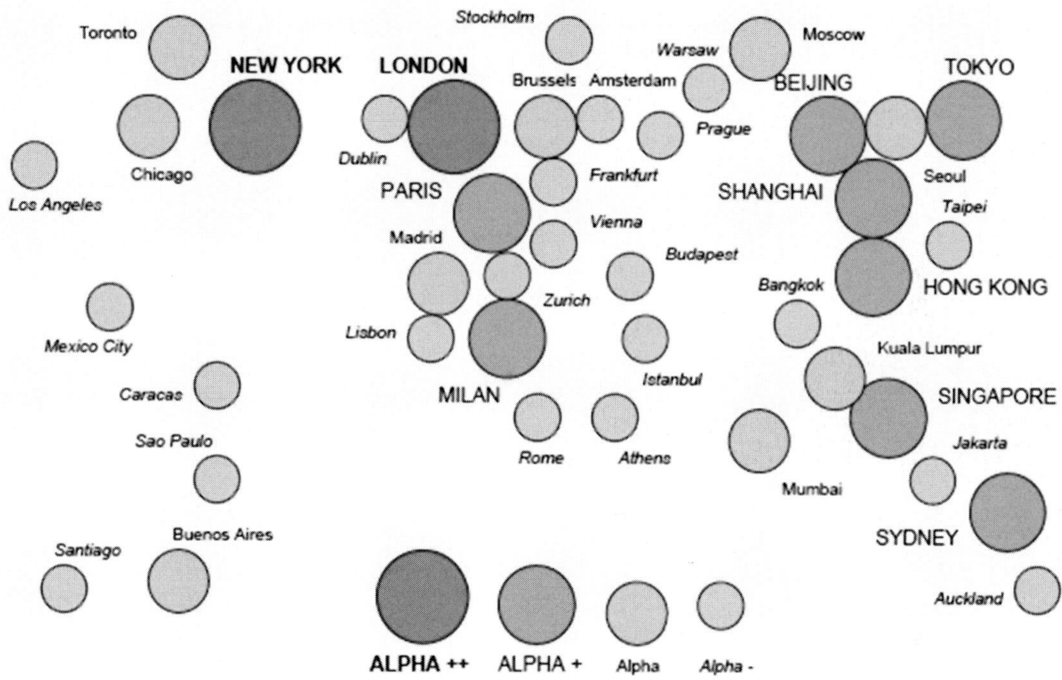

Plate 48 **The world according to GaWC.** The Globalization and World Cities project directed by Professor Peter Taylor at Loughborough University, England, is the best source for detailed information about the global city system. Many of their data-rich reports are illustrated with cartograms like this. The GaWC website (www.lboro.ac.uk/gawc/) is an extraordinary and highly recommended resource on contemporary world urbanization.

Plate 49 **The City of Silk, Kuwait.** Approved by the government of Kuwait in 2008, the Madinat al-Hareer (City of Silk) development will include a business and conference center and associated hotels, spas, and parks. Designed by architect Eric R. Kuhne of the CivicArts research and development group, the project will cover 2,500 square kilometers, house 700,000, and cost US$100 billion (estimated).

Plate 50 **The persistence of urban slums: Mumbai, India**. As some sectors of cities in the age of globalization reach new heights of innovation and productivity, other sectors are stuck in slum conditions as bad as those described by Friedrich Engels in the early days of the Industrial Revolution. From Mexico City and Rio de Janeiro to Lagos, Nigeria, and Manila, Philippines, as many as 1 billion people worldwide live in conditions such as these with incomes of $1 per day or less, and the struggle against gross social inequality continues as an important part of the urban agenda for reform.

Illustration Credits

Every effort has been made to contact copyright holders for their permission to reprint plates in this book. The publishers would be grateful to hear from any copyright holder who is not here acknowledged and will undertake to rectify any errors or omissions in future editions of this book. Following is copyright information for the plates that appear in this book.

1 **The demographic S curves of urbanization.** Copyright © 2010 Michael Brestel.

2 **"A View of the City of Babylon." Maurice Bardin.** Copyright © by the Oriental Institute, University of Chicago. Used by permission.

3 **A view of ancient Athens.** Copyright © Dien-Jen Ru. Used by permission of the artist.

4 **A walled medieval city: Carcassonne, France.** Postcard ca. 1900. Public domain.

5 **The nineteenth-century industrial city.** Augustus Welby Pugin, *Contrasts: Or a Parallel Between the Noble Edifices of the Middle Ages and Corresponding Buildings of the Present Day: Showing the Present Decay of Taste* (London: Charles Dolman, 1841). Public domain.

6 **A modern downtown of the 1920s: San Francisco's Market Street, ca. 1925** Unknown photographer, from the collection of Frederic Stout. Public domain.

7 **Levittown, New York, 1947.** Copyright © Levittown, New York Public Library. Used by permission of the New York Public Library.

8 **The auto-centered metropolis, 1922.** Security Pacific Collection, Los Angeles Public Library. Copyright © Los Angeles Public Library. Used by permission of Los Angeles Public Library.

9 **Sprawl suburbia.** Photograph by David Shankbone and placed in the public domain through online public license. Used by permission of the photographer.

10 **"Music in the Street, Music in the Parlor."** Unknown artist, 1868, *The Illustrated News*. Public domain.

11 **"The Hearth-Stone of the Poor."** Sol Eytinge, Jr., 1876, *Harper's Weekly*. Public domain.

12 **"City Sketches."** C.A. Barry, 1855 *Ballou's Pictorial Drawing-Room Companion*. Public domain.

13 **"Bicycles and Tricycles – How They Are Made."** Unknown artist, 1887, *Frank Leslie's Illustrated Newspaper*. Public domain.

14 **"Burning Mills, Oswego, New York."** George N. Barnard, 1853, *Oswego Daily Times*. Used by permission of George Eastman House.

15 **"A Rainy Day on Broadway."** Edward Anthony, 1859. Used by permission of George Eastman House.

16 **"Bandits' Roost, 39½ Mulberry Street."** Jacob Riis, New York, 1889. The Jacob A. Riis collection # 101 Copyright © Museum of the City of New York. Used by permission of the Museum of the City of New York.

17 **Ellis Island Immigrant Portraits.** Lewis Hine, ca. 1910. In Grove S. Dow, *Social Problems of Today* (New York: Thomas Y. Crowell, 1925). Public domain.

18 **"The Steerage."** Alfred Stieglitz, 1907. Copyright © Museum of Modern Art, New York. Used by permission of the Museum of Modern Art, New York.

19 **Untitled (roadside sign).** Dorothea Lange, 1939. Farm Security Administration, US Department of Agriculture. Printed in C.E. Lively and Conrad Tauber, *Rural Migration in the United States* (Washington, DC: Works Progress Administration, 1939). Public domain.

20 **"Ford Plant, Detroit."** Charles Sheeler, 1927. Copyright © Museum of Modern Art, New York. Used by permission of the Museum of Modern Art, New York.

21 **"The Critic."** Weegee (Arthur Fellig), 1943. Copyright © 1994, International Center of Photography, New York. Bequest of Wilma Wilcox. Used by permission of the International Center of Photography, New York.

22 **Still from *The Crowd.*** King Vidor, 1926. Museum of Modern Art, New York, Film Stills Archive. Copyright © Museum of Modern Art, New York. Used by permission of the Museum of Modern Art, New York.

23 **Still from *Metropolis.*** Fritz Lang, 1929. Museum of Modern Art, New York, Film Stills Archive. Copyright © Museum of Modern Art, New York. Used by permission of the Museum of Modern Art, New York.

24 **New urbanist neo-traditionalism: street in Seaside, Florida.** Copyright © Duany – Plater-Zyberk. Used by permission of Duany – Plater-Zyberk.

25 **Iconic architecture: Guggenheim Museum, Bilbao, Spain.** Copyright © Sydney Pollack, by permission. Courtesy of Gehry Partners, LLP.

26 **The Mall has it all: North Garden, Mall of America, Minneapolis.** Copyright © Mall of America/ Simon MOA Management Co., Inc. Used by permission of Mall of America.

27 **The persistence of tradition.** Copyright © Paul V. Turner. Used by permission of the photographer.

28 **The persistence of poverty and decay.** Copyright © by Ken Alexander. Used by permission of the photographer.

29 **The streets belong to the kids.** Copyright © StoutFoto, San Francisco. Used by permission of the photographer.

30 **The streets belong to the people.** Copyright © 1925 Frans Masereel, DACS 2010. Used by permission of the copyright holder.

31 **Urban pride: architecture as symbolic power – Burj Khalifa, Dubai.** Copyright © Emaar Properties. Courtesy of Burj Khalifa, Armani Hotels and Resorts. Used by permission.

32 **Urban terror: World Trade Center, New York, September 11, 2001.** Copyright © Wanda McCormick, photographer, Readio.com. Used by permission of the photographer.

33 **Central Park, New York, 1863**. Copyright © Museum of the City of New York. Used by permission of the Museum of the City of New York.

34 **Arturo Soria y Mata, plan for a linear city around Madrid, 1894**. From Arturo Soria y Mata, *The Linear City*, reprinted in Richard LeGates and Frederic Stout (eds), *Early Urban Planning 1870–1940* (London: Routledge/Thoemmes, 1998).

35 **Ebenezer Howard's plan for a Garden City, 1898**. From Ebenezer Howard, *Garden Cities of To-morrow*, reprinted in Richard LeGates and Frederic Stout (eds), *Early Urban Planning 1870–1940* (London: Routledge/Thoemmes, 1998).

36 **Plan for Welwyn Garden City, 1909**. Royal Town Planning Institute. Public domain.

37 **Le Corbusier's "Plan Voisin" for a City of 3 Million People, 1925**. Le Corbusier, *Urbanisme* (London: John Rodher, 1929). Public domain.

38 **Plan for Radburn, New Jersey, 1929**. Clarence Stein and Henry Wright architects. Public domain.

39 **Frank Lloyd Wright's plan for Broadacre City, 1935**. Frank Lloyd Wright, *When Democracy Builds* (Chicago, IL: University of Chicago Press 1945).

40 **Paseo del Rio, San Antonio Texas**. Copyright © Alexander Garvin. Used by permission of Alexander Garvin.

41 **Quincy Market, Boston, Massachusetts.** Copyright © Alexander Garvin. Used by permission of Alexander Garvin.

42 **Peter Calthorpe, "The Crossings," Mountain View, California.** Copyright © Peter Calthorpe Associates. Used by permission of Peter Calthorpe.
43 **Strøget. A pedestrian-only street in Copenhagen, Denmark.** Copyright © Jan Gehl Architects. Used by permission.
44 **An Asian Megacity: Shanghai, China**. Copyright © Peter Bialobrzeski. From Peter Bialobrzeski, *Neon Tigers: Photographs of Asian Megacities* (Ostfildern, Germany: Hatje Cantz, 2004). Used by permission of Redux Pictures, 12th Floor, 116 E. 16th Street, New York, NY 10003.
45 **Work in a contemporary Chinese Factory.** Photograph by Edward Burtynsky from Jennifer Baichwal's film *Manufactured Landscapes* (Mercury Films, 2006). Used by permission.
46 **Work at Google.com.** Copyright © by Google.com. Used by permission.
47 **Visualizing the digital interconnections of the global cities network.** Copyright © by Stephen Eick of VisTracks LLC. Used by permission.
48 **The world according to GaWC.** Copyright © Globalization and World Cities research group, Loughborough University, 1998 and 2008. Used by permission of GaWC.
49 **The City of Silk, Kuwait.** Copyright © CivicArts, Eric R. Kuhne & Associates. Used by permission.
50 **The persistence of urban slums: Mumbai, India.** Copyright © Sarah Lane of Sarahlane.com. Used by permission.

PART TITLE ILLUSTRATIONS

Prologue Lisa Ryan, 1998. Copyright © Lisa Ryan. Used by permission of Lisa Ryan.
PART 1 Hartmann Schedel, "Hierosolima" (Jerusalem). *Liber Cronicarum. (Nuremberg, 1493). Public domain.*
PART 2 Gustave Doré, "London Traffic," 1872. Public domain.
PART 3 Frederic Stout and Lisa Ryan, Geography collage 1999. Copyright © Frederic Stout and Lisa Ryan. Used by permission of Frederic Stout and Lisa Ryan.
PART 4 Thomas Nast, "Who Stole the People's Money?" *Harpers Weekly* August 19, 1871. Public domain.
PART 5 Robert Owen, *The Crisis* (London: J. Eamonson, 1832). Public domain.
PART 6 Postwar British cartoon. Public domain.
PART 7 Sketch of Las Ramblas, Barcelona, Spain, by Allan Jacobs. Copyright © MIT Press. Used by permission.
PART 8 Arcology by Paola Soleri. Copyright © The Cosanti Foundation. Used by permission of the Cosanti Foundation.

Copyright Information

Every effort has been made to contact copyright holders for their permission to reprint selections in this book. The publishers would be grateful to hear from any copyright holder who is not here acknowledged and will undertake to rectify any errors or omissions in future editions of this book. Following is copyright information for the selections that appear in this book.

PROLOGUE

PROLOGUE LeGates, Richard T., "Prologue: How to Study Cities." Copyright © 2010 Richard T. LeGates. Reprinted by permission of the author.

1 THE EVOLUTION OF CITIES

DAVIS Davis, Kingsley, "The Urbanization of the Human Population," *Scientific American* (September 1965). Copyright © 1965 Scientific American, Inc. All rights reserved. Reprinted by permission of *Scientific American.*

CHILDE Childe, V. Gordon, "The Urban Revolution," *Town Planning Review*, 21 (April 1950), 3–17. Reprinted by permission of the *Town Planning Review.*

KITTO Kitto, H.D.F., "The Polis," from Kitto, *The Greeks* (London: Penguin, 1951; revised edition, 1957). Copyright © 1951, 1957, H.D.F. Kitto. Reprinted by permission of Penguin Books Ltd.

ENGELS Engels, Friedrich, "The Great Towns," from Engels, *The Condition of the Working Class in England in 1844* (1845). Public domain.

WARNER Warner, Sam Bass, "Evolution and Transformation: The American Industrial Metropolis, 1840–1940." Copyright © 2009 Sam Bass Warner. Specially commissioned for inclusion in *The City Reader*, fifth edition, and printed by permission of the author.

JACKSON Jackson, Kenneth T., "The Drive-in Culture of Contemporary America," from Jackson, *Crabgrass Frontier: The Suburbanization of the United States* (New York: Oxford University Press, 1985). Copyright © 1985 Oxford University Press, Inc. Reprinted by permission of Oxford University Press, Inc.

FISHMAN Fishman, Robert, "Beyond Suburbia: The Rise of the Technoburb," from Fishman, *Bourgeois Utopias: The Rise and Fall of Suburbia* (New York: Basic Books, 1987). Copyright © 1987 Basic Books, Inc. Reprinted by permission of Basic Books, a division of Perseus Books Group.

2 URBAN CULTURE AND SOCIETY

MUMFORD Mumford, Lewis, "What is a City?" *Architectural Record*, 82 (November 1937). Copyright © 1937 McGraw-Hill, Inc. All rights reserved. Reprinted by permission of *Architectural Record*, www.architecturalrecord.com.

WIRTH Wirth, Louis, "Urbanism as a Way of Life," *American Journal of Sociology*, 44(1) (July 1938). Copyright © 1938 University of Chicago Press. Reprinted by permission of University of Chicago Press.

JACOBS Jacobs, Jane, "The Uses of Sidewalks: Safety," from Jacobs, *The Death and Life of Great American Cities* (New York: Random House, 1961). Copyright © 1961, 1989 Jane Jacobs. Published by Random House and Jonathan Cape. Reprinted by permission of Random House Inc. and The Random House Group Limited.

DU BOIS Du Bois, W.E.B., "The Negro Problems of Philadelphia," "The Question of Earning a Living," and "Color Prejudice," from Du Bois, *The Philadelphia Negro* (Philadelphia, PA: University of Pennsylvania Press, 1899). Public domain.

WILSON Wilson, William Julius, "From Institutional to Jobless Ghettos," from Wilson, *When Work Disappears: The World of the New Urban Poor* (New York: Knopf, 1996). Copyright © 1996 William Julius Wilson. Reprinted by permission of the author.

ANDERSON Anderson, Elijah, "The Code of the Street" and "Decent and Street Families," from Anderson, *Code of the Street: Decency, Violence, and the Moral Life of the Inner City* (New York: Norton, 1999). Copyright © 1999 Elijah Anderson. Reprinted by permission of W.W. Norton and Company.

PUTNAM Putnam, Robert D., "Bowling Alone: America's Declining Social Capital," *Journal of Democracy* 6(1) (January, 1995), 65–78. Copyright © Robert D. Putnam. Reprinted by permission of the author.

FLORIDA Florida, Richard, "The Creative Class," from Florida, *The Rise of the Creative Class: And How It's Transforming Work, Leisure, Community and Everyday Life* (New York: Basic Books, 2002). Copyright © 2002 Richard Florida. Reprinted by permission of the author.

STOUT Stout, Frederic, "Visions of a New Reality: The City and the Emergence of Modern Visual Culture." Copyright © 1999 Frederic Stout. Reprinted by permission of the author.

3 URBAN SPACE

BURGESS Burgess, Ernest W., "The Growth of the City: An Introduction to a Research Project," from Robert E. Park, Ernest W. Burgess, and Roderick D. McKenzie (eds), *The City* (Chicago, IL: University of Chicago Press, 1925). Copyright © 1925 University of Chicago Press. Reprinted by permission of the University of Chicago Press.

DEAR Dear, Michael, "The Los Angeles School of Urbanism: An Intellectual History." Copyright © 2003 Michael Dear. Reprinted by permission of the author.

SPAIN Spain, Daphne, "What Happened to Gender Relations on the Way from Chicago to Los Angeles?" *City and Community*, 1(2) (June 2002), 155–169. Copyright © 2002 Blackwell Publishing Co. Reprinted by permission of Blackwell Publishing.

MADANIPOUR Madanipour, Ali, "Social Exclusion and Space," from Ali Madanipour, Goran Cars, and Judith Allen (eds), *Social Exclusion in European Cities: Processes, Experiences, and Responses* (London: Jessica Kingsley Publishers, 1998). Copyright © Jessica Kingsley Publishers. Reprinted by permission of Jessica Kingsley Publishers.

DAVIS Davis, Mike, "Fortress L.A.," from Davis, *City of Quartz: Excavating the Future in Los Angeles* (London: Verso, 1990). Copyright © 1990 Verso. Reprinted by permission of Verso.

JACKSON Jackson, John Brinckerhoff, "The Almost Perfect Town," *Landscape*, 2(1) (Autumn, 1952). Reprinted by permission of the author.

BRUEGMANN Bruegmann, Robert, "The Causes of Sprawl," from Bruegmann, *Sprawl: A Compact History* (Chicago, IL: University of Chicago Press, 2005). Copyright © 2005 University of Chicago Press. Reprinted by permission of the University of Chicago Press.

4 URBAN POLITICS, GOVERNANCE, AND ECONOMICS

HARVEY Harvey, David, "Contested Cities: Spatial Process and Spatial Form," from Nick Jewson and Susanne MacGregor (eds), *Transforming Cities* (London: Routledge, 1997). Copyright © 1997 Routledge. Reprinted by permission of Taylor & Francis Books, UK.

ARNSTEIN Arnstein, Sherry, "A Ladder of Citizen Participation," *Journal of the American Institute of Planners*, 8(3) (July, 1969). Copyright © 1969 Journal of the American Planning Association. Reprinted by permission of the *Journal of the American Planning Association*.

MOLOTCH Molotch, Harvey, "The City as a Growth Machine: Towards a Political Economy of Place," *American Journal of Sociology*, 82 (September, 1976), 309–332. Copyright © 1976 American Journal of Sociology. Reprinted by permission of *American Journal of Sociology*.

WILSON and KELLING Wilson, James Q. and Kelling, George L., "Broken Windows," *Atlantic Monthly*, 249(3) (March, 1982). Copyright © 1982 by James Q. Wilson. Reprinted by permission of James Q. Wilson and *Atlantic Monthly*.

THOMPSON Thompson, Wilbur, "The City as a Distorted Price System," *Psychology Today* 2(3) (August, 1968). Copyright © 1968 Psychology Today. Reprinted by permission of *Psychology Today*.

PORTER Porter, Michael, "The Competitive Advantage of the Inner City," *Harvard Business Review*, (May/June, 1995). Copyright © 1995 President and Fellows of Harvard College. All rights reserved. Reprinted by permission of the President and Fellows of Harvard College.

ORFIELD Orfield, Myron, "Metropolitics and Fiscal Equity." Copyright © 2009 Myron Orfield. Specially commissioned for inclusion in *The City Reader, fifth edition* Printed by permission of the author.

5 URBAN PLANNING HISTORY AND VISIONS

OLMSTED Olmsted, Frederick Law, "Public Parks and the Enlargement of Towns," *American Social Science Association* (1870). Public domain.

HOWARD Howard, Ebenezer, "Author's Introduction" and "The Town–Country Magnet," from Howard, *Garden Cities of To-morrow* (1898/1902). Public domain.

LE CORBUSIER Le Corbusier, "A Contemporary City," from Le Corbusier, *The City of Tomorrow and its Planning*. (New York: Dover, 1987). Translated from the 8th French edition of *Urbanisme* by Frederick Etchells. Original copyright © 1929. Copyright © 1971 by MIT Press. Reprinted by permission of MIT Press.

WRIGHT Wright, Frank Lloyd, "Broadacre City: A New Community Plan," from *Architectural Record*, 77 (April, 1935). Copyright © 1935, 1943, 1999, 2002 Frank Lloyd Wright Foundation, Scottsdale, Arizona. Reprinted by permission of the Frank Lloyd Wright Foundation.

WORLD COMMISSION ON ENVIRONMENT AND DEVELOPMENT "Towards Sustainable Development," from WCED, *Our Common Future* (New York: Oxford University Press, 1987). Copyright © 1987 World Commission on Environment and Development. Reprinted by permission of Oxford University Press.

CONGRESS FOR THE NEW URBANISM "Charter of the New Urbanism." Copyright © 1993 Congress for the New Urbanism. Reprinted by permission of the Congress for the New Urbanism.

CALTHORPE and FULTON Calthorpe, Peter and Fulton, William, "Designing the Region" and "Designing the Region is Designing the Neighborhood," from Calthorpe and Fulton, *The Regional City: Planning for the End of Sprawl* (Washington, DC: Island Press, 2001). Copyright © Island Press. Reprinted by permission of Island Press, Washington, DC.